STUDENT'S SOLUTIONS MANUAL

JUDITH A. PENNA

INTRODUCTORY ALGEBRA

TWELFTH EDITION

Marvin L. Bittinger

Indiana University Purdue University Indianapolis

Judith A. Beecher

Barbara L. Johnson

Indiana University Purdue University Indianapolis

PEARSON

Boston Columbus Indianapolis New York San Francisco Upper Saddle River
Amsterdam Cape Town Dubai London Madrid Milan Munich Paris Montreal Toronto
Delhi Mexico City São Paulo Sydney Hong Kong Seoul Singapore Taipei Tokyo

Publishing as Pearson, 75 Arlington Street, Boston, MA 02116.

ISBN-13: 978-0-321-92299-1
ISBN-10: 0-321-92299-9

1 2 3 4 5 6 EBM 17 16 15 14

PEARSON

www.pearsonhighered.com

Contents

Chapter R

Prealgebra Review

Exercise Set R.1

RC1. If one number is a factor of another, then the GCF is the smaller of the two numbers.

RC3. If one number is a factor of another, then the LCM is the larger of the two numbers.

1. We first find some factorizations:

$20 = 1 \cdot 20,\ 20 = 2 \cdot 10,\ 20 = 4 \cdot 5$

The factors of 20 are 1, 2, 4, 5, 10, and 20.

3. We first find some factorizations:

$72 = 1 \cdot 72,\ 72 = 2 \cdot 36,\ 72 = 3 \cdot 24,$

$72 = 4 \cdot 18,\ 72 = 6 \cdot 12,\ 72 = 8 \cdot 9$

The factors of 72 are 1, 2, 3, 4, 6, 8, 9, 12, 18, 24, 36 and 72.

5. $15 = 3 \cdot 5$

Both factors are prime, so we have the prime factorization of 15.

7. $49 = 7 \cdot 7$

Both factors are prime, so we have the prime factorization of 49.

9. We begin by factoring 18 in any way that we can and continue factoring until each factor is prime.

$18 = 2 \cdot 9 = 2 \cdot 3 \cdot 3$

11. We begin by factoring 40 in any way that we can and continue factoring until each factor is prime.

$40 = 4 \cdot 10 = 2 \cdot 2 \cdot 2 \cdot 5$

13. We begin by factoring 90 in any way that we can and continue factoring until each factor is prime.

$90 = 2 \cdot 45 = 2 \cdot 9 \cdot 5 = 2 \cdot 3 \cdot 3 \cdot 5$

15. We go through the table of primes until we find a prime that divides 210. We continue dividing by that prime until it is not possible to do so any longer. We continue this process until each factor is prime.

$210 = 2 \cdot 105 = 2 \cdot 3 \cdot 35 = 2 \cdot 3 \cdot 5 \cdot 7$

17. We go through the table of primes until we find a prime that divides 91. The first such prime is 7.

$91 = 7 \cdot 13$

Both factors are prime, so this is the prime factorization.

19. We first find the prime factorization of each number.

$36 = 2 \cdot 2 \cdot 3 \cdot 3$

$48 = 2 \cdot 2 \cdot 2 \cdot 2 \cdot 3$

Each number has two factors of 2 and one factor of 3. The GCF is $2 \cdot 2 \cdot 3$, or 12.

21. We first find the prime factorization of each number.

13 is prime.

$52 = 2 \cdot 2 \cdot 13$

Each number has one factor of 13. The GCF is 13.

23. We first find the prime factorization of each number.

$15 = 3 \cdot 5$

$28 = 2 \cdot 2 \cdot 7$

There is no common prime factor, so the GCF is 1.

25. We first find the prime factorization of each number.

$54 = 2 \cdot 3 \cdot 3 \cdot 3$

$180 = 2 \cdot 2 \cdot 3 \cdot 3 \cdot 5$

Each number has one factor of 2 and two factors of 3. The GCF is $2 \cdot 3 \cdot 3$, or 18.

27. We first find the prime factorization of each number.

$18 = 2 \cdot 3 \cdot 3$

$66 = 2 \cdot 3 \cdot 11$

Each number has one factor of 2 and one factor of 3. The GCF is $2 \cdot 3$, or 6.

29. We first find the prime factorization of each number.

$15 = 3 \cdot 5$

$40 = 2 \cdot 2 \cdot 2 \cdot 5$

$60 = 2 \cdot 2 \cdot 3 \cdot 5$

Each number has a factor of 5. The GCF is 5.

31. We first find the prime factorization of each number.

$70 = 2 \cdot 5 \cdot 7$

$105 = 3 \cdot 5 \cdot 7$

$350 = 2 \cdot 5 \cdot 5 \cdot 7$

Each number has one factor of 5 and one factor of 7. The GCF is $5 \cdot 7$, or 35.

33. a) We find the prime factorizations:

$24 = 2 \cdot 2 \cdot 2 \cdot 3$

$36 = 2 \cdot 2 \cdot 3 \cdot 3$

b) We write 2 as a factor three times (the greatest number of times it occurs in any one factorization). We write 3 as a factor two times (the greatest number of times it occurs in any one factorization). The LCM is $2 \cdot 2 \cdot 2 \cdot 3 \cdot 3$, or 72.

35. $3 = 3$ (3 is prime)
$15 = 3 \cdot 5$

The LCM is $3 \cdot 5$, or 15.

37. $30 = 2 \cdot 3 \cdot 5$

$40 = 2 \cdot 2 \cdot 2 \cdot 5$

The LCM is $2 \cdot 2 \cdot 2 \cdot 3 \cdot 5$, or 120.

39. 13 and 23 are both prime. The LCM is $13 \cdot 23$, or 299.

41. $18 = 2 \cdot 3 \cdot 3$

$30 = 2 \cdot 3 \cdot 5$

The LCM is $2 \cdot 3 \cdot 3 \cdot 5$, or 90.

43. $30 = 2 \cdot 3 \cdot 5$

$36 = 2 \cdot 2 \cdot 3 \cdot 3$

The LCM is $2 \cdot 2 \cdot 3 \cdot 3 \cdot 5$, or 180.

45. $12 = 2 \cdot 2 \cdot 3$

$28 = 2 \cdot 2 \cdot 7$

The LCM is $2 \cdot 2 \cdot 3 \cdot 7$, or 84.

47. $24 = 2 \cdot 2 \cdot 2 \cdot 3$

$36 = 2 \cdot 2 \cdot 3 \cdot 3$

$12 = 2 \cdot 2 \cdot 3$

The LCM is $2 \cdot 2 \cdot 2 \cdot 3 \cdot 3$, or 72.

49. $5 = 5$ (5 is prime)
$12 = 2 \cdot 2 \cdot 3$
$15 = 3 \cdot 5$

The LCM is $2 \cdot 2 \cdot 3 \cdot 5$, or 60.

51. $6 = 2 \cdot 3$
$12 = 2 \cdot 2 \cdot 3$
$18 = 2 \cdot 3 \cdot 3$

The LCM is $2 \cdot 2 \cdot 3 \cdot 3$, or 36.

53. Jupiter: $12 = 2 \cdot 2 \cdot 3$
Saturn: $30 = 2 \cdot 3 \cdot 5$

The LCM is $2 \cdot 2 \cdot 3 \cdot 5$, or 60.

Jupiter and Saturn will appear in the same direction in the night sky as seen from the earth every 60 years.

55. Saturn: $30 = 2 \cdot 3 \cdot 5$
Uranus: $84 = 2 \cdot 2 \cdot 3 \cdot 7$

The LCM is $2 \cdot 2 \cdot 3 \cdot 5 \cdot 7$, or 420.

Saturn and Uranus will appear in the same direction in the night sky as seen from the earth every 420 years.

57. The smallest number of strands that can be used is the LCM of 10 and 3.

$10 = 2 \cdot 5$

$3 = 3$

$\text{LCM} = 2 \cdot 5 \cdot 3 = 30$

The smallest number of strands that can be used is 30.

59. $8 = 2 \cdot 2 \cdot 2$
$12 = 2 \cdot 2 \cdot 3$

The greatest number of times 2 occurs in the factorizations of 8 and 12 is three times. The greatest number of times 3 occurs is once. Thus, the LCM must contain exactly three factors of 2 and one factor of 3. The LCM is $2 \cdot 2 \cdot 2 \cdot 3$.

a) No; $2 \cdot 2 \cdot 3 \cdot 3$ is not a multiple of 8.

b) No; $2 \cdot 2 \cdot 2 \cdot 3 \cdot 5$ is a multiple of both 8 and 12, but it is not the least common multiple.

c) No; $2 \cdot 3 \cdot 3$ is not a multiple of 8 or 12.

d) Yes; $2 \cdot 2 \cdot 2 \cdot 3$ is a multiple of both 8 and 12, and it is the smallest such multiple.

Exercise Set R.2

RC1. $9 + 0 = 9$ illustrates the identity property of 0. The correct answer is (d).

RC3. $12 = 2 \cdot 2 \cdot 3$ and $15 = 3 \cdot 5$, so the LCD of $\frac{5}{12}$ and $\frac{1}{15}$ is $2 \cdot 2 \cdot 3 \cdot 5$. The correct answer is (c).

RC5. $\frac{31}{31}$ is equivalent of 1, so the correct answer is (e).

1.
$$\begin{aligned} \frac{3}{4} &= \frac{3}{4} \cdot 1 && \text{Identity property of 1} \\ &= \frac{3}{4} \cdot \frac{3}{3} && \text{Using } \frac{3}{3} \text{ for 1} \\ &= \frac{9}{12} && \text{Multiplying numerators and denominators} \end{aligned}$$

3.
$$\begin{aligned} \frac{3}{5} &= \frac{3}{5} \cdot 1 && \text{Identity property of 1} \\ &= \frac{3}{5} \cdot \frac{20}{20} && \text{Using } \frac{20}{20} \text{ for 1} \\ &= \frac{60}{100} && \text{Multiplying numerators and denominators} \end{aligned}$$

5.
$$\begin{aligned} \frac{13}{20} &= \frac{13}{20} \cdot 1 && \text{Identity property of 1} \\ &= \frac{13}{20} \cdot \frac{8}{8} && \text{Using } \frac{8}{8} \text{ for 1} \\ &= \frac{104}{160} \end{aligned}$$

7. We will use $\frac{3}{3}$ for 1 since $24 = 8 \cdot 3$.

$$\frac{7}{8} = \frac{7}{8} \cdot 1 = \frac{7}{8} \cdot \frac{3}{3} = \frac{21}{24}$$

9. We will use $\frac{4}{4}$ for 1 since $16 = 4 \cdot 4$.

$$\frac{5}{4} = \frac{5}{4} \cdot 1 = \frac{5}{4} \cdot \frac{4}{4} = \frac{20}{16}$$

11. We will use $\frac{23}{23}$ for 1 since $437 = 19 \cdot 23$.

$$\frac{17}{19} = \frac{17}{19} \cdot 1 = \frac{17}{19} \cdot \frac{23}{23} = \frac{391}{437}$$

13. $\frac{18}{27} = \frac{2 \cdot 9}{3 \cdot 9}$ Factoring numerator and denominator

$= \frac{2}{3} \cdot \frac{9}{9}$ Factoring the fractional expression

$= \frac{2}{3} \cdot 1$

$= \frac{2}{3}$ Identity property of 1

15. $\frac{56}{14} = \frac{4 \cdot 14}{1 \cdot 14}$ Factoring and inserting a factor of 1 in the denominator

$= \frac{4}{1} \cdot \frac{14}{14}$ Factoring the fractional expression

$= \frac{4}{1} \cdot 1$

$= 4$ Identity property of 1

17. $\frac{6}{42} = \frac{1 \cdot 6}{7 \cdot 6}$ Factoring and inserting a factor of 1 in the numerator

$= \frac{1}{7} \cdot \frac{6}{6}$

$= \frac{1}{7} \cdot 1$

$= \frac{1}{7}$

19. $\frac{56}{7} = \frac{8 \cdot 7}{1 \cdot 7} = \frac{8}{1} \cdot \frac{7}{7} = \frac{8}{1} \cdot 1 = 8$

21. $\frac{19}{76} = \frac{1 \cdot 19}{4 \cdot 19}$ Factoring and inserting a factor of 1 in the numerator

$= \frac{1 \cdot \cancel{19}}{4 \cdot \cancel{19}}$ Removing a factor of 1: $\frac{19}{19} = 1$

$= \frac{1}{4}$

23. $\frac{100}{20} = \frac{5 \cdot 20}{1 \cdot 20}$ Factoring and inserting a factor of 1 in the denominator

$= \frac{5 \cdot \cancel{20}}{1 \cdot \cancel{20}}$ Removing a factor of 1: $\frac{20}{20} = 1$

$= \frac{5}{1}$

$= 5$ Simplifying

25. $\frac{425}{525} = \frac{17 \cdot 25}{21 \cdot 25}$ Factoring the numerator and the denominator

$= \frac{17 \cdot \cancel{25}}{21 \cdot \cancel{25}}$ Removing a factor of 1: $\frac{25}{25} = 1$

$= \frac{17}{21}$

27. $\frac{2600}{1400} = \frac{2 \cdot 13 \cdot 100}{2 \cdot 7 \cdot 100}$ Factoring

$= \frac{13 \cdot \cancel{2} \cdot \cancel{100}}{7 \cdot \cancel{2} \cdot \cancel{100}}$ Removing a factor of 1: $\frac{2 \cdot 100}{2 \cdot 100} = 1$

$= \frac{13}{7}$

29. $\frac{8 \cdot x}{6 \cdot x} = \frac{2 \cdot 4 \cdot x}{2 \cdot 3 \cdot x}$ Factoring

$= \frac{4 \cdot \cancel{2} \cdot \cancel{x}}{3 \cdot \cancel{2} \cdot \cancel{x}}$ Removing a factor of 1: $\frac{2 \cdot x}{2 \cdot x} = 1$

$= \frac{4}{3}$

31. $\frac{1}{3} \cdot \frac{1}{4} = \frac{1 \cdot 1}{3 \cdot 4}$ Multiplying numerators and denominators

$= \frac{1}{12}$

33. $\frac{15}{4} \cdot \frac{3}{4} = \frac{15 \cdot 3}{4 \cdot 4} = \frac{45}{16}$

35. $\frac{1}{3} + \frac{1}{3} = \frac{1+1}{3}$ Adding numerators; keeping the same denominator

$= \frac{2}{3}$

37. $\frac{4}{9} + \frac{13}{18} = \frac{4}{9} \cdot \frac{2}{2} + \frac{13}{18}$ LCD is 18.

$= \frac{8}{18} + \frac{13}{18}$

$= \frac{21}{18}$

$= \frac{7 \cdot \cancel{3}}{6 \cdot \cancel{3}} = \frac{7}{6}$ Simplifying

39. $\frac{3}{10} + \frac{8}{15} = \frac{3}{10} \cdot \frac{3}{3} + \frac{8}{15} \cdot \frac{2}{2}$ LCD is 30.

$= \frac{9}{30} + \frac{16}{30}$

$= \frac{25}{30}$

$= \frac{5 \cdot \cancel{5}}{6 \cdot \cancel{5}} = \frac{5}{6}$ Simplifying

41. $\frac{7}{30} + \frac{5}{12} = \frac{7}{30} \cdot \frac{2}{2} + \frac{5}{12} \cdot \frac{5}{5}$ LCD is 60.

$= \frac{14}{60} + \frac{25}{60}$

$= \frac{39}{60}$

$= \frac{\cancel{3} \cdot 13}{\cancel{3} \cdot 20} = \frac{13}{20}$ Simplifying

43. $\dfrac{5}{4}-\dfrac{3}{4}=\dfrac{2}{4}$
$=\dfrac{1\cdot\cancel{2}}{2\cdot\cancel{2}}=\dfrac{1}{2}$

45. $\dfrac{11}{12}-\dfrac{3}{8}=\dfrac{11}{12}\cdot\dfrac{2}{2}-\dfrac{3}{8}\cdot\dfrac{3}{3}$ LCD is 24.
$=\dfrac{22}{24}-\dfrac{9}{24}$
$=\dfrac{13}{24}$

47. $\dfrac{11}{12}-\dfrac{2}{5}=\dfrac{11}{12}\cdot\dfrac{5}{5}-\dfrac{2}{5}\cdot\dfrac{12}{12}$ LCD is 60.
$=\dfrac{55}{60}-\dfrac{24}{60}$
$=\dfrac{31}{60}$

49. $5\cdot\dfrac{2}{3}=\dfrac{5\cdot 2}{3}=\dfrac{10}{3}$

51. $\dfrac{8}{9}\div\dfrac{4}{15}=\dfrac{8}{9}\cdot\dfrac{15}{4}=\dfrac{8\cdot 15}{9\cdot 4}=\dfrac{2\cdot\cancel{4}\cdot\cancel{3}\cdot 5}{\cancel{3}\cdot 3\cdot\cancel{4}}=\dfrac{10}{3}$

53. $\dfrac{1}{8}\div\dfrac{1}{4}=\dfrac{1}{8}\cdot\dfrac{4}{1}=\dfrac{1\cdot 4}{8\cdot 1}=\dfrac{1\cdot\cancel{4}}{2\cdot\cancel{4}\cdot 1}=\dfrac{1}{2}$

55. $\dfrac{\frac{13}{12}}{\frac{39}{5}}=\dfrac{13}{12}\div\dfrac{39}{5}=\dfrac{13}{12}\cdot\dfrac{5}{39}=\dfrac{13\cdot 5}{12\cdot 39}=\dfrac{\cancel{13}\cdot 5}{12\cdot 3\cdot\cancel{13}}=\dfrac{5}{36}$

57. $100\div\dfrac{1}{5}=\dfrac{100}{1}\div\dfrac{1}{5}=\dfrac{100}{1}\cdot\dfrac{5}{1}=\dfrac{100\cdot 5}{1\cdot 1}=\dfrac{500}{1}=500$

59. $\dfrac{3}{4}\div 10=\dfrac{3}{4}\cdot\dfrac{1}{10}=\dfrac{3\cdot 1}{4\cdot 10}=\dfrac{3}{40}$

61. $1000-\dfrac{1}{100}=1000\cdot\dfrac{100}{100}-\dfrac{1}{100}$ LCD is 100.
$=\dfrac{100,000}{100}-\dfrac{1}{100}$
$=\dfrac{99,999}{100}$

63. $30\div\dfrac{1}{30}=30\cdot\dfrac{30}{1}=900$

65. We begin by factoring 28 in any way that we can and continue factoring until each factor is prime.

$28=4\cdot 7=2\cdot 2\cdot 7$

67. We begin by factoring 1000 in any way that we can and continue factoring until each factor is prime.

$1000=4\cdot 250=2\cdot 2\cdot 2\cdot 125=2\cdot 2\cdot 2\cdot 5\cdot 25=$
$2\cdot 2\cdot 2\cdot 5\cdot 5\cdot 5$

69. We first find the prime factorization of each number.

$28=2\cdot 2\cdot 7$
$42=2\cdot 3\cdot 7$

Each number has one factor of 2 and one factor of 7. The GCF is $2\cdot 7$, or 14.

71. We first find the prime factorization of each number.

$248=2\cdot 2\cdot 2\cdot 31$
$400=2\cdot 2\cdot 2\cdot 2\cdot 5\cdot 5$

Each number has three factors of 2. The GCF is $2\cdot 2\cdot 2$, or 8.

73. $16=2\cdot 2\cdot 2\cdot 2$
$24=2\cdot 2\cdot 2\cdot 3$

The LCM $=2\cdot 2\cdot 2\cdot 2\cdot 3$, or 48.

75. $48=2\cdot 2\cdot 2\cdot 2\cdot 3$
$64=2\cdot 2\cdot 2\cdot 2\cdot 2\cdot 2$
$96=2\cdot 2\cdot 2\cdot 2\cdot 2\cdot 3$
The LCM $=2\cdot 2\cdot 2\cdot 2\cdot 2\cdot 2\cdot 3$, or 192.

77. $\dfrac{192}{256}=\dfrac{3\cdot\cancel{64}}{4\cdot\cancel{64}}=\dfrac{3}{4}$

79. $\dfrac{64\cdot a\cdot b}{16\cdot a\cdot b}=\dfrac{4\cdot\cancel{16}\cdot\cancel{a}\cdot\cancel{b}}{1\cdot\cancel{16}\cdot\cancel{a}\cdot\cancel{b}}=\dfrac{4}{1}=4$

81. $\dfrac{36\cdot(2\cdot h)}{8\cdot(9\cdot h)}=\dfrac{9\cdot 4\cdot 2\cdot h}{2\cdot 4\cdot 9\cdot h}=1$

Exercise Set R.3

RC1. In 0.3057, the digit 5 is in the thousandths place.

RC3. In 490.4, the digit 9 is in the tens place.

RC5. In 0.678, the digit 7 is in the hundredths place.

1. 5.3. (1 place) $5.3=\dfrac{53}{10}$ (1 zero)

3. 0.67. (2 places) $0.67=\dfrac{67}{100}$ (2 zeros)

5. 2.0007. (4 places) $2.0007=\dfrac{20,007}{10,000}$ (4 zeros)

7. 7889.8. (1 place) $7889.8=\dfrac{78,898}{10}$ (1 zero)

9. $\dfrac{1}{10}$ (1 zero) 0.1. (1 place) $\dfrac{1}{10}=0.1$

11. $\dfrac{1}{10,000}$ 0.0001. $\dfrac{1}{10,000} = 0.0001$

$\uparrow$ 4 zeros, 4 places

13. $\dfrac{9999}{1000}$ 9.999. $\dfrac{9999}{1000} = 9.999$

$\uparrow$ 3 zeros, 3 places

15. $\dfrac{4578}{10,000}$ 0.4578. $\dfrac{4578}{10,000} = 0.4578$

$\uparrow$ 4 zeros, 4 places

17.

$$\begin{array}{r} {\scriptstyle 1\ \ 1}\quad \\ 415.78 \\ +\ \ 29.16 \\ \hline 444.94 \end{array}$$

19.

$$\begin{array}{r} {\scriptstyle 1}\qquad\quad \\ 234.000 \\ +156.617 \\ \hline 390.617 \end{array}$$

21. $85 + 67.95 + 2.774$

We have:

$$\begin{array}{r} {\scriptstyle 1\ 1\ 1}\qquad \\ 85.\qquad \\ 67.95\ \ \\ +\ \ 2.774 \\ \hline 155.724 \end{array}$$

23. $17.95 + 16.99 + 28.85$

We have:

$$\begin{array}{r} {\scriptstyle 2\ 2\ 1} \\ 17.95 \\ 16.99 \\ +28.85 \\ \hline 63.79 \end{array}$$

25.

$$\begin{array}{r} {\scriptstyle 10\ 10}\quad \\ {\scriptstyle 7\ \ \not 0\ \ \not 0\ \ 10} \\ 7\not 8.\not 1\ \not 1\ \not 0 \\ -45.876 \\ \hline 32.234 \end{array}$$

27.

$$\begin{array}{r} {\scriptstyle 16}\qquad \\ {\scriptstyle 7\ \not 6\ 9\ 10} \\ 3\not 8.\not 7\not 0\not 0 \\ -11.865 \\ \hline 26.835 \end{array}$$

29.

$$\begin{array}{r} {\scriptstyle 16}\qquad \\ {\scriptstyle 4\ \not 6\ 18}\quad \\ \not 5\ \not 7.\not 8\ 6 \\ -\ \ 9.95 \\ \hline 47.91 \end{array}$$

31.

$$\begin{array}{r} {\scriptstyle 2\ 9\ 9\ 9\ 10} \\ \not 3.\not 0\not 0\not 0\not 0 \\ -1.0807 \\ \hline 1.9193 \end{array}$$

33.

$$\begin{array}{rl} {\scriptstyle 2\ 3}\quad & \\ 7.34 & \longleftarrow \text{2 decimal places} \\ \times\qquad 1.8 & \longleftarrow \text{1 decimal place} \\ \hline 5872 & \\ 7340 & \\ \hline 13.212 & \longleftarrow \text{3 decimal places} \end{array}$$

35.

$$\begin{array}{rl} {\scriptstyle 5}\ \ \\ {\scriptstyle 1}\ \ \\ 0.86 & \longleftarrow \text{2 decimal places} \\ \times\quad 0.93 & \longleftarrow \text{2 decimal places} \\ \hline 258 & \\ 7740 & \\ \hline 0.7998 & \longleftarrow \text{4 decimal places} \end{array}$$

37.

$$\begin{array}{rl} 17.95 & \longleftarrow \text{2 decimal places} \\ \times\qquad 10 & \longleftarrow \text{0 decimal places} \\ \hline 179.50 & \longleftarrow \text{2 decimal places} \end{array}$$

39.

$$\begin{array}{rl} {\scriptstyle 1\ 2}\ \ \\ {\scriptstyle 4\ 5}\ \ \\ 0.457 & \longleftarrow \text{3 decimal places} \\ \times\quad 3.08 & \longleftarrow \text{2 decimal places} \\ \hline 3656 & \\ 137100 & \\ \hline 1.40756 & \longleftarrow \text{5 decimal places} \end{array}$$

41.

$$\begin{array}{rl} {\scriptstyle 5\ 3\ 1}\ \ \\ {\scriptstyle 5\ 3\ 1}\ \ \\ 3.642 & \longleftarrow \text{3 decimal places} \\ \times\quad 0.99 & \longleftarrow \text{2 decimal places} \\ \hline 32778 & \\ 327780 & \\ \hline 3.60558 & \longleftarrow \text{5 decimal places} \end{array}$$

43. Place the decimal point in the quotient directly above the decimal point in the dividend. Then divide as whole numbers.

$$\begin{array}{r} 2.3 \\ 72\overline{)165.6} \\ 144\ \ \\ \hline 216 \\ 216 \\ \hline 0 \end{array}$$

45.

$$\begin{array}{r} 5.2 \\ 8.5_{\wedge}\overline{)44.2_{\wedge}0} \\ 425\ \ \\ \hline 170 \\ 170 \\ \hline 0 \end{array}$$

47.

$$\begin{array}{r} 0.023 \\ 9.9_{\wedge}\overline{)0.2_{\wedge}277} \\ 198\ \ \\ \hline 297 \\ 297 \\ \hline 0 \end{array}$$

49.
$$\begin{array}{r}
18.75 \\
0.64_\wedge\overline{)\,12.00_\wedge00} \\
\underline{64} \\
560 \\
\underline{512} \\
480 \\
\underline{448} \\
320 \\
\underline{320} \\
0
\end{array}$$

51.
$$\begin{array}{r}
660. \\
1.05_\wedge\overline{)\,693.00_\wedge} \\
\underline{630} \\
630 \\
\underline{630} \\
0 \\
\underline{0} \\
0
\end{array}$$

53.
$$\begin{array}{r}
0.68 \\
8.6_\wedge\overline{)\,5.8_\wedge48} \\
\underline{516} \\
688 \\
\underline{688} \\
0
\end{array}$$

55.
$$\begin{array}{r}
0.34375 \\
32\overline{)\,11.00000} \\
\underline{96} \\
140 \\
\underline{128} \\
120 \\
\underline{96} \\
240 \\
\underline{224} \\
160 \\
\underline{160} \\
0
\end{array}$$

Decimal notation for $\frac{11}{32}$ is 0.34375.

57.
$$\begin{array}{r}
1.1818 \\
11\overline{)\,13.0000} \\
\underline{11} \\
20 \\
\underline{11} \\
90 \\
\underline{88} \\
20 \\
\underline{11} \\
90 \\
\underline{88} \\
2
\end{array}$$

Since 2 and 9 alternate as remainders, the sequence of digits following the decimal point in the quotient repeats. Thus, decimal notation for $\frac{13}{11}$ is 1.1818..., or $1.\overline{18}$.

59.
$$\begin{array}{r}
0.55 \\
9\overline{)\,5.00} \\
\underline{45} \\
50 \\
\underline{45} \\
5
\end{array}$$

The number 5 repeats as a remainder, so the digit 5 will repeat in the quotient. Thus, decimal notation for $\frac{5}{9}$ is 0.55..., or $0.\overline{5}$.

61.
$$\begin{array}{r}
2.11 \\
9\overline{)\,19.00} \\
\underline{18} \\
10 \\
\underline{9} \\
1
\end{array}$$

The number 1 repeats as a remainder, so the digit 1 will repeat in the quotient. Thus, decimal notation for $\frac{19}{9}$ is 2.11..., or $2.\overline{1}$.

63. 745.06534

Round to the nearest hundredth: The digit in the hundredths place is 6. The next digit to the right, 5, is 5 or higher, so we round up: 745.07

Round to the nearest tenth: The digit in the tenths place is 0. The next digit to the right, 6, is 5 or higher, so we round up: 745.1

Round to the nearest one: The digit in the ones place is 5. The next digit to the right, 0, is less than 5, so we round down: 745

Round to the nearest ten: The digit in the tens place is 4. The next digit to the right, 5, is 5 or higher, so we round up: 750

Round to the nearest hundred: The digit in the hundreds place is 7. The next digit to the right, 4, is less than 5, so we round down: 700

65. 6780.50568

Round to the nearest hundredth: The digit in the hundredths place is 0. The next digit to the right, 5, is 5 or higher, so we round up: 6780.51

Round to the nearest tenth: The digit in the tenths place is 5. The next digit to the right, 0, is less than 5, so we round down: 6780.5

Round to the nearest one: The digit in the ones place is 0. The next digit to the right, 5, is 5 or higher, so we round up: 6781

Round to the nearest ten: The digit in the tens place is 8. The next digit to the right, 0, is less than 5, so we round down: 6780

Round to the nearest hundred: The digit in the hundreds place is 7. The next digit to the right, 8, is 5 or higher, so we round up: 6800

67. \$17.988

Round to the nearest cent (nearest hundredth): The digit in the hundredths place is 8. The next digit to the right, 8, is 5 or higher, so we round up: \$17.99

Round to the nearest dollar (nearest one): The digit in the ones place is 7. The next digit to the right, 9, is 5 or higher, so we round up: \$18

69. \$346.075

Round to the nearest cent (nearest hundredth): The digit in the hundredths place is 7. The next digit to the right, 5, is 5 or higher, so we round up: \$346.08

Round to the nearest dollar (nearest one): The digit in the ones place is 6. The next digit to the right, 0, is less than 5, so we round down: \$346

71. \$16.95

The digit in the ones place is 6. The next digit to the right, 9, is 5 or higher, so we round up: \$17

73. \$189.50

The digit in the ones place is 9. The next digit to the right, 5, is 5 or higher, so we round up: \$190

75.

$$\begin{array}{r}
0.2\,8\,5\,7\,1 \\
7\,\overline{)\,2.0\,0\,0\,0\,0\,0} \\
\underline{1\,4} \\
6\,0 \\
\underline{5\,6} \\
4\,0 \\
\underline{3\,5} \\
5\,0 \\
\underline{4\,9} \\
1\,0 \\
\underline{7} \\
3
\end{array}$$

$\frac{2}{7} \approx 0.28571$

We round to the nearest:

ten-thousandth: 0.2857

thousandth: 0.286

hundredth: 0.29

tenth: 0.3

one: 0

77.

$$\begin{array}{r}
0.5\,8\,9\,7\,4 \\
3\,9\,\overline{)\,2\,3.0\,0\,0\,0\,0\,0} \\
\underline{1\,9\,5} \\
3\,5\,0 \\
\underline{3\,1\,2} \\
3\,8\,0 \\
\underline{3\,5\,1} \\
2\,9\,0 \\
\underline{2\,7\,3} \\
1\,7\,0 \\
\underline{1\,5\,6} \\
1\,4
\end{array}$$

$\frac{23}{39} \approx 0.58974$

We round to the nearest:

ten-thousandth: 0.5897

thousandth: 0.590

hundredth: 0.59

tenth: 0.6

one: 1

79. $\frac{7}{8} + \frac{5}{32} = \frac{7}{8} \cdot \frac{4}{4} + \frac{5}{32}$, LCD is 32.

$$= \frac{28}{32} + \frac{5}{32}$$
$$= \frac{33}{32}$$

81. $\frac{15}{16} \cdot \frac{11}{12} = \frac{15 \cdot 11}{16 \cdot 12}$

$$= \frac{\cancel{3} \cdot 5 \cdot 11}{16 \cdot \cancel{3} \cdot 4}$$
$$= \frac{55}{64}$$

83. The LCD is 210.

$$\frac{9}{70} + \frac{8}{15} = \frac{9}{70} \cdot \frac{3}{3} + \frac{8}{15} \cdot \frac{14}{14}$$
$$= \frac{27}{210} + \frac{112}{210}$$
$$= \frac{139}{210}$$

85. The LCD is 1000.

$$\frac{9}{10} + \frac{1}{100} + \frac{113}{1000} = \frac{9}{10} \cdot \frac{100}{100} + \frac{1}{100} \cdot \frac{10}{10} + \frac{113}{1000}$$
$$= \frac{900}{1000} + \frac{10}{1000} + \frac{113}{1000}$$
$$= \frac{1023}{1000}$$

87. We begin by factoring 208 in any way we can and continue factoring until each factor is prime.

$$208 = 2 \cdot 104 = 2 \cdot 4 \cdot 26 = 2 \cdot 2 \cdot 2 \cdot 2 \cdot 13$$

89. We first find the prime factorization of each number.

$72 = 2 \cdot 2 \cdot 2 \cdot 3 \cdot 3$

$150 = 2 \cdot 3 \cdot 5 \cdot 5$

Each number has one factor of 2 and one factor of 3. The GCF is $2 \cdot 3$, or 6.

91.
$$
\begin{array}{r}
0.7\,1\,4\,2\,8\,5 \\
7\,\overline{)\,5.0\,0\,0\,0\,0\,0} \\
\underline{4\,9} \\
1\,0 \\
\underline{7} \\
3\,0 \\
\underline{2\,8} \\
2\,0 \\
\underline{1\,4} \\
6\,0 \\
\underline{5\,6} \\
4\,0 \\
\underline{3\,5} \\
5
\end{array}
$$

The number 5 appears again, so the sequence of digits in the quotient will repeat. Then decimal notation for $\frac{5}{7}$ is $0.\overline{714285}$.

93.
$$
\begin{array}{r}
0.6\,4\,2\,8\,5\,7\,1 \\
1\,4\,\overline{)\,9.0\,0\,0\,0\,0\,0\,0} \\
\underline{8\,4} \\
6\,0 \\
\underline{5\,6} \\
4\,0 \\
\underline{2\,8} \\
1\,2\,0 \\
\underline{1\,1\,2} \\
8\,0 \\
\underline{7\,0} \\
1\,0\,0 \\
\underline{9\,8} \\
2\,0 \\
\underline{1\,4} \\
6
\end{array}
$$

The number 6 repeats as a remainder so the sequence of digits 428571 will repeat. Then decimal notation for $\frac{9}{14}$ is $0.6\overline{428571}$.

Exercise Set R.4

RC1. $500\% = 5$; the correct answer is (g).

RC3. $\frac{1}{400} = 0.0025 = 0.25\%$; the correct answer is (c).

RC5. $50\% = \frac{1}{2}$; the correct answer is (h).

RC7. $\frac{1}{10} = 0.1 = 10\%$; the correct answer is (d).

1. 13% 0.13.

Move the decimal point 2 places to the left.

$13\% = 0.13$

3. 35.4% 0.35.4

Move the decimal point 2 places to the left.

$35.4\% = 0.354$

20.1% 0.20.1

Move the decimal point 2 places to the left.

$20.1\% = 0.201$

5. 63% 0.63.

Move the decimal point 2 places to the left.

$63\% = 0.63$

7. 94.1% 0.94.1

Move the decimal point 2 places to the left.

$94.1\% = 0.941$

9. 1% 0.01.

Move the decimal point 2 places to the left.

$1\% = 0.01$

11. 0.61% 0.00.61

Move the decimal point 2 places to the left.

$0.61\% = 0.0061$

13. 240% 2.40.

Move the decimal point 2 places to the left.

$240\% = 2.4$

15. 3.25% 0.03.25

Move the decimal point 2 places to the left.

$3.25\% = 0.0325$

17. $93\% = 93 \times \frac{1}{100} = \frac{93}{100}$

$6\% = 6 \times \frac{1}{100} = \frac{6}{100}$

$1\% = 1 \times \frac{1}{100} = \frac{1}{100}$

19. $39\% = 39 \times \frac{1}{100}$ Replacing % with $\times\frac{1}{100}$

$= \frac{39}{100}$

21. $88\% = 88 \times \frac{1}{100}$ Replacing % with $\times\frac{1}{100}$

$= \frac{88}{100}$

23. $60\% = 60 \times \frac{1}{100}$ Replacing % with $\times\frac{1}{100}$

$= \frac{60}{100}$

25. $28.9\% = 28.9 \times \frac{1}{100}$ Replacing % with $\times\frac{1}{100}$

$= \frac{28.9}{100}$

$= \frac{28.9}{100} \cdot \frac{10}{10}$ Multiplying by 1 to get a whole number in the numerator

$= \frac{289}{1000}$

27. $110\% = 110 \times \frac{1}{100} = \frac{110}{100}$

29. $0.042\% = 0.042 \times \frac{1}{100} = \frac{0.042}{100} = \frac{0.042}{100} \times \frac{1000}{1000} = \frac{42}{100,000}$

31. $250\% = 250 \times \frac{1}{100} = \frac{250}{100}$

33. $3.47\% = 3.47 \times \frac{1}{100} = \frac{3.47}{100} = \frac{3.47}{100} \times \frac{100}{100} = \frac{347}{10,000}$

35. 0.214 0.21.4 (arrow)

Move the decimal point 2 places to the right.

$0.214 = 21.4\%$

37. 0.14 0.14. (arrow)

Move the decimal point 2 places to the right.

$0.14 = 14\%$

39. 0.99 0.99. (arrow)

Move the decimal point 2 places to the right.

$0.99 = 99\%$

41. 1 1.00. (arrow)

Move the decimal point 2 places to the right.

$1 = 100\%$

43. 0.0047 0.00.47 (arrow)

Move the decimal point 2 places to the right.

$0.0047 = 0.47\%$

45. 0.072 0.07.2 (arrow)

Move the decimal point 2 places to the right.

$0.072 = 7.2\%$

47. 9.2 9.20. (arrow)

Move the decimal point 2 places to the right.

$9.2 = 920\%$

49. 0.0068 0.00.68 (arrow)

Move the decimal point 2 places to the right.

$0.0068 = 0.68\%$

51. $\frac{1}{6} = 0.16\overline{6} = 16.\overline{6}\%$, or $16\frac{2}{3}\%$

53. $\frac{13}{20} = 0.65 = 65\%$

55. $\frac{29}{100} = 0.29 = 29\%$

57. $\frac{8}{10} = 0.8 = 80\%$

59. $\frac{3}{5} = 0.6 = 60\%$

61. $\frac{2}{3} = 0.66\overline{6} = 66.\overline{6}\%$, or $66\frac{2}{3}\%$

63. $\frac{7}{4} = 1.75 = 175\%$

65. $\frac{3}{4} = 0.75 = 75\%$

67. $\frac{2}{5} = \frac{2}{5} \cdot \frac{20}{20} = \frac{40}{100} = 40\%$

69.
$$\begin{array}{r} 2.25 \\ 4\overline{)9.00} \\ 8 \\ \hline 1\,0 \\ 8 \\ \hline 2\,0 \\ 2\,0 \\ \hline 0 \end{array}$$

Decimal notation for $\frac{9}{4}$ is 2.25.

71.
$$\begin{array}{r l} 2\,3.\,4\,5\,8 & \leftarrow \text{3 decimal places} \\ \times \quad 7.\,0\,3 & \leftarrow \text{2 decimal places} \\ \hline 7\,0\,3\,7\,4 & \\ 1\,6\,4\,2\,0\,6\,0\,0 & \\ \hline 1\,6\,4.\,9\,0\,9\,7\,4 & \leftarrow \text{5 decimal places} \end{array}$$

73. 0.0123. (arrow) 4 places $0.0123 = \frac{123}{10,000}$ ↑ 4 zeros

75. $\frac{87}{24} = \frac{3 \cdot 29}{3 \cdot 8} = \frac{3}{3} \cdot \frac{29}{8} = \frac{29}{8}$

77. 18.234

The digit in the hundredths place is 3. The next digit to the right, 4, is less than 5, so we round down: 18.23.

79. $34.3\% = 34.3 \times \frac{1}{100} = \frac{34.3}{100} = \frac{34.3}{100} \cdot \frac{10}{10} = \frac{343}{1000}$

81. $18\% + 14\% = 0.18 + 0.14 = 0.32 = 32\%$

83. $1 - 30\% = 1 - 0.3 = 0.7 = 70\%$

85. $1 + 5\% = 100\% + 5\% = 105\%$

87. $3(1 + 15\%) = 3(1 + 0.15) = 3(1.15) = 3.45 = 345\%$

89. $\frac{100\%}{40} = \frac{1}{40} = 0.025 = 2.5\%$

Exercise Set R.5

RC1. Since we always do calculations within parentheses first, the addition should be done first.

RC3. There are no parentheses or exponential expressions. Thus, we would do the multiplication first.

1. $\underbrace{5 \times 5 \times 5 \times 5}_{\text{4 factors}} = 5^4$

3. $\underbrace{10 \cdot 10 \cdot 10}_{\text{3 factors}} = 10^3$

5. $\underbrace{10 \times 10 \times 10 \times 10 \times 10 \times 10}_{\text{6 factors}} = 10^6$

7. $7^2 = 7 \cdot 7 = 49$

9. $9^5 = 9 \cdot 9 \cdot 9 \cdot 9 \cdot 9 = 59,049$

11. $1^4 = 1 \cdot 1 \cdot 1 \cdot 1 = 1$

13. $(2.3)^2 = (2.3)(2.3) = 5.29$

15. $(0.2)^3 = (0.2)(0.2)(0.2) = 0.008$

17. $(20.4)^2 = (20.4)(20.4) = 416.16$

19. $\left(\frac{3}{8}\right)^2 = \left(\frac{3}{8}\right)\left(\frac{3}{8}\right) = \frac{9}{64}$

21. $5^3 = 5 \cdot 5 \cdot 5 = 125$

23. $\left(\frac{3}{2}\right)^3 = \frac{3}{2} \cdot \frac{3}{2} \cdot \frac{3}{2} = \frac{27}{8}$

25. $1000 \times (1.02)^3 = 1000 \times (1.02 \times 1.02 \times 1.02) =$
$1000 \times 1.061208 = 1061.208$

27. $9 + 2 \times 8 = 9 + 16$ Multiplying
$= 25$ Adding

29. $9(8) + 7(6) = 72 + 42$ Multiplying
$= 114$ Adding

31. $39 - 4 \times 2 + 2 = 39 - 8 + 2$ Multiplying
$= 31 + 2$ Subtracting
$= 33$ Adding

33. $9 \div 3 + 16 \div 8 = 3 + 2$ Dividing
$= 5$ Adding

35. $(6 \cdot 3)^2 = 18^2$ Multiplying within parentheses
$= 324$

37. $4 \cdot 5^2 = 4 \cdot 25$ Evaluating the exponential expression
$= 100$

39. $(8 + 2)^3 = 10^3$ Adding within parentheses
$= 1000$ Evaluating the exponential expression

41. $6 + 4^2 = 6 + 16$ Evaluating the exponential expression
$= 22$ Adding

43. $4^3 \div 8 - 4$
$= 64 \div 8 - 4$ Evaluating the exponential expression
$= 8 - 4$ Dividing
$= 4$ Subtracting

45. $120 - 3^3 \cdot 4 \div 6$
$= 120 - 27 \cdot 4 \div 6$ Evaluating the exponential expression
$= 120 - 108 \div 6$ Multiplying
$= 120 - 18$ Dividing
$= 102$ Subtracting

47. $6[9 + (3 + 4)] = 6[9 + 7]$ Adding inside the parentheses
$= 6[16]$ Adding inside the brackets
$= 96$ Multiplying

49. $8 + (7 + 9) = 8 + 16$ Adding inside the parentheses
$= 24$ Adding

51. $15(4 + 2) = 15(6)$ Adding inside the parentheses
$= 90$ Multiplying

53. $12 - (8 - 4) = 12 - 4$ Subtracting inside the parentheses
$= 8$ Subtracting

55. $1000 \div 100 \div 10 = 10 \div 10$ Dividing in order from left to right
$= 1$ Doing the second division

57. $2000 \div \frac{3}{50} \cdot \frac{3}{2}$
$= \frac{2000}{1} \cdot \frac{50}{3} \cdot \frac{3}{2}$ Dividing
$= \frac{100,000}{3} \cdot \frac{3}{2}$ Completing the division
$= \frac{100,000 \cdot 3}{3 \cdot 2}$ Multiplying
$= \frac{\cancel{2} \cdot 50,000 \cdot \cancel{3}}{1 \cdot \cancel{3} \cdot \cancel{2}}$
$= 50,000$ Simplifying

59. We multiply and divide in order from left to right.

$$\begin{aligned}75 \div 15 \cdot 4 \cdot 8 \div 32 &= 5 \cdot 4 \cdot 8 \div 32\\ &= 20 \cdot 8 \div 32\\ &= 160 \div 32\\ &= 5\end{aligned}$$

61. We multiply and divide in order from left to right.

$$\begin{aligned}16 \cdot 5 \div 80 \div 12 \cdot 36 \cdot 9 &= 80 \div 80 \div 12 \cdot 36 \cdot 9\\ &= 1 \div 12 \cdot 36 \cdot 9\\ &= \frac{1}{12} \cdot 36 \cdot 9\\ &= \frac{36}{12} \cdot 9\\ &= 3 \cdot 9\\ &= 27\end{aligned}$$

63. We will do the calculations in the numerator and in the denominator and then divide the results.

$$\begin{aligned}\frac{80-6^2}{9^2+3^2} &= \frac{80-36}{81+9}\\ &= \frac{44}{90}\\ &= \frac{22 \cdot \not{2}}{45 \cdot \not{2}}\\ &= \frac{22}{45}\end{aligned}$$

65. $$\begin{aligned}\frac{3(6+7)-5\cdot 4}{6\cdot 7+8(4-1)} &= \frac{3\cdot 13-5\cdot 4}{6\cdot 7+8\cdot 3}\\ &= \frac{39-20}{42+24}\\ &= \frac{19}{66}\end{aligned}$$

67.

$8 \cdot 2 - (12 - 0) \div 3 - (5 - 2)$	
$= 8 \cdot 2 - 12 \div 3 - 3$	Subtracting inside the parentheses
$= 16 - 4 - 3$	Multiplying and dividing in order from left to right
$= 12 - 3$	Subtracting in order
$= 9$	from left to right

69. $\frac{5}{16} = 0.3125 \quad 0.31.25$

Move the decimal point two places to the right.

$\frac{5}{16} = 0.3125 = 31.25\%$

71. $\frac{9}{2001} = \frac{\not{3} \cdot 3}{\not{3} \cdot 667} = \frac{3}{667}$

73. We begin by factoring 48 in any way that we can and continue factoring until each factor is prime.

$48 = 2 \cdot 24 = 2 \cdot 4 \cdot 6 = 2 \cdot 2 \cdot 2 \cdot 2 \cdot 3$

75. $\frac{10^5}{10^3} = \frac{\not{10} \cdot \not{10} \cdot \not{10} \cdot 10 \cdot 10}{\not{10} \cdot \not{10} \cdot \not{10} \cdot 1} = 10 \cdot 10 = 10^2$

77. $5^4 \cdot 5^2 = (5 \cdot 5 \cdot 5 \cdot 5) \cdot (5 \cdot 5) = 5^6$

Chapter R Vocabulary Reinforcement

1. In the expression 8^2, 8 is called the base.
2. The LCM of 15 and 40 is 120.
3. To factor a number means to express the number as a product.
4. For the fraction $\frac{4}{13}$, 4 is called the numerator.
5. $417 \cdot 1 = 417$ illustrates the identity property of 1.
6. If a natural number, other than 1, is not prime, it is called a composite number.
7. The fraction expression $\frac{2}{13}$ is equivalent to the fraction expression $\frac{14}{91}$.
8. $24 + 0 = 24$ illustrates the identity property of 0.
9. A natural number that has exactly two different factors, itself and 1, is called a prime number.
10. The GCF of 15 and 40 is 5.
11. $\frac{11}{3}$ is the reciprocal of $\frac{3}{11}$.
12. In the expression 7^5, 5 is called the exponent.

Chapter R Concept Reinforcement

1. The least common multiple of two numbers is a multiple of both numbers, so the LCM is always equal to or larger than the larger number. The given statement is true.
2. The statement is false. To convert from decimal notation to percent notation, we move the decimal point two places to the *right* and write the percent symbol.
3. The statement is true. The number 1 is not prime, because it does not have two *different* factors.
4. The statement is false. For example, the GCF of 8 and 16 is 8 which is not smaller than 8.
5. $100\% = 1$, so a percent that is greater than 100% is greater than 1. The given statement is true.
6. The statement is false. Since a composite number is not a prime number, the prime factorization of a composite number cannot contain that number.

Chapter R Review Exercises

1. We begin by factoring 92 in any way that we can and continue factoring until each factor is prime.

$92 = 2 \cdot 46 = 2 \cdot 2 \cdot 23$

2. We begin by factoring 1400 in any way that we can and continue factoring until each factor is prime.

$1400 = 14 \cdot 100 = 2 \cdot 7 \cdot 4 \cdot 25 = 2 \cdot 7 \cdot 2 \cdot 2 \cdot 5 \cdot 5$, or $2 \cdot 2 \cdot 2 \cdot 5 \cdot 5 \cdot 7$

3. We first find the prime factorization of each number.

$20 = 2 \cdot 2 \cdot 5$

$50 = 2 \cdot 5 \cdot 5$

Each number has one factor of 2 and one factor of 5. The GCF is $2 \cdot 5$, or 10.

4. We first find the prime factorization of each number.

$18 = 2 \cdot 3 \cdot 3$

$45 = 3 \cdot 3 \cdot 5$

$900 = 2 \cdot 2 \cdot 3 \cdot 3 \cdot 5 \cdot 5$

Each number has two factors of 3. The GCF is $3 \cdot 3$, or 9.

5. $13 = 13$ (13 is prime.)

$32 = 2 \cdot 2 \cdot 2 \cdot 2 \cdot 2$

The LCM is $2 \cdot 2 \cdot 2 \cdot 2 \cdot 2 \cdot 13$, or 416.

6. $5 = 5$ (5 is prime.)

$18 = 2 \cdot 3 \cdot 3$

$45 = 3 \cdot 3 \cdot 5$

The LCM is $2 \cdot 3 \cdot 3 \cdot 5$, or 90.

7. $\frac{2}{5} = \frac{2}{5} \cdot 1 = \frac{2}{5} \cdot \frac{6}{6} = \frac{12}{30}$

8. $\frac{12}{23} = \frac{12}{23} \cdot 1 = \frac{12}{23} \cdot \frac{8}{8} = \frac{96}{184}$

9. We will use $\frac{8}{8}$ for 1 since $64 = 8 \cdot 8$.

$\frac{5}{8} = \frac{5}{8} \cdot 1 = \frac{5}{8} \cdot \frac{8}{8} = \frac{40}{64}$

10. We will use $\frac{7}{7}$ for 1 since $84 = 12 \cdot 7$.

$\frac{13}{12} = \frac{13}{12} \cdot 1 = \frac{13}{12} \cdot \frac{7}{7} = \frac{91}{84}$

11. $\frac{20}{48} = \frac{4 \cdot 5}{4 \cdot 12} = \frac{4}{4} \cdot \frac{5}{12} = 1 \cdot \frac{5}{12} = \frac{5}{12}$

12.
$$\begin{aligned}\frac{1020}{1820} &= \frac{2 \cdot 2 \cdot 3 \cdot 5 \cdot 17}{2 \cdot 2 \cdot 5 \cdot 7 \cdot 13} \\ &= \frac{\not{2} \cdot \not{2} \cdot 3 \cdot \not{5} \cdot 17}{\not{2} \cdot \not{2} \cdot \not{5} \cdot 7 \cdot 13} \\ &= \frac{3 \cdot 17}{7 \cdot 13} \\ &= \frac{51}{91}\end{aligned}$$

13. The LCD is 36.
$$\begin{aligned}\frac{4}{9} + \frac{5}{12} &= \frac{4}{9} \cdot \frac{4}{4} + \frac{5}{12} \cdot \frac{3}{3} \\ &= \frac{16}{36} + \frac{15}{36} \\ &= \frac{31}{36}\end{aligned}$$

14. $\frac{3}{4} \div 3 = \frac{3}{4} \cdot \frac{1}{3} = \frac{\not{3} \cdot 1}{4 \cdot \not{3}} = \frac{1}{4}$

15. The LCD is 15.
$$\begin{aligned}\frac{2}{3} - \frac{1}{15} &= \frac{2}{3} \cdot \frac{5}{5} - \frac{1}{15} \\ &= \frac{10}{15} - \frac{1}{15} = \frac{9}{15} \\ &= \frac{\not{3} \cdot 3}{\not{3} \cdot 5} = \frac{3}{5}\end{aligned}$$

16. $\frac{9}{10} \cdot \frac{16}{5} = \frac{9 \cdot 16}{10 \cdot 5} = \frac{9 \cdot \not{2} \cdot 8}{\not{2} \cdot 5 \cdot 5} = \frac{72}{25}$

17. The LCD is 144.
$$\begin{aligned}\frac{11}{18} + \frac{13}{16} &= \frac{11}{18} \cdot \frac{8}{8} + \frac{13}{16} \cdot \frac{9}{9} \\ &= \frac{88}{144} + \frac{117}{144} \\ &= \frac{205}{144}\end{aligned}$$

18. The LCD is 72.
$$\begin{aligned}\frac{35}{36} + \frac{23}{24} &= \frac{35}{36} \cdot \frac{2}{2} + \frac{23}{24} \cdot \frac{3}{3} \\ &= \frac{70}{72} + \frac{69}{72} \\ &= \frac{139}{72}\end{aligned}$$

19. The LCD is 54.
$$\begin{aligned}\frac{25}{27} + \frac{17}{18} &= \frac{25}{27} \cdot \frac{2}{2} + \frac{17}{18} \cdot \frac{3}{3} \\ &= \frac{50}{54} + \frac{51}{54} \\ &= \frac{101}{54}\end{aligned}$$

20. The LCD is 84.
$$\begin{aligned}\frac{29}{42} + \frac{17}{28} &= \frac{29}{42} \cdot \frac{2}{2} + \frac{17}{28} \cdot \frac{3}{3} \\ &= \frac{58}{84} + \frac{51}{84} \\ &= \frac{109}{84}\end{aligned}$$

21. The LCD is 72.
$$\begin{aligned}\frac{35}{36} - \frac{19}{24} &= \frac{35}{36} \cdot \frac{2}{2} - \frac{19}{24} \cdot \frac{3}{3} \\ &= \frac{70}{72} - \frac{57}{72} \\ &= \frac{13}{72}\end{aligned}$$

22. The LCD is 144.

$$\begin{aligned}\frac{13}{16}-\frac{11}{18}&=\frac{13}{16}\cdot\frac{9}{9}-\frac{11}{18}\cdot\frac{8}{8}\\&=\frac{117}{144}-\frac{88}{144}\\&=\frac{29}{144}\end{aligned}$$

23. The LCD is 84.

$$\begin{aligned}\frac{29}{42}-\frac{17}{28}&=\frac{29}{42}\cdot\frac{2}{2}-\frac{17}{28}\cdot\frac{3}{3}\\&=\frac{58}{84}-\frac{51}{84}=\frac{7}{84}\\&=\frac{\not{7}\cdot 1}{\not{7}\cdot 12}=\frac{1}{12}\end{aligned}$$

24. The LCD is 180.

$$\begin{aligned}\frac{11}{36}-\frac{1}{20}&=\frac{11}{36}\cdot\frac{5}{5}-\frac{1}{20}\cdot\frac{9}{9}\\&=\frac{55}{180}-\frac{9}{180}=\frac{46}{180}\\&=\frac{\not{2}\cdot 23}{\not{2}\cdot 90}=\frac{23}{90}\end{aligned}$$

25. 17.97. (2 places) $\quad 17.97=\dfrac{1797}{100}$ (2 zeros)

26. $\dfrac{2337}{10,000}$ (4 zeros) $\quad$ 0.2337. (4 places) $\quad \dfrac{2337}{10,000}=0.2337$

27.

$$\begin{array}{r}\scriptstyle 1\ 1\quad 1\\ 2\,3\,4\,4.5\,6\\ +\quad 9\,8.3\,4\,5\\ \hline 2\,4\,4\,2.9\,0\,5\end{array}$$

28.

$$\begin{array}{r}\scriptstyle 1\ 1\ 1\quad\ \ \\ 6.0\,4\\ 7\,8.\\ +\quad 1.9\,8\,9\,8\\ \hline 8\,6.0\,2\,9\,8\end{array}$$

29.

$$\begin{array}{r}\scriptstyle 1\ 10\ 3\ 9\ 10\\ \not{2}\,\not{0}.\not{4}\,\not{0}\,\not{0}\\ -\,1\,1.0\,5\,8\\ \hline 9.3\,4\,2\end{array}$$

30.

$$\begin{array}{r}\scriptstyle 12\\ \scriptstyle 8\ 9\ 2\ 12\\ 7\,8\,\not{9}.\not{0}\,\not{3}\,\not{2}\\ -\,6\,5\,5.7\,6\,8\\ \hline 1\,3\,3.2\,6\,4\end{array}$$

31.

$$\begin{array}{r}1\,7.9\,5\\ \times\quad 2\,4\\ \hline 7\,1\,8\,0\\ 3\,5\,9\,0\,0\\ \hline 4\,3\,0.8\,0\end{array}$$

32.

$$\begin{array}{r}5\,6.9\,5\\ \times\quad 1.9\,4\\ \hline 2\,2\,7\,8\,0\\ 5\,1\,2\,5\,5\,0\\ 5\,6\,9\,5\,0\,0\\ \hline 1\,1\,0.4\,8\,3\,0\end{array}$$

33.

$$\begin{array}{r}5\,5.6\\ 2.8_{\wedge}\overline{)1\,5\,5.6_{\wedge}8}\\ 1\,4\,0\\ \hline 1\,5\,6\\ 1\,4\,0\\ \hline 1\,6\,8\\ 1\,6\,8\\ \hline 0\end{array}$$

34.

$$\begin{array}{r}0.4\,5\\ 5\,2\overline{)2\,3.4\,0}\\ 2\,0\,8\\ \hline 2\,6\,0\\ 2\,6\,0\\ \hline 0\end{array}$$

35.

$$\begin{array}{r}1.5\,8\,3\\ 1\,2\overline{)1\,9.0\,0\,0}\\ 1\,2\\ \hline 7\,0\\ 6\,0\\ \hline 1\,0\,0\\ 9\,6\\ \hline 4\,0\\ 3\,6\\ \hline 4\end{array}$$

The number 4 repeats as a remainder, so the digit 3 will repeat in the quotient. Thus, decimal notation for $\dfrac{19}{12}$ is $1.5833\ldots$, or $1.58\overline{3}$.

36. Round 34.067 to the nearest tenth. The digit in the tenths place is 0. The next digit to the right, 6, is 5 or higher, so we round up: 34.1.

37. 10.5% $\quad$ 0.10.5

Move the decimal point 2 places to the left.

10.5% = 0.105

38. $35.7\%=35.7\times\dfrac{1}{100}=\dfrac{35.7}{100}=\dfrac{35.7}{100}\times\dfrac{10}{10}=\dfrac{357}{1000}$

39. 0.0502 $\quad$ 0.05.02

Move the decimal point 2 places to the right.

0.0502 = 5.02%

40. $\dfrac{396}{1000}=0.396=39.6\%$

41. $\frac{5}{8} = 0.625 = 62.5\%$

42. $\frac{29}{25} = 1.16 = 116\%$

43. $\underbrace{6 \cdot 6 \cdot 6}_{\text{3 factors}} = 6^3$

44. $(1.06)^2 = (1.06)(1.06) = 1.1236$

45.
$$\begin{aligned} & 120 - 6^2 \div 4 + 8 & \\ &= 120 - 36 \div 4 + 8 && \text{Evaluating the exponential expression} \\ &= 120 - 9 + 8 && \text{Dividing} \\ &= 111 + 8 && \text{Subtracting} \\ &= 119 && \text{Adding} \end{aligned}$$

46. We multiply and divide in order from left to right.
$$\begin{aligned} 64 \div 16 \cdot 32 \div 48 \div 12 \cdot 18 &= 4 \cdot 32 \div 48 \div 12 \cdot 18 \\ &= 128 \div 48 \div 12 \cdot 18 \\ &= \frac{128}{48} \div 12 \cdot 18 \\ &= \frac{128}{48} \cdot \frac{1}{12} \cdot 18 \\ &= \frac{128 \cdot 1}{48 \cdot 12} \cdot 18 \\ &= \frac{128 \cdot 18}{48 \cdot 12} \\ &= \frac{\cancel{2} \cdot \cancel{2} \cdot \cancel{2} \cdot \cancel{2} \cdot \cancel{2} \cdot \cancel{2} \cdot 2 \cdot 2 \cdot \cancel{3} \cdot \cancel{3}}{\cancel{2} \cdot \cancel{2} \cdot \cancel{2} \cdot \cancel{2} \cdot \cancel{3} \cdot \cancel{2} \cdot \cancel{2} \cdot \cancel{3} \cdot 1} \\ &= 4 \end{aligned}$$

47.
$$\begin{aligned} (120 - 6^2) \div 4 + 8 &= (120 - 36) \div 4 + 8 \\ &= 84 \div 4 + 8 \\ &= 21 + 8 \\ &= 29 \end{aligned}$$

48. We will multiply and divide in order from left to right.
$$\begin{aligned} & 64 \cdot 16 \div 32 \div 48 \div 12 \cdot 18 \\ &= 1024 \div 32 \div 48 \div 12 \cdot 18 \\ &= 32 \div 48 \div 12 \cdot 18 \\ &= \frac{32}{48} \div 12 \cdot 18 \\ &= \frac{32}{48} \cdot \frac{1}{12} \cdot 18 \\ &= \frac{32 \cdot 1}{48 \cdot 12} \cdot 18 \\ &= \frac{32 \cdot 18}{48 \cdot 12} \\ &= \frac{576}{576} \\ &= 1 \end{aligned}$$

49. $(120 - 6^2) \div (4 + 8) = (120 - 36) \div 12 = 84 \div 12 = 7$

50.
$$\begin{aligned} & 8^2 \cdot 2^4 \div 2^2 \cdot 8 \div 48 \div 12 \cdot 18 \\ &= 64 \cdot 16 \div 4 \cdot 8 \div 48 \div 12 \cdot 18 \\ &= 1024 \div 4 \cdot 8 \div 48 \div 12 \cdot 18 \\ &= 256 \cdot 8 \div 48 \div 12 \cdot 18 \\ &= 2048 \div 48 \div 12 \cdot 18 \\ &= \frac{2048}{48} \div 12 \cdot 18 \\ &= \frac{2048}{48} \cdot \frac{1}{12} \cdot 18 \\ &= \frac{2048 \cdot 1}{48 \cdot 12} \cdot 18 \\ &= \frac{2048 \cdot 18}{48 \cdot 12} \\ &= \frac{\cancel{4} \cdot \cancel{16} \cdot 32 \cdot 2 \cdot \cancel{3} \cdot \cancel{3}}{\cancel{3} \cdot \cancel{16} \cdot \cancel{3} \cdot \cancel{4} \cdot 1} \\ &= 64 \end{aligned}$$

51.
$$\begin{aligned} \frac{4(18-8) + 7 \cdot 9}{9^2 - 8^2} &= \frac{4(10) + 7 \cdot 9}{81 - 64} \\ &= \frac{40 + 63}{17} \\ &= \frac{103}{17} \end{aligned}$$

Chapter R Discussion and Writing Exercises

1. Canceling is possible only when the numerator and denominator of a fraction have common factors.

2. The parentheses are not necessary. Using the rules for order of operations, the multiplication will be performed first regardless of whether it is in parentheses.

3. One approach would be to express 0.1 as $\frac{1}{10}$ and divide as follows:

$5.367 \div \frac{1}{10} = 5.367 \cdot 10 = 53.67$

Chapter R Test

1. We begin by factoring 300 in any way that we can and continue factoring until each factor is prime.

$300 = 3 \cdot 100 = 3 \cdot 10 \cdot 10 = 3 \cdot 2 \cdot 5 \cdot 2 \cdot 5$, or $2 \cdot 2 \cdot 3 \cdot 5 \cdot 5$

2. We first find the prime factorization of each number.

$42 = 2 \cdot 3 \cdot 7$

$56 = 2 \cdot 2 \cdot 2 \cdot 7$

$140 = 2 \cdot 2 \cdot 5 \cdot 7$

Each number has one factor of 2 and one factor of 7. The GCF is $2 \cdot 7$, or 14.

3. $15 = 3 \cdot 5$

$24 = 2 \cdot 2 \cdot 2 \cdot 3$

$60 = 2 \cdot 2 \cdot 3 \cdot 5$

The LCM is $2 \cdot 2 \cdot 2 \cdot 3 \cdot 5$, or 120.

4. $\frac{3}{7}=\frac{3}{7}\cdot 1=\frac{3}{7}\cdot\frac{7}{7}=\frac{21}{49}$

5. We will use $\frac{3}{3}$ for 1 since $48=16\cdot 3$.

$\frac{11}{16}=\frac{11}{16}\cdot 1=\frac{11}{16}\cdot\frac{3}{3}=\frac{33}{48}$

6. $\frac{16}{24}=\frac{2\cdot 8}{3\cdot 8}=\frac{2}{3}\cdot\frac{8}{8}=\frac{2}{3}\cdot 1=\frac{2}{3}$

7. $\frac{925}{1525}=\frac{25\cdot 37}{25\cdot 61}=\frac{25}{25}\cdot\frac{37}{61}=1\cdot\frac{37}{61}=\frac{37}{61}$

8. $\frac{10}{27}\div\frac{8}{3}=\frac{10}{27}\cdot\frac{3}{8}=\frac{10\cdot 3}{27\cdot 8}=\frac{\not{2}\cdot 5\cdot\not{3}}{\not{3}\cdot 9\cdot\not{2}\cdot 4}=\frac{5}{9\cdot 4}=\frac{5}{36}$

9. The LCD is 40.

$\frac{9}{10}-\frac{5}{8}=\frac{9}{10}\cdot\frac{4}{4}-\frac{5}{8}\cdot\frac{5}{5}=\frac{36}{40}-\frac{25}{40}=\frac{11}{40}$

10. The LCD is 36.

$\frac{11}{12}+\frac{17}{18}=\frac{11}{12}\cdot\frac{3}{3}+\frac{17}{18}\cdot\frac{2}{2}=\frac{33}{36}+\frac{34}{36}=\frac{67}{36}$

11. $\frac{10}{27}\cdot\frac{3}{8}=\frac{10\cdot 3}{27\cdot 8}=\frac{\not{2}\cdot 5\cdot\not{3}}{\not{3}\cdot 9\cdot\not{2}\cdot 4}=\frac{5}{36}$

12. 6.78. (2 places) $\quad 6.78=\frac{678}{100}$ (2 zeros)

13. $\frac{1895}{1000}$ (3 zeros) $\quad$ 1.895. (3 places) $\quad \frac{1895}{1000}=1.895$

14.
$$\begin{array}{r} {}^{1\,1\,1} \\ 7.14 \\ 89. \\ +\ 2.8787 \\ \hline 99.0187 \end{array}$$

15.
$$\begin{array}{r} {}^{7\ 9\ 9\ 9\ 10} \\ 1\not{8}\not{0}\not{0}.\not{0}\not{0} \\ -\quad 3.42 \\ \hline 1796.58 \end{array}$$

16.
$$\begin{array}{r} 123.6 \\ \times\ 3.52 \\ \hline 2472 \\ 61800 \\ 370800 \\ \hline 435.072 \end{array}$$

17.
$$\begin{array}{r} 1.6 \\ 7.2_{\wedge}\overline{)\,11.5_{\wedge}2} \\ 72 \\ \hline 432 \\ 432 \\ \hline 0 \end{array}$$

18.
$$\begin{array}{r} 2.09 \\ 11\overline{)\,23.00} \\ 22 \\ \hline 100 \\ 99 \\ \hline 1 \end{array}$$

The number 1 repeats as a remainder, so the sequence of digits following the decimal point in the quotient repeats. Thus, decimal notation for $\frac{23}{11}$ is 2.0909..., or $2.\overline{09}$.

19. Round 234.7284 to the nearest tenth. The digit in the tenths place is 7. The next digit to the right, 2, is less than 5, so we round down: 234.7.

20. Round 234.7284 to the nearest thousandth. The digit in the thousandths place is 8. The next digit to the right, 4, is less than 5, so we round down: 234.728.

21. 0.7% $\quad$ 0.00.7

Move the decimal point 2 places to the left.

$0.7\%=0.007$

22. $91\%=91\times\frac{1}{100}=\frac{91}{100}$

23. $\frac{11}{25}=0.44=44\%$

24. $5^4=5\cdot 5\cdot 5\cdot 5=625$

25. $(1.2)^2=1.2\cdot 1.2=1.44$

26.
$$\begin{aligned} 200-2^3+5\times 10 &= 200-8+5\times 10 \\ &= 200-8+50 \\ &= 192+50 \\ &= 242 \end{aligned}$$

27. $8000\div 0.16\div 2.5=50,000\div 2.5=20,000$

28. 5.4% $\quad$ 0.05.4

Move the decimal point two places to the left.

$5.4\%=0.054$

29. $\frac{12}{1000}=0.012=1.2\%$

30. $\frac{38xy}{2x}=\frac{\not{2}\cdot 19\cdot\not{x}\cdot y}{\not{2}\cdot\not{x}\cdot 1}=19y$

Chapter 1

Introduction to Real Numbers and Algebraic Expressions

Exercise Set 1.1

RC1. $3/q$ involves division. (See page 45 in the text.)

RC3. $3 \cdot q$ involves multiplication. (See page 45 in the text.)

1. Substitute 56 for x: $56-24=32$, so it takes Abigail 32 min to get to work if it takes Jayden 56 min.

Substitute 93 for x: $93-24=69$, so it takes Abigail 69 min to get to work if it takes Jayden 93 min.

Substitute 105 for x: $105-24=81$, so it takes Abigail 81 min to get to work if it takes Jayden 105 min.

3. Substitute 65 for r and 4 for t, and carry out the multiplication:

$$d = rt = 65 \cdot 4 = 260 \text{ mi}$$

5. Since a square is a rectangle with all sides the same length, we can use the formula for the area of a rectangle on page 45 in the text. We substitute 24 in. for both l and w.

$$\begin{aligned} A &= lw \\ &= (24 \text{ in.})(24 \text{ in.}) \\ &= 24 \cdot 24 \cdot \text{in.} \cdot \text{in.} \\ &= 576 \text{ in}^2 \end{aligned}$$

7. Substitute 45 m for b and 86 m for h, and carry out the multiplication:

$$\begin{aligned} A = \frac{1}{2}bh &= \frac{1}{2}(45 \text{ m})(86 \text{ m}) \\ &= \frac{1}{2}(45)(86)(\text{m})(\text{m}) \\ &= 1935 \text{ m}^2 \end{aligned}$$

9. $8x = 8 \cdot 7 = 56$

11. $\frac{a}{b} = \frac{24}{3} = 8$

13. $\frac{3p}{q} = \frac{3 \cdot 2}{6} = \frac{6}{6} = 1$

15. $\frac{x+y}{5} = \frac{10+20}{5} = \frac{30}{5} = 6$

17. $\frac{x-y}{8} = \frac{20-4}{8} = \frac{16}{8} = 2$

19. $b+7$, or $7+b$

21. $c-12$

23. $a+b$, or $b+a$

25. $x \div y$, or $\frac{x}{y}$, or x/y, or $x \cdot \frac{1}{y}$

27. $x+w$, or $w+x$

29. $n-m$

31. $2z$

33. $3m$, or $m \cdot 3$

35. $4a+6$, or $6+4a$

37. $xy-8$

39. $2t-5$

41. $3n+11$, or $11+3n$

43. $4x+3y$, or $3y+4x$

45. A 5% increase in s is represented by $0.05s$, so we have $s+0.05s$.

47. The distance traveled is the product of the speed and the time. Thus, Danielle traveled $65t$ miles.

49. $\$50 - x$

51. $\$12.50n$

53. We use the list of primes. The first prime that is a factor of 108 is 2.

$$108 = 2 \cdot 54$$

We keep dividing by 2 until it is no longer possible to do so.

$$108 = 2 \cdot 2 \cdot 27$$

Now we do the same thing for the next prime, 3.

$$108 = 2 \cdot 2 \cdot 3 \cdot 3 \cdot 3$$

This is the prime factorization of 108.

55.
$$\begin{aligned} \frac{3}{8} + \frac{5}{14} &= \frac{3}{8} \cdot \frac{7}{7} + \frac{5}{14} \cdot \frac{4}{4} \\ &= \frac{21}{56} + \frac{20}{56} \\ &= \frac{41}{56} \end{aligned}$$

57.
$$\begin{array}{r} 1.03 \\ \times\ 0.05 \\ \hline 0.0515 \end{array}$$

59.
$$\begin{aligned} 16 &= 2 \cdot 2 \cdot 2 \cdot 2 \\ 24 &= 2 \cdot 2 \cdot 2 \cdot 3 \\ 32 &= 2 \cdot 2 \cdot 2 \cdot 2 \cdot 2 \end{aligned}$$

The LCM is $2 \cdot 2 \cdot 2 \cdot 2 \cdot 2 \cdot 3$, or 96.

61. $\dfrac{a-2b+c}{4b-a} = \dfrac{20-2\cdot10+5}{4\cdot10-20}$
$= \dfrac{20-20+5}{40-20}$
$= \dfrac{0+5}{20}$
$= \dfrac{5}{20} = \dfrac{\not{5}\cdot 1}{\not{5}\cdot 4}$
$= \dfrac{1}{4}$

63. $\dfrac{12-c}{c+12b} = \dfrac{12-12}{12+12\cdot 1} = \dfrac{0}{12+12} = \dfrac{0}{24} = 0$

Exercise Set 1.2

RC1. $-2\frac{5}{7}$ is $\frac{5}{7}$ of the way from -2 to -3. It is point H.

RC3. -2.25 is one-fourth of the way from -2 to -3. It is point J.

RC5. $|-4| = 4$; this is point B.

RC7. K lies to the left of B on the number line, so the statement is true.

RC9. E lies to the right of C on the number line, so the statement is false.

1. The integer 24 corresponds to 24° above zero; the integer -2 corresponds to 2° below zero.

3. The integer 3,600,000,000 corresponds to a temperature of 3,600,000,000°F; the integer -460 corresponds to a temperature of 460°F below zero.

5. The integer 1454 corresponds to a height of 1454 ft; the integer -55 corresponds to a depth of 55 ft below ground level.

7. The number $\frac{10}{3}$ can be named $3\frac{1}{3}$, or $3.3\overline{3}$. The graph is $\frac{1}{3}$ of the way from 3 to 4.

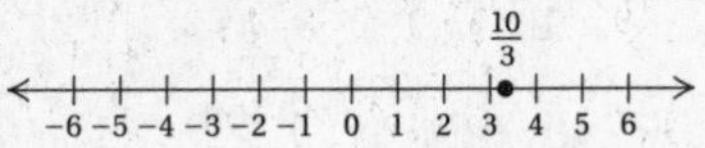

9. The graph of -5.2 is $\frac{2}{10}$ of the way from -5 to -6.

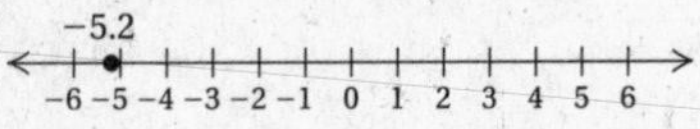

11. The graph of $-4\frac{2}{5}$ is $\frac{2}{5}$ of the way from -4 to -5.

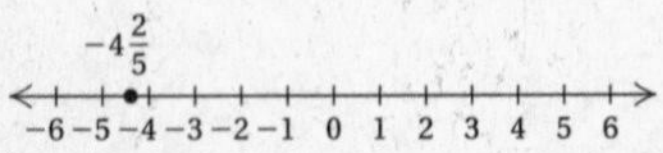

13. We first find decimal notation for $\frac{7}{8}$. Since $\frac{7}{8}$ means $7\div8$, we divide.

$$\begin{array}{r} 0.8\,7\,5 \\ 8\,\overline{)\,7.0\,0\,0} \\ \underline{6\,4} \\ 6\,0 \\ \underline{5\,6} \\ 4\,0 \\ \underline{4\,0} \\ 0 \end{array}$$

Thus $\frac{7}{8} = 0.875$, so $-\frac{7}{8} = -0.875$.

15. $\frac{5}{6}$ means $5\div6$, so we divide.

$$\begin{array}{r} 0.8\,3\,3\ldots \\ 6\,\overline{)\,5.0\,0\,0} \\ \underline{4\,8} \\ 2\,0 \\ \underline{1\,8} \\ 2\,0 \\ \underline{1\,8} \\ 2 \end{array}$$

We have $\frac{5}{6} = 0.8\overline{3}$.

17. First we find decimal notation for $\frac{7}{6}$. Since $\frac{7}{6}$ means $7\div6$, we divide.

$$\begin{array}{r} 1.1\,6\,6\ldots \\ 6\,\overline{)\,7.0\,0\,0} \\ \underline{6} \\ 1\,0 \\ \underline{6} \\ 4\,0 \\ \underline{3\,6} \\ 4\,0 \\ \underline{3\,6} \\ 4 \end{array}$$

Thus $\frac{7}{6} = 1.1\overline{6}$, so $-\frac{7}{6} = -1.1\overline{6}$.

19. $\frac{2}{3}$ means $2\div3$, so we divide.

$$\begin{array}{r} 0.6\,6\,6\ldots \\ 3\,\overline{)\,2.0\,0\,0} \\ \underline{1\,8} \\ 2\,0 \\ \underline{1\,8} \\ 2\,0 \\ \underline{1\,8} \\ 2 \end{array}$$

We have $\frac{2}{3} = 0.\overline{6}$.

21. $\frac{1}{10}$ means $1\div10$, so we divide.

$$\begin{array}{r} 0.1 \\ 1\,0\,\overline{)\,1.0} \\ \underline{1\,0} \\ 0 \end{array}$$

We have $\frac{1}{10} = 0.1$

23. We first find decimal notation for $\frac{1}{2}$. Since $\frac{1}{2}$ means $1 \div 2$, we divide.

$$\begin{array}{r} 0.5 \\ 2\overline{)1.0} \\ \underline{1\,0} \\ 0 \end{array}$$

Thus $\frac{1}{2} = 0.5$, so $-\frac{1}{2} = -0.5$

25. $\frac{4}{25}$ means $4 \div 25$, so we divide.

$$\begin{array}{r} 0.1\,6 \\ 2\,5\overline{)4.0\,0} \\ \underline{2\,5} \\ 1\,5\,0 \\ \underline{1\,5\,0} \\ 0 \end{array}$$

We have $\frac{4}{25} = 0.16$.

27. Since 8 is to the right of 0, we have $8 > 0$.

29. Since -8 is to the left of 3, we have $-8 < 3$.

31. Since -8 is to the left of 8, we have $-8 < 8$.

33. Since -8 is to the left of -5, we have $-8 < -5$.

35. Since -5 is to the right of -11, we have $-5 > -11$.

37. Since 2.14 is to the right of 1.24, we have $2.14 > 1.24$.

39. Since -12.88 is to the left of -6.45, we have $-12.88 < -6.45$.

41. $-\frac{1}{2} = -\frac{1}{2} \cdot \frac{3}{3} = -\frac{3}{6}$

$-\frac{2}{3} = -\frac{2}{3} \cdot \frac{2}{2} = -\frac{4}{6}$

Since $-\frac{3}{6}$ is to the right of $-\frac{4}{6}$, then $-\frac{1}{2}$ is to the right of $-\frac{2}{3}$, and we have $-\frac{1}{2} > -\frac{2}{3}$.

43. Since $-\frac{2}{3}$ is to the left of $\frac{1}{3}$, we have $-\frac{2}{3} < \frac{1}{3}$.

45. Convert to decimal notation $\frac{5}{12} = 0.4166\ldots$ and $\frac{11}{25} = 0.44$. Since $0.4166\ldots$ is to the left of 0.44, $\frac{5}{12} < \frac{11}{25}$.

47. $x < -6$ has the same meaning as $-6 > x$.

49. $y \geq -10$ has the same meaning as $-10 \leq y$.

51. $-5 \leq -6$ is false since neither $-5 < -6$ nor $-5 = -6$ is true.

53. $4 \geq 4$ is true since $4 = 4$ is true.

55. $-3 \geq -11$ is true since $-3 > -11$ is true.

57. $0 \geq 8$ is false since neither $0 > 8$ nor $0 = 8$ is true.

59. The distance of -3 from 0 is 3, so $|-3| = 3$.

61. The distance of 11 from 0 is 11, so $|11| = 11$.

63. The distance of $-\frac{2}{3}$ from 0 is $\frac{2}{3}$, so $\left|-\frac{2}{3}\right| = \frac{2}{3}$.

65. The distance of $\frac{0}{4}$ from 0 is $\frac{0}{4}$, or 0, so $\left|\frac{0}{4}\right| = 0$.

67. The distance of -2.65 from 0 is 2.65, so $|-2.65| = 2.65$.

69. 110% 1.10.

Move the decimal point 2 places to the left.

110% = 1.1

71. $\frac{13}{25} = 0.52 = 52\%$

73. $3^4 = 3 \cdot 3 \cdot 3 \cdot 3 = 81$

75.
$$\begin{aligned} 3(7+2^3) &= 3(7+8) \\ &= 3(15) \\ &= 45 \end{aligned}$$

77. $\frac{2}{3} = 0.\overline{6}$, $-\frac{1}{7} = -0.\overline{142857}$, $\frac{1}{3} = 0.\overline{3}$, $-\frac{2}{7} = -0.\overline{285714}$, $-\frac{2}{3} = -0.\overline{6}$, $\frac{2}{5} = 0.4$, $-\frac{1}{3} = -0.\overline{3}$, $-\frac{2}{5} = -0.4$, $\frac{9}{8} = 1.125$

Listing from least to greatest, we have

$-\frac{2}{3}, -\frac{2}{5}, -\frac{1}{3}, -\frac{2}{7}, -\frac{1}{7}, \frac{1}{3}, \frac{2}{5}, \frac{2}{3}, \frac{9}{8}$.

79. $0.\overline{9} = 3(0.\overline{3}) = 3 \cdot \frac{1}{3} = \frac{1}{1}$

81. First consider $0.\overline{5}$.

$0.\overline{5} = 0.\overline{3} \cdot \frac{5}{3} = \frac{1}{3} \cdot \frac{5}{3} = \frac{5}{9}$

Then, $5.\overline{5} = 5 + 0.\overline{5} = 5 + \frac{5}{9} = 5\frac{5}{9} = \frac{50}{9}$.

Exercise Set 1.3

RC1. To add $7+2$, start at 0, move right to 7, and then move 2 units right. The sum is 9.

RC3. To add $4+(-6)$, start at 0, move right to 4, and then move 6 units left. The sum is -2.

1. $2+(-9)$ The absolute values are 2 and 9. The difference is $9-2$, or 7. The negative number has the larger absolute value, so the answer is negative. $2+(-9) = -7$

3. $-11+5$ The absolute values are 11 and 5. The difference is $11-5$, or 6. The negative number has the larger absolute value, so the answer is negative. $-11+5 = -6$

5. $-8+8$ A negative and a positive number. The numbers have the same absolute value. The sum is 0. $-8+8 = 0$

7. $-3+(-5)$ Two negatives. Add the absolute values, getting 8. Make the answer negative. $-3+(-5) = -8$

9. $-7+0$ One number is 0. The answer is the other number. $-7+0=-7$

11. $0+(-27)$ One number is 0. The answer is the other number. $0+(-27)=-27$

13. $17+(-17)$ A negative and a positive number. The numbers have the same absolute value. The sum is 0. $17+(-17)=0$

15. $-17+(-25)$ Two negatives. Add the absolute values, getting 42. Make the answer negative. $-17+(-25)=-42$

17. $18+(-18)$ A positive and a negative number. The numbers have the same absolute value. The sum is 0. $18+(-18)=0$

19. $-28+28$ A negative and a positive number. The numbers have the same absolute value. The sum is 0. $-28+28=0$

21. $8+(-5)$ The absolute values are 8 and 5. The difference is $8-5$, or 3. The positive number has the larger absolute value, so the answer is positive. $8+(-5)=3$

23. $-4+(-5)$ Two negatives. Add the absolute values, getting 9. Make the answer negative. $-4+(-5)=-9$

25. $13+(-6)$ The absolute values are 13 and 6. The difference is $13-6$, or 7. The positive number has the larger absolute value, so the answer is positive. $13+(-6)=7$

27. $-25+25$ A negative and a positive number. The numbers have the same absolute value. The sum is 0. $-25+25=0$

29. $53+(-18)$ The absolute values are 53 and 18. The difference is $53-18$, or 35. The positive number has the larger absolute value, so the answer is positive. $53+(-18)=35$

31. $-8.5+4.7$ The absolute values are 8.5 and 4.7. The difference is $8.5-4.7$, or 3.8. The negative number has the larger absolute value, so the answer is negative. $-8.5+4.7=-3.8$

33. $-2.8+(-5.3)$ Two negatives. Add the absolute values, getting 8.1. Make the answer negative. $-2.8+(-5.3)=-8.1$

35. $-\frac{3}{5}+\frac{2}{5}$ The absolute values are $\frac{3}{5}$ and $\frac{2}{5}$. The difference is $\frac{3}{5}-\frac{2}{5}$, or $\frac{1}{5}$. The negative number has the larger absolute value, so the answer is negative. $-\frac{3}{5}+\frac{2}{5}=-\frac{1}{5}$

37. $-\frac{2}{9}+\left(-\frac{5}{9}\right)$ Two negatives. Add the absolute values, getting $\frac{7}{9}$. Make the answer negative. $-\frac{2}{9}+\left(-\frac{5}{9}\right)=-\frac{7}{9}$

39. $-\frac{5}{8}+\frac{1}{4}$ The absolute values are $\frac{5}{8}$ and $\frac{1}{4}$. The difference is $\frac{5}{8}-\frac{2}{8}$, or $\frac{3}{8}$. The negative number has the larger absolute value, so the answer is negative. $-\frac{5}{8}+\frac{1}{4}=-\frac{3}{8}$

41. $-\frac{5}{8}+\left(-\frac{1}{6}\right)$ Two negatives. Add the absolute values, getting $\frac{15}{24}+\frac{4}{24}$, or $\frac{19}{24}$. Make the answer negative. $-\frac{5}{8}+\left(-\frac{1}{6}\right)=-\frac{19}{24}$

43. $-\frac{3}{8}+\frac{5}{12}$ The absolute values are $\frac{3}{8}$ and $\frac{5}{12}$. The difference is $\frac{10}{24}-\frac{9}{24}$, or $\frac{1}{24}$. The positive number has the larger absolute value, so the answer is positive. $-\frac{3}{8}+\frac{5}{12}=\frac{1}{24}$

45. $-\frac{1}{6}+\frac{7}{10}$ The absolute values are $\frac{1}{6}$ and $\frac{7}{10}$. The difference is $\frac{21}{30}-\frac{5}{30}=\frac{16}{30}=\frac{\not{2}\cdot 8}{\not{2}\cdot 15}=\frac{8}{15}$. The positive number has the larger absolute value, so the answer is positive. $-\frac{1}{6}+\frac{7}{10}=\frac{8}{15}$

47. $\frac{7}{15}+\left(-\frac{1}{9}\right)$ The absolute values are $\frac{7}{15}$ and $\frac{1}{9}$. The difference is $\frac{21}{45}-\frac{5}{45}=\frac{16}{45}$. The positive number has the larger absolute value, so the answer is positive. $\frac{7}{15}+\left(-\frac{1}{9}\right)=\frac{16}{45}$

49. $76+(-15)+(-18)+(-6)$

a) Add the negative numbers: $-15+(-18)+(-6)=-39$

b) Add the results: $76+(-39)=37$

51. $-44+\left(-\frac{3}{8}\right)+95+\left(-\frac{5}{8}\right)$

a) Add the negative numbers: $-44+\left(-\frac{3}{8}\right)+\left(-\frac{5}{8}\right)=-45$

b) Add the results: $-45+95=50$

53. The additive inverse of 24 is -24 because $24+(-24)=0$.

55. The additive inverse of -26.9 is 26.9 because $-26.9+26.9=0$.

57. If $x=8$, then $-x=-8$. (The opposite of 8 is -8.)

59. If $x=-\frac{13}{8}$ then $-x=-\left(-\frac{13}{8}\right)=\frac{13}{8}$. (The opposite of $-\frac{13}{8}$ is $\frac{13}{8}$.)

61. If $x=-43$ then $-(-x)=-(-(-43))=-43$. (The opposite of the opposite of -43 is -43.)

63. If $x=\frac{4}{3}$ then $-(-x)=-\left(-\frac{4}{3}\right)=\frac{4}{3}$. (The opposite of the opposite of $\frac{4}{3}$ is $\frac{4}{3}$.)

65. $-(-24)=24$ (The opposite of -24 is 24.)

67. $-\left(-\frac{3}{8}\right)=\frac{3}{8}$ (The opposite of $-\frac{3}{8}$ is $\frac{3}{8}$.)

69. Let E = the elevation of Mauna Kea above sea level.

Elevation above sea level is total height plus elevation below sea level.

$$\underbrace{\text{Elevation above sea level}}_{E} \ \underbrace{\text{is}}_{=} \ \underbrace{\text{total height}}_{33,480} \ \underbrace{\text{plus}}_{+} \ \underbrace{\text{elevation below sea level.}}_{(-19,684)}$$

We carry out the addition.

$$E = 33,480 + (-19,684) = 13,796$$

The elevation of Mauna Kea is 13,796 ft above sea level.

71. Let T = the final temperature. We will express the rise in temperature as a positive number and a decrease in the temperature as a negative number.

$$\underbrace{\text{Final temperature}}_{T} \ \underbrace{\text{is}}_{=} \ \underbrace{\text{original temperature}}_{32} \ \underbrace{\text{plus}}_{+}$$

$$\underbrace{\text{rise in temperature}}_{15} \ \underbrace{\text{plus}}_{+} \ \underbrace{\text{decrease in temperature.}}_{(-50)}$$

We add from left to right.

$$\begin{aligned} T &= 32 + 15 + (-50) \\ &= 47 + (-50) \\ &= -3 \end{aligned}$$

The final temperature was $-3°$F.

73. Let S = the sum of the profits and losses. We add the five numbers in the bar graph to find S.

$S = \$10,500 + (-\$8300) + (-\$4100) + (\$6200) = \$4300$

The sum of the profits and losses is \$4300. That is, there was a profit of \$4300.

75. Let B = the new balance in the account at the end of August. We will express the payments as positive numbers and the original balance and the amount of the new charge as negative numbers.

$$\underbrace{\text{New balance}}_{B} \ \underbrace{\text{is}}_{=} \ \underbrace{\text{original balance}}_{-470} \ \underbrace{\text{plus}}_{+} \ \underbrace{\text{amount of first payment}}_{45} \ \underbrace{\text{plus}}_{+}$$

$$\underbrace{\text{amount of new charge}}_{(-160)} \ \underbrace{\text{plus}}_{+} \ \underbrace{\text{amount of second payment.}}_{500}$$

We add from left to right.

$$\begin{aligned} B &= -470 + 45 + (-160) + 500 \\ &= -425 + (-160) + 500 \\ &= -585 + 500 \\ &= -85 \end{aligned}$$

The balance in the account is $-\$85$. Lyle owes \$85.

77. $5a - 2b = 5 \cdot 9 - 2 \cdot 3 = 45 - 6 = 39$

79. First we find decimal notation for $\frac{1}{12}$ by dividing.

$$\begin{array}{r} 0.0\,8\,3\,3 \\ 12\,\overline{)\,1.0\,0\,0\,0} \\ \underline{9\,6} \\ 4\,0 \\ \underline{3\,6} \\ 4\,0 \\ \underline{3\,6} \\ 4 \end{array}$$

We have $\frac{1}{12} = 0.08\overline{3}$, so $-\frac{1}{12} = -0.08\overline{3}$.

81. 0 is 0 units from 0, so $|0| = 0$.

83. When x is positive, the opposite of x, $-x$, is negative, so $-x$ is negative for all positive numbers x.

85. If a is positive, $-a$ is negative. Thus $-a + b$, the sum of two negative numbers, is negative. The correct answer is B.

Exercise Set 1.4

RC1. $18 - 6 = 18 + (-6)$; the correct answer is (c).

RC3. $-18 = 6 = -18 + (-6)$; the correct answer is (d).

1. $2 - 9 = 2 + (-9) = -7$

3. $-8 - (-2) = -8 + 2 = -6$

5. $-11 - (-11) = -11 + 11 = 0$

7. $12 - 16 = 12 + (-16) = -4$

9. $20 - 27 = 20 + (-27) = -7$

11. $-9 - (-3) = -9 + 3 = -6$

13. $-40 - (-40) = -40 + 40 = 0$

15. $7 - (-7) = 7 + 7 = 14$

17. $8 - (-3) = 8 + 3 = 11$

19. $-6 - 8 = -6 + (-8) = -14$

21. $-4 - (-9) = -4 + 9 = 5$

23. $-6 - (-5) = -6 + 5 = -1$

25. $8 - (-10) = 8 + 10 = 18$

27. $-5 - (-2) = -5 + 2 = -3$

29. $-7 - 14 = -7 + (-14) = -21$

31. $0 - (-5) = 0 + 5 = 5$

33. $-8 - 0 = -8 + 0 = -8$

35. $7 - (-5) = 7 + 5 = 12$

37. $2 - 25 = 2 + (-25) = -23$

39. $-42 - 26 = -42 + (-26) = -68$

41. $-71 - 2 = -71 + (-2) = -73$

43. $24-(-92)=24+92=116$

45. $-50-(-50)=-50+50=0$

47. $-\frac{3}{8}-\frac{5}{8}=-\frac{3}{8}+\left(-\frac{5}{8}\right)=-\frac{8}{8}=-1$

49. $\frac{3}{4}-\frac{2}{3}=\frac{3}{4}+\left(-\frac{2}{3}\right)=\frac{9}{12}+\left(-\frac{8}{12}\right)=\frac{1}{12}$

51. $-\frac{3}{4}-\frac{2}{3}=-\frac{3}{4}+\left(-\frac{2}{3}\right)=-\frac{9}{12}+\left(-\frac{8}{12}\right)=-\frac{17}{12}$

53. $-\frac{5}{8}-\left(-\frac{3}{4}\right)=-\frac{5}{8}+\frac{3}{4}=-\frac{5}{8}+\frac{6}{8}=\frac{1}{8}$

55. $6.1-(-13.8)=6.1+13.8=19.9$

57. $-2.7-5.9=-2.7+(-5.9)=-8.6$

59. $0.99-1=0.99+(-1)=-0.01$

61. $-79-114=-79+(-114)=-193$

63. $0-(-500)=0+500=500$

65. $-2.8-0=-2.8+0=-2.8$

67. $7-10.53=7+(-10.53)=-3.53$

69. $\frac{1}{6}-\frac{2}{3}=\frac{1}{6}+\left(-\frac{2}{3}\right)=\frac{1}{6}+\left(-\frac{4}{6}\right)=-\frac{3}{6}$, or $-\frac{1}{2}$

71. $-\frac{4}{7}-\left(-\frac{10}{7}\right)=-\frac{4}{7}+\frac{10}{7}=\frac{6}{7}$

73. $-\frac{7}{10}-\frac{10}{15}=-\frac{7}{10}+\left(-\frac{10}{15}\right)=-\frac{21}{30}+\left(-\frac{20}{30}\right)=-\frac{41}{30}$

75. $\frac{1}{5}-\frac{1}{3}=\frac{1}{5}+\left(-\frac{1}{3}\right)=\frac{3}{15}+\left(-\frac{5}{15}\right)=-\frac{2}{15}$

77. $\frac{5}{12}-\frac{7}{16}=\frac{5}{12}+\left(-\frac{7}{16}\right)=\frac{20}{48}+\left(-\frac{21}{48}\right)=-\frac{1}{48}$

79. $-\frac{2}{15}-\frac{7}{12}=-\frac{2}{15}+\left(-\frac{7}{12}\right)=-\frac{8}{60}+\left(-\frac{35}{60}\right)=-\frac{43}{60}$

81. $18-(-15)-3-(-5)+2=18+15+(-3)+5+2=37$

83. $-31+(-28)-(-14)-17=(-31)+(-28)+14+(-17)=-62$

85. $-34-28+(-33)-44=(-34)+(-28)+(-33)+(-44)=-139$

87. $-93-(-84)-41-(-56)=(-93)+84+(-41)+56=6$

89. $-5.4-(-30.9)+30.8+40.2-(-12)=$
$-5.4+30.9+30.8+40.2+12=108.5$

91. $-\frac{7}{12}+\frac{3}{4}-\left(-\frac{5}{8}\right)-\frac{13}{24}=-\frac{7}{12}+\frac{3}{4}+\frac{5}{8}+\left(-\frac{13}{24}\right)=$
$-\frac{28}{48}+\frac{36}{48}+\frac{30}{48}+\left(-\frac{26}{48}\right)=\frac{12}{48}=\frac{\cancel{12}\cdot 1}{4\cdot\cancel{12}}=\frac{1}{4}$

93. Let $D=$ the difference in elevations.

Difference in elevations	is	higher elevation	minus	lower elevation.
↓	↓	↓	↓	↓
D	$=$	$29,035$	$-$	(-1348)

We carry out the subtraction.

$$D=29,035-(-1348)=29,035+1348=30,383$$

The difference in elevations is 30,383 ft.

95. Let $A=$ the amount owed.

Amount owed	is	amount of charge	minus	amount of return.
↓	↓	↓	↓	↓
A	$=$	476.89	$-$	128.95

We subtract.

$$A=476.89-128.95=347.94$$

Francisca owes \$347.94.

97. Let $D=$ the difference in elevations.

Difference in elevations	is	higher elevation	minus	lower elevation.
↓	↓	↓	↓	↓
D	$=$	3776	$-$	(-4)

We carry out the subtraction.

$$D=3776-(-4)=3776+4=3780$$

Fujiyama is 3780 m higher than Hachirogata.

99. Let $D=$ the difference in the elevations.

Difference in elevations	is	higher elevation	minus	lower elevation.
↓	↓	↓	↓	↓
D	$=$	-131	$-$	(-512)

We carry out the subtraction.

$$D=-131-(-512)=-131+512=381$$

Lake Assal is 381 ft lower than the Valdes Peninsula.

101. Let $T=$ the difference in the temperatures.

Difference in temperatures	is	higher temperature	minus	lower temperature.
↓	↓	↓	↓	↓
T	$=$	840	$-$	(-290)

We carry out the subtraction.

$$T=840-(-290)=840+290=1130$$

The difference between the temperatures is 1130°F.

103. $y+7$, or $7+y$

105. $a-h$

107. $r+s$, or $s+r$

109. False. $3 - 0 = 3, 0 - 3 = -3, 3 - 0 \neq 0 - 3$

111. True

113. True by definition of opposites.

Chapter 1 Mid-Chapter Review

1. The statement is true. See page 54 in the text.

2. The statement is false. If $a > b$, then a lies to the *right* of b on the number line.

3. The statement is true. See page 58 in the text.

4. The statement is false. We translate 7 less than y as $y - 7$. The expression $7 - y$ represents y less than 7.

5. $-x = -(-4) = 4;$
$-(-x) = -(-(-4)) = -(4) = -4$

6. $5 - 13 = 5 + (-13) = -8$

7. $-6 - 7 = -6 + (-7) = -13$

8. $\frac{3m}{n} = \frac{3 \cdot 8}{6} = \frac{3 \cdot 2 \cdot 4}{3 \cdot 2 \cdot 1} = \frac{3 \cdot 2}{3 \cdot 2} \cdot \frac{4}{1} = 4$

9. $\frac{a+b}{2} = \frac{5+17}{2} = \frac{22}{2} = 11$

10. Let y represent the number. We have $3y$.

11. Let n represent the number. We have $n - 5$.

12. The integer 450 corresponds to a \$450 deposit. The integer -79 corresponds to writing a check for \$79.

13. -3.5 is halfway between -4 and -3 on the number line.

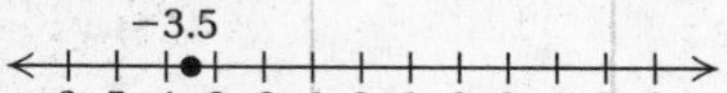

14. We first find decimal notation for $\frac{4}{5}$. Since $\frac{4}{5}$ means $4 \div 5$, we divide.

```
    0.8
5 ) 4.0
    4 0
      0
```

Thus $\frac{4}{5} = 0.8$, so $-\frac{4}{5} = -0.8$.

15. $\frac{7}{3}$ means $7 \div 3$, so we divide.

```
    2.3 3
3 ) 7.0 0
    6
    1 0
      9
      1 0
        9
        1
```

We have $\frac{7}{3} = 2.\overline{3}$.

16. Since -5 is to the left of -3, we have $-5 < -3$.

17. Since -9.9 is to the right of -10.1, we have $-9.9 > -10.1$.

18. $-8 \geq -5$ is false because neither $-8 > -5$ nor $-8 = -5$ is true.

19. $-4 \leq -4$ is true because $-4 = -4$ is true.

20. $5 > y$ has the same meaning as $y < 5$.

21. $t \leq -3$ has the same meaning as $-3 \geq t$.

22. The distance of 15.6 from 0 is 15.6, so $|15.6| = 15.6$.

23. The distance of -18 from 0 is 18, so $|-18| = 18$.

24. The distance of 0 from 0 is 0, so $|0| = 0$.

25. The distance of $-\frac{12}{5}$ from 0 is $\frac{12}{5}$, so $\left|-\frac{12}{5}\right| = \frac{12}{5}$.

26. The opposite of -5.6 is 5.6 because $-5.6 + 5.6 = 0$.

27. The opposite of $\frac{7}{4}$ is $-\frac{7}{4}$ because $\frac{7}{4} + \left(-\frac{7}{4}\right) = 0$.

28. The opposite of 0 is 0 because $0 + 0 = 0$.

29. The opposite of -49 is 49 because $-49 + 49 = 0$.

30. $-x = -(-19) = 19$

31. $-(-x) = -(-2.3) = 2.3$

32. $7 + (-9)$ The absolute values are 7 and 9. The difference is $9 - 7$, or 2. The negative number has the larger absolute value, so the answer is negative. $7 + (-9) = -2$

33. $-\frac{3}{8} + \frac{1}{4}$ The absolute values are $\frac{3}{8}$ and $\frac{1}{4}$. The difference is $\frac{3}{8} - \frac{1}{4}$, or $\frac{3}{8} - \frac{2}{8}$, or $\frac{1}{8}$. The negative number has the larger absolute value, so the answer is negative.
$-\frac{3}{8} + \frac{1}{4} = -\frac{1}{8}$

34. $3.6 + (-3.6)$ A negative and a positive number. The numbers have the same absolute value. The sum is 0. $3.6 + (-3.6) = 0$

35. $-8 + (-9)$ Two negative numbers. Add the absolute values, getting 17. Make the answer negative.
$-8 + (-9) = -17$

36. $\frac{2}{3} + \left(-\frac{9}{8}\right)$ The absolute values are $\frac{2}{3}$ and $\frac{9}{8}$. The difference is $\frac{9}{8} - \frac{2}{3}$, or $\frac{27}{24} - \frac{16}{24}$, or $\frac{11}{24}$. The negative number has the larger absolute value, so the answer is negative.
$\frac{2}{3} + \left(-\frac{9}{8}\right) = -\frac{11}{24}$

37. $-4.2 + (-3.9)$ Two negative numbers. Add the absolute values, getting 8.1. Make the answer negative.
$-4.2 + (-3.9) = -8.1$

38. $-14+5$ The absolute values are 14 and 5. The difference is $14-5$, or 9. The negative number has the larger absolute value, so the answer is negative. $-14+5=-9$

39. $19+(-21)$ The absolute values are 19 and 21. The difference is $21-19$, or 2. The negative number has the larger absolute value, so the answer is negative.

$19+(-21)=-2$

40. $-4.1-6.3=-4.1+(-6.3)=-10.4$

41. $5-(-11)=5+11=16$

42. $-\frac{1}{4}-\left(-\frac{3}{5}\right)=-\frac{1}{4}+\frac{3}{5}=-\frac{5}{20}+\frac{12}{20}=\frac{7}{20}$

43. $12-24=12+(-24)=-12$

44. $-8-(-4)=-8+4=-4$

45. $-\frac{1}{2}-\frac{5}{6}=-\frac{1}{2}+\left(-\frac{5}{6}\right)=-\frac{3}{6}+\left(-\frac{5}{6}\right)=-\frac{8}{6}=-\frac{4}{3}$

46. $12.3-14.1=12.3+(-14.1)=-1.8$

47. $6-(-7)=6+7=13$

48.
$$\begin{aligned}16-(-9)-20-(-4)&=16+9+(-20)+4\\&=25+(-20)+4\\&=5+4\\&=9\end{aligned}$$

49.
$$\begin{aligned}-4+(-10)-(-3)-12&=-4+(-10)+3+(-12)\\&=-14+3+(-12)\\&=-11+(-12)\\&=-23\end{aligned}$$

50.
$$\begin{aligned}17-(-25)+15-(-18)&=17+25+15+18\\&=42+15+18\\&=57+18\\&=75\end{aligned}$$

51.
$$\begin{aligned}-9+(-3)+16-(-10)&=-9+(-3)+16+10\\&=-12+16+10\\&=4+10\\&=14\end{aligned}$$

52. Let T = the difference in the temperatures.

Difference in temperature	is	higher temperature	minus	lower temperature.
↓	↓	↓	↓	↓
T	$=$	25	$-$	(-8)

We carry out the subtraction.

$T=25-(-8)=25+8=33$

The difference in temperatures is 33°C.

53. Let P = the sum of the gains and losses.

$P=56.12+(-1.18)+1.22+(-1.36)=54.80$

The value of the stock at the end of the day was $54.80.

54. Answers may vary. Three examples are $\frac{6}{13}$, -23.8, and $\frac{43}{5}$. These are rational numbers because they can be named in the form $\frac{a}{b}$, where a and b are integers and b is not 0. They are not integers, however, because they are not whole numbers or the opposites of whole numbers.

55. Answers may vary. Three examples are π, $-\sqrt{7}$, and 0.31311311131111.... Rational numbers can be named as described in Exercise 54 above. Real numbers that are not rational are irrational. Decimal notation for rational numbers either terminates or repeats. Decimal notation for irrational numbers neither terminates nor repeats.

56. Answers may vary. Think of adding two negative numbers on the number line. We start to the left of 0 and move to the left, so the result is also to the left of 0, or negative.

57. Yes; consider $m-(-n)$ where both m and n are positive. (Note that, if n is positive, $-n$ is negative.) Then

$m-(-n)=m+n$. Now $m+n$, the sum of two positive numbers, is positive.

Exercise Set 1.5

RC1. To multiply a positive number and a negative number, multiply their absolute values. The answer is negative.

RC3. The product of an even number of negative numbers is positive.

RC5. $-3^2=-3\cdot 3=-9$

RC7. $-\left(\frac{1}{2}\right)^2=-\left(\frac{1}{4}\right)=-\frac{1}{4}$

1. -8

3. -24

5. -72

7. 16

9. 42

11. -120

13. -238

15. 1200

17. 98

19. -72

21. -12.4

23. 30

25. 21.7

27. $\frac{2}{3}\cdot\left(-\frac{3}{5}\right)=-\left(\frac{2\cdot 3}{3\cdot 5}\right)=-\left(\frac{2}{5}\cdot\frac{3}{3}\right)=-\frac{2}{5}$

29. $-\frac{3}{8}\cdot\left(-\frac{2}{9}\right)=\frac{3\cdot 2}{8\cdot 9}=\frac{3\cdot 2\cdot 1}{4\cdot 2\cdot 3\cdot 3}=\frac{3\cdot 2}{3\cdot 2}\cdot\frac{1}{4\cdot 3}=\frac{1}{12}$

31. -17.01

33. $7 \cdot (-4) \cdot (-3) \cdot 5 = 7 \cdot 12 \cdot 5 = 7 \cdot 60 = 420$

35. $-\frac{2}{3} \cdot \frac{1}{2} \cdot \left(-\frac{6}{7}\right) = -\frac{2}{6} \cdot \left(-\frac{6}{7}\right) = \frac{2 \cdot 6}{7 \cdot 6} = \frac{2}{7} \cdot \frac{6}{6} = \frac{2}{7}$

37. $-3 \cdot (-4) \cdot (-5) = 12 \cdot (-5) = -60$

39. $-2 \cdot (-5) \cdot (-3) \cdot (-5) = 10 \cdot 15 = 150$

41. $-4 \cdot (-1.8) \cdot 7 = (7.2) \cdot 7 = 50.4$

43. $-\frac{1}{9} \cdot \left(-\frac{2}{3}\right) \cdot \left(\frac{5}{7}\right) = \frac{2}{27} \cdot \frac{5}{7} = \frac{10}{189}$

45. $4 \cdot (-4) \cdot (-5) \cdot (-12) = -16 \cdot (60) = -960$

47. $0.07 \cdot (-7) \cdot 6 \cdot (-6) = 0.07 \cdot 6 \cdot (-7) \cdot (-6) = 0.42 \cdot (42) = 17.64$

49. $\left(-\frac{5}{6}\right)\left(\frac{1}{8}\right)\left(-\frac{3}{7}\right)\left(-\frac{1}{7}\right) = \left(-\frac{5}{48}\right)\left(\frac{3}{49}\right) = -\frac{5 \cdot 3}{16 \cdot 3 \cdot 49} =$
$-\frac{5}{16 \cdot 49} \cdot \frac{3}{3} = -\frac{5}{784}$

51. 0, The product of 0 and any real number is 0.

53. $(-8)(-9)(-10) = 72(-10) = -720$

55. $(-6)(-7)(-8)(-9)(-10) = 42 \cdot 72 \cdot (-10) = 3024 \cdot (-10) =$
$-30,240$

57. $(-1)^{12}$
$= (-1)(-1)(-1)(-1)(-1)(-1)(-1)(-1)(-1)(-1)(-1)(-1)$
$= 1 \cdot 1 \cdot 1 \cdot 1 \cdot 1 \cdot 1 = 1$

59. For $x = 4$:

$(-x)^2 = (-4)^2 = 16$

$-x^2 = -(4)^2 = -(16) = -16$

For $x = -4$:

$(-x)^2 = [-(-4)]^2 = [4]^2 = 16$

$-x^2 = -(-4)^2 = -(16) = -16$

61. For $y = \frac{2}{5}$:

$(-y)^2 = \left(-\frac{2}{5}\right)^2 = \frac{4}{25}$

$-y^2 = -\left(\frac{2}{5}\right)^2 = -\frac{4}{25}$

For $y = -\frac{2}{5}$:

$(-y)^2 = \left[-\left(-\frac{2}{5}\right)\right]^2 = \left(\frac{2}{5}\right)^2 = \frac{4}{25}$

$-y^2 = -\left(-\frac{2}{5}\right)^2 = -\frac{4}{25}$

63. For $t = 3$:

$-(-t)^2 = -(-3)^2 = -9$

$-t^2 = -(3)^2 = -9$

For $t = -3$:

$-(-t)^2 = -[-(-3)]^2 = -[3]^2 = -9$

$-t^2 = -(-3)^2 = -9$

65. For $x = 7$:

$(-3x)^2 = (-3 \cdot 7)^2$ Substituting
$= (-21)^2$ Multiplying inside the parentheses
$= (-21)(-21)$ Evaluating the power
$= 441$

$-3x^2 = -3(7)^2$ Substituting
$= -3 \cdot 49$ Evaluating the power
$= -147$

For $x = -7$:

$(-3x)^2 = [-3(-7)]^2 = 21^2 = 441$

$-3x^2 = -3(-7)^2 = -3(49) = -147$

67. When $x = 2$: $5x^2 = 5(2)^2$ Substituting
$= 5 \cdot 4$ Evaluating the power
$= 20$

When $x = -2$: $5x^2 = 5(-2)^2$ Substituting
$= 5 \cdot 4$ Evaluating the power
$= 20$

69. When $x = 1$, $-2x^3 = -2 \cdot 1^3 = -2 \cdot 1 = -2$.
When $x = -1$, $-2x^3 = -2(-1)^3 = -2(-1) = 2$.

71. This is a multistep problem. First we find the number of degrees the temperature dropped. Since it dropped 3°C each minute for 8 minutes we have a drop d given by

$d = 8 \cdot (-3) = -24.$

Now let $T =$ the temperature at 11:08 AM.

$T = 0 + (-24) = -24$

The temperature was -24°C at 11:08 AM.

73. Let $w =$ the total weight change. Since Dave's weight decreases 2 lb each week for 10 weeks we have

$w = 10 \cdot (-2) = -20.$

Thus, the total weight change is -20 lb.

75. This is a multistep problem. First we find the total decrease in price. Since it decreased \$1.38 each hour for 8 hours we have a decrease in price d given by

$d = 8(-\$1.38) = -\$11.04.$

Now let $P =$ the price of the stock after 8 hours.

$P = \$23.75 + (-\$11.04) = \$12.71$

After 8 hours the price of the stock was \$12.71.

77. This is a multistep problem. First we find the total distance the diver rises. Since the diver rises 7 meters each minute for 9 minutes, the total distance d the diver rises is given by

$d = 9 \cdot 7 = 63.$

Now let $E =$ the diver's elevation after 9 minutes.

$E = -95 + 63 = -32$

The diver's elevation is -32 m, or 32 m below the surface.

79. This is a multistep problem. First we find the total drop in temperature. Since it dropped 6°F per hour for 4 hr, the total drop in temperature d is given by

$d = 4(-6°\text{F}) = -24°\text{F}$.

Now let T = the temperature at the end of the 4-hr period.

$T = 62°\text{F} + (-24°\text{F}) = 38°\text{F}$

At the end of the 4-hr period the temperature was 38°F.

81. $\frac{x-2y}{3} = \frac{20-2\cdot 7}{3} = \frac{20-14}{3} = \frac{6}{3} = 2$

83. $-\frac{1}{2} - \left(-\frac{1}{6}\right) = -\frac{1}{2} + \frac{1}{6} = -\frac{3}{6} + \frac{1}{6} = -\frac{2}{6} = -\frac{1}{3}$

85. $31 - (-13) = 31 + 13 = 44$

87. Since -10 is to the right of -12, it is true that $-10 > -12$.

89. Since 4 is to the right of -8, it is false that $4 < -8$.

91. If a is positive and b is negative, then ab is negative and thus $-ab$ is positive. The correct answer is A.

93. The largest quotient is found by dividing the largest number, 10, by the smallest positive number, $\frac{1}{5}$.

$10 \div \frac{1}{5} = 10 \cdot \frac{5}{1} = 50$

The smallest quotient is found by dividing the smallest negative number, -5, by the smallest positive number, $\frac{1}{5}$.

$-5 \div \frac{1}{5} = -5 \cdot \frac{5}{1} = -25$

Exercise Set 1.6

RC1. The numbers 4 and -4 are called <u>opposites</u> of each other.

RC3. The additive inverse, or opposite, of a number is what we add to the number to get <u>0</u>.

1. $48 \div (-6) = -8$ Check: $-8(-6) = 48$

3. $\frac{28}{-2} = -14$ Check: $-14(-2) = 28$

5. $\frac{-24}{8} = -3$ Check: $-3 \cdot 8 = -24$

7. $\frac{-36}{-12} = 3$ Check: $3(-12) = -36$

9. $\frac{-72}{9} = -8$ Check: $-8 \cdot 9 = -72$

11. $-100 \div (-50) = 2$ Check: $2(-50) = -100$

13. $-108 \div 9 = -12$ Check: $9(-12) = -108$

15. $\frac{200}{-25} = -8$ Check: $-8(-25) = 200$

17. Not defined

19. $\frac{0}{-2.6} = 0$ Check: $0(-2.6) = 0$

21. The reciprocal of $\frac{15}{7}$ is $\frac{7}{15}$ because $\frac{15}{7} \cdot \frac{7}{15} = 1$.

23. The reciprocal of $-\frac{47}{13}$ is $-\frac{13}{47}$ because $\left(-\frac{47}{13}\right)\cdot\left(-\frac{13}{47}\right) = 1$.

25. The reciprocal of 13 is $\frac{1}{13}$ because $13 \cdot \frac{1}{13} = 1$.

27. The reciprocal of -32 is $-\frac{1}{32}$ because $-32 \cdot \left(-\frac{1}{32}\right) = 1$.

29. The reciprocal of $-\frac{1}{7.1}$ is -7.1 because $\left(-\frac{1}{7.1}\right)(-7.1) = 1$.

31. The reciprocal of $\frac{1}{9}$ is $\frac{9}{1}$, or 9, because $\frac{1}{9} \cdot 9 = 1$.

33. The reciprocal of $\frac{1}{4y}$ is $4y$ because $\frac{1}{4y} \cdot 4y = 1$.

35. The reciprocal of $\frac{2a}{3b}$ is $\frac{3b}{2a}$ because $\frac{2a}{3b} \cdot \frac{3b}{2a} = 1$.

37. $4 \cdot \frac{1}{17}$

39. $8 \cdot \left(-\frac{1}{13}\right)$

41. $13.9 \cdot \left(-\frac{1}{1.5}\right)$

43. $\frac{2}{3} \cdot \left(-\frac{5}{4}\right)$

45. $x \cdot y$

47. $(3x+4)\left(\frac{1}{5}\right)$

49. $\frac{3}{4} \div \left(-\frac{2}{3}\right) = \frac{3}{4} \cdot \left(-\frac{3}{2}\right) = -\frac{9}{8}$

51. $-\frac{5}{4} \div \left(-\frac{3}{4}\right) = -\frac{5}{4} \cdot \left(-\frac{4}{3}\right) = \frac{20}{12} = \frac{5\cdot 4}{3\cdot 4} = \frac{5}{3}$

53. $-\frac{2}{7} \div \left(-\frac{4}{9}\right) = -\frac{2}{7} \cdot \left(-\frac{9}{4}\right) = \frac{18}{28} = \frac{9\cdot 2}{14\cdot 2} = \frac{9}{14}$

55. $-\frac{3}{8} \div \left(-\frac{8}{3}\right) = -\frac{3}{8} \cdot \left(-\frac{3}{8}\right) = \frac{9}{64}$

57. $-\frac{5}{6} \div \frac{2}{3} = -\frac{5}{6} \cdot \frac{3}{2} = -\frac{15}{12} = -\frac{3\cdot 5}{3\cdot 4} = -\frac{5}{4}$

59. $-\frac{9}{4} \div \frac{5}{12} = -\frac{9}{4} \cdot \frac{12}{5} = -\frac{108}{20} = -\frac{4\cdot 27}{4\cdot 5} = -\frac{27}{5}$

61. $\frac{-11}{-13} = \frac{11}{13}$ The opposite of a number divided by the opposite of another number is the quotient of the two numbers.

63. $-6.6 \div 3.3 = -2$ Do the long division. Make the answer negative.

65. $\frac{48.6}{-3} = -16.2$ Do the long division. Make the answer negative.

67. $-\frac{12.5}{5} = -2.5$ Do the long division. Make the answer negative.

69. $11.25 \div (-9) = -1.25$ Do the long division. Make the answer negative.

71. $\frac{-9}{17-17} = \frac{-9}{0}$ Division by 0 is not defined.

73. Amount of increase, in millions: $102 - 71 = 31$

$\frac{31}{71} \approx 0.44 = 44\%$

The percent increase is about 44%.

75. Amount of decrease, in billions: $168 - 212 = -44$

$\frac{-44}{212} \approx -0.21 = -21\%$

The percent decrease is about -21%.

77. $\frac{1}{4} - \frac{1}{2} = \frac{1}{4} + \left(-\frac{1}{2}\right) = \frac{1}{4} + \left(-\frac{2}{4}\right) = -\frac{1}{4}$

79. $35 \cdot (-1.2) = -42$

81. $13.4 + (-4.9)$ The absolute values are 13.4 and 4.9. The difference is $13.4 - 4.9$, or 8.5. The positive number has the larger absolute value, so the answer is positive.

$13.4 + (-4.9) = 8.5$

83. First we find decimal notation for $\frac{1}{11}$. Since $\frac{1}{11}$ means $1 \div 11$, we divide.

$$\begin{array}{r} 0.0\,9\,0\,9\ldots \\ 11\,\overline{)\,1.0\,0\,0\,0} \\ 9\,9 \\ \hline 1\,0\,0 \\ 9\,9 \\ \hline 1 \end{array}$$

Thus $\frac{1}{11} = 0.\overline{09}$, so $-\frac{1}{11} = -0.\overline{09}$.

85. $\frac{15}{4}$ means $15 \div 4$, so we divide.

$$\begin{array}{r} 3.7\,5 \\ 4\,\overline{)\,1\,5.0\,0} \\ 1\,2 \\ \hline 3\,0 \\ 2\,8 \\ \hline 2\,0 \\ 2\,0 \\ \hline 0 \end{array}$$

We have $\frac{15}{4} = 3.75$.

87. The reciprocal of -10.5 is $\frac{1}{-10.5}$.

The reciprocal of $\frac{1}{-10.5} = -10.5$.

We see that the reciprocal of the reciprocal is the original number.

89. $-a$ is positive and b is negative, so $\frac{-a}{b}$ is the quotient of a positive and a negative number and, thus, is negative.

91. a is negative and $-b$ is positive, so $\frac{a}{-b}$ is the quotient of a negative number and a positive number and, thus, is negative. Then $-\left(\frac{a}{-b}\right)$ is the opposite of a negative number and, thus, is positive.

93. $-a$ and $-b$ are both positive, so $\frac{-a}{-b}$ is the quotient of two positive numbers and, thus, is positive. Then $-\left(\frac{-a}{-b}\right)$ is the opposite of a positive number and, thus, is negative.

Exercise Set 1.7

RC1. $8 \cdot \left(\frac{1}{2} \cdot 9\right) = \left(8 \cdot \frac{1}{2}\right) \cdot 9$ illustrates the associative law of multiplication. The correct answer is (g).

RC3. $5(6-3) = 5 \cdot 6 - 5 \cdot 3$ illustrates the distributive law of multiplication over subtraction. The correct answer is (f).

RC5. $3 + 0 = 3$ illustrates the identity property of 0. The correct answer is (d).

RC7. $8 + \left(\frac{1}{2} + 9\right) = \left(8 + \frac{1}{2}\right) + 9$ illustrates the associative law of addition. The correct answer is (b).

1. Note that $5y = 5 \cdot y$. We multiply by 1, using y/y as an equivalent expression for 1:

$$\frac{3}{5} = \frac{3}{5} \cdot 1 = \frac{3}{5} \cdot \frac{y}{y} = \frac{3y}{5y}$$

3. Note that $15x = 3 \cdot 5x$. We multiply by 1, using $5x/5x$ as an equivalent expression for 1:

$$\frac{2}{3} = \frac{2}{3} \cdot 1 = \frac{2}{3} \cdot \frac{5x}{5x} = \frac{10x}{15x}$$

5. Note that $x^2 = x \cdot x$. We multiply by 1, using x/x as an equivalent expression for 1.

$$\frac{2}{x} = \frac{2}{x} \cdot 1 = \frac{2}{x} \cdot \frac{x}{x} = \frac{2x}{x^2}$$

7.
$$\begin{aligned} -\frac{24a}{16a} &= -\frac{3 \cdot 8a}{2 \cdot 8a} \\ &= -\frac{3}{2} \cdot \frac{8a}{8a} \\ &= -\frac{3}{2} \cdot 1 && \left(\frac{8a}{8a} = 1\right) \\ &= -\frac{3}{2} && \text{Identity property of 1} \end{aligned}$$

9.
$$\begin{aligned} -\frac{42ab}{36ab} &= -\frac{7 \cdot 6ab}{6 \cdot 6ab} \\ &= -\frac{7}{6} \cdot \frac{6ab}{6ab} \\ &= -\frac{7}{6} \cdot 1 && \left(\frac{6ab}{6ab} = 1\right) \\ &= -\frac{7}{6} && \text{Identity property of 1} \end{aligned}$$

11. $\frac{20st}{15t} = \frac{4s \cdot 5t}{3 \cdot 5t}$
$= \frac{4s}{3} \cdot \frac{5t}{5t}$
$= \frac{4s}{3} \cdot 1 \quad \left(\frac{5t}{5t} = 1\right)$
$= \frac{4s}{3}$ Identity property of 1

13. $8 + y$, commutative law of addition

15. nm, commutative law of multiplication

17. $xy + 9$, commutative law of addition
$9 + yx$, commutative law of multiplication

19. $c + ab$, commutative law of addition
$ba + c$, commutative law of multiplication

21. $(a + b) + 2$, associative law of addition

23. $8(xy)$, associative law of multiplication

25. $a + (b + 3)$, associative law of addition

27. $(3a)b$, associative law of multiplication

29. a) $(a + b) + 2 = a + (b + 2)$, associative law of addition
b) $(a + b) + 2 = (b + a) + 2$, commutative law of addition
c) $(a + b) + 2 = (b + a) + 2$ Using the commutative law first,
$= b + (a + 2)$ then the associative law
There are other correct answers.

31. a) $5 + (v + w) = (5 + v) + w$, associative law of addition
b) $5 + (v + w) = 5 + (w + v)$, commutative law of addition
c) $5 + (v + w) = 5 + (w + v)$ Using the commutative law first,
$= (5 + w) + v$ then the associative law
There are other correct answers.

33. a) $(xy)3 = x(y3)$, associative law of multiplication
b) $(xy)3 = (yx)3$, commutative law of multiplication
c) $(xy)3 = (yx)3$ Using the commutative law first,
$= y(x3)$ then the associative law
There are other correct answers.

35. a) $7(ab) = (7a)b$
b) $7(ab) = (7a)b = b(7a)$
c) $7(ab) = 7(ba) = (7b)a$
There are other correct answers.

37. $2(b + 5) = 2 \cdot b + 2 \cdot 5 = 2b + 10$

39. $7(1 + t) = 7 \cdot 1 + 7 \cdot t = 7 + 7t$

41. $6(5x + 2) = 6 \cdot 5x + 6 \cdot 2 = 30x + 12$

43. $7(x + 4 + 6y) = 7 \cdot x + 7 \cdot 4 + 7 \cdot 6y = 7x + 28 + 42y$

45. $7(x - 3) = 7 \cdot x - 7 \cdot 3 = 7x - 21$

47. $-3(x - 7) = -3 \cdot x - (-3) \cdot 7 = -3x - (-21) = -3x + 21$

49. $\frac{2}{3}(b - 6) = \frac{2}{3} \cdot b - \frac{2}{3} \cdot 6 = \frac{2}{3}b - 4$

51. $7.3(x - 2) = 7.3 \cdot x - 7.3 \cdot 2 = 7.3x - 14.6$

53. $-\frac{3}{5}(x - y + 10) = -\frac{3}{5} \cdot x - \left(-\frac{3}{5}\right) \cdot y + \left(-\frac{3}{5}\right) \cdot 10 =$
$-\frac{3}{5}x - \left(-\frac{3}{5}y\right) + (-6) = -\frac{3}{5}x + \frac{3}{5}y - 6$

55. $-9(-5x - 6y + 8) = -9(-5x) - (-9)6y + (-9)8$
$= 45x - (-54y) + (-72) = 45x + 54y - 72$

57. $-4(x - 3y - 2z) = -4 \cdot x - (-4)3y - (-4)2z$
$= -4x - (-12y) - (-8z) = -4x + 12y + 8z$

59. $3.1(-1.2x + 3.2y - 1.1) = 3.1(-1.2x) + (3.1)3.2y - 3.1(1.1)$
$= -3.72x + 9.92y - 3.41$

61. $4x + 3z$ Parts are separated by plus signs. The terms are $4x$ and $3z$.

63. $7x + 8y - 9z = 7x + 8y + (-9z)$ Separating parts with plus signs
The terms are $7x$, $8y$, and $-9z$.

65. $2x + 4 = 2 \cdot x + 2 \cdot 2 = 2(x + 2)$

67. $30 + 5y = 5 \cdot 6 + 5 \cdot y = 5(6 + y)$

69. $14x + 21y = 7 \cdot 2x + 7 \cdot 3y = 7(2x + 3y)$

71. $14t - 7 = 7 \cdot 2t - 7 \cdot 1 = 7(2t - 1)$

73. $8x - 24 = 8 \cdot x - 8 \cdot 3 = 8(x - 3)$

75. $18a - 24b = 6 \cdot 3a - 6 \cdot 4b = 6(3a - 4b)$

77. $-4y + 32 = -4 \cdot y - 4(-8) = -4(y - 8)$
We could also factor this expression as follows:
$-4y + 32 = 4(-y) + 4 \cdot 8 = 4(-y + 8)$

79. $5x + 10 + 15y = 5 \cdot x + 5 \cdot 2 + 5 \cdot 3y = 5(x + 2 + 3y)$

81. $16m - 32n + 8 = 8 \cdot 2m - 8 \cdot 4n + 8 \cdot 1 = 8(2m - 4n + 1)$

83. $12a + 4b - 24 = 4 \cdot 3a + 4 \cdot b - 4 \cdot 6 = 4(3a + b - 6)$

85. $8x + 10y - 22 = 2 \cdot 4x + 2 \cdot 5y - 2 \cdot 11 = 2(4x + 5y - 11)$

87. $ax - a = a \cdot x - a \cdot 1 = a(x - 1)$

89. $ax - ay - az = a \cdot x - a \cdot y - a \cdot z = a(x - y - z)$

91. $-18x + 12y + 6 = -6 \cdot 3x - 6(-2y) - 6(-1) = -6(3x - 2y - 1)$
We could also factor this expression as follows:
$-18x + 12y + 6 = 6(-3x) + 6 \cdot 2y + 6 \cdot 1 = 6(-3x + 2y + 1)$

93. $\frac{2}{3}x - \frac{5}{3}y + \frac{1}{3} = \frac{1}{3} \cdot 2x - \frac{1}{3} \cdot 5y + \frac{1}{3} \cdot 1 =$
$\frac{1}{3}(2x - 5y + 1)$

95. $36x - 6y + 18z = 6 \cdot 6x - 6 \cdot y + 6 \cdot 3z = 6(6x - y + 3z)$

97. $9a + 10a = (9 + 10)a = 19a$

99. $10a - a = 10a - 1 \cdot a = (10 - 1)a = 9a$

101. $2x + 9z + 6x = 2x + 6x + 9z = (2 + 6)x + 9z = 8x + 9z$

103. $7x + 6y^2 + 9y^2 = 7x + (6 + 9)y^2 = 7x + 15y^2$

105.
$$\begin{aligned} & 41a + 90 - 60a - 2 = 41a - 60a + 90 - 2 \\ & = (41 - 60)a + (90 - 2) \\ & = -19a + 88 \end{aligned}$$

107.
$$\begin{aligned} & 23 + 5t + 7y - t - y - 27 \\ = & 23 - 27 + 5t - 1 \cdot t + 7y - 1 \cdot y \\ = & (23 - 27) + (5 - 1)t + (7 - 1)y \\ = & -4 + 4t + 6y, \text{ or } 4t + 6y - 4 \end{aligned}$$

109. $\frac{1}{2}b + \frac{1}{2}b = \left(\frac{1}{2} + \frac{1}{2}\right)b = 1b = b$

111. $2y + \frac{1}{4}y + y = 2y + \frac{1}{4}y + 1 \cdot y = \left(2 + \frac{1}{4} + 1\right)y = 3\frac{1}{4}y$, or $\frac{13}{4}y$

113. $11x - 3x = (11 - 3)x = 8x$

115. $6n - n = (6 - 1)n = 5n$

117. $y - 17y = (1 - 17)y = -16y$

119.
$$\begin{aligned} & -8 + 11a - 5b + 6a - 7b + 7 \\ = & 11a + 6a - 5b - 7b - 8 + 7 \\ = & (11 + 6)a + (-5 - 7)b + (-8 + 7) \\ = & 17a - 12b - 1 \end{aligned}$$

121. $9x + 2y - 5x = (9 - 5)x + 2y = 4x + 2y$

123. $11x + 2y - 4x - y = (11 - 4)x + (2 - 1)y = 7x + y$

125. $2.7x + 2.3y - 1.9x - 1.8y = (2.7 - 1.9)x + (2.3 - 1.8)y = 0.8x + 0.5y$

127.
$$\begin{aligned} & \frac{13}{2}a + \frac{9}{5}b - \frac{2}{3}a - \frac{3}{10}b - 42 \\ = & \left(\frac{13}{2} - \frac{2}{3}\right)a + \left(\frac{9}{5} - \frac{3}{10}\right)b - 42 \\ = & \left(\frac{39}{6} - \frac{4}{6}\right)a + \left(\frac{18}{10} - \frac{3}{10}\right)b - 42 \\ = & \frac{35}{6}a + \frac{15}{10}b - 42 \\ = & \frac{35}{6}a + \frac{3}{2}b - 42 \end{aligned}$$

129. $18 - (-20) = 18 + 20 = 38$

131. $-\frac{4}{15} \cdot (-15) = \frac{4 \cdot \cancel{15}}{\cancel{15} \cdot 1} = 4$

133. $\frac{2}{7} \div \left(-\frac{7}{2}\right) = \frac{2}{7} \cdot \left(-\frac{2}{7}\right) = -\frac{4}{49}$

135. $-\frac{1}{2} + \frac{3}{2} = \frac{2}{2} = 1$

137. $9w = 9 \cdot 20 = 180$

139. Since -43 is to the left of -40, it is true that $-43 < -40$.

141. $-6 \le -6$ is true since $-6 = -6$ is true.

143. No; for any replacement other than 5 the two expressions do not have the same value. For example, let $t = 2$. Then $3 \cdot 2 + 5 = 6 + 5 = 11$, but $3 \cdot 5 + 2 = 15 + 2 = 17$.

145. Yes; commutative law of addition

147.
$$\begin{aligned} & q + qr + qrs + qrst && \text{There are no like terms.} \\ = & q \cdot 1 + q \cdot r + q \cdot rs + q \cdot rst \\ = & q(1 + r + rs + rst) && \text{Factoring} \end{aligned}$$

Exercise Set 1.8

RC1. Multiplication

RC3. Subtraction

RC5. Division

1. $-(2x + 7) = -2x - 7$ Changing the sign of each term

3. $-(8 - x) = -8 + x$ Changing the sign of each term

5. $-4a + 3b - 7c$

7. $-6x + 8y - 5$

9. $-3x + 5y + 6$

11. $8x + 6y + 43$

13. $9x - (4x + 3) = 9x - 4x - 3$ Removing parentheses by changing the sign of every term

$= 5x - 3$ Collecting like terms

15. $2a - (5a - 9) = 2a - 5a + 9 = -3a + 9$

17. $2x + 7x - (4x + 6) = 2x + 7x - 4x - 6 = 5x - 6$

19. $2x - 4y - 3(7x - 2y) = 2x - 4y - 21x + 6y = -19x + 2y$

21.
$$\begin{aligned} & 15x - y - 5(3x - 2y + 5z) \\ = & 15x - y - 15x + 10y - 25z && \text{Multiplying each term in parentheses by } -5 \\ = & 9y - 25z \end{aligned}$$

23. $(3x + 2y) - 2(5x - 4y) = 3x + 2y - 10x + 8y = -7x + 10y$

25.
$$\begin{aligned} & (12a - 3b + 5c) - 5(-5a + 4b - 6c) \\ = & 12a - 3b + 5c + 25a - 20b + 30c \\ = & 37a - 23b + 35c \end{aligned}$$

27.
$$\begin{aligned} 9 - 2(5 - 4) & = 9 - 2 \cdot 1 && \text{Computing } 5 - 4 \\ & = 9 - 2 && \text{Computing } 2 \cdot 1 \\ & = 7 \end{aligned}$$

29. $8[7 - 6(4 - 2)] = 8[7 - 6(2)] = 8[7 - 12] = 8[-5] = -40$

31.
$$\begin{aligned} & [4(9 - 6) + 11] - [14 - (6 + 4)] \\ = & [4(3) + 11] - [14 - 10] \\ = & [12 + 11] - [14 - 10] \\ = & 23 - 4 \\ = & 19 \end{aligned}$$

33. $[10(x+3)-4]+[2(x-1)+6]$
$=[10x+30-4]+[2x-2+6]$
$=[10x+26]+[2x+4]$
$=10x+26+2x+4$
$=12x+30$

35. $[7(x+5)-19]-[4(x-6)+10]$
$=[7x+35-19]-[4x-24+10]$
$=[7x+16]-[4x-14]$
$=7x+16-4x+14$
$=3x+30$

37. $3\{[7(x-2)+4]-[2(2x-5)+6]\}$
$=3\{[7x-14+4]-[4x-10+6]\}$
$=3\{[7x-10]-[4x-4]\}$
$=3\{7x-10-4x+4\}$
$=3\{3x-6\}$
$=9x-18$

39. $4\{[5(x-3)+2]-3[2(x+5)-9]\}$
$=4\{[5x-15+2]-3[2x+10-9]\}$
$=4\{[5x-13]-3[2x+1]\}$
$=4\{5x-13-6x-3\}$
$=4\{-x-16\}$
$=-4x-64$

41. $8-2\cdot 3-9=8-6-9$ Multiplying
$=2-9$ Doing all additions and subtractions in order from left to right
$=-7$

43. $(8-2)\div(3-9)=6\div -6$ Subtracting inside the parentheses
$=-1$ Dividing

45. $[(-24)\div(-3)]\div\left(-\frac{1}{2}\right)=8\div\left(-\frac{1}{2}\right)=8\cdot(-2)=-16$

47. $16\cdot(-24)+50=-384+50=-334$

49. $2^4+2^3-10=16+8-10=24-10=14$

51. $5^3+26\cdot 71-(16+25\cdot 3)=5^3+26\cdot 71-(16+75)=$
$5^3+26\cdot 71-91=125+26\cdot 71-91=125+1846-91=$
$1971-91=1880$

53. $4\cdot 5-2\cdot 6+4=20-12+4=8+4=12$

55. $4^3/8=64/8=8$

57. $8(-7)+6(-5)=-56-30=-86$

59. $19-5(-3)+3=19+15+3=34+3=37$

61. $9\div(-3)+16\div 8=-3+2=-1$

63. $-4^2+6=-16+6=-10$

65. $-8^2-3=-64-3=-67$

67. $12-20^3=12-8000=-7988$

69. $2\cdot 10^3-5000=2\cdot 1000-5000=2000-5000=-3000$

71. $6[9-(3-4)]=6[9-(-1)]=6[9+1]=6[10]=60$

73. $-1000\div(-100)\div 10=10\div 10=1$

75. $8-(7-9)=8-(-2)=8+2=10$

77. $\frac{10-6^2}{9^2+3^2}=\frac{10-36}{81+9}=\frac{-26}{90}=-\frac{13}{45}$

79. $\frac{3(6-7)-5\cdot 4}{6\cdot 7-8(4-1)}=\frac{3(-1)-5\cdot 4}{42-8\cdot 3}=\frac{-3-20}{42-24}=-\frac{23}{18}$

81. $\frac{|2^3-3^2|+|12\cdot 5|}{-32\div(-16)\div(-4)}=\frac{|8-9|+|12\cdot 5|}{-32\div(-16)\div(-4)}=$
$\frac{|-1|+|60|}{2\div(-4)}=\frac{1+60}{-\frac{1}{2}}=\frac{61}{-\frac{1}{2}}=61(-2)=-122$

83. $\frac{x-y}{y}=\frac{38-2}{2}=\frac{36}{2}=18$

85. $|-0.4|=0.4$ because -0.4 is 0.4 units from 0.

87. The reciprocal of -9 is $-\frac{1}{9}$ because $-9\left(-\frac{1}{9}\right)=1$.

89. $5-30=5+(-30)=-25$

91. $-5-(-30)=-5+30=25$

93. $\{x-[f-(f-x)]+[x-f]\}-3x$
$=\{x-[f-f+x]+[x-f]\}-3x$
$=\{x-[x]+[x-f]\}-3x$
$=\{x-x+x-f\}-3x=x-f-3x=-2x-f$

95. a) $x^2+3=7^2+3=49+3=52;$
$x^2+3=(-7)^2+3=49+3=52;$
$x^2+3=(-5.013)^2+3=25.130169+3=28.130169$

b) $1-x^2=1-5^2=1-25=-24;$
$1-x^2=1-(-5)^2=1-25=-24;$
$1-x^2=1-(-10.455)^2=1-109.307025=$
-108.307025

97. $\frac{-15+20+50+(-82)+(-7)+(-2)}{6}=\frac{-36}{6}=-6$

Chapter 1 Vocabulary Reinforcement

1. The set of integers is $\{\ldots,-5,-4,-3,-2,-1,0,1,2,3,\ldots\}$.

2. Two numbers whose sum is 0 are called additive inverses of each other.

3. The commutative law of addition says that $a+b=b+a$ for any real numbers a and b.

4. The identity property of 1 states that for any real number a, $a\cdot 1=1\cdot a=a$.

5. The associative law of addition says that $a+(b+c)=(a+b)+c$ for any real numbers a, b, and c.

6. Two numbers whose product is 1 are called multiplicative inverses of each other.

7. The equation $y+0=y$ illustrates the identity property of 0.

Chapter 1 Concept Reinforcement

1. True; the set of integers is composed of the set of whole numbers and the set of opposites of the natural numbers.

2. True; see page 81 in the text.

3. False; the product of a number and its multiplicative inverse is 1.

4. False; $a < b$ has the same meaning as $b > a$.

Chapter 1 Study Guide

1. $2a + b = 2(-1) + 16 = -2 + 16 = 14$

2. Since -6 is to the left of -3, we have $-6 < -3$.

3. The number $-\frac{5}{4}$ is negative, so we make it positive.
$$\left|-\frac{5}{4}\right| = \frac{5}{4}$$

4. $-5.6 + (-2.9)$ Two negative numbers. Add the absolute values, getting 8.5. Make the answer negative.
$-5.6 + (-2.9) = -8.5$

5. $7 - 9 = 7 + (-9) = -2$

6. $-8(-7) = 56$

7. $-48 \div 6 = -8$

8. $-\frac{3}{4} \div \left(-\frac{5}{3}\right) = -\frac{3}{4} \cdot \left(-\frac{3}{5}\right) = \frac{9}{20}$

9. $\frac{45y}{27y} = \frac{5 \cdot 9y}{3 \cdot 9y} = \frac{5}{3} \cdot \frac{9y}{9y} = \frac{5}{3} \cdot 1 = \frac{5}{3}$

10. $5(x + 3y - 4z) = 5 \cdot x + 5 \cdot 3y - 5 \cdot 4z$
$= 5x + 15y - 20z$

11. $27x + 9y - 36z = 9 \cdot 3x + 9 \cdot y - 9 \cdot 4z = 9(3x + y - 4z)$

12. $6a - 4b - a + 2b = 6a - a - 4b + 2b$
$= (6 - 1)a + (-4 + 2)b$
$= 5a - 2b$

13. $8a - b - (4a + 3b) = 8a - b - 4a - 3b = 4a - 4b$

14. $75 \div (-15) + 24 \div 8$
$= -5 + 3$ Dividing
$= -2$ Adding

Chapter 1 Review Exercises

1. Substitute 17 for x and 5 for y and carry out the computation.
$$\frac{x - y}{3} = \frac{17 - 5}{3} = \frac{12}{3} = 4$$

2. $19\%x$, $0.19x$

3. The integer 620 corresponds to earning \$620; the integer -125 corresponds to receiving a speeding ticket for \$125.

4. The distance of -38 from 0 is 38, so $|-38| = 38$.

5. The distance of 126 from 0 is 126, so $|126| = 126$.

6. The graph of -2.5 is halfway between -3 and -2.

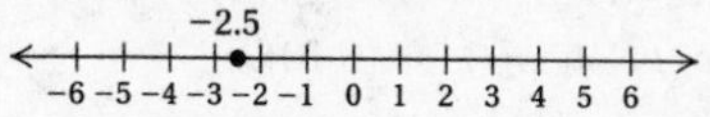

7. The graph of $\frac{8}{9}$ is $\frac{8}{9}$ of the way from 0 to 1.

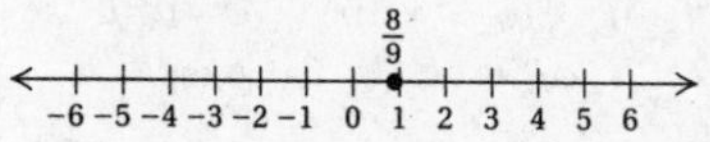

8. Since -3 is to the left of 10, we have $-3 < 10$.

9. Since -1 is to the right of -6, we have $-1 > -6$.

10. Since 0.126 is to the right of -12.6, we have $0.126 > -12.6$.

11. $-\frac{2}{3} = -\frac{2}{3} \cdot \frac{10}{10} = -\frac{20}{30}$
$-\frac{1}{10} = -\frac{1}{10} \cdot \frac{3}{3} = -\frac{3}{30}$
Since $-\frac{20}{30}$ is to the left of $-\frac{3}{30}$, then $-\frac{2}{3}$ is to the left of $-\frac{1}{10}$ and we have $-\frac{2}{3} < -\frac{1}{10}$.

12. $x > -3$ has the same meaning as $-3 < x$.

13. $-9 \leq 11$ is true since $-9 < 11$ is true.

14. $-11 \geq -3$ is false since neither $-11 > -3$ nor $-11 = -3$ is true.

15. The opposite of 3.8 is -3.8 because $3.8 + (-3.8) = 0$.

16. The opposite of $-\frac{3}{4}$ is $\frac{3}{4}$ because $-\frac{3}{4} + \frac{3}{4} = 0$.

17. The reciprocal of $\frac{3}{8}$ is $\frac{8}{3}$ because $\frac{3}{8} \cdot \frac{8}{3} = 1$.

18. The reciprocal of -7 is $-\frac{1}{7}$ because $-7 \cdot \left(-\frac{1}{7}\right) = 1$.

19. If $x = -34$, then $-x = -(-34) = 34$.

20. If $x = 5$, then $-(-x) = -(-5) = 5$.

21. $4 + (-7)$
The absolute values are 4 and 7. The difference is $7 - 4$, or 3. The negative number has the larger absolute value, so the answer is negative. $4 + (-7) = -3$

22. $6 + (-9) + (-8) + 7$
a) Add the negative numbers: $-9 + (-8) = -17$
b) Add the positive numbers: $6 + 7 = 13$
c) Add the results: $-17 + 13 = -4$

23. $-3.8 + 5.1 + (-12) + (-4.3) + 10$

a) Add the negative numbers: $-3.8 + (-12) + (-4.3) = -20.1$

b) Add the positive numbers: $5.1 + 10 = 15.1$

c) Add the results: $-20.1 + 15.1 = -5$

24. $$\begin{aligned} -3-(-7)+7-10 &= -3+7+7+(-10) \\ &= 4+7+(-10) \\ &= 11+(-10) \\ &= 1 \end{aligned}$$

25. $-\frac{9}{10} - \frac{1}{2} = -\frac{9}{10} - \frac{5}{10} = -\frac{9}{10} + \left(-\frac{5}{10}\right) = -\frac{14}{10} = -\frac{7\cdot 2}{5\cdot 2} = -\frac{7}{5}\cdot\frac{2}{2} = -\frac{7}{5}$

26. $-3.8 - 4.1 = -3.8 + (-4.1) = -7.9$

27. $-9\cdot(-6) = 54$

28. $-2.7(3.4) = -9.18$

29. $\frac{2}{3}\cdot\left(-\frac{3}{7}\right) = -\left(\frac{2\cdot 3}{3\cdot 7}\right) = -\left(\frac{2}{7}\cdot\frac{3}{3}\right) = -\frac{2}{7}$

30. $3\cdot(-7)\cdot(-2)\cdot(-5) = -21\cdot 10 = -210$

31. $35 \div (-5) = -7$ Check: $-7\cdot(-5) = 35$

32. $-5.1 \div 1.7 = -3$ Check: $-3\cdot(1.7) = -5.1$

33. $-\frac{3}{11} \div -\frac{4}{11} = -\frac{3}{11}\cdot\left(-\frac{11}{4}\right) = \frac{3\cdot 11}{11\cdot 4} = \frac{3}{4}\cdot\frac{11}{11} = \frac{3}{4}$

34. $$\begin{aligned} 2(-3.4-12.2)-8(-7) &= 2(-15.6)-8(-7) \\ &= -31.2+56 \\ &= 24.8 \end{aligned}$$

35. $$\begin{aligned} \frac{-12(-3)-2^3-(-9)(-10)}{3\cdot 10+1} &= \frac{-12(-3)-8-(-9)(-10)}{30+1} \\ &= \frac{36-8-90}{31} \\ &= \frac{28-90}{31} \\ &= \frac{-62}{31} \\ &= -2 \end{aligned}$$

36. $$\begin{aligned} -16\div 4-30\div(-5) &= -4-(-6) \\ &= -4+6 \\ &= 2 \end{aligned}$$

37. $\frac{-4[7-(10-13)]}{|-2(8)-4|} = \frac{-4[7-(-3)]}{|-16-4|} = \frac{-4[7+3]}{|-20|} = \frac{-4[10]}{20} = \frac{-40}{20} = -2$

38. Let t = the total gain or loss. We represent the gains as positive numbers and the loss as a negative number. We add the gains and the loss to find t.

$t = 5 + (-12) + 15 = -7 + 15 = 8$

There is a total gain of 8 yd.

39. Let a = Chang's total assets after he borrows \$2500.

Total assets	is	Initial assets	minus	Amount of loan
↓	↓	↓	↓	↓
a	$=$	2140	$-$	2500

We carry out the subtraction.

$a = 2140 - 2500 = -360$

Chang's total assets were −\$360.

40. First we multiply to find the total drop d in the price:

$d = 8(-\$1.63) = -\$13.04$

Now we add this number to the opening price to find the price p after 8 hr:

$p = \$17.68 + (-\$13.04) = \$4.64$

After 8 hr the price of the stock was \$4.64 per share.

41. Yuri spent the \$68 in his account plus an additional \$64.65, so he spent a total of \$68 + \$64.65, or \$132.65, on seven equally-priced tee shirts. Then each tee shirt cost $\frac{\$132.65}{7}$, or \$18.95.

42. $5(3x-7) = 5\cdot 3x - 5\cdot 7 = 15x - 35$

43. $-2(4x-5) = -2\cdot 4x - (-2)(5) = -8x - (-10) = -8x + 10$

44. $10(0.4x+1.5) = 10\cdot 0.4x + 10\cdot 1.5 = 4x + 15$

45. $-8(3-6x) = -8\cdot 3 - (-8)(6x) = -24 - (-48x) = -24 + 48x$

46. $2x - 14 = 2\cdot x - 2\cdot 7 = 2(x-7)$

47. $-6x + 6 = -6\cdot x - 6(-1) = -6(x-1)$

The expression can also be factored as follows:

$-6x + 6 = 6(-x) + 6\cdot 1 = 6(-x+1)$

48. $5x + 10 = 5\cdot x + 5\cdot 2 = 5(x+2)$

49. $-3x + 12y - 12 = -3\cdot x - 3(-4y) - 3\cdot 4 = -3(x - 4y + 4)$

We could also factor this expression as follows:

$-3x + 12y - 12 = 3(-x) + 3\cdot 4y + 3(-4) = 3(-x + 4y - 4)$

50. $$\begin{aligned} 11a+2b-4a-5b &= 11a-4a+2b-5b \\ &= (11-4)a+(2-5)b \\ &= 7a-3b \end{aligned}$$

51. $$\begin{aligned} 7x-3y-9x+8y &= 7x-9x-3y+8y \\ &= (7-9)x+(-3+8)y \\ &= -2x+5y \end{aligned}$$

52. $$\begin{aligned} 6x+3y-x-4y &= 6x-x+3y-4y \\ &= (6-1)x+(3-4)y \\ &= 5x-y \end{aligned}$$

53. $$\begin{aligned} -3a+9b+2a-b &= -3a+2a+9b-b \\ &= (-3+2)a+(9-1)b \\ &= -a+8b \end{aligned}$$

54. $2a - (5a - 9) = 2a - 5a + 9 = -3a + 9$

55. $3(b+7)-5b = 3b+21-5b = -2b+21$

56. $3[11-3(4-1)] = 3[11-3\cdot 3] = 3[11-9] = 3\cdot 2 = 6$

57. $2[6(y-4)+7] = 2[6y-24+7] = 2[6y-17] = 12y-34$

58.
$$\begin{aligned}&[8(x+4)-10]-[3(x-2)+4]\\ &= [8x+32-10]-[3x-6+4]\\ &= 8x+22-[3x-2]\\ &= 8x+22-3x+2\\ &= 5x+24\end{aligned}$$

59.
$$\begin{aligned}&5\{[6(x-1)+7]-[3(3x-4)+8]\}\\ &= 5\{[6x-6+7]-[9x-12+8]\}\\ &= 5\{6x+1-[9x-4]\}\\ &= 5\{6x+1-9x+4\}\\ &= 5\{-3x+5\}\\ &= -15x+25\end{aligned}$$

60. $18x-6y+30 = 6\cdot 3x-6\cdot y+6\cdot 5 = 6(3x-y+5)$

Answer D is correct.

61. $mn+5 = nm+5$ by the commutative law of multiplication.

$mn+5 = 5+mn$ by the commutative law of addition.

$mn+5 = 5+mn = 5+nm$ by the commutative laws of addition and multiplication.

Thus we can eliminate answers A, C, and D. It is not possible to write $mn+5$ as $5n+m$ using the commutative and associative laws, so answer B is correct.

62.
$$\begin{aligned}-\left|\frac{7}{8}-\left(-\frac{1}{2}\right)-\frac{3}{4}\right| &= -\left|\frac{7}{8}+\frac{1}{2}-\frac{3}{4}\right|\\ &= -\left|\frac{7}{8}+\frac{4}{8}-\frac{6}{8}\right|\\ &= -\left|\frac{11}{8}-\frac{6}{8}\right|\\ &= -\left|\frac{5}{8}\right|\\ &= -\frac{5}{8}\end{aligned}$$

63.
$$\begin{aligned}&(|2.7-3|+3^2-|-3|)\div(-3)\\ &= (|2.7-3|+9-|-3|)\div(-3)\\ &= (|-0.3|+9-|-3|)\div(-3)\\ &= (0.3+9-3)\div(-3)\\ &= (9.3-3)\div(-3)\\ &= 6.3\div(-3)\\ &= -2.1\end{aligned}$$

64.
$$\underbrace{2000-1990}_{10} + \underbrace{1980-1970}_{10} + \ldots + \underbrace{20-10}_{10}$$
$$\downarrow \qquad\qquad \downarrow \qquad\qquad \downarrow$$
$$10 \quad + \quad 10 \quad + \ldots + \quad 10$$

Counting by 10's from 10 through 2000 gives us 2000/10, or 200, numbers in the expression. There are 200/2, or 100, pairs of numbers in the expression. Each pair is a difference that is equivalent to 10. Thus, the expression is equal to $100\cdot 10$, or 1000.

65. Note that the sum of the lengths of the three horizontal segments at the top of the figure, two of which are not labeled and one of which is labeled b, is equivalent to the length of the horizontal segment at the bottom of the figure, a. Then the perimeter is $a+b+b+a+a+a$, or $4a+2b$.

Chapter 1 Discussion and Writing Exercises

1. The sum of each pair of opposites such as -50 and 50, -49 and 49, and so on is 0. The sum of these sums and the remaining integer 0 is 0.

2. The product of an even number of negative numbers is positive, so $(-7)^8$ is positive. The product of an odd number of negative numbers is negative, so $(-7)^{11}$ is negative.

3. Consider $\dfrac{a}{b} = q$ where a and b are both negative numbers. Then $q\cdot b = a$, so q must be a positive number.

4. Consider $\dfrac{a}{b} = q$ where a is a negative number and b is a positive number. Then $q\cdot b = a$, so q must be a negative number.

5. We use the distributive law when we collect like terms even though we might not always write this step.

6. Jake expects the calculator to multiply 2 and 3 first and then divide 18 by that product. This procedure does not follow the rules for order of operations.

Chapter 1 Test

1. Substitute 10 for x and 5 for y and carry out the computations.
$$\frac{3x}{y} = \frac{3\cdot 10}{5} = \frac{30}{5} = 6$$

2. Using x for "some number," we have $x-9$.

3. Since -3 is the right of -8 on the number line, we have $-3 > -8$.

4. Since $-\frac{1}{2}$ is to the left of $-\frac{1}{8}$ on the number line, we have $-\frac{1}{2} < -\frac{1}{8}$.

5. Since -0.78 is to the right of -0.87 on the number line, we have $-0.78 > -0.87$.

6. $-2 > x$ has the same meaning as $x < -2$.

7. $-13 \le -3$ is true since $-13 < -3$ is true.

8. The distance of -7 from 0 is 7, so $|-7| = 7$.

9. The distance of $\frac{9}{4}$ from 0 is $\frac{9}{4}$, so $\left|\frac{9}{4}\right| = \frac{9}{4}$.

10. The distance of -2.7 from 0 is 2.7, so $|-2.7| = 2.7$.

11. The opposite of $\frac{2}{3}$ is $-\frac{2}{3}$ because $\frac{2}{3}+\left(-\frac{2}{3}\right) = 0$.

12. The opposite of -1.4 is 1.4 because $-1.4 + 1.4 = 0$.

13. The reciprocal of -2 is $-\frac{1}{2}$ because $-2\left(-\frac{1}{2}\right) = 1$.

14. The reciprocal of $\frac{4}{7}$ is $\frac{7}{4}$ because $\frac{4}{7} \cdot \frac{7}{4} = 1$.

15. If $x = -8$, then $-x = -(-8) = 8$.

16. $3.1 - (-4.7) = 3.1 + 4.7 = 7.8$

17.
$$\begin{aligned} -8 + 4 + (-7) + 3 &= -4 + (-7) + 3 \\ &= -11 + 3 \\ &= -8 \end{aligned}$$

18.
$$\begin{aligned} -\frac{1}{5} + \frac{3}{8} &= -\frac{1}{5} \cdot \frac{8}{8} + \frac{3}{8} \cdot \frac{5}{5} \\ &= -\frac{8}{40} + \frac{15}{40} \\ &= \frac{7}{40} \end{aligned}$$

19. $2 - (-8) = 2 + 8 = 10$

20. $3.2 - 5.7 = 3.2 + (-5.7) = -2.5$

21.
$$\begin{aligned} \frac{1}{8} - \left(-\frac{3}{4}\right) &= \frac{1}{8} + \frac{3}{4} \\ &= \frac{1}{8} + \frac{3}{4} \cdot \frac{2}{2} \\ &= \frac{1}{8} + \frac{6}{8} \\ &= \frac{7}{8} \end{aligned}$$

22. $4 \cdot (-12) = -48$

23. $-\frac{1}{2} \cdot \left(-\frac{3}{8}\right) = \frac{3}{16}$

24. $-45 \div 5 = -9$ Check: $-9 \cdot 5 = -45$

25. $-\frac{3}{5} \div \left(-\frac{4}{5}\right) = -\frac{3}{5} \cdot \left(-\frac{5}{4}\right) = \frac{3 \cdot 5}{5 \cdot 4} = \frac{3 \cdot \cancel{5}}{\cancel{5} \cdot 4} = \frac{3}{4}$

26. $4.864 \div (-0.5) = -9.728$

27.
$$\begin{aligned} -2(16) - |2(-8) - 5^3| &= -2(16) - |2(-8) - 125| \\ &= -2(16) - |-16 - 125| \\ &= -2(16) - |-141| \\ &= -2(16) - 141 \\ &= -32 - 141 \\ &= -173 \end{aligned}$$

28. $-20 \div (-5) + 36 \div (-4) = 4 + (-9) = -5$

29. Let $P =$ the number of points by which the market has changed over the five week period.

$$\underbrace{\text{Total change}}_{\displaystyle P} \; = \; \underbrace{\text{Week 1 change}}_{\displaystyle -13} \; + \; \underbrace{\text{Week 2 change}}_{\displaystyle (-16)} \; + \; \underbrace{\text{Week 3 change}}_{\displaystyle 36} \; + \; \underbrace{\text{Week 4 change}}_{\displaystyle (-11)} \; + \; \underbrace{\text{Week 5 change}}_{\displaystyle 19}$$

We carry out the computation.

$$\begin{aligned} P &= -13 + (-16) + 36 + (-11) + 19 \\ &= -29 + 36 + (-11) + 19 \\ &= 7 + (-11) + 19 \\ &= -4 + 19 \\ &= 15 \end{aligned}$$

The market rose 15 points.

30. Let $D =$ the difference in the elevations.

$$\underbrace{\text{Difference in elevations}}_{\displaystyle D} \; \underset{\displaystyle =}{\text{is}} \; \underbrace{\text{Higher elevation}}_{\displaystyle 2229} \; \underset{\displaystyle -}{\text{minus}} \; \underbrace{\text{Lower elevation}}_{\displaystyle (-15)}$$

We carry out the subtraction.

$$D = 2229 - (-15) = 2229 + 15 = 2244$$

The difference in elevations is 2244 m.

31. First we multiply to find the total decrease d in the population.

$$d = 6 \cdot 420 = 2520$$

The population decreased by 2520 over the six year period.

Now we subtract to find the new population p.

$$18,600 - 2520 = 16,080$$

After 6 yr the population was 16,080.

32. First we subtract to find the total drop in temperature t.

$$t = 16°\text{C} - (-17°\text{C}) = 16°\text{C} + 17°\text{C} = 33°\text{C}$$

Then we divide to find by how many degrees d the temperature dropped each minute in the 44 minutes from 11:08 A.M. to 11:52 A.M.

$$d = 33 \div 44 = 0.75$$

The temperature changed about $-0.75°$C each minute.

33. $3(6 - x) = 3 \cdot 6 - 3 \cdot x = 18 - 3x$

34. $-5(y - 1) = -5 \cdot y - (-5)(1) = -5y - (-5) = -5y + 5$

35. $12 - 22x = 2 \cdot 6 - 2 \cdot 11x = 2(6 - 11x)$

36. $7x + 21 + 14y = 7 \cdot x + 7 \cdot 3 + 7 \cdot 2y = 7(x + 3 + 2y)$

37.
$$\begin{aligned} 6 + 7 - 4 - (-3) &= 6 + 7 + (-4) + 3 \\ &= 13 + (-4) + 3 \\ &= 9 + 3 \\ &= 12 \end{aligned}$$

38. $5x - (3x - 7) = 5x - 3x + 7 = 2x + 7$

39.
$$\begin{aligned} 4(2a - 3b) + a - 7 &= 8a - 12b + a - 7 \\ &= 9a - 12b - 7 \end{aligned}$$

40. $$\begin{aligned} & 4\{3[5(y-3)+9]+2(y+8)\} \\ &= 4\{3[5y-15+9]+2y+16\} \\ &= 4\{3[5y-6]+2y+16\} \\ &= 4\{15y-18+2y+16\} \\ &= 4\{17y-2\} \\ &= 68y-8 \end{aligned}$$

41. $256 \div (-16) \div 4 = -16 \div 4 = -4$

42. $$\begin{aligned} 2^3 - 10[4-(-2+18)3] &= 2^3 - 10[4-(16)3] \\ &= 2^3 - 10[4-48] \\ &= 2^3 - 10[-44] \\ &= 8 - 10[-44] \\ &= 8 + 440 \\ &= 448 \end{aligned}$$

43. Choices A, C, and D are true because $-5 = -5$ is true. Choice B is not true because -5 is not to the left of -5 on the number line. The correct answer is B.

44. $$\begin{aligned} & |-27-3(4)| - |-36| + |-12| \\ &= |-27-12| - |-36| + |-12| \\ &= |-39| - |-36| + |-12| \\ &= 39 - 36 + 12 \\ &= 3 + 12 \\ &= 15 \end{aligned}$$

45. $$\begin{aligned} & a - \{3a - [4a - (2a - 4a)]\} \\ &= a - \{3a - [4a - (-2a)]\} \\ &= a - \{3a - [4a + 2a]\} \\ &= a - \{3a - 6a\} \\ &= a - \{-3a\} \\ &= a + 3a \\ &= 4a \end{aligned}$$

46. The perimeter is equivalent to the perimeter of a square with sides x along with four additional segments of length y. We have $x + x + x + x + y + y + y + y = 4x + 4y$.

Chapter 2

Solving Equations and Inequalities

Exercise Set 2.1

RC1. The first step in solving $9 = x - 4$ is to add 4 on both sides. The correct answer is (f).

RC3. The first step in solving $x - 3 = 9$ is to add 3 on both sides. The correct answer is (e).

1.
$$\begin{array}{c|l} x+17 = 32 & \text{Writing the equation} \\ \hline 15+17 \ ?\ 32 & \text{Substituting 15 for } x \\ 32 & \text{TRUE} \end{array}$$

Since the left-hand and right-hand sides are the same, 15 is a solution of the equation.

3.
$$\begin{array}{c|l} x-7 = 12 & \text{Writing the equation} \\ \hline 21-7 \ ?\ 12 & \text{Substituting 21 for } x \\ 14 & \text{FALSE} \end{array}$$

Since the left-hand and right-hand sides are not the same, 21 is not a solution of the equation.

5.
$$\begin{array}{c|l} 6x = 54 & \text{Writing the equation} \\ \hline 6(-7) \ ?\ 54 & \text{Substituting} \\ -42 & \text{FALSE} \end{array}$$

-7 is not a solution of the equation.

7.
$$\begin{array}{c|l} \frac{x}{6} = 5 & \text{Writing the equation} \\ \hline \frac{30}{6} \ ?\ 5 & \text{Substituting} \\ 5 & \text{TRUE} \end{array}$$

5 is a solution of the equation.

9.
$$\begin{array}{c|l} 5x+7 = 107 & \\ \hline 5\cdot 20+7 \ ?\ 107 & \text{Substituting} \\ 100+7 & \\ 107 & \text{TRUE} \end{array}$$

20 is a solution of the equation.

11.
$$\begin{array}{c|l} 7(y-1) = 63 & \\ \hline 7(-10-1) \ ?\ 63 & \text{Substituting} \\ 7(-11) & \\ -77 & \text{FALSE} \end{array}$$

-10 is not a solution of the equation.

13.
$$\begin{aligned} x+2 &= 6 \\ x+2-2 &= 6-2 \quad \text{Subtracting 2 on both sides} \\ x &= 4 \quad \text{Simplifying} \end{aligned}$$

Check:
$$\begin{array}{c|c} x+2 = 6 & \\ \hline 4+2 \ ?\ 6 & \\ 6 & \text{TRUE} \end{array}$$

The solution is 4.

15.
$$\begin{aligned} x+15 &= -5 \\ x+15-15 &= -5-15 \quad \text{Subtracting 15 on both sides} \\ x &= -20 \end{aligned}$$

Check:
$$\begin{array}{c|c} x+15 = -5 & \\ \hline -20+15 \ ?\ -5 & \\ -5 & \text{TRUE} \end{array}$$

The solution is -20.

17.
$$\begin{aligned} x+6 &= -8 \\ x+6-6 &= -8-6 \\ x &= -14 \end{aligned}$$

Check:
$$\begin{array}{c|c} x+6 = -8 & \\ \hline -14+6 \ ?\ -8 & \\ -8 & \text{TRUE} \end{array}$$

The solution is -14.

19.
$$\begin{aligned} x+16 &= -2 \\ x+16-16 &= -2-16 \\ x &= -18 \end{aligned}$$

Check:
$$\begin{array}{c|c} x+16 = -2 & \\ \hline -18+16 \ ?\ -2 & \\ -2 & \text{TRUE} \end{array}$$

The solution is -18.

21.
$$\begin{aligned} x-9 &= 6 \\ x-9+9 &= 6+9 \\ x &= 15 \end{aligned}$$

Check:
$$\begin{array}{c|c} x-9 = 6 & \\ \hline 15-9 \ ?\ 6 & \\ 6 & \text{TRUE} \end{array}$$

The solution is 15.

23.
$$\begin{aligned} x-7 &= -21 \\ x-7+7 &= -21+7 \\ x &= -14 \end{aligned}$$

Check:
$$\begin{array}{c|c} x-7 = -21 & \\ \hline -14-7 \ ?\ -21 & \\ -21 & \text{TRUE} \end{array}$$

The solution is -14.

25.
$$\begin{aligned} 5+t &= 7 \\ -5+5+t &= -5+7 \\ t &= 2 \end{aligned}$$

Check:
$$\begin{array}{c|c} 5+t = 7 & \\ \hline 5+2 \ ?\ 7 & \\ 7 & \text{TRUE} \end{array}$$

The solution is 2.

27.
$$-7+y=13$$
$$7+(-7)+y=7+13$$
$$y=20$$

Check: $-7+y=13$

$-7+20 \ ? \ 13$

$13 \mid$ TRUE

The solution is 20.

29.
$$-3+t=-9$$
$$3+(-3)+t=3+(-9)$$
$$t=-6$$

Check: $-3+t=-9$

$-3+(-6) \ ? \ -9$

$-9 \mid$ TRUE

The solution is -6.

31.
$$x+\frac{1}{2}=7$$
$$x+\frac{1}{2}-\frac{1}{2}=7-\frac{1}{2}$$
$$x=6\frac{1}{2}$$

Check: $x+\frac{1}{2}=7$

$6\frac{1}{2}+\frac{1}{2} \ ? \ 7$

$7 \mid$ TRUE

The solution is $6\frac{1}{2}$.

33.
$$12=a-7.9$$
$$12+7.9=a-7.9+7.9$$
$$19.9=a$$

Check: $12=a-7.9$

$12 \ ? \ 19.9-7.9$

$\mid 12$ TRUE

The solution is 19.9.

35.
$$r+\frac{1}{3}=\frac{8}{3}$$
$$r+\frac{1}{3}-\frac{1}{3}=\frac{8}{3}-\frac{1}{3}$$
$$r=\frac{7}{3}$$

Check: $r+\frac{1}{3}=\frac{8}{3}$

$\frac{7}{3}+\frac{1}{3} \ ? \ \frac{8}{3}$

$\frac{8}{3} \mid$ TRUE

The solution is $\frac{7}{3}$.

37.
$$m+\frac{5}{6}=-\frac{11}{12}$$
$$m+\frac{5}{6}-\frac{5}{6}=-\frac{11}{12}-\frac{5}{6}$$
$$m=-\frac{11}{12}-\frac{5}{6}\cdot\frac{2}{2}$$
$$m=-\frac{11}{12}-\frac{10}{12}$$
$$m=-\frac{21}{12}=-\frac{\not{3}\cdot 7}{\not{3}\cdot 4}$$
$$m=-\frac{7}{4}$$

Check: $m+\frac{5}{6}=-\frac{11}{12}$

$-\frac{7}{4}+\frac{5}{6} \ ? \ -\frac{11}{12}$

$-\frac{21}{12}+\frac{10}{12} \mid$

$-\frac{11}{12} \mid$ TRUE

The solution is $-\frac{7}{4}$.

39.
$$x-\frac{5}{6}=\frac{7}{8}$$
$$x-\frac{5}{6}+\frac{5}{6}=\frac{7}{8}+\frac{5}{6}$$
$$x=\frac{7}{8}\cdot\frac{3}{3}+\frac{5}{6}\cdot\frac{4}{4}$$
$$x=\frac{21}{24}+\frac{20}{24}$$
$$x=\frac{41}{24}$$

Check: $x-\frac{5}{6}=\frac{7}{8}$

$\frac{41}{24}-\frac{5}{6} \ ? \ \frac{7}{8}$

$\frac{41}{24}-\frac{20}{24} \mid \frac{21}{24}$

$\frac{21}{24} \mid$ TRUE

The solution is $\frac{41}{24}$.

41.
$$-\frac{1}{5}+z=-\frac{1}{4}$$
$$\frac{1}{5}-\frac{1}{5}+z=\frac{1}{5}-\frac{1}{4}$$
$$z=\frac{1}{5}\cdot\frac{4}{4}-\frac{1}{4}\cdot\frac{5}{5}$$
$$z=\frac{4}{20}-\frac{5}{20}$$
$$z=-\frac{1}{20}$$

Check:

$$\begin{array}{c|c} -\frac{1}{5}+z & -\frac{1}{4} \\ \hline -\frac{1}{5}+\left(-\frac{1}{20}\right) \ ? & -\frac{1}{4} \\ -\frac{4}{20}+\left(-\frac{1}{20}\right) & -\frac{5}{20} \\ -\frac{5}{20} & \end{array} \quad \text{TRUE}$$

The solution is $-\frac{1}{20}$.

43.
$$\begin{aligned} x+2.3 &= 7.4 \\ x+2.3-2.3 &= 7.4-2.3 \\ x &= 5.1 \end{aligned}$$

Check:

$$\begin{array}{c|c} x+2.3 & 7.4 \\ \hline 5.1+2.3 \ ? & 7.4 \\ 7.4 & \end{array} \quad \text{TRUE}$$

The solution is 5.1.

45.
$$\begin{aligned} 7.6 &= x-4.8 \\ 7.6+4.8 &= x-4.8+4.8 \\ 12.4 &= x \end{aligned}$$

Check:

$$\begin{array}{c|c} 7.6 & x-4.8 \\ \hline 7.6 \ ? & 12.4-4.8 \\ & 7.6 \end{array} \quad \text{TRUE}$$

The solution is 12.4.

47.
$$\begin{aligned} -9.7 &= -4.7+y \\ 4.7+(-9.7) &= 4.7+(-4.7)+y \\ -5 &= y \end{aligned}$$

Check:

$$\begin{array}{c|c} -9.7 & -4.7+y \\ \hline -9.7 \ ? & -4.7+(-5) \\ & -9.7 \end{array} \quad \text{TRUE}$$

The solution is -5.

49.
$$\begin{aligned} 5\frac{1}{6}+x &= 7 \\ -5\frac{1}{6}+5\frac{1}{6}+x &= -5\frac{1}{6}+7 \\ x &= -\frac{31}{6}+\frac{42}{6} \\ x &= \frac{11}{6}, \text{ or } 1\frac{5}{6} \end{aligned}$$

Check:

$$\begin{array}{c|c} 5\frac{1}{6}+x & 7 \\ \hline 5\frac{1}{6}+1\frac{5}{6} \ ? & 7 \\ 7 & \end{array} \quad \text{TRUE}$$

The solution is $\frac{11}{6}$, or $1\frac{5}{6}$.

51.
$$\begin{aligned} q+\frac{1}{3} &= -\frac{1}{7} \\ q+\frac{1}{3}-\frac{1}{3} &= -\frac{1}{7}-\frac{1}{3} \\ q &= -\frac{1}{7}\cdot\frac{3}{3}-\frac{1}{3}\cdot\frac{7}{7} \\ q &= -\frac{3}{21}-\frac{7}{21} \\ q &= -\frac{10}{21} \end{aligned}$$

Check:

$$\begin{array}{c|c} q+\frac{1}{3} & -\frac{1}{7} \\ \hline -\frac{10}{21}+\frac{1}{3} \ ? & -\frac{1}{7} \\ -\frac{10}{21}+\frac{7}{21} & -\frac{3}{21} \\ -\frac{3}{21} & \end{array} \quad \text{TRUE}$$

The solution is $-\frac{10}{21}$.

53. $\frac{2}{3}\div\left(-\frac{4}{9}\right)=\frac{2}{3}\cdot\left(-\frac{9}{4}\right)=-\frac{2\cdot 9}{3\cdot 4}=-\frac{\cancel{2}\cdot\cancel{3}\cdot 3}{\cancel{3}\cdot\cancel{2}\cdot 2}=-\frac{3}{2}$

55.
$$\begin{aligned} -\frac{2}{3}-\left(-\frac{5}{8}\right) &= -\frac{2}{3}+\frac{5}{8} \\ &= -\frac{2}{3}\cdot\frac{8}{8}+\frac{5}{8}\cdot\frac{3}{3} \\ &= -\frac{16}{24}+\frac{15}{24} \\ &= -\frac{1}{24} \end{aligned}$$

57. The translation is $\$83 - x$.

59.
$$\begin{aligned} x+\frac{4}{5} &= -\frac{2}{3}-\frac{4}{15} \\ x+\frac{4}{5} &= -\frac{2}{3}\cdot\frac{5}{5}-\frac{4}{15} \quad \text{Adding on the right side} \\ x+\frac{4}{5} &= -\frac{10}{15}-\frac{4}{15} \\ x+\frac{4}{5} &= -\frac{14}{15} \\ x+\frac{4}{5}-\frac{4}{5} &= -\frac{14}{15}-\frac{4}{5} \\ x &= -\frac{14}{15}-\frac{4}{5}\cdot\frac{3}{3} \\ x &= -\frac{14}{15}-\frac{12}{15} \\ x &= -\frac{26}{15} \end{aligned}$$

The solution is $-\frac{26}{15}$.

61.
$$\begin{aligned} 16+x-22 &= -16 \\ x-6 &= -16 \quad \text{Adding on the left side} \\ x-6+6 &= -16+6 \\ x &= -10 \end{aligned}$$

The solution is -10.

63. $x+3=3+x$
$x+3-3=3+x-3$
$x=x$

$x=x$ is true for all real numbers. Thus the solution is all real numbers.

Exercise Set 2.2

RC1. To solve $3=-\frac{1}{12}x$, we would first multiply both sides by -12. The correct answer is (f).

RC3. To solve $12x=-6$, we would first divide both sides by 12. The correct answer is (a).

1. $6x=36$

$\frac{6x}{6}=\frac{36}{6}$ Dividing by 6 on both sides

$1\cdot x=6$ Simplifying

$x=6$ Identity property of 1

Check: $6x=36$; $6\cdot 6 \stackrel{?}{} 36$; 36 TRUE

The solution is 6.

3. $5y=45$

$\frac{5y}{5}=\frac{45}{5}$ Dividing by 5 on both sides

$1\cdot y=9$ Simplifying

$y=9$ Identity property of 1

Check: $5y=45$; $5\cdot 9 \stackrel{?}{} 45$; 45 TRUE

The solution is 9.

5. $84=7x$

$\frac{84}{7}=\frac{7x}{7}$ Dividing by 7 on both sides

$12=1\cdot x$

$12=x$

Check: $84=7x$; $84 \stackrel{?}{} 7\cdot 12$; 84 TRUE

The solution is 12.

7. $-x=40$

$-1\cdot x=40$

$\frac{-1\cdot x}{-1}=\frac{40}{-1}$

$1\cdot x=-40$

$x=-40$

Check: $-x=40$; $-(-40) \stackrel{?}{} 40$; 40 TRUE

The solution is -40.

9. $-1=-z$

$-1=-1\cdot z$

$\frac{-1}{-1}=\frac{-1\cdot z}{-1}$

$1=1\cdot z$

$1=z$

Check: $-1=-z$; $-1 \stackrel{?}{} -(1)$; -1 TRUE

The solution is 1.

11. $7x=-49$

$\frac{7x}{7}=\frac{-49}{7}$

$1\cdot x=-7$

$x=-7$

Check: $7x=-49$; $7(-7) \stackrel{?}{} -49$; -49 TRUE

The solution is -7.

13. $-12x=72$

$\frac{-12x}{-12}=\frac{72}{-12}$

$1\cdot x=-6$

$x=-6$

Check: $-12x=72$; $-12(-6) \stackrel{?}{} 72$; 72 TRUE

The solution is -6.

15. $-21x=-126$

$\frac{-21x}{-21}=\frac{-126}{-21}$

$1\cdot x=6$

$x=6$

Check: $-21x=-126$; $-21\cdot 6 \stackrel{?}{} -126$; -126 TRUE

The solution is 6.

17. $\frac{t}{7}=-9$

$7\cdot\frac{1}{7}t=7\cdot(-9)$

$1\cdot t=-63$

$t=-63$

Check: $\frac{t}{7}=-9$; $\frac{-63}{7} \stackrel{?}{} -9$; -9 TRUE

The solution is -63.

19.
$$\frac{n}{-6} = 8$$
$$-6\left(\frac{n}{-6}\right) = -6 \cdot 8$$
$$1 \cdot n = -48$$
$$n = -48$$

Check: $\frac{n}{-6} = 8$

$\frac{-48}{-6}$? 8

8 TRUE

The solution is -48.

21.
$$\frac{3}{4}x = 27$$
$$\frac{4}{3} \cdot \frac{3}{4}x = \frac{4}{3} \cdot 27$$
$$1 \cdot x = \frac{4 \cdot \not{3} \cdot 3 \cdot 3}{\not{3} \cdot 1}$$
$$x = 36$$

Check: $\frac{3}{4}x = 27$

$\frac{3}{4} \cdot 36$? 27

27 TRUE

The solution is 36.

23.
$$-\frac{2}{3}x = 6$$
$$-\frac{3}{2}\left(-\frac{2}{3}x\right) = -\frac{3}{2} \cdot 6$$
$$1 \cdot x = -\frac{3 \cdot \not{2} \cdot 3}{\not{2} \cdot 1}$$
$$x = -9$$

Check: $-\frac{2}{3}x = 6$

$-\frac{2}{3}(-9)$? 6

6 TRUE

The solution is -9.

25.
$$\frac{-t}{3} = 7$$
$$3 \cdot \frac{1}{3} \cdot (-t) = 3 \cdot 7$$
$$-t = 21$$
$$-1 \cdot (-1 \cdot t) = -1 \cdot 21$$
$$1 \cdot t = -21$$
$$t = -21$$

Check: $\frac{-t}{3} = 7$

$\frac{-(-21)}{3}$? 7

$\frac{21}{3}$

7 TRUE

The solution is -21.

27.
$$-\frac{m}{3} = \frac{1}{5}$$
$$-\frac{1}{3} \cdot m = \frac{1}{5}$$
$$-3 \cdot \left(-\frac{1}{3} \cdot m\right) = -3 \cdot \frac{1}{5}$$
$$m = -\frac{3}{5}$$

Check: $-\frac{m}{3} = \frac{1}{5}$

$-\frac{-\frac{3}{5}}{3}$? $\frac{1}{5}$

$-\left(-\frac{3}{5} \div 3\right)$

$-\left(-\frac{3}{5} \cdot \frac{1}{3}\right)$

$-\left(-\frac{1}{5}\right)$

$\frac{1}{5}$ TRUE

The solution is $-\frac{3}{5}$.

29.
$$-\frac{3}{5}r = \frac{9}{10}$$
$$-\frac{5}{3} \cdot \left(-\frac{3}{5}r\right) = -\frac{5}{3} \cdot \frac{9}{10}$$
$$1 \cdot r = -\frac{\not{5} \cdot \not{3} \cdot 3}{\not{3} \cdot \not{5} \cdot 2}$$
$$r = -\frac{3}{2}$$

Check: $-\frac{3}{5}r = \frac{9}{10}$

$-\frac{3}{5} \cdot \left(-\frac{3}{2}\right)$? $\frac{9}{10}$

$\frac{9}{10}$ TRUE

The solution is $-\frac{3}{2}$.

31.
$$-\frac{3}{2}r = -\frac{27}{4}$$
$$-\frac{2}{3}\cdot\left(-\frac{3}{2}r\right) = -\frac{2}{3}\cdot\left(-\frac{27}{4}\right)$$
$$1\cdot r = \frac{\not{2}\cdot\not{3}\cdot 3\cdot 3}{\not{3}\cdot\not{2}\cdot 2}$$
$$r = \frac{9}{2}$$

Check: $-\frac{3}{2}r = -\frac{27}{4}$

$$-\frac{3}{2}\cdot\frac{9}{2} \;?\; -\frac{27}{4}$$
$$-\frac{27}{4} \quad \text{TRUE}$$

The solution is $\frac{9}{2}$.

33.
$$6.3x = 44.1$$
$$\frac{6.3x}{6.3} = \frac{44.1}{6.3}$$
$$1\cdot x = 7$$
$$x = 7$$

Check: $6.3x = 44.1$

$$6.3\cdot 7 \;?\; 44.1$$
$$44.1 \quad \text{TRUE}$$

The solution is 7.

35.
$$-3.1y = 21.7$$
$$\frac{-3.1y}{-3.1} = \frac{21.7}{-3.1}$$
$$1\cdot y = -7$$
$$y = -7$$

Check: $-3.1y = 21.7$

$$-3.1(-7) \;?\; 21.7$$
$$21.7 \quad \text{TRUE}$$

The solution is -7.

37.
$$38.7m = 309.6$$
$$\frac{38.7m}{38.7} = \frac{309.6}{38.7}$$
$$1\cdot m = 8$$
$$m = 8$$

Check: $38.7m = 309.6$

$$38.7\cdot 8 \;?\; 309.6$$
$$309.6 \quad \text{TRUE}$$

The solution is 8.

39.
$$-\frac{2}{3}y = -10.6$$
$$-\frac{3}{2}\cdot\left(-\frac{2}{3}y\right) = -\frac{3}{2}\cdot(-10.6)$$
$$1\cdot y = \frac{31.8}{2}$$
$$y = 15.9$$

Check: $-\frac{2}{3}y = -10.6$

$$-\frac{2}{3}\cdot(15.9) \;?\; -10.6$$
$$-\frac{31.8}{3}$$
$$-10.6 \quad \text{TRUE}$$

The solution is 15.9.

41.
$$\frac{-x}{5} = 10$$
$$5\cdot\frac{-x}{5} = 5\cdot 10$$
$$-x = 50$$
$$-1\cdot(-x) = -1\cdot 50$$
$$x = -50$$

Check: $\frac{-x}{5} = 10$

$$\frac{-(-50)}{5} \;?\; 10$$
$$\frac{50}{5}$$
$$10 \quad \text{TRUE}$$

The solution is -50.

43.
$$-\frac{t}{2} = 7$$
$$2\cdot\left(-\frac{t}{2}\right) = 2\cdot 7$$
$$-t = 14$$
$$-1\cdot(-t) = -1\cdot 14$$
$$t = -14$$

Check: $-\frac{t}{2} = 7$

$$-\frac{-14}{2} \;?\; 7$$
$$-(-7)$$
$$7 \quad \text{TRUE}$$

The solution is -14.

45. $3x + 4x = (3+4)x = 7x$

47. $-4x + 11 - 6x + 18x = (-4 - 6 + 18)x + 11 = 8x + 11$

49. $3x - (4 + 2x) = 3x - 4 - 2x = x - 4$

51. $8y - 6(3y + 7) = 8y - 18y - 42 = -10y - 42$

53. The translation is $8r$ miles.

55.
$$-0.2344m = 2028.732$$
$$\frac{-0.2344m}{-0.2344} = \frac{2028.732}{-0.2344}$$
$$1\cdot m = -8655$$
$$m = -8655$$

The solution is -8655.

57. For all x, $0 \cdot x = 0$. There is no solution to $0 \cdot x = 9$.

59.
$$\begin{aligned} 2|x| &= -12 \\ \frac{2|x|}{2} &= \frac{-12}{2} \\ 1 \cdot |x| &= -6 \\ |x| &= -6 \end{aligned}$$

Absolute value cannot be negative. The equation has no solution.

61.
$$\begin{aligned} 3x &= \frac{b}{a} \\ \frac{1}{3} \cdot 3x &= \frac{1}{3} \cdot \frac{b}{a} \\ x &= \frac{b}{3a} \end{aligned}$$

The solution is $\frac{b}{3a}$.

63.
$$\begin{aligned} \frac{a}{b}x &= 4 \\ \frac{b}{a} \cdot \frac{a}{b}x &= \frac{b}{a} \cdot 4 \\ x &= \frac{4b}{a} \end{aligned}$$

The solution is $\frac{4b}{a}$.

Exercise Set 2.3

RC1. The LCM of the denominators is 10, so multiplying by 10 on both sides will clear the fractions. The correct answer is (d).

RC3. The LCM of the denominators is 12, so multiplying by 12 on both sides will clear the fractions. The correct answer is (c).

RC5. The LCM of the denominators is 35, so multiplying by 35 on both sides will clear the fractions. The correct answer is (b).

1.
$$\begin{aligned} 5x + 6 &= 31 \\ 5x + 6 - 6 &= 31 - 6 && \text{Subtracting 6 on both sides} \\ 5x &= 25 && \text{Simplifying} \\ \frac{5x}{5} &= \frac{25}{5} && \text{Dividing by 5 on both sides} \\ x &= 5 && \text{Simplifying} \end{aligned}$$

Check:
$$\begin{array}{r|l} 5x + 6 & 31 \\ \hline 5 \cdot 5 + 6 \ ? & 31 \\ 25 + 6 & \\ 31 & \qquad \text{TRUE} \end{array}$$

The solution is 5.

3.
$$\begin{aligned} 8x + 4 &= 68 \\ 8x + 4 - 4 &= 68 - 4 && \text{Subtracting 4 on both sides} \\ 8x &= 64 && \text{Simplifying} \\ \frac{8x}{8} &= \frac{64}{8} && \text{Dividing by 8 on both sides} \\ x &= 8 && \text{Simplifying} \end{aligned}$$

Check:
$$\begin{array}{r|l} 8x + 4 & 68 \\ \hline 8 \cdot 8 + 4 \ ? & 68 \\ 64 + 4 & \\ 68 & \qquad \text{TRUE} \end{array}$$

The solution is 8.

5.
$$\begin{aligned} 4x - 6 &= 34 \\ 4x - 6 + 6 &= 34 + 6 && \text{Adding 6 on both sides} \\ 4x &= 40 \\ \frac{4x}{4} &= \frac{40}{4} && \text{Dividing by 4 on both sides} \\ x &= 10 \end{aligned}$$

Check:
$$\begin{array}{r|l} 4x - 6 & 34 \\ \hline 4 \cdot 10 - 6 \ ? & 34 \\ 40 - 6 & \\ 34 & \qquad \text{TRUE} \end{array}$$

The solution is 10.

7.
$$\begin{aligned} 3x - 9 &= 33 \\ 3x - 9 + 9 &= 33 + 9 \\ 3x &= 42 \\ \frac{3x}{3} &= \frac{42}{3} \\ x &= 14 \end{aligned}$$

Check:
$$\begin{array}{r|l} 3x - 9 & 33 \\ \hline 3 \cdot 14 - 9 \ ? & 33 \\ 42 - 9 & \\ 33 & \qquad \text{TRUE} \end{array}$$

The solution is 14.

9.
$$\begin{aligned} 7x + 2 &= -54 \\ 7x + 2 - 2 &= -54 - 2 \\ 7x &= -56 \\ \frac{7x}{7} &= \frac{-56}{7} \\ x &= -8 \end{aligned}$$

Check:
$$\begin{array}{r|l} 7x + 2 & -54 \\ \hline 7(-8) + 2 \ ? & -54 \\ -56 + 2 & \\ -54 & \qquad \text{TRUE} \end{array}$$

The solution is -8.

11.
$$\begin{aligned} -45 &= 6y + 3 \\ -45 - 3 &= 6y + 3 - 3 \\ -48 &= 6y \\ \frac{-48}{6} &= \frac{6y}{6} \\ -8 &= y \end{aligned}$$

Check:
$$\begin{array}{r|l} -45 & 6y + 3 \\ \hline -45 \ ? & 6(-8) + 3 \\ & -48 + 3 \\ & -45 \qquad \text{TRUE} \end{array}$$

The solution is -8.

13.
$$-4x+7=35$$
$$-4x+7-7=35-7$$
$$-4x=28$$
$$\frac{-4x}{-4}=\frac{28}{-4}$$
$$x=-7$$

Check: $-4x+7=35$

$-4(-7)+7 \ ? \ 35$

$28+7$

35 TRUE

The solution is -7.

15.
$$\frac{5}{4}x-18=-3$$
$$\frac{5}{4}x-18+18=-3+18$$
$$\frac{5}{4}x=15$$
$$\frac{4}{5}\cdot\frac{5}{4}x=\frac{4}{5}\cdot 15$$
$$x=12$$

Check: $\frac{5}{4}x-18=-3$

$\frac{5}{4}\cdot 12-18 \ ? \ -3$

$15-18$

-3 TRUE

The solution is 12.

17. $5x+7x=72$

$12x=72$ Collecting like terms

$\frac{12x}{12}=\frac{72}{12}$ Dividing by 12 on both sides

$x=6$

Check: $5x+7x=72$

$5\cdot 6+7\cdot 6 \ ? \ 72$

$30+42$

72 TRUE

The solution is 6.

19. $8x+7x=60$

$15x=60$ Collecting like terms

$\frac{15x}{15}=\frac{60}{15}$ Dividing by 15 on both sides

$x=4$

Check: $8x+7x=60$

$8\cdot 4+7\cdot 4 \ ? \ 60$

$32+28$

60 TRUE

The solution is 4.

21. $4x+3x=42$
$$7x=42$$
$$\frac{7x}{7}=\frac{42}{7}$$
$$x=6$$

Check: $4x+3x=42$

$4\cdot 6+3\cdot 6 \ ? \ 42$

$24+18$

42 TRUE

The solution is 6.

23. $-6y-3y=27$
$$-9y=27$$
$$\frac{-9y}{-9}=\frac{27}{-9}$$
$$y=-3$$

Check: $-6y-3y=27$

$-6(-3)-3(-3) \ ? \ 27$

$18+9$

27 TRUE

The solution is -3.

25. $-7y-8y=-15$
$$-15y=-15$$
$$\frac{-15y}{-15}=\frac{-15}{-15}$$
$$y=1$$

Check: $-7y-8y=-15$

$-7\cdot 1-8\cdot 1 \ ? \ -15$

$-7-8$

-15 TRUE

The solution is 1.

27.
$$x+\frac{1}{3}x=8$$
$$\left(1+\frac{1}{3}\right)x=8$$
$$\frac{4}{3}x=8$$
$$\frac{3}{4}\cdot\frac{4}{3}x=\frac{3}{4}\cdot 8$$
$$x=6$$

Check: $x+\frac{1}{3}x=8$

$6+\frac{1}{3}\cdot 6 \ ? \ 8$

$6+2$

8 TRUE

The solution is 6.

29. $10.2y - 7.3y = -58$

$2.9y = -58$

$\dfrac{2.9y}{2.9} = \dfrac{-58}{2.9}$

$y = -20$

Check: $10.2y - 7.3y = -58$

$10.2(-20) - 7.3(-20)$	? -58
$-204 + 146$	
-58	TRUE

The solution is -20.

31. $8y - 35 = 3y$

$8y = 3y + 35$ Adding 35 and simplifying

$8y - 3y = 35$ Subtracting $3y$ and simplifying

$5y = 35$ Collecting like terms

$\dfrac{5y}{5} = \dfrac{35}{5}$ Dividing by 5

$y = 7$

Check: $8y - 35 = 3y$

$8 \cdot 7 - 35$	? $3 \cdot 7$
$56 - 35$	21
21	TRUE

The solution is 7.

33. $8x - 1 = 23 - 4x$

$8x + 4x = 23 + 1$ Adding 1 and $4x$ and simplifying

$12x = 24$ Collecting like terms

$\dfrac{12x}{12} = \dfrac{24}{12}$ Dividing by 12

$x = 2$

Check: $8x - 1 = 23 - 4x$

$8 \cdot 2 - 1$	? $23 - 4 \cdot 2$
$16 - 1$	$23 - 8$
15	15 TRUE

The solution is 2.

35. $2x - 1 = 4 + x$

$2x - x = 4 + 1$ Adding 1 and $-x$

$x = 5$ Collecting like terms

Check: $2x - 1 = 4 + x$

$2 \cdot 5 - 1$	? $4 + 5$
$10 - 1$	9
9	TRUE

The solution is 5.

37. $6x + 3 = 2x + 11$

$6x - 2x = 11 - 3$

$4x = 8$

$\dfrac{4x}{4} = \dfrac{8}{4}$

$x = 2$

Check: $6x + 3 = 2x + 11$

$6 \cdot 2 + 3$	? $2 \cdot 2 + 11$
$12 + 3$	$4 + 11$
15	15 TRUE

The solution is 2.

39. $5 - 2x = 3x - 7x + 25$

$5 - 2x = -4x + 25$

$4x - 2x = 25 - 5$

$2x = 20$

$\dfrac{2x}{2} = \dfrac{20}{2}$

$x = 10$

Check: $5 - 2x = 3x - 7x + 25$

$5 - 2 \cdot 10$	? $3 \cdot 10 - 7 \cdot 10 + 25$
$5 - 20$	$30 - 70 + 25$
-15	$-40 + 25$
	-15 TRUE

The solution is 10.

41. $4 + 3x - 6 = 3x + 2 - x$

$3x - 2 = 2x + 2$ Collecting like terms on each side

$3x - 2x = 2 + 2$

$x = 4$

Check: $4 + 3x - 6 = 3x + 2 - x$

$4 + 3 \cdot 4 - 6$	? $3 \cdot 4 + 2 - 4$
$4 + 12 - 6$	$12 + 2 - 4$
$16 - 6$	$14 - 4$
10	10 TRUE

The solution is 4.

43. $4y - 4 + y + 24 = 6y + 20 - 4y$

$5y + 20 = 2y + 20$

$5y - 2y = 20 - 20$

$3y = 0$

$y = 0$

Check: $4y - 4 + y + 24 = 6y + 20 - 4y$

$4 \cdot 0 - 4 + 0 + 24$	? $6 \cdot 0 + 20 - 4 \cdot 0$
$0 - 4 + 0 + 24$	$0 + 20 - 0$
20	20 TRUE

The solution is 0.

45. $\frac{7}{2}x + \frac{1}{2}x = 3x + \frac{3}{2} + \frac{5}{2}x$

The least common multiple of all the denominators is 2. We multiply by 2 on both sides.

$$\begin{aligned}
2\left(\frac{7}{2}x + \frac{1}{2}x\right) &= 2\left(3x + \frac{3}{2} + \frac{5}{2}x\right)\\
2 \cdot \frac{7}{2}x + 2 \cdot \frac{1}{2}x &= 2 \cdot 3x + 2 \cdot \frac{3}{2} + 2 \cdot \frac{5}{2}x\\
7x + x &= 6x + 3 + 5x\\
8x &= 11x + 3\\
8x - 11x &= 3\\
-3x &= 3\\
\frac{-3x}{-3} &= \frac{3}{-3}\\
x &= -1
\end{aligned}$$

Check: $\frac{7}{2}x + \frac{1}{2}x = 3x + \frac{3}{2} + \frac{5}{2}x$

$\frac{7}{2}(-1) + \frac{1}{2}(-1)$?	$3(-1) + \frac{3}{2} + \frac{5}{2}(-1)$
$-\frac{7}{2} - \frac{1}{2}$	$-3 + \frac{3}{2} - \frac{5}{2}$
-4	$-\frac{8}{2}$
	-4 TRUE

The solution is -1.

47. $\frac{2}{3} + \frac{1}{4}t = \frac{1}{3}$

The least common multiple of all the denominators is 12. We multiply by 12 on both sides.

$$\begin{aligned}
12\left(\frac{2}{3} + \frac{1}{4}t\right) &= 12 \cdot \frac{1}{3}\\
12 \cdot \frac{2}{3} + 12 \cdot \frac{1}{4}t &= 12 \cdot \frac{1}{3}\\
8 + 3t &= 4\\
3t &= 4 - 8\\
3t &= -4\\
\frac{3t}{3} &= \frac{-4}{3}\\
t &= -\frac{4}{3}
\end{aligned}$$

Check: $\frac{2}{3} + \frac{1}{4}t = \frac{1}{3}$

$\frac{2}{3} + \frac{1}{4}\left(-\frac{4}{3}\right)$?	$\frac{1}{3}$
$\frac{2}{3} - \frac{1}{3}$	
$\frac{1}{3}$	TRUE

The solution is $-\frac{4}{3}$.

49. $\frac{2}{3} + 3y = 5y - \frac{2}{15}$, LCM is 15

$$\begin{aligned}
15\left(\frac{2}{3} + 3y\right) &= 15\left(5y - \frac{2}{15}\right)\\
15 \cdot \frac{2}{3} + 15 \cdot 3y &= 15 \cdot 5y - 15 \cdot \frac{2}{15}\\
10 + 45y &= 75y - 2\\
10 + 2 &= 75y - 45y\\
12 &= 30y\\
\frac{12}{30} &= \frac{30y}{30}\\
\frac{2}{5} &= y
\end{aligned}$$

Check: $\frac{2}{3} + 3y = 5y - \frac{2}{15}$

$\frac{2}{3} + 3 \cdot \frac{2}{5}$?	$5 \cdot \frac{2}{5} - \frac{2}{15}$
$\frac{2}{3} + \frac{6}{5}$	$2 - \frac{2}{15}$
$\frac{10}{15} + \frac{18}{15}$	$\frac{30}{15} - \frac{2}{15}$
$\frac{28}{15}$	$\frac{28}{15}$ TRUE

The solution is $\frac{2}{5}$.

51. $\frac{5}{3} + \frac{2}{3}x = \frac{25}{12} + \frac{5}{4}x + \frac{3}{4}$, LCM is 12

$$\begin{aligned}
12\left(\frac{5}{3} + \frac{2}{3}x\right) &= 12\left(\frac{25}{12} + \frac{5}{4}x + \frac{3}{4}\right)\\
12 \cdot \frac{5}{3} + 12 \cdot \frac{2}{3}x &= 12 \cdot \frac{25}{12} + 12 \cdot \frac{5}{4}x + 12 \cdot \frac{3}{4}\\
20 + 8x &= 25 + 15x + 9\\
20 + 8x &= 15x + 34\\
20 - 34 &= 15x - 8x\\
-14 &= 7x\\
\frac{-14}{7} &= \frac{7x}{7}\\
-2 &= x
\end{aligned}$$

Check: $\frac{5}{3} + \frac{2}{3}x = \frac{25}{12} + \frac{5}{4}x + \frac{3}{4}$

$\frac{5}{3} + \frac{2}{3}(-2)$?	$\frac{25}{12} + \frac{5}{4}(-2) + \frac{3}{4}$
$\frac{5}{3} - \frac{4}{3}$	$\frac{25}{12} - \frac{5}{2} + \frac{3}{4}$
$\frac{1}{3}$	$\frac{25}{12} - \frac{30}{12} + \frac{9}{12}$
	$\frac{4}{12}$
	$\frac{1}{3}$ TRUE

The solution is -2.

53.

$$2.1x + 45.2 = 3.2 - 8.4x$$

Greatest number of decimal places is 1

$$10(2.1x + 45.2) = 10(3.2 - 8.4x)$$

Multiplying by 10 to clear decimals

$$10(2.1x) + 10(45.2) = 10(3.2) - 10(8.4x)$$
$$21x + 452 = 32 - 84x$$
$$21x + 84x = 32 - 452$$
$$105x = -420$$
$$\frac{105x}{105} = \frac{-420}{105}$$
$$x = -4$$

Check: $2.1x + 45.2 = 3.2 - 8.4x$

$$\begin{array}{r|l} 2.1(-4) + 45.2 \ ? & 3.2 - 8.4(-4) \\ -8.4 + 45.2 & 3.2 + 33.6 \\ 36.8 & 36.8 \qquad \text{TRUE} \end{array}$$

The solution is -4.

55.

$$1.03 - 0.62x = 0.71 - 0.22x$$

Greatest number of decimal places is 2

$$100(1.03 - 0.62x) = 100(0.71 - 0.22x)$$

Multiplying by 100 to clear decimals

$$100(1.03) - 100(0.62x) = 100(0.71) - 100(0.22x)$$
$$103 - 62x = 71 - 22x$$
$$32 = 40x$$
$$\frac{32}{40} = \frac{40x}{40}$$
$$\frac{4}{5} = x, \text{ or}$$
$$0.8 = x$$

Check: $1.03 - 0.62x = 0.71 - 0.22x$

$$\begin{array}{r|l} 1.03 - 0.62(0.8) \ ? & 0.71 - 0.22(0.8) \\ 1.03 - 0.496 & 0.71 - 0.176 \\ 0.534 & 0.534 \qquad \text{TRUE} \end{array}$$

The solution is $\frac{4}{5}$, or 0.8.

57.

$$\frac{2}{7}x - \frac{1}{2}x = \frac{3}{4}x + 1, \text{ LCM is } 28$$
$$28\left(\frac{2}{7}x - \frac{1}{2}x\right) = 28\left(\frac{3}{4}x + 1\right)$$
$$28 \cdot \frac{2}{7}x - 28 \cdot \frac{1}{2}x = 28 \cdot \frac{3}{4}x + 28 \cdot 1$$
$$8x - 14x = 21x + 28$$
$$-6x = 21x + 28$$
$$-6x - 21x = 28$$
$$-27x = 28$$
$$x = -\frac{28}{27}$$

Check: $\frac{2}{7}x - \frac{1}{2}x = \frac{3}{4}x + 1$

$$\begin{array}{r|l} \frac{2}{7}\left(-\frac{28}{27}\right) - \frac{1}{2}\left(-\frac{28}{27}\right) \ ? & \frac{3}{4}\left(-\frac{28}{27}\right) + 1 \\ -\frac{8}{27} + \frac{14}{27} & -\frac{21}{27} + 1 \\ \frac{6}{27} & \frac{6}{27} \qquad \text{TRUE} \end{array}$$

The solution is $-\frac{28}{27}$.

59. $3(2y - 3) = 27$

$$6y - 9 = 27 \qquad \text{Using a distributive law}$$
$$6y = 27 + 9 \qquad \text{Adding 9}$$
$$6y = 36$$
$$y = 6 \qquad \text{Dividing by 6}$$

Check: $3(2y - 3) = 27$

$$\begin{array}{r|l} 3(2 \cdot 6 - 3) \ ? & 27 \\ 3(12 - 3) & \\ 3 \cdot 9 & \\ 27 & \qquad \text{TRUE} \end{array}$$

The solution is 6.

61.

$$40 = 5(3x + 2)$$
$$40 = 15x + 10 \qquad \text{Using a distributive law}$$
$$40 - 10 = 15x$$
$$30 = 15x$$
$$2 = x$$

Check: $40 = 5(3x + 2)$

$$\begin{array}{r|l} 40 \ ? & 5(3 \cdot 2 + 2) \\ & 5(6 + 2) \\ & 5 \cdot 8 \\ & 40 \qquad \text{TRUE} \end{array}$$

The solution is 2.

63.

$$-23 + y = y + 25$$
$$-y - 23 + y = -y + y + 25$$
$$-23 = 25 \qquad \text{FALSE}$$

The equation has no solution.

65.

$$-23 + x = x - 23$$
$$-x - 23 + x = -x + x - 23$$
$$-23 = -23 \qquad \text{TRUE}$$

All real numbers are solutions.

67. $2(3 + 4m) - 9 = 45$

$$6 + 8m - 9 = 45$$
$$8m - 3 = 45 \qquad \text{Collecting like terms}$$
$$8m = 45 + 3$$
$$8m = 48$$
$$m = 6$$

Check: $2(3+4m)-9=45$

$$\begin{array}{r|l} 2(3+4\cdot 6)-9 & ?\ 45 \\ 2(3+24)-9 & \\ 2\cdot 27-9 & \\ 54-9 & \\ 45 & \quad \text{TRUE} \end{array}$$

The solution is 6.

69. $5r-(2r+8)=16$

$5r-2r-8=16$

$3r-8=16$ Collecting like terms

$3r=16+8$

$3r=24$

$r=8$

Check: $5r-(2r+8)=16$

$$\begin{array}{r|l} 5\cdot 8-(2\cdot 8+8) & ?\ 16 \\ 40-(16+8) & \\ 40-24 & \\ 16 & \quad \text{TRUE} \end{array}$$

The solution is 8.

71. $6-2(3x-1)=2$

$6-6x+2=2$

$8-6x=2$

$8-2=6x$

$6=6x$

$1=x$

Check: $6-2(3x-1)=2$

$$\begin{array}{r|l} 6-2(3\cdot 1-1) & ?\ 2 \\ 6-2(3-1) & \\ 6-2\cdot 2 & \\ 6-4 & \\ 2 & \quad \text{TRUE} \end{array}$$

The solution is 1.

73. $5(d+4)=7(d-2)$

$5d+20=7d-14$

$20+14=7d-5d$

$34=2d$

$17=d$

Check: $5(d+4)=7(d-2)$

$$\begin{array}{r|l} 5(17+4) & ?\ 7(17-2) \\ 5\cdot 21 & 7\cdot 15 \\ 105 & 105 \quad \text{TRUE} \end{array}$$

The solution is 17.

75. $8(2t+1)=4(7t+7)$

$16t+8=28t+28$

$16t-28t=28-8$

$-12t=20$

$t=-\frac{20}{12}$

$t=-\frac{5}{3}$

Check: $8(2t+1)=4(7t+7)$

$$\begin{array}{r|l} 8\left(2\left(-\frac{5}{3}\right)+1\right) & ?\ 4\left(7\left(-\frac{5}{3}\right)+7\right) \\ 8\left(-\frac{10}{3}+1\right) & 4\left(-\frac{35}{3}+7\right) \\ 8\left(-\frac{7}{3}\right) & 4\left(-\frac{14}{3}\right) \\ -\frac{56}{3} & -\frac{56}{3} \quad \text{TRUE} \end{array}$$

The solution is $-\frac{5}{3}$.

77. $5x+5-7x=15-12x+10x-10$

$-2x+5=5-2x$ Collecting like terms

$2x-2x+5=2x+5-2x$ Adding $2x$

$5=5$ TRUE

All real numbers are solutions.

79. $22x-5-15x+3=10x-4-3x+11$

$7x-2=7x+7$ Collecting like terms

$-7x+7x-2=-7x+7x+7$

$-2=7$ FALSE

The equation has no solution.

81. $3(r-6)+2=4(r+2)-21$

$3r-18+2=4r+8-21$

$3r-16=4r-13$

$13-16=4r-3r$

$-3=r$

Check: $3(r-6)+2=4(r+2)-21$

$$\begin{array}{r|l} 3(-3-6)+2 & ?\ 4(-3+2)-21 \\ 3(-9)+2 & 4(-1)-21 \\ -27+2 & -4-21 \\ -25 & -25 \quad \text{TRUE} \end{array}$$

The solution is -3.

83. $19-(2x+3)=2(x+3)+x$

$19-2x-3=2x+6+x$

$16-2x=3x+6$

$16-6=3x+2x$

$10=5x$

$2=x$

Check: $19-(2x+3)=2(x+3)+x$

$$\begin{array}{r|l} 19-(2\cdot 2+3) & ?\ 2(2+3)+2 \\ 19-(4+3) & 2\cdot 5+2 \\ 19-7 & 10+2 \\ 12 & 12 \quad \text{TRUE} \end{array}$$

The solution is 2.

85.
$$\begin{aligned}
2[4-2(3-x)]-1 &= 4[2(4x-3)+7]-25\\
2[4-6+2x]-1 &= 4[8x-6+7]-25\\
2[-2+2x]-1 &= 4[8x+1]-25\\
-4+4x-1 &= 32x+4-25\\
4x-5 &= 32x-21\\
-5+21 &= 32x-4x\\
16 &= 28x\\
\frac{16}{28} &= x\\
\frac{4}{7} &= x
\end{aligned}$$

The check is left to the student.

The solution is $\frac{4}{7}$.

87.
$$\begin{aligned}
11-4(x+1)-3 &= 11+2(4-2x)-16\\
11-4x-4-3 &= 11+8-4x-16\\
4-4x &= 3-4x\\
4x+4-4x &= 4x+3-4x\\
4 &= 3 \qquad \text{FALSE}
\end{aligned}$$

The equation has no solution.

89.
$$\begin{aligned}
22x-1-12x &= 5(2x-1)+4\\
22x-1-12x &= 10x-5+4\\
10x-1 &= 10x-1\\
-10x+10x-1 &= -10x+10x-1\\
-1 &= -1 \qquad \text{TRUE}
\end{aligned}$$

All real numbers are solutions.

91.
$$\begin{aligned}
0.7(3x+6) &= 1.1-(x+2)\\
2.1x+4.2 &= 1.1-x-2\\
10(2.1x+4.2) &= 10(1.1-x-2) \quad \text{Clearing decimals}\\
21x+42 &= 11-10x-20\\
21x+42 &= -10x-9\\
21x+10x &= -9-42\\
31x &= -51\\
x &= -\frac{51}{31}
\end{aligned}$$

The check is left to the student.

The solution is $-\frac{51}{31}$.

93. Do the long division. The answer is negative.

```
           6 . 5
  3. 4^ ) 2 2. 1^0
          2 0 4
            1 7 0
            1 7 0
                0
```

$-22.1 \div 3.4 = -6.5$

95. $7x-21-14y = 7\cdot x - 7\cdot 3 - 7\cdot 2y = 7(x-3-2y)$

97.
$$\begin{aligned}
-3+2(-5)^2(-3)-7 &= -3+2(25)(-3)-7\\
&= -3+50(-3)-7\\
&= -3-150-7\\
&= -153-7\\
&= -160
\end{aligned}$$

99. $23(2x-4)-15(10-3x) = 46x-92-150+45x = 91x-242$

101. First we multiply to remove the parentheses.
$$\begin{aligned}
\frac{2}{3}\left(\frac{7}{8}-4x\right)-\frac{5}{8} &= \frac{3}{8}\\
\frac{7}{12}-\frac{8}{3}x-\frac{5}{8} &= \frac{3}{8}, \text{ LCM is 24}\\
24\left(\frac{7}{12}-\frac{8}{3}x-\frac{5}{8}\right) &= 24\cdot\frac{3}{8}\\
24\cdot\frac{7}{12}-24\cdot\frac{8}{3}x-24\cdot\frac{5}{8} &= 9\\
14-64x-15 &= 9\\
-1-64x &= 9\\
-64x &= 10\\
x &= -\frac{10}{64}\\
x &= -\frac{5}{32}
\end{aligned}$$

The solution is $-\frac{5}{32}$.

103.
$$\begin{aligned}
\frac{4-3x}{7} &= \frac{2+5x}{49}-\frac{x}{14}\\
98\left(\frac{4-3x}{7}\right) &= 98\left(\frac{2+5x}{49}-\frac{x}{14}\right), \text{ LCM is 98}\\
\frac{98(4-3x)}{7} &= 98\left(\frac{2+5x}{49}\right)-98\cdot\frac{x}{14}\\
14(4-3x) &= 2(2+5x)-7x\\
56-42x &= 4+10x-7x\\
56-42x &= 4+3x\\
56-42x+42x &= 4+3x+42x\\
56 &= 4+45x\\
56-4 &= 4+45x-4\\
52 &= 45x\\
\frac{52}{45} &= x
\end{aligned}$$

The solution is $\frac{52}{45}$.

Exercise Set 2.4

RC1.
$$\begin{aligned}
4x &= 7y\\
\frac{4x}{4} &= \frac{7y}{4}\\
x &= \frac{7y}{4}, \text{ or } \frac{7}{4}y
\end{aligned}$$

The correct answer is (b).

RC3.
$$xz + 4y = w$$
$$xz = w - 4y$$
$$\frac{xz}{z} = \frac{w-4y}{z}$$
$$x = \frac{w-4y}{z}$$
The correct answer is (d).

1. We substitute 344 for r and 24 for f.
$$w = \frac{r}{f}$$
$$w = \frac{344}{24}$$
$$w = 14.\overline{3}, \text{ or } 14\frac{1}{3}$$
The wavelength is $14\frac{1}{3}$ meters per cycle.

3. We substitute 84 for c and 8 for w and calculate D.
$$D = \frac{c}{w} = \frac{84}{8} = 10.5$$
The calorie density is 10.5 calories per oz.

5. a) We substitute 75 for r and 4.5 for t.
$$d = rt$$
$$d = 75(4.5) = 337.5$$
The car will travel 337.5 mi.

b)
$$d = rt$$
$$\frac{d}{r} = \frac{rt}{r}$$
$$\frac{d}{r} = t$$

7. a) We substitute 21,345 for n and calculate f.
$$f = \frac{21,345}{15} = 1423$$
There are 1423 full-time equivalent students.

b)
$$f = \frac{n}{15}$$
$$15 \cdot f = 15 \cdot \frac{n}{15}$$
$$15f = n$$

9.
$$y = 5x$$
$$\frac{y}{5} = \frac{5x}{5}$$
$$\frac{y}{5} = x$$

11.
$$a = bc$$
$$\frac{a}{b} = \frac{bc}{b}$$
$$\frac{a}{b} = c$$

13.
$$n = m + 11$$
$$n - 11 = m + 11 - 11$$
$$n - 11 = m$$

15.
$$y = x - \frac{3}{5}$$
$$y + \frac{3}{5} = x - \frac{3}{5} + \frac{3}{5}$$
$$y + \frac{3}{5} = x$$

17.
$$y = 13 + x$$
$$y - 13 = 13 + x - 13$$
$$y - 13 = x$$

19.
$$y = x + b$$
$$y - b = x + b - b$$
$$y - b = x$$

21.
$$y = 5 - x$$
$$y - 5 = 5 - x - 5$$
$$y - 5 = -x$$
$$-1 \cdot (y-5) = -1 \cdot (-x)$$
$$-y + 5 = x, \text{ or}$$
$$5 - y = x$$

23.
$$y = a - x$$
$$y - a = a - x - a$$
$$y - a = -x$$
$$-1 \cdot (y-a) = -1 \cdot (-x)$$
$$-y + a = x, \text{ or}$$
$$a - y = x$$

25.
$$8y = 5x$$
$$\frac{8y}{8} = \frac{5x}{8}$$
$$y = \frac{5x}{8}, \text{ or } \frac{5}{8}x$$

27.
$$By = Ax$$
$$\frac{By}{A} = \frac{Ax}{A}$$
$$\frac{By}{A} = x$$

29.
$$W = mt + b$$
$$W - b = mt + b - b$$
$$W - b = mt$$
$$\frac{W-b}{m} = \frac{mt}{m}$$
$$\frac{W-b}{m} = t$$

31.
$$y = bx + c$$
$$y - c = bx + c - c$$
$$y - c = bx$$
$$\frac{y-c}{b} = \frac{bx}{b}$$
$$\frac{y-c}{b} = x$$

33. $A = bh$

$\dfrac{A}{b} = \dfrac{bh}{b}$ Dividing by b

$\dfrac{A}{b} = h$

35. $P = 2l + 2w$

$P - 2l = 2l + 2w - 2l$ Subtracting $2l$

$P - 2l = 2w$

$\dfrac{P-2l}{2} = \dfrac{2w}{2}$ Dividing by 2

$\dfrac{P-2l}{2} = w$, or

$\dfrac{1}{2}P - l = w$

37. $A = \dfrac{a+b}{2}$

$2A = a + b$ Multiplying by 2

$2A - b = a$ Subtracting b

39. $A = \dfrac{a+b+c}{3}$

$3A = a + b + c$ Multiplying by 3

$3A - a - c = b$ Subtracting a and c

41. $A = at + b$

$A - b = at$ Subtracting b

$\dfrac{A-b}{a} = t$ Dividing by a

43. $Ax + By = c$

$Ax = c - By$ Subtracting By

$\dfrac{Ax}{A} = \dfrac{c - By}{A}$ Dividing by A

$x = \dfrac{c - By}{A}$

45. $F = ma$

$\dfrac{F}{m} = \dfrac{ma}{m}$ Dividing by m

$\dfrac{F}{m} = a$

47. $E = mc^2$

$\dfrac{E}{m} = \dfrac{mc^2}{m}$ Dividing by m

$\dfrac{E}{m} = c^2$

49. $v = \dfrac{3k}{t}$

$tv = t \cdot \dfrac{3k}{t}$ Multiplying by t

$tv = 3k$

$\dfrac{tv}{v} = \dfrac{3k}{v}$ Dividing by v

$t = \dfrac{3k}{v}$

51. $\dfrac{3x - 2y}{y} = \dfrac{3 \cdot 6 - 2 \cdot 2}{2} = \dfrac{18-4}{2} = \dfrac{14}{2} = 7$

53. $-45.8 - (-32.6) = -45.8 + 32.6 = -13.2$

55. $87\dfrac{1}{2} - 123 = \dfrac{175}{2} - \dfrac{246}{2} = -\dfrac{71}{2} = -35\dfrac{1}{2}$

57. $0.082 + (-9.407) = -9.325$

59. $2y - 3 + y = 8 - 5y$

$3y - 3 = 8 - 5y$ Collecting like terms

$3y + 5y = 8 + 3$ Adding $5y$ and 3

$8y = 11$ Simplifying

$y = \dfrac{11}{8}$ Dividing by 8

The solution is $\dfrac{11}{8}$.

61. $2(5x + 6) = x - 15$

$10x + 12 = x - 15$ Removing parentheses

$10x - x = -15 - 12$ Subtracting x and 12

$9x = -27$ Simplifying

$x = -3$ Dividing by 9

The solution is -3.

63.

$$H = \frac{2}{a-b}$$
$$(a-b)H = (a-b)\left(\frac{2}{a-b}\right)$$
$$Ha - Hb = 2$$
$$Ha - Hb - Ha = 2 - Ha$$
$$-Hb = 2 - Ha$$
$$-1(-Hb) = -1(2 - Ha)$$
$$Hb = -2 + Ha$$
$$\frac{Hb}{H} = \frac{-2+Ha}{H}$$
$$b = \frac{-2+Ha}{H}, \text{ or } \frac{Ha-2}{H}, \text{ or } a - \frac{2}{H}$$

$$H = \frac{2}{a-b}$$
$$(a-b)H = (a-b) \cdot \frac{2}{a-b}$$
$$Ha - Hb = 2$$
$$Ha - Hb + Hb = 2 + Hb$$
$$Ha = 2 + Hb$$
$$\frac{Ha}{H} = \frac{2+Hb}{H}$$
$$a = \frac{2+Hb}{H}, \text{ or } \frac{2}{H} + b$$

65. $A = lw$

When l and w both double, we have

$2l \cdot 2w = 4lw = 4A$,

so A quadruples.

67. $A = \frac{1}{2}bh$

When b increases by 4 units we have

$$\frac{1}{2}(b+4)h = \frac{1}{2}bh + 2h = A + 2h,$$

so A increases by $2h$ units.

Chapter 2 Mid-Chapter Review

1. The solution of $3 - x = 4x$ is $\frac{3}{5}$; the solution of $5x = -3$ is $-\frac{3}{5}$. The equations have different solutions, so they are not equivalent. The given statement is false.

2. True; see page 127 in the text.

3. True; see page 132 in the text.

4. False; see page 144 in the text.

5.
$$\begin{aligned} x+5 &= -3 \\ x+5-5 &= -3-5 \\ x+0 &= -8 \\ x &= -8 \end{aligned}$$

6.
$$\begin{aligned} -6x &= 42 \\ \frac{-6x}{-6} &= \frac{42}{-6} \\ 1 \cdot x &= -7 \\ x &= -7 \end{aligned}$$

7.
$$\begin{aligned} 5y+z &= t \\ 5y+z-z &= t-z \\ 5y &= t-z \\ \frac{5y}{5} &= \frac{t-z}{5} \\ y &= \frac{t-z}{5} \end{aligned}$$

8.
$$\begin{aligned} x+5 &= 11 \\ x+5-5 &= 11-5 \\ x &= 6 \end{aligned}$$

The solution is 6.

9.
$$\begin{aligned} x+9 &= -3 \\ x+9-9 &= -3-9 \\ x &= -12 \end{aligned}$$

The solution is -12.

10.
$$\begin{aligned} 8 &= t+1 \\ 8-1 &= t+1-1 \\ 7 &= t \end{aligned}$$

The solution is 7.

11.
$$\begin{aligned} -7 &= y+3 \\ -7-3 &= y+3-3 \\ -10 &= y \end{aligned}$$

The solution is -10.

12.
$$\begin{aligned} x-6 &= 14 \\ x-6+6 &= 14+6 \\ x &= 20 \end{aligned}$$

The solution is 20.

13.
$$\begin{aligned} y-7 &= -2 \\ y-7+7 &= -2+7 \\ y &= 5 \end{aligned}$$

The solution is 5.

14.
$$\begin{aligned} -\frac{3}{2}+z &= -\frac{3}{4} \\ -\frac{3}{2}+z+\frac{3}{2} &= -\frac{3}{4}+\frac{3}{2} \\ z &= -\frac{3}{4}+\frac{6}{4} \\ z &= \frac{3}{4} \end{aligned}$$

The solution is $\frac{3}{4}$.

15.
$$\begin{aligned} -3.3 &= -1.9+t \\ -3.3+1.9 &= -1.9+t+1.9 \\ -1.4 &= t \end{aligned}$$

The solution is -1.4.

16.
$$\begin{aligned} 7x &= 42 \\ \frac{7x}{7} &= \frac{42}{7} \\ x &= 6 \end{aligned}$$

The solution is 6.

17.
$$\begin{aligned} 17 &= -t \\ 17 &= -1 \cdot t \\ \frac{17}{-1} &= \frac{-1 \cdot t}{-1} \\ -17 &= t \end{aligned}$$

The solution is -17.

18.
$$\begin{aligned} 6x &= -54 \\ \frac{6x}{6} &= \frac{-54}{6} \\ x &= -9 \end{aligned}$$

The solution is -9.

19.
$$\begin{aligned} -5y &= -85 \\ \frac{-5y}{-5} &= \frac{-85}{-5} \\ y &= 17 \end{aligned}$$

The solution is 17.

20.
$$\begin{aligned} \frac{x}{7} &= 3 \\ \frac{1}{7} \cdot x &= 3 \\ 7 \cdot \frac{1}{7}x &= 7 \cdot 3 \\ x &= 21 \end{aligned}$$

The solution is 21.

21.
$$\begin{aligned} \frac{2}{3}x &= 12 \\ \frac{3}{2}\cdot\frac{2}{3}x &= \frac{3}{2}\cdot 12 \\ x &= \frac{3\cdot\cancel{2}\cdot 6}{\cancel{2}\cdot 1} \\ x &= 18 \end{aligned}$$
The solution is 18.

22.
$$\begin{aligned} -\frac{t}{5} &= 3 \\ -\frac{1}{5}\cdot t &= 3 \\ -5\left(-\frac{1}{5}\cdot t\right) &= -5\cdot 3 \\ t &= -15 \end{aligned}$$
The solution is −15.

23.
$$\begin{aligned} \frac{3}{4}x &= -\frac{9}{8} \\ \frac{4}{3}\cdot\frac{3}{4}x &= \frac{4}{3}\left(-\frac{9}{8}\right) \\ x &= -\frac{\cancel{4}\cdot\cancel{3}\cdot 3}{\cancel{3}\cdot 2\cdot\cancel{4}} \\ x &= -\frac{3}{2} \end{aligned}$$
The solution is $-\frac{3}{2}$.

24.
$$\begin{aligned} 3x+2 &= 5 \\ 3x+2-2 &= 5-2 \\ 3x &= 3 \\ \frac{3x}{3} &= \frac{3}{3} \\ x &= 1 \end{aligned}$$
The solution is 1.

25.
$$\begin{aligned} 5x+4 &= -11 \\ 5x+4-4 &= -11-4 \\ 5x &= -15 \\ \frac{5x}{5} &= \frac{-15}{5} \\ x &= -3 \end{aligned}$$
The solution is −3.

26.
$$\begin{aligned} 6x-7 &= 2 \\ 6x-7+7 &= 2+7 \\ 6x &= 9 \\ \frac{6x}{6} &= \frac{9}{6} \\ x &= \frac{9}{6} = \frac{3\cdot\cancel{3}}{2\cdot\cancel{3}} \\ x &= \frac{3}{2} \end{aligned}$$
The solution is $\frac{3}{2}$.

27.
$$\begin{aligned} -4x-9 &= -5 \\ -4x-9+9 &= -5+9 \\ -4x &= 4 \\ \frac{-4x}{-4} &= \frac{4}{-4} \\ x &= -1 \end{aligned}$$
The solution is −1.

28.
$$\begin{aligned} 6x+5x &= 33 \\ 11x &= 33 \\ \frac{11x}{11} &= \frac{33}{11} \\ x &= 3 \end{aligned}$$
The solution is 3.

29.
$$\begin{aligned} -3y-4y &= 49 \\ -7y &= 49 \\ \frac{-7y}{-7} &= \frac{49}{-7} \\ y &= -7 \end{aligned}$$
The solution is −7.

30.
$$\begin{aligned} 3x-4 &= 12-x \\ 3x-4+x &= 12-x+x \\ 4x-4 &= 12 \\ 4x-4+4 &= 12+4 \\ 4x &= 16 \\ \frac{4x}{4} &= \frac{16}{4} \\ x &= 4 \end{aligned}$$
The solution is 4.

31.
$$\begin{aligned} 5-6x &= 9-8x \\ 5-6x+8x &= 9-8x+8x \\ 5+2x &= 9 \\ 5+2x-5 &= 9-5 \\ 2x &= 4 \\ \frac{2x}{2} &= \frac{4}{2} \\ x &= 2 \end{aligned}$$
The solution is 2.

32.
$$4y-\frac{3}{2}=\frac{3}{4}+2y$$
$$4\left(4y-\frac{3}{2}\right)=4\left(\frac{3}{4}+2y\right)\quad\text{Clearing fractions}$$
$$4\cdot 4y-4\cdot\frac{3}{2}=4\cdot\frac{3}{4}+4\cdot 2y$$
$$16y-6=3+8y$$
$$16y-6-8y=3+8y-8y$$
$$8y-6=3$$
$$8y-6+6=3+6$$
$$8y=9$$
$$\frac{8y}{8}=\frac{9}{8}$$
The solution is $\frac{9}{8}$.

33.
$$\frac{4}{5}+\frac{1}{6}t=\frac{1}{10}$$
$$30\left(\frac{4}{5}+\frac{1}{6}t\right)=30\cdot\frac{1}{10}\quad\text{Clearing fractions}$$
$$30\cdot\frac{4}{5}+30\cdot\frac{1}{6}t=\frac{30}{10}$$
$$24+5t=3$$
$$24+5t-24=3-24$$
$$5t=-21$$
$$\frac{5t}{5}=\frac{-21}{5}$$
$$t=-\frac{21}{5}$$
The solution is $-\frac{21}{5}$.

34.
$$0.21n-1.05=2.1-0.14n$$
$$100(0.21n-1.05)=100(2.1-0.14n)$$
Clearing decimals
$$100(0.21n)-100(1.05)=100(2.1)-100(0.14n)$$
$$21n-105=210-14n$$
$$21n-105+14n=210-14n+14n$$
$$35n-105=210$$
$$35n-105+105=210+105$$
$$35n=315$$
$$\frac{35n}{35}=\frac{315}{35}$$
$$n=9$$
The solution is 9.

35.
$$5(3y-1)=-35$$
$$15y-5=-35$$
$$15y-5+5=-35+5$$
$$15y=-30$$
$$\frac{15y}{15}=\frac{-30}{15}$$
$$y=-2$$
The solution is -2.

36.
$$7-2(5x+3)=1$$
$$7-10x-6=1$$
$$1-10x=1$$
$$1-10x-1=1-1$$
$$-10x=0$$
$$\frac{-10x}{-10}=\frac{0}{-10}$$
$$x=0$$
The solution is 0.

37.
$$-8+t=t-8$$
$$-8+t-t=t-8-t$$
$$-8=-8$$
We have an equation that is true for all real numbers. Thus, all real numbers are solutions.

38.
$$z+12=-12+z$$
$$z+12-z=-12+z-z$$
$$12=-12$$
We have a false equation. There are no solutions.

39.
$$4(3x+2)=5(2x-1)$$
$$12x+8=10x-5$$
$$12x+8-10x=10x-5-10x$$
$$2x+8=-5$$
$$2x+8-8=-5-8$$
$$2x=-13$$
$$\frac{2x}{2}=\frac{-13}{2}$$
$$x=-\frac{13}{2}$$
The solution is $-\frac{13}{2}$.

40.
$$8x-6-2x=3(2x-4)+6$$
$$6x-6=6x-12+6$$
$$6x-6=6x-6$$
$$6x-6-6x=6x-6-6x$$
$$-6=-6$$
We have an equation that is true for all real numbers. Thus, all real numbers are solutions.

41.
$$A=4b$$
$$\frac{A}{4}=\frac{4b}{4}$$
$$\frac{A}{4}=b$$

42.
$$y=x-1.5$$
$$y+1.5=x-1.5+1.5$$
$$y+1.5=x$$

43.
$$\begin{aligned} n &= s - m \\ n - s &= s - m - s \\ n - s &= -m \\ -1(n - s) &= -1(-m) \\ -n + s &= m, \text{ or} \\ s - n &= m \end{aligned}$$

44.
$$\begin{aligned} 4t &= 9w \\ \frac{4t}{4} &= \frac{9w}{4} \\ t &= \frac{9w}{4} \end{aligned}$$

45.
$$\begin{aligned} B &= at - c \\ B + c &= at - c + c \\ B + c &= at \\ \frac{B + c}{a} &= \frac{at}{a} \\ \frac{B + c}{a} &= t \end{aligned}$$

46.
$$\begin{aligned} M &= \frac{x + y + z}{2} \\ 2 \cdot M &= 2\left(\frac{x + y + z}{2}\right) \\ 2M &= x + y + z \\ 2M - x - z &= x + y + z - x - z \\ 2M - x - z &= y \end{aligned}$$

47. Equivalent expressions have the same value for all possible replacements for the variable(s). Equivalent equations have the same solution(s).

48. The equations are not equivalent because they do not have the same solutions. Although 5 is a solution of both equations, -5 is a solution of $x^2 = 25$ but not of $x = 5$.

49. For an equation $x + a = b$, add the opposite of a (or subtract a) on both sides of the equation.

50. It appears that the student added $\frac{1}{3}$ on the right side of the equation rather than subtracting $\frac{1}{3}$.

51. For an equation $ax = b$, multiply by $1/a$ (or divide by a) on both sides of the equation.

52. Answers may vary. A walker who knows how far and how long she walks each day wants to know her average speed each day.

Exercise Set 2.5

RC1. $\underbrace{13}$ is $\underbrace{82\%}$ of $\underbrace{\text{what number?}}$

$$13 = 82\% \cdot b$$

The correct answer is (d).

RC3. $\underbrace{82}$ is $\underbrace{\text{what percent}}$ of $\underbrace{13?}$

$$82 = p \cdot 13$$

The correct answer is (e).

RC5. $\underbrace{13}$ is $\underbrace{\text{what percent}}$ of $\underbrace{82?}$

$$13 = p \cdot 82$$

The correct answer is (f).

1. ***Translate.***

$\underbrace{\text{What percent}}$ of $\underbrace{180}$ is $\underbrace{36?}$

$$p \cdot 180 = 36$$

Solve. We divide by 36 on both sides and convert the answer to percent notation.

$$\begin{aligned} p \cdot 180 &= 36 \\ \frac{p \cdot 180}{180} &= \frac{36}{180} \\ p &= 0.2 \\ p &= 20\% \end{aligned}$$

Thus, 36 is 20% of 180. The answer is 20%.

3. ***Translate.***

45 is 30% of what?

$$45 = 30\% \cdot b$$

Solve. We solve the equation.

$$\begin{aligned} 45 &= 30\% \cdot b \\ 45 &= 0.3b \quad \text{Converting to decimal notation} \\ \frac{45}{0.3} &= \frac{0.3b}{0.3} \\ 150 &= b \end{aligned}$$

Thus, 45 is 30% of 150. The answer is 150.

5. ***Translate.***

What is 65% of 840?

$$a = 65\% \cdot 840$$

Solve. We convert 65% to decimal notation and multiply.

$$\begin{aligned} a &= 65\% \cdot 840 \\ a &= 0.65 \times 840 \\ a &= 546 \end{aligned}$$

Thus, 546 is 65% of 840. The answer is 546.

7. ***Translate.***

30 is $\underbrace{\text{what percent}}$ of 125?

$$30 = p \cdot 125$$

Solve. We solve the equation.

$$30 = p \cdot 125$$
$$\frac{30}{125} = \frac{p \cdot 125}{125}$$
$$0.24 = p$$
$$24\% = p$$

Thus, 30 is 24% of 125. The answer is 24%.

9. ***Translate.***

$$\underbrace{12\%}\ \text{of}\ \underbrace{\text{what number}}\ \text{is}\ 0.3?$$
$$12\% \cdot b = 0.3$$

Solve. We solve the equation.

$$12\% \cdot b = 0.3$$
$$0.12b = 0.3 \quad \text{Converting to decimal notation}$$
$$\frac{0.12b}{0.12} = \frac{0.3}{0.12}$$
$$b = 2.5$$

Thus, 12% of 2.5 is 0.3. The answer is 2.5.

11. ***Translate.***

$$\underbrace{2}\ \text{is}\ \underbrace{\text{what percent}}\ \text{of}\ \underbrace{40?}$$
$$2 = p \cdot 40$$

Solve. We divide by 40 on both sides and convert the answer to percent notation.

$$2 = p \cdot 40$$
$$\frac{2}{40} = \frac{p \cdot 40}{40}$$
$$0.05 = p$$
$$5\% = p$$

Thus, 2 is 5% of 40. The answer is 5%.

13. ***Translate.***

$$\underbrace{\text{What percent}}\ \text{of}\ 68\ \text{is}\ 17?$$
$$p \cdot 68 = 17$$

Solve. We divide by 68 on both sides and then convert to percent notation.

$$p \cdot 68 = 17$$
$$p = \frac{17}{68}$$
$$p = 0.25 = 25\%$$

The answer is 25%.

15. ***Translate.***

$$\text{What is } 35\% \text{ of } 240?$$
$$a = 35\% \cdot 240$$

Solve. We convert 35% to decimal notation and multiply.

$$a = 35\% \cdot 240$$
$$a = 0.35 \cdot 240$$
$$a = 84$$

The answer is 84.

17. ***Translate.***

$$\underbrace{\text{What percent}}\ \text{of}\ 125\ \text{is}\ 30?$$
$$p \cdot 125 = 30$$

Solve. We divide by 125 on both sides and then convert to percent notation.

$$p \cdot 125 = 30$$
$$p = \frac{30}{125}$$
$$p = 0.24 = 24\%$$

The answer is 24%.

19. ***Translate.***

$$\underbrace{\text{What percent}}\ \text{of}\ 300\ \text{is}\ 48?$$
$$p \cdot 300 = 48$$

Solve. We divide by 300 on both sides and then convert to percent notation.

$$p \cdot 300 = 48$$
$$p = \frac{48}{300}$$
$$p = 0.16 = 16\%$$

The answer is 16%.

21. ***Translate.***

$$14\ \text{is}\ 30\%\ \text{of}\ \underbrace{\text{what number}?}$$
$$14 = 30\% \cdot b$$

Solve. We solve the equation.

$$14 = 0.3b \quad (30\% = 0.3)$$
$$\frac{14}{0.3} = b$$
$$46.\overline{6} = b$$

The answer is $46.\overline{6}$, or $46\frac{2}{3}$, or $\frac{140}{3}$.

23. ***Translate.***

$$\text{What is } 2\% \text{ of } 40?$$
$$a = 2\% \cdot 40$$

Solve. We convert 2% to decimal notation and multiply.

$$a = 2\% \cdot 40$$
$$a = 0.02 \cdot 40$$
$$a = 0.8$$

The answer is 0.8.

25. ***Translate.***

$$0.8\ \text{is}\ 16\%\ \text{of}\ \underbrace{\text{what number}?}$$
$$0.8 = 16\% \cdot b$$

Solve. We solve the equation.

$$0.8 = 0.16b \quad (16\% = 0.16)$$
$$\frac{0.8}{0.16} = b$$
$$5 = b$$

The answer is 5.

27. ***Translate.***

$$\underbrace{54}\ \text{is}\ 135\%\ \text{of}\ \underbrace{\text{what number}}?$$

$$54 = 135\% \cdot b$$

Solve. We solve the equation.

$$54 = 1.35b \quad (135\% = 1.35)$$

$$\frac{54}{1.35} = b$$

$$40 = b$$

The answer is 40.

29. First we translate.

What is 9.1% of 8909?

$$a = 9.1\% \cdot 8909$$

Solve. We convert 9.1% to decimal notation and multiply.

$$a = 9.1\% \cdot 8909 = 0.091 \cdot 8909 \approx 811$$

The projected population of South America is about 811 million.

31. First we translate.

What is 59.2% of 8909?

$$a = 59.2\% \cdot 8909$$

Solve. We convert 59.2% to decimal notation and multiply.

$$a = 59.2\% \cdot 8909 = 0.592 \cdot 8909 \approx 5274$$

The projected population of Asia is about 5274 million, or 5.274 billion.

33. First we translate.

What is 19.8% of 8909?

$$a = 19.8\% \cdot 8909$$

Solve. We convert 19.8% to decimal notation and multiply.

$$a = 19.8\% \cdot 8909 = 0.198 \cdot 8909 \approx 1764$$

The projected population of Africa is about 1764 million, or 1.764 billion.

35. First we reword and translate.

What is 41% of 2360

$$a = 41\% \cdot 2360$$

Solve. We convert 41% to decimal notation and multiply.

$$a = 41\% \cdot 2360 = 0.41 \cdot 2360 \approx 968$$

The net revenue from adult eBook sales was about \$968 million in 2011.

37. First we reword and translate.

What is 3.4% of \$6500?

$$a = 3.4\% \cdot 6500$$

Solve. We convert 3.4% to decimal notation and multiply.

$$a = 3.4\% \cdot 6500 = 0.034 \cdot 6500 = 221$$

Sarah will pay \$221 in interest.

39. First we reword and translate.

49,854,000 is $\underbrace{\text{what percent}}$ of 237,400,000?

$$49{,}854{,}000 = p \cdot 237{,}400{,}000$$

Solve. We divide by 237,400,000 on both sides and convert to percent notation.

$$49{,}854{,}000 = p \cdot 237{,}400{,}000$$

$$\frac{49{,}854{,}000}{237{,}400{,}000} = p$$

$$0.21 = p$$

$$21\% = p$$

21% of adults age 18 and older in the United States have at least one tattoo.

41. a) First we reword and translate.

$\underbrace{\text{What percent}}$ of \$12 is \$1.50?

$$p \cdot 12 = 1.50$$

Solve. We divide by 12 on both sides and convert to percent notation.

$$p \cdot 12 = 1.50$$

$$\frac{p \cdot 12}{12} = \frac{1.50}{12}$$

$$p = 0.125$$

$$p = 12.5\%$$

The tip was 12.5% of the cost of the meal.

b) We add to find the total cost of the meal, including tip:

$$\$12 + \$1.50 = \$13.50$$

43. a) First we reword and translate.

15% of what is \$4.65?

$$15\% \cdot b = 4.65$$

Solve. We solve the equation.

$$15\% \cdot b = 4.65$$

$$0.15 \cdot b = 4.65$$

$$\frac{0.15 \cdot b}{0.15} = \frac{4.65}{0.15}$$

$$b = 31$$

The cost of the meal before the tip was \$31.

b) We add to find the total cost of the meal, including tip:

$$\$31 + \$4.65 = \$35.65$$

45. First we reword and translate.

15.1% of what is 12,959?

$$15.1\% \cdot b = 12{,}959$$

Solve. We solve the equation.

$$15.1\% \cdot b = 12,959$$
$$0.151 \cdot b = 12,959$$
$$\frac{0.151 \cdot b}{0.151} = \frac{12,959}{0.151}$$
$$b \approx 85,821$$

The total acreage of Portland is about 85,821 acres.

47. We subtract to find the increase, in millions.

$3.1 - 1.7 = 1.4$

The increase was 1.4 million.

Now we find the percent of increase.

$$\underbrace{1.4}_{1.4}\ \underbrace{\text{is}}_{=}\ \underbrace{\text{what percent}}_{p}\ \underbrace{\text{of}}_{\cdot}\ \underbrace{1.7?}_{1.7}$$

We divide by 1.7 on both sides and then convert to percent notation.

$$1.4 = p \cdot 1.7$$
$$\frac{1.4}{1.7} = p$$
$$0.824 \approx p$$
$$82.4\% \approx p$$

The percent of increase is about 82.4%.

49. We subtract to find the decrease, in billions of dollars.

$48.6 - 22.8 = 25.8$

Now we find the percent decrease.

$$\underbrace{25.8}_{25.8}\ \underbrace{\text{is}}_{=}\ \underbrace{\text{what percent}}_{p}\ \underbrace{\text{of}}_{\cdot}\ \underbrace{48.6?}_{48.6}$$

We divide by 48.6 on both sides and convert to percent notation.

$$25.8 = p \cdot 48.6$$
$$\frac{25.8}{48.6} = p$$
$$0.531 \approx p$$
$$53.1\% \approx p$$

The percent decrease was about 53.1%.

51. First we subtract to find the increase.

$29,396 - 19,162 = 10,234$

Now we find the percent increase.

$$\underbrace{10,234}_{10,234}\ \underbrace{\text{is}}_{=}\ \underbrace{\text{what percent}}_{p}\ \underbrace{\text{of}}_{\cdot}\ \underbrace{19,162?}_{19,162}$$

We divide by 19,162 on both sides and convert to percent notation.

$$10,234 = p \cdot 19,162$$
$$\frac{10,234}{19,162} = p$$
$$0.534 \approx p$$
$$53.4\% \approx p$$

The percent increase was about 53.4%.

53. First we subtract to find the decrease, in millions.

$2.6 - 2.1 = 0.5$

Now we find the percent decrease.

$$\underbrace{0.5}_{0.5}\ \underbrace{\text{is}}_{=}\ \underbrace{\text{what percent}}_{p}\ \underbrace{\text{of}}_{\cdot}\ \underbrace{2.6?}_{2.6}$$

We divide by 2.6 on both sides and convert to percent notation.

$$0.5 = p \cdot 2.6$$
$$\frac{0.5}{2.6} = p$$
$$0.192 \approx p$$
$$19.2\% \approx p$$

The percent decrease is about 19.2%.

55. $3(4 + q) = 3 \cdot 4 + 3 \cdot q = 12 + 3q$

57. $\dfrac{75yw}{40y} = \dfrac{5 \cdot 15 \cdot y \cdot w}{5 \cdot 8 \cdot y} = \dfrac{5 \cdot y}{5 \cdot y} \cdot \dfrac{15 \cdot w}{8} = \dfrac{15w}{8}$

59.
$$\begin{aligned} -2[3 - 5(7 - 2)] &= -2[3 - 5(5)] \\ &= -2[3 - 25] \\ &= -2[-22] \\ &= 44 \end{aligned}$$

61. Since 6 ft = 6 × 1 ft = 6 × 12 in. = 72 in., we can express 6 ft 4 in. as 72 in. + 4 in., or 76 in.

Translate. We reword the problem.

$$\underbrace{96.1\%}_{96.1\%}\ \underbrace{\text{of}}_{\cdot}\ \underbrace{\text{what}}_{b}\ \underbrace{\text{is}}_{=}\ \underbrace{76\text{ in.}?}_{76}$$

Solve. We solve the equation.

$$96.1\% \cdot b = 76$$
$$0.961 \cdot b = 76$$
$$\frac{0.961 \cdot b}{0.961} = \frac{76}{0.961}$$
$$b \approx 79$$

Note that 79 in. = 72 in. + 7 in. = 6 ft 7 in.

Jaraan's final adult height will be about 6 ft 7 in.

Exercise Set 2.6

RC1. Familiarize yourself with the problem situation.

RC3. Solve the equation.

RC5. State the answer to the problem clearly.

1. *Familiarize*. Let m = the number of Medals of Honor awarded for valor in the Civil War.

Translate.

$$\underbrace{\text{Medals for World War II}}_{464}\ \underbrace{\text{plus}}_{+}\ \underbrace{\text{Excess Civil War medals}}_{1058}\ \underbrace{\text{is}}_{=}\ \underbrace{\text{Medals for Civil War}}_{m}$$

Solve. We carry out the addition.

$$464 + 1058 = m$$
$$1522 = m$$

Check. $1522 - 1058 = 464$, the number of medals for World War II, so the answer checks.

State. 1522 Medals of Honor were awarded for valor in the Civil War.

3. ***Familiarize***. Using the labels on the drawing in the text, we let x = the length of the shorter piece, in inches, and $3x$ = the length of the longer piece, in inches.

Translate. We reword the problem.

$$\underbrace{\text{The length of the shorter piece}}_{x} \quad \underbrace{\text{plus}}_{+} \quad \underbrace{\text{the length of the longer piece}}_{3x} \quad \underbrace{\text{is}}_{=} \quad \underbrace{240 \text{ ft.}}_{240}$$

Solve. We solve the equation.

$$x + 3x = 240$$
$$4x = 240 \quad \text{Collecting like terms}$$
$$\frac{4x}{4} = \frac{240}{4}$$
$$x = 60$$

If x is 60, then $3x = 3 \cdot 60$, or 180.

Check. 180 is three times 60, and $60 + 180 = 240$. The answer checks.

State. The lengths of the pieces are 60 in. and 180 in.

5. ***Familiarize***. Let s = the ridership on the public transit system in San Diego in the first quarter of 2012, in millions.

Translate.

$$\underbrace{\text{Ridership in San Diego}}_{s} \quad \underbrace{\text{plus}}_{+} \quad \underbrace{\text{Excess Boston ridership}}_{77.4} \quad \underbrace{\text{is}}_{=} \quad \underbrace{\text{Ridership in Boston}}_{99.2}$$

Solve. We subtract 77.4 on both sides.

$$s + 77.4 = 99.2$$
$$s + 77.4 - 77.4 = 99.2 - 77.4$$
$$s = 21.8$$

Check. $21.8 + 77.4 = 99.2$, the ridership in Boston, so the answer checks.

State. The ridership on the public transit system in San Diego in the first quarter of 2012 was 21.8 million.

7. ***Familiarize***. Let d = the runner's distance from the finish, in miles. Then $2d$ = the distance from the start.

Translate.

$$\underbrace{\text{Distance from start}}_{2d} \quad \underbrace{\text{plus}}_{+} \quad \underbrace{\text{Distance from finish}}_{d} \quad \underbrace{\text{is}}_{=} \quad \underbrace{\text{Total distance}}_{13.1}$$

Solve.

$$2d + d = 13.1$$
$$3d = 13.1$$
$$d = \frac{13.1}{3}$$
$$d \approx 4.37$$

Check. $2(4.37) + 4.37 = 8.74 + 4.37 = 13.11 \approx 13.1$, the length of the course, so the answer checks.

State. The runner was about 4.37 mi from the finish.

9. ***Familiarize.*** Let x = the smaller number and $x + 1$ = the larger number.

Translate. We reword the problem.

$$\underbrace{\text{First number}}_{x} \; + \; \underbrace{\text{second number}}_{(x+1)} \; \underbrace{\text{is}}_{=} \; \underbrace{2409}_{2409}$$

Solve. We solve the equation.

$$x + (x + 1) = 2409$$
$$2x + 1 = 2409 \quad \text{Collecting like terms}$$
$$2x + 1 - 1 = 2409 - 1 \quad \text{Subtracting 1}$$
$$2x = 2408$$
$$\frac{2x}{2} = \frac{2408}{2} \quad \text{Dividing by 2}$$
$$x = 1204$$

If x is 1204, then $x + 1$ is 1205.

Check. 1204 and 1205 are consecutive integers, and their sum is 2409. The answer checks.

State. The apartment numbers are 1204 and 1205.

11. ***Familiarize.*** Let a = the first number. Then $a + 1$ = the second number, and $a + 2$ = the third number.

Translate. We reword the problem.

$$\underbrace{\text{First number}}_{a} \; + \; \underbrace{\text{second number}}_{(a+1)} \; + \; \underbrace{\text{third number}}_{(a+2)} \; \underbrace{\text{is}}_{=} \; \underbrace{126}_{126}$$

Solve. We solve the equation.

$$a + (a + 1) + (a + 2) = 126$$
$$3a + 3 = 126 \quad \text{Collecting like terms}$$
$$3a + 3 - 3 = 126 - 3$$
$$3a = 123$$
$$\frac{3a}{3} = \frac{123}{3}$$
$$a = 41$$

If a is 41, then $a + 1$ is 42 and $a + 2$ is 43.

Check. 41, 42, and 43 are consecutive integers, and their sum is 126. The answer checks.

State. The numbers are 41, 42, and 43.

13. ***Familiarize.*** Let x = the first odd integer. Then $x + 2$ = the next odd integer and $(x + 2) + 2$, or $x + 4$ = the third odd integer.

Translate. We reword the problem.

$$\underbrace{\text{First odd integer}} + \underbrace{\text{second odd integer}} + \underbrace{\text{third odd integer}} \text{ is } \underbrace{189}$$
$$x + (x+2) + (x+4) = 189$$

Solve. We solve the equation.

$x+(x+2)+(x+4)=189$

$3x+6=189$ Collecting like terms

$3x+6-6=189-6$

$3x=183$

$\dfrac{3x}{3}=\dfrac{183}{3}$

$x=61$

If x is 61, then $x+2$ is 63 and $x+4$ is 65.

Check. 61, 63, and 65 are consecutive odd integers, and their sum is 189. The answer checks.

State. The integers are 61, 63, and 65.

15. ***Familiarize.*** Let h = the height of the photo, in inches. Then $3h+2$ = the width.

Translate. We used the formula $P=2l+2w$.

$292=2\cdot h+2(3h+2)$

Solve.

$292=2\cdot h+2(3h+2)$

$292=2h+6h+4$

$292=8h+4$

$288=8h$

$36=h$

If $h=36$, then $3h+2=3\cdot 36+2=108+2=110$.

Check. $2\cdot 36+2\cdot 110=72+220=292$, the perimeter, so the answer checks.

State. The dimensions of the photo are 36 in. × 110 in.

17. ***Familiarize.*** Let p = the regular price of a 3-lb bag of coffee beans.

Translate.

$$\underbrace{\text{Regular price}} \text{ minus } 30\% \text{ of } \underbrace{\text{regular price}} \text{ is } \underbrace{\text{sale price}}$$
$$p - 30\% \cdot p = 44.10$$

Solve.

$p-30\%\cdot p=44.10$

$p-0.3p=44.10$

$0.7p=44.10$

$p=\dfrac{44.10}{0.7}$

$p=63$

Check. 30% of \$63 is \$18.90 and \$63 − \$18.90 = \$44.10, the sale price, so the answer checks.

State. The regular price of a bag of coffee beans is \$63.

19. ***Familiarize.*** Let w = the price of the wallet itself.

Translate.

$$\underbrace{\text{Price of wallet}} \text{ plus } 7\% \text{ of } \underbrace{\text{price of wallet}} \text{ is } \underbrace{\text{price with tax}}$$
$$w + 7\% \cdot w = 26.70$$

Solve.

$w+7\%\cdot w=26.70$

$w+0.07w=26.70$

$1.07w=26.70$

$w=\dfrac{26.70}{1.07}$

$w\approx 24.95$

Check. 7% of \$24.95 is about \$1.75 and \$24.95 + \$1.75 = \$26.70, the price with tax, so the answer checks.

State. The wallet itself cost \$24.95.

21. ***Familiarize.*** Let n = the number of visits required for a total parking cost of \$27.00. The parking cost for each $1\frac{1}{2}$ hour visit is \$1.50 for the first hour plus \$1.00 for part of a second hour, or \$2.50. Then the total parking cost for n visits is $2.50n$ dollars.

Translate. We reword the problem.

$$\underbrace{\text{Total parking cost}} \text{ is } \$27.00.$$
$$2.50n = 27.00$$

Solve. We solve the equation.

$2.5n=27$

$10(2.5n)=10(27)$ Clearing the decimal

$25n=270$

$\dfrac{25n}{25}=\dfrac{270}{25}$

$n=10.8$

If the total parking cost is \$27.00 for 10.8 visits, then the cost will be more than \$27.00 for 11 or more visits.

Check. The parking cost for 10 visits is \$2.50(10), or \$25, and the parking cost for 11 visits is \$2.50(11), or \$27.50. Since 11 is the smallest number for which the parking cost exceeds \$27.00, the answer checks.

State. The minimum number of weekly visits for which it is worthwhile to buy a parking pass is 11.

23. ***Familiarize.*** Let x = the measure of the first angle. Then $3x$ = the measure of the second angle, and $x+40$ = the measure of the third angle. Recall that the sum of measures of the angles of a triangle is 180°.

Translate.

$$\underbrace{\text{Measure of first angle}} + \underbrace{\text{measure of second angle}} + \underbrace{\text{measure of third angle}} \text{ is } \underbrace{180.}$$
$$x + 3x + (x+40) = 180$$

Solve. We solve the equation.

$$\begin{aligned} x+3x+(x+40) &= 180 \\ 5x+40 &= 180 \\ 5x+40-40 &= 180-40 \\ 5x &= 140 \\ \frac{5x}{5} &= \frac{140}{5} \\ x &= 28 \end{aligned}$$

Possible answers for the angle measures are as follows:

First angle: $x = 28°$

Second angle: $3x = 3(28) = 84°$

Third angle: $x + 40 = 28 + 40 = 68°$

Check. Consider 28°, 84°, and 68°. The second angle is three times the first, and the third is 40° more than the first. The sum, 28° + 84° + 68°, is 180°. These numbers check.

State. The measures of the angles are 28°, 84°, and 68°.

25. ***Familiarize.*** Using the labels on the drawing in the text, we let x = the measure of the first angle, $x + 5$ = the measure of the second angle, and $3x + 10$ = the measure of the third angle. Recall that the sum of measures of the angles of a triangle is 180°.

Translate.

$$\underbrace{\text{Measure of first angle}}_{x} \ \underset{+}{+} \ \underbrace{\text{measure of second angle}}_{(x+5)} \ \underset{+}{+} \ \underbrace{\text{measure of third angle}}_{(3x+10)} \ \underset{=}{\text{is}} \ \underset{180}{180.}$$

Solve. We solve the equation.

$$\begin{aligned} x+(x+5)+(3x+10) &= 180 \\ 5x+15 &= 180 \\ 5x+15-15 &= 180-15 \\ 5x &= 165 \\ \frac{5x}{5} &= \frac{165}{5} \\ x &= 33 \end{aligned}$$

Possible answers for the angle measures are as follows:

First angle: $x = 33°$

Second angle: $x + 5 = 33 + 5 = 38°$

Third angle: $3x + 10 = 3(33) + 10 = 109°$

Check. The second angle is 5° more than the first, and the third is 10° more than 3 times the first. The sum, 33° + 38° + 109°, is 180°. The numbers check.

State. The measures of the angles are 33°, 38°, and 109°.

27. ***Familiarize***. Let a = the amount Sarah invested. The investment grew by 28% of a, or $0.28a$.

Translate.

$$\underbrace{\text{Amount invested}}_{a} \ \underset{+}{\text{plus}} \ \underbrace{\text{amount of growth}}_{0.28a} \ \underset{=}{\text{is}} \ \underset{448}{\$448.}$$

Solve. We solve the equation.

$$\begin{aligned} a+0.28a &= 448 \\ 1.28a &= 448 \\ a &= 350 \end{aligned}$$

Check. 28% of \$350 is 0.28(\$350), or \$98, and \$350 + \$98 = \$448. The answer checks.

State. Sarah invested \$350.

29. ***Familiarize***. Let b = the balance on the credit card at the beginning of the month. The balance grew by 2% of b, or $0.02b$.

Translate.

$$\underbrace{\text{Original balance}}_{b} \ \underset{+}{\text{plus}} \ \underbrace{\text{amount of growth}}_{0.02b} \ \underset{=}{\text{is}} \ \underset{870}{\$870.}$$

Solve. We solve the equation.

$$\begin{aligned} b+0.02b &= 870 \\ 1.02b &= 870 \\ b &\approx \$852.94 \end{aligned}$$

Check. 2% of \$852.94 is 0.02(\$852.94), or \$17.06, and \$852.94 + \$17.06 = \$870. The answer checks.

State. The balance at the beginning of the month was \$852.94.

31. ***Familiarize***. The total cost is the initial charge plus the mileage charge. Let d = the distance, in miles, that one can travel for \$39.50. The mileage charge is the cost per mile times the number of miles traveled or $2d$.

Translate.

$$\underbrace{\text{Initial charge}}_{3.50} \ \underset{+}{\text{plus}} \ \underbrace{\text{mileage charge}}_{2d} \ \underset{=}{\text{is}} \ \underset{39.50}{\$39.50.}$$

Solve. We solve the equation.

$$\begin{aligned} 3.50+2d &= 39.50 \\ 2d &= 36 \\ d &= 18 \end{aligned}$$

Check. An 18-mi taxi ride would cost $\$3.50 + \$2 \cdot 18$, or \$3.50 + \$36, or \$39.50. The answer checks.

State. One can travel 18 mi for \$39.50.

33. ***Familiarize***. Let c = the cost of the meal before the tip. We know that the cost of the meal before the tip plus the tip, 15% of the cost, is the total cost, \$44.39.

Translate.

$$\underbrace{\text{Cost of meal}}_{c} \ \underset{+}{\text{plus}} \ \underset{15\%c}{\text{tip}} \ \underset{=}{\text{is}} \ \underset{44.39}{\$44.39}$$

Solve. We solve the equation.

$$c + 15\%c = 44.39$$
$$c + 0.15c = 44.39$$
$$1c + 0.15c = 44.39$$
$$1.15c = 44.39$$
$$\frac{1.15c}{1.15} = \frac{44.39}{1.15}$$
$$c = 38.60$$

Check. We find 15% of \$38.60 and add it to \$38.60: 15% × \$38.60 = 0.15 × \$38.60 = \$5.79 and \$38.60+\$5.79 = \$44.39. The answer checks.

State. The cost of the meal before the tip was added was \$38.60.

35. ***Familiarize.*** Let s = the third score. Then $s - 7$ = the second score.

Translate. We add the three scores and divide by 3 to find the average.

$$\frac{67 + (s - 7) + s}{3} = 84$$

Solve.

$$\frac{67 + (s - 7) + s}{3} = 84$$
$$\frac{60 + 2s}{3} = 84$$
$$3\left(\frac{60 + 2s}{3}\right) = 3 \cdot 84$$
$$60 + 2s = 252$$
$$60 + 2s - 60 = 252 - 60$$
$$2s = 192$$
$$s = 96$$

If $s = 96$, then $s - 7 = 89$.

Check. We find the average of 67, 89, and 96.

$$\frac{67 + 89 + 96}{3} = \frac{252}{3} = 84$$

The answer checks.

State. Mariana's scores on the second and third tests were 89 and 96, respectively.

37. ***Familiarize.*** Let y = the number.

Translate. We reword the problem.

$$\underbrace{\text{Two times a number}}_{2 \cdot y}\ \underbrace{\text{plus}}_{+}\ \underbrace{16}_{16}\ \underbrace{\text{is}}_{=}\ \underbrace{\tfrac{2}{3}\text{ of the number}}_{\tfrac{2}{3} \cdot y}$$

$$2 \cdot y + 16 = \frac{2}{3} \cdot y$$

Solve. We solve the equation.

$$2y + 16 = \frac{2}{3}y$$
$$3(2y + 16) = 3 \cdot \frac{2}{3}y \quad \text{Clearing the fraction}$$
$$6y + 48 = 2y$$
$$48 = -4y \quad \text{Subtracting 6y}$$
$$-12 = y \quad \text{Dividing by } -8$$

Check. We double −12 and get −24. Adding 16, we get −8. Also, $\frac{2}{3}(-12) = -8$. The answer checks.

State. The number is −12.

39. $$-\frac{4}{5} - \frac{3}{8} = -\frac{4}{5} + \left(-\frac{3}{8}\right) = -\frac{32}{40} + \left(-\frac{15}{40}\right) = -\frac{47}{40}$$

41. $$-\frac{4}{5} \cdot \frac{3}{8} = -\frac{4 \cdot 3}{5 \cdot 8} = -\frac{4 \cdot 3}{5 \cdot 2 \cdot 4} = -\frac{\not{4} \cdot 3}{5 \cdot 2 \cdot \not{4}} = -\frac{3}{10}$$

43. $$\frac{1}{10} \div \left(-\frac{1}{100}\right) = \frac{1}{10} \cdot \left(-\frac{100}{1}\right) = -\frac{1 \cdot 100}{10 \cdot 1} = -\frac{\not{1} \cdot \not{10} \cdot 10}{\not{10} \cdot \not{1} \cdot 1} = -\frac{10}{1} = -10$$

45. $-25.6(-16) = 409.6$

47. $-25.6 + (-16) = -41.6$

49. $12 + yz = yz + 12$ Commutatiave law of addition
$= zy + 12$ Commutative law of multiplication
$12 + yz = 12 + zy$ Commutative law of multiplication.

51. ***Familiarize.*** Let a = the original number of apples. Then $\frac{1}{3}a$, $\frac{1}{4}a$, $\frac{1}{8}a$, and $\frac{1}{5}a$ are given to four people, respectively. The fifth and sixth people get 10 apples and 1 apple, respectively.

Translate. We reword the problem.

$$\underbrace{\text{The total number of apples}}_{\frac{1}{3}a + \frac{1}{4}a + \frac{1}{8}a + \frac{1}{5}a + 10 + 1}\ \underbrace{\text{is}}_{=}\ \underbrace{a}_{a}$$

$$\frac{1}{3}a + \frac{1}{4}a + \frac{1}{8}a + \frac{1}{5}a + 10 + 1 = a$$

Solve. We solve the equation.

$$\frac{1}{3}a + \frac{1}{4}a + \frac{1}{8}a + \frac{1}{5}a + 10 + 1 = a, \text{ LCD is } 120$$
$$120\left(\frac{1}{3}a + \frac{1}{4}a + \frac{1}{8}a + \frac{1}{5}a + 11\right) = 120 \cdot a$$
$$40a + 30a + 15a + 24a + 1320 = 120a$$
$$109a + 1320 = 120a$$
$$1320 = 11a$$
$$120 = a$$

Check. If the original number of apples was 120, then the first four people got $\frac{1}{3} \cdot 120$, $\frac{1}{4} \cdot 120$, $\frac{1}{8} \cdot 120$, and $\frac{1}{5} \cdot 120$, or 40, 30, 15, and 24 apples, respectively. Adding all the apples we get 40 + 30 + 15 + 24 + 10 + 1, or 120. The result checks.

State. There were originally 120 apples in the basket.

53. Divide the largest triangle into three triangles, each with a vertex at the center of the circle and with height x as shown.

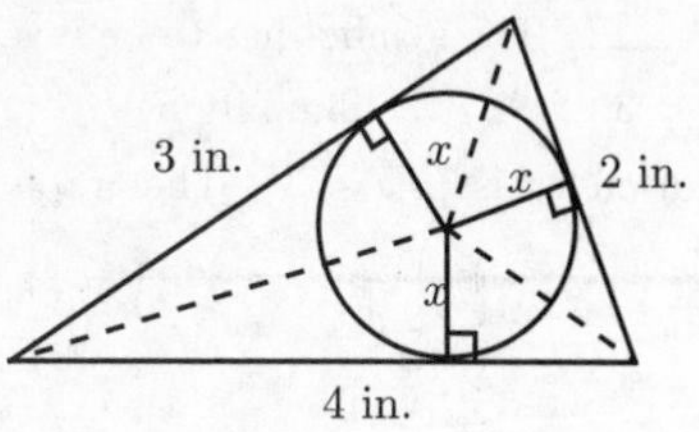

Then the sum of the areas of the three smaller triangles is the area of the original triangle. We have:

$$\frac{1}{2}\cdot 3x+\frac{1}{2}\cdot 2x+\frac{1}{2}\cdot 4x=2.9047$$
$$2\left(\frac{1}{2}\cdot 3x+\frac{1}{2}\cdot 2x+\frac{1}{2}\cdot 4x\right)=2(2.9047)$$
$$3x+2x+4x=5.8094$$
$$9x=5.8094$$
$$x\approx 0.65$$

Thus, x is about 0.65 in.

Exercise Set 2.7

RC1.
$$x+10\geq 12$$
$$x+10-10\geq 12-10$$
$$x\geq 2$$

We see that $x+10\geq 12$ and $x\leq 2$ are not equivalent.

RC3.
$$-\frac{3}{4}y<6$$
$$-\frac{4}{3}\left(-\frac{3}{4}y\right)>-\frac{4}{3}\cdot 6$$
$$y>-8$$

We see that $-\frac{3}{4}y<6$ and $y>-8$ are equivalent.

1. $x>-4$

a) Since $4>-4$ is true, 4 is a solution.

b) Since $0>-4$ is true, 0 is a solution.

c) Since $-4>-4$ is false, -4 is not a solution.

d) Since $6>-4$ is true, 6 is a solution.

e) Since $5.6>-4$ is true, 5.6 is a solution.

3. $x\geq 6.8$

a) Since $-6\geq 6.8$ is false, -6 is not a solution.

b) Since $0\geq 6.8$ is false, 0 is not a solution.

c) Since $6\geq 6.8$ is false, 6 is not a solution.

d) Since $8\geq 6.8$ is true, 8 is a solution.

e) Since $-3\frac{1}{2}\geq 6.8$ is false, $-3\frac{1}{2}$ is not a solution.

5. The solutions of $x>4$ are those numbers greater than 4. They are shown on the graph by shading all points to the right of 4. The open circle at 4 indicates that 4 is not part of the graph.

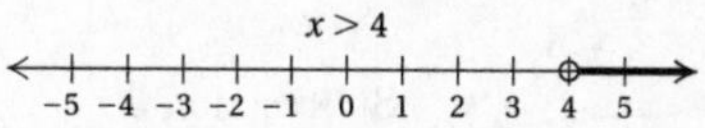

7. The solutions of $t<-3$ are those numbers less than -3. They are shown on the graph by shading all points to the left of -3. The open circle at -3 indicates that -3 is not part of the graph.

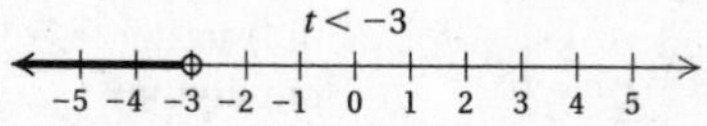

9. The solutions of $m\geq -1$ are are shown by shading the point for -1 and all points to the right of -1. The closed circle at -1 indicates that -1 is part of the graph.

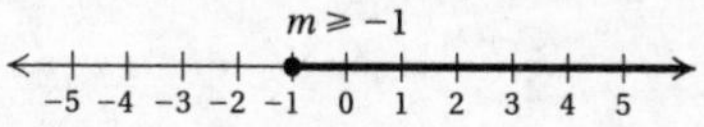

11. In order to be a solution of the inequality $-3<x\leq 4$, a number must be a solution of both $-3<x$ and $x\leq 4$. The solution set is graphed as follows:

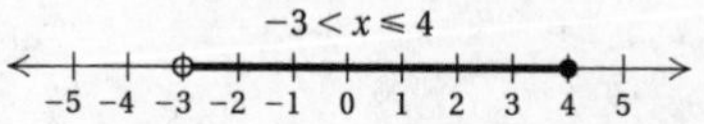

The open circle at -3 means that -3 is not part of the graph. The closed circle at 4 means that 4 is part of the graph.

13. In order to be a solution of the inequality $0<x<3$, a number must be a solution of both $0<x$ and $x<3$. The solution set is graphed as follows:

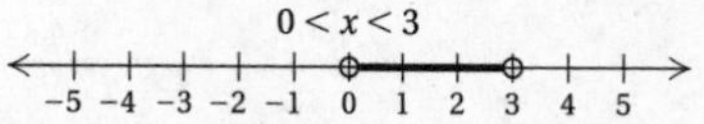

The open circles at 0 and at 3 mean that 0 and 3 are not part of the graph.

15.
$$x+7>2$$
$$x+7-7>2-7 \quad \text{Subtracting 7}$$
$$x>-5 \quad \text{Simplifying}$$

The solution set is $\{x|x>-5\}$.

The graph is as follows:

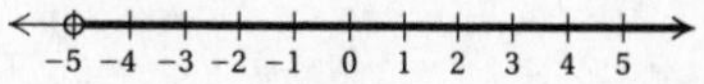

17.
$$x+8\leq -10$$
$$x+8-8\leq -10-8 \quad \text{Subtracting 8}$$
$$x\leq -18 \quad \text{Simplifying}$$

The solution set is $\{x|x\leq -18\}$.

The graph is as follows:

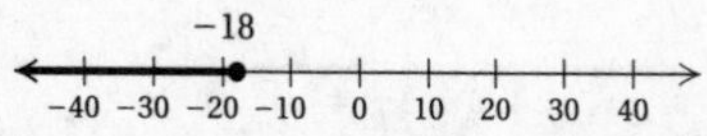

19.
$$\begin{aligned} y-7 &> -12 \\ y-7+7 &> -12+7 && \text{Adding 7} \\ y &> -5 && \text{Simplifying} \end{aligned}$$
The solution set is $\{y|y>-5\}$.

21.
$$\begin{aligned} 2x+3 &> x+5 \\ 2x+3-3 &> x+5-3 && \text{Subtracting 3} \\ 2x &> x+2 && \text{Simplifying} \\ 2x-x &> x+2-x && \text{Subtracting } x \\ x &> 2 && \text{Simplifying} \end{aligned}$$
The solution set is $\{x|x>2\}$.

23.
$$\begin{aligned} 3x+9 &\le 2x+6 \\ 3x+9-9 &\le 2x+6-9 && \text{Subtracting 9} \\ 3x &\le 2x-3 && \text{Simplifying} \\ 3x-2x &\le 2x-3-2x && \text{Subtracting } 2x \\ x &\le -3 && \text{Simplifying} \end{aligned}$$
The solution set is $\{x|x\le -3\}$.

25.
$$\begin{aligned} 5x-6 &< 4x-2 \\ 5x-6+6 &< 4x-2+6 \\ 5x &< 4x+4 \\ 5x-4x &< 4x+4-4x \\ x &< 4 \end{aligned}$$
The solution set is $\{x|x<4\}$.

27.
$$\begin{aligned} -9+t &> 5 \\ -9+t+9 &> 5+9 \\ t &> 14 \end{aligned}$$
The solution set is $\{t|t>14\}$.

29.
$$\begin{aligned} y+\frac{1}{4} &\le \frac{1}{2} \\ y+\frac{1}{4}-\frac{1}{4} &\le \frac{1}{2}-\frac{1}{4} \\ y &\le \frac{2}{4}-\frac{1}{4} && \text{Obtaining a common denominator} \\ y &\le \frac{1}{4} \end{aligned}$$
The solution set is $\left\{y\middle|y\le \frac{1}{4}\right\}$.

31.
$$\begin{aligned} x-\frac{1}{3} &> \frac{1}{4} \\ x-\frac{1}{3}+\frac{1}{3} &> \frac{1}{4}+\frac{1}{3} \\ x &> \frac{3}{12}+\frac{4}{12} && \text{Obtaining a common denominator} \\ x &> \frac{7}{12} \end{aligned}$$
The solution set is $\left\{x\middle|x>\frac{7}{12}\right\}$.

33.
$$\begin{aligned} 5x &< 35 \\ \frac{5x}{5} &< \frac{35}{5} && \text{Dividing by 5} \\ x &< 7 \end{aligned}$$
The solution set is $\{x|x<7\}$. The graph is as follows:

35.
$$\begin{aligned} -12x &> -36 \\ \frac{-12x}{-12} &< \frac{-36}{-12} && \text{Dividing by } -12 \end{aligned}$$
↑ The symbol has to be reversed.
$$x < 3 \qquad \text{Simplifying}$$
The solution set is $\{x|x<3\}$. The graph is as follows:

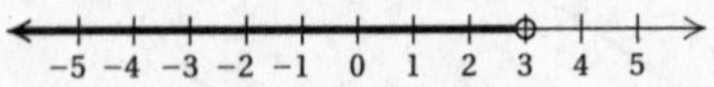

37.
$$\begin{aligned} 5y &\ge -2 \\ \frac{5y}{5} &\ge \frac{-2}{5} && \text{Dividing by 5} \\ y &\ge -\frac{2}{5} \end{aligned}$$
The solution set is $\left\{y\middle|y\ge -\frac{2}{5}\right\}$.

39.
$$\begin{aligned} -2x &\le 12 \\ \frac{-2x}{-2} &\ge \frac{12}{-2} && \text{Dividing by } -2 \end{aligned}$$
↑ The symbol has to be reversed.
$$x \ge -6 \qquad \text{Simplifying}$$
The solution set is $\{x|x\ge -6\}$.

41.
$$\begin{aligned} -4y &\ge -16 \\ \frac{-4y}{-4} &\le \frac{-16}{-4} && \text{Dividing by } -4 \end{aligned}$$
↑ The symbol has to be reversed.
$$y \le 4 \qquad \text{Simplifying}$$
The solution set is $\{y|y\le 4\}$.

43.
$$\begin{aligned} -3x &< -17 \\ \frac{-3x}{-3} &> \frac{-17}{-3} && \text{Dividing by } -3 \end{aligned}$$
↑ The symbol has to be reversed.
$$x > \frac{17}{3} \qquad \text{Simplifying}$$
The solution set is $\left\{x\middle|x>\frac{17}{3}\right\}$.

45.
$$\begin{aligned} -2y &> \frac{1}{7} \\ -\frac{1}{2}\cdot(-2y) &< -\frac{1}{2}\cdot\frac{1}{7} \end{aligned}$$
↑ The symbol has to be reversed.
$$y < -\frac{1}{14}$$
The solution set is $\left\{y\middle|y< -\frac{1}{14}\right\}$.

47.
$$\begin{aligned} -\frac{6}{5} &\le -4x \\ -\frac{1}{4}\cdot\left(-\frac{6}{5}\right) &\ge -\frac{1}{4}\cdot(-4x) \\ \frac{6}{20} &\ge x \\ \frac{3}{10} &\ge x, \text{ or } x \le \frac{3}{10} \end{aligned}$$
The solution set is $\left\{x\middle|\frac{3}{10}\ge x\right\}$, or $\left\{x\middle|x\le\frac{3}{10}\right\}$.

49. $4+3x<28$
$-4+4+3x<-4+28$ Adding -4
$3x<24$ Simplifying
$\frac{3x}{3}<\frac{24}{3}$ Dividing by 3
$x<8$

The solution set is $\{x|x<8\}$.

51. $3x-5\le 13$
$3x-5+5\le 13+5$ Adding 5
$3x\le 18$
$\frac{3x}{3}\le\frac{18}{3}$ Dividing by 3
$x\le 6$

The solution set is $\{x|x\le 6\}$.

53. $13x-7<-46$
$13x-7+7<-46+7$
$13x<-39$
$\frac{13x}{13}<\frac{-39}{13}$
$x<-3$

The solution set is $\{x|x<-3\}$.

55. $30>3-9x$
$30-3>3-9x-3$ Subtracting 3
$27>-9x$
$\frac{27}{-9}<\frac{-9x}{-9}$ Dividing by -9
↑ The symbol has to be reversed.
$-3<x$

The solution set is $\{x|-3<x\}$, or $\{x|x>-3\}$.

57. $4x+2-3x\le 9$
$x+2\le 9$ Collecting like terms
$x+2-2\le 9-2$
$x\le 7$

The solution set is $\{x|x\le 7\}$.

59. $-3<8x+7-7x$
$-3<x+7$ Collecting like terms
$-3-7<x+7-7$
$-10<x$

The solution set is $\{x|-10<x\}$, or $\{x|x>-10\}$.

61. $6-4y>4-3y$
$6-4y+4y>4-3y+4y$ Adding $4y$
$6>4+y$
$-4+6>-4+4+y$ Adding -4
$2>y$, or $y<2$

The solution set is $\{y|2>y\}$, or $\{y|y<2\}$.

63. $5-9y\le 2-8y$
$5-9y+9y\le 2-8y+9y$
$5\le 2+y$
$-2+5\le -2+2+y$
$3\le y$, or $y\ge 3$

The solution set is $\{y|3\le y\}$, or $\{y|y\ge 3\}$.

65. $19-7y-3y<39$
$19-10y<39$ Collecting like terms
$-19+19-10y<-19+39$
$-10y<20$
$\frac{-10y}{-10}>\frac{20}{-10}$
↑ The symbol has to be reversed.
$y>-2$

The solution set is $\{y|y>-2\}$.

67. $0.9x+19.3>5.3-2.6x$
$10(0.9x+19.3)>10(5.3-2.6x)$ Multiplying by 10 to clear decimals
$9x+193>53-26x$
$35x>-140$ Adding $26x$ and subtracting 193
$x>-4$ Dividing by 35

The solution set is $\{x|x>-4\}$.

69. $\frac{x}{3}-2\le 1$
$3\left(\frac{x}{3}-2\right)\le 3\cdot 1$ Multiplying by 3 to to clear the fraction
$x-6\le 3$ Simplifying
$x\le 9$ Adding 6

The solution set is $\{x|x\le 9\}$.

71. $\frac{y}{5}+1\le\frac{2}{5}$
$5\left(\frac{y}{5}+1\right)\le 5\cdot\frac{2}{5}$ Clearing fractions
$y+5\le 2$
$y\le -3$ Subtracting 5

The solution set is $\{y|y\le -3\}$.

73. $3(2y-3)<27$
$6y-9<27$ Removing parentheses
$6y<36$ Adding 9
$y<6$ Dividing by 6

The solution set is $\{y|y<6\}$.

75. $2(3+4m)-9\ge 45$
$6+8m-9\ge 45$ Removing parentheses
$8m-3\ge 45$ Collecting like terms
$8m\ge 48$ Adding 3
$m\ge 6$ Dividing by 8

The solution set is $\{m|m\ge 6\}$.

77. $8(2t+1)>4(7t+7)$
$16t+8>28t+28$
$16t-28t>28-8$
$-12t>20$
$t<-\frac{20}{12}$ Dividing by -12 and reversing the symbol
$t<-\frac{5}{3}$

The solution set is $\left\{t\middle|t<-\frac{5}{3}\right\}$.

79. $3(r-6)+2 < 4(r+2)-21$
$3r-18+2 < 4r+8-21$
$3r-16 < 4r-13$
$-16+13 < 4r-3r$
$-3 < r$, or $r > -3$

The solution set is $\{r|r > -3\}$.

81. $0.8(3x+6) \geq 1.1-(x+2)$
$2.4x+4.8 \geq 1.1-x-2$
$10(2.4x+4.8) \geq 10(1.1-x-2)$ Clearing decimals
$24x+48 \geq 11-10x-20$
$24x+48 \geq -10x-9$ Collecting like terms
$24x+10x \geq -9-48$
$34x \geq -57$
$x \geq -\frac{57}{34}$

The solution set is $\left\{x\middle|x \geq -\frac{57}{34}\right\}$.

83. $\frac{5}{3}+\frac{2}{3}x < \frac{25}{12}+\frac{5}{4}x+\frac{3}{4}$

The number 12 is the least common multiple of all the denominators. We multiply by 12 on both sides.

$$12\left(\frac{5}{3}+\frac{2}{3}x\right) < 12\left(\frac{25}{12}+\frac{5}{4}x+\frac{3}{4}\right)$$
$$12\cdot\frac{5}{3}+12\cdot\frac{2}{3}x < 12\cdot\frac{25}{12}+12\cdot\frac{5}{4}x+12\cdot\frac{3}{4}$$
$$20+8x < 25+15x+9$$
$$20+8x < 34+15x$$
$$20-34 < 15x-8x$$
$$-14 < 7x$$
$$-2 < x, \text{ or } x > -2$$

The solution set is $\{x|x > -2\}$.

85. $-\frac{3}{4}+\frac{1}{8}$ One negative and one positive number. Find the difference of the absolute values. Then make the answer negative, since the negative number has the larger absolute value.

$-\frac{3}{4}+\frac{1}{8} = -\frac{6}{8}+\frac{1}{8} = -\frac{5}{8}$

87. $-2.3-7.1 = -2.3+(-7.1) = -9.4$

89. $5-3^2+(8-2)^2\cdot 4 = 5-3^2+6^2\cdot 4$
$= 5-9+36\cdot 4$
$= 5-9+144$
$= -4+144$
$= 140$

91. $5(2x-4)-3(4x+1) = 10x-20-12x-3 =$
$-2x-23$

93. $|x| < 3$

a) Since $|0| = 0$ and $0 < 3$ is true, 0 is a solution.

b) Since $|-2| = 2$ and $2 < 3$ is true, -2 is a solution.

c) Since $|-3| = 3$ and $3 < 3$ is false, -3 is not a solution.

d) Since $|4| = 4$ and $4 < 3$ is false, 4 is not a solution.

e) Since $|3| = 3$ and $3 < 3$ is false, 3 is not a solution.

f) Since $|1.7| = 1.7$ and $1.7 < 3$ is true, 1.7 is a solution.

g) Since $|-2.8| = 2.8$ and $2.8 < 3$ is true, -2.8 is a solution.

95. $x+3 < 3+x$
$x-x < 3-3$ Subtracting x and 3
$0 < 0$

We get a false inequality. There is no solution.

Exercise Set 2.8

RC1. If r is at most q, then r must be less than or equal to q, so we have $r \leq q$.

RC3. $r < q$

RC5. If q exceeds r, then q is greater than r, or r is less than q, so we have $r < q$.

1. $n \geq 7$

3. $w > 2$ kg

5. 90 mph $< s <$ 110 mph

7. $w \leq 20$ hr

9. $c \geq \$3.20$

11. $x > 8$

13. $y \leq -4$

15. $n \geq 1300$

17. $w \leq 500$ L

19. $3x+2 < 13$, or $2+3x < 13$

21. ***Familiarize***. Let s represent the score on the fourth test.

Translate.

$\underbrace{\text{The average score}}$ $\underbrace{\text{is at least}}$ 80.

$$\frac{82+76+78+s}{4} \quad \geq \quad 80$$

Solve.

$$\frac{82+76+78+s}{4} \geq 80$$
$$4\left(\frac{82+76+78+s}{4}\right) \geq 4\cdot 80$$
$$82+76+78+s \geq 320$$
$$236+s \geq 320$$
$$s \geq 84$$

Check. As a partial check we show that the average is at least 80 when the fourth test score is 84.

$$\frac{82+76+78+84}{4} = \frac{320}{4} = 80$$

State. Xavier will get at least a B if the score on the fourth test is at least 84. The solution set is $\{s|s \geq 84\}$.

23. ***Familiarize.*** We use the formula for converting Celsius temperatures to Fahrenheit temperatures, $F = \frac{9}{5}C + 32$.

Translate.

$$\underbrace{\text{Fahrenheit temperature}}_{\downarrow\atop \frac{9}{5}C+32}\ \underbrace{\text{is less than}}_{\downarrow\atop <}\ \underset{\downarrow\atop 1945.4}{1945.4.}$$

Solve.

$$\begin{aligned}\frac{9}{5}C + 32 &< 1945.4\\ \frac{9}{5}C &< 1913.4\\ \frac{5}{9}\cdot\frac{9}{5}C &< \frac{5}{9}(1913.4)\\ C &< 1063\end{aligned}$$

Check. As a partial check we can show that the Fahrenheit temperature is less than 1945.4°for a Celsius temperature less than 1063°and is greater than 1945.4°for a Celsius temperature greater than 1063°.

$$F = \frac{9}{5}\cdot 1062 + 32 = 1943.6 < 1945.4$$

$$F = \frac{9}{5}\cdot 1064 + 32 = 1947.2 > 1945.4$$

State. Gold stays solid for temperatures less than 1063°C. The solution set is $\{C|C < 1063°\}$.

25. ***Familiarize.*** $R = -0.075t + 3.85$

In the formula R represents the world record and t represents the years since 1930. When $t = 0$ (1930), the record was $-0.075\cdot 0 + 3.85$, or 3.85 minutes. When $t = 2$ (1932), the record was $-0.075(2) + 3.85$, or 3.7 minutes. For what values of t will $-0.075t + 3.85$ be less than 3.5?

Translate. The record is to be less than 3.5. We have the inequality

$$R < 3.5.$$

To find the t values which satisfy this condition we substitute $-0.075t + 3.85$ for R.

$$-0.075t + 3.85 < 3.5$$

Solve.

$$\begin{aligned}-0.075t + 3.85 &< 3.5\\ -0.075t &< 3.5 - 3.85\\ -0.075t &< -0.35\\ t &> \frac{-0.35}{-0.075}\\ t &> 4\frac{2}{3}\end{aligned}$$

Check. With inequalities it is impossible to check each solution. But we can check to see if the solution set we obtained seems reasonable.

When $t = 4\frac{1}{2}$, $R = -0.075(4.5) + 3.85$, or 3.5125.

When $t = 4\frac{2}{3}$, $R = -0.075\left(\frac{14}{3}\right) + 3.85$, or 3.5.

When $t = 4\frac{3}{4}$, $R = -0.075(4.75) + 3.85$, or 3.49375.

Since $r = 3.5$ when $t = 4\frac{2}{3}$ and R decreases as t increases, R will be less than 3.5 when t is greater than $4\frac{2}{3}$.

State. The world record will be less than 3.5 minutes more than $4\frac{2}{3}$ years after 1930. If we let Y = the year, then the solution set is $\{Y|Y \geq 1935\}$.

27. ***Familiarize.*** Let c = the number of copies Myra has made. The total cost of the copies is the setup fee of \$5 plus \$4 times the number of copies, or $\$4\cdot c$.

Translate.

$$\underbrace{\text{Setup fee}}_{\downarrow\atop 5}\ \underset{\downarrow\atop +}{\text{plus}}\ \underbrace{\text{copying cost}}_{\downarrow\atop 4c}\ \underbrace{\text{cannot exceed}}_{\downarrow\atop \leq}\ \underset{\downarrow\atop 65}{\$65.}$$

Solve. We solve the inequality.

$$\begin{aligned}5 + 4c &\leq 65\\ 4c &\leq 60\\ c &\leq 15\end{aligned}$$

Check. As a partial check, we show that Myra can have 15 copies made and not exceed her \$65 budget.

$$\$5 + \$4\cdot 15 = 5 + 60 = \$65$$

State. Myra can have 15 or fewer copies made and stay within her budget.

29. ***Familiarize.*** As in the drawing in the text, we let L = the length of the envelope. Recall that the area of a rectangle is the product of the length and the width.

Translate.

$$\underbrace{\text{Length}}_{\downarrow\atop L}\ \underset{\downarrow\atop \cdot}{\text{times}}\ \underbrace{\text{width}}_{\downarrow\atop 3\frac{1}{2}}\ \underbrace{\text{is at least}}_{\downarrow\atop \geq}\ \underbrace{17\frac{1}{2}\text{ in}^2}_{\downarrow\atop 17\frac{1}{2}}$$

Solve.

$$\begin{aligned}L\cdot 3\frac{1}{2} &\geq 17\frac{1}{2}\\ L\cdot\frac{7}{2} &\geq \frac{35}{2}\\ L\cdot\frac{7}{2}\cdot\frac{2}{7} &\geq \frac{35}{2}\cdot\frac{2}{7}\\ L &\geq 5\end{aligned}$$

The solution set is $\{L|L \geq 5\}$.

Check. We can obtain a partial check by substituting a number greater than or equal to 5 in the inequality. For example, when $L = 6$:

$$L\cdot 3\frac{1}{2} = 6\cdot 3\frac{1}{2} = 6\cdot\frac{7}{2} = 21 \geq 17\frac{1}{2}$$

The result appears to be correct.

State. Lengths of 5 in. or more will satisfy the constraints. The solution set is $\{L|L \geq 5\text{ in.}\}$.

31. ***Familiarize.*** Let m represent the length of a telephone call, in minutes.

Translate.

$$\underbrace{\text{\$0.75 charge}}_{\downarrow \atop 0.75} \quad \underbrace{\text{plus}}_{\downarrow \atop +} \quad \underbrace{\text{charge for time used}}_{\downarrow \atop 0.45m} \quad \underbrace{\text{is at least}}_{\downarrow \atop \geq} \quad \underbrace{\text{\$3.00.}}_{\downarrow \atop 3}$$

Solve. We solve the inequality.

$$\begin{aligned} 0.75 + 0.45m &\geq 3 \\ 0.45m &\geq 2.25 \\ m &\geq 5 \end{aligned}$$

Check. As a partial check, we can show that if a call lasts 5 minutes it costs at least \$3.00:

$$\$0.75 + \$0.45(5) = \$0.75 + \$2.25 = \$3.00.$$

State. Simon's calls last at least 5 minutes each.

33. ***Familiarize.*** Let $c =$ the number of courses for which Angelica registers. Her total tuition is the \$35 registration fee plus \$375 times the number of courses for which she registers, or $\$375 \cdot c$.

Translate.

$$\underbrace{\text{Registration fee}}_{\downarrow \atop 35} \quad \underbrace{\text{plus}}_{\downarrow \atop +} \quad \underbrace{\text{fee for courses}}_{\downarrow \atop 375 \cdot c} \quad \underbrace{\text{cannot exceed}}_{\downarrow \atop \leq} \quad \underbrace{\text{\$1000.}}_{\downarrow \atop 1000}$$

Solve. We solve the inequality.

$$\begin{aligned} 35 + 375c &\leq 1000 \\ 375c &\leq 965 \\ c &\leq 2.5\overline{73} \end{aligned}$$

Check. Although the solution set of the inequality is all numbers less than or equal to $2.5\overline{73}$, since c represents the number of courses for which Angelica registers, we round down to 2. If she registers for 2 courses, her tuition is $\$35 + \$375 \cdot 2$, or \$785 which does not exceed \$1000. If she registers for 3 courses, her tuition is $\$35 + \$375 \cdot 3$, or \$1160 which exceeds \$1000.

State. Angelica can register for at most 2 courses.

35. ***Familiarize.*** Let $s =$ the number of servings of fruits or vegetables Dale eats on Saturday.

Translate.

$$\underbrace{\text{Average number of fruit or vegetable servings}}_{\downarrow \atop \frac{4+6+7+4+6+4+s}{7}} \quad \underbrace{\text{is at least}}_{\downarrow \atop \geq} \quad \underbrace{5.}_{\downarrow \atop 5}$$

Solve. We first multiply by 7 to clear the fraction.

$$\begin{aligned} 7\left(\frac{4+6+7+4+6+4+s}{7}\right) &\geq 7 \cdot 5 \\ 4+6+7+4+6+4+s &\geq 35 \\ 31 + s &\geq 35 \\ s &\geq 4 \end{aligned}$$

Check. As a partial check, we show that Dale can eat 4 servings of fruits or vegetables on Saturday and average at least 5 servings per day for the week:

$$\frac{4+6+7+4+6+4+4}{7} = \frac{35}{7} = 5$$

State. Dale should eat at least 4 servings of fruits or vegetables on Saturday.

37. ***Familiarize.*** We first make a drawing. We let l represent the length, in feet.

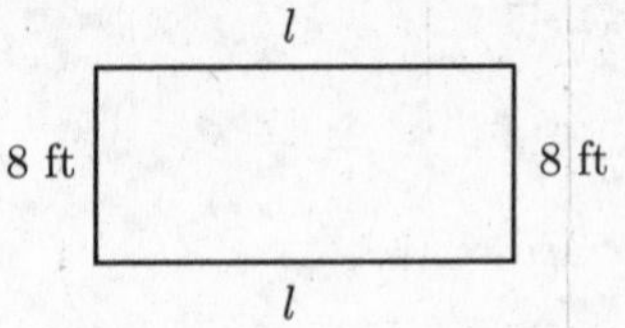

The perimeter is $P = 2l + 2w$, or $2l + 2 \cdot 8$, or $2l + 16$.

Translate. We translate to 2 inequalities.

$$\underbrace{\text{The perimeter}}_{\downarrow \atop 2l+16} \quad \underbrace{\text{is at least}}_{\downarrow \atop \geq} \quad \underbrace{\text{200 ft.}}_{\downarrow \atop 200}$$

$$\underbrace{\text{The perimeter}}_{\downarrow \atop 2l+16} \quad \underbrace{\text{is at most}}_{\downarrow \atop \leq} \quad \underbrace{\text{200 ft.}}_{\downarrow \atop 200}$$

Solve. We solve each inequality.

$$\begin{aligned} 2l + 16 &\geq 200 & \qquad 2l + 16 &\leq 200 \\ 2l &\geq 184 & 2l &\leq 184 \\ l &\geq 92 & l &\leq 92 \end{aligned}$$

Check. We check to see if the solutions seem reasonable.

When $l = 91$ ft, $P = 2 \cdot 91 + 16$, or 198 ft.

When $l = 92$ ft, $P = 2 \cdot 92 + 16$, or 200 ft.

When $l = 93$ ft, $P = 2 \cdot 93 + 16$, or 202 ft.

From these calculations, it appears that the solutions are correct.

State. Lengths greater than or equal to 92 ft will make the perimeter at least 200 ft. Lengths less than or equal to 92 ft will make the perimeter at most 200 ft.

39. ***Familiarize.*** Using the label on the drawing in the text, we let L represent the length.

The area is the length times the width, or $4L$.

Translate.

$$\underbrace{\text{Area}}_{\downarrow \atop 4L} \quad \underbrace{\text{is less than}}_{\downarrow \atop <} \quad \underbrace{86 \text{ cm}^2.}_{\downarrow \atop 86}$$

Solve.

$$\begin{aligned} 4L &< 86 \\ L &< 21.5 \end{aligned}$$

Check. We check to see if the solution seems reasonable.

When $L = 22$, the area is $22 \cdot 4$, or 88 cm^2.

When $L = 21.5$, the area is $21.5(4)$, or 86 cm^2.

When $L = 21$, the area is $21 \cdot 4$, or 84 cm^2.

From these calculations, it would appear that the solution is correct.

State. The area will be less than 86 cm^2 for lengths less than 21.5 cm.

41. ***Familiarize***. Let v = the blue book value of the car. Since the car was repaired, we know that \$8500 does not exceed $0.8v$ or, in other words, $0.8v$ is at least \$8500.

Translate.

$$\underbrace{\text{80\% of the blue book value}}_{\displaystyle 0.8v} \ \underbrace{\text{is at least}}_{\displaystyle \geq} \ \underbrace{\$8500.}_{\displaystyle 8500}$$

Solve.

$$0.8v \geq 8500$$
$$v \geq \frac{8500}{0.8}$$
$$v \geq 10,625$$

Check. As a partial check, we show that 80% of \$10,625 is at least \$8500:

$$0.8(\$10,625) = \$8500$$

State. The blue book value of the car was at least \$10,625.

43. ***Familiarize***. Let r = the amount of fat in a serving of the regular peanut butter, in grams. If reduced fat peanut butter has at least 25% less fat than regular peanut butter, then it has at most 75% as much fat as the regular peanut butter.

Translate.

$$\underbrace{\text{12 g of fat}}_{\displaystyle 12} \ \underbrace{\text{is at most}}_{\displaystyle \leq} \ \underbrace{\text{75\%}}_{\displaystyle 0.75} \ \underbrace{\text{of}}_{\displaystyle \cdot} \ \underbrace{\text{the amount of fat in regular peanut butter.}}_{\displaystyle r}$$

Solve.

$$12 \leq 0.75r$$
$$16 \leq r$$

Check. As a partial check, we show that 12 g of fat does not exceed 75% of 16 g of fat:

$$0.75(16) = 12$$

State. Regular peanut butter contains at least 16 g of fat per serving.

45. ***Familiarize***. Let h = the height of the triangle, in ft. Recall that the formula for the area of a triangle with base b and height h is $A = \frac{1}{2}bh$.

Translate.

$$\underbrace{\text{Area}}_{\displaystyle \frac{1}{2}\left(1\frac{1}{2}\right)h} \ \underbrace{\text{is at least}}_{\displaystyle \geq} \ \underbrace{3 \text{ ft}^2.}_{\displaystyle 3}$$

Solve. We solve the inequality.

$$\frac{1}{2}\left(1\frac{1}{2}\right)h \geq 3$$
$$\frac{1}{2} \cdot \frac{3}{2} \cdot h \geq 3$$
$$\frac{3}{4}h \geq 3$$
$$h \geq \frac{4}{3} \cdot 3$$
$$h \geq 3$$

Check. As a partial check, we show that the area of the triangle is 3 ft^2 when the height is 4 ft.

$$\frac{1}{2}\left(1\frac{1}{2}\right)(4) = \frac{1}{2} \cdot \frac{3}{2} \cdot \frac{4}{1} = 3$$

State. The height should be at least 4 ft.

47. ***Familiarize***. Let w = the number of weeks after July 1. After w weeks the water level has dropped $\frac{2}{3}w$ ft.

Translate.

$$\underbrace{\text{Original depth}}_{\displaystyle 25} \ \underbrace{\text{minus}}_{\displaystyle -} \ \underbrace{\text{drop in water level}}_{\displaystyle \frac{2}{3}w} \ \underbrace{\text{does not exceed}}_{\displaystyle \leq} \ \underbrace{\text{21 ft.}}_{\displaystyle 21}$$

Solve. We solve the inequality.

$$25 - \frac{2}{3}w \leq 21$$
$$-\frac{2}{3}w \leq -4$$
$$w \geq -\frac{3}{2}(-4)$$
$$w \geq 6$$

Check. As a partial check we show that the water level is 21 ft 6 weeks after July 1.

$$25 - \frac{2}{3} \cdot 6 = 25 - 4 = 21 \text{ ft}$$

Since the water level continues to drop during the weeks after July 1, the answer seems reasonable.

State. The water level will not exceed 21 ft for dates at least 6 weeks after July 1.

49. ***Familiarize***. The average number of calls per week is the sum of the calls for the three weeks divided by the number of weeks, 3. We let c represent the number of calls made during the third week.

Translate. The average of the three weeks is given by

$$\frac{17 + 22 + c}{3}.$$

Since the average must be at least 20, this means that it must be greater than or equal to 20. Thus, we can translate the problem to the inequality

$$\frac{17 + 22 + c}{3} \geq 20.$$

Solve. We first multiply by 3 to clear the fraction.

$$3\left(\frac{17+22+c}{3}\right) \geq 3\cdot 20$$
$$17+22+c \geq 60$$
$$39+c \geq 60$$
$$c \geq 21$$

Check. Suppose c is a number greater than or equal to 21. Then by adding 17 and 22 on both sides of the inequality we get

$$17+22+c \geq 17+22+21$$
$$17+22+c \geq 60$$

so

$$\frac{17+22+c}{3} \geq \frac{60}{3}, \text{ or } 20.$$

State. 21 calls or more will maintain an average of at least 20 for the three-week period.

51. $-13+x=27$

$$-13+x+13=27+13$$
$$x=40$$

The solution is 40.

53. $4a-3=45$

$$4a-3+3=45+3$$
$$4a=48$$
$$\frac{4a}{4}=\frac{48}{4}$$
$$a=12$$

The solution is 12.

55. $-\frac{1}{2}+x=x-\frac{1}{2}$

$$-\frac{1}{2}+x+\frac{1}{2}=x-\frac{1}{2}+\frac{1}{2}$$
$$x=x$$

We get an equation that is true for all values of x. All real numbers are solutions of the equation.

57. $\underbrace{\text{What percent}}$ of 200 is 15?

$$\downarrow \quad \downarrow \quad \downarrow \quad \downarrow \quad \downarrow$$
$$p \quad \cdot \quad 200 \quad = \quad 15$$

Solve. We divide by 200 on both sides and convert to percent notation.

$$p\cdot 200=15$$
$$p=\frac{15}{200}$$
$$p=0.075$$
$$p=7.5\%$$

59. 25 is 2% of $\underbrace{\text{what number?}}$

$$\downarrow \quad \downarrow \quad \downarrow \quad \downarrow \quad \downarrow$$
$$25 \quad = \quad 2\% \quad \cdot \quad b$$

Solve.

$$25=2\%\cdot b$$
$$25=0.02\cdot b$$
$$\frac{25}{0.02}=\frac{0.02\cdot b}{0.02}$$
$$1250=b$$

61. ***Familiarize.*** We use the formula $F=\frac{9}{5}C+32$.

Translate. We are interested in temperatures such that $5° < F < 15°$. Substituting for F, we have:

$$5<\frac{9}{5}C+32<15$$

Solve.

$$5<\frac{9}{5}C+32<15$$
$$5\cdot 5<5\left(\frac{9}{5}C+32\right)<5\cdot 15$$
$$25<9C+160<75$$
$$-135<9C<-85$$
$$-15<C<-9\frac{4}{9}$$

Check. The check is left to the student.

State. Green ski wax works best for temperatures between $-15°\text{C}$ and $-9\frac{4}{9}°\text{C}$.

63. ***Familiarize.*** Let f = the fat content of a serving of regular nacho cheese tortilla chips, in grams. A product that contains 60% less fat than another product has 40% of the fat content of that product. If the reduced-fat tortilla chips cannot be labeled lowfat, then they contain at least 3 g of fat.

Translate.

40%	of	the fat content of regular tortilla chips	is at least	3 grams of fat
↓	↓	↓	↓	↓
0.4	·	f	$\geq$	3

Solve.

$$0.4f \geq 3$$
$$f \geq 7.5$$

Check. As a partial check, we show that 40% of 7.5 g is not less than 3 g.

$$0.4(7.5)=3$$

State. A serving of regular nacho cheese tortilla chips contains at least 7.5 g of fat.

Chapter 2 Vocabulary Reinforcement

1. Any replacement for the variable that makes an equation true is called a solution of the equation.

2. The addition principle for equations states that for any real numbers a, b, and c, $a=b$ is equivalent to $a+c=b+c$.

3. The multiplication principle states that for any real numbers a, b, and c, $a=b$ is equivalent to $a\cdot c=b\cdot c$.

4. An inequality is a number sentence with $<$, $\leq$, $>$, or $\geq$ as its verb.

5. Equations with the same solution are called equivalent equations.

Chapter 2 Concept Reinforcement

1. True; see page 144 in the text.

2. True; for any number n, $n \geq n$ is true because $n = n$ is true.

3. False; the solution set of $2x - 7 \leq 11$ is $\{x|x \leq 9\}$; the solution set of $x < 2$ is $\{x|x < 2\}$. The inequalities do not have the same solution set, so they are not equivalent.

4. True; if $x > y$, then $-1 \cdot x < -1 \cdot y$ (reversing the inequality symbol), or $-x < -y$.

Chapter 2 Study Guide

1.
$$\begin{aligned} 4(x-3) &= 6(x+2) \\ 4x - 12 &= 6x + 12 \\ 4x - 12 - 6x &= 6x + 12 - 6x \\ -2x - 12 &= 12 \\ -2x - 12 + 12 &= 12 + 12 \\ -2x &= 24 \\ \frac{-2x}{-2} &= \frac{24}{-2} \\ x &= -12 \end{aligned}$$

The solution is -12.

2.
$$\begin{aligned} 4 + 3y - 7 &= 3 + 3(y-2) \\ 4 + 3y - 7 &= 3 + 3y - 6 \\ 3y - 3 &= -3 + 3y \\ 3y - 3 - 3y &= -3 + 3y - 3y \\ -3 &= -3 \end{aligned}$$

Every real number is a solution of the equation $-3 = -3$, so all real numbers are solutions of the original equation.

3.
$$\begin{aligned} 4(x-3) + 7 &= -5 + 4x + 10 \\ 4x - 12 + 7 &= -5 + 4x + 10 \\ 4x - 5 &= 5 + 4x \\ 4x - 5 - 4x &= 5 + 4x - 4x \\ -5 &= 5 \end{aligned}$$

We get a false equation, so the original equation has no solution.

4.
$$\begin{aligned} A &= \frac{1}{2}bh \\ 2 \cdot A &= 2 \cdot \frac{1}{2}bh \\ 2A &= bh \\ \frac{2A}{h} &= \frac{bh}{h} \\ \frac{2A}{h} &= b \end{aligned}$$

5. Graph: $x > 1$

The solutions of $x > 1$ are all numbers greater than 1. We shade all points to the right of 1 and use an open circle at 1 to indicate that 1 is not part of the graph.

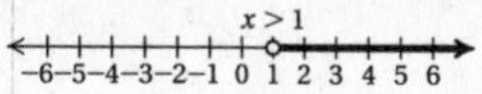

6. Graph: $x \leq -1$

The solutions of $x \leq -1$ are all numbers less than or equal to -1. We shade all points to the left of -1 and use a closed circle at -1 to indicate that -1 is part of the graph.

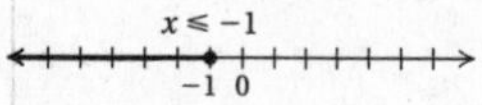

7.
$$\begin{aligned} 6y + 5 &> 3y - 7 \\ 6y + 5 - 3y &> 3y - 7 - 3y \\ 3y + 5 &> -7 \\ 3y + 5 - 5 &> -7 - 5 \\ 3y &> -12 \\ \frac{3y}{3} &> \frac{-12}{3} \\ y &> -4 \end{aligned}$$

The solution set is $\{y|y > -4\}$.

Chapter 2 Review Exercises

1.
$$\begin{aligned} x + 5 &= -17 \\ x + 5 - 5 &= -17 - 5 \\ x &= -22 \end{aligned}$$

The solution is -22.

2.
$$\begin{aligned} n - 7 &= -6 \\ n - 7 + 7 &= -6 + 7 \\ n &= 1 \end{aligned}$$

The solution is 1.

3.
$$\begin{aligned} x - 11 &= 14 \\ x - 11 + 11 &= 14 + 11 \\ x &= 25 \end{aligned}$$

The solution is 25.

4.
$$\begin{aligned} y - 0.9 &= 9.09 \\ y - 0.9 + 0.9 &= 9.09 + 0.9 \\ y &= 9.99 \end{aligned}$$

The solution is 9.99.

5.
$$\begin{aligned} -\frac{2}{3}x &= -\frac{1}{6} \\ -\frac{3}{2} \cdot \left(-\frac{2}{3}x\right) &= -\frac{3}{2} \cdot \left(-\frac{1}{6}\right) \\ 1 \cdot x &= \frac{\cancel{3} \cdot 1}{2 \cdot 2 \cdot \cancel{3}} \\ x &= \frac{1}{4} \end{aligned}$$

The solution is $\frac{1}{4}$.

6. $-8x = -56$

$$\frac{-8x}{-8} = \frac{-56}{-8}$$

$$x = 7$$

The solution is 7.

7. $$-\frac{x}{4} = 48$$

$$4 \cdot \frac{1}{4} \cdot (-x) = 4 \cdot 48$$

$$-x = 192$$

$$-1 \cdot (-1 \cdot x) = -1 \cdot 192$$

$$x = -192$$

The solution is -192.

8. $15x = -35$

$$\frac{15x}{15} = \frac{-35}{15}$$

$$x = -\frac{\cancel{5} \cdot 7}{3 \cdot \cancel{5}}$$

$$x = -\frac{7}{3}$$

The solution is $-\frac{7}{3}$.

9. $$\frac{4}{5}y = -\frac{3}{16}$$

$$\frac{5}{4} \cdot \frac{4}{5}y = \frac{5}{4} \cdot \left(-\frac{3}{16}\right)$$

$$y = -\frac{15}{64}$$

The solution is $-\frac{15}{64}$.

10. $$5 - x = 13$$

$$5 - x - 5 = 13 - 5$$

$$-x = 8$$

$$-1 \cdot (-1 \cdot x) = -1 \cdot 8$$

$$x = -8$$

The solution is -8.

11. $$\frac{1}{4}x - \frac{5}{8} = \frac{3}{8}$$

$$\frac{1}{4}x - \frac{5}{8} + \frac{5}{8} = \frac{3}{8} + \frac{5}{8}$$

$$\frac{1}{4}x = 1$$

$$4 \cdot \frac{1}{4}x = 4 \cdot 1$$

$$x = 4$$

The solution is 4.

12. $$5t + 9 = 3t - 1$$

$$5t + 9 - 3t = 3t - 1 - 3t$$

$$2t + 9 = -1$$

$$2t + 9 - 9 = -1 - 9$$

$$2t = -10$$

$$\frac{2t}{2} = \frac{-10}{2}$$

$$t = -5$$

The solution is -5.

13. $$7x - 6 = 25x$$

$$7x - 6 - 7x = 25x - 7x$$

$$-6 = 18x$$

$$\frac{-6}{18} = \frac{18x}{18}$$

$$-\frac{\cancel{6} \cdot 1}{3 \cdot \cancel{6}} = x$$

$$-\frac{1}{3} = x$$

The solution is $-\frac{1}{3}$.

14. $$14y = 23y - 17 - 10$$

$$14y = 23y - 27 \quad \text{Collecting like terms}$$

$$14y - 23y = 23y - 27 - 23y$$

$$-9y = -27$$

$$\frac{-9y}{-9} = \frac{-27}{-9}$$

$$y = 3$$

The solution is 3.

15. $$0.22y - 0.6 = 0.12y + 3 - 0.8y$$

$$0.22y - 0.6 = -0.68y + 3 \quad \text{Collecting like terms}$$

$$0.22y - 0.6 + 0.68y = -0.68y + 3 + 0.68y$$

$$0.9y - 0.6 = 3$$

$$0.9y - 0.6 + 0.6 = 3 + 0.6$$

$$0.9y = 3.6$$

$$\frac{0.9y}{0.9} = \frac{3.6}{0.9}$$

$$y = 4$$

The solution is 4.

16.
$$\frac{1}{4}x - \frac{1}{8}x = 3 - \frac{1}{16}x$$
$$\frac{2}{8}x - \frac{1}{8}x = 3 - \frac{1}{16}x$$
$$\frac{1}{8}x = 3 - \frac{1}{16}x$$
$$\frac{1}{8}x + \frac{1}{16}x = 3 - \frac{1}{16}x + \frac{1}{16}x$$
$$\frac{2}{16}x + \frac{1}{16}x = 3$$
$$\frac{3}{16}x = 3$$
$$\frac{16}{3} \cdot \frac{3}{16}x = \frac{16}{3} \cdot 3$$
$$x = \frac{16 \cdot \cancel{3}}{\cancel{3} \cdot 1}$$
$$x = 16$$
The solution is 16.

17.
$$14y + 17 + 7y = 9 + 21y + 8$$
$$21y + 17 = 21y + 17$$
$$21y + 17 - 21y = 21y + 17 - 21y$$
$$17 = 17 \qquad \text{TRUE}$$
All real numbers are solutions.

18.
$$4(x + 3) = 36$$
$$4x + 12 = 36$$
$$4x + 12 - 12 = 36 - 12$$
$$4x = 24$$
$$\frac{4x}{4} = \frac{24}{4}$$
$$x = 6$$
The solution is 6.

19.
$$3(5x - 7) = -66$$
$$15x - 21 = -66$$
$$15x - 21 + 21 = -66 + 21$$
$$15x = -45$$
$$\frac{15x}{15} = \frac{-45}{15}$$
$$x = -3$$
The solution is -3.

20.
$$8(x - 2) - 5(x + 4) = 20 + x$$
$$8x - 16 - 5x - 20 = 20 + x$$
$$3x - 36 = 20 + x$$
$$3x - 36 - x = 20 + x - x$$
$$2x - 36 = 20$$
$$2x - 36 + 36 = 20 + 36$$
$$2x = 56$$
$$\frac{2x}{2} = \frac{56}{2}$$
$$x = 28$$
The solution is 28.

21.
$$-5x + 3(x + 8) = 16$$
$$-5x + 3x + 24 = 16$$
$$-2x + 24 = 16$$
$$-2x + 24 - 24 = 16 - 24$$
$$-2x = -8$$
$$\frac{-2x}{-2} = \frac{-8}{-2}$$
$$x = 4$$
The solution is 4.

22.
$$6(x - 2) - 16 = 3(2x - 5) + 11$$
$$6x - 12 - 16 = 6x - 15 + 11$$
$$6x - 28 = 6x - 4$$
$$6x - 28 - 6x = 6x - 4 - 6x$$
$$-28 = -4 \qquad \text{False}$$
There is no solution.

23. Since $-3 \le 4$ is true, -3 is a solution.

24. Since $7 \le 4$ is false, 7 is not a solution.

25. Since $4 \le 4$ is true, 4 is a solution.

26.
$$y + \frac{2}{3} \ge \frac{1}{6}$$
$$y + \frac{2}{3} - \frac{2}{3} \ge \frac{1}{6} - \frac{2}{3}$$
$$y \ge \frac{1}{6} - \frac{4}{6}$$
$$y \ge -\frac{3}{6}$$
$$y \ge -\frac{1}{2}$$
The solution set is $\left\{y \middle| y \ge -\frac{1}{2}\right\}$.

27.
$$9x \ge 63$$
$$\frac{9x}{9} \ge \frac{63}{9}$$
$$x \ge 7$$
The solution set is $\{x|x \ge 7\}$.

28.
$$2 + 6y > 14$$
$$2 + 6y - 2 > 14 - 2$$
$$6y > 12$$
$$\frac{6y}{6} > \frac{12}{6}$$
$$y > 2$$
The solution set is $\{y|y > 2\}$.

29.
$$\begin{aligned} 7-3y &\ge 27+2y \\ 7-3y-2y &\ge 27+2y-2y \\ 7-5y &\ge 27 \\ 7-5y-7 &\ge 27-7 \\ -5y &\ge 20 \\ \frac{-5y}{-5} &\le \frac{20}{-5} \quad \text{Reversing the inequality symbol} \\ y &\le -4 \end{aligned}$$
The solution set is $\{y|y \le -4\}$.

30.
$$\begin{aligned} 3x+5 &< 2x-6 \\ 3x+5-2x &< 2x-6-2x \\ x+5 &< -6 \\ x+5-5 &< -6-5 \\ x &< -11 \end{aligned}$$
The solution set is $\{x|x < -11\}$.

31.
$$\begin{aligned} -4y &< 28 \\ \frac{-4y}{-4} &> \frac{28}{-4} \quad \text{Reversing the inequality symbol} \\ y &> -7 \end{aligned}$$
The solution set is $\{y|y > -7\}$.

32.
$$\begin{aligned} 4-8x &< 13+3x \\ 4-8x-3x &< 13+3x-3x \\ 4-11x &< 13 \\ 4-11x-4 &< 13-4 \\ -11x &< 9 \\ \frac{-11x}{-11} &> \frac{9}{-11} \quad \text{Reversing the inequality symbol} \\ x &> -\frac{9}{11} \end{aligned}$$
The solution set is $\left\{x\middle|x > -\frac{9}{11}\right\}$.

33.
$$\begin{aligned} -4x &\le \frac{1}{3} \\ -\frac{1}{4}\cdot(-4x) &\ge -\frac{1}{4}\cdot\frac{1}{3} \quad \text{Reversing the inequality symbol} \\ x &\ge -\frac{1}{12} \end{aligned}$$
The solution set is $\left\{x\middle|x \ge -\frac{1}{12}\right\}$.

34.
$$\begin{aligned} 4x-6 &< x+3 \\ 4x-6-x &< x+3-x \\ 3x-6 &< 3 \\ 3x-6+6 &< 3+6 \\ 3x &< 9 \\ \frac{3x}{3} &< \frac{9}{3} \\ x &< 3 \end{aligned}$$
The solution set is $\{x|x < 3\}$. The graph is as follows:

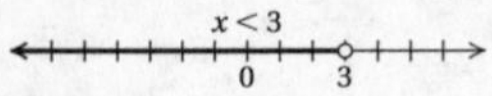

35. In order to be a solution of $-2 < x \le 5$, a number must be a solution of both $-2 < x$ and $x \le 5$. The solution set is graphed as follows:

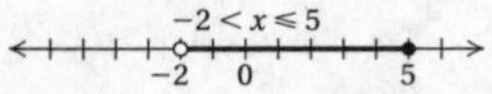

36. The solutions of $y > 0$ are those numbers greater than 0. The graph is as follows:

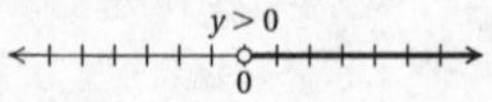

37.
$$\begin{aligned} C &= \pi d \\ \frac{C}{\pi} &= \frac{\pi d}{\pi} \\ \frac{C}{\pi} &= d \end{aligned}$$

38.
$$\begin{aligned} V &= \frac{1}{3}Bh \\ 3\cdot V &= 3\cdot\frac{1}{3}Bh \\ 3V &= Bh \\ \frac{3V}{h} &= \frac{Bh}{h} \\ \frac{3V}{h} &= B \end{aligned}$$

39.
$$\begin{aligned} A &= \frac{a+b}{2} \\ 2\cdot A &= 2\cdot\left(\frac{a+b}{2}\right) \\ 2A &= a+b \\ 2A-b &= a+b-b \\ 2A-b &= a \end{aligned}$$

40.
$$\begin{aligned} y &= mx+b \\ y-b &= mx+b-b \\ y-b &= mx \\ \frac{y-b}{m} &= \frac{mx}{m} \\ \frac{y-b}{m} &= x \end{aligned}$$

41. ***Familiarize***. Let $w =$ the width, in miles. Then $w+90 =$ the length. Recall that the perimeter P of a rectangle with length l and width w is given by $P = 2l + 2w$.

Translate. Substitute 1280 for P and $w+90$ for l in the formula above.
$$\begin{aligned} P &= 2l+2w \\ 1280 &= 2(w+90)+2w \end{aligned}$$

Solve. We solve the equation.

$$\begin{aligned} 1280 &= 2(w+90)+2w \\ 1280 &= 2w+180+2w \\ 1280 &= 4w+180 \\ 1280-180 &= 4w+180-180 \\ 1100 &= 4w \\ \frac{1100}{4} &= \frac{4w}{4} \\ 275 &= w \end{aligned}$$

If $w = 275$, then $w + 90 = 275 + 90 = 365$.

Check. The length, 365 mi, is 90 mi more than the width, 275 mi. The perimeter is $2 \cdot 365$ mi $+ 2 \cdot 275$ mi $= 730$ mi $+ 550$ mi $= 1280$ mi. The answer checks.

State. The length is 365 mi, and the width is 275 mi.

42. ***Familiarize.*** Let x = the number on the first marker. Then $x+1$ = the number on the second marker.

Translate.

First number plus second number is 691.

$$x + (x+1) = 691$$

Solve. We solve the equation.

$$\begin{aligned} x+(x+1) &= 691 \\ 2x+1 &= 691 \\ 2x+1-1 &= 691-1 \\ 2x &= 690 \\ \frac{2x}{2} &= \frac{690}{2} \\ x &= 345 \end{aligned}$$

If $x = 345$, then $x + 1 = 345 + 1 = 346$.

Check. 345 and 346 are consecutive integers and $345 + 346 = 691$. The answer checks.

State. The numbers on the markers are 345 and 346.

43. ***Familiarize.*** Let c = the cost of the entertainment center in February.

Translate.

Cost in February plus \$332 is Cost in June

$$c + 332 = 2449$$

Solve. We solve the equation.

$$\begin{aligned} c+332 &= 2449 \\ c+332-332 &= 2449-332 \\ c &= 2117 \end{aligned}$$

Check. \$2117 + \$332 = \$2449, so the answer checks.

State. The entertainment center cost \$2117 in February.

44. ***Familiarize.*** Let a = the number of subscriptions Ty sold.

Translate.

Commission per subscription times number sold is Total commission

$$4 \times a = 108$$

Solve. We solve the equation.

$$\begin{aligned} 4\cdot a &= 108 \\ \frac{4\cdot a}{4} &= \frac{108}{4} \\ a &= 27 \end{aligned}$$

Check. $\$4 \cdot 27 = \$108$, so the answer checks.

State. Ty sold 27 magazine subscriptions.

45. ***Familiarize.*** Let x = the measure of the first angle. Then $x + 50$ = the measure of the second angle, and $2x - 10$ = the measure of the third angle. Recall that the sum of measures of the angles of a triangle is 180°.

Translate.

Measure of first angle + measure of second angle + measure of third angle is 180°.

$$x + (x+50) + (2x-10) = 180$$

Solve. We solve the equation.

$$\begin{aligned} x+(x+50)+(2x-10) &= 180 \\ 4x+40 &= 180 \\ 4x+40-40 &= 180-40 \\ 4x &= 140 \\ \frac{4x}{4} &= \frac{140}{4} \\ x &= 35 \end{aligned}$$

If $x = 35$, then $x + 50 = 35 + 50 = 85$ and $2x - 10 = 2 \cdot 35 - 10 = 70 - 10 = 60$.

Check. The measure of the second angle is 50° more than the measure of the first angle, and the measure of the third angle is 10° less than twice the measure of the first angle. The sum of the measure is 35° + 85° + 60° = 180°. The answer checks.

State. The measures of the angles are 35°, 85°, and 60°.

46. *Translate.*

What number is 20% of 75?

$$a = 20\% \cdot 75$$

Solve. We convert 20% to decimal notation and multiply.

$$\begin{aligned} a &= 20\%\cdot 75 \\ a &= 0.2\cdot 75 \\ a &= 15 \end{aligned}$$

Thus, 15 is 20% of 75.

47. *Translate.*

15 is what percent of 80?

$$15 = p \cdot 80$$

Solve. We solve the equation.

$$15 = p \cdot 80$$
$$\frac{15}{80} = \frac{p \cdot 80}{80}$$
$$0.1875 = p$$
$$18.75\% = p$$

Thus, 15 is 18.75% of 80.

48. *Translate.*

$$\underbrace{18 \text{ is } 3\% \text{ of what number?}}$$
$$18 = 3\% \cdot b$$

Solve. We solve the equation.

$$18 = 3\% \cdot b$$
$$18 = 0.03 \cdot b$$
$$\frac{18}{0.03} = \frac{0.03 \cdot b}{0.03}$$
$$600 = b$$

Thus, 18 is 3% of 600.

49. We subtract to find the increase.

$$164,440 - 87,872 = 76,568$$

The increase is 76,568.

Now we find the percent increase.

$$76,568 \text{ is } \underbrace{\text{what percent}} \text{ of } 87,872?$$
$$76,568 = p \cdot 87,872$$

We divide by 87,872 on both sides and then convert to percent notation.

$$76,568 = p \cdot 87,872$$
$$\frac{76,568}{87,872} = \frac{p \cdot 87,872}{87,872}$$
$$0.871 \approx p$$
$$87.1\% \approx p$$

The percent increase is about 87.1%.

50. We subtract to find the decrease, in billions.

$$102.4 - 73.5 = 28.9$$

Now we find the percent decrease.

$$28.9 \text{ is } \underbrace{\text{what percent}} \text{ of } 102.4?$$
$$28.9 = p \cdot 102.4$$

We divide by 102.4 on both sides and then convert to percent notation.

$$28.9 = p \cdot 102.4$$
$$\frac{28.9}{102.4} = \frac{p \cdot 102.4}{102.4}$$
$$0.282 \approx p$$
$$28.2\% \approx p$$

The percent decrease is about 28.2%.

51. ***Familiarize.*** Let $p =$ the price before the reduction.

Translate.

$$\underbrace{\text{Price before reduction}} \text{ minus } 30\% \text{ of price is } \$154.$$
$$p - 30\% \cdot p = 154$$

Solve. We solve the equation.

$$p - 30\% \cdot p = 154$$
$$p - 0.3p = 154$$
$$0.7p = 154$$
$$\frac{0.7p}{0.7} = \frac{154}{0.7}$$
$$p = 220$$

Check. 30% of \$220 is $0.3 \cdot \$220 = \$66$ and $\$220 - \$66 = \$154$, so the answer checks.

State. The price before the reduction was \$220.

52. ***Familiarize.*** Let $s =$ the previous salary.

Translate.

$$\underbrace{\text{Previous salary}} \text{ plus } 8\% \text{ of } \underbrace{\text{previous salary}} \text{ is } \$78,300.$$
$$s + 8\% \cdot s = 78,300$$

Solve. We solve the equation.

$$s + 8\% \cdot s = 78,300$$
$$s + 0.08s = 78,300$$
$$1.08s = 78,300$$
$$\frac{1.08s}{1.08} = \frac{78,300}{1.08}$$
$$s = 72,500$$

Check. 8% of $\$72,500 = 0.08 \cdot \$72,500 = \$5800$ and $\$72,500 + \$5800 = \$78,300$, so the answer checks.

State. The previous salary was \$72,500.

53. ***Familiarize.*** Let $a =$ the amount the organization actually owes. This is the price of the supplies without sales tax added. Then the incorrect amount is $a + 5\%$ of a, or $a + 0.05a$, or $1.05a$.

Translate.

$$\underbrace{\text{Incorrect amount}} \text{ is } \$145.90.$$
$$1.05a = 145.90$$

Solve. We solve the equation.

$$1.05a = 145.90$$
$$\frac{1.05a}{1.05} = \frac{145.90}{1.05}$$
$$a \approx 138.95$$

Check. 5% of \$138.95 is $0.05 \cdot \$138.95 \approx \$6.95$, and $\$138.95 + \$6.95 = \$145.90$, so the answer checks.

State. The organization actually owes \$138.95.

54. ***Familiarize.*** Let s represent the score on the next test.

Translate.

$$\underbrace{\text{The average score}}\ \underbrace{\text{is at least}}\ 80.$$

$$\frac{71+75+82+86+s}{5} \geq 80$$

Solve.

$$\frac{71+75+82+86+s}{5} \geq 80$$

$$5\left(\frac{71+75+82+86+s}{5}\right) \geq 5\cdot 80$$

$$71+75+82+86+s \geq 400$$

$$314+s \geq 400$$

$$s \geq 86$$

Check. As a partial check we show that the average is at least 80 when the next test score is 86.

$$\frac{71+75+82+86+86}{5} = \frac{400}{5} = 80$$

State. The lowest grade Noah can get on the next test and have an average test score of 80 is 86.

55. ***Familiarize.*** Let w represent the width of the rectangle, in cm. The perimeter is given by $P = 2l+2w$, or $2\cdot 43+2w$, or $86+2w$.

Translate.

$$\underbrace{\text{The perimeter}}\ \underbrace{\text{is greater than}}\ \underbrace{120\text{ cm}}.$$

$$86+2w \quad > \quad 120$$

Solve.

$$86+2w > 120$$

$$2w > 34$$

$$w > 17$$

Check. We check to see if the solution seems reasonable.

When $w = 16$ cm, $P = 2\cdot 43+2\cdot 16$, or 118 cm.

When $w = 17$ cm, $P = 2\cdot 43+2\cdot 17$, or 120 cm.

When $w = 18$ cm, $P = 2\cdot 43+2\cdot 18$, or 122 cm.

It appears that the solution is correct.

State. The solution set is $\{w|w > 17\text{ cm}\}$.

56.

$$4(3x-5)+6 = 8+x$$

$$12x-20+6 = 8+x$$

$$12x-14 = 8+x$$

$$12x-14-x = 8+x-x$$

$$11x-14 = 8$$

$$11x-14+14 = 8+14$$

$$11x = 22$$

$$\frac{11x}{11} = \frac{22}{11}$$

$$x = 2$$

The solution is 2. This is between 1 and 5, so the correct answer is C.

57.

$$3x+4y = P$$

$$3x+4y-3x = P-3x$$

$$4y = P-3x$$

$$\frac{4y}{4} = \frac{P-3x}{4}$$

$$y = \frac{P-3x}{4}$$

Answer A is correct.

58.

$$2|x|+4 = 50$$

$$2|x| = 46$$

$$|x| = 23$$

The solutions are the numbers whose distance from 0 is 23. Those numbers are -23 and 23.

59. $|3x| = 60$

The solutions are the values of x for which the distance of $3\cdot x$ from 0 is 60. Then we have:

$$3x = -60 \quad or \quad 3x = 60$$

$$x = -20 \quad or \quad x = 20$$

The solutions are -20 and 20.

60.

$$y = 2a-ab+3$$

$$y-3 = 2a-ab$$

$$y-3 = a(2-b)$$

$$\frac{y-3}{2-b} = a$$

Chapter 2 Discussion and Writing Exercises

1. The end result is the same either way. If s is the original salary, the new salary after a 5% raise followed by an 8% raise is $1.08(1.05s)$. If the raises occur in the opposite order, the new salary is $1.05(1.08s)$. By the commutative and associate laws of multiplication, we see that these are equal. However, it would be better to receive the 8% raise first, because this increase yields a higher salary initially than a 5% raise.

2. No; Erin paid 75% of the original price and was offered credit for 125% of this amount, not to be used on sale items. Now 125% of 75% is 93.75%, so Erin would have a credit of 93.75% of the original price. Since this credit can be applied only to nonsale items, she has less purchasing power than if the amount she paid were refunded and she could spend it on sale items.

3. The inequalities are equivalent by the multiplication principle for inequalities. If we multiply both sides of one inequality by -1, the other inequality results.

4. For any pair of numbers, their relative position on the number line is reversed when both are multiplied by the same negative number. For example, -3 is to the left of 5 on the number line ($-3 < 5$), but 12 is to the right of -20. That is, $-3(-4) > 5(-4)$.

5. Answers may vary. Fran is more than 3 years older than Todd.

6. Let n represent "a number." Then "five more than a number" translates to $n+5$, or $5+n$, and "five is more than a number" translates to $5>n$.

Chapter 2 Test

1.
$$x+7=15$$
$$x+7-7=15-7$$
$$x=8$$
The solution is 8.

2.
$$t-9=17$$
$$t-9+9=17+9$$
$$t=26$$
The solution is 26.

3.
$$3x=-18$$
$$\frac{3x}{3}=\frac{-18}{3}$$
$$x=-6$$
The solution is -6.

4.
$$-\frac{4}{7}x=-28$$
$$-\frac{7}{4}\cdot\left(-\frac{4}{7}x\right)=-\frac{7}{4}\cdot(-28)$$
$$x=\frac{7\cdot\cancel{4}\cdot 7}{\cancel{4}\cdot 1}$$
$$x=49$$
The solution is 49.

5.
$$3t+7=2t-5$$
$$3t+7-2t=2t-5-2t$$
$$t+7=-5$$
$$t+7-7=-5-7$$
$$t=-12$$
The solution is -12.

6.
$$\frac{1}{2}x-\frac{3}{5}=\frac{2}{5}$$
$$\frac{1}{2}x-\frac{3}{5}+\frac{3}{5}=\frac{2}{5}+\frac{3}{5}$$
$$\frac{1}{2}x=1$$
$$2\cdot\frac{1}{2}x=2\cdot 1$$
$$x=2$$
The solution is 2.

7.
$$8-y=16$$
$$8-y-8=16-8$$
$$-y=8$$
$$-1\cdot(-1\cdot y)=-1\cdot 8$$
$$y=-8$$
The solution is -8.

8.
$$-\frac{2}{5}+x=-\frac{3}{4}$$
$$-\frac{2}{5}+x+\frac{2}{5}=-\frac{3}{4}+\frac{2}{5}$$
$$x=-\frac{15}{20}+\frac{8}{20}$$
$$x=-\frac{7}{20}$$
The solution is $-\frac{7}{20}$.

9.
$$3(x+2)=27$$
$$3x+6=27$$
$$3x+6-6=27-6$$
$$3x=21$$
$$\frac{3x}{3}=\frac{21}{3}$$
$$x=7$$
The solution is 7.

10.
$$-3x-6(x-4)=9$$
$$-3x-6x+24=9$$
$$-9x+24=9$$
$$-9x+24-24=9-24$$
$$-9x=-15$$
$$\frac{-9x}{-9}=\frac{-15}{-9}$$
$$x=\frac{\cancel{3}\cdot 5}{\cancel{3}\cdot 3}$$
$$x=\frac{5}{3}$$
The solution is $\frac{5}{3}$.

11. We multiply by 10 to clear the decimals.
$$0.4p+0.2=4.2p-7.8-0.6p$$
$$10(0.4p+0.2)=10(4.2p-7.8-0.6p)$$
$$4p+2=42p-78-6p$$
$$4p+2=36p-78$$
$$4p+2-36p=36p-78-36p$$
$$-32p+2=-78$$
$$-32p+2-2=-78-2$$
$$-32p=-80$$
$$\frac{-32p}{-32}=\frac{-80}{-32}$$
$$p=\frac{5\cdot\cancel{16}}{2\cdot\cancel{16}}$$
$$p=\frac{5}{2}$$
The solution is $\frac{5}{2}$.

12. $$\begin{aligned} 4(3x-1)+11 &= 2(6x+5)-8 \\ 12x-4+11 &= 12x+10-8 \\ 12x+7 &= 12x+2 \\ 12x+7-12x &= 12x+2-12x \\ 7 &= 2 \qquad \text{FALSE} \end{aligned}$$

There are no solutions.

13. $$\begin{aligned} -2+7x+6 &= 5x+4+2x \\ 7x+4 &= 7x+4 \\ 7x+4-7x &= 7x+4-7x \\ 4 &= 4 \qquad \text{TRUE} \end{aligned}$$

All real numbers are solutions.

14. $$\begin{aligned} x+6 &\le 2 \\ x+6-6 &\le 2-6 \\ x &\le -4 \end{aligned}$$

The solution set is $\{x|x \le -4\}$.

15. $$\begin{aligned} 14x+9 &> 13x-4 \\ 14x+9-13x &> 13x-4-13x \\ x+9 &> -4 \\ x+9-9 &> -4-9 \\ x &> -13 \end{aligned}$$

The solution set is $\{x|x > -13\}$.

16. $$\begin{aligned} 12x &\le 60 \\ \frac{12x}{12} &\le \frac{60}{12} \\ x &\le 5 \end{aligned}$$

The solution set is $\{x|x \le 5\}$.

17. $$\begin{aligned} -2y &\ge 26 \\ \frac{-2y}{-2} &\le \frac{26}{-2} \qquad \text{Reversing the inequality symbol} \\ y &\le -13 \end{aligned}$$

The solution set is $\{y|y \le -13\}$.

18. $$\begin{aligned} -4y &\le -32 \\ \frac{-4y}{-4} &\ge \frac{-32}{-4} \qquad \text{Reversing the inequality symbol} \\ y &\ge 8 \end{aligned}$$

The solution set is $\{y|y \ge 8\}$.

19. $$\begin{aligned} -5x &\ge \frac{1}{4} \\ -\frac{1}{5}\cdot(-5x) &\le -\frac{1}{5}\cdot\frac{1}{4} \qquad \text{Reversing the inequality symbol} \\ x &\le -\frac{1}{20} \end{aligned}$$

The solution set is $\left\{x\middle|x \le -\frac{1}{20}\right\}$.

20. $$\begin{aligned} 4-6x &> 40 \\ 4-6x-4 &> 40-4 \\ -6x &> 36 \\ \frac{-6x}{-6} &< \frac{36}{-6} \qquad \text{Reversing the inequality symbol} \\ x &< -6 \end{aligned}$$

The solution set is $\{x|x < -6\}$.

21. $$\begin{aligned} 5-9x &\ge 19+5x \\ 5-9x-5x &\ge 19+5x-5x \\ 5-14x &\ge 19 \\ 5-14x-5 &\ge 19-5 \\ -14x &\ge 14 \\ \frac{-14x}{-14} &\le \frac{14}{-14} \qquad \text{Reversing the inequality symbol} \\ x &\le -1 \end{aligned}$$

The solution set is $\{x|x \le -1\}$.

22. The solutions of $y \le 9$ are shown by shading the point for 9 and all points to the left of 9. The closed circle at 9 indicates that 9 is part of the graph.

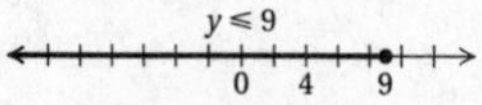

23. $$\begin{aligned} 6x-3 &< x+2 \\ 6x-3-x &< x+2-x \\ 5x-3 &< 2 \\ 5x-3+3 &< 2+3 \\ 5x &< 5 \\ \frac{5x}{5} &< \frac{5}{5} \\ x &< 1 \end{aligned}$$

The solution set is $\{x|x < 1\}$. The graph is as follows:

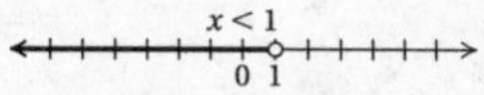

24. In order to be a solution of the inequality $-2 \le x \le 2$, a number must be a solution of both $-2 \le x$ and $x \le 2$. The solution set is graphed as follows:

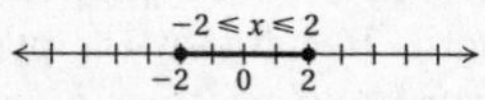

25. *Translate.*

What number is 24% of 75?

$$a = 24\% \cdot 75$$

Solve. We convert 24% to decimal notation and multiply.

$$\begin{aligned} a &= 24\% \cdot 75 \\ a &= 0.24 \cdot 75 \\ a &= 18 \end{aligned}$$

Thus, 18 is 24% of 75.

26. *Translate.*

$$\begin{array}{ccccc} 15.84 & \text{is} & \underbrace{\text{what percent}} & \text{of} & 96? \\ \downarrow & \downarrow & \downarrow & \downarrow & \downarrow \\ 15.84 & = & p & \cdot & 96 \end{array}$$

Solve.

$$15.84 = p \cdot 96$$
$$\frac{15.84}{96} = \frac{p \cdot 96}{96}$$
$$0.165 = p$$
$$16.5\% = p$$

Thus, 15.84 is 16.5% of 96.

27. *Translate.*

$$\begin{array}{ccccc} 800 & \text{is} & 2\% & \text{of} & \underbrace{\text{what number}}? \\ \downarrow & \downarrow & \downarrow & \downarrow & \downarrow \\ 800 & = & 2\% & \cdot & b \end{array}$$

Solve.

$$800 = 2\% \cdot b$$
$$800 = 0.02 \cdot b$$
$$\frac{800}{0.02} = \frac{0.02 \cdot b}{0.02}$$
$$40,000 = b$$

Thus, 800 is 2% of 40,000.

28. We subtract to find the increase.

$$29.2 - 18.2 = 11$$

Now we find the percent of increase.

$$\begin{array}{ccccc} 11 & \text{is} & \underbrace{\text{what percent}} & \text{of} & 18.2? \\ \downarrow & \downarrow & \downarrow & \downarrow & \downarrow \\ 11 & = & p & \cdot & 18.2 \end{array}$$

We divide by 18.2 on both sides and then convert to percent notation.

$$11 = p \cdot 18.2$$
$$\frac{11}{18.2} = \frac{p \cdot 18.2}{18.2}$$
$$0.604 \approx p$$
$$60.4\% \approx p$$

The percent increase is about 60.4%.

29. ***Familiarize.*** Let w = the width of the photograph, in cm. Then $w + 4$ = the length. Recall that the perimeter P of a rectangle with length l and width w is given by $P = 2l + 2w$.

Translate. We substitute 36 for P and $w + 4$ for l in the formula above.

$$P = 2l + 2w$$
$$36 = 2(w + 4) + 2w$$

Solve. We solve the equation.

$$36 = 2(w + 4) + 2w$$
$$36 = 2w + 8 + 2w$$
$$36 = 4w + 8$$
$$36 - 8 = 4w + 8 - 8$$
$$28 = 4w$$
$$\frac{28}{4} = \frac{4w}{4}$$
$$7 = w$$

If $w = 7$, then $w + 4 = 7 + 4 = 11$.

Check. The length, 11 cm, is 4 cm more than the width, 7 cm. The perimeter is $2 \cdot 11$ cm $+ 2 \cdot 7$ cm $= 22$ cm $+ 14$ cm $= 36$ cm. The answer checks.

State. The width is 7 cm, and the length is 11 cm.

30. ***Familiarize.*** Let t = the total cost of raising a child to age 17.

Translate.

$$\begin{array}{ccccc} \underbrace{\text{Cost for child care and K-12 education}} & \text{is} & 18\% & \text{of} & \underbrace{\text{Total cost}} \\ \downarrow & \downarrow & \downarrow & \downarrow & \downarrow \\ 41,500 & = & 18\% & \cdot & t \end{array}$$

Solve.

$$41,500 = 18\% \cdot t$$
$$41,500 = 0.18t$$
$$\frac{41,500}{0.18} = \frac{0.18t}{0.18}$$
$$230,556 \approx t$$

Check. 18% of \$230,556 is about \$41,500, so the answer checks.

State. The total cost of raising a child to age 17 is about \$230,556.

31. ***Familiarize.*** Let x = the first integer. Then $x + 1$ = the second and $x + 2$ = the third.

Translate.

$$\begin{array}{cccccccc} \underbrace{\text{First integer}} & \text{plus} & \underbrace{\text{second integer}} & \text{plus} & \underbrace{\text{third integer}} & \text{is} & 7530. \\ \downarrow & \downarrow & \downarrow & \downarrow & \downarrow & \downarrow & \downarrow \\ x & + & (x + 1) & + & (x + 2) & = & 7530 \end{array}$$

Solve.

$$x + (x + 1) + (x + 2) = 7530$$
$$3x + 3 = 7530$$
$$3x + 3 - 3 = 7530 - 3$$
$$3x = 7527$$
$$\frac{3x}{3} = \frac{7527}{3}$$
$$x = 2509$$

If $x = 2509$, then $x + 1 = 2510$ and $x + 2 = 2511$.

Check. The numbers 2509, 2510, and 2511 are consecutive integers and $2509 + 2510 + 2511 = 7530$. The answer checks.

State. The integers are 2509, 2510, and 2511.

32. ***Familiarize.*** Let $x=$ the amount originally invested. Using the formula for simple interest, $I=Prt$, the interest earned in one year will be $x\cdot 5\%\cdot 1$, or $5\%x$.

Translate.

$$\underbrace{\text{Amount invested}}_{\downarrow\; x}\;\; \underset{\downarrow\; +}{\text{plus}}\;\; \underset{\downarrow\; 5\%x}{\text{interest}}\;\; \underset{\downarrow\; =}{\text{is}}\;\; \underbrace{\text{amount after 1 year.}}_{\downarrow\; 924}$$

Solve. We solve the equation.

$$\begin{aligned} x+5\%x &= 924\\ x+0.05x &= 924\\ 1.05x &= 924\\ \frac{1.05x}{1.05} &= \frac{924}{1.05}\\ x &= 880 \end{aligned}$$

Check. 5% of \$880 is $0.05\cdot \$880=\$44$ and $\$880+\$44=\$924$, so the answer checks.

State. \$880 was originally invested.

33. ***Familiarize.*** Using the labels on the drawing in the text, we let $x=$ the length of the shorter piece, in meters, and $x+2=$ the length of the longer piece.

Translate.

$$\underbrace{\text{Length of shorter piece}}_{\downarrow\; x}\;\; \underset{\downarrow\; +}{\text{plus}}\;\; \underbrace{\text{length of longer piece}}_{\downarrow\; (x+2)}\;\; \underset{\downarrow\; =}{\text{is}}\;\; \underbrace{8\text{ m}}_{\downarrow\; 8}.$$

Solve. We solve the equation.

$$\begin{aligned} x+(x+2) &= 8\\ 2x+2 &= 8\\ 2x+2-2 &= 8-2\\ 2x &= 6\\ \frac{2x}{2} &= \frac{6}{2}\\ x &= 3 \end{aligned}$$

If $x=3$, then $x+2=3+2=5$.

Check. One piece is 2 m longer than the other and the sum of the lengths is 3 m+5 m, or 8 m. The answer checks.

State. The lengths of the pieces are 3 m and 5 m.

34. ***Familiarize.*** Let $l=$ the length of the rectangle, in yd. The perimeter is given by $P=2l+2w$, or $2l+2\cdot 96$, or $2l+192$.

Translate.

$$\underbrace{\text{The perimeter}}_{\downarrow\; 2l+192}\;\; \underbrace{\text{is at least}}_{\downarrow\; \geq}\;\; \underbrace{540\text{ yd}}_{\downarrow\; 540}.$$

Solve.

$$\begin{aligned} 2l+192 &\geq 540\\ 2l &\geq 348\\ l &\geq 174 \end{aligned}$$

Check. We check to see if the solution seems reasonable.

When $l=174$ yd, $P=2\cdot 174+2\cdot 96$, or 540 yd.

When $l=175$ yd, $P=2\cdot 175+2\cdot 96$, or 542 yd.

It appears that the solution is correct.

State. For lengths that are at least 174 yd, the perimeter will be at least 540 yd. The solution set can be expressed as $\{l|l\geq 174\text{ yd}\}$.

35. ***Familiarize.*** Let $s=$ the amount Jason spends in the sixth month.

Translate.

$$\underbrace{\text{Average spending}}_{\downarrow\; \frac{98+89+110+85+83+s}{6}}\;\; \underbrace{\text{is no more than}}_{\downarrow\; \leq}\;\; \underset{\downarrow\; 95}{\$95.}$$

Solve.

$$\begin{aligned} \frac{98+89+110+85+83+s}{6} &\leq 95\\ 6\left(\frac{98+89+110+85+83+s}{6}\right) &\leq 6\cdot 95\\ 98+89+110+85+83+s &\leq 570\\ 465+s &\leq 570\\ s &\leq 105 \end{aligned}$$

Check. As a partial check we show that the average spending is \$95 when Jason spends \$105 in the sixth month.

$$\frac{98+89+110+85+83+105}{6}=\frac{570}{6}=95$$

State. Jason can spend no more than \$105 in the sixth month. The solution set can be expressed as $\{s|s\leq \$105\}$.

36. ***Familiarize.*** Let $c=$ the number of copies made. For 3 months, the rental charge is $3\cdot \$225$, or \$675. Expressing 3.2¢ as \$0.032, the charge for the copies is given by $\$0.032\cdot c$.

Translate.

$$\underbrace{\text{Rental charge}}_{\downarrow\; 675}\;\; \underset{\downarrow\; +}{\text{plus}}\;\; \underbrace{\text{copy charge}}_{\downarrow\; 0.032c}\;\; \underbrace{\text{is no more than}}_{\downarrow\; \leq}\;\; \underset{\downarrow\; 4500}{\$4500.}$$

Solve.

$$\begin{aligned} 675+0.032c &\leq 4500\\ 0.032c &\leq 3825\\ c &\leq 119{,}531 \end{aligned}$$

Check. We check to see if the solution seems reasonable.

When $c=119,530$, the total cost is $\$675+\$0.032(119,530)$, or about \$4499.96.

When $c=119,532$, the total cost is $\$675+\$0.032(119,532)$, or about \$4500.02.

It appears that the solution is correct.

State. No more than 119,531 copies can be made. The solution set can be expressed as $\{c|c\leq 119,531\}$.

37. $A = 2\pi rh$

$$\frac{A}{2\pi h} = \frac{2\pi rh}{2\pi h}$$

$$\frac{A}{2\pi h} = r$$

38. $y = 8x + b$

$$y - b = 8x + b - b$$

$$y - b = 8x$$

$$\frac{y-b}{8} = \frac{8x}{8}$$

$$\frac{y-b}{8} = x$$

39. We subtract to find the increase, in millions.

$$70.3 - 40.4 = 29.9$$

Now we find the percent increase.

$$\underbrace{29.9}\ \text{is}\ \underbrace{\text{what percent}}\ \text{of}\ 40.4?$$

$$29.9 = p \cdot 40.4$$

We divide by 40.4 on both sides and then convert to percent notation.

$$29.9 = p \cdot 40.4$$

$$\frac{29.9}{40.4} = \frac{p \cdot 40.4}{40.4}$$

$$0.74 \approx p$$

$$74\% \approx p$$

The percent increase is about 74%. Answer D is correct.

40. $c = \dfrac{1}{a-d}$

$$(a-d) \cdot c = a - d \cdot \left(\frac{1}{a-d}\right)$$

$$ac - dc = 1$$

$$ac - dc - ac = 1 - ac$$

$$-dc = 1 - ac$$

$$\frac{-dc}{-c} = \frac{1-ac}{-c}$$

$$d = \frac{1-ac}{-c}$$

Since $\dfrac{1-ac}{-c} = \dfrac{-1}{-1} \cdot \dfrac{1-ac}{-c} = \dfrac{-1(1-ac)}{-1(-c)} = \dfrac{-1+ac}{c}$, or $\dfrac{ac-1}{c}$, we can also express the result as $d = \dfrac{ac-1}{c}$.

41. $3|w| - 8 = 37$

$$3|w| = 45$$

$$|w| = 15$$

The solutions are the numbers whose distance from 0 is 15. They are -15 and 15.

42. ***Familiarize.*** Let t = the number of tickets given away.

Translate. We add the number of tickets given to the five people.

$$\frac{1}{3}t + \frac{1}{4}t + \frac{1}{5}t + 8 + 5 = t$$

Solve.

$$\frac{1}{3}t + \frac{1}{4}t + \frac{1}{5}t + 8 + 5 = t$$

$$\frac{20}{60}t + \frac{15}{60}t + \frac{12}{60}t + 8 + 5 = t$$

$$\frac{47}{60}t + 13 = t$$

$$13 = t - \frac{47}{60}t$$

$$13 = \frac{60}{60}t - \frac{47}{60}t$$

$$13 = \frac{13}{60}t$$

$$\frac{60}{13} \cdot 13 = \frac{60}{13} \cdot \frac{13}{60}t$$

$$60 = t$$

Check. $\dfrac{1}{3} \cdot 60 = 20$, $\dfrac{1}{4} \cdot 60 = 15$, $\dfrac{1}{5} \cdot 60 = 12$; then $20 + 15 + 12 + 8 + 5 = 60$. The answer checks.

State. 60 tickets were given away.

Cumulative Review Chapters 1 - 2

1. $\dfrac{y-x}{4} = \dfrac{12-6}{4} = \dfrac{6}{4} = \dfrac{\cancel{2} \cdot 3}{\cancel{2} \cdot 2} = \dfrac{3}{2}$

2. $\dfrac{3x}{y} = \dfrac{3 \cdot 5}{4} = \dfrac{15}{4}$

3. $x - 3 = 3 - 3 = 0$

4. $2w - 4$

5. Since -4 is to the right of -6, we have $-4 > -6$.

6. Since 0 is to the right of -5, we have $0 > -5$.

7. Since -8 is to the left of 7, we have $-8 < 7$.

8. The opposite of $\dfrac{2}{5}$ is $-\dfrac{2}{5}$ because $\dfrac{2}{5} + \left(-\dfrac{2}{5}\right) = 0$.

The reciprocal of $\dfrac{2}{5}$ is $\dfrac{5}{2}$ because $\dfrac{2}{5} \cdot \dfrac{5}{2} = 1$.

9. The distance of 3 from 0 is 3, so $|3| = 3$.

10. The distance of $-\dfrac{3}{4}$ from 0 is $\dfrac{3}{4}$, so $\left|-\dfrac{3}{4}\right| = \dfrac{3}{4}$.

11. The distance of 0 from 0 is 0, so $|0| = 0$.

12. $-6.7 + 2.3$

One negative number and one positive number. The absolute values are 6.7 and 2.3. The difference of the absolute values is $6.7 - 2.3 = 4.4$. The negative number has the larger absolute value, so the sum is negative.

$$-6.7 + 2.3 = -4.4$$

13. $-\dfrac{1}{6} - \dfrac{7}{3} = -\dfrac{1}{6} + \left(-\dfrac{7}{3}\right) = -\dfrac{1}{6} + \left(-\dfrac{14}{6}\right) = -\dfrac{15}{6} = -\dfrac{\cancel{3} \cdot 5}{2 \cdot \cancel{3}} = -\dfrac{5}{2}$

14. $-\frac{5}{8}\left(-\frac{4}{3}\right)=\frac{5\cdot 4}{8\cdot 3}=\frac{5\cdot \cancel{4}}{2\cdot \cancel{4}\cdot 3}=\frac{5}{6}$

15. $(-7)(5)(-6)(-0.5)=-35(3)=-105$

16. $81\div(-9)=-9$

17. $-10.8\div 3.6=-3$

18. $-\frac{4}{5}\div-\frac{25}{8}=-\frac{4}{5}\cdot-\frac{8}{25}=\frac{4\cdot 8}{5\cdot 25}=\frac{32}{125}$

19. $5(3x+5y+2z)=5\cdot 3x+5\cdot 5y+5\cdot 2z=$
$15x+25y+10z$

20. $4(-3x-2)=4(-3x)-4\cdot 2=-12x-8$

21. $-6(2y-4x)=-6\cdot 2y-(-6)(4x)=-12y-(-24x)=$
$-12y+24x$

22. $64+18x+24y=2\cdot 32+2\cdot 9x+2\cdot 12y=2(32+9x+12y)$

23. $16y-56=8\cdot 2y-8\cdot 7=8(2y-7)$

24. $5a-15b+25=5\cdot a-5\cdot 3b+5\cdot 5=5(a-3b+5)$

25.
$$\begin{aligned}9b+18y+6b+4y&=9b+6b+18y+4y\\&=(9+6)b+(18+4)y\\&=15b+22y\end{aligned}$$

26.
$$\begin{aligned}3y+4+6z+6y&=3y+6y+4+6z\\&=(3+6)y+4+6z\\&=9y+4+6z\end{aligned}$$

27.
$$\begin{aligned}-4d-6a+3a-5d+1&=-4d-5d-6a+3a+1\\&=(-4-5)d+(-6+3)a+1\\&=-9d-3a+1\end{aligned}$$

28.
$$\begin{aligned}3.2x+2.9y-5.8x-8.1y&=3.2x-5.8x+2.9y-8.1y\\&=(3.2-5.8)x+(2.9-8.1)y\\&=-2.6x-5.2y\end{aligned}$$

29. $7-2x-(-5x)-8=7-2x+5x-8=-1+3x$

30. $-3x-(-x+y)=-3x+x-y=-2x-y$

31. $-3(x-2)-4x=-3x+6-4x=-7x+6$

32. $10-2(5-4x)=10-10+8x=8x$

33.
$$\begin{aligned}&[3(x+6)-10]-[5-2(x-8)]\\&=[3x+18-10]-[5-2x+16]\\&=[3x+8]-[21-2x]\\&=3x+8-21+2x\\&=5x-13\end{aligned}$$

34.
$$\begin{aligned}x+1.75&=6.25\\x+1.75-1.75&=6.25-1.75\\x&=4.5\end{aligned}$$
The solution is 4.5.

35.
$$\begin{aligned}\frac{5}{2}y&=\frac{2}{5}\\\frac{2}{5}\cdot\frac{5}{2}y&=\frac{2}{5}\cdot\frac{2}{5}\\y&=\frac{4}{25}\end{aligned}$$
The solution is $\frac{4}{25}$.

36.
$$\begin{aligned}-2.6+x&=8.3\\-2.6+x+2.6&=8.3+2.6\\x&=10.9\end{aligned}$$
The solution is 10.9.

37.
$$\begin{aligned}4\tfrac{1}{2}+y&=8\tfrac{1}{3}\\4\tfrac{1}{2}+y-4\tfrac{1}{2}&=8\tfrac{1}{3}-4\tfrac{1}{2}\\y&=8\tfrac{2}{6}-4\tfrac{3}{6}\\y&=7\tfrac{8}{6}-4\tfrac{3}{6}\quad\left(8\tfrac{2}{6}=7+1\tfrac{2}{6}=7+\tfrac{8}{6}=7\tfrac{8}{6}\right)\\y&=3\tfrac{5}{6}\end{aligned}$$
The solution is $3\frac{5}{6}$.

38.
$$\begin{aligned}-\frac{3}{4}x&=36\\-\frac{4}{3}\left(-\frac{3}{4}x\right)&=-\frac{4}{3}\cdot 36\\x&=-\frac{4\cdot 36}{3}=-\frac{4\cdot\cancel{3}\cdot 12}{\cancel{3}\cdot 1}\\x&=-48\end{aligned}$$
The solution is -48.

39.
$$\begin{aligned}\frac{2}{5}x&=-\frac{3}{20}\\\frac{5}{2}\cdot\frac{2}{5}x&=\frac{5}{2}\left(-\frac{3}{20}\right)\\x&=-\frac{3\cdot\cancel{5}}{2\cdot 4\cdot\cancel{5}}\\x&=-\frac{3}{8}\end{aligned}$$
The solution is $-\frac{3}{8}$.

40.
$$\begin{aligned}5.8x&=-35.96\\\frac{5.8x}{5.8}&=\frac{-35.96}{5.8}\\x&=-6.2\end{aligned}$$
The solution is -6.2.

41.
$$\begin{aligned}-4x+3&=15\\-4x+3-3&=15-3\\-4x&=12\\\frac{-4x}{-4}&=\frac{12}{-4}\\x&=-3\end{aligned}$$
The solution is -3.

42.
$$\begin{aligned} -3x+5 &= -8x-7 \\ -3x+5+8x &= -8x-7+8x \\ 5x+5 &= -7 \\ 5x+5-5 &= -7-5 \\ 5x &= -12 \\ \frac{5x}{5} &= \frac{-12}{5} \\ x &= -\frac{12}{5} \end{aligned}$$

The solution is $-\frac{12}{5}$.

43.
$$\begin{aligned} 4y-4+y &= 6y+20-4y \\ 5y-4 &= 2y+20 \\ 5y-4-2y &= 2y+20-2y \\ 3y-4 &= 20 \\ 3y-4+4 &= 20+4 \\ 3y &= 24 \\ \frac{3y}{3} &= \frac{24}{3} \\ y &= 8 \end{aligned}$$

The solution is 8.

44.
$$\begin{aligned} -3(x-2) &= -15 \\ -3x+6 &= -15 \\ -3x+6-6 &= -15-6 \\ -3x &= -21 \\ \frac{-3x}{-3} &= \frac{-21}{-3} \\ x &= 7 \end{aligned}$$

The solution is 7.

45. First we will multiply by the least common multiple of all the denominators to clear the fractions.
$$\begin{aligned} \frac{1}{3}x-\frac{5}{6} &= \frac{1}{2}+2x \\ 6\left(\frac{1}{3}x-\frac{5}{6}\right) &= 6\left(\frac{1}{2}+2x\right) \\ 6\cdot\frac{1}{3}x-6\cdot\frac{5}{6} &= 6\cdot\frac{1}{2}+6\cdot 2x \\ 2x-5 &= 3+12x \\ 2x-5-12x &= 3+12x-12x \\ -10x-5 &= 3 \\ -10x-5+5 &= 3+5 \\ -10x &= 8 \\ \frac{-10x}{-10} &= \frac{8}{-10} \\ x &= -\frac{8}{10} = -\frac{\cancel{2}\cdot 4}{\cancel{2}\cdot 5} \\ x &= -\frac{4}{5} \end{aligned}$$

The solution is $-\frac{4}{5}$.

46. First we will multiply by 10 to clear the decimals.
$$\begin{aligned} -3.7x+6.2 &= -7.3x-5.8 \\ 10(-3.7x+6.2) &= 10(-7.3x-5.8) \\ -37x+62 &= -73x-58 \\ -37x+62+73x &= -73x-58+73x \\ 36x+62 &= -58 \\ 36x+62-62 &= -58-62 \\ 36x &= -120 \\ \frac{36x}{36} &= \frac{-120}{36} \\ x &= -\frac{10\cdot\cancel{12}}{3\cdot\cancel{12}} \\ x &= -\frac{10}{3} \end{aligned}$$

The solution is $-\frac{10}{3}$.

47.
$$\begin{aligned} 4(x+2) &= 4(x-2)+16 \\ 4x+8 &= 4x-8+16 \\ 4x+8 &= 4x+8 \\ 4x+8-4x &= 4x+8-4x \\ 8 &= 8 \qquad \text{TRUE} \end{aligned}$$

All real numbers are solutions.

48.
$$\begin{aligned} 0(x+3)+4 &= 0 \\ 0+4 &= 0 \\ 4 &= 0 \qquad \text{FALSE} \end{aligned}$$

There is no solution.

49.
$$\begin{aligned} 3x-1 &< 2x+1 \\ 3x-1-2x &< 2x+1-2x \\ x-1 &< 1 \\ x-1+1 &< 1+1 \\ x &< 2 \end{aligned}$$

The solution set is $\{x|x<2\}$.

50.
$$\begin{aligned} 3y+7 &> 5y+13 \\ 3y+7-5y &> 5y+13-5y \\ -2y+7 &> 13 \\ -2y+7-7 &> 13-7 \\ -2y &> 6 \\ \frac{-2y}{-2} &< \frac{6}{-2} \qquad \text{Reversing the inequality symbol} \\ y &< -3 \end{aligned}$$

The solution set is $\{y|y<-3\}$.

51.
$$\begin{aligned}
5-y &\le 2y-7\\
5-y-2y &\le 2y-7-2y\\
5-3y &\le -7\\
5-3y-5 &\le -7-5\\
-3y &\le -12\\
\frac{-3y}{-3} &\ge \frac{-12}{-3} \quad \text{Reversing the inequality symbol}\\
y &\ge 4
\end{aligned}$$
The solution set is $\{y|y \ge 4\}$.

52.
$$\begin{aligned}
H &= 65-m\\
H-65 &= 65-m-65\\
H-65 &= -m\\
-1(H-65) &= -1\cdot(-1\cdot m)\\
-H+65 &= m, \text{ or}\\
65-H &= m
\end{aligned}$$

53.
$$\begin{aligned}
I &= Prt\\
\frac{I}{Pr} &= \frac{Prt}{Pr}\\
\frac{I}{Pr} &= t
\end{aligned}$$

54. *Translate.*

$$\begin{array}{ccccccc}
\underbrace{\text{What number}} & \text{is} & 24\% & \text{of} & 105?\\
\downarrow & \downarrow & \downarrow & \downarrow & \downarrow\\
a & = & 24\% & \cdot & 105
\end{array}$$

Solve. We convert 24% to decimal notation and multiply.
$$\begin{aligned}
a &= 24\%\cdot 105\\
a &= 0.24\cdot 105\\
a &= 25.2
\end{aligned}$$
Thus, 25.2 is 24% of 105.

55. *Translate.*

$$\begin{array}{ccccc}
39.6 & \text{is} & \underbrace{\text{what percent}} & \text{of} & 88?\\
\downarrow & \downarrow & \downarrow & \downarrow & \downarrow\\
39.6 & = & p & \cdot & 88
\end{array}$$

Solve. We solve the equation.
$$\begin{aligned}
39.6 &= p\cdot 88\\
\frac{39.6}{88} &= \frac{p\cdot 88}{88}\\
0.45 &= p\\
45\% &= p
\end{aligned}$$
Thus, 39.6 is 45% of 88.

56. ***Translate.***

$$\begin{array}{ccccc}
\$163.35 & \text{is} & 45\% & \text{of} & \underbrace{\text{what number}}?\\
\downarrow & \downarrow & \downarrow & \downarrow & \downarrow\\
163.35 & = & 45\% & \cdot & b
\end{array}$$

Solve.
$$\begin{aligned}
163.35 &= 45\%\cdot b\\
163.35 &= 0.45\cdot b\\
\frac{163.35}{0.45} &= \frac{0.45\cdot b}{0.45}\\
363 &= b
\end{aligned}$$
Thus, \$163.35 is 45% of \$363.

57. ***Familiarize.*** Let $p =$ the price before the reduction.

Translate.

$$\begin{array}{cccccc}
\underline{\text{Price before reduction}} & \text{minus} & 25\% & \text{of} & \text{price} & \text{is} & \$18.45.\\
\downarrow & \downarrow & \downarrow & \downarrow & \downarrow & \downarrow & \downarrow\\
p & - & 25\% & \cdot & p & = & 18.45
\end{array}$$

Solve. We solve the equation.
$$\begin{aligned}
p-25\%\cdot p &= 18.45\\
p-0.25p &= 18.45\\
0.75p &= 18.45\\
\frac{0.75p}{0.75p} &= \frac{18.45}{0.75}\\
p &= 24.6
\end{aligned}$$

Check. 25% of \$24.60 is $0.25\cdot\$24.60 = \$6.15$ and $\$24.60 - \$6.15 = \$18.45$, so the answer checks.

State. The price before the reduction was \$24.60.

58. ***Familiarize.*** Let $m =$ the amount Melinda paid for her rollerblades. Then $m+17 =$ the amount Susan paid for hers.

Translate.

$$\begin{array}{ccccc}
\underbrace{\begin{matrix}\text{Amount}\\ \text{Melinda paid}\end{matrix}} & \text{plus} & \underbrace{\begin{matrix}\text{amount}\\ \text{Susan paid}\end{matrix}} & \text{is} & \$107.\\
\downarrow & \downarrow & \downarrow & \downarrow & \downarrow\\
m & + & (m+17) & = & 107
\end{array}$$

Solve.
$$\begin{aligned}
m+(m+17) &= 107\\
2m+17 &= 107\\
2m+17-17 &= 107-17\\
2m &= 90\\
\frac{2m}{2} &= \frac{90}{2}\\
m &= 45
\end{aligned}$$

The exercise asks only for the amount Melinda paid, but we also find the amount Susan paid so that we can check the answer.

If $m = 45$, then $m+17 = 45+17 = 62$.

Check. \$62 is \$17 more than \$45, and $\$45 + \$62 = \$107$. The answer checks.

State. Melinda paid \$45 for her rollerblades.

59. ***Familiarize.*** Let $x =$ the amount originally invested. Using the formula for simple interest, $I = Prt$, the interest earned in one year will be $x\cdot 8\%\cdot 1$, or $8\%x$.

Translate.

$$\underbrace{\text{Amount invested}}_{\downarrow\; x} \quad \underset{+}{\text{plus}} \quad \underset{8\%x}{\text{interest}} \quad \underset{=}{\text{is}} \quad \underbrace{\text{amount after 1 year.}}_{\downarrow\; 1134}$$

Solve.

$$x + 8\%x = 1134$$
$$x + 0.08x = 1134$$
$$1.08x = 1134$$
$$\frac{1.08x}{1.08} = \frac{1134}{1.08}$$
$$x = 1050$$

Check. 8% of \$1050 is 0.08·\$1050 = \$84 and \$1050+\$84 = \$1134, so the answer checks.

State. \$1050 was originally invested.

60. ***Familiarize.*** Let l = the length of the first piece of wire, in meters. Then $l+3$ = the length of the second piece and $\frac{4}{5}l$ = the length of the third piece.

Translate.

$$\underbrace{\text{Length of first piece}}_{\downarrow\; l} \quad \underset{+}{\text{plus}} \quad \underbrace{\text{length of second piece}}_{\downarrow\; (l+3)} \quad \underset{+}{\text{plus}} \quad \underbrace{\text{length of third piece}}_{\downarrow\; \frac{4}{5}l} \quad \underset{=}{\text{is}} \quad \underbrace{143 \text{ m}}_{\downarrow\; 143}.$$

Solve.

$$l + (l+3) + \frac{4}{5}l = 143$$
$$\frac{5}{5}l + \frac{5}{5}l + 3 + \frac{4}{5}l = 143$$
$$\frac{14}{5}l + 3 = 143$$
$$\frac{14}{5}l + 3 - 3 = 143 - 3$$
$$\frac{14}{5}l = 140$$
$$\frac{5}{14}\cdot\frac{14}{5}l = \frac{5}{14}\cdot 140$$
$$l = \frac{5\cdot 10\cdot \cancel{14}}{\cancel{14}\cdot 1}$$
$$l = 50$$

If $l = 50$, then $l + 3 = 50 + 3 = 53$ and $\frac{4}{5}l = \frac{4}{5}\cdot 50 = 40$.

Check. The second piece is 3 m longer than the first piece and the third piece is $\frac{4}{5}$ as long as the first. Also, 50 m + 53 m + 40 m = 143 m. The answer checks.

State. The lengths of the pieces are 50 m, 53 m, and 40 m.

61. ***Familiarize.*** Let s = Nadia's score on the fourth test.

Translate.

$$\underbrace{\text{The average score}}_{\downarrow\; \frac{82+76+78+s}{4}} \quad \underbrace{\text{is at least}}_{\downarrow\; \geq} \quad \underset{80}{80.}$$

Solve.

$$\frac{82+76+78+s}{4} \geq 80$$
$$4\left(\frac{82+76+78+s}{4}\right) \geq 4\cdot 80$$
$$82+76+78+s \geq 320$$
$$236 + s \geq 320$$
$$s \geq 84$$

Check. As a partial check we show that the average is at least 80 when the fourth test score is 84.

$$\frac{82+76+78+84}{4} = \frac{320}{4} = 80$$

State. Scores greater than or equal to 84 will earn Nadia at least a B. The solution set is $\{s|s \geq 84\}$.

62. $-125 \div 25 \cdot 625 \div 5 = -5 \cdot 625 \div 5$
$= -3125 \div 5$
$= -625$

Answer C is correct.

63. ***Familiarize.*** Let s = the salary at the beginning of the year. After a 4% increase the new salary is $s + 4\%s$, or $s + 0.04s$, or $1.04s$. Then after a 3% cost-of-living adjustment the final salary is $1.04s + 3\% \cdot 1.04s$, or $1.04s + 0.03 \cdot 1.04s$, or $1.04s + 0.0312s$, or $1.0712s$.

Translate.

$$\underbrace{\text{Final salary}}_{\downarrow\; 1.0712s} \quad \underset{=}{\text{is}} \quad \underset{48,418.24}{\$48,418.24.}$$

Solve.

$$1.0712s = 48,418.24$$
$$\frac{1.0712s}{1.0712} = \frac{48,418.24}{1.0712}$$
$$s = 45,200$$

Check. 4% of \$45,200 is 0.04 · \$45,200 = \$1808 and \$45,200 + \$1808 = \$47,008. Then 3% of \$47,008 is 0.03 · \$47,008 = \$1410.24 and \$47,008 + \$1410.24 = \$48,418.24. The answer checks.

State. At the beginning of the year the salary was \$45,200.

64. First we subtract to find the amount of the reduction.

9 in. − 6.3 in. = 2.7 in.

Translate.

$$\underbrace{2.7 \text{ in.}}_{\downarrow\; 2.7} \quad \underset{=}{\text{is}} \quad \underbrace{\text{what percent}}_{\downarrow\; p} \quad \underset{\cdot}{\text{of}} \quad \underbrace{9 \text{ in.}}_{\downarrow\; 9}?$$

Solve.

$$2.7 = p \cdot 9$$
$$\frac{2.7}{9} = \frac{p\cdot 9}{9}$$
$$0.3 = p$$
$$30\% = p$$

The drawing should be reduced 30%.

65. $4|x| - 13 = 3$

$$4|x| = 16$$

$$|x| = 4$$

The solutions are the numbers whose distance from 0 is 4. They are -4 and 4.

66. First we multiply by 28 to clear the fractions.

$$\frac{2+5x}{4} = \frac{11}{28} + \frac{8x+3}{7}$$

$$28\left(\frac{2+5x}{4}\right) = 28\left(\frac{11}{28} + \frac{8x+3}{7}\right)$$

$$\frac{28(2+5x)}{4} = 28 \cdot \frac{11}{28} + \frac{28(8x+3)}{7}$$

$$7(2+5x) = 11 + 4(8x+3)$$

$$14 + 35x = 11 + 32x + 12$$

$$14 + 35x = 32x + 23$$

$$14 + 3x = 23$$

$$3x = 9$$

$$x = 3$$

The solution is 3.

67. $$p = \frac{2}{m+Q}$$

$$(m+Q) \cdot p = (m+Q) \cdot \frac{2}{m+Q}$$

$$mp + Qp = 2$$

$$Qp = 2 - mp$$

$$Q = \frac{2 - mp}{p}$$

Chapter 3

Graphs of Linear Equations

Exercise Set 3.1

RC1. The statement is true. See page 218 in the text.

RC3. The statement is false. See page 215 in the text.

1. $(2,5)$ is 2 units right and 5 units up.

$(-1,3)$ is 1 unit left and 3 units up.

$(3,-2)$ is 3 units right and 2 units down.

$(-2,-4)$ is 2 units left and 4 units down.

$(0,4)$ is 0 units left or right and 4 units up.

$(0,-5)$ is 0 units left or right and 5 units down.

$(5,0)$ is 5 units right and 0 units up or down.

$(-5,0)$ is 5 units left and 0 units up or down.

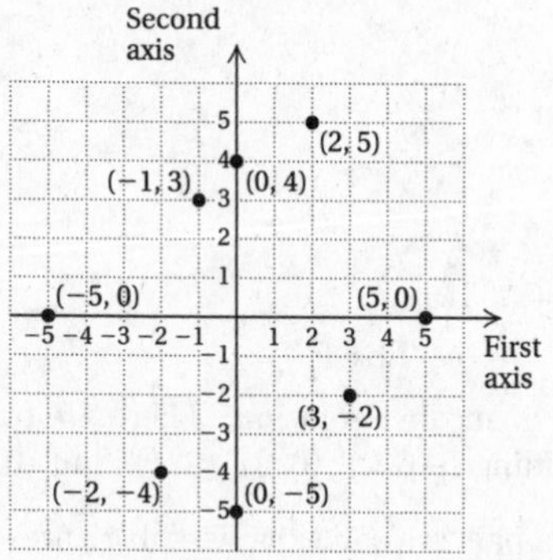

3. Since the first coordinate is negative and the second coordinate positive, the point $(-5,3)$ is located in quadrant II.

5. Since the first coordinate is positive and the second coordinate negative, the point $(100,-1)$ is in quadrant IV.

7. Since both coordinates are negative, the point $(-6,-29)$ is in quadrant III.

9. Since one of the coordinates is 0, the point $(3.8,0)$ lies on an axis. It is not in a quadrant.

11. Since the first coordinate is negative and the second coordinate is positive, the point $\left(-\frac{1}{3},\frac{15}{7}\right)$ is in quadrant II.

13. Since the first coordinate is positive and the second coordinate is negative, the point $\left(12\frac{7}{8},-1\frac{1}{2}\right)$ is in quadrant IV.

15.

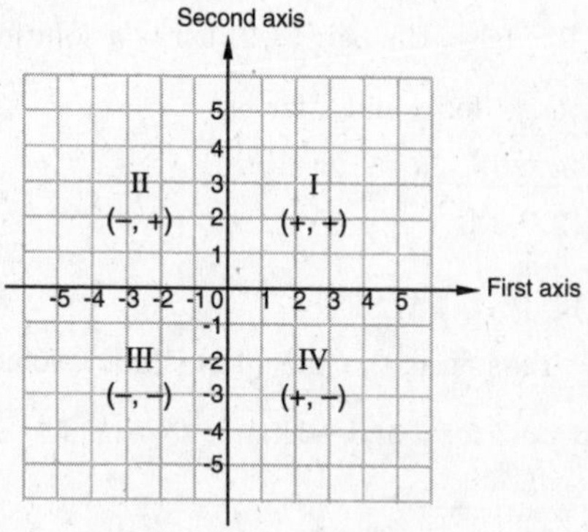

If the first coordinate is negative and the second coordinate is positive, the point is in quadrant II.

17. See the figure in Exercise 15.

If the first coordinate is positive, then the point must be in either quadrant I or quadrant IV.

19. If the first and second coordinates are equal, they must either be both positive or both negative. The point must be in either quadrant I (both positive) or quadrant III (both negative).

21.

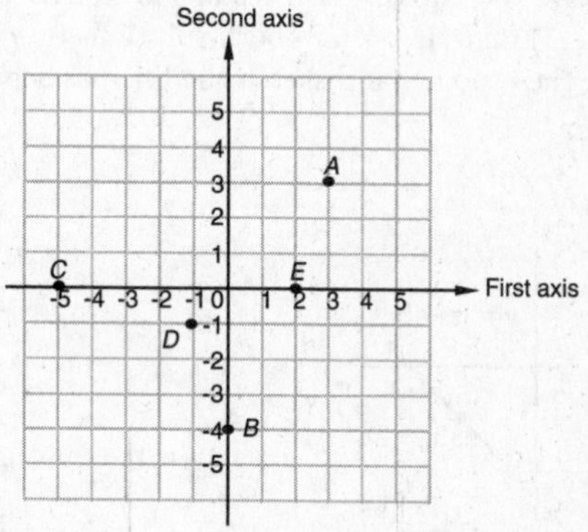

Point A is 3 units right and 3 units up. The coordinates of A are $(3,3)$.

Point B is 0 units left or right and 4 units down. The coordinates of B are $(0,-4)$.

Point C is 5 units left and 0 units up or down. The coordinates of C are $(-5,0)$.

Point D is 1 unit left and 1 unit down. The coordinates of D are $(-1,-1)$.

Point E is 2 units right and 0 units up or down. The coordinates of E are $(2,0)$.

23. We substitute 2 for x and 9 for y (alphabetical order of variables).

$$\begin{array}{l|l} \multicolumn{2}{c}{y = 3x - 1} \\ \hline 9 \ ? & 3 \cdot 2 - 1 \\ & 6 - 1 \\ & 5 \qquad \text{FALSE} \end{array}$$

Since $9 = 5$ is false, the pair $(2, 9)$ is not a solution.

25. We substitute 4 for x and 2 for y.

$$\begin{array}{r|l} \multicolumn{2}{c}{2x + 3y = 12} \\ \hline 2 \cdot 4 + 3 \cdot 2 & ? \ 12 \\ 8 + 6 & \\ 14 & \qquad \text{FALSE} \end{array}$$

Since $14 = 12$ is false, the pair (4,2) is not a solution.

27. We substitute 3 for a and -1 for b.

$$\begin{array}{r|l} \multicolumn{2}{c}{3a - 4b = 13} \\ \hline 3 \cdot 3 - 4(-1) & ? \ 13 \\ 9 + 4 & \\ 13 & \qquad \text{TRUE} \end{array}$$

Since $13 = 13$ is true, the pair $(3, -1)$ is a solution.

29. To show that a pair is a solution, we substitute, replacing x with the first coordinate and y with the second coordinate in each pair.

$$\begin{array}{l|l} \multicolumn{2}{c}{y = x - 5} \\ \hline -1 \ ? & 4 - 5 \\ & -1 \qquad \text{TRUE} \end{array} \qquad \begin{array}{l|l} \multicolumn{2}{c}{y = x - 5} \\ \hline -4 \ ? & 1 - 5 \\ & -4 \qquad \text{TRUE} \end{array}$$

In each case the substitution results in a true equation. Thus, $(4, -1)$ and $(1, -4)$ are both solutions of $y = x - 5$. We graph these points and sketch the line passing through them.

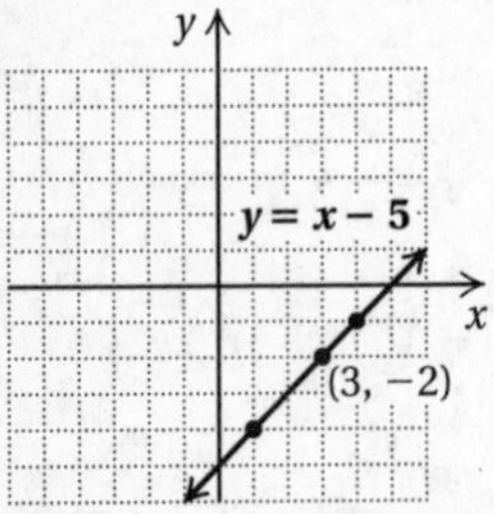

The line appears to pass through $(3, -2)$ also. We check to determine if $(3, -2)$ is a solution of $y = x - 5$.

$$\begin{array}{l|l} \multicolumn{2}{c}{y = x - 5} \\ \hline -2 \ ? & 3 - 5 \\ & -2 \qquad \text{TRUE} \end{array}$$

Thus, $(3, -2)$ is another solution. There are other correct answers, including $(-1, -6)$, $(2, -3)$, $(0, -5)$, $(5, 0)$, and $(6, 1)$.

31. To show that a pair is a solution, we substitute, replacing x with the first coordinate and y with the second coordinate in each pair.

$$\begin{array}{l|l} \multicolumn{2}{c}{y = \frac{1}{2}x + 3} \\ \hline 5 \ ? & \frac{1}{2} \cdot 4 + 3 \\ & 2 + 3 \\ & 5 \qquad \text{TRUE} \end{array} \qquad \begin{array}{l|l} \multicolumn{2}{c}{y = \frac{1}{2}x + 3} \\ \hline 2 \ ? & \frac{1}{2}(-2) + 3 \\ & -1 + 3 \\ & 2 \qquad \text{TRUE} \end{array}$$

In each case the substitution results in a true equation. Thus, $(4, 5)$ and $(-2, 2)$ are both solutions of $y = \frac{1}{2}x + 3$. We graph these points and sketch the line passing through them.

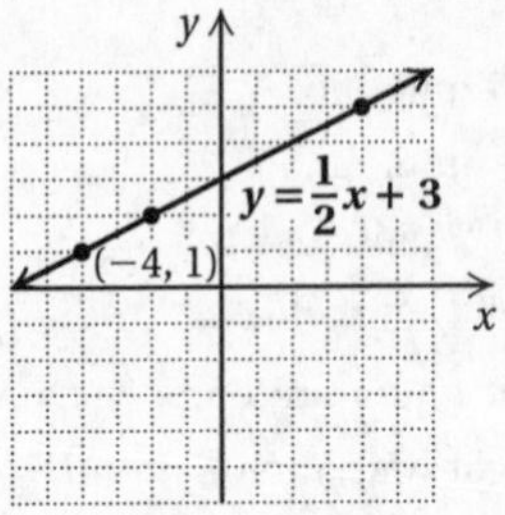

The line appears to pass through $(-4, 1)$ also. We check to determine if $(-4, 1)$ is a solution of $y = \frac{1}{2}x + 3$.

$$\begin{array}{l|l} \multicolumn{2}{c}{y = \frac{1}{2}x + 3} \\ \hline 1 \ ? & \frac{1}{2}(-4) + 3 \\ & -2 + 3 \\ & 1 \qquad \text{TRUE} \end{array}$$

Thus, $(-4, 1)$ is another solution. There are other correct answers, including $(-6, 0)$, $(0, 3)$, $(2, 4)$, and $(6, 6)$.

33. To show that a pair is a solution, we substitute, replacing x with the first coordinate and y with the second coordinate in each pair.

$$\begin{array}{r|l} \multicolumn{2}{c}{4x - 2y = 10} \\ \hline 4 \cdot 0 - 2(-5) & ? \ 10 \\ 10 & \qquad \text{TRUE} \end{array}$$

$$\begin{array}{r|l} \multicolumn{2}{c}{4x - 2y = 10} \\ \hline 4 \cdot 4 - 2 \cdot 3 & ? \ 10 \\ 16 - 6 & \\ 10 & \qquad \text{TRUE} \end{array}$$

In each case the substitution results in a true equation. Thus, $(0, -5)$ and $(4, 3)$ are both solutions of $4x - 2y = 10$. We graph these points and sketch the line passing through them.

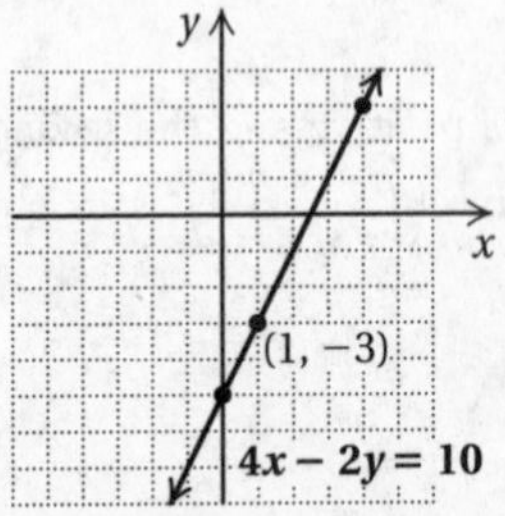

The line appears to pass through $(1, -3)$ also. We check to determine if $(1, -3)$ is a solution of $4x - 2y = 10$.

$4x - 2y = 10$	
$4 \cdot 1 - 2(-3)$	? 10
$4 + 6$	
10	TRUE

Thus, $(1, -3)$ is another solution. There are other correct answers, including $(2, -1)$, $(3, 1)$, and $(5, 5)$.

35. $y = x + 1$

The equation is in the form $y = mx + b$. The y-intercept is $(0, 1)$. We find five other pairs.

When $x = -2$, $y = -2 + 1 = -1$.

When $x = -1$, $y = -1 + 1 = 0$.

When $x = 1$, $y = 1 + 1 = 2$.

When $x = 2$, $y = 2 + 1 = 3$.

When $x = 3$, $y = 3 + 1 = 4$.

x	y
-2	-1
-1	0
0	1
1	2
2	3
3	4

Plot these points, draw the line they determine, and label the graph $y = x + 1$.

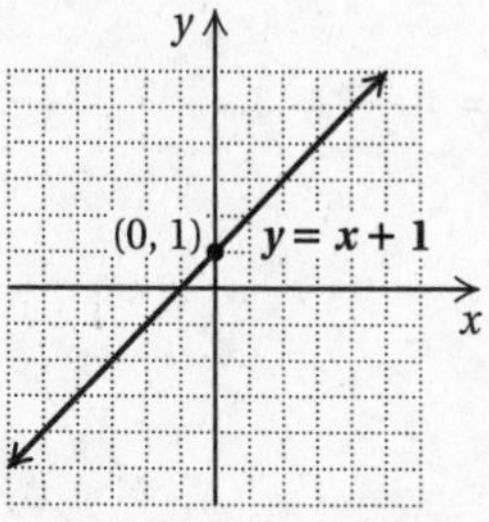

37. $y = x$

The equation is equivalent to $y = x + 0$. The y-intercept is $(0, 0)$. We find five other points.

When $x = -2$, $y = -2$.

When $x = -1$, $y = -1$.

When $x = 1$, $y = 1$.

When $x = 2$, $y = 2$.

When $x = 3$, $y = 3$.

x	y
-2	-2
-1	-1
0	0
1	1
2	2
3	3

Plot these points, draw the line they determine, and label the graph $y = x$.

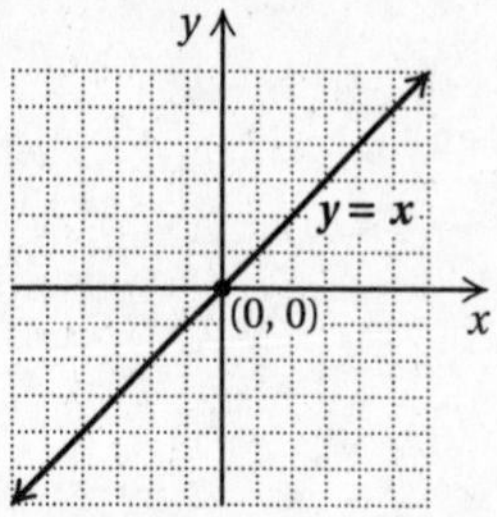

39. $y = \frac{1}{2}x$

The equation is equivalent to $y = \frac{1}{2}x + 0$. The y-intercept is $(0, 0)$. We find two other points.

When $x = -2$, $y = \frac{1}{2}(-2) = -1$.

When $x = 4$, $y = \frac{1}{2} \cdot 4 = 2$.

x	y
-2	-1
0	0
4	2

Plot these points, draw the line they determine, and label the graph $y = \frac{1}{2}x$.

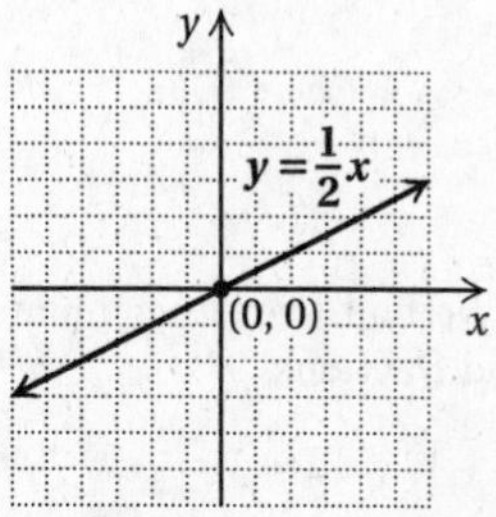

41. $y = x - 3$

The equation is equivalent to $y = x + (-3)$. The y-intercept is $(0, -3)$. We find two other points.

When $x = -2$, $y = -2 - 3 = -5$.

When $x = 4$, $y = 4 - 3 = 1$.

x	y
-2	-5
0	-3
4	1

Plot these points, draw the line they determine, and label the graph $y = x - 3$.

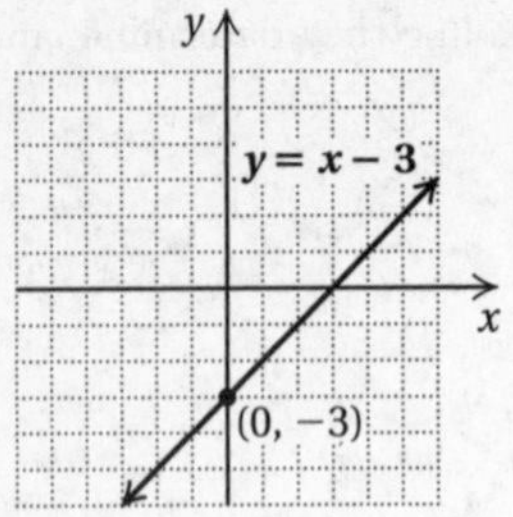

43. $y = 3x - 2 = 3x + (-2)$

The y-intercept is $(0, -2)$. We find two other points.

When $x = -1$, $y = 3(-1) - 2 = -3 - 2 = -5$.

When $x = 2$, $y = 3 \cdot 2 - 2 = 6 - 2 = 4$.

x	y
-1	-5
0	-2
2	4

Plot these points, draw the line they determine, and label the graph $y = 3x - 2$.

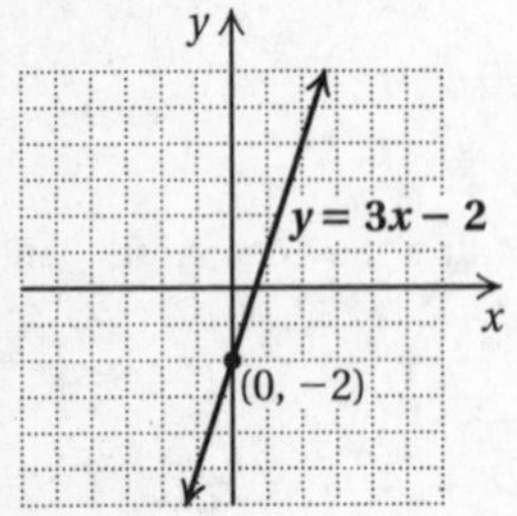

45. $y = \frac{1}{2}x + 1$

The y-intercept is $(0, 1)$. We find two other points using multiples of 2 for x to avoid fractions.

When $x = -4$, $y = \frac{1}{2}(-4) + 1 = -2 + 1 = -1$.

When $x = 4$, $y = \frac{1}{2} \cdot 4 + 1 = 2 + 1 = 3$.

x	y
-4	-1
0	1
4	3

Plot these points, draw the line they determine, and label the graph $y = \frac{1}{2}x + 1$.

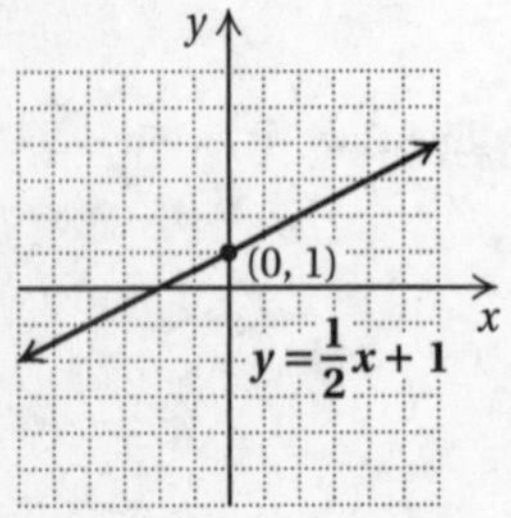

47. $x + y = -5$

$y = -x - 5$

$y = -x + (-5)$

The y-intercept is $(0, -5)$. We find two other points.

When $x = -4$, $y = -(-4) - 5 = 4 - 5 = -1$.

When $x = -1$, $y = -(-1) - 5 = 1 - 5 = -4$.

x	y
-4	-1
0	-5
-1	-4

Plot these points, draw the line they determine, and label the graph $x + y = -5$.

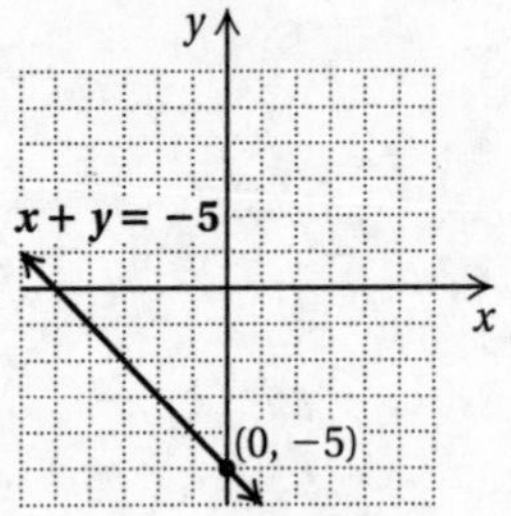

49. $y = \frac{5}{3}x - 2 = \frac{5}{3}x + (-2)$

The y-intercept is $(0, -2)$. We find two other points using multiples of 3 for x to avoid fractions.

When $x = -3$, $y = \frac{5}{3}(-3) - 2 = -5 - 2 = -7$.

When $x = 3$, $y = \frac{5}{3} \cdot 3 - 2 = 5 - 2 = 3$.

x	y
-3	-7
0	-2
3	3

Plot these points, draw the line they determine, and label the graph $y = \frac{5}{3}x - 2$.

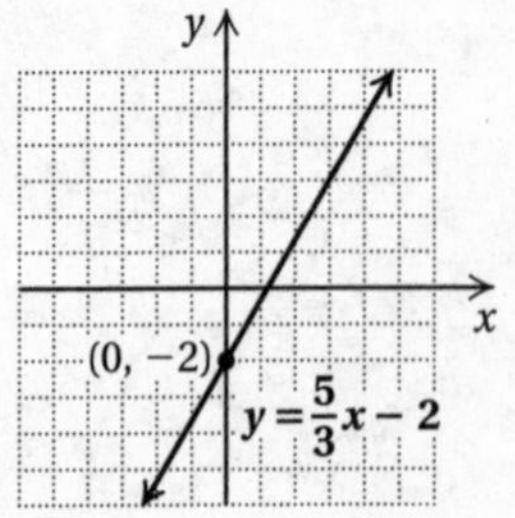

51. $x + 2y = 8$

$2y = -x + 8$

$y = -\frac{1}{2}x + 4$

The y-intercept is $(0, 4)$. We find two other points using multiples of 2 for x to avoid fractions.

When $x = -2$, $y = -\frac{1}{2}(-2) + 4 = 1 + 4 = 5$.

When $x = 4$, $y = -\frac{1}{2} \cdot 4 + 4 = -2 + 4 = 2$.

x	y
-2	5
0	4
4	2

Plot these points, draw the line they determine, and label the graph $x + 2y = 8$.

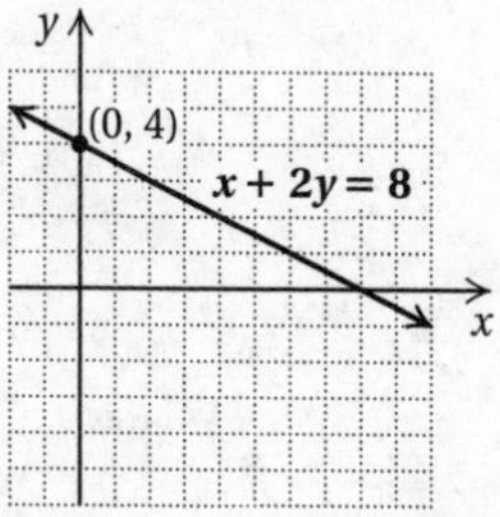

53. $y = \frac{3}{2}x + 1$

The y-intercept is $(0, 1)$. We find two other points using multiples of 2 for x to avoid fractions.

When $x = -4$, $y = \frac{3}{2}(-4) + 1 = -6 + 1 = -5$.

When $x = 2$, $y = \frac{3}{2} \cdot 2 + 1 = 3 + 1 = 4$.

x	y
-4	-5
0	1
2	4

Plot these points, draw the line they determine, and label the graph $y = \frac{3}{2}x + 1$.

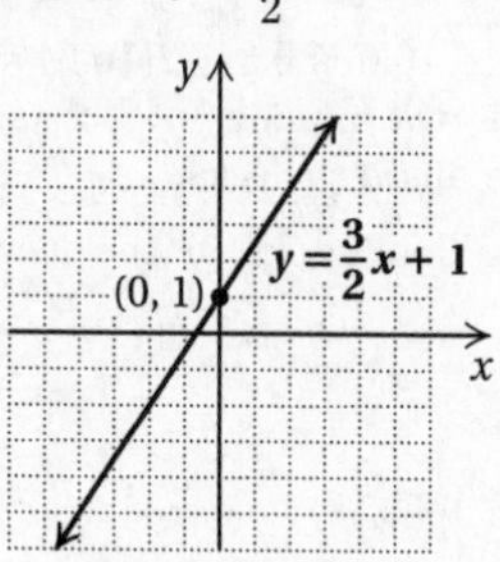

55. $8x - 2y = -10$

$-2y = -8x - 10$

$y = 4x + 5$

The y-intercept is $(0, 5)$. We find two other points.

When $x = -2$, $y = 4(-2) + 5 = -8 + 5 = -3$.

When $x = -1$, $y = 4(-1) + 5 = -4 + 5 = 1$.

x	y
-2	-3
-1	1
0	5

Plot these points, draw the line they determine, and label the graph $8x - 2y = -10$.

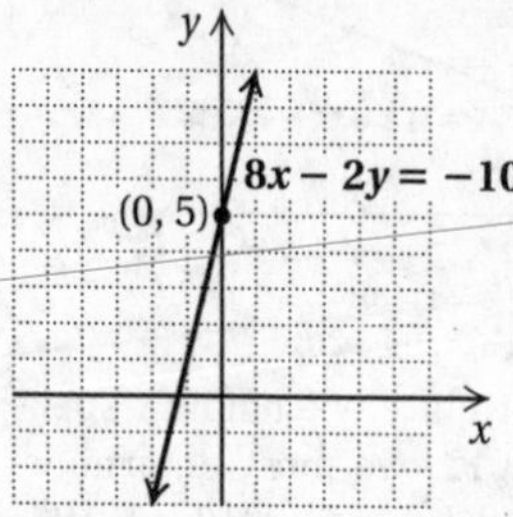

57. $8y + 2x = -4$

$8y = -2x - 4$

$y = -\frac{1}{4}x - \frac{1}{2}$

$y = -\frac{1}{4}x + \left(-\frac{1}{2}\right)$

The y-intercept is $\left(0, -\frac{1}{2}\right)$. We find two other points.

When $x = -2$, $y = -\frac{1}{4}(-2) - \frac{1}{2} = \frac{1}{2} - \frac{1}{2} = 0$.

When $x = 2$, $y = -\frac{1}{4} \cdot 2 - \frac{1}{2} = -\frac{1}{2} - \frac{1}{2} = -1$.

x	y
-2	0
0	$-\frac{1}{2}$
2	-1

Plot these points, draw the line they determine, and label the graph $8y + 2x = -4$.

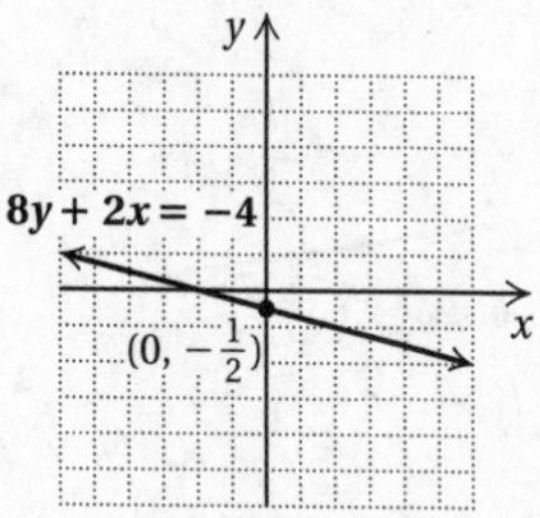

59. a) We substitute 0, 4 $(2014 - 2010 = 4)$, and 5 $(2015 - 2010 = 5)$ for t and then calculate A.

If $t = 0$, then $A = 12.83 \cdot 0 + 68.38 = \$68.38$ billion.

If $t = 4$, then $A = 12.83 \cdot 4 + 68.38 = 51.32 + 68.38 = \$119.7$ billion.

If $t = 5$, then $A = 12.83 \cdot 5 + 68.38 = 64.15 + 68.38 = \$132.53$ billion.

b) We plot the three ordered pairs we found in part (a). Note that negative t- and A-values have no meaning in this problem.

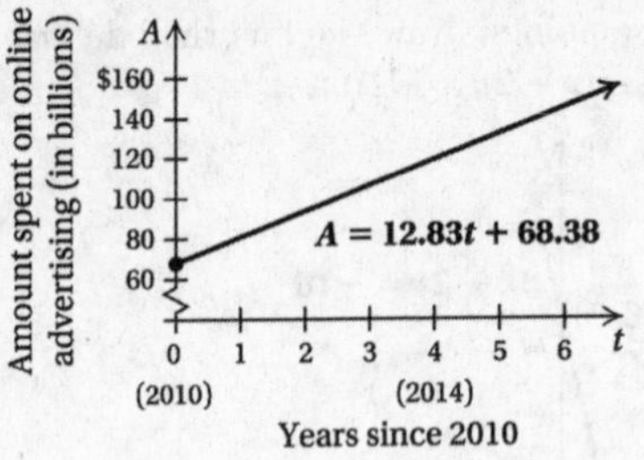

To use the graph to estimate the amount spent for online advertising in 2012, we must determine the A-value that is paired with $t = 2$ $(2012 - 2010 = 2)$. We locate 2 on the t-axis, go up to the graph, and then find the value on the A-axis that corresponds to this point. It appears that about $94 billion was spent for online advertising in 2012.

c) Substitute 158.19 for A and then solve for t.

$$A = 12.83t + 68.38$$
$$158.19 = 12.83t + 68.38$$
$$89.81 = 12.83t$$
$$7 = t$$

About 7 yr after 2010, or in 2017, online advertising spending will be $158.19 billion.

61. a) We substitute 0, 7 $(2007-2000 = 7)$, and 12 $(2012-2000 = 12)$ for t and then calculate S.

If $t = 0$, $S = -0.125 \cdot 0 + 6.898 = 6.898$ million.

If $t = 7$, $S = -0.125 \cdot 7 + 6.898 = -0.875 + 6.898 = 6.023$ million.

If $t = 12$, $S = -0.125 \cdot 12 + 6.898 = -1.5 + 6.898 = 5.398$ million.

b) Plot the three ordered pairs we found in part (a). Note that negative t- and S-values have no meaning in this problem.

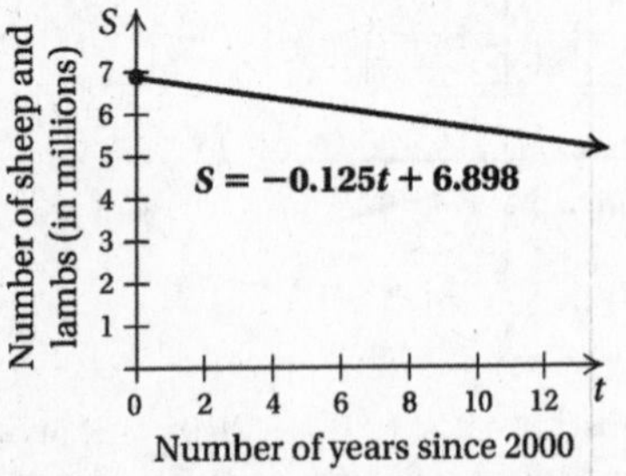

To use the graph to estimate the number of sheep and lambs in 2010, we must determine the S-value that is paired with 2010, or $t = 10$. We locate 10 on the t-axis, go up to the graph, and then find the value on the S-axis that corresponds to this point. It appears that there will be about 5.5 million sheep and lambs in 2010.

c) Substitute 4.898 for S and then solve for t.

$$S = -0.125t + 6.898$$
$$4.898 = -0.125t + 6.898$$
$$-2 = -0.125t$$
$$16 = t$$

There will be 4.898 million sheep and lambs 16 years after 2000, or in 2016.

63. The distance of -12 from 0 is 12, so $|-12| = 12$.

65. The distance of 0 from 0 is 0, so $|0| = 0$.

67. $2x - 14 = 29$

$$2x = 43$$
$$x = \frac{43}{2}$$

The solution is $\frac{43}{2}$.

69. $-10 = 1.2y + 2$

$$-12 = 1.2y$$
$$-10 = y$$

The solution is -10.

71. ***Familiarize.*** Let b = the number of books in the New York Public Library, in millions.

Translate.

Books in Library of Congress	are	0.3	more than	2	times	books in New York Public Library
↓	↓	↓	↓	↓	↓	↓
33.5	=	0.3	+	2	·	b

Solve.

$$33.5 = 0.3 + 2 \cdot b$$
$$33.2 = 2b$$
$$16.6 = b$$

Check. 0.3 more than twice 16.6 is $0.3 + 2(16.6) = 0.3 + 33.2 = 33.5$. This is the number of books in the Library of Congress, so the answer checks.

State. There are 16.6 million books in the New York Public Library.

73.

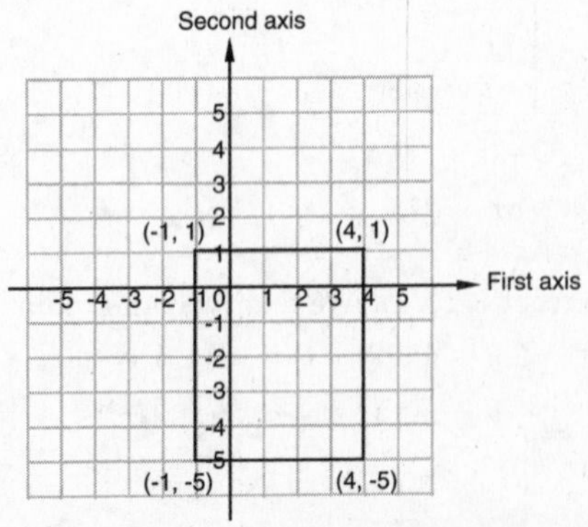

The coordinates of the fourth vertex are $(-1, -5)$.

75. Answers may vary.

We select eight points such that the sum of the coordinates for each point is 6.

$(-1,7)$	$-1+7=6$
$(0,6)$	$0+6=6$
$(1,5)$	$1+5=6$
$(2,4)$	$2+4=6$
$(3,3)$	$3+3=6$
$(4,2)$	$4+2=6$
$(5,1)$	$5+1=6$
$(6,0)$	$6+0=6$

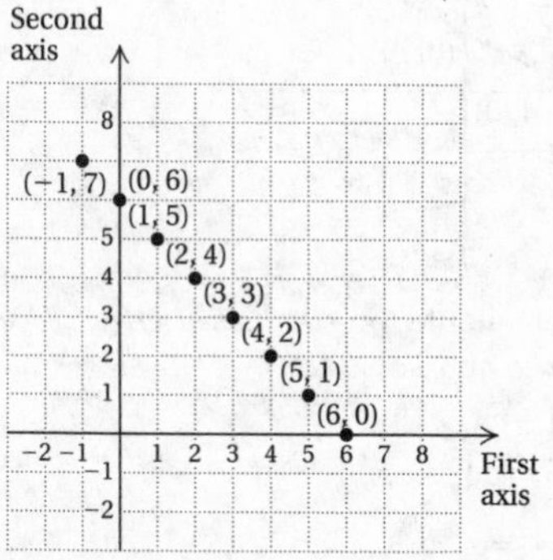

77.

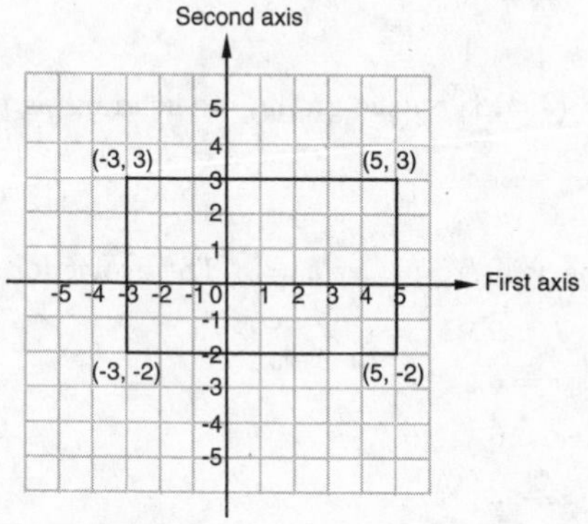

The length is 8 linear units, and the width is 5 linear units.

$P = 2l + 2w$

$P = 2 \cdot 8 + 2 \cdot 5 = 16 + 10 = 26$ linear units

Exercise Set 3.2

RC1. The graph of $y = -3$ is a <u>horizontal</u> line with $(0, -3)$ as its <u>y-intercept</u>.

RC3. To find the x-intercept, let $\underline{y = 0}$.

RC5. To find the y-intercept, let $\underline{x = 0}$.

1. (a) The graph crosses the y-axis at $(0, 5)$, so the y-intercept is $(0, 5)$.

(b) The graph crosses the x- axis at $(2, 0)$, so the x-intercept is $(2, 0)$.

3. (a) The graph crosses the y-axis at $(0, -4)$, so the y-intercept is $(0, -4)$.

(b) The graph crosses the x-axis at $(3, 0)$, so the x-intercept is $(3, 0)$.

5. $3x + 5y = 15$

(a) To find the y-intercept, let $x = 0$. This is the same as covering up the x-term and then solving.

$$5y = 15$$
$$y = 3$$

The y-intercept is $(0, 3)$.

(b) To find the x-intercept, let $y = 0$. This is the same as covering up the y-term and then solving.

$$3x = 15$$
$$x = 5$$

The x-intercept is $(5, 0)$.

7. $7x - 2y = 28$

(a) To find the y-intercept, let $x = 0$. This is the same as covering up the x-term and then solving.

$$-2y = 28$$
$$y = -14$$

The $y-$intercept is $(0, -14)$.

(b) To find the x-intercept, let $y = 0$. This is the same as covering up the y-term and then solving.

$$7x = 28$$
$$x = 4$$

The x-intercept is $(4, 0)$.

9. $-4x + 3y = 10$

(a) To find the y-intercept, let $x = 0$. This is the same as covering up the x-term and then solving.

$$3y = 10$$
$$y = \frac{10}{3}$$

The y-intercept is $\left(0, \frac{10}{3}\right)$.

(b) To find the x-intercept, let $y = 0$. This is the same as covering up the y-term and then solving.

$$-4x = 10$$
$$x = -\frac{5}{2}$$

The x-intercept is $\left(-\frac{5}{2}, 0\right)$.

11. $6x - 3 = 9y$

$6x - 9y = 3$ Writing the equation in the form $Ax + By = C$

(a) To find the y-intercept, let $x = 0$. This is the same as covering up the x-term and then solving.

$$-9y = 3$$
$$y = -\frac{1}{3}$$

The y-intercept is $\left(0, -\frac{1}{3}\right)$.

(b) To find the x-intercept, let $y = 0$. This is the same as covering up the y-term and then solving.

$$6x = 3$$
$$x = \frac{1}{2}$$

The x-intercept is $\left(\frac{1}{2}, 0\right)$.

13. $x + 3y = 6$

To find the x-intercept, let $y = 0$. Then solve for x.

$$x + 3y = 6$$
$$x + 3 \cdot 0 = 6$$
$$x = 6$$

Thus, $(6, 0)$ is the x-intercept.

To find the y-intercept, let $x = 0$. Then solve for y.

$$x + 3y = 6$$
$$0 + 3y = 6$$
$$3y = 6$$
$$y = 2$$

Thus, $(0, 2)$ is the y-intercept.

Plot these points and draw the line.

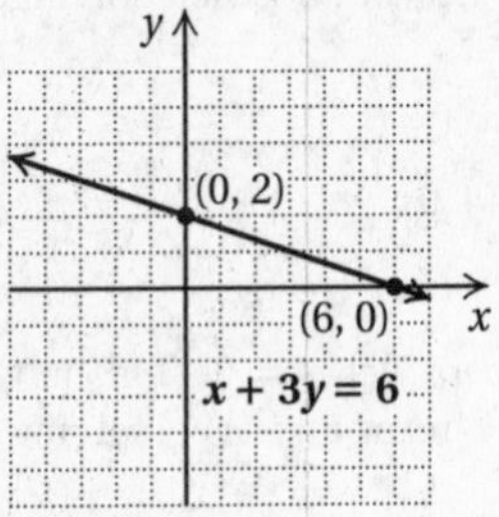

A third point should be used as a check. We substitute any value for x and solve for y.

We let $x = 3$. Then

$$x + 3y = 6$$
$$3 + 3y = 6$$
$$3y = 3$$
$$y = 1$$

The point $(3, 1)$ is on the graph, so the graph is probably correct.

15. $-x + 2y = 4$

To find the x-intercept, let $y = 0$. Then solve for x.

$$-x + 2y = 4$$
$$-x + 2 \cdot 0 = 4$$
$$-x = 4$$
$$x = -4$$

Thus, $(-4, 0)$ is the x-intercept.

To find the y-intercept, let $x = 0$. Then solve for y.

$$-x + 2y = 4$$
$$-0 + 2y = 4$$
$$2y = 4$$
$$y = 2$$

Thus, $(0, 2)$ is the y-intercept.

Plot these points and draw the line.

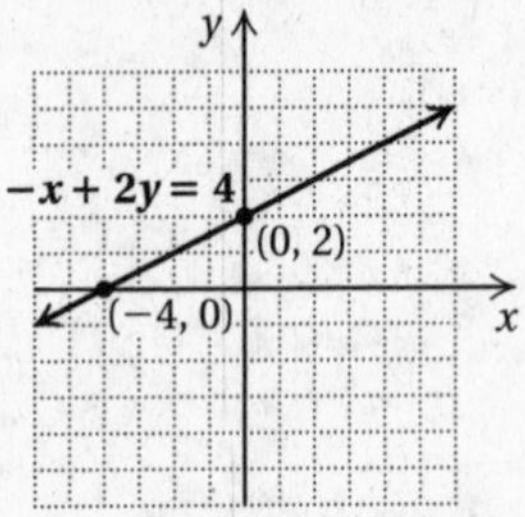

A third point should be used as a check. We substitute any value for x and solve for y.

We let $x = 4$. Then

$$-x + 2y = 4$$
$$-4 + 2y = 4$$
$$2y = 8$$
$$y = 4$$

The point $(4, 4)$ is on the graph, so the graph is probably correct.

17. $3x + y = 6$

To find the x-intercept, let $y = 0$. Then solve for x.

$$3x + y = 6$$
$$3x + 0 = 6$$
$$3x = 6$$
$$x = 2$$

Thus, $(2, 0)$ is the x-intercept.

To find the y-intercept, let $x = 0$. Then solve for y.

$$3x + y = 6$$
$$3 \cdot 0 + y = 6$$
$$y = 6$$

Thus, $(0, 6)$ is the y-intercept.

Plot these points and draw the line.

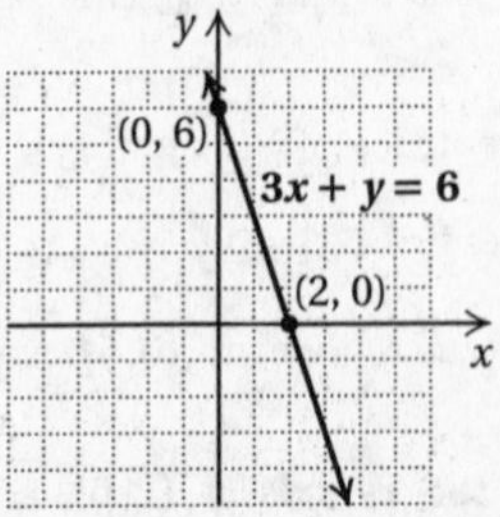

A third point should be used as a check. We substitute any value for x and solve for y.

We let $x = 1$. Then

$$3x + y = 6$$
$$3 \cdot 1 + y = 6$$
$$3 + y = 6$$
$$y = 3$$

The point $(1, 3)$ is on the graph, so the graph is probably correct.

19. $2y - 2 = 6x$

To find the x-intercept, let $y = 0$. Then solve for x.

$$2y - 2 = 6x$$
$$2 \cdot 0 - 2 = 6x$$
$$-2 = 6x$$
$$-\frac{1}{3} = x$$

Thus, $\left(-\frac{1}{3}, 0\right)$ is the x-intercept.

To find the y-intercept, let $x = 0$. Then solve for y.

$$2y - 2 = 6x$$
$$2y - 2 = 6 \cdot 0$$
$$2y - 2 = 0$$
$$2y = 2$$
$$y = 1$$

Thus, $(0, 1)$ is the y-intercept.

It is helpful to plot another point since the intercepts are so close together. This point can also serve as a check.

We let $x = 1$. Then

$$2y - 2 = 6x$$
$$2y - 2 = 6 \cdot 1$$
$$2y - 2 = 6$$
$$2y = 8$$
$$y = 4$$

Plot the point $(1, 4)$ and the intercepts and draw the line.

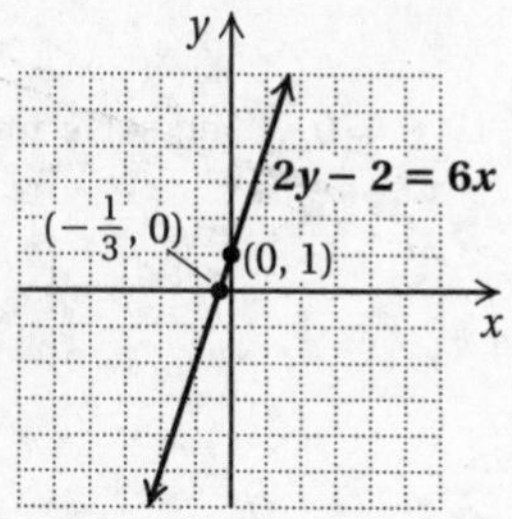

21. $3x - 9 = 3y$

To find the x-intercept, let $y = 0$. Then solve for x.

$$3x - 9 = 3y$$
$$3x - 9 = 3 \cdot 0$$
$$3x - 9 = 0$$
$$3x = 9$$
$$x = 3$$

Thus, $(3, 0)$ is the x-intercept.

To find the y-intercept, let $x = 0$. Then solve for y.

$$3x - 9 = 3y$$
$$3 \cdot 0 - 9 = 3y$$
$$-9 = 3y$$
$$-3 = y$$

Thus, $(0, -3)$ is the y-intercept.

Plot these points and draw the line.

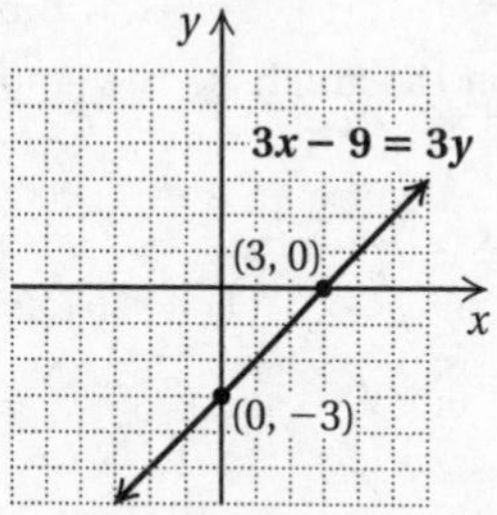

A third point should be used as a check. We substitute any value for x and solve for y.

We let $x = 1$. Then

$$3x - 9 = 3y$$
$$3 \cdot 1 - 9 = 3y$$
$$3 - 9 = 3y$$
$$-6 = 3y$$
$$-2 = y$$

The point $(1, -2)$ is on the graph, so the graph is probably correct.

23. $2x - 3y = 6$

To find the x-intercept, let $y = 0$. Then solve for x.

$$2x - 3y = 6$$
$$2x - 3 \cdot 0 = 6$$
$$2x = 6$$
$$x = 3$$

Thus, $(3, 0)$ is the x-intercept.

To find the y-intercept, let $x = 0$. Then solve for y.

$$2x - 3y = 6$$
$$2 \cdot 0 - 3y = 6$$
$$-3y = 6$$
$$y = -2$$

Thus, $(0, -2)$ is the y-intercept.

Plot these points and draw the line.

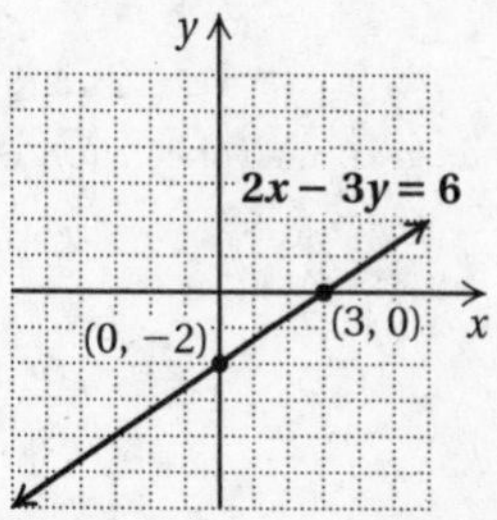

A third point should be used as a check. We substitute any value for x and solve for y.

We let $x = -3$.

$$2x - 3y = 6$$
$$2(-3) - 3y = 6$$
$$-6 - 3y = 6$$
$$-3y = 12$$
$$y = -4$$

The point $(-3, -4)$ is on the graph, so the graph is probably correct.

25. $4x + 5y = 20$

To find the x-intercept, let $y = 0$. Then solve for x.

$$4x + 5y = 20$$
$$4x + 5 \cdot 0 = 20$$
$$4x = 20$$
$$x = 5$$

Thus, $(5, 0)$ is the x-intercept.

To find the y-intercept, let $x = 0$. Then solve for y.

$$4x + 5y = 20$$
$$4 \cdot 0 + 5y = 20$$
$$5y = 20$$
$$y = 4$$

Thus, $(0, 4)$ is the y-intercept.

Plot these points and draw the graph.

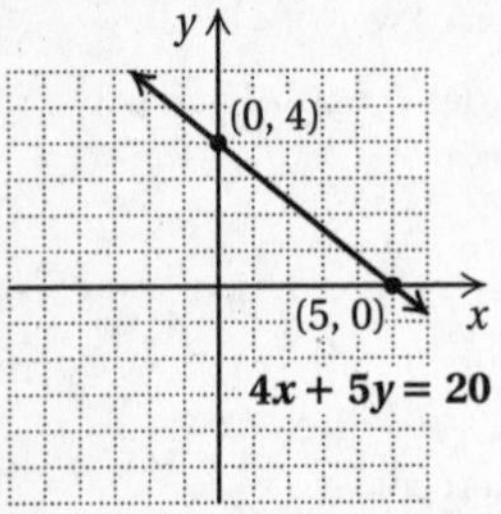

A third point should be used as a check. We substitute any value for x and solve for y.

We let $x = 4$. Then

$$4x + 5y = 20$$
$$4 \cdot 4 + 5y = 20$$
$$16 + 5y = 20$$
$$5y = 4$$
$$y = \frac{4}{5}$$

The point $\left(4, \frac{4}{5}\right)$ is on the graph, so the graph is probably correct.

27. $2x + 3y = 8$

To find the x-intercept, let $y = 0$. Then solve for x.

$$2x + 3y = 8$$
$$2x + 3 \cdot 0 = 8$$
$$2x = 8$$
$$x = 4$$

Thus, $(4, 0)$ is the x-intercept.

To find the y-intercept, let $x = 0$. Then solve for y.

$$2x + 3y = 8$$
$$2 \cdot 0 + 3y = 8$$
$$3y = 8$$
$$y = \frac{8}{3}$$

Thus, $\left(0, \frac{8}{3}\right)$ is the y-intercept.

Plot these points and draw the graph.

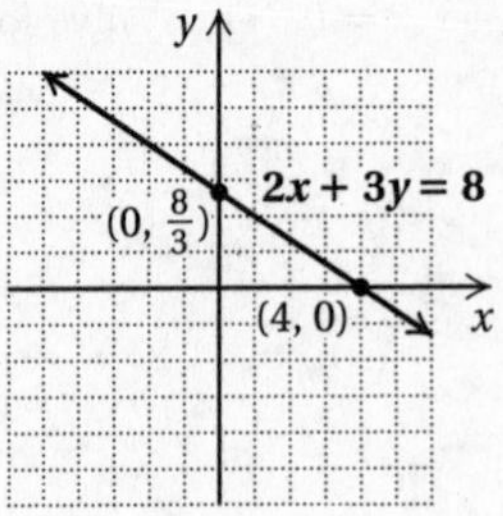

A third point should be used as a check.

We let $x = 1$. Then

$$2x + 3y = 8$$
$$2 \cdot 1 + 3y = 8$$
$$2 + 3y = 8$$
$$3y = 6$$
$$y = 2$$

The point $(1, 2)$ is on the graph, so the graph is probably correct.

29. $3x + 4y = 5$

To find the x-intercept, let $y = 0$. Then solve for x.

$$3x + 4y = 5$$
$$3x + 4 \cdot 0 = 5$$
$$3x = 5$$
$$x = \frac{5}{3}$$

Thus, $\left(\frac{5}{3}, 0\right)$ is the x-intercept.

To find the y-intercept, let $x = 0$. Then solve for y.

$$3x + 4y = 5$$
$$3 \cdot 0 + 4y = 5$$
$$4y = 5$$
$$y = \frac{5}{4}$$

Thus, $\left(0, \frac{5}{4}\right)$ is the y-intercept.

It is helpful to plot another point since the intercepts are so close together. This point can also serve as a check.

We let $x = 3$. Then

$$3x + 4y = 5$$
$$3 \cdot 3 + 4y = 5$$
$$9 + 4y = 5$$
$$4y = -4$$
$$y = -1$$

Plot the point $(3, -1)$ and the intercepts and draw the line.

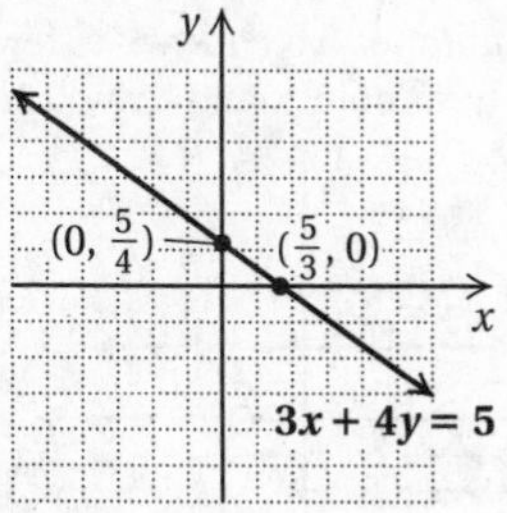

31. $3x - 2 = y$

To find the x-intercept, let $y = 0$. Then solve for x.

$$3x - 2 = y$$
$$3x - 2 = 0$$
$$3x = 2$$
$$x = \frac{2}{3}$$

Thus, $\left(\frac{2}{3}, 0\right)$ is the x-intercept.

To find the y-intercept, let $x = 0$. Then solve for y.

$$3x - 2 = y$$
$$3 \cdot 0 - 2 = y$$
$$-2 = y$$

Thus, $(0, -2)$ is the y-intercept.

Plot these points and draw the line.

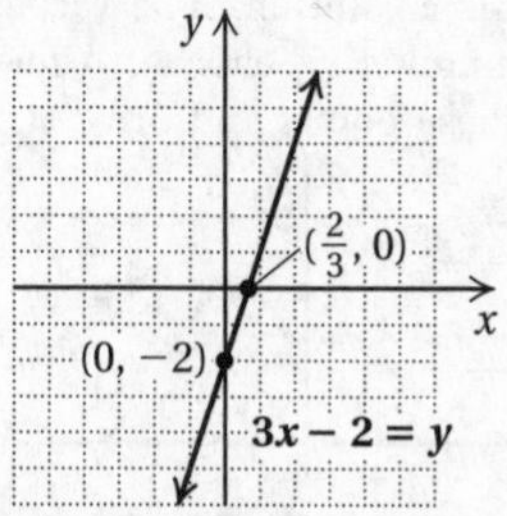

A third point should be used as a check.

We let $x = 2$. Then

$$3x - 2 = y$$
$$3 \cdot 2 - 2 = y$$
$$6 - 2 = y$$
$$4 = y$$

The point $(2, 4)$ is on the graph, so the graph is probably correct.

33. $6x - 2y = 12$

To find the x-intercept, let $y = 0$. Then solve for x.

$$6x - 2y = 12$$
$$6x - 2 \cdot 0 = 12$$
$$6x = 12$$
$$x = 2$$

Thus, $(2, 0)$ is the x-intercept.

To find the y-intercept, let $x = 0$. Then solve for y.

$$6x - 2y = 12$$
$$6 \cdot 0 - 2y = 12$$
$$-2y = 12$$
$$y = -6$$

Thus, $(0, -6)$ is the y-intercept.

Plot these points and draw the line.

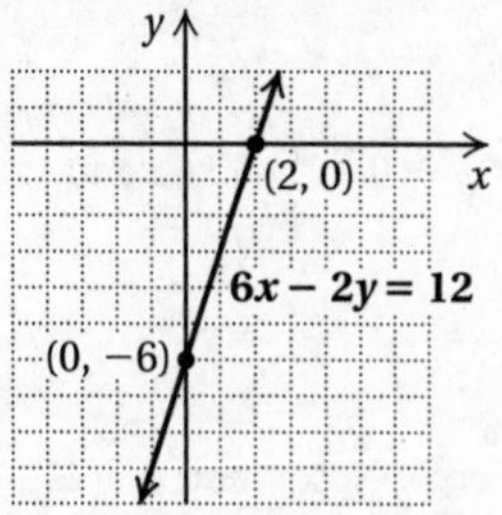

We use a third point as a check.

We let $x = 1$. Then

$$6x - 2y = 12$$
$$6 \cdot 1 - 2y = 12$$
$$6 - 2y = 12$$
$$-2y = 6$$
$$y = -3$$

The point $(1, -3)$ is on the graph, so the graph is probably correct.

35. $y = -3 - 3x$

To find the x-intercept, let $y = 0$. Then solve for x.

$$y = -3 - 3x$$
$$0 = -3 - 3x$$
$$3x = -3$$
$$x = -1$$

Thus, $(-1, 0)$ is the x-intercept.

To find the y-intercept, let $x = 0$. Then solve for y.

$$y = -3 - 3x$$
$$y = -3 - 3 \cdot 0$$
$$y = -3$$

Thus, $(0, -3)$ is the y-intercept.

Plot these points and draw the graph.

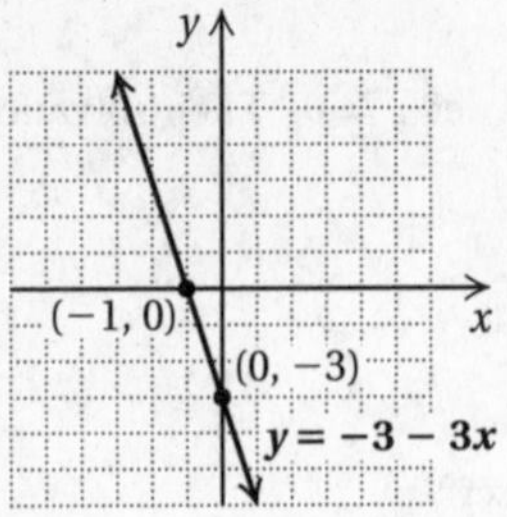

We use a third point as a check.

We let $x = -2$. Then

$$y = -3 - 3x$$
$$y = -3 - 3 \cdot (-2)$$
$$y = -3 + 6$$
$$y = 3$$

The point $(-2, 3)$ is on the graph, so the graph is probably correct.

37. $y - 3x = 0$

To find the x-intercept, let $y = 0$. Then solve for x.

$$0 - 3x = 0$$
$$-3x = 0$$
$$x = 0$$

Thus, $(0, 0)$ is the x-intercept. Note that this is also the y-intercept.

In order to graph the line, we will find a second point.

When $x = 1$, $y - 3 \cdot 1 = 0$

$$y - 3 = 0$$
$$y = 3$$

Plot the points and draw the graph.

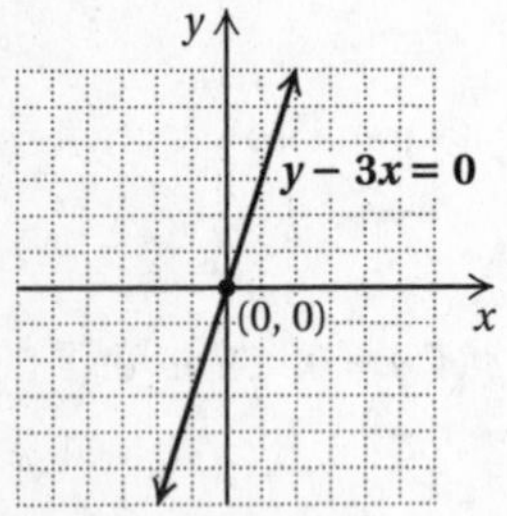

We use a third point as a check.

We let $x = -1$. Then

$$y - 3(-1) = 0$$
$$y + 3 = 0$$
$$y = -3$$

The point $(-1, -3)$ is on the graph, so the graph is probably correct.

39. $x = -2$

Any ordered pair $(-2, y)$ is a solution. The variable x must be -2, but y can be any number we choose. A few solutions are listed below. Plot these points and draw the line.

x	y
-2	-2
-2	0
-2	4

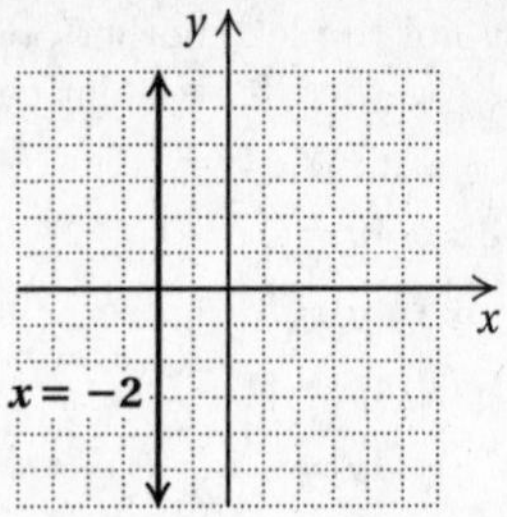

41. $y = 2$

Any ordered pair $(x, 2)$ is a solution. The variable y must be 2, but x can be any number we choose. A few solutions are listed below. Plot these points and draw the line.

x	y
-3	2
0	2
2	2

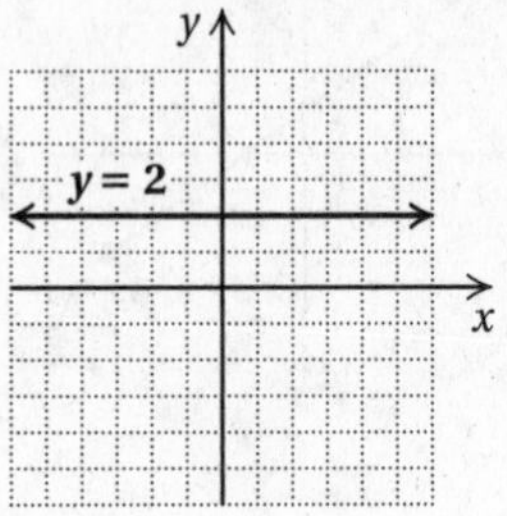

43. $x = 2$

Any ordered pair $(2, y)$ is a solution. The variable x must be 2, but y can be any number we choose. A few solutions are listed below. Plot these points and draw the line.

x	y
2	-1
2	4
2	5

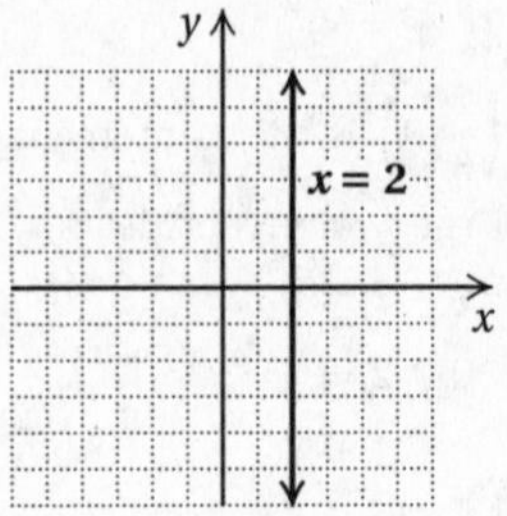

45. $y = 0$

Any ordered pair $(x, 0)$ is a solution. The variable y must be 0, but x can be any number we choose. A few solutions are listed below. Plot these points and draw the line.

x	y
-5	0
-1	0
3	0

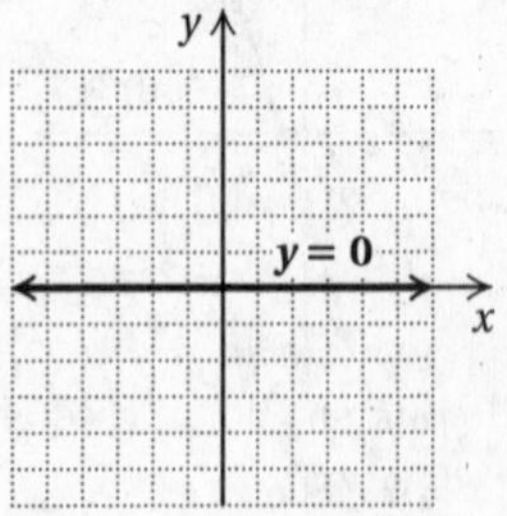

47. $x = \frac{3}{2}$

Any ordered pair $\left(\frac{3}{2}, y\right)$ is a solution. The variable x must be $\frac{3}{2}$, but y can be any number we choose. A few solutions are listed below. Plot these points and draw the line.

x	y
$\frac{3}{2}$	-2
$\frac{3}{2}$	0
$\frac{3}{2}$	4

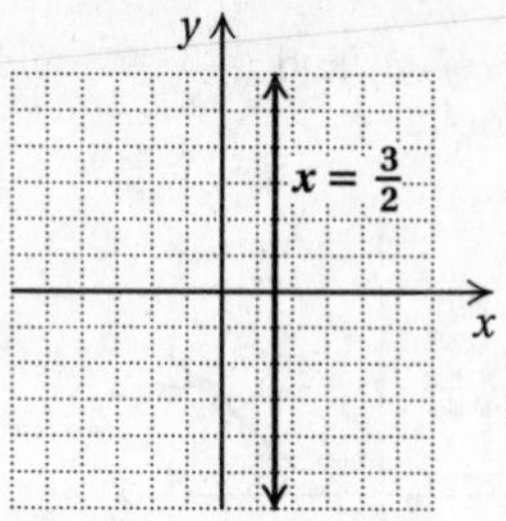

49. $3y = -5$

$y = -\frac{5}{3}$ Solving for y

Any ordered pair $\left(x, -\frac{5}{3}\right)$ is a solution. A few solutions are listed below. Plot these points and draw the line.

x	y
-3	$-\frac{5}{3}$
0	$-\frac{5}{3}$
2	$-\frac{5}{3}$

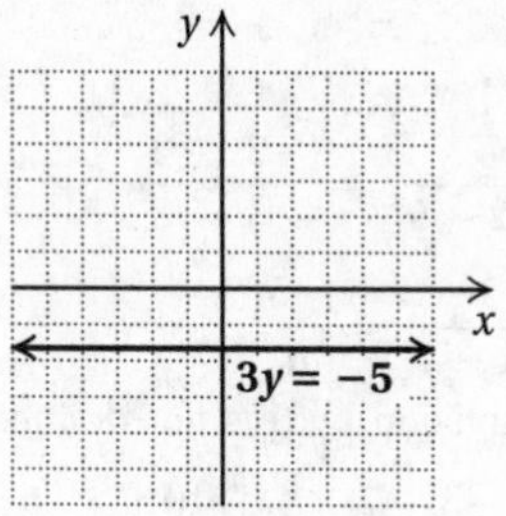

51. $4x + 3 = 0$

$4x = -3$

$x = -\frac{3}{4}$ Solving for x

Any ordered pair $\left(-\frac{3}{4}, y\right)$ is a solution. A few solutions are listed below. Plot these points and draw the line.

x	y
$-\frac{3}{4}$	-2
$-\frac{3}{4}$	0
$-\frac{3}{4}$	3

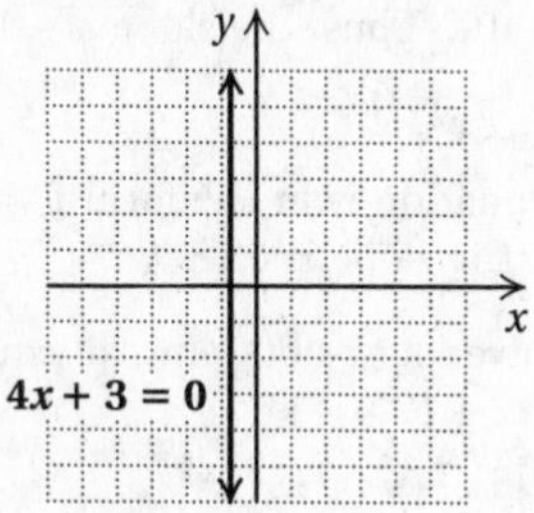

53. $48 - 3y = 0$

$-3y = -48$

$y = 16$ Solving for y

Any ordered pair $(x, 16)$ is a solution. A few solutions are listed below. Plot these points and draw the line.

x	y
-4	16
0	16
2	16

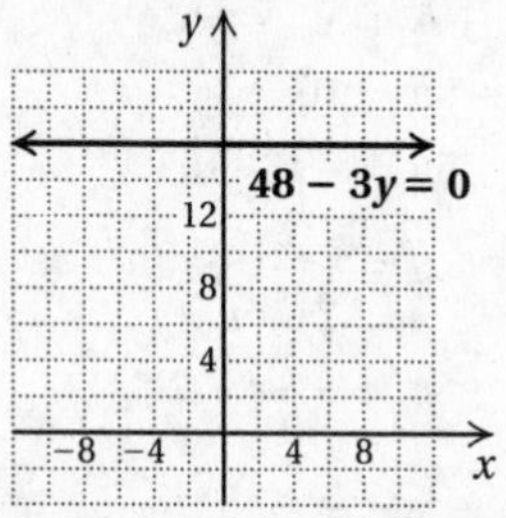

55. Note that every point on the horizontal line passing through $(0, -1)$ has -1 as the y-coordinate. Thus, the equation of the line is $y = -1$.

57. Note that every point on the vertical line passing through $(4, 0)$ has 4 as the x-coordinate. Thus, the equation of the line is $x = 4$.

59.
$$\begin{aligned} x + (x-1) &< (x+2) - (x+1) \\ 2x - 1 &< x + 2 - x - 1 \\ 2x - 1 &< 1 \\ 2x &< 2 \\ x &< 1 \end{aligned}$$

The solution set is $\{x|x < 1\}$.

61.
$$\begin{aligned} \frac{2x}{7} - 4 &\le -2 \\ \frac{2x}{7} &\le 2 \\ \frac{7}{2} \cdot \frac{2x}{7} &\le \frac{7}{2} \cdot 2 \\ x &\le 7 \end{aligned}$$

The solution set is $\{x|x \le 7\}$.

63. A line parallel to the x-axis has an equation of the form $y = b$. Since the y-coordinate of one point on the line is -4, then $b = -4$ and the equation is $y = -4$.

65. Substitute -4 for x and 0 for y.

$$\begin{aligned} 3(-4) + k &= 5 \cdot 0 \\ -12 + k &= 0 \\ k &= 12 \end{aligned}$$

Exercise Set 3.3

RC1. The slope of a horizontal line is 0. The correct answer is (d).

RC3. The change in x is called the run. The correct answer is (b).

RC5. Slope is rise/run. The correct answer is (c).

1. We consider (x_1, y_1) to be $(-3, 5)$ and (x_2, y_2) to be $(4, 2)$.

$$m = \frac{y_2 - y_1}{x_2 - x_1} = \frac{2 - 5}{4 - (-3)} = \frac{-3}{7} = -\frac{3}{7}$$

3. We can choose any two points. We consider (x_1, y_1) to be $(-3, -1)$ and (x_2, y_2) to be $(0, 1)$.

$$m = \frac{y_2 - y_1}{x_2 - x_1} = \frac{1 - (-1)}{0 - (-3)} = \frac{2}{3}$$

5. We can choose any two points. We consider (x_1, y_1) to be $(-4, -2)$ and (x_2, y_2) to be $(4, 4)$.

$$m = \frac{y_2 - y_1}{x_2 - x_1} = \frac{4 - (-2)}{4 - (-4)} = \frac{6}{8} = \frac{3}{4}$$

7. We consider (x_1, y_1) to be $(-4, -2)$ and (x_2, y_2) to be $(3, -2)$.

$$m = \frac{y_2 - y_1}{x_2 - x_1} = \frac{-2 - (-2)}{3 - (-4)} = \frac{0}{7} = 0$$

9. We plot $(-2,4)$ and $(3,0)$ and draw the line containing these points.

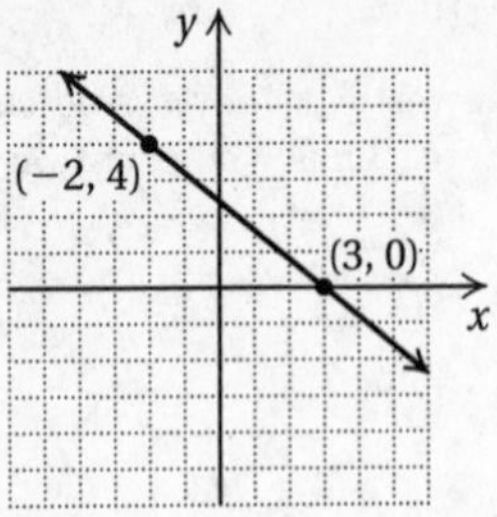

To find the slope, consider (x_1,y_1) to be $(-2,4)$ and (x_2,y_2) to be $(3,0)$.

$$m=\frac{y_2-y_1}{x_2-x_1}=\frac{0-4}{3-(-2)}=\frac{-4}{5}=-\frac{4}{5}$$

11. We plot $(-4,0)$ and $(-5,-3)$ and draw the line containing these points.

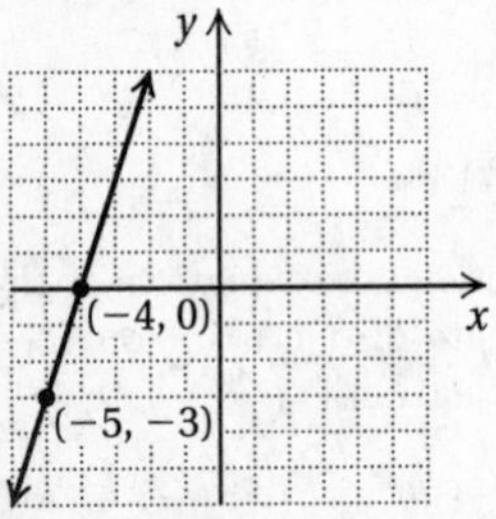

To find the slope, consider (x_1,y_1) to be $(-4,0)$ and (x_2,y_2) to be $(-5,-3)$.

$$m=\frac{y_2-y_1}{x_2-x_1}=\frac{-3-0}{-5-(-4)}=\frac{-3}{-1}=3$$

13. We plot $(-4,1)$ and $(2,-3)$ and draw the line containing these points.

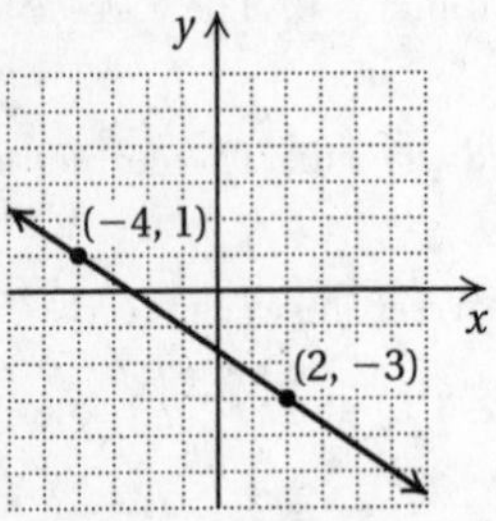

To find the slope, consider (x_1,y_1) to be $(-4,1)$ and (x_2,y_2) to be $(2,-3)$.

$$m=\frac{y_2-y_1}{x_2-x_1}=\frac{-3-1}{2-(-4)}=\frac{-4}{6}=-\frac{2}{3}$$

15. We plot $(5,3)$ and $(-3,-4)$ and draw the line containing these points.

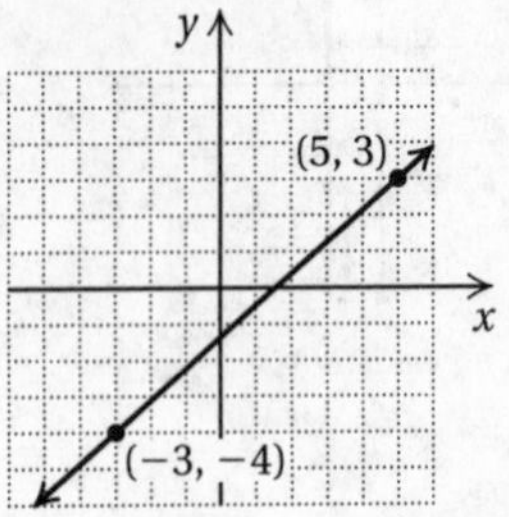

To find the slope, consider (x_1,y_1) to be $(5,3)$ and (x_2,y_2) to be $(-3,-4)$.

$$m=\frac{y_2-y_1}{x_2-x_1}=\frac{-4-3}{-3-5}=\frac{-7}{-8}=\frac{7}{8}$$

17. $m=\dfrac{-\frac{1}{2}-\frac{3}{2}}{2-5}=\dfrac{-2}{-3}=\dfrac{2}{3}$

19. $m=\dfrac{-2-3}{4-4}=\dfrac{-5}{0}$

Since division by 0 is not defined, the slope is not defined.

21. $m=\dfrac{-3-7}{15-(-11)}=\dfrac{-10}{26}=-\dfrac{5}{13}$

23. $m=\dfrac{\frac{3}{11}-\frac{3}{11}}{\frac{5}{4}-\left(-\frac{1}{2}\right)}=\dfrac{0}{\frac{7}{4}}=0$

25. $y=-10x=-10x+0$

The equation can be written in the form $y=mx+b$, where $m=-10$. Thus, the slope is -10.

27. $y=3.78x-4$

The equation is in the form $y=mx+b$, where $m=3.78$. Thus, the slope is 3.78.

29. We solve for y, obtaining an equation of the form $y=mx+b$.

$$\begin{aligned}3x-y&=4\\-y&=-3x+4\\-1(-y)&=-1(-3x+4)\\y&=3x-4\end{aligned}$$

The slope is 3.

31. We solve for y, obtaining an equation of the form $y=mx+b$.

$$\begin{aligned}x+5y&=10\\5y&=-x+10\\y&=\frac{1}{5}(-x+10)\\y&=-\frac{1}{5}x+2\end{aligned}$$

The slope is $-\dfrac{1}{5}$.

33. We solve for y, obtaining an equation of the form $y = mx + b$.

$$3x + 2y = 6$$
$$2y = -3x + 6$$
$$y = \frac{1}{2}(-3x + 6)$$
$$y = -\frac{3}{2}x + 3$$

The slope is $-\frac{3}{2}$.

35. The graph of $x = \frac{2}{15}$ is a vertical line, so the slope is not defined.

37. $y = 2 - x = -1 \cdot x + 2$

The equation can be written in the form $y = mx+b$, where $m = -1$. Thus, the slope is -1.

39. We solve for y, obtaining an equation of the form $y = mx + b$.

$$9x = 3y + 5$$
$$9x - 5 = 3y$$
$$\frac{1}{3}(9x - 5) = y$$
$$3x - \frac{5}{3} = y$$

The slope is 3.

41. We solve for y, obtaining an equation of the form $y = mx + b$.

$$5x - 4y + 12 = 0$$
$$5x + 12 = 4y$$
$$\frac{1}{4}(5x + 12) = y$$
$$\frac{5}{4}x + 3 = y$$

The slope is $\frac{5}{4}$.

43. $y = 4$

The equation can be thought of as $y = 0 \cdot x + 4$, so the slope is 0.

45.
$$x = \frac{3}{4}y - 2$$
$$x + 2 = \frac{3}{4}y$$
$$\frac{4}{3}(x + 2) = \frac{4}{3} \cdot \frac{3}{4}y$$
$$\frac{4}{3}x + \frac{8}{3} = y$$

The slope is $\frac{4}{3}$.

47.
$$\frac{2}{3}y = -\frac{7}{4}x$$
$$\frac{3}{2} \cdot \frac{2}{3}y = \frac{3}{2}\left(-\frac{7}{4}x\right)$$
$$y = -\frac{21}{8}x$$

The slope is $-\frac{21}{8}$.

49. $m = \frac{\text{rise}}{\text{run}} = \frac{2.4}{8.2} = \frac{2.4}{8.2} \cdot \frac{10}{10} = \frac{24}{82}$

$= \frac{\not{2} \cdot 12}{\not{2} \cdot 41} = \frac{12}{41}$

51. $m = \frac{\text{rise}}{\text{run}} = \frac{56}{258} = \frac{\not{2} \cdot 28}{\not{2} \cdot 129} = \frac{28}{129}$

53. Grade $= \frac{8\frac{1}{2}}{280} = \frac{8.5}{280} \approx 0.030 = 3.0\%$

The grade meets the rapid-transit rail standards.

55. Rate of change $= m = \frac{34 - 26.58}{2015 - 2009}$

$= \frac{7.42}{6}$

≈ 1.24 million children per year

57. Rate of change $= m = \frac{343,829 - 484,949}{2010 - 2000}$

$= \frac{-141,120}{10}$

$\approx -14,100$ people per year

59. We use the data points (2006, 358,000,000) and (2011, 511,000,000).

Rate of change $= m = \frac{511,000,000 - 358,000,000}{2011 - 2006}$

$= \frac{153,000,000}{5}$

$\approx 30,600,000$ lb per year

61.
$$2x - 11 = 4$$
$$2x = 15$$
$$x = \frac{15}{2}$$

The solution is $\frac{15}{2}$.

63. $\frac{1}{3}p - p = \frac{1}{3}p - 1 \cdot p = \left(\frac{1}{3} - 1\right)p = -\frac{2}{3}p$

65. Note that the sum of the coordinates of each point on the graph is 5. Thus, we have $x + y = 5$, or $y = -x + 5$.

67. Note that each y-coordinate is 2 more than the corresponding x-coordinate. Thus, we have $y = x + 2$.

Exercise Set 3.4

RC1. The y-intercept is $(0, -2)$. Using two points on the graph, we find that the slope is $-\frac{1}{3}$. Thus, the equation is $y = -\frac{1}{3}x - 2$, so answer (e) is correct.

RC3. The y-intercept is $(0, -2)$. Using two points on the graph, we find that the slope is -3. Thus, the equation is
$y = -3x - 2$, so answer (a) is correct.

1. $y = -4x - 9$

The equation is already in the form $y = mx + b$. The slope is -4 and the y-intercept is $(0, -9)$.

3. $y = 1.8x$

We can think of $y = 1.8x$ as $y = 1.8x + 0$. The slope is 1.8 and the y-intercept is $(0, 0)$.

5. We solve for y.

$$-8x - 7y = 21$$
$$-7y = 8x + 21$$
$$y = -\frac{1}{7}(8x + 21)$$
$$y = -\frac{8}{7}x - 3$$

The slope is $-\frac{8}{7}$ and the y-intercept is $(0, -3)$.

7. We solve for y.

$$4x = 9y + 7$$
$$4x - 7 = 9y$$
$$\frac{1}{9}(4x - 7) = y$$
$$\frac{4}{9}x - \frac{7}{9} = y$$

The slope is $\frac{4}{9}$ and the y-intercept is $\left(0, -\frac{7}{9}\right)$.

9. We solve for y.

$$-6x = 4y + 2$$
$$-6x - 2 = 4y$$
$$\frac{1}{4}(-6x - 2) = y$$
$$-\frac{3}{2}x - \frac{1}{2} = y$$

The slope is $-\frac{3}{2}$ and the y-intercept is $\left(0, -\frac{1}{2}\right)$.

11. $y = -17$

We can think of $y = -17$ as $y = 0x - 17$. The slope is 0 and the y-intercept is $(0, -17)$.

13. We substitute -7 for m and -13 for b in the equation $y = mx + b$.

$$y = -7x - 13$$

15. We substitute 1.01 for m and -2.6 for b in the equation $y = mx + b$.

$$y = 1.01x - 2.6$$

17. We substitute 0 for m and -5 for b in the equation $y = mx + b$.

$$y = 0 \cdot x - 5, \text{ or } y = -5$$

19. We know the slope is -2, so the equation is $y = -2x + b$. Using the point $(-3, 0)$, we substitute -3 for x and 0 for y in $y = -2x + b$. Then we solve for b.

$$y = -2x + b$$
$$0 = -2(-3) + b$$
$$0 = 6 + b$$
$$-6 = b$$

Thus, we have the equation $y = -2x - 6$.

21. We know the slope is $\frac{3}{4}$, so the equation is $y = \frac{3}{4}x + b$. Using the point $(2, 4)$, we substitute 2 for x and 4 for y in $y = \frac{3}{4}x + b$. Then we solve for b.

$$y = \frac{3}{4}x + b$$
$$4 = \frac{3}{4} \cdot 2 + b$$
$$4 = \frac{3}{2} + b$$
$$\frac{5}{2} = b$$

Thus, we have the equation $y = \frac{3}{4}x + \frac{5}{2}$.

23. We know the slope is 1, so the equation is $y = 1 \cdot x + b$, or $y = x + b$. Using the point $(2, -6)$, we substitute 2 for x and -6 for y in $y = x + b$. Then we solve for y.

$$y = x + b$$
$$-6 = 2 + b$$
$$-8 = b$$

Thus, we have the equation $y = x - 8$.

25. We substitute -3 for m and 3 for b in the equation $y = mx + b$.

$$y = -3x + 3$$

27. $(12, 16)$ and $(1, 5)$

First we find the slope.

$$m = \frac{16 - 5}{12 - 1} = \frac{11}{11} = 1$$

Thus, $y = 1 \cdot x + b$, or $y = x + b$. We can use either point to find b. We choose $(1, 5)$. Substitute 1 for x and 5 for y in $y = x + b$.

$$y = x + b$$
$$5 = 1 + b$$
$$4 = b$$

Thus, the equation is $y = x + 4$.

29. $(0, 4)$ and $(4, 2)$

First we find the slope.

$$m = \frac{4-2}{0-4} = \frac{2}{-4} = -\frac{1}{2}$$

Thus, $y = -\frac{1}{2}x + b$. One of the given points is the y-intercept $(0, 4)$. Thus, we substitute 4 for b in $y = -\frac{1}{2}x + b$. The equation is $y = -\frac{1}{2}x + 4$.

31. $(3, 2)$ and $(1, 5)$

First we find the slope.

$$m = \frac{2-5}{3-1} = \frac{-3}{2} = -\frac{3}{2}$$

Thus, $y = -\frac{3}{2}x + b$. We can use either point to find b. We choose $(3, 2)$. Substitute 3 for x and 2 for y in $y = -\frac{3}{2}x + b$.

$$y = -\frac{3}{2}x + b$$
$$2 = -\frac{3}{2} \cdot 3 + b$$
$$2 = -\frac{9}{2} + b$$
$$\frac{13}{2} = b$$

Thus, the equation is $y = -\frac{3}{2}x + \frac{13}{2}$.

33. $\left(4, -\frac{2}{5}\right)$ and $\left(4, \frac{2}{5}\right)$

First we find the slope.

$$m = \frac{-\frac{2}{5} - \frac{2}{5}}{4-4} = \frac{-\frac{4}{5}}{0}$$

The slope is not defined. Thus, we have a vertical line. Since the first coordinates of the given points are 4, the equation of the line is $x = 4$.

35. $(-4, 5)$ and $(-2, -3)$

First we find the slope.

$$m = \frac{5-(-3)}{-4-(-2)} = \frac{8}{-2} = -4$$

Thus, $y = -4x + b$. We can use either point to find b. We choose $(-4, 5)$. Substitute -4 for x and 5 for y in $y = -4x + b$.

$$y = -4x + b$$
$$5 = -4(-4) + b$$
$$5 = 16 + b$$
$$-11 = b$$

Thus, the equation is $y = -4x - 11$.

37. $\left(-2, \frac{1}{4}\right)$ and $\left(3, \frac{1}{4}\right)$

First we find the slope.

$$m = \frac{\frac{1}{4} - \frac{1}{4}}{-2-3} = \frac{0}{-5} = 0$$

We could observe here that, because the slope is 0, we have a horizontal line. Both $y-$ coordinates are $\frac{1}{4}$, so the equation is $y = \frac{1}{4}$.

We could also use one of the points to find that the y-intercept is $\frac{1}{4}$ and then write the equation, $y = 0 \cdot x + \frac{1}{4}$, or $y = \frac{1}{4}$.

39. a) First we find the slope.

$$m = \frac{68.8 - 35.0}{12 - 0} = \frac{33.8}{12} \approx 2.82$$

We see from the graph that the y-intercept is $(0, 35.0)$, so the equation of the line is $S = 2.82t + 35$.

b) The rate of change is the slope, \$2.82 billion per year.

c) In 2015, $t = 2015 - 2000$, or 15.

$S = 2.82(15) + 35 = 42.3 + 35 = \$77.3$ billion.

41.
$$3x - 4(9 - x) = 17$$
$$3x - 36 + 4x = 17$$
$$7x - 36 = 17$$
$$7x = 53$$
$$x = \frac{53}{7}$$

The solution is $\frac{53}{7}$.

43.
$$4(a-3) + 6 = 21 - \frac{1}{2}a$$
$$4a - 12 + 6 = 21 - \frac{1}{2}a$$
$$4a - 6 = 21 - \frac{1}{2}a$$
$$2(4a - 6) = 2\left(21 - \frac{1}{2}a\right)$$
$$8a - 12 = 42 - a$$
$$9a - 12 = 42$$
$$9a = 54$$
$$a = 6$$

The solution is 6.

45.
$$40(2x - 7) = 50(4 - 6x)$$
$$80x - 280 = 200 - 300x$$
$$380x - 280 = 200$$
$$380x = 480$$
$$x = \frac{480}{380}$$
$$x = \frac{24}{19}$$

The solution is $\frac{24}{19}$.

47. What is 30% of 12?

$$\downarrow \quad \downarrow \quad \downarrow \quad \downarrow \quad \downarrow$$
$$a \quad = \quad 30\% \quad \cdot \quad 12$$

We convert 30% to decimal notation and multiply.

$a = 30\% \cdot 12$

$a = 0.3 \cdot 12$

$a = 3.6$

49. $\underbrace{\text{What percent}}$ of 50 is 2.5?

$$\begin{array}{ccccc} \downarrow & \downarrow & \downarrow & \downarrow & \downarrow \\ p & \cdot & 50 & = & 2.5 \end{array}$$

We solve the equation.

$p \cdot 50 = 2.5$

$p = \frac{2.5}{50} = 0.05$

$p = 5\%$

51. First find the slope of $3x - y + 4 = 0$.

$3x - y + 4 = 0$

$3x + 4 = y$

The slope is 3.

Thus, $y = 3x + b$. Using the point $(2, -3)$, we substitute 2 for x and -3 for y in $y = 3x + b$. Then we solve for b.

$y = 3x + b$

$-3 = 3 \cdot 2 + b$

$-3 = 6 + b$

$-9 = b$

Thus, the equation is $y = 3x - 9$.

53. First find the slope of $3x - 2y = 8$.

$3x - 2y = 8$

$-2y = -3x + 8$

$y = \frac{3}{2}x - 4$

The slope is $\frac{3}{2}$.

Then find the y-intercept of $2y + 3x = -4$.

$2y + 3x = -4$

$2y = -3x - 4$

$y = -\frac{3}{2}x - 2$

The y-intercept is $(0, -2)$.

Finally, write the equation of the line with slope $\frac{3}{2}$ and y-intercept $(0, -2)$.

$y = mx + b$

$y = \frac{3}{2}x + (-2)$

$y = \frac{3}{2}x - 2$

Chapter 3 Mid-Chapter Review

1. $\left|-\frac{3}{4}\right| = \frac{3}{4}$; $\left|-\frac{5}{2}\right| = \frac{5}{2}$; since $\frac{5}{2} > \frac{3}{4}$, a slope of $-\frac{5}{2}$ is steeper than a slope of $-\frac{3}{4}$. Thus, the given statement is false.

2. True; $\frac{d-b}{c-a}$ is $\frac{\text{rise}}{\text{run}}$, or slope.

3. True; when $x = 0$, we have $y = \frac{C}{B}$.

4. False; in quadrant IV, the first coordinate is positive and the second coordinate is negative.

5. a) The y-intercept is $(0, -3)$.

b) The x-intercept is $(-3, 0)$.

c) The slope is $\frac{-3-0}{0-(-3)} = \frac{-3}{3} = -1$.

d) The equation of the line in $y = mx + b$ form is $y = -1 \cdot x + (-3)$, or $y = -x - 3$.

6. a) The x-intercept is $(c, 0)$.

b) The y-intercept is $(0, d)$.

c) The slope is $\frac{d-0}{0-c} = \frac{d}{-c} = -\frac{d}{c}$.

d) The equation of the line in $y = mx + b$ form is $y = -\frac{d}{c}x + d$.

7.

$$\begin{array}{c|c} -2q - 7p & 19 \\ \hline -2(-5) - 7 \cdot 8 \ ? & 19 \\ 10 - 56 & \\ -46 & \quad \text{FALSE} \end{array}$$

$(8, -5)$ is not a solution.

8.

$$\begin{array}{c|c} 6y & -3x + 1 \\ \hline 6 \cdot \frac{2}{3} \ ? & -3(-1) + 1 \\ 4 & 3 + 1 \\ & 4 \quad \text{TRUE} \end{array}$$

$\left(-1, \frac{2}{3}\right)$ is a solution.

9. To find the x-intercept, let $y = 0$. Then solve for x.

$-3x + 2y = 18$

$-3x + 2 \cdot 0 = 18$

$-3x = 18$

$x = -6$

The x-intercept is $(-6, 0)$.

To find the y-intercept, let $x = 0$. Then solve for y.

$$-3x + 2y = 18$$
$$-3 \cdot 0 + 2y = 18$$
$$2y = 18$$
$$y = 9$$

The y-intercept is $(0, 9)$.

10. To find the x-intercept, let $y = 0$. Then solve for x.

$$x - \frac{1}{2} = 10y$$
$$x - \frac{1}{2} = 10 \cdot 0$$
$$x - \frac{1}{2} = 0$$
$$x = \frac{1}{2}$$

The x-intercept is $\left(\frac{1}{2}, 0\right)$.

To find the y-intercept, let $x = 0$. Then solve for y.

$$x - \frac{1}{2} = 10y$$
$$0 - \frac{1}{2} = 10y$$
$$-\frac{1}{2} = 10y$$
$$-\frac{1}{20} = y$$

The y-intercept is $\left(0, -\frac{1}{20}\right)$.

11. Graph: $-2x + y = -3$.

$$-2x + y = -3$$
$$y = 2x - 3$$

The y-intercept is $(0, -3)$.

We find the x-intercept.

$$-2x + 0 = -3$$
$$-2x = -3$$
$$x = \frac{3}{2}$$

The x-intercept is $\left(\frac{3}{2}, 0\right)$.

We find a third point as a check. Let $x = 1$.

$$y = 2 \cdot 1 - 3 = 2 - 3 = -1$$

A third point is $(1, -1)$. We plot this point and the intercepts and draw the line.

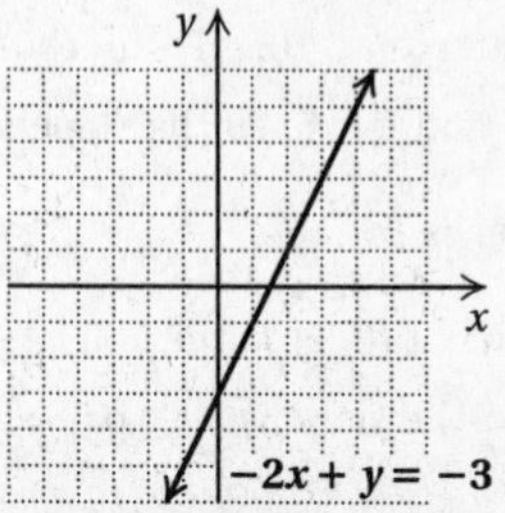

12. Graph: $y = -\frac{3}{2}$.

This is an equation of the form $y = b$ with $b = -\frac{3}{2}$. Its graph is a horizontal line with y-intercept $(0, b)$, or $\left(0, -\frac{3}{2}\right)$.

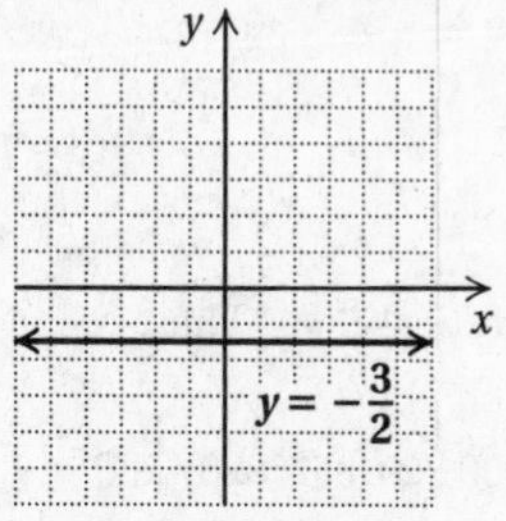

13. Graph: $y = -x + 4$.

The y-intercept is $(0, 4)$.

We find the x-intercept.

$$0 = -x + 4$$
$$x = 4$$

The x-intercept is $(4, 0)$.

We find a third point as a check. Let $x = 3$.

$$y = -3 + 4 = 1$$

A third point is $(3, 1)$. We plot this point and the intercepts and draw the line.

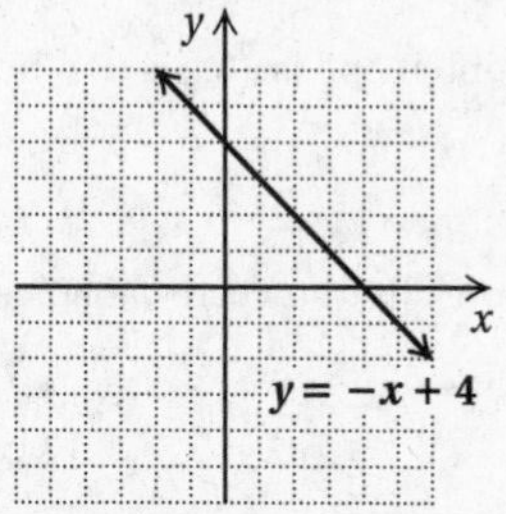

14. Graph: $x = 0$.

This is an equation of the form $x = a$ with $a = 0$. Its graph is a vertical line with x-intercept $(0, a)$, or $(0, 0)$.

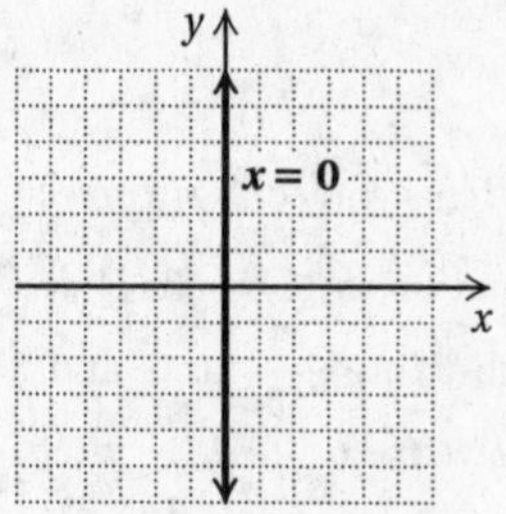

15. $m = \dfrac{4 - (-6)}{-2 - \frac{1}{4}} = \dfrac{10}{-2\frac{1}{4}} = \dfrac{10}{-\frac{9}{4}} = 10\left(-\frac{4}{9}\right) = -\dfrac{40}{9}$

16. $m = \dfrac{-3 - 3}{6 - (-6)} = \dfrac{-6}{12} = -\dfrac{1}{2}$

17. $y = 0.728$, or $y = 0 \cdot x + 0.728$

The slope is 0.

18. $$13x - y = -5$$
$$13x = y - 5$$
$$13x + 5 = y$$

The slope is 13.

19. $$12x + 7 = 0$$
$$12x = -7$$
$$x = -\frac{7}{12}$$

This is the equation of a vertical line. Thus, the slope is not defined.

20. We use the data points (2000, 20,851,820) and (2010, 25,145,561).

Rate of change $= m = \dfrac{25,145,561 - 20,851,820}{2010 - 2000} = \dfrac{4,293,741}{10} \approx 429,400$ people per year

21. $y = -1$, or $y = 0 \cdot x - 1$

The slope is 0 and the y-intercept is $(0, -1)$. Choice D is correct.

22. $x = 1$

This is the equation of a vertical line with x-intercept $(1, 0)$. The slope of a vertical line is not defined. Thus, choice C is correct.

23. $y = -x - 1$

The slope is -1 and the y-intercept is $(0, -1)$. Choice B is correct.

24. $y = x - 1$

The slope is 1 so either A or E is the correct choice. We find the x-intercept.

$$0 = x - 1$$
$$1 = x$$

The x-intercept is $(1, 0)$. Thus, choice E is correct.

25. $y = x + 1$

The slope is 1 so either A or E is the correct choice. We find the x-intercept.

$$0 = x + 1$$
$$-1 = x$$

The x-intercept is $(-1, 0)$. Thus choice A is correct.

26. The slope is -3 so the equation is $y = -3x + b$. Using the point $\left(-\frac{1}{3}, 3\right)$, we substitute $-\frac{1}{3}$ for x and 3 for y in $y = -3x + b$. Then we solve for b.

$$y = -3x + b$$
$$3 = -3 \cdot -\frac{1}{3} + b$$
$$3 = 1 + b$$
$$2 = b$$

The equation is $y = -3x + 2$.

27. $\left(\frac{1}{2}, 6\right)$ and $\left(\frac{1}{2}, -6\right)$

First we find the slope.

$$m = \frac{6 - (-6)}{\frac{1}{2} - \frac{1}{2}} = \frac{12}{0}$$

The slope is not defined. Thus, we have a vertical line with x-coordinate $\frac{1}{2}$. The equation is $x = \frac{1}{2}$.

28. $(3, -4)$ and $(-7, -2)$

First we find the slope.

$$m = \frac{-2 - (-4)}{-7 - 3} = \frac{2}{-10} = -\frac{1}{5}$$

Thus $y = -\frac{1}{5}x + b$. We use the point $(3, -4)$ to find b.

$$y = -\frac{1}{5}x + b$$
$$-4 = -\frac{1}{5} \cdot 3 + b$$
$$-4 = -\frac{3}{5} + b$$
$$-\frac{20}{5} + \frac{3}{5} = b$$
$$-\frac{17}{5} = b$$

The equation is $y = -\frac{1}{5}x - \frac{17}{5}$.

29. $(3, -4)$ and $(2, -4)$

First we find the slope.

$$m = \frac{-4 - (-4)}{3 - 2} = \frac{0}{1} = 0$$

The slope is 0, so we have a horizontal line with y-coordinate -4. The equation is $y = -4$.

30. No; an equation $x = a$, $a \neq 0$, does not have a y-intercept.

31. Most would probably say that the second equation would be easier to graph because it has been solved for y. This makes it more efficient to find the y-value that corresponds to a given x-value.

32. $A = 0$. If the line is horizontal, then regardless of the value of x, the value of y remains constant. Thus, Ax must be 0 and, hence, $A = 0$.

33. Any ordered pair $(7, y)$ is a solution of $x = 7$. Thus, all points on the graph are 7 units to the right of the y-axis, so they lie on a vertical line.

Exercise Set 3.5

RC1. To move from the y-intercept, $(0, -1)$ to the point $(-5, -5)$, we go down 4 and left 5, so the slope is $\frac{-4}{-5}$, or $\frac{4}{5}$. The correct answer is (c).

RC3. To move from the y-intercept $(0, 1)$ to the point $(-5, 5)$, we go up 4 and left 5, so the slope is $\frac{4}{-5}$, or $-\frac{4}{5}$. The correct answer is (b).

1. Slope $\frac{2}{5}$; y-intercept $(0, 1)$

We plot $(0, 1)$ and from there move up 2 units and right 5 units. This locates the point $(5, 3)$. We plot $(5, 3)$ and draw a line passing through $(0, 1)$ and $(5, 3)$.

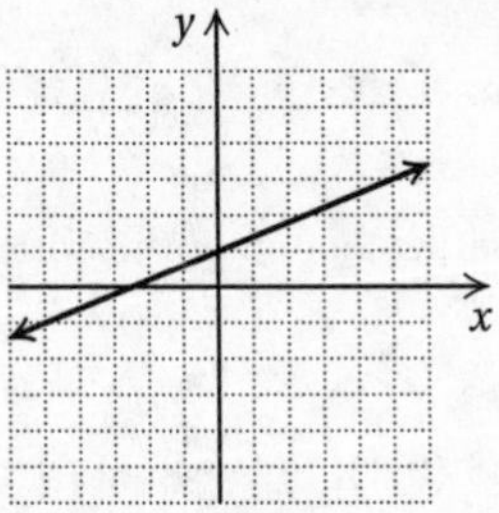

3. Slope $\frac{5}{3}$; y-intercept $(0, -2)$

We plot $(0, -2)$ and from there move up 5 units and right 3 units. This locates the point $(3, 3)$. We plot $(3, 3)$ and draw a line passing through $(0, -2)$ and $(3, 3)$.

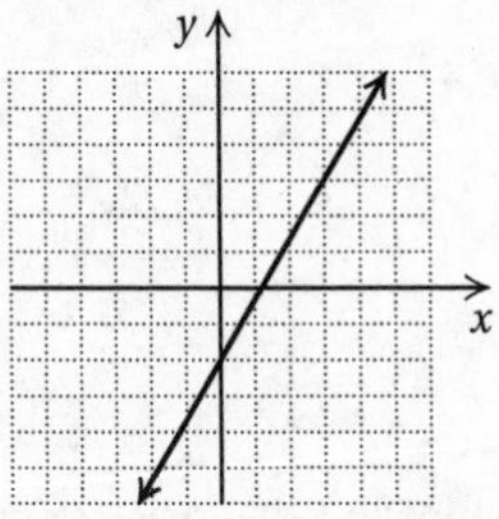

5. Slope $-\frac{3}{4}$; y-intercept $(0, 5)$

We plot $(0, 5)$. We can think of the slope as $\frac{-3}{4}$, so from $(0, 5)$ we move down 3 units and right 4 units. This locates the point $(4, 2)$. We plot $(4, 2)$ and draw a line passing through $(0, 5)$ and $(4, 2)$.

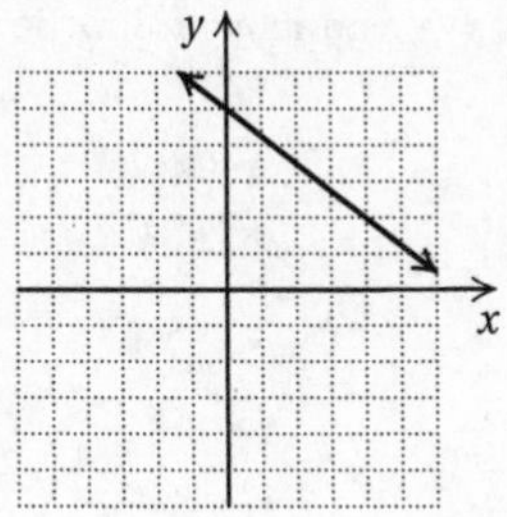

7. Slope $-\frac{1}{2}$; y-intercept $(0, 3)$

We plot $(0, 3)$. We can think of the slope as $\frac{-1}{2}$, so from $(0, 3)$ we move down 1 unit and right 2 units. This locates the point $(2, 2)$. We plot $(2, 2)$ and draw a line passing through $(0, 3)$ and $(2, 2)$

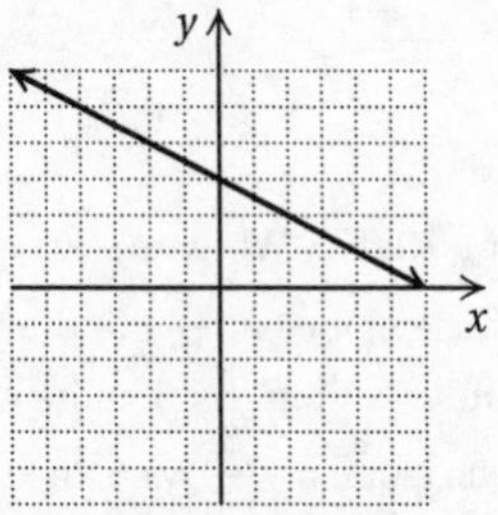

9. Slope 2; y-intercept $(0, -4)$

We plot $(0, -4)$. We can think of the slope as $\frac{2}{1}$, so from $(0, -4)$ we move up 2 units and right 1 unit. This locates the point $(1, -2)$. We plot $(1, -2)$ and draw a line passing through $(0, -4)$ and $(1, -2)$.

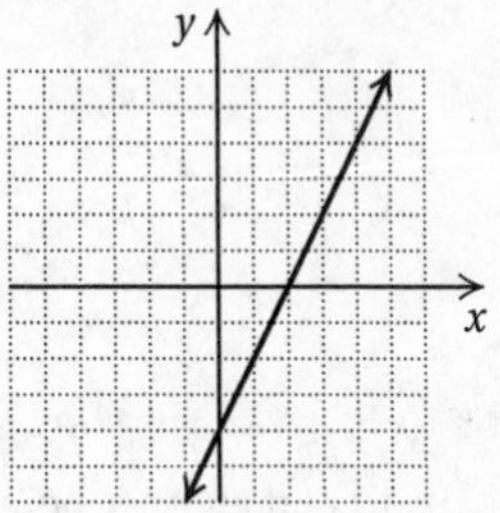

11. Slope -3; y-intercept $(0, 2)$

We plot $(0, 2)$. We can think of the slope as $\frac{-3}{1}$, so from $(0, 2)$ we move down 3 units and right 1 unit. This locates the point $(1, -1)$. We plot $(1, -1)$ and draw a line passing through $(0, 2)$ and $(1, -1)$.

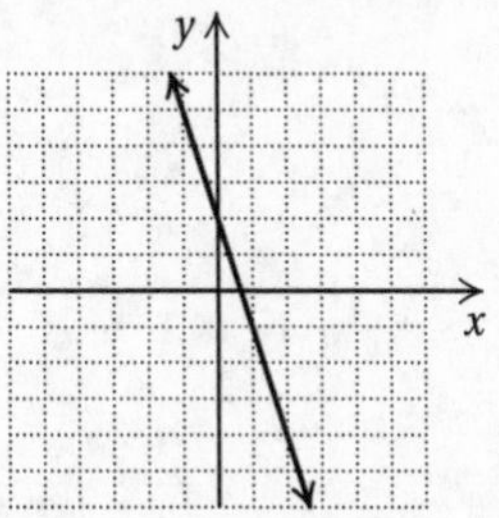

13. $y = \frac{3}{5}x + 2$

First we plot the y-intercept $(0, 2)$. We can start at the y-intercept and use the slope, $\frac{3}{5}$, to find another point. We move up 3 units and right 5 units to get a new point $(5, 5)$. Thinking of the slope as $\frac{-3}{-5}$ we can start at $(0, 2)$ and move down 3 units and left 5 units to get another point $(-5, -1)$.

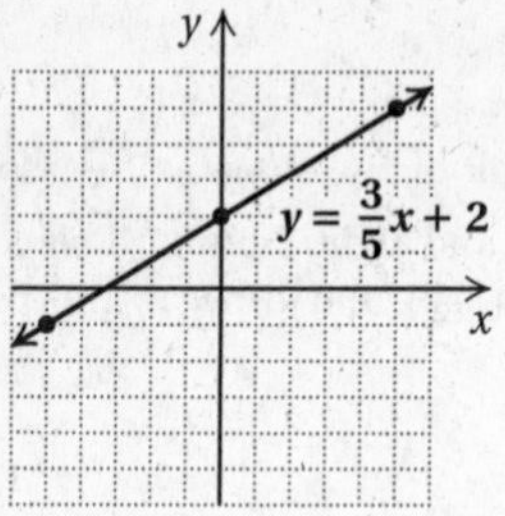

15. $y = -\frac{3}{5}x + 1$

First we plot the y-intercept $(0, 1)$. We can start at the y-intercept and, thinking of the slope as $\frac{-3}{5}$, find another point by moving down 3 units and right 5 units to the point $(5, -2)$. Thinking of the slope as $\frac{3}{-5}$ we can start at $(0, 1)$ and move up 3 units and left 5 units to get another point $(-5, 4)$.

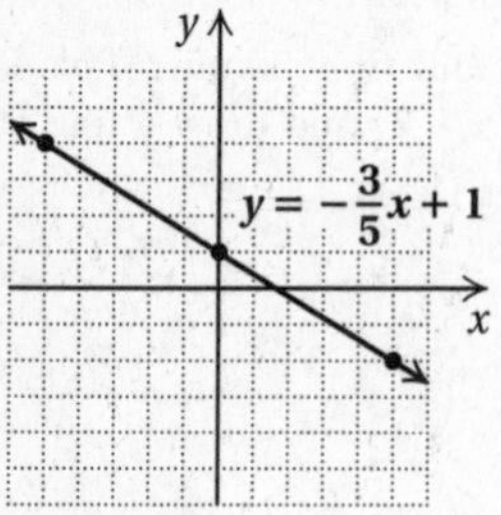

17. $y = \frac{5}{3}x + 3$

First we plot the y-intercept $(0, 3)$. We can start at the y-intercept and use the slope, $\frac{5}{3}$, to find another point. We move up 5 units and right 3 units to get a new point $(3, 8)$. Thinking of the slope as $\frac{-5}{-3}$ we can start at $(0, 3)$ and move down 5 units and left 3 units to get another point $(-3, -2)$.

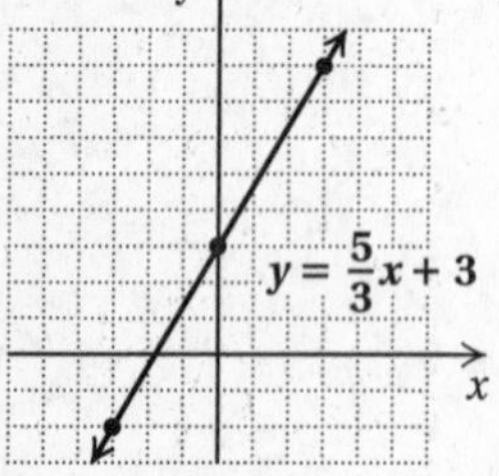

19. $y = -\frac{3}{2}x - 2$

First we plot the y-intercept $(0, -2)$. We can start at the y-intercept and, thinking of the slope as $\frac{-3}{2}$, find another point by moving down 3 units and right 2 units to the point $(2, -5)$. Thinking of the slope as $\frac{3}{-2}$ we can start at $(0, -2)$ and move up 3 units and left 2 units to get another point $(-2, 1)$.

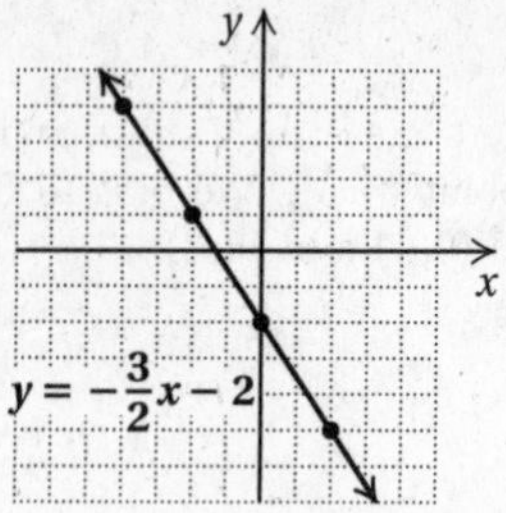

21. We first rewrite the equation in slope-intercept form.

$$2x + y = 1$$
$$y = -2x + 1$$

Now we plot the y-intercept $(0, 1)$. We can start at the y-intercept and, thinking of the slope as $\frac{-2}{1}$, find another point by moving down 2 units and right 1 unit to the point $(1, -1)$. In a similar manner, we can move from the point $(1, -1)$ to find a third point $(2, -3)$.

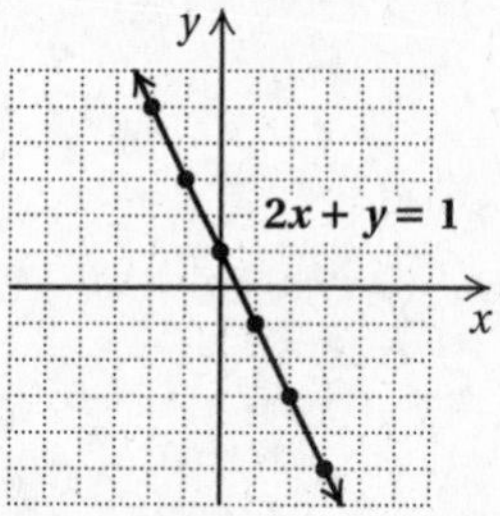

23. We first rewrite the equation in slope-intercept form.

$$3x - y = 4$$
$$-y = -3x + 4$$
$$y = 3x - 4 \quad \text{Multiplying by } -1$$

Now we plot the y-intercept $(0, -4)$. We can start at the y-intercept and, thinking of the slope as $\frac{3}{1}$, find another point by moving up 3 units and right 1 unit to the point $(1, -1)$. In a similar manner, we can move from the point $(1, -1)$ to find a third point $(2, 2)$.

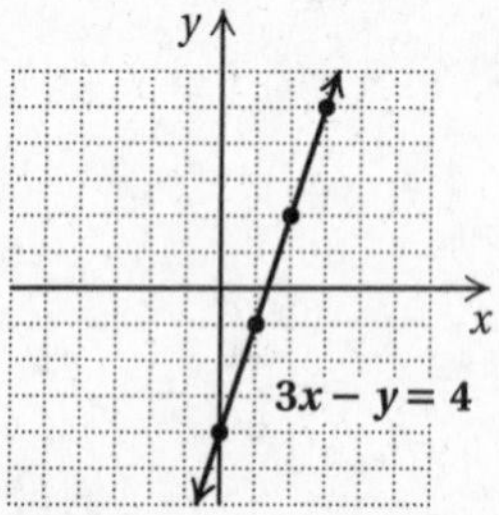

25. We first rewrite the equation in slope-intercept form.

$$2x + 3y = 9$$
$$3y = -2x + 9$$
$$y = \frac{1}{3}(-2x + 9)$$
$$y = -\frac{2}{3}x + 3$$

Now we plot the y-intercept $(0, 3)$. We can start at the y-intercept and, thinking of the slope as $\frac{-2}{3}$, find another point by moving down 2 units and right 3 units to the point $(3, 1)$. Thinking of the slope as $\frac{2}{-3}$ we can start at $(0, 3)$ and move up 2 units and left 3 units to get another point $(-3, 5)$.

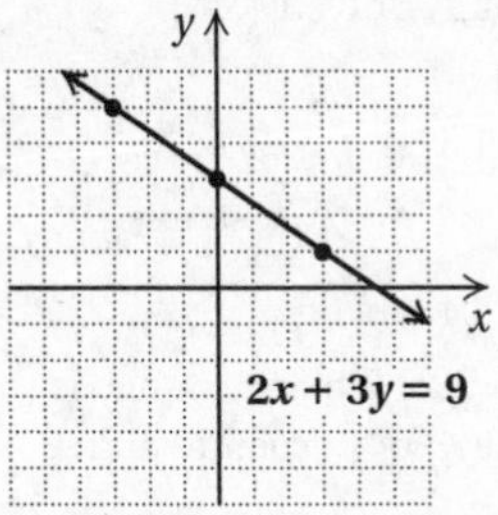

27. We first rewrite the equation in slope-intercept form.

$$x - 4y = 12$$
$$-4y = -x + 12$$
$$y = -\frac{1}{4}(-x + 12)$$
$$y = \frac{1}{4}x - 3$$

Now we plot the y-intercept $(0, -3)$. We can start at the y-intercept and use the slope, $\frac{1}{4}$, to find another point. We move up 1 unit and right 4 units to the point $(4, -2)$. Thinking of the slope as $\frac{-1}{-4}$ we can start at $(0, -3)$ and move down 1 unit and left 4 units to get another point $(-4, -4)$.

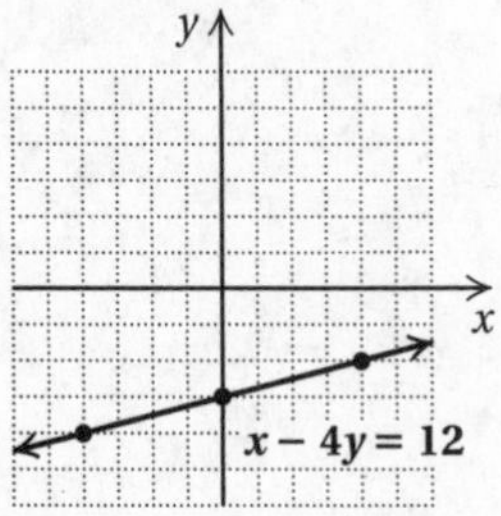

29. We first rewrite the equation in slope-intercept form.

$$x + 2y = 6$$
$$2y = -x + 6$$
$$y = \frac{1}{2}(-x + 6)$$
$$y = -\frac{1}{2}x + 3$$

Now we plot the y-intercept $(0, 3)$. We can start at the y-intercept and, thinking of the slope as $\frac{-1}{2}$, find another point by moving down 1 unit and right 2 units to the point $(2, 2)$. Thinking of the slope as $\frac{1}{-2}$ we can start at $(0, 3)$ and move up 1 unit and left 2 units to get another point $(-2, 4)$.

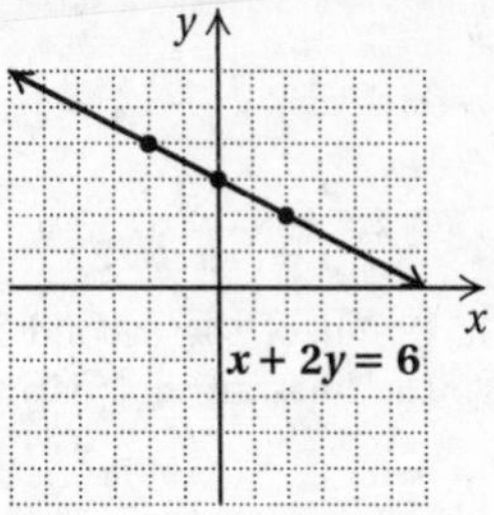

31. $m = \frac{y_2 - y_1}{x_2 - x_1} = \frac{7 - (-6)}{8 - (-2)} = \frac{13}{10}$

33. $m = \frac{y_2 - y_1}{x_2 - x_1} = \frac{4.6 - (-2.3)}{14.5 - 4.5} = \frac{6.9}{10} = \frac{69}{100}$, or 0.69

35. $m = \frac{y_2 - y_1}{x_2 - x_1} = \frac{-6 - (-6)}{8 - (-2)} = \frac{0}{10} = 0$

37. $8 - (-11) + 23 = 8 + 11 + 23 = 19 + 23 = 42$

39.
$$\begin{aligned} -10 - (-30) + 5 - 2 &= -10 + 30 + 5 + (-2) \\ &= 20 + 5 + (-2) \\ &= 25 + (-2) \\ &= 23 \end{aligned}$$

41. First we plot $(-3, 1)$. Then, thinking of the slope as $\frac{2}{1}$, from $(-3, 1)$ we move up 2 units and right 1 unit to locate the point $(-2, 3)$. We plot $(-2, 3)$ and draw a line passing through $(-3, 1)$ and $(-2, 3)$.

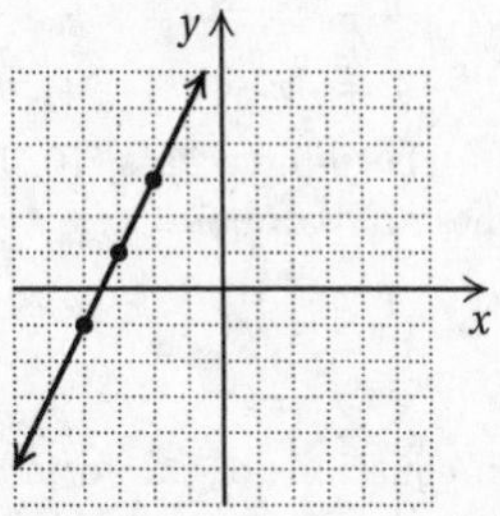

Exercise Set 3.6

RC1. Lines (b) and (e) have the same slope and different y-intercepts, so they are parallel.

RC3. The slopes of lines (c) and (d) are reciprocals, so they are perpendicular.

1. 1. The first equation is already solved for y:

$$y = x + 4$$

2. We solve the second equation for y:

$$y - x = -3$$
$$y = x - 3$$

The slope of each line is 1. The y-intercepts, $(0, 4)$ and $(0, -3)$, are different. The lines are parallel.

3. We solve each equation for y:

1. $y+3=6x$
$y=6x-3$

2. $-6x-y=2$
$-y=6x+2$
$y=-6x-2$

The slope of the first line is 6 and of the second is -6. Since the slopes are different, the lines are not parallel.

5. We solve each equation for y:

1. $10y+32x=16.4$
$10y=-32x+16.4$
$y=-3.2x+1.64$

2. $y+3.5=0.3125x$
$y=0.3125x-3.5$

The slope of the first line is -3.2 and of the second is 0.3125. Since the slopes are different, the lines are not parallel.

7. 1. The first equation is already solved for y:
$y=2x+7$

2. We solve the second equation for y:
$5y+10x=20$
$5y=-10x+20$
$y=-2x+4$

The slope of the first line is 2 and of the second is -2. Since the slopes are different, the lines are not parallel.

9. We solve each equation for y:

1. $3x-y=-9$
$3x+9=y$

2. $2y-6x=-2$
$2y=6x-2$
$y=3x-1$

The slope of each line is 3. The y-intercepts, $(0,9)$ and $(0,-1)$ are different. The lines are parallel.

11. $x=3$,
$x=4$

These are vertical lines with equations of the form $x=p$ and $x=q$, where $p\neq q$. Thus, they are parallel.

13. The graph of $x=-6$ is a vertical line, and the graph of $y=6$ is a horizontal line, so the lines are not parallel.

15. 1. The first equation is already solved for y:
$y=-4x+3$

2. We solve the second equation for y:
$4y+x=-1$
$4y=-x-1$
$y=-\frac{1}{4}x-\frac{1}{4}$

The slopes are -4 and $-\frac{1}{4}$. Their product is $-4\left(-\frac{1}{4}\right)=1$. Since the product of the slopes is not -1, the lines are not perpendicular.

17. We solve each equation for y:

1. $x+y=6$
$y=-x+6$

2. $4y-4x=12$
$4y=4x+12$
$y=x+3$

The slopes are -1 and 1. Their product is $-1\cdot 1=-1$. The lines are perpendicular.

19. 1. The first equation is already solved for y:
$y=-0.3125x+11$

2. We solve the second equation for y:
$y-3.2x=-14$
$y=3.2x-14$

The slopes are -0.3125 and 3.2. Their product is $-0.3125(3.2)=-1$. The lines are perpendicular.

21. 1. The first equation is already solved for y:
$y=-x+8$

2. We solve the second equation for y:
$x-y=-1$
$x+1=y$

The slopes are -1 and 1. Their product is $-1\cdot 1=-1$. The lines are perpendicular.

23. We solve each equation for y:

1. $\frac{3}{8}x-\frac{y}{2}=1$
$8\left(\frac{3}{8}x-\frac{y}{2}\right)=8\cdot 1$
$8\cdot\frac{3}{8}x-8\cdot\frac{y}{2}=8$
$3x-4y=8$
$-4y=-3x+8$
$y=\frac{3}{4}x-2$

2. $\frac{4}{3}x-y+1=0$
$\frac{4}{3}x+1=y$

The slopes are $\frac{3}{4}$ and $\frac{4}{3}$. Their product is $\frac{3}{4}\left(\frac{4}{3}\right)=1$. Since the product of the slopes is not -1, the lines are not perpendicular.

25. $x=0$,
$y=-2$

The first line is vertical and the second is horizontal, so the lines are perpendicular.

27. These are both horizontal lines, so they are not perpendicular.

29. We solve each equation for y:

1. $3y + 21 = 2x$

$3y = 2x - 21$

$y = \frac{2}{3}x - 7$

2. $3y = 2x + 24$

$y = \frac{2}{3}x + 8$

The slope of each line is $\frac{2}{3}$. The y-intercepts, $(0, -7)$ and $(0, 8)$, are different. The lines are parallel.

31. We solve each equation for y:

1. $3y = 2x - 21$

$y = \frac{2}{3}x - 7$

2. $2y - 16 = 3x$

$2y = 3x + 16$

$y = \frac{3}{2}x + 8$

The slopes, $\frac{2}{3}$ and $\frac{3}{2}$, are different so the lines are not parallel. The product of the slopes is $\frac{2}{3} \cdot \frac{3}{2} = 1 \neq -1$, so the lines are not perpendicular. Thus, the lines are neither parallel nor perpendicular.

33.

$2x - 15y = -17$	
$2 \cdot 1 - 15(-1)$?	-17
$2 + 15$	
17	FALSE

$(1, -1)$ is not a solution of the equation.

35. $-40x + 5y = 80$

To find the x-intercept, let $y = 0$. Then solve for x.

$$-40x + 5 \cdot 0 = 80$$
$$-40x = 80$$
$$x = -2$$

The x-intercept is $(-2, 0)$.

To find the y-intercept, let $x = 0$. Then solve for y.

$$-40 \cdot 0 + 5y = 80$$
$$5y = 80$$
$$y = 16$$

The y-intercept is $(0, 16)$.

37. First we find the slope of the given line:

$$y - 3x = 4$$
$$y = 3x + 4$$

The slope is 3.

Then we use the slope-intercept equation to write the equation of a line with slope 3 and y-intercept $(0, 6)$:

$$y = mx + b$$
$y = 3x + 6$ Substituting 3 for m and 6 for b

39. First we find the slope of the given line:

$$3y - x = 0$$
$$3y = x$$
$$y = \frac{1}{3}x$$

The slope is $\frac{1}{3}$.

We can find the slope of the line perpendicular to the given line by taking the reciprocal of $\frac{1}{3}$ and changing the sign. We get -3.

Then we use the slope-intercept equation to write the equation of a line with slope -3 and y-intercept $(0, 2)$:

$$y = mx + b$$
$y = -3x + 2$ Substituting -3 for m and 2 for b

41. First we find the slope of the given line:

$$4x - 8y = 12$$
$$-8y = -4x + 12$$
$$y = \frac{1}{2}x - \frac{3}{2}$$

The slope is $\frac{1}{2}$, so the equation is $y = \frac{1}{2}x + b$. Substitute -2 for x and 0 for y and solve for b.

$$y = \frac{1}{2}x + b$$
$$0 = \frac{1}{2}(-2) + b$$
$$0 = -1 + b$$
$$1 = b$$

Thus, the equation is $y = \frac{1}{2}x + 1$.

43. We find the slope of each line:

1. $4y = kx - 6$

$y = \frac{k}{4}x - \frac{3}{2}$

2. $5x + 20y = 12$

$20y = -5x + 12$

$y = -\frac{1}{4}x + \frac{3}{5}$

The slopes are $\frac{k}{4}$ and $-\frac{1}{4}$. If the lines are perpendicular, the product of their slopes is -1.

$$\frac{k}{4}\left(-\frac{1}{4}\right) = -1$$
$$-\frac{k}{16} = -1$$
$$k = 16$$

45. First we find the equation of A, a line containing the points $(1, -1)$ and $(4, 3)$:

The slope is $\frac{3-(-1)}{4-1} = \frac{4}{3}$, so the equation is $y = \frac{4}{3}x + b$. Use either point to find b. We choose $(1, -1)$.

$$y = \frac{4}{3}x + b$$
$$-1 = \frac{4}{3} \cdot 1 + b$$
$$-1 = \frac{4}{3} + b$$
$$-\frac{7}{3} = b$$

Thus, the equation of line A is $y = \frac{4}{3}x - \frac{7}{3}$.

The slope of A is $\frac{4}{3}$. Since A and B are perpendicular we find the slope of B by taking the reciprocal of $\frac{4}{3}$ and changing the sign. We get $-\frac{3}{4}$, so the equation is $y = -\frac{3}{4}x + b$. We use the point $(1, -1)$ to find b.

$$y = -\frac{3}{4}x + b$$
$$-1 = -\frac{3}{4} \cdot 1 + b$$
$$-1 = -\frac{3}{4} + b$$
$$-\frac{1}{4} = b$$

Thus, the equation of line B is $y = -\frac{3}{4}x - \frac{1}{4}$.

Exercise Set 3.7

RC1. $=$

RC3. x-intercept

RC5. $<$

RC7. half-plane

RC9. false

1. We use alphabetical order to replace x by -3 and y by -5.

$$\begin{array}{r|l} -x - 3y < 18 & \\ \hline -(-3) - 3(-5) \ ?\ 18 & \\ 3 + 15 & \\ 18 & \quad \text{FALSE} \end{array}$$

Since $18 < 18$ is false, $(-3, -5)$ is not a solution.

3. We use alphabetical order to replace x by 1 and y by -10.

$$\begin{array}{r|l} 7y - 9x \le -3 & \\ \hline 7(-10) - 9 \cdot 1 \ ?\ -3 & \\ -70 - 9 & \\ -79 & \quad \text{TRUE} \end{array}$$

Since $-79 \le -3$ is true, $(1, -10)$ is a solution.

5. Graph $x > 2y$.

First graph the line $x = 2y$, or $y = \frac{1}{2}x$. Two points on the line are $(0, 0)$ and $(4, 2)$. We draw a dashed line since the inequality symbol is $>$. Then we pick a test point that is not on the line. We try $(-2, 1)$.

$$\begin{array}{r|l} x > 2y & \\ \hline -2 \ ?\ 2 \cdot 1 & \\ & 2 \quad \text{FALSE} \end{array}$$

We see that $(-2, 1)$ is not a solution of the inequality, so we shade the points in the region that does not contain $(-2, 1)$.

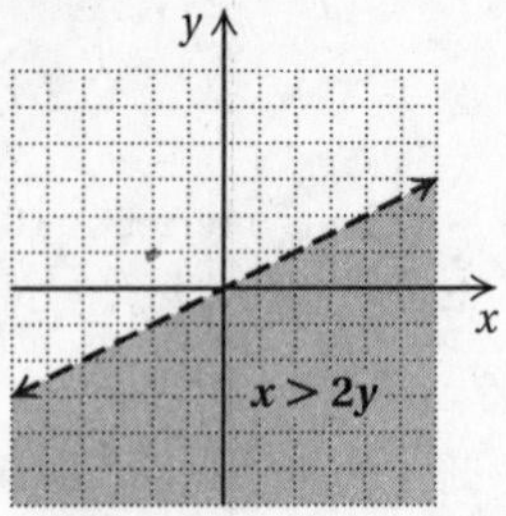

7. Graph $y \le x - 3$.

First graph the line $y = x - 3$. The intercepts are $(0, -3)$ and $(3, 0)$. We draw a solid line since the inequality symbol is $\le$. Then we pick a test point that is not on the line. We try $(0, 0)$.

$$\begin{array}{r|l} y \le x - 3 & \\ \hline 0 \ ?\ 0 - 3 & \\ & -3 \quad \text{FALSE} \end{array}$$

We see that $(0, 0)$ is not a solution of the inequality, so we shade the region that does not contain $(0, 0)$.

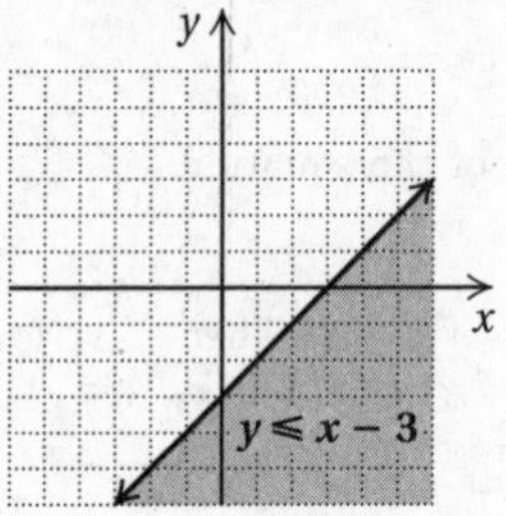

9. Graph $x + y \le 3$.

First graph the line $x + y = 3$. The intercepts are $(0, 3)$ and $(3, 0)$. We draw a solid line since the inequality symbol is $\le$. Then we test the point $(0, 0)$.

$$\begin{array}{r|l} x + y \le 3 & \\ \hline 0 + 0 \ ?\ 3 & \\ 0 & \quad \text{TRUE} \end{array}$$

Since $(0, 0)$ is a solution of the inequality, we shade the region that contains $(0, 0)$.

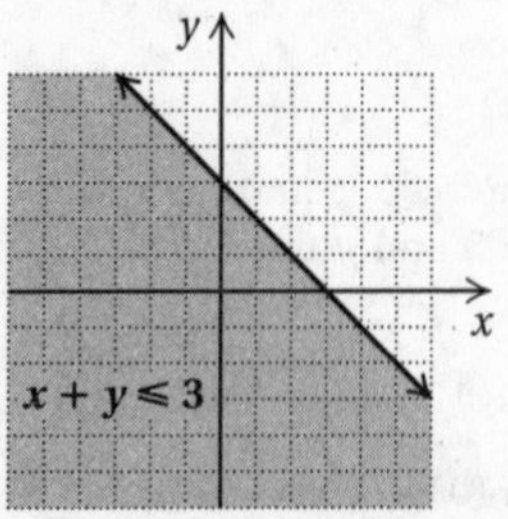

11. Graph $y > x - 2$.

First graph the line $y = x - 2$. The intercepts are $(0, -2)$ and $(2, 0)$. We draw a dashed line since the inequality symbol is $>$. Then we test the point $(0,0)$.

$$\begin{array}{c|l} y > x-2 & \\ \hline 0 \;?\; 0-2 & \\ & -2 \quad \text{TRUE} \end{array}$$

Since $(0,0)$ is a solution of the inequality, we shade the region that contains $(0,0)$.

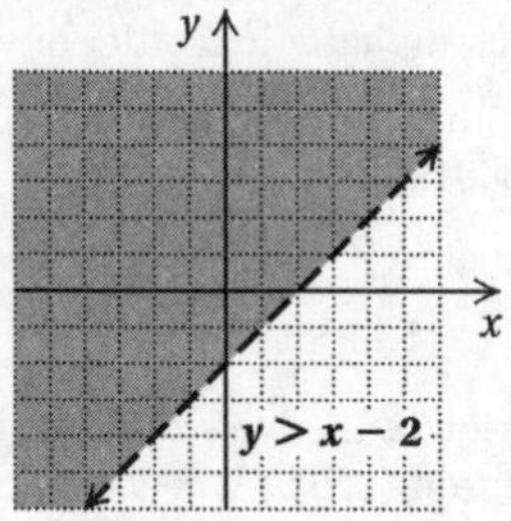

13. Graph $x - y > 7$.

First graph the line $x - y = 7$. The intercepts are $(0,-7)$ and $(7,0)$. We draw a dashed line since the inequality symbol is $>$. Then we test the point $(0,0)$.

$$\begin{array}{c|l} x-y > 7 & \\ \hline 0-0 \;?\; 7 & \\ 0 & \text{FALSE} \end{array}$$

Since $(0,0)$ is not a solution of the inequality, we shade the region that does not contain $(0,0)$.

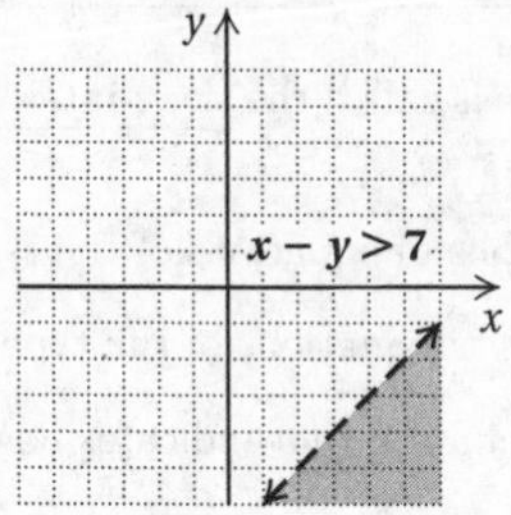

15. Graph $y \geq 4x - 1$.

First graph the line $y = 4x - 1$. The intercepts are $\left(\frac{1}{4}, 0\right)$ and $(0,-1)$. We draw a solid line since the inequality symbol is $\geq$. Then we test the point $(0,0)$.

$$\begin{array}{c|l} y \geq 4x-1 & \\ \hline 0 \;?\; 4\cdot 0-1 & \\ & -1 \quad \text{TRUE} \end{array}$$

Since $(0,0)$ is a solution of the inequality, we shade the half-plane that contains $(0,0)$.

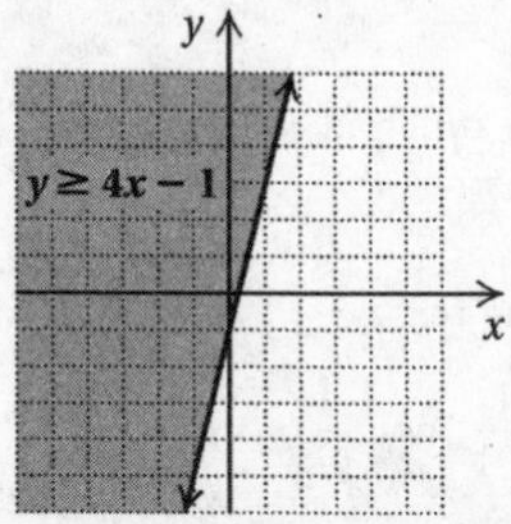

17. Graph $y \geq 1 - 2x$.

First graph the line $y = 1 - 2x$. The intercepts are $(0,1)$ and $\left(\frac{1}{2}, 0\right)$. We draw a solid line since the inequality symbol is $\geq$. Then we test the point $(0,0)$.

$$\begin{array}{c|l} y \geq 1-2x & \\ \hline 0 \;?\; 1-2\cdot 0 & \\ & 1 \quad \text{FALSE} \end{array}$$

Since $(0,0)$ is not a solution of the inequality, we shade the region that does not contain $(0,0)$.

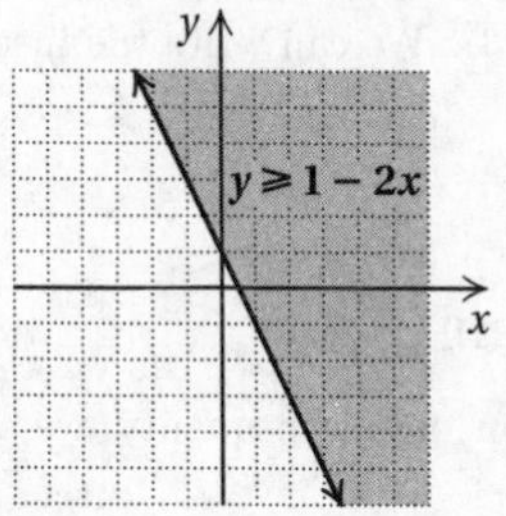

19. Graph $2x + 3y \leq 12$.

First graph the line $2x+3y = 12$. The intercepts are $(0,4)$ and $(6,0)$. We draw a solid line since the inequality symbol is $\leq$. Then we test the point $(0,0)$.

$$\begin{array}{c|l} 2x+3y \leq 12 & \\ \hline 2\cdot 0+3\cdot 0 \;?\; 12 & \\ 0 & \text{TRUE} \end{array}$$

Since $(0,0)$ is a solution of the inequality, we shade the region containing $(0,0)$.

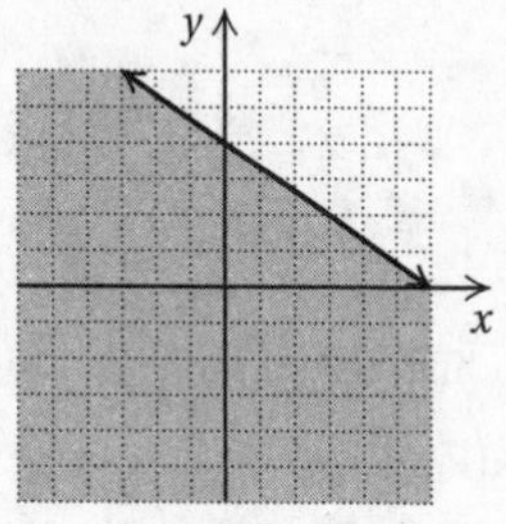

2x + 3y ≤ 12

21. Graph $y \leq 3$.

First graph the line $y = 3$ using a solid line since the inequality symbol is $\leq$. Then pick a test point that is not on the line. We choose $(1,-2)$. We can write the inequality as $0x + y \leq 3$.

$$\begin{array}{c|l} 0x+y \leq 3 & \\ \hline 0\cdot 1+(-2) \;?\; 3 & \\ -2 & \text{TRUE} \end{array}$$

Since $(1,-2)$ is a solution of the inequality, we shade the region containing $(1,-2)$.

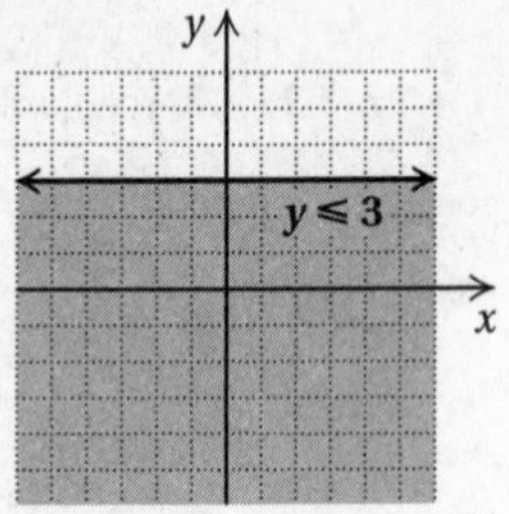

23. Graph $x \geq -1$.

Graph the line $x = -1$ using a solid line since the inequality symbol is $\geq$. Then pick a test point that is not on the line. We choose $(2,3)$. We can write the inequality as $x + 0y \geq -1$.

$$\begin{array}{c|c} \multicolumn{2}{c}{x + 0y \geq -1} \\ \hline 2 + 0 \cdot 3 \; ? \; -1 & \\ 2 & \text{TRUE} \end{array}$$

Since $(2,3)$ is a solution of the inequality, we shade the region containing $(2,3)$.

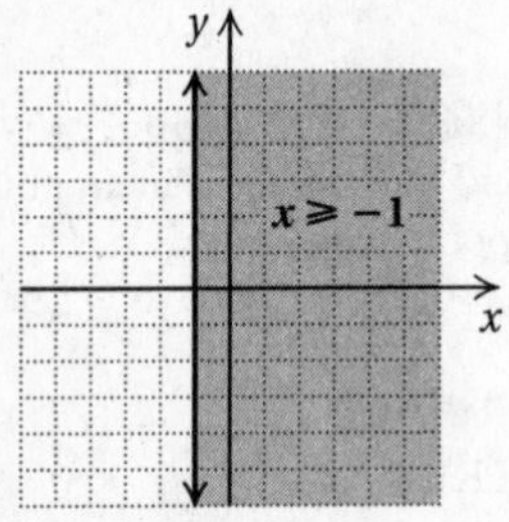

25. First we solve each equation for y:

1. $5y + 50 = 4x$ $\quad$ 2. $5y = 4x + 15$

$$5y = 4x - 50 \qquad y = \frac{4}{5}x + 3$$

$$y = \frac{4}{5}x - 10$$

The slope of each line is $\frac{4}{5}$. The y-intercepts, $(0,-10)$ and $(0,3)$, are different. The lines are parallel.

27. First we solve each equation for y:

1. $5y + 50 = 4x$ $\quad$ 2. $4y = 5x + 12$

$$5y = 4x - 50 \qquad y = \frac{5}{4}x + 3$$

$$y = \frac{4}{5}x - 10$$

The slope, $\frac{4}{5}$ and $\frac{5}{4}$, are different, so the lines are not parallel. The product of the slopes is $\frac{4}{5} \cdot \frac{5}{4} = 1 \neq -1$, so the lines are not perpendicular. Thus, the lines are neither parallel nor perpendicular.

29. The c children weigh $35c$ kg, and the a adults weigh $75a$ kg. Together, the children and adults weigh $35c + 75a$ kg. When this total is more than 1000 kg the elevator is overloaded, so we have $35c + 75a > 1000$. (Of course, c and a would also have to be nonnegative, but we will not deal with nonnegativity constraints here.)

To graph $35c + 75a > 1000$, we first graph $35c + 75a = 1000$ using a dashed line. Two points on the line are $(4,20)$ and $(11,5)$. (We are using alphabetical order of variables.) Then we test the point $(0,0)$.

$$\begin{array}{c|c} \multicolumn{2}{c}{35c + 75a > 1000} \\ \hline 35 \cdot 0 + 75 \cdot 0 \; ? \; 1000 & \\ 0 & \text{FALSE} \end{array}$$

Since $(0,0)$ is not a solution of the inequality, we shade the region that does not contain $(0,0)$.

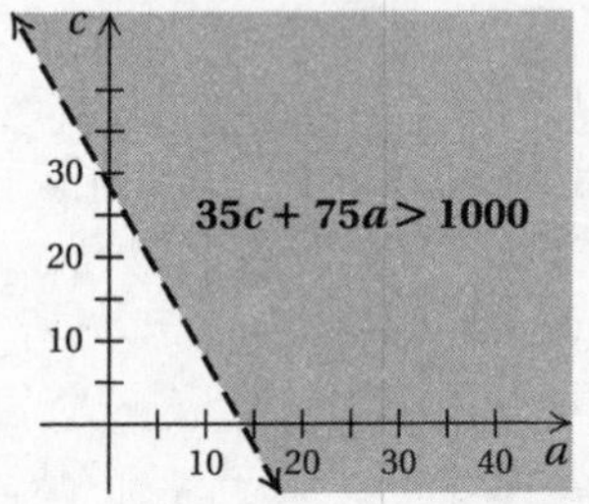

Chapter 3 Vocabulary Reinforcement

1. The equation $y = mx + b$ is called the slope-intercept equation.
2. Horizontal lines are graphs of equations of the type $y = b$.
3. Vertical lines are graphs of equations of the type $x = a$.
4. The slope of a line is a number that indicates how the line slants.
5. The x-intercept of a line, if it exists, indicates where the line crosses the x-axis, and thus will always have 0 as the second coordinate.
6. The y-intercept of a line, if it exists, indicates where the line crosses the y-axis, and thus will always have 0 as the first coordinate.

Chapter 3 Concept Reinforcement

1. True; $(0,0)$ is a solution of $y = mx$, so the x- and y-intercepts are both $(0,0)$.
2. False; parallel lines have different y-intercepts.
3. False; $0 > 0$ is false.
4. True; see page 215 in the text.
5. False; when $y = 0$, $x = \frac{C}{A}$, not $\frac{A}{C}$.

Chapter 3 Study Guide

1. Point F is 2 units right and 4 units up. Its coordinates are $(2, 4)$.

Point G is 2 units left and 0 units up or down. Its coordinates are $(-2, 0)$.

Point H is 3 units left and 5 units down. Its coordinates are $(-3, -5)$.

2.
$$x + 2y = 8$$
$$2y = -x + 8$$
$$y = -\frac{1}{2}x + 4$$

The y-intercept is $(0, 4)$. We find two other points.

When $x = 2$, $y = -\frac{1}{2} \cdot 2 + 4 = -1 + 4 = 3$.

When $x = -2$, $y = -\frac{1}{2}(-2) + 4 = 1 + 4 = 5$.

We plot $(0, 4)$, $(2, 3)$, and $(-2, 5)$ and draw the line.

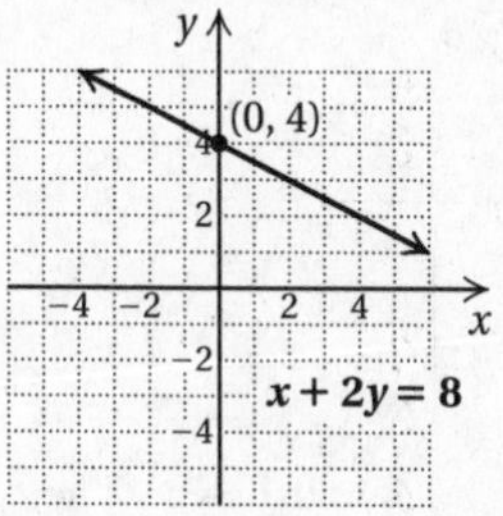

3. To find the x-intercept, we let $y = 0$ and solve for x.
$$y - 2x = -4$$
$$0 - 2x = -4$$
$$-2x = -4$$
$$x = 2$$

The x-intercept is $(2, 0)$.

To find the y-intercept, we let $x = 0$ and solve for y.
$$y - 2x = -4$$
$$y - 2 \cdot 0 = -4$$
$$y = -4$$

The y-intercept is $(0, -4)$.

We find a third point as a check. Let $x = 4$.
$$y - 2 \cdot 4 = -4$$
$$y - 8 = -4$$
$$y = 4$$

A third point is $(4, 4)$. We plot this point and the intercepts and draw the graph.

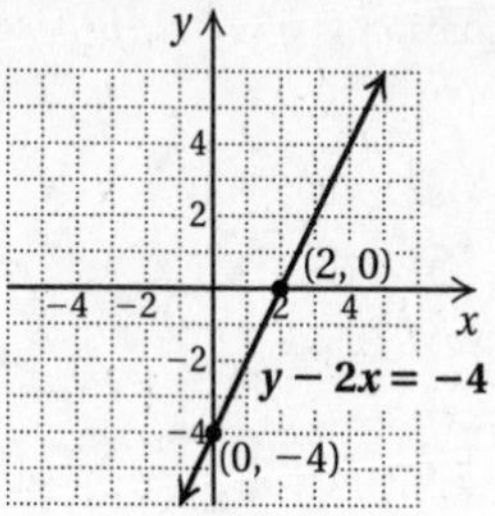

4. Graph: $y = -\frac{5}{2}$.

The graph is a horizontal line with y-intercept $\left(0, -\frac{5}{2}\right)$.

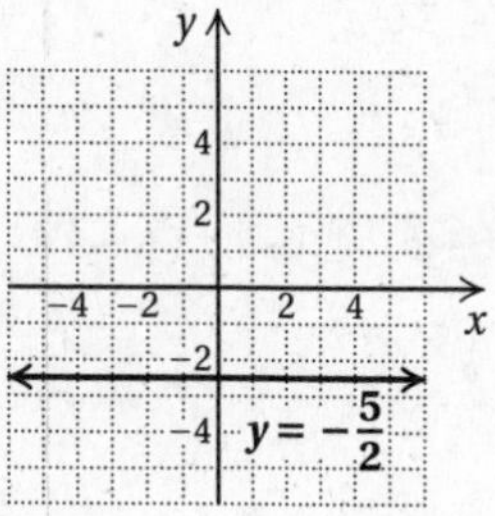

5. Graph: $x = 2$.

The graph is a vertical line with x-intercept $(2, 0)$.

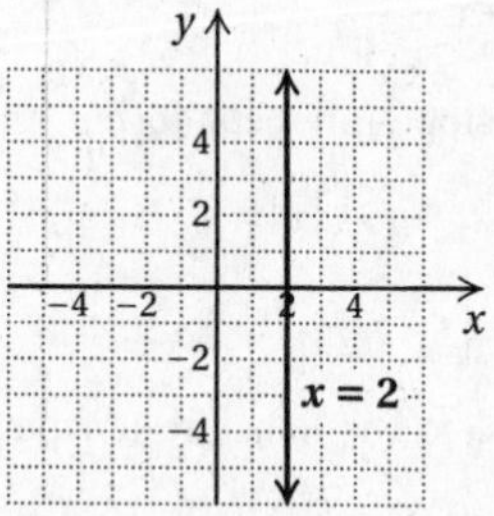

6. $m = \dfrac{20 - 14}{-8 - (-8)} = \dfrac{6}{0}$

The slope is not defined.

7. $m = \dfrac{20 - (-1)}{16 - 2} = \dfrac{21}{14} = \dfrac{3}{2}$

8. $m = \dfrac{2.8 - 2.8}{0.5 - 1.5} = \dfrac{0}{-1} = 0$

9. $x = 0.25$ is the equation of a vertical line. The slope is not defined.

10.
$$7y + 14x = -28$$
$$7y = -14x - 28$$
$$y = -2x - 4$$

The slope is -2.

11. $y = -5$, or $y = 0 \cdot x - 5$

The slope is 0.

12. The slope is 6, so the equation is $y = 6x + b$. Use the point $(-1, 1)$ to find b.

$$y = 6x + b$$
$$1 = 6(-1) + b$$
$$1 = -6 + b$$
$$7 = b$$

The equation is $y = 6x + 7$.

13. First we find the slope.

$$m = \frac{-2-(-3)}{1-7} = \frac{1}{-6} = -\frac{1}{6}$$

The equation is $y = -\frac{1}{6}x + b$. We use one of the points to find b.

$$y = -\frac{1}{6}x + b$$
$$-2 = -\frac{1}{6} \cdot 1 + b \quad \text{Using } (1, -2)$$
$$-2 = -\frac{1}{6} + b$$
$$-\frac{12}{6} + \frac{1}{6} = b$$
$$-\frac{11}{6} = b$$

The equation is $y = -\frac{1}{6}x - \frac{11}{6}$.

14. Write the equations in slope-intercept form.

1. $4y = -x - 12$ 2. $y - 4x = \frac{1}{2}$

$y = -\frac{1}{4}x - 3$ $y = 4x + \frac{1}{2}$

The slopes are different, so the lines are not parallel.

The product of the slopes is $-\frac{1}{4} \cdot 4$, or -1, so the lines are perpendicular.

15. Write the equations in slope-intercept form.

1. $2y - x = -4$ 2. $x - 2y = -12$

$2y = x - 4$ $-2y = -x - 12$

$y = \frac{1}{2}x - 2$ $y = \frac{1}{2}x + 6$

The slopes are same, $\frac{1}{2}$. The y-intercepts, $(0, -2)$ and $(0, 6)$, are different. The lines are parallel.

16. Graph: $y - 3x \le -3$.

We first graph the line $y - 3x = -3$. The intercepts are $(1, 0$ and $(0, -3)$. We draw a solid line because the inequality symbol is $\le$. We test the point $(0, 0)$.

$$\begin{array}{c|c} \multicolumn{2}{c}{y - 3x \le -3} \\ \hline 0 - 3 \cdot 0 \ ? & -3 \\ 0 & \end{array} \quad \text{FALSE}$$

We see that $(0, 0)$ is not a solution, so we shade the half-plane that does not contain $(0, 0)$.

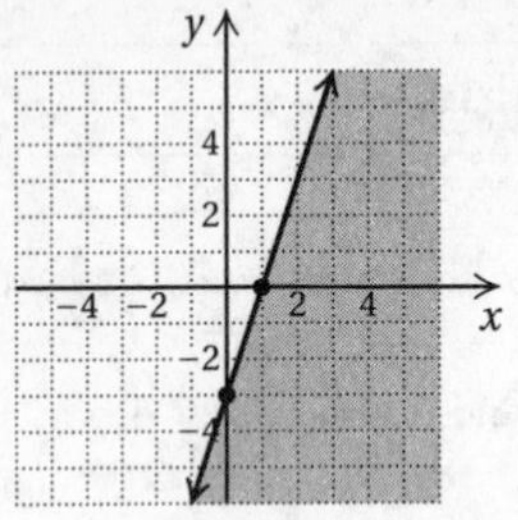

$y - 3x \le -3$

Chapter 3 Review Exercises

1. $(2, 5)$ is 2 units right and 5 units up. See the graph following Exercise 3 below.

2. $(0, -3)$ is 0 units right or left and 3 units down. See the graph following Exercise 3 below.

3. $(-4, -2)$ is 4 units left and 2 units down. See the graph below.

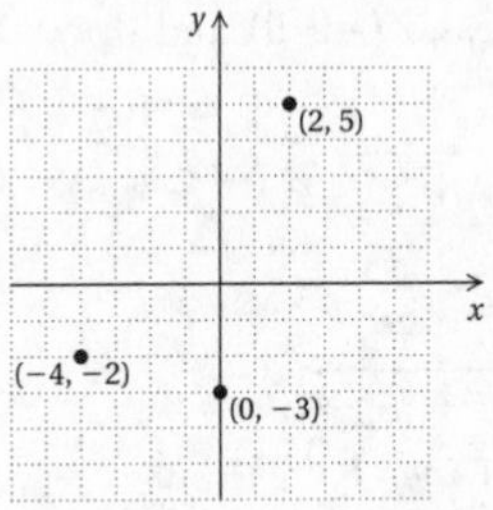

4. Point A is 5 units left and 1 unit down. The coordinates of A are $(-5, -1)$.

5. Point B is 2 units left and 5 units up. The coordinates of B are $(-2, 5)$.

6. Point C is 3 units right and 0 units up or down. The coordinates of C are $(3, 0)$.

7. Since the first coordinate is positive and the second coordinate is negative, the point $(3, -8)$ is in quadrant IV.

8. Since both coordinates are negative, the point $(-20, -14)$ is in quadrant III.

9. Since both coordinates are positive, the point $(4.9, 1.3)$ is in quadrant I.

10. We substitute 2 for x and -6 for y.

$$\begin{array}{c|c} \multicolumn{2}{c}{2y - x = 10} \\ \hline 2(-6) - 2 \ ? & 10 \\ -12 - 2 & \\ -14 & \end{array} \quad \text{FALSE}$$

Since $-14 = 10$ is false, the pair $(2, -6)$ is not a solution.

11. We substitute 0 for x and 5 for y.

$$\begin{array}{c|c} \multicolumn{2}{c}{2y - x = 10} \\ \hline 2 \cdot 5 - 0 \ ? & 10 \\ 10 - 0 & \\ 10 & \end{array} \quad \text{TRUE}$$

Since $10 = 10$ is true, the pair $(0, 5)$ is a solution.

12. To show that a pair is a solution, we substitute, replacing x with the first coordinate and y with the second coordinate in each pair.

$2x - y = 3$		
$2 \cdot 0 - (-3)$	? 3	
$0 + 3$		
3		TRUE

$2x - y = 3$		
$2 \cdot 2 - 1$	? 3	
$4 - 1$		
3		TRUE

In each case the substitution results in a true equation. Thus, $(0, -3)$ and $(2, 1)$ are both solutions of $2x - y = 3$. We graph these points and sketch the line passing through them.

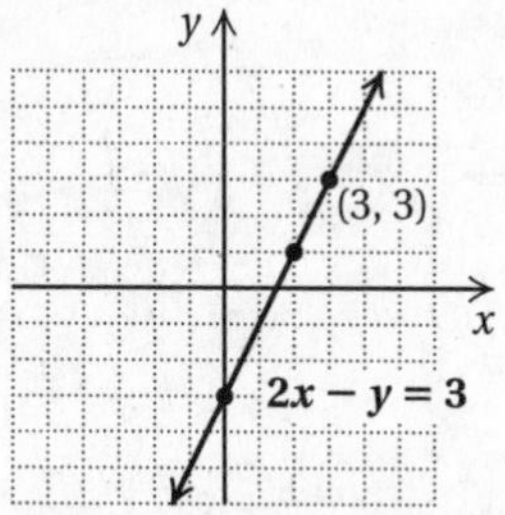

The line appears to pass through $(3, 3)$ also. We check to determine if $(3, 3)$ is a solution of $2x - y = 3$.

$2x - y = 3$		
$2 \cdot 3 - 3$	? 3	
$6 - 3$		
3		TRUE

Thus, $(3, 3)$ is another solution. There are other correct answers, including $(-1, -5)$ and $(4, 5)$.

13. $y = 2x - 5$

The y-intercept is $(0, -5)$. We find two other points.

When $x = 2$, $y = 2 \cdot 2 - 5 = 4 - 5 = -1$.

When $x = 4$, $y = 2 \cdot 4 - 5 = 8 - 5 = 3$.

x	y
0	-5
2	-1
4	3

Plot these points, draw the line they determine, and label the graph $y = 2x - 5$.

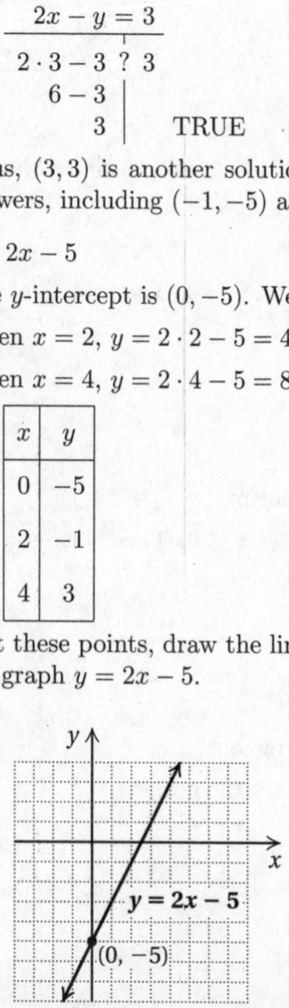

14. $y = -\frac{3}{4}x$

The equation is equivalent to $y = -\frac{3}{4}x + 0$. The y-intercept is $(0, 0)$. We find two other points.

When $x = -4$, $y = -\frac{3}{4}(-4) = 3$.

When $x = 4$, $y = -\frac{3}{4} \cdot 4 = -3$.

x	y
-4	3
0	0
4	-3

Plot these points, draw the line they determine, and label the graph $y = -\frac{3}{4}x$.

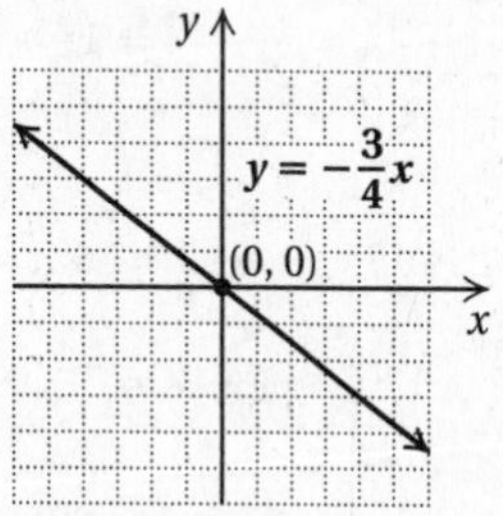

15. $y = -x + 4$

The y-intercept is $(0, 4)$. We find two other points.

When $x = -1$, $y = -(-1) + 4 = 1 + 4 = 5$.

When $x = 4$, $y = -4 + 4 = 0$.

x	y
-1	5
0	4
4	0

Plot these points, draw the line they determine, and label the graph $y = -x + 4$.

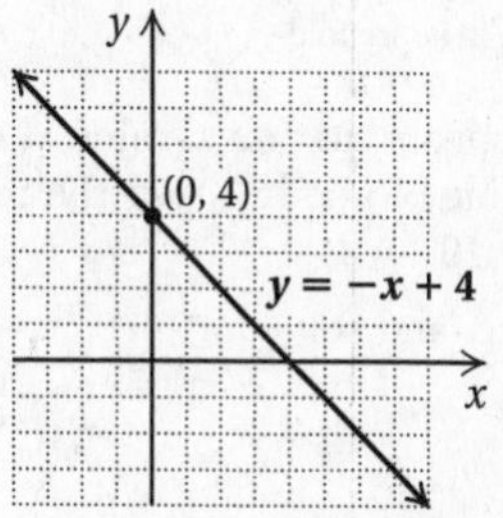

16. $y = 3 - 4x$, or $y = -4x + 3$

The y-intercept is $(0, 3)$. We find two other points.

When $x = 1$, $y = -4 \cdot 1 + 3 = -4 + 3 = -1$.

When $x = 2$, $y = -4 \cdot 2 + 3 = -8 + 3 = -5$.

x	y
0	3
1	-1
2	-5

Plot these points, draw the line they determine, and label the graph $y = 3 - 4x$.

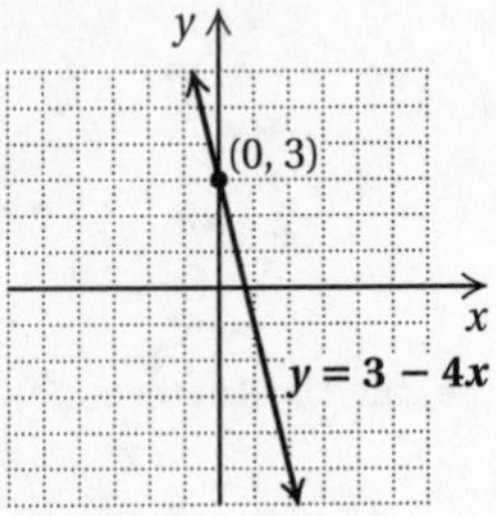

17. a) When $n = 1$, $S = \frac{3}{2} \cdot 1 + 13 = \frac{3}{2} + 13 = 1\frac{1}{2} + 13 = 14\frac{1}{2}$ ft^3.

When $n = 2$, $S = \frac{3}{2} \cdot 2 + 13 = 3 + 13 = 16$ ft^3.

When $n = 5$, $S = \frac{3}{2} \cdot 5 + 13 = \frac{15}{2} + 13 = 7\frac{1}{2} + 13 = 20\frac{1}{2}$ ft^3.

When $n = 10$, $S = \frac{3}{2} \cdot 10 + 13 = 15 + 13 = 28$ ft^3.

b) We plot the points found in part (a): $\left(1, 14\frac{1}{2}\right)$, $(2, 16)$, $\left(5, 20\frac{1}{2}\right)$ and $(10, 28)$. Then we draw the graph.

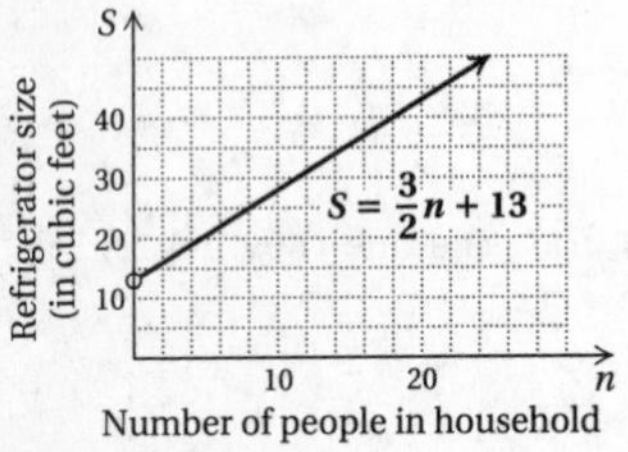

From the graph, it appears that an x-value of 4 corresponds to the S-value of about 19, so the recommended size is about 19.

c) We substitute 22 for S and solve for n.

$$22 = \frac{3}{2}n + 13$$
$$9 = \frac{3}{2}n$$
$$\frac{2}{3} \cdot 9 = \frac{2}{3} \cdot \frac{3}{2}n$$
$$6 = n$$

A 22-ft^3 refrigerator is recommended for a household of 6 residents.

18. $x - 2y = 6$

To find the x-intercept, let $y = 0$. Then solve for x.

$$x - 2y = 6$$
$$x - 2 \cdot 0 = 6$$
$$x = 6$$

Thus, $(6, 0)$ is the x-intercept.

To find the y-intercept, let $x = 0$. Then solve for y.

$$x - 2y = 6$$
$$0 - 2y = 6$$
$$-2y = 6$$
$$y = -3$$

Thus, $(0, -3)$ is the y-intercept.

Plot these points and draw the graph.

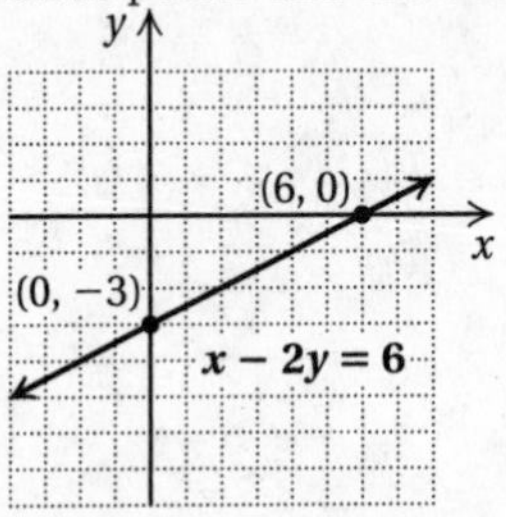

We use a third point as a check.

We let $x = -2$. Then

$$x - 2y = 6$$
$$-2 - 2y = 6$$
$$-2y = 8$$
$$y = -4.$$

The point $(-2, -4)$ is on the graph, so the graph is probably correct.

19. $5x - 2y = 10$

To find the x-intercept, let $y = 0$. Then solve for x.

$$5x - 2y = 10$$
$$5x - 2 \cdot 0 = 10$$
$$5x = 10$$
$$x = 2$$

Thus, $(2, 0)$ is the x-intercept.

To find the y-intercept, let $x = 0$. Then solve for y.

$$5x - 2y = 10$$
$$5 \cdot 0 - 2y = 10$$
$$-2y = 10$$
$$y = -5$$

Thus, $(0, -5)$ is the y-intercept.

Plot these points and draw the graph.

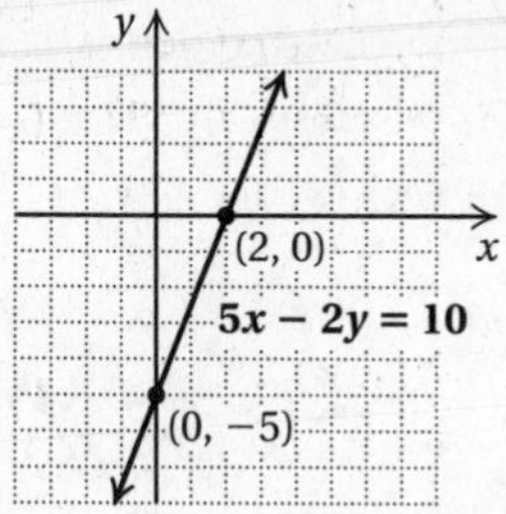

We use a third point as a check.

We let $x = 4$. Then

$$\begin{aligned} 5x - 2y &= 10 \\ 5 \cdot 4 - 2y &= 10 \\ 20 - 2y &= 10 \\ -2y &= -10 \\ y &= 5. \end{aligned}$$

The point $(4, 5)$ is on the graph, so the graph is probably correct.

20. $y = 3$

Any ordered pair $(x, 3)$ is a solution. The variable y must be 3, but x can be any number we choose. A few solutions are listed below. Plot these points and draw the line.

x	y
-3	3
0	3
2	3

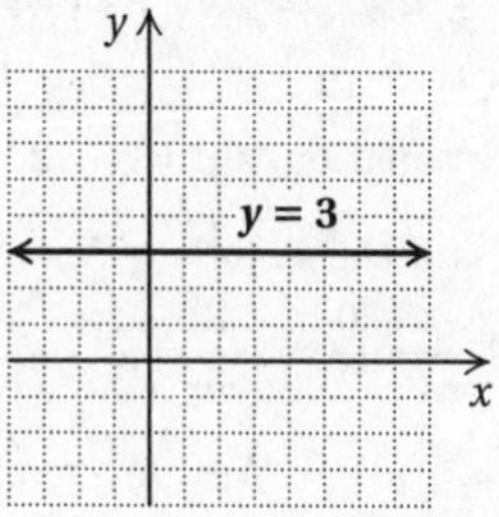

21. $5x - 4 = 0$

$$\begin{aligned} 5x &= 4 \\ x &= \frac{4}{5} \quad \text{Solving for } x \end{aligned}$$

Any ordered pair $\left(\frac{4}{5}, y\right)$ is a solution. A few solutions are listed below. Plot these points and draw the graph.

x	y
$\frac{4}{5}$	-3
$\frac{4}{5}$	0
$\frac{4}{5}$	2

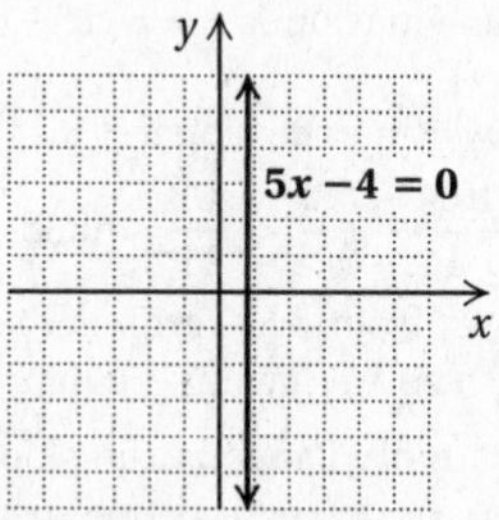

22. We can choose any two points. We consider (x_1, y_1) to be $(-3, 1)$ and (x_2, y_2) to be $(3, 3)$.

$$m = \frac{y_2 - y_1}{x_2 - x_1} = \frac{3 - 1}{3 - (-3)} = \frac{2}{6} = \frac{1}{3}$$

23. We can choose any two points. We consider (x_1, y_1) to be $(3, 1)$ and (x_2, y_2) to be $(-3, 3)$.

$$m = \frac{3 - 1}{-3 - 3} = \frac{2}{-6} = -\frac{1}{3}$$

24. We plot $(-5, -2)$ and $(5, 4)$ and draw the line containing those points.

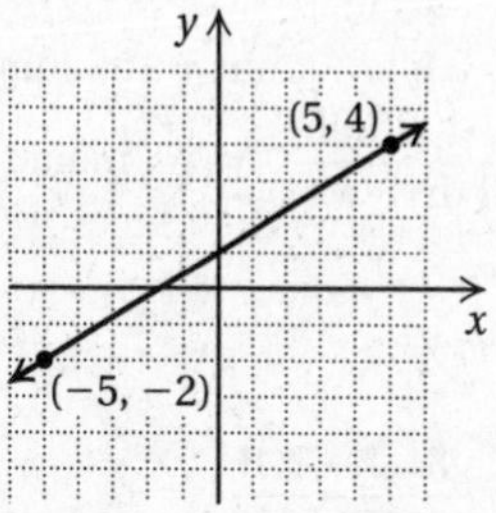

To find the slope, consider (x_1, y_1) to be $(-5, -2)$ and (x_2, y_2) to be $(5, 4)$.

$$m = \frac{y_2 - y_1}{x_2 - x_1} = \frac{4 - (-2)}{5 - (-5)} = \frac{6}{10} = \frac{3}{5}$$

25. We plot $(-5, 5)$ and $(4, -4)$ and draw the line containing those points.

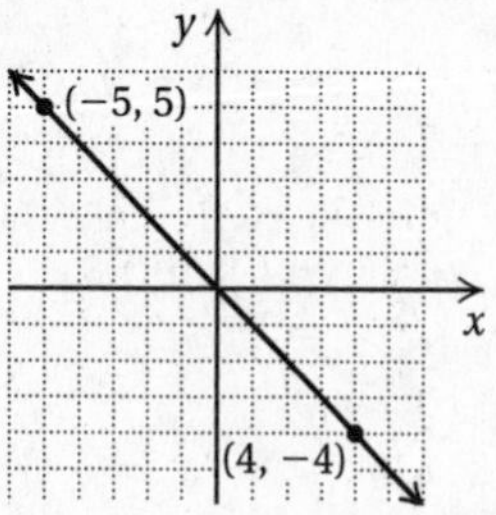

To find the slope, consider (x_1, y_1) to be $(4, -4)$ and (x_2, y_2) to be $(-5, 5)$.

$$m = \frac{y_2 - y_1}{x_2 - x_1} = \frac{5 - (-4)}{-5 - 4} = \frac{9}{-9} = -1$$

26. $y = -\frac{5}{8}x - 3$

The equation is in the form $y = mx + b$, where $m = -\frac{5}{8}$. Thus, the slope is $-\frac{5}{8}$.

27. We solve for y, obtaining an equation of the form $y = mx + b$.

$$\begin{aligned} 2x - 4y &= 8 \\ -4y &= -2x + 8 \\ -\frac{1}{4}(-4y) &= -\frac{1}{4}(-2x + 8) \\ y &= \frac{1}{2}x - 2 \end{aligned}$$

The slope is $\frac{1}{2}$.

28. The graph of $x = -2$ is a vertical line, so the slope is not defined.

29. $y = 9$, or $y = 0 \cdot x + 9$

The slope is 0.

30. 5:30 P.M. is 2.5 hr after 3:00 P.M. In this time the number of driveways plowed was $13 - 7$, or 6.

a) Rate of change $= \dfrac{6 \text{ driveways}}{2.5 \text{ hr}} = 2.4$ driveways per hour

b) $2.5 \text{ hr} = 2.5 \times 1 \text{ hr} = 2.5 \times 60 \text{ min} = 150 \text{ min}$

Rate of change $= \dfrac{150 \text{ min}}{6 \text{ driveways}} = 25$ minutes per driveway

31. Grade $= \dfrac{315}{4500} = 0.07 = 7\%$

32.
$$\begin{aligned}\text{Rate of change} = m &= \frac{31.4 - 7.4}{2011 - 2001}\\ &= \frac{24}{10}\\ &= \$2.4 \text{ billion per year}\end{aligned}$$

33. $y = -9x + 46$

The equation is in the form $y = mx + b$. The slope is -9 and the y-intercept is $(0, 46)$.

34. We solve for y.
$$\begin{aligned}x + y &= 9\\ y &= -x + 9\end{aligned}$$
The slope is -1 and the y-intercept is $(0, 9)$.

35. We solve for y.
$$\begin{aligned}3x - 5y &= 4\\ -5y &= -3x + 4\\ -\frac{1}{5}(-5y) &= -\frac{1}{5}(-3x + 4)\\ y &= \frac{3}{5}x - \frac{4}{5}\end{aligned}$$
The slope is $\frac{3}{5}$ and the y-intercept is $\left(0, -\frac{4}{5}\right)$.

36. We substitute -2.8 for m and 19 for b in the equation $y = mx + b$.
$$y = -2.8x + 19$$

37. We substitute $\frac{5}{8}$ for m and $-\frac{7}{8}$ for b in the equation $y = mx + b$.
$$y = \frac{5}{8}x - \frac{7}{8}$$

38. We know the slope is 3, so the equation is $y = 3x + b$. Using the point $(1, 2)$, we substitute 1 for x and 2 for y in $y = 3x + b$. Then we solve for b.
$$\begin{aligned}y &= 3x + b\\ 2 &= 3 \cdot 1 + b\\ 2 &= 3 + b\\ -1 &= b\end{aligned}$$
Thus, we have the equation $y = 3x - 1$.

39. We know the slope is $\frac{2}{3}$, so the equation is $y = \frac{2}{3}x + b$. Using the point $(-2, -5)$, we substitute -2 for x and -5 for y in $y = \frac{2}{3}x + b$. Then we solve for b.
$$\begin{aligned}y &= \frac{2}{3}x + b\\ -5 &= \frac{2}{3}(-2) + b\\ -5 &= -\frac{4}{3} + b\\ -\frac{11}{3} &= b\end{aligned}$$
Thus, we have the equation $y = \frac{2}{3}x - \frac{11}{3}$.

40. The slope is -2 and the y-intercept is $(0, -4)$, so we have the equation $y = -2x - 4$.

41. First we find the slope.
$$m = \frac{1 - 7}{-1 - 5} = \frac{-6}{-6} = 1$$
Thus, $y = 1 \cdot x + b$, or $y = x + b$. We can use either point to find b. We choose $(5, 7)$. Substitute 5 for x and 7 for y in $y = x + b$.
$$\begin{aligned}y &= x + b\\ 7 &= 5 + b\\ 2 &= b\end{aligned}$$
Thus, the equation is $y = x + 2$.

42. First we find the slope.
$$m = \frac{-3 - 0}{-4 - 2} = \frac{-3}{-6} = \frac{1}{2}$$
Thus, $y = \frac{1}{2}x + b$. We can use either point to find b. We choose $(2, 0)$. Substitute 2 for x and 0 for y in $y = \frac{1}{2}x + b$.
$$\begin{aligned}y &= \frac{1}{2}x + b\\ 0 &= \frac{1}{2} \cdot 2 + b\\ 0 &= 1 + b\\ -1 &= b\end{aligned}$$
Thus, the equation is $y = \frac{1}{2}x - 1$.

43. a) First we find the slope.
$$m = \frac{10.4 - 6.2}{11 - 0} = \frac{4.2}{11} \approx 0.38$$
We see from the graph that the y-intercept is $(0, 6.2)$, so we have the equation $y = 0.38x + 6.2$.

b) The rate of change is the slope, an increase of 0.38 million, or 380,000, people per year.

c) In 2016, $x = 2016 - 2000 = 16$.

$y = 0.38(16) + 6.2 = 6.08 + 6.2 = 12.28$ million, or 12,280,000 people

44. Slope -1, y-intercept $(0,4)$

We plot $(0,4)$. We can think of the slope as $\dfrac{-1}{1}$, so from $(0,4)$ we move down 1 unit and right 1 unit. This locates the point $(1,3)$. We plot $(1,3)$ and draw a line passing through $(0,4)$ and $(1,3)$.

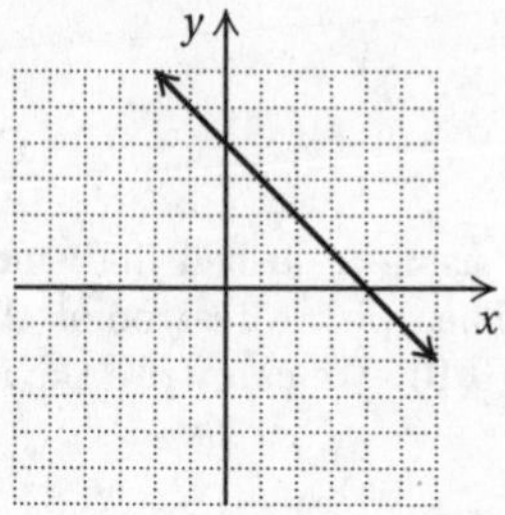

45. Slope $\dfrac{5}{3}$, y-intercept $(0,-3)$.

Plot $(0,-3)$ and from there move up 5 units and right 3 units. This locates the point $(3,2)$. We plot $(3,2)$ and draw a line passing through $(0,-3)$ and $(3,2)$.

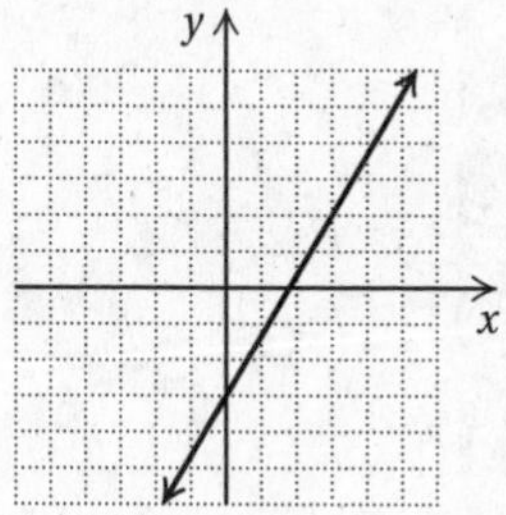

46. $y = -\dfrac{3}{5}x + 2$

First we plot the y-intercept $(0,2)$. We can start at the y-intercept and, thinking of the slope as $\dfrac{-3}{5}$, find another point by moving down 3 units and right 5 units to the point $(5,-1)$. Thinking of the slope as $\dfrac{3}{-5}$ we can start at $(0,2)$ and move up 3 units and left 5 units to get another point $(-5,5)$.

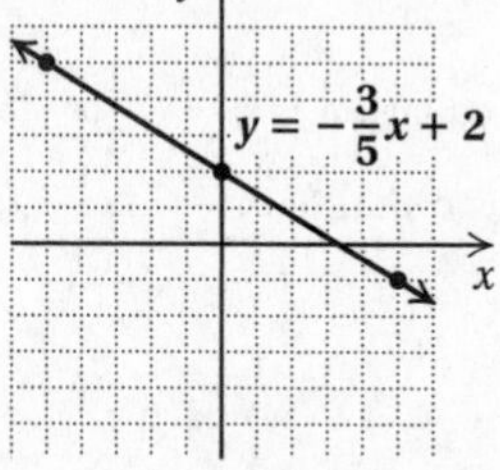

47. First we rewrite the equation in slope-intercept form.

$$2y - 3x = 6$$
$$2y = 3x + 6$$
$$y = \frac{1}{2}(3x + 6)$$
$$y = \frac{3}{2}x + 3$$

Now we plot the y-intercept $(0,3)$. We can start at the y-intercept and use the slope, $\dfrac{3}{2}$, to find another point. We move up 3 units and right 2 units to the point $(2,6)$. Thinking of the slope as $\dfrac{-3}{-2}$ we can start at $(0,3)$ and move down 3 units and left 2 units to get another point, $(-2,0)$.

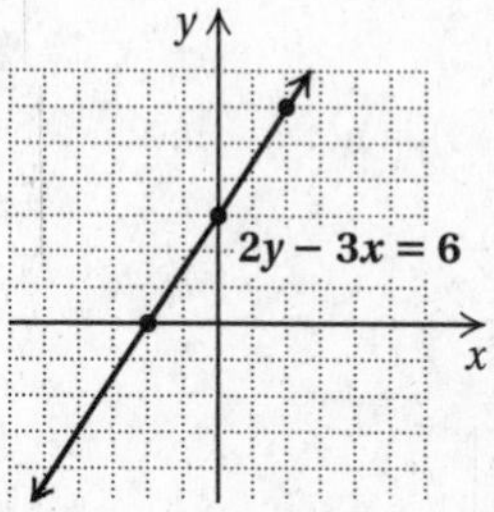

48. First we solve each equation for y:

1. $4x + y = 6$
   $y = -4x + 6$
2. $4x + y = 8$
   $y = -4x + 8$

The slope of each line is -4. The y-intercepts, $(0,6)$ and $(0,8)$, are different. The lines are parallel.

49. We solve the first equation for y.

$$2x + y = 10$$
$$y = -2x + 10$$

The second equation is already solved for y.

$$y = \frac{1}{2}x - 4$$

The slopes, -2 and $\dfrac{1}{2}$, are not the same so the lines are not parallel. The product of the slopes is $-2 \cdot \dfrac{1}{2} = -1$, so the lines are perpendicular.

50. First we solve each equation for y:

1. $x + 4y = 8$
   $4y = -x + 8$
   $y = \frac{1}{4}(-x + 8)$
   $y = -\frac{1}{4}x + 2$
2. $x = -4y - 10$
   $x + 10 = -4y$
   $-\frac{1}{4}(x + 10) = y$
   $-\frac{1}{4}x - \frac{5}{2} = y$

The slope of each line is $-\dfrac{1}{4}$. The y-intercepts, $(0,2)$ and $\left(0, -\dfrac{5}{2}\right)$, are different. The lines are parallel.

51. First we solve each equation for y:

1. $3x - y = 6$
   $-y = -3x + 6$
   $y = -1(-3x + 6)$
   $y = 3x - 6$
2. $3x + y = 8$
   $y = -3x + 8$

The slopes, 3 and -3, are not the same so the lines are not parallel. The product of the slopes is $3(-3) = -9 \neq -1$, so the lines are not perpendicular. Thus, the lines are neither parallel nor perpendicular.

52. $x - 2y > 1$

$0 - 2 \cdot 0 \ ? \ 1$

$0 \ |$ FALSE

Since $0 > 1$ is false, $(0, 0)$ is not a solution.

53. $x - 2y > 1$

$1 - 2 \cdot 3 \ ? \ 1$

$1 - 6 \ |$

$-5 \ |$ FALSE

Since $-5 > 1$ is false, $(1, 3)$ is not a solution.

54. $x - 2y > 1$

$4 - 2(-1) \ ? \ 1$

$4 + 2 \ |$

$6 \ |$ TRUE

Since $6 > 1$ is true, $(4, -1)$ is a solution.

55. Graph $x < y$.

First graph the line $x = y$, or $y = x$. Two points on the line are $(0, 0)$ and $(3, 3)$. We draw a dashed line since the inequality symbol is $<$. Then we pick a test point that is not on the line. We try $(1, 2)$.

$x < y$

$1 \ ? \ 2$ TRUE

We see that $(1, 2)$ is a solution of the inequality, so we shade the region that contains $(1, 2)$.

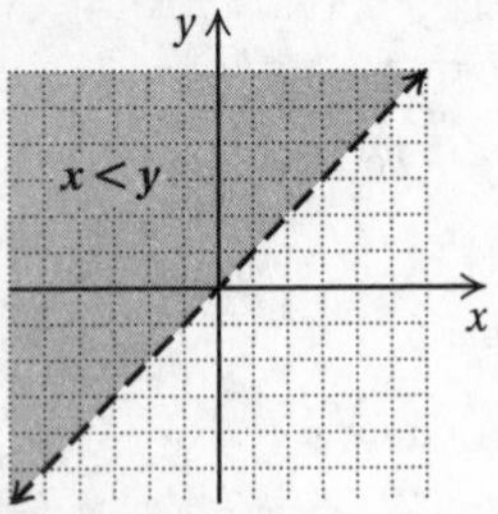

56. Graph $x + 2y \geq 4$.

First graph the line $x + 2y = 4$. The intercepts are $(0, 2)$ and $(4, 0)$. We draw a solid line since the inequality symbol is $\geq$. Then we test the point $(0, 0)$.

$x + 2y \geq 4$

$0 + 2 \cdot 0 \ ? \ 1$

$0 \ |$ FALSE

Since $(0, 0)$ is not a solution of the inequality, we shade the region that does not contain $(0, 0)$.

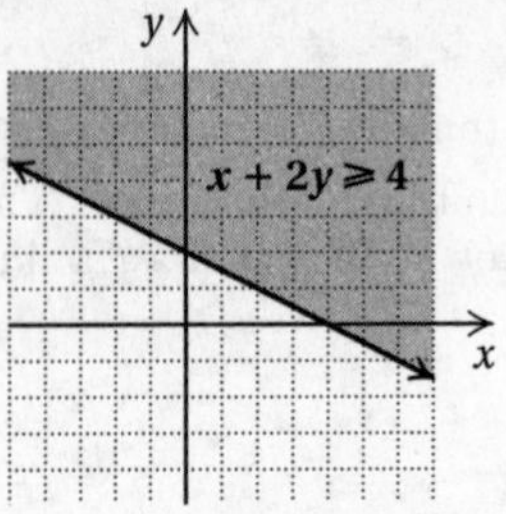

57. Graph $x > -2$.

Graph the line $x = -2$ using a dashed line since the inequality symbol is $>$. Then pick a test point that is not on the line. We choose $(0, 0)$. We can write the inequality as $x + 0y > -2$.

$x + 0y > -2$

$0 + 0 \cdot 0 \ ? \ -2$

$0 \ |$ TRUE

Since $(0, 0)$ is a solution of the inequality, we shade the region containing $(0, 0)$.

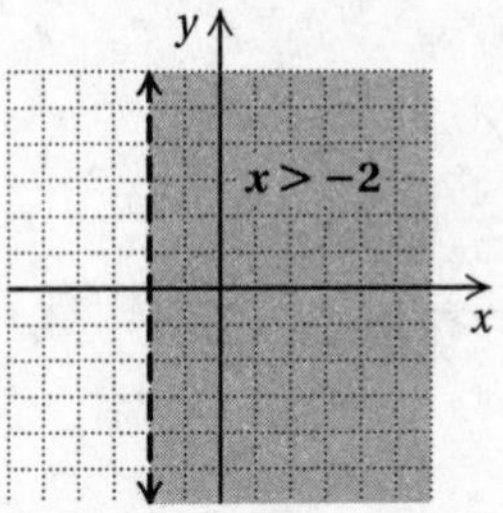

58. We write the equations in slope-intercept form.

1. $-x + \frac{1}{2}y = -2$ 2. $2y + x - 8 = 0$

$\frac{1}{2}y = x - 2$ $2y = -x + 8$

$y = 2x - 4$ $y = -\frac{1}{2}x + 4$

The slopes are different, so the lines are not parallel. The product of the slopes is $2\left(-\frac{1}{2}\right) = -1$, so the lines are perpendicular. Answer D is correct.

59. The slope is $-\frac{8}{3}$, so the equation is $y = -\frac{8}{3}x + b$.

We use the point $(-3, 8)$ to find b.

$$y = -\frac{8}{3}x + b$$

$$8 = -\frac{8}{3}(-3) + b$$

$$8 = 8 + b$$

$$0 = b$$

The equation is $y = -\frac{8}{3}x + 0$, or $y = -\frac{8}{3}x$. Answer C is correct.

60. We plot the given points. We see that the fourth vertex is $(-2,-3)$.

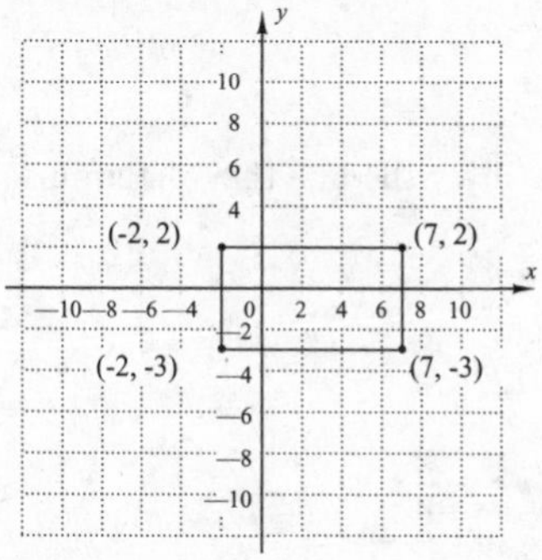

The length of the rectangle is 9 units and the width is 5 units.

$A = l \cdot w$

$A = (9 \text{ units})(5 \text{ units}) = 45$ square units

$P = 2l + 2w$

$P = 2 \cdot 9 \text{ units} + 2 \cdot 5 \text{ units}$

$P = 18 \text{ units} + 10 \text{ units} = 28 \text{ units}$

61. a) From Telluride to Station St. Sophia, the ascent is $10,550 \text{ ft} - 8725 = 1825$ ft. From Station St. Sophia to Mountain View, the descent is $10,550 \text{ ft} - 9500 \text{ ft} = 1050$ ft. The total ascent and descent is $1825 \text{ ft} + 1050 \text{ ft} = 2875$ ft. The time that elapses from 11:55 A.M. to 12:07 P.M. is 12 minutes Then the average rate of ascent and descent is

$$\frac{2875 \text{ ft}}{12 \text{ min}} = 239.58\overline{3} \text{ feet per minute.}$$

b) The average rate of ascent and descent in minutes per foot is

$$\frac{12 \text{ min}}{2875 \text{ ft}} \approx 0.004 \text{ minutes per foot.}$$

Chapter 3 Discussion and Writing Exercises

1. If one equation represents a vertical line (that is, is of the form $x = a$) and the other represents a horizontal line (that is, is of the form $y = b$), then the graphs are perpendicular. If neither line is of one of the forms above, then solve each equation for y in order to determine the slope of each. Then, if the product of the slopes is -1, the graphs are perpendicular.

2. If $b > 0$, then the y-intercept of $y = mx + b$ is on the positive y-axis and the graph of $y = mx + b$ lies "above" the origin. Using $(0, 0)$ as a test point, we have the false inequality $0 > b$ so the region above $y = mx + b$ is shaded.

If $b = 0$, the line $y = mx + b$ or $y = mx$, passes through the origin. Testing a point above the line, such as $(1, m+1)$, we have the true inequality $m + 1 > m$ so the region above the line is shaded.

If $b < 0$, then the y-intercept of $y = mx + b$ is on the negative y-axis and the graph of $y = mx + b$ lies "below" the origin. Using $(0, 0)$ as a test point we get the true inequality $0 > b$ so the region above $y = mx + b$ is shaded.

Thus, we see that in any case the graph of any inequality of the form $y > mx + b$ is always shaded above the line $y = mx + b$.

3. The y-intercept is the point at which the graph crosses the y-axis. Since a point on the y-axis is neither left nor right of the origin, the first or x-coordinate of the point is 0.

4. The graph of $x < 1$ on a number line consists of the points in the set $\{x|x < 1\}$.

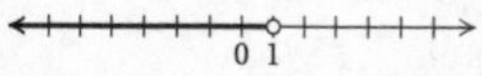

The graph of $x < 1$ on a plane consists of the points, or ordered pairs, in the set $\{(x, y)|x + 0 \cdot y < 1\}$. This is the set of ordered pairs with first coordinate less than 1.

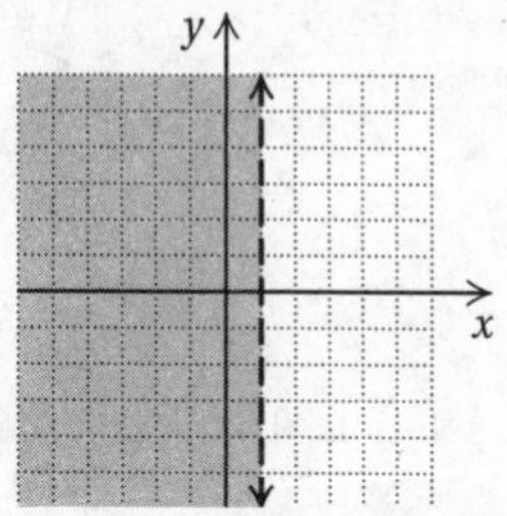

5. First plot the y-intercept, $(0, 2458)$. Then, thinking of the slope as $\frac{37}{100}$, plot a second point on the line by moving up 37 units and right 100 units from the y-intercept and plot a third point by moving down 37 units and left 100 units. Finally, draw a line through the three points.

6. If the equations are of the form $x = p$ and $x = q$, where $p \neq q$, then the graphs are parallel vertical lines. If neither equation is of the form $x = p$, then solve each for y in order to determine the slope and y-intercept of each. If the slopes are the same and the y-intercepts are different, the lines are parallel.

Chapter 3 Test

1. Since the first coordinate is negative and the second coordinate is positive, the point $\left(-\frac{1}{2}, 7\right)$ is in quadrant II.

2. Since both coordinates are negative, the point $(-5, -6)$ is in quadrant III.

3. Point A is 5 units left and 1 unit up. The coordinates of A are $(-5, 1)$.

4. Point B is 0 units left or right and 4 units down. The coordinates of B are $(0, -4)$.

5.

$$\begin{array}{r|l} y - 2x = 5 & \\ \hline -3 - 2(-4) \ ? & 5 \\ -3 + 8 & \\ 5 & \text{TRUE} \end{array}$$

$y - 2x = 5$	
$3 - 2(-1)$	? 5
$3 + 2$	
5	TRUE

In each case we get a true equation, so $(-4, -3)$ and $(-1, 3)$ are solutions of $y - 2x = 5$. We plot these points and draw the line passing through them.

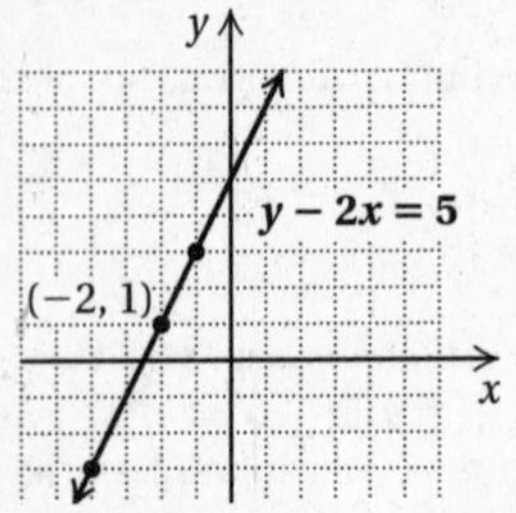

The line appears to pass through $(-2, 1)$. We check to determine if $(-2, 1)$ is a solution of $y - 2x = 5$.

$y - 2x = 5$	
$1 - 2(-2)$	? 5
$1 + 4$	
5	TRUE

Thus, $(-2, 1)$ is another solution. There are other correct answers, including $(-5, -5)$, $(-3, -1)$, and $(0, 5)$.

6. $y = 2x - 1$

The y-intercept is $(0, -1)$. We find two other points.

When $x = -2$, $y = 2(-2) - 1 = -4 - 1 = -5$.

When $x = 3$, $y = 2 \cdot 3 - 1 = 6 - 1 = 5$.

x	y
-2	-5
0	-1
3	5

Plot these points, draw the line they determine, and label the graph $y = 2x - 1$.

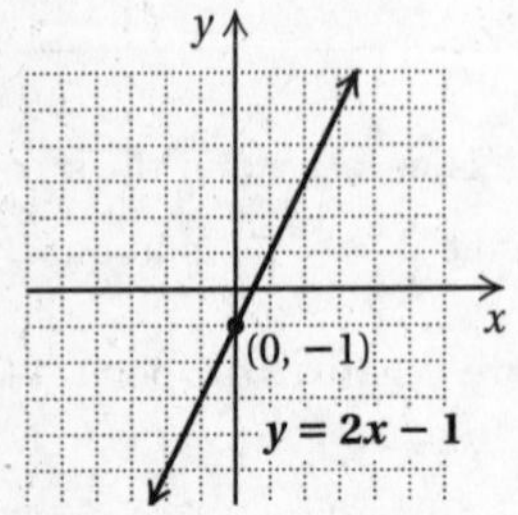

7. $y = -\frac{3}{2}x$, or $y = -\frac{3}{2}x + 0$

The y-intercept is $(0, 0)$. We find two other points.

When $x = -2$, $y = -\frac{3}{2}(-2) = 3$.

When $x = 2$, $y = -\frac{3}{2} \cdot 2 = -3$.

x	y
-2	3
0	0
2	-3

Plot these points, draw the line they determine, and label the graph $y = -\frac{3}{2}x$.

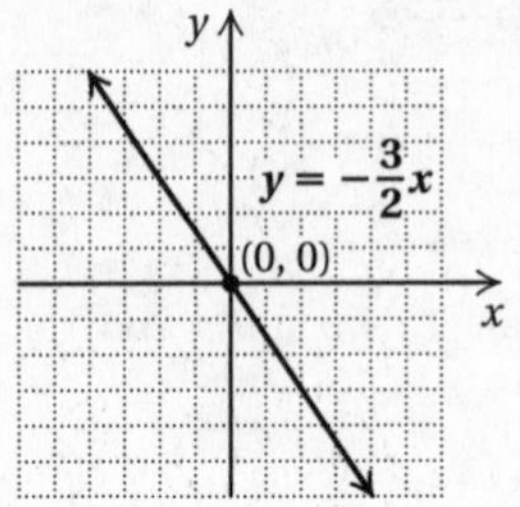

8. $2x - 4y = -8$

To find the x-intercept, let $y = 0$. Then solve for x.

$$2x - 4y = -8$$
$$2x - 4 \cdot 0 = -8$$
$$2x = -8$$
$$x = -4$$

Thus, $(-4, 0)$ is the x-intercept.

To find the y-intercept, let $x = 0$. Then solve for y.

$$2x - 4y = -8$$
$$2 \cdot 0 - 4y = -8$$
$$-4y = -8$$
$$y = 2$$

Thus, $(0, 2)$ is the y-intercept.

Plot these points and draw the graph.

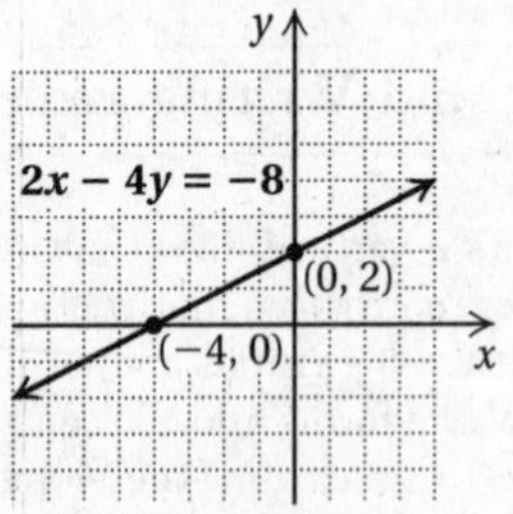

We use a third point as a check.

We let $x = 4$. Then

$$2x - 4y = -8$$
$$2 \cdot 4 - 4y = -8$$
$$8 - 4y = -8$$
$$-4y = -16$$
$$y = 4$$

The point $(4, 4)$ is on the graph, so the graph is probably correct.

9. $2x - y = 3$

To find the x-intercept, let $y = 0$. Then solve for x.

$$2x - y = 3$$
$$2x - 0 = 3$$
$$2x = 3$$
$$x = \frac{3}{2}$$

Thus, $\left(\frac{3}{2}, 0\right)$ is the x-intercept.

To find the y-intercept, let $x = 0$. Then solve for y.

$$2x - y = 3$$
$$2 \cdot 0 - y = 3$$
$$-y = 3$$
$$y = -3$$

Thus, $(0, -3)$ is the y-intercept.

Plot these points and draw the graph.

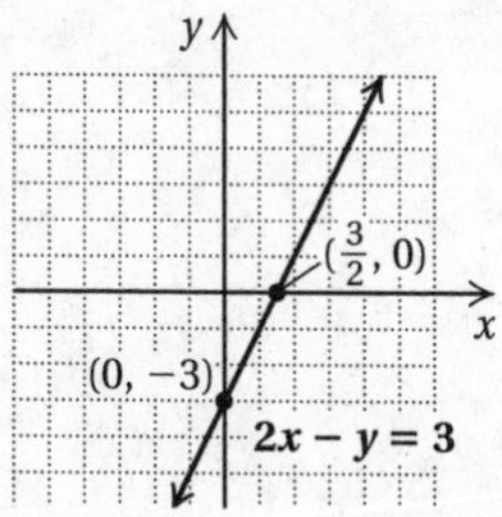

We use a third point as a check.

We let $x = 3$. Then

$$2x - y = 3$$
$$2 \cdot 3 - y = 3$$
$$6 - y = 3$$
$$-y = -3$$
$$y = 3$$

The point $(3, 3)$ is on the line, so the graph is probably correct.

10. $2x + 8 = 0$

$$2x = -8$$
$$x = -4 \quad \text{Solving for } x$$

Any ordered pair $(-4, y)$ is a solution. A few solutions are listed below. Plot these points and draw the graph.

x	y
-4	-3
-4	0
-4	2

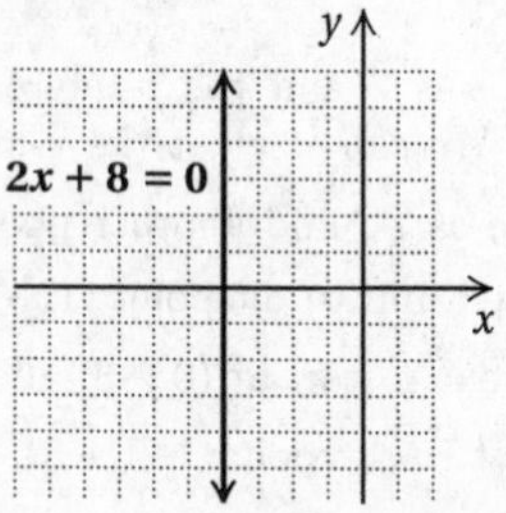

11. $y = 5$

Any ordered pair $(x, 5)$ is a solution. A few solutions are listed below. Plot these points and draw the graph.

x	y
-2	5
0	5
3	5

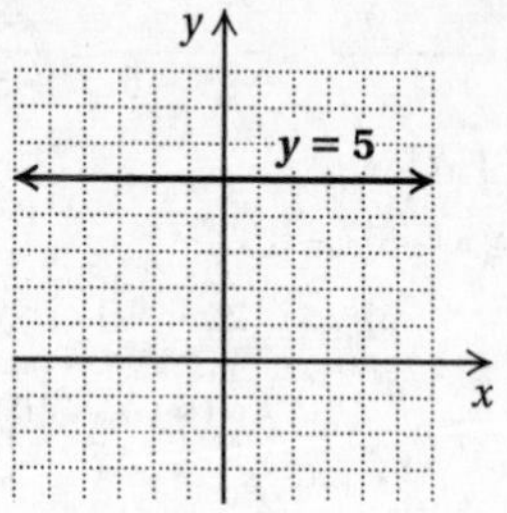

12. a) In 2007, $t = 0$, and $C = 606 \cdot 0 + 8593 = \$8593$.

In 2009, $t = 2009 - 2007 = 2$, and $C = 606 \cdot 2 + 8593 = 1212 + 8593 = \$9805$.

In 2012, $t = 2012 - 2007 = 5$, and $C = 606 \cdot 5 + 8593 = 3030 + 8593 = \$11,623$.

b) We plot the points found in part (a), $(0, 8593)$, $(2, 9805)$, and $(5, 11,623)$. Then we draw the line passing through these points.

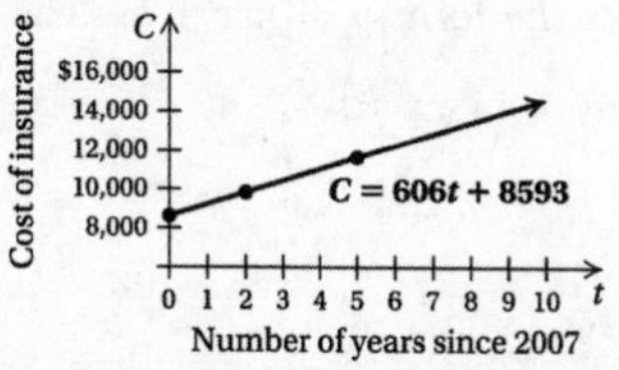

c) We substitute 15,259 for C and solve for t.

$$C = 606t + 8593$$
$$15,259 = 606t + 8593$$
$$6666 = 606t$$
$$11 = t$$

We predict that the cost of health insurance will be \$15,259 11 years after 2007, or in 2018.

13. The time that elapses from 1:00 P.M. to 5:00 P.M. is 4 hours.

$$\text{Rate} = \frac{450 \text{ miles} - 100 \text{ miles}}{4 \text{ hours}} = \frac{350 \text{ miles}}{4 \text{ hours}} =$$
87.5 miles per hour

14. We can choose any two points. We consider (x_1, y_1) to be $(2, 4)$ and (x_2, y_2) to be $(5, -2)$.

$$m = \frac{y_2 - y_1}{x_2 - x_1} = \frac{-2 - 4}{5 - 2} = \frac{-6}{3} = -2$$

15. We plot $(-3, 1)$ and $(5, 4)$ and draw the line containing these points.

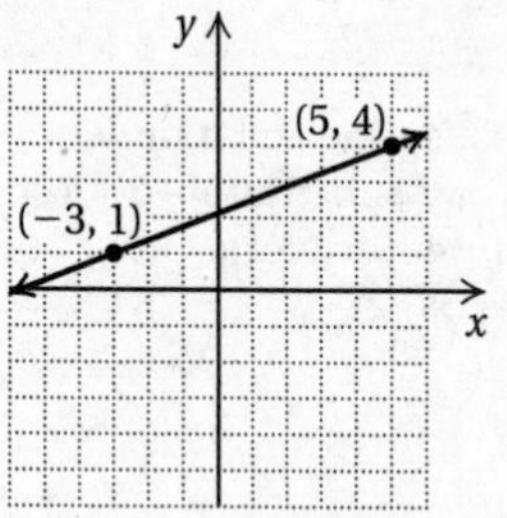

To find the slope, consider (x_1, y_1) to be $(5, 4)$ and (x_2, y_2) to be $(-3, 1)$.

$$m = \frac{y_2 - y_1}{x_2 - x_1} = \frac{1-4}{-3-5} = \frac{-3}{-8} = \frac{3}{8}$$

16. a) We solve for y.

$$2x - 5y = 10$$
$$-5y = -2x + 10$$
$$y = -\frac{1}{5}(-2x + 10)$$
$$y = \frac{2}{5}x - 2$$

The slope is $\frac{2}{5}$.

b) The graph of $x = -2$ is a vertical line, so the slope is not defined.

17. Slope $= \frac{-54}{1080} = -\frac{1}{20}$, or -0.05

18. $y = 2x - \frac{1}{4}$

The equation is in the form $y = mx + b$. The slope is 2 and the y-intercept is $\left(0, -\frac{1}{4}\right)$.

19. We solve for y.

$$-4x + 3y = -6$$
$$3y = 4x - 6$$
$$y = \frac{1}{3}(4x - 6)$$
$$y = \frac{4}{3}x - 2$$

The slope is $\frac{4}{3}$ and the y-intercept is $(0, -2)$.

20. We substitute 1.8 for m and -7 for b in the equation $y = mx + b$.

$$y = 1.8x - 7$$

21. We substitute $-\frac{3}{8}$ for m and $-\frac{1}{8}$ for b in the equation $y = mx + b$.

$$y = -\frac{3}{8}x - \frac{1}{8}$$

22. We know the slope is 1, so the equation is $y = 1 \cdot x + b$, or $y = x + b$. Using the point $(3, 5)$, we substitute 3 for x and 5 for y in $y = x + b$. Then we solve for b.

$$y = x + b$$
$$5 = 3 + b$$
$$2 = b$$

Thus, the equation is $y = x + 2$.

23. We know the slope is -3, so the equation is $y = -3x + b$. Using the point $(-2, 0)$, we substitute -2 for x and 0 for y in $y = -3x + b$. Then we solve for b.

$$y = -3x + b$$
$$0 = -3(-2) + b$$
$$0 = 6 + b$$
$$-6 = b$$

Thus, the equation is $y = -3x - 6$.

24. First we find the slope.

$$m = \frac{-2-1}{2-1} = \frac{-3}{1} = -3$$

Thus, $y = -3x + b$. We can use either point to find b. We choose $(1, 1)$. Substitute 1 for both x and y in $y = -3x + b$.

$$y = -3x + b$$
$$1 = -3 \cdot 1 + b$$
$$1 = -3 + b$$
$$4 = b$$

Thus, the equation is $y = -3x + 4$.

25. First we find the slope.

$$m = \frac{-1-(-3)}{4-(-4)} = \frac{2}{8} = \frac{1}{4}$$

Thus, $y = \frac{1}{4}x + b$. We can use either point to find b. We choose $(4, -1)$. Substitute 4 for x and -1 for y in $y = \frac{1}{4}x + b$.

$$y = \frac{1}{4}x + b$$
$$-1 = \frac{1}{4} \cdot 4 + b$$
$$-1 = 1 + b$$
$$-2 = b$$

Thus, the equation is $y = \frac{1}{4}x - 2$.

26. Slope $-\frac{3}{2}$, y-intercept $(0, 1)$

We plot $(0, 1)$. We can think of the slope as $\frac{-3}{2}$, so from $(0, 1)$ we move down 3 units and right 2 units. This locates the point $(2, -2)$. We plot $(2, -2)$ and draw a line passing through $(0, 1)$ and $(2, -2)$.

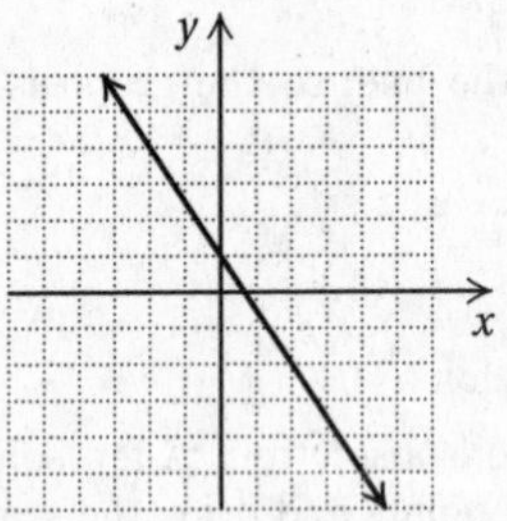

27. $y = 2x - 3$

The slope is 2 and the y-intercept is $(0, -3)$. We plot $(0, -3)$. We can start at the y-intercept and, thinking of the slope as $\frac{2}{1}$, find another point by moving up 2 units and right 1 unit to the point $(1, -1)$. Thinking of the slope as $\frac{-2}{-1}$ we can start at $(0, -3)$ and move down 2 units and left 1 unit to the point $(-1, -5)$.

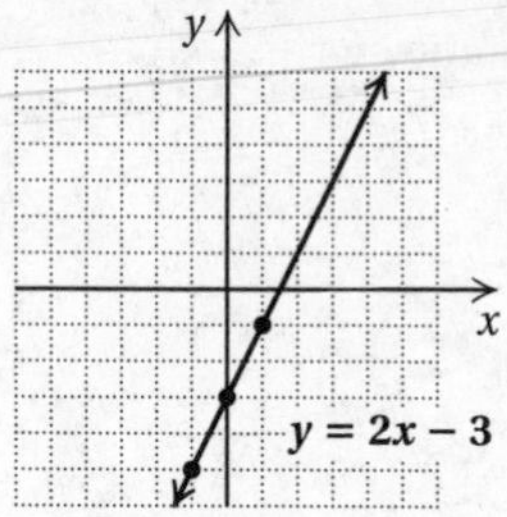

28. First we solve each equation for y:

1. $2x + y = 8$ $\quad$ 2. $2x + y = 4$

$y = -2x + 8$ $\quad$ $y = -2x + 4$

The slope of each line is -2. The y-intercepts, $(0, 8)$ and $(0, 4)$ are different. The lines are parallel.

29. We solve the first equation for y.

$$2x + 5y = 2$$
$$5y = -2x + 2$$
$$y = \frac{1}{5}(-2x + 2)$$
$$y = -\frac{2}{5}x + \frac{2}{5}$$

The second equation is in the form $y = mx + b$:

$$y = 2x + 4$$

The slopes, $-\frac{2}{5}$ and 2, are not the same, so the lines are not parallel. The product of the slopes is $-\frac{2}{5} \cdot 2 = -\frac{4}{5} \neq -1$, so the lines are not perpendicular. Thus, the lines are neither parallel nor perpendicular.

30. First we solve each equation for y:

1. $x + 2y = 8$ $\quad$ 2. $-2x + y = 8$

$2y = -x + 8$ $\quad$ $y = 2x + 8$

$$y = \frac{1}{2}(-x + 8)$$
$$y = -\frac{1}{2}x + 4$$

The slopes, $-\frac{1}{2}$ and 2, are not the same, so the lines are not parallel. The product of the slope is $-\frac{1}{2} \cdot 2 = -1$, so the lines are perpendicular.

31.

$3y - 2x < -2$	
$3 \cdot 0 - 2 \cdot 0$?	-2
0	FALSE

Since $0 < -2$ is false, $(0, 0)$ is not a solution.

32.

$3y - 2x < -2$	
$3(-10) - 2(-4)$?	-2
$-30 + 8$	
-22	TRUE

Since $-22 < -2$ is true, $(-4, -10)$ is a solution.

33. Graph $y > x - 1$.

First graph the line $y = x - 1$. Two points on the line are $(0, -1)$ and $(4, 3)$. We draw a dashed line since the inequality symbol is $>$. Then we test a point that is not on the line. We try $(0, 0)$.

$y > x - 1$	
0 ?	$0 - 1$
	-1 TRUE

Since $(0, 0)$ is a solution of the inequality, we shade the region containing $(0, 0)$.

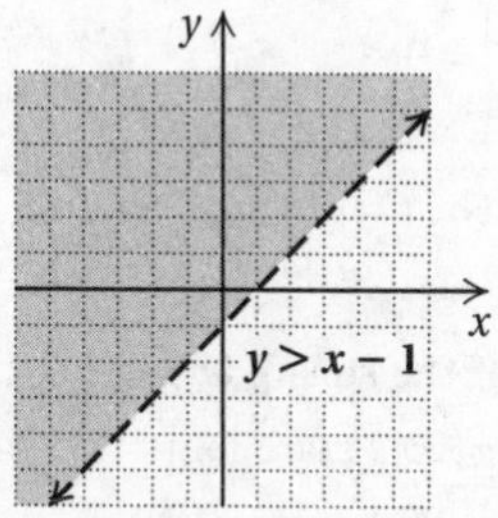

34. Graph $2x - y \geq 4$.

First graph the line $2x - y = 4$. The intercepts are $(0, -4)$ and $(2, 0)$. We draw a solid line since the inequality symbol is $\geq$. Then we test a point that is not on the line. We try $(0, 0)$.

$2x - y \geq 4$	
$2 \cdot 0 - 0$?	4
0	FALSE

Since $(0, 0)$ is not a solution of the inequality, we shade the half-plane that does not contain $(0, 0)$.

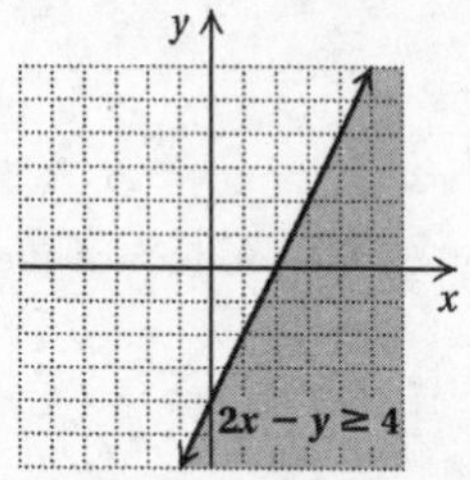

35. First we solve each equation for y.

1. $15x + 21y = 7$

$$21y = -15x + 7$$
$$y = \frac{1}{21}(-15x + 7)$$
$$y = -\frac{5}{7}x + \frac{1}{3}$$

2. $35y + 14 = -25x$

$$35y = -25x - 14$$
$$y = \frac{1}{35}(-25x - 14)$$
$$y = -\frac{5}{7}x - \frac{2}{5}$$

The lines have the same slope, $-\frac{5}{7}$. The y-intercepts, $\left(0, \frac{1}{3}\right)$ and $\left(0, -\frac{2}{5}\right)$ are different. Thus, the lines are parallel. Answer A is correct.

36. We plot the given points. We see that the other vertices of the square are $(-3, 4)$ and $(2, -1)$.

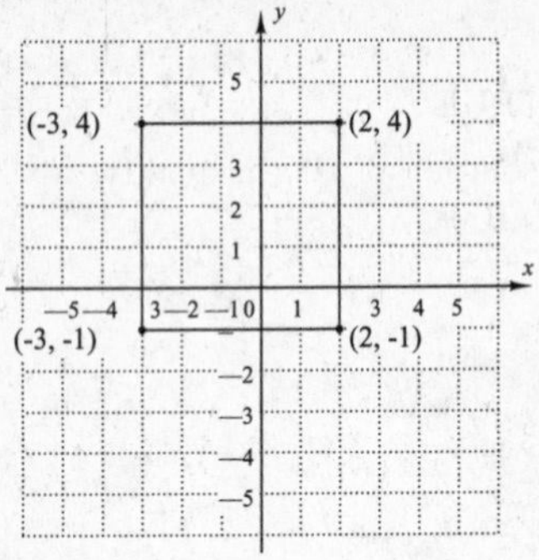

The length of the sides of the square is 5 units.

$$A = s^2 = (5\text{ units})^2 = 25\text{ square units}$$
$$P = 4s = 4 \cdot 5\text{ units} = 20\text{ linear units}$$

37. First solve each equation for y.

$$3x + 7y = 14$$
$$7y = -3x + 14$$
$$y = \frac{1}{7}(-3x + 14)$$
$$y = -\frac{3}{7}x + 2$$

$$ky - 7x = -3$$
$$ky = 7x - 3$$
$$y = \frac{1}{k}(7x - 3)$$
$$y = \frac{7}{k}x - \frac{3}{k}$$

If the lines are perpendicular, the product of their slopes is -1.

$$-\frac{3}{7} \cdot \frac{7}{k} = -1$$
$$-\frac{3 \cdot \cancel{7}}{\cancel{7} \cdot k} = -1$$
$$-\frac{3}{k} = -1$$
$$k\left(-\frac{3}{k}\right) = k(-1)$$
$$-3 = -k$$
$$3 = k$$

Cumulative Review Chapters 1 - 3

1. $\dfrac{2m - n}{5} = \dfrac{2(-1) - 2}{5} = \dfrac{-2 - 2}{5} = \dfrac{-4}{5} = -\dfrac{4}{5}$

2. $-\dfrac{2}{3}(x - 6y + 3) = -\dfrac{2}{3} \cdot x - \dfrac{2}{3}(-6y) - \dfrac{2}{3} \cdot 3$

$= -\dfrac{2}{3}x + 4y - 2$

3. $18w - 24 + 9y = 3 \cdot 6w - 3 \cdot 8 + 3 \cdot 3y = 3(6w - 8 + 3y)$

4. First we find decimal notation for $\dfrac{7}{9}$.

$$\begin{array}{r} 0.7\,7\ldots \\ 9\,\overline{)\,7.0\,0} \\ \underline{6\,3} \\ 7\,0 \\ \underline{6\,3} \\ 7 \end{array}$$

$\dfrac{7}{9} = 0.\overline{7}$, so $-\dfrac{7}{9} = -0.\overline{7}$.

5. The distance of $-2\frac{1}{5}$ from 0 is $2\frac{1}{5}$, so $\left|-2\frac{1}{5}\right| = 2\frac{1}{5}$.

6. The opposite of 8.17 is -8.17 because $8.17 + (-8.17) = 0$.

7. The reciprocal of $-\dfrac{8}{7}$ is $-\dfrac{7}{8}$ because $-\dfrac{8}{7}\left(-\dfrac{7}{8}\right) = 1$.

8. $2x - 5y + (-3x) + 4y = 2x + (-3x) - 5y + 4y = -x - y$

9. $-2.6 + (-0.4)$

Two negative numbers. Add the absolute values, getting 3. Make the answer negative.

$-2.6 + (-0.4) = -3$

10.
$$\begin{aligned} 3 - [81 \div (1 + 2^3)] &= 3 - [81 \div (1 + 8)] \\ &= 3 - [81 \div 9] \\ &= 3 - 9 \\ &= -6 \end{aligned}$$

11. $\dfrac{5}{18} \div \left(-\dfrac{5}{12}\right) = \dfrac{5}{18} \cdot \left(-\dfrac{12}{5}\right) = -\dfrac{5 \cdot 12}{18 \cdot 5} = -\dfrac{\cancel{5} \cdot 2 \cdot \cancel{6}}{3 \cdot \cancel{6} \cdot \cancel{5}} = -\dfrac{2}{3}$

12.
$$\begin{aligned} 6(x + 4) - 5[x - (2x - 3)] &= 6(x + 4) - 5[x - 2x + 3] \\ &= 6(x + 4) - 5[-x + 3] \\ &= 6x + 24 + 5x - 15 \\ &= 11x + 9 \end{aligned}$$

13. $\left(-\dfrac{1}{2}\right)(-1.1)(4.8) = 0.55(4.8) = 2.64$

14.
$$\begin{aligned} -20 + 30 \div 10 \cdot 6 &= -20 + 3 \cdot 6 \\ &= -20 + 18 \\ &= -2 \end{aligned}$$

15.
$$\frac{4}{9}y = -36$$
$$\frac{9}{4} \cdot \frac{4}{9}y = \frac{9}{4}(-36)$$
$$y = -\frac{9 \cdot 36}{4} = -\frac{9 \cdot \cancel{4} \cdot 9}{\cancel{4} \cdot 1}$$
$$y = -81$$

The solution is -81.

16.
$$-8 + w = w + 7$$
$$-8 + w - w = w + 7 - w$$
$$-8 = 7$$

We get a false equation. There is no solution.

17.
$$7.5 - 2x = 0.5x$$
$$10(7.5 - 2x) = 10(0.5x) \quad \text{Clearing decimals}$$
$$75 - 20x = 5x$$
$$75 = 25x \quad \text{Adding } 20x$$
$$3 = x \quad \text{Dividing by 3}$$
The solution is 3.

18.
$$4(x+2) = 4(x-2) + 16$$
$$4x + 8 = 4x - 8 + 16$$
$$4x + 8 = 4x + 8$$
$$4x + 8 - 4x = 4x + 8 - 4x$$
$$8 = 8$$
We get an equation that is true for all real numbers. All real numbers are solutions of the original equation.

19. $2(x+2) \geq 5(2x+3)$
$$2x + 4 \geq 10x + 15$$
$$-8x + 4 \geq 15 \quad \text{Subtracting } 10x$$
$$-8x \geq 11 \quad \text{Subtracting 4}$$
$$x \leq -\frac{11}{8} \quad \text{Dividing by } -8 \text{ and reversing the inequality symbol}$$
The solution set is $\left\{x \middle| x \leq -\frac{11}{8}\right\}$.

20.
$$x - \frac{5}{6} = \frac{1}{2}$$
$$6\left(x - \frac{5}{6}\right) = 6 \cdot \frac{1}{2} \quad \text{Clearing fractions}$$
$$6x - 5 = 3$$
$$6x = 8$$
$$x = \frac{8}{6} = \frac{4}{3}$$
The solution is $\frac{4}{3}$.

21.
$$9x - 12y = -3$$
$$-12y = -9x - 3$$
$$-\frac{1}{12}(-12y) = -\frac{1}{12}(-9x - 3)$$
$$y = \frac{3}{4}x + \frac{1}{4}$$
The slope is $\frac{3}{4}$.

22. $x = -\frac{15}{16}$ is the equation of a vertical line. Thus, the slope is not defined.

23. The slope is -20, so the equation is $y = -20x + b$. Use the point $(-8, -2)$ to find b.
$$y = -20x + b$$
$$-2 = -20(-8) + b$$
$$-2 = 160 + b$$
$$-162 = b$$
The equation is $y = -20x - 162$.

24. The graph of $x = -5$ is a vertical line and the graph of $y = \frac{2}{7}$ is a horizontal line, so the lines are perpendicular.

25.
$$A = \frac{1}{2}h(b+c)$$
$$2A = h(b+c) \quad \text{Multiplying by 2}$$
$$\frac{2A}{b+c} = h \quad \text{Dividing by } b+c$$

26. The first coordinate is positive and the second is negative, so $(3, -1)$ is in quadrant IV.

27. To find the x-intercept, substitute 0 for y and solve for x.
$$2x - 7y = 21$$
$$2x - 7 \cdot 0 = 21$$
$$2x = 21$$
$$x = \frac{21}{2}$$
The x-intercept is $\left(\frac{21}{2}, 0\right)$.

To find the y-intercept, substitute 0 for x and solve for y.
$$2x - 7y = 21$$
$$2 \cdot 0 - 7y = 21$$
$$-7y = 21$$
$$y = -3$$
The y-intercept is $(0, -3)$.

28. 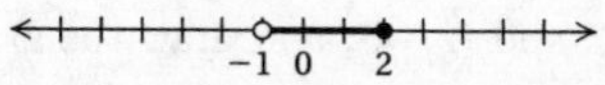

29. Graph: $2x + 5y = 10$.

Find the x-intercept:
$$2x + 5 \cdot 0 = 10$$
$$2x = 10$$
$$x = 5$$
The x-intercept is $(5, 0)$.

Find the y-intercept:
$$2 \cdot 0 + 5y = 10$$
$$5y = 10$$
$$y = 2$$
The y-intercept is $(0, 2)$.

Find a third point as a check. Let $x = -5$.
$$2(-5) + 5y = 10$$
$$-10 + 5y = 10$$
$$5y = 20$$
$$y = 4$$
A third point is $(-5, 4)$. Plot this point and the intercepts and draw the graph.

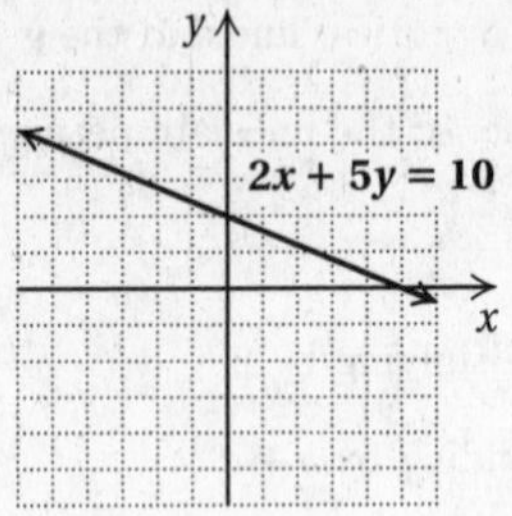

30. Graph: $y = -2$.

This is the equation of a horizontal line with y-intercept $(0, -2)$.

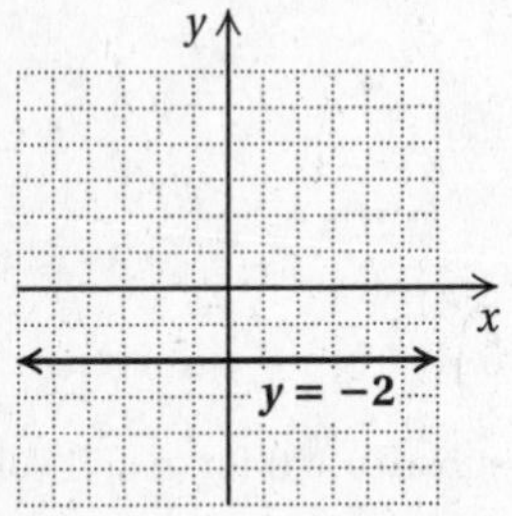

31. Graph: $y = -2x + 1$.

We see from the equation that the y-intercept is $(0, 1)$. Find two more points on the graph.

If $x = -1$, $y = -2(-1) + 1 = 2 + 1 = 3$.

If $x = 2$, $y = -2 \cdot 2 + 1 = -4 + 1 = -3$.

We plot $(0, 1)$, $(-1, 3)$, and $(2, -3)$ and draw the graph.

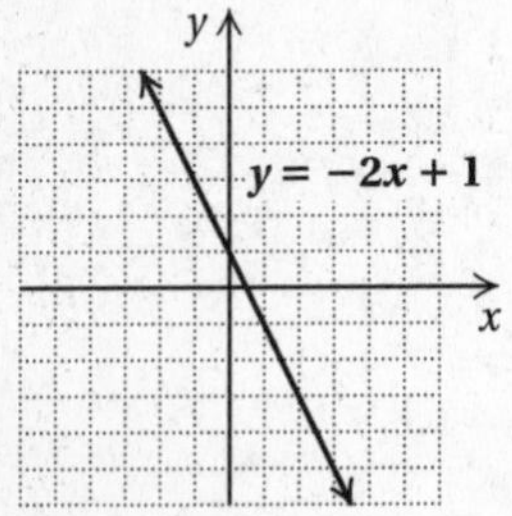

32. Graph: $3y + 6 = 2x$.

Find the x-intercept:

$$\begin{aligned} 3 \cdot 0 + 6 &= 2x \\ 6 &= 2x \\ 3 &= x \end{aligned}$$

The x-intercept is $(3, 0)$.

Find the y-intercept:

$$\begin{aligned} 3y + 6 &= 2 \cdot 0 \\ 3y + 6 &= 0 \\ 3y &= -6 \\ y &= -2 \end{aligned}$$

The y-intercept is $(0, -2)$.

Find a third point as a check. Let $x = -3$.

$$\begin{aligned} 3y + 6 &= 2(-3) \\ 3y + 6 &= -6 \\ 3y &= -12 \\ y &= -4 \end{aligned}$$

A third point is $(-3, -4)$. Plot this point and the intercepts and draw the graph.

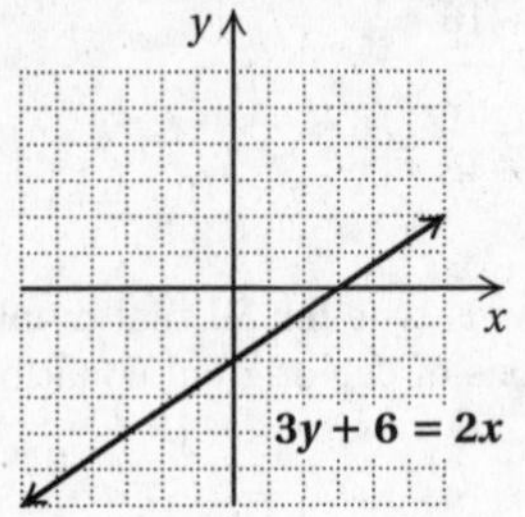

33. Graph: $y = -\frac{3}{2}x$.

When $x = -2$, $y = -\frac{3}{2}(-2) = 3$.

When $x = 0$, $y = -\frac{3}{2} \cdot 0 = 0$.

When $x = 2$, $y = -\frac{3}{2} \cdot 2 = -3$.

We plot the points $(-2, 3)$, $(0, 0)$, and $(2, -3)$ and draw the graph.

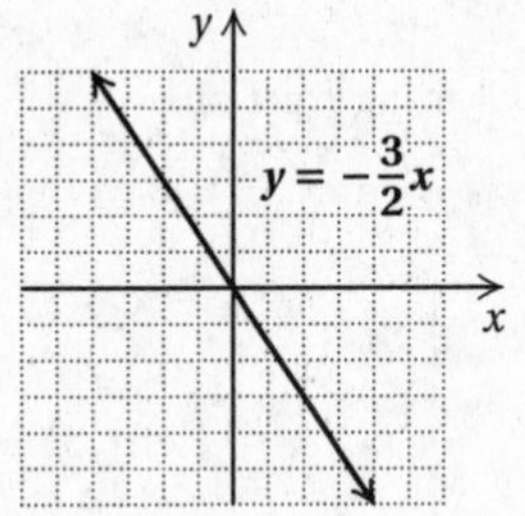

34. Graph: $x = 4.5$.

The graph is the equation of a vertical line with x-intercept $(4.5, 0)$.

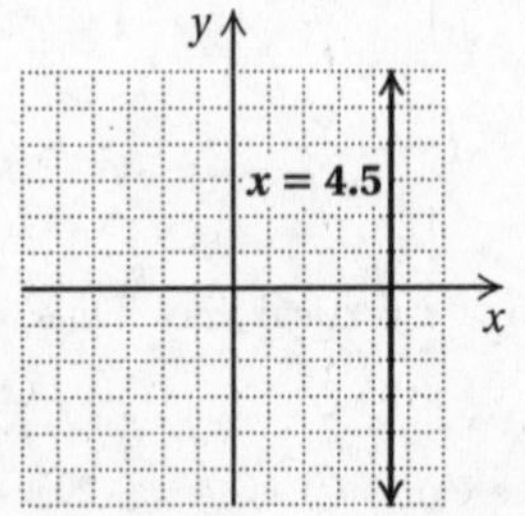

35. We use the points $(-2, 3)$ and $(2, 2)$ to find the slope.

$$m = \frac{2 - 3}{2 - (-2)} = \frac{-1}{4} = -\frac{1}{4}$$

36. ***Familiarize***. Let x = the number, in millions, of Americans with O-negative blood. Then $x + 94.2$ = the number of Americans with O-positive blood.

$$\underbrace{\text{O-negative Americans}} \quad \text{plus} \quad \underbrace{\text{O-positive Americans}} \quad \text{is} \quad \underbrace{\text{Total O-positive or O-negative Americans.}}$$

$$x \quad + \quad (x + 94.2) \quad = \quad 134.6$$

Solve.

$$\begin{aligned} x + (x + 94.2) &= 134.6 \\ 2x + 94.2 &= 134.6 \\ 2x &= 40.4 \\ x &= 20.2 \end{aligned}$$

Check. If 20.2 million Americans have O-negative blood, then 20.2 + 94.2, or 114.4 million have O-positive blood. Adding 20.2 and 114.4, we get 134.6 million, the number of Americans with either O-positive or O-negative blood. The answer checks.

State. 20.2 million Americans have O-negative blood.

37. ***Familiarize***. Let h = the number of hours Cory can work on the fifth day without violating his contract.

$$\underbrace{\text{Number of hours worked}} \quad \underbrace{\text{cannot exceed}} \quad \underbrace{\text{40 hr.}}$$

$$7 + 10 + 9 + 6 + h \quad \le \quad 40$$

Solve.

$$\begin{aligned} 7 + 10 + 9 + 6 + h &\le 40 \\ 32 + h &\le 40 \\ h &\le 8 \end{aligned}$$

Check. If Cory works 8 hr on the fifth day, then he works 7+10+9+6+8, or 40 hr. This does not violate the contract. If he works 9 hr on the fifth day, he works 7+10+9+6+9, or 41 hr, and he violates his contract. We have a partial check.

State. The number of hours Cory can work on the fifth day is given by $\{h | h \le 8 \text{ hr}\}$.

38. ***Familiarize***. Let w = the length of the first piece of wire. Then $w + 3$ = the length of the second piece, and $\frac{4}{5}w$ = the length of the third piece.

Translate.

$$\underbrace{\text{Length of first piece}} \quad \text{plus} \quad \underbrace{\text{length of second piece}} \quad \text{plus} \quad \underbrace{\text{length of third piece}} \quad \text{is} \quad \underbrace{\text{total length}}$$

$$w \quad + \quad (w + 3) \quad + \quad \frac{4}{5}w \quad = \quad 143$$

Solve.

$$\begin{aligned} w + (w + 3) + \frac{4}{5}w &= 143 \\ \frac{14}{5}w + 3 &= 143 \\ \frac{14}{5}w &= 140 \\ \frac{5}{14} \cdot \frac{14}{5}w &= \frac{5}{14} \cdot 140 = \frac{5 \cdot 10 \cdot \cancel{14}}{\cancel{14} \cdot 1} \\ w &= 50 \end{aligned}$$

If $w = 50$, then $w + 3 = 50 + 3 = 53$, and $\frac{4}{5}w = \frac{4}{5} \cdot 50 = 40$.

Check. The length of the second piece of wire, 53 m is 3 m more than 50 m, the length of the first piece. The length of the third piece, 40 m, is $\frac{4}{5}$ of 50 m, the length of the first piece. The answer checks.

State. The first piece should be 50 m, the second 53 m, and the third 40 m.

39. $1000 \div 100 \cdot 10 - 10 = 10 \cdot 10 - 10 = 100 - 10 = 90$

Answer A is correct.

40. $m = \dfrac{-7 - 3}{2 - (-4)} = \dfrac{-10}{6} = -\dfrac{5}{3}$

Answer D is correct.

41.

$$\begin{aligned} 4|x| - 13 &= 3 \\ 4|x| &= 16 \\ |x| &= 4 \end{aligned}$$

The numbers -4 and 4 are 4 units from 0. They are the solutions.

42.

$$\frac{2 + 5x}{4} = \frac{11}{28} + \frac{8x + 3}{7}, \quad \text{LCD} = 28$$

$$\begin{aligned} 28\left(\frac{2 + 5x}{4}\right) &= 28\left(\frac{11}{28} + \frac{8x + 3}{7}\right) \\ 7(2 + 5x) &= 11 + 4(8x + 3) \\ 14 + 35x &= 11 + 32x + 12 \\ 14 + 35x &= 23 + 32x \\ 14 + 3x &= 23 \\ 3x &= 9 \\ x &= 3 \end{aligned}$$

The solution is 3.

43.

$$\begin{aligned} p &= \frac{2}{m + Q} \\ p(m + Q) &= 2 \\ pm + pQ &= 2 \\ pQ &= 2 - pm \\ Q &= \frac{2 - pm}{p}, \text{ or } \frac{2}{p} - m \end{aligned}$$

Chapter 4

Polynomials: Operations

Exercise Set 4.1

RC1. $y^1 = y$; answer (c) is correct.

RC3. $y^1 \cdot y^1 = y^{1+1} = y^2$; answer (e) is correct.

RC5. $\dfrac{y^8}{y^9} = y^{8-9} = y^{-1} = \dfrac{1}{y}$; answer (d) is correct.

1. 3^4 means $3 \cdot 3 \cdot 3 \cdot 3$.

3. $(-1.1)^5$ means $(-1.1)(-1.1)(-1.1)(-1.1)(-1.1)$.

5. $\left(\dfrac{2}{3}\right)^4$ means $\left(\dfrac{2}{3}\right)\left(\dfrac{2}{3}\right)\left(\dfrac{2}{3}\right)\left(\dfrac{2}{3}\right)$.

7. $(7p)^2$ means $(7p)(7p)$.

9. $8k^3$ means $8 \cdot k \cdot k \cdot k$.

11. $-6y^4$ means $-6 \cdot y \cdot y \cdot y \cdot y$.

13. $a^0 = 1,\ a \neq 0$

15. $b^1 = b$

17. $\left(\dfrac{2}{3}\right)^0 = 1$

19. $(-7.03)^1 = -7.03$

21. $8.38^0 = 1$

23. $(ab)^1 = ab$

25. $ab^0 = a \cdot b^0 = a \cdot 1 = a$

27. $m^3 = 3^3 = 3 \cdot 3 \cdot 3 = 27$

29. $p^1 = 19^1 = 19$

31. $-x^4 = -(-3)^4 = -(-3)(-3)(-3)(-3) = -81$

33. $x^4 = 4^4 = 4 \cdot 4 \cdot 4 \cdot 4 = 256$

35.
$$\begin{aligned} y^2 - 7 &= 10^2 - 7 \\ &= 100 - 7 \quad \text{Evaluating the power} \\ &= 93 \quad \text{Subtracting} \end{aligned}$$

37.
$$\begin{aligned} 161 - b^2 &= 161 - 5^2 \\ &= 161 - 25 \quad \text{Evaluating the power} \\ &= 136 \quad \text{Subtracting} \end{aligned}$$

39.
$$\begin{aligned} x^1 + 3 &= 7^1 + 3 \\ &= 7 + 3 \quad (7^1 = 7) \\ &= 10 \end{aligned}$$

$$\begin{aligned} x^0 + 3 &= 7^0 + 3 \\ &= 1 + 3 \quad (7^0 = 1) \\ &= 4 \end{aligned}$$

41.
$$\begin{aligned} A = \pi r^2 &\approx 3.14 \times (34\text{ ft})^2 \\ &\approx 3.14 \times 1156\text{ ft}^2 \quad \text{Evaluating the power} \\ &\approx 3629.84\text{ ft}^2 \end{aligned}$$

43. $3^{-2} = \dfrac{1}{3^2} = \dfrac{1}{9}$

45. $10^{-3} = \dfrac{1}{10^3} = \dfrac{1}{1000}$

47. $a^{-3} = \dfrac{1}{a^3}$

49. $\dfrac{1}{8^{-2}} = 8^2 = 64$

51. $\dfrac{1}{y^{-4}} = y^4$

53. $5z^{-4} = 5 \cdot z^{-4} = 5 \cdot \dfrac{1}{z^4} = \dfrac{5}{z^4}$

55. $xy^{-2} = x \cdot y^{-2} = x \cdot \dfrac{1}{y^2} = \dfrac{x}{y^2}$

57. $\dfrac{1}{4^3} = 4^{-3}$

59. $\dfrac{1}{x^3} = x^{-3}$

61. $\dfrac{1}{a^5} = a^{-5}$

63. $2^4 \cdot 2^3 = 2^{4+3} = 2^7$

65. $9^{17} \cdot 9^{21} = 9^{17+21} = 9^{38}$

67. $x^4 \cdot x = x^{4+1} = x^5$

69. $x^{14} \cdot x^3 = x^{14+3} = x^{17}$

71. $(3y)^4(3y)^8 = (3y)^{4+8} = (3y)^{12}$

73. $(7y)^1(7y)^{16} = (7y)^{1+16} = (7y)^{17}$

75. $3^{-5} \cdot 3^8 = 3^{-5+8} = 3^3$

77. $x^{-2} \cdot x = x^{-2+1} = x^{-1} = \dfrac{1}{x}$

79. $x^{-7} \cdot x^{-6} = x^{-7+(-6)} = x^{-13} = \dfrac{1}{x^{13}}$

81. $a^{11} \cdot a^{-3} \cdot a^{-18} = a^{11+(-3)+(-18)} = a^{-10} = \dfrac{1}{a^{10}}$

83. $(x^4y^7)(x^2y^8) = (x^4x^2)(y^7y^8) = x^{4+2}y^{7+8} = x^6y^{15}$

85. $(s^2t^3)(st^4) = (s^2s)(t^3t^4) = s^{2+1}t^{3+4} = s^3t^7$

87. $\dfrac{7^5}{7^2} = 7^{5-2} = 7^3$

89. $\dfrac{y^9}{y} = y^{9-1} = y^8$

91. $\dfrac{16^2}{16^8} = 16^{2-8} = 16^{-6} = \dfrac{1}{16^6}$

93. $\dfrac{m^6}{m^{12}} = m^{6-12} = m^{-6} = \dfrac{1}{m^6}$

95. $\dfrac{(8x)^6}{(8x)^{10}} = (8x)^{6-10} = (8x)^{-4} = \dfrac{1}{(8x)^4}$

97. $\dfrac{x}{x^{-1}} = x^{1-(-1)} = x^2$

99. $\dfrac{z^{-6}}{z^{-2}} = z^{-6-(-2)} = z^{-4} = \dfrac{1}{z^4}$

101. $\dfrac{x^{-5}}{x^{-8}} = x^{-5-(-8)} = x^3$

103. $\dfrac{m^{-9}}{m^{-9}} = m^{-9-(-9)} = m^0 = 1$

105. $\dfrac{a^5b^3}{a^2b} = \dfrac{a^5}{a^2}\cdot\dfrac{b^3}{b} = a^{5-2}b^{3-1} = a^3b^2$

107. $5^2 = 5\cdot 5 = 25$

$5^{-2} = \dfrac{1}{5^2} = \dfrac{1}{25}$

$\left(\dfrac{1}{5}\right)^2 = \dfrac{1}{5}\cdot\dfrac{1}{5} = \dfrac{1}{25}$

$\left(\dfrac{1}{5}\right)^{-2} = \dfrac{1}{\left(\dfrac{1}{5}\right)^2} = \dfrac{1}{\dfrac{1}{25}} = 1\cdot\dfrac{25}{1} = 25$

$-5^2 = -(5)(5) = -25$

$(-5)^2 = (-5)(-5) = 25$

$-\left(-\dfrac{1}{5}\right)^2 = -\left(-\dfrac{1}{5}\right)\left(-\dfrac{1}{5}\right) = -\dfrac{1}{25}$

$\left(-\dfrac{1}{5}\right)^{-2} = \dfrac{1}{\left(-\dfrac{1}{5}\right)^2} = \dfrac{1}{\dfrac{1}{25}} = 1\cdot\dfrac{25}{1} = 25$

109. ***Familiarize***. Let x = the length of the shorter piece. Then $2x$ = the length of the longer piece.

Translate.

Length of shorter piece plus length of longer piece is 12 in.

$$x + 2x = 12$$

Solve.

$x + 2x = 12$

$3x = 12$

$\dfrac{3x}{3} = \dfrac{12}{3}$

$x = 4$

If $x = 4$, $2x = 2\cdot 4 = 8$.

Check. The longer piece, 8 in., is twice as long as the shorter piece, 4 in. Also, 4 in. + 8 in. = 12 in., the total length of the sandwich. The answer checks.

State. The lengths of the pieces are 4 in. and 8 in.

111. Total: $1800 + 1500 + 700 = 4000$ lb

For the peanuts:

1800 is what percent of 4000?

$1800 = p \cdot 4000$

For the cashews:

1500 is what percent of 4000?

$1500 = p \cdot 4000$

For the almonds:

700 is what percent of 4000?

$700 = p \cdot 4000$

To solve each equation we divide both sides by 4000 and convert to percent notation.

$1800 = p\cdot 4000$

$0.45 = p$

$45\% = p$

Peanuts are 45% of the total.

$1500 = p\cdot 4000$

$0.375 = p$

$37.5\% = p$

Cashews are 37.5% of the total.

$700 = p\cdot 4000$

$0.175 = p$

$17.5\% = p$

Almonds are 17.5% of the total.

113. $2x - 4 - 5x + 8 = x - 3$

$-3x + 4 = x - 3$

$-4x + 4 = -3$

$-4x = -7$

$x = \dfrac{7}{4}$

The solution is $\dfrac{7}{4}$.

115. $-6(2-x) + 10(5x-7) = 10$

$-12 + 6x + 50x - 70 = 10$

$56x - 82 = 10$ Collecting like terms

$56x - 82 + 82 = 10 + 82$ Adding 82

$56x = 92$

$\dfrac{56x}{56} = \dfrac{92}{56}$ Dividing by 56

$x = \dfrac{23}{14}$

The solution is $\dfrac{23}{14}$.

117. Let $y_1 = (x+1)^2$ and $y_2 = x^2+1$. A graph of the equations or a table of values shows that $(x+1)^2 = x^2+1$ is not correct.

119. Let $y_1 = (5x)^0$ and $y_2 = 5x^0$. A graph of the equations or a table of values shows that $(5x)^0 = 5x^0$ is not correct.

121. $(y^{2x})(y^{3x}) = y^{2x+3x} = y^{5x}$

123. $\dfrac{a^{6t}(a^{7t})}{a^{9t}} = \dfrac{a^{6t+7t}}{a^{9t}} = \dfrac{a^{13t}}{a^{9t}} = a^{13t-9t} = a^{4t}$

125. $\dfrac{(0.8)^5}{(0.8)^3(0.8)^2} = \dfrac{(0.8)^5}{(0.8)^{3+2}} = \dfrac{(0.8)^5}{(0.8)^5} = 1$

127. Since the bases are the same, the expression with the larger exponent is larger. Thus, $3^5 > 3^4$.

129. Since the exponents are the same, the expression with the larger base is larger. Thus, $4^3 < 5^3$.

131. $\dfrac{1}{-z^4} = \dfrac{1}{-(-10)^4} = \dfrac{1}{-(-10)(-10)(-10)(-10)} =$

$\dfrac{1}{-10,000} = -\dfrac{1}{10,000}$

133. No; for example, $(3+4)^2 = 7^2 = 49$, but $3^2+4^2 = 9+16 = 25$.

Exercise Set 4.2

RC1. To raise a power to a power, multiply the exponents.

RC3. To convert a number less than 1 to scientific notation, move the decimal point to the right.

1. $(2^3)^2 = 2^{3\cdot 2} = 2^6$

3. $(5^2)^{-3} = 5^{2(-3)} = 5^{-6} = \dfrac{1}{5^6}$

5. $(x^{-3})^{-4} = x^{(-3)(-4)} = x^{12}$

7. $(a^{-2})^9 = a^{-2\cdot 9} = a^{-18} = \dfrac{1}{a^{18}}$

9. $(t^{-3})^{-6} = t^{(-3)(-6)} = t^{18}$

11. $(t^4)^{-3} = t^{4(-3)} = t^{-12} = \dfrac{1}{t^{12}}$

13. $(x^{-2})^{-4} = x^{-2(-4)} = x^8$

15. $(ab)^3 = a^3b^3$ Raising each factor to the third power

17. $(ab)^{-3} = a^{-3}b^{-3} = \dfrac{1}{a^3b^3}$

19. $(mn^2)^{-3} = m^{-3}(n^2)^{-3} = m^{-3}n^{2(-3)} =$

$m^{-3}n^{-6} = \dfrac{1}{m^3n^6}$

21. $(4x^3)^2 = 4^2(x^3)^2$ Raising each factor to the second power

$= 16x^6$

23. $(3x^{-4})^2 = 3^2(x^{-4})^2 = 3^2x^{-4\cdot 2} = 9x^{-8} = \dfrac{9}{x^8}$

25. $(x^4y^5)^{-3} = (x^4)^{-3}(y^5)^{-3} = x^{4(-3)}y^{5(-3)} =$

$x^{-12}y^{-15} = \dfrac{1}{x^{12}y^{15}}$

27. $(x^{-6}y^{-2})^{-4} = (x^{-6})^{-4}(y^{-2})^{-4} = x^{(-6)(-4)}y^{(-2)(-4)} =$

$x^{24}y^8$

29. $(a^{-2}b^7)^{-5} = (a^{-2})^{-5}(b^7)^{-5} = a^{10}b^{-35} = \dfrac{a^{10}}{b^{35}}$

31. $(5r^{-4}t^3)^2 = 5^2(r^{-4})^2(t^3)^2 = 25r^{-4\cdot 2}t^{3\cdot 2} =$

$25r^{-8}t^6 = \dfrac{25t^6}{r^8}$

33. $(a^{-5}b^7c^{-2})^3 = (a^{-5})^3(b^7)^3(c^{-2})^3 =$

$a^{-5\cdot 3}b^{7\cdot 3}c^{-2\cdot 3} = a^{-15}b^{21}c^{-6} = \dfrac{b^{21}}{a^{15}c^6}$

35. $(3x^3y^{-8}z^{-3})^2 = 3^2(x^3)^2(y^{-8})^2(z^{-3})^2 =$

$9x^6y^{-16}z^{-6} = \dfrac{9x^6}{y^{16}z^6}$

37. $(-4x^3y^{-2})^2 = (-4)^2(x^3)^2(y^{-2})^2 = 16x^6y^{-4} = \dfrac{16x^6}{y^4}$

39. $(-a^{-3}b^{-2})^{-4} = (-1\cdot a^{-3}b^{-2})^{-4} =$

$(-1)^{-4}(a^{-3})^{-4}(b^{-2})^{-4} = \dfrac{1}{(-1)^4}\cdot a^{12}b^8 = \dfrac{a^{12}b^8}{1} = a^{12}b^8$

41. $\left(\dfrac{y^3}{2}\right)^2 = \dfrac{(y^3)^2}{2^2} = \dfrac{y^6}{4}$

43. $\left(\dfrac{a^2}{b^3}\right)^4 = \dfrac{(a^2)^4}{(b^3)^4} = \dfrac{a^8}{b^{12}}$

45. $\left(\dfrac{y^2}{2}\right)^{-3} = \dfrac{(y^2)^{-3}}{2^{-3}} = \dfrac{y^{-6}}{2^{-3}} = \dfrac{\frac{1}{y^6}}{\frac{1}{2^3}} = \dfrac{1}{y^6}\cdot\dfrac{2^3}{1} = \dfrac{8}{y^6}$

47. $\left(\dfrac{7}{x^{-3}}\right)^2 = \dfrac{7^2}{(x^{-3})^2} = \dfrac{49}{x^{-6}} = 49x^6$

49. $\left(\dfrac{x^2y}{z}\right)^3 = \dfrac{(x^2)^3y^3}{z^3} = \dfrac{x^6y^3}{z^3}$

51. $\left(\dfrac{a^2b}{cd^3}\right)^{-2} = \dfrac{(a^2)^{-2}b^{-2}}{c^{-2}(d^3)^{-2}} = \dfrac{a^{-4}b^{-2}}{c^{-2}d^{-6}} = \dfrac{\frac{1}{a^4}\cdot\frac{1}{b^2}}{\frac{1}{c^2}\cdot\frac{1}{d^6}} =$

$\dfrac{\frac{1}{a^4b^2}}{\frac{1}{c^2d^6}} = \dfrac{1}{a^4b^2}\cdot\dfrac{c^2d^6}{1} = \dfrac{c^2d^6}{a^4b^2}$

53. 2.8,000,000,000. (10 places)

Large number, so the exponent is positive.

$28,000,000,000 = 2.8\times 10^{10}$

55. 9.07,000,000,000,000,000.
↑ 17 places

Large number, so the exponent is positive.

$907,000,000,000,000,000 = 9.07 \times 10^{17}$

57. 0.000003.04
↑ 6 places

Small number, so the exponent is negative.

$0.00000304 = 3.04 \times 10^{-6}$

59. 0.00000001.8
↑ 8 places

Small number, so the exponent is negative.

$0.000000018 = 1.8 \times 10^{-8}$

61. 1.00,000,000,000.
↑ 11 places

Large number, so the exponent is positive.

$100,000,000,000 = 1.0 \times 10^{11} = 10^{11}$

63. 419,854,000

4.19,854,000.
↑ 8 places

Large number, so the exponent is positive.

$419,854,000 = 4.19854 \times 10^{8}$

65. 0.0000006.8
↑ 7 places

Small number, so the exponent is negative.

$0.00000068 = 6.8 \times 10^{-7}$

67. 8.74×10^{7}

Positive exponent, so the answer is a large number.

8.7400000.
↑ 7 places

$8.74 \times 10^{7} = 87,400,000$

69. 5.704×10^{-8}

Negative exponent, so the answer is a small number.

0.00000005.704
↑ 8 places

$5.704 \times 10^{-8} = 0.00000005704$

71. $10^{7} = 1 \times 10^{7}$

Positive exponent, so the answer is a large number.

1.0000000.
↑ 7 places

$10^{7} = 10,000,000$

73. $10^{-5} = 1 \times 10^{-5}$

Negative exponent, so the answer is a small number.

0.00001.
↑ 5 places

$10^{-5} = 0.00001$

75. $(3 \times 10^{4})(2 \times 10^{5}) = (3 \cdot 2) \times (10^{4} \cdot 10^{5})$
$= 6 \times 10^{9}$

77. $(5.2 \times 10^{5})(6.5 \times 10^{-2}) = (5.2 \cdot 6.5) \times (10^{5} \cdot 10^{-2})$
$= 33.8 \times 10^{3}$

The answer at this stage is 33.8×10^{3} but this is not scientific notation since 33.8 is not a number between 1 and 10. We convert 33.8 to scientific notation and simplify.

$33.8 \times 10^{3} = (3.38 \times 10^{1}) \times 10^{3} = 3.38 \times (10^{1} \times 10^{3}) = 3.38 \times 10^{4}$

The answer is 3.38×10^{4}.

79. $(9.9 \times 10^{-6})(8.23 \times 10^{-8}) = (9.9 \cdot 8.23) \times (10^{-6} \cdot 10^{-8})$
$= 81.477 \times 10^{-14}$

The answer at this stage is 81.477×10^{-14}. We convert 81.477 to scientific notation and simplify.

$81.477 \times 10^{-14} = (8.1477 \times 10^{1}) \times 10^{-14} = 8.1477 \times (10^{1} \times 10^{-14}) = 8.1477 \times 10^{-13}$.

The answer is 8.1477×10^{-13}.

81. $\frac{8.5 \times 10^{8}}{3.4 \times 10^{-5}} = \frac{8.5}{3.4} \times \frac{10^{8}}{10^{-5}}$
$= 2.5 \times 10^{8-(-5)}$
$= 2.5 \times 10^{13}$

83. $(3.0 \times 10^{6}) \div (6.0 \times 10^{9}) = \frac{3.0 \times 10^{6}}{6.0 \times 10^{9}}$
$= \frac{3.0}{6.0} \times \frac{10^{6}}{10^{9}}$
$= 0.5 \times 10^{6-9}$
$= 0.5 \times 10^{-3}$

The answer at this stage is 0.5×10^{-3}. We convert 0.5 to scientific notation and simplify.

$0.5 \times 10^{-3} = (5.0 \times 10^{-1}) \times 10^{-3} = 5.0 \times (10^{-1} \times 10^{-3}) = 5.0 \times 10^{-4}$

85. $\frac{7.5 \times 10^{-9}}{2.5 \times 10^{12}} = \frac{7.5}{2.5} \times \frac{10^{-9}}{10^{12}}$
$= 3.0 \times 10^{-9-12}$
$= 3.0 \times 10^{-21}$

87. There are 60 seconds in one minute and 60 minutes in one hour, so there are 60(60), or 3600 seconds in one hour. There are 24 hours in one day and 365 days in one year, so there are 3600(24)(365), or 31,536,000 seconds in one year.

$4,200,000 \times 31,536,000$
$= (4.2 \times 10^{6}) \times (3.1536 \times 10^{7})$
$= (4.2 \times 3.1536) \times (10^{6} \times 10^{7})$
$\approx 13.25 \times 10^{13}$
$\approx (1.325 \times 10) \times 10^{13}$
$\approx 1.325 \times (10 \times 10^{13})$
$\approx 1.325 \times 10^{14}$

About 1.325×10^{14} cubic feet of water is discharged from the Amazon River in 1 yr.

89. $\dfrac{1.908 \times 10^{24}}{6 \times 10^{21}} = \dfrac{1.908}{6} \times \dfrac{10^{24}}{10^{21}}$

$$= 0.318 \times 10^3$$
$$= (3.18 \times 10^{-1}) \times 10^3$$
$$= 3.18 \times (10^{-1} \times 10^3)$$
$$= 3.18 \times 10^2$$

The mass of Jupiter is 3.18×10^2 times the mass of Earth.

91. 10 billion trillion $= 1 \times 10 \times 10^9 \times 10^{12}$

$$= 1 \times 10^{22}$$

There are 1×10^{22} stars in the known universe.

93. $\dfrac{2.5 \times 10^{18}}{2 \times 10^9} = 1.25 \times 10^9$ bytes per person per day

95. $\dfrac{1.55 \times 10^{-3}}{5 \times 10^{-6}} = 0.31 \times 10^3$

$$= (3.1 \times 10^{-1}) \times 10^3$$
$$= 3.1 \times 10^2 \text{ sheets}$$

97. First we divide the distance from the earth to the moon by 3 days to find the number of miles per day the space vehicle travels. Note that $240,000 = 2.4 \times 10^5$.

$$\frac{2.4 \times 10^5}{3} = 0.8 \times 10^5 = 8 \times 10^4$$

The space vehicle travels 8×10^4 miles per day. Now divide the distance from the earth to Mars by 8×10^4 to find how long it will take the space vehicle to reach Mars. Note that $35,000,000 = 3.5 \times 10^7$.

$$\frac{3.5 \times 10^7}{8 \times 10^4} = 0.4375 \times 10^3 = 4.375 \times 10^2$$

It takes 4.375×10^2 days for the space vehicle to travel from the earth to Mars.

99. $y = x - 5$

The equation is equivalent to $y = x + (-5)$. The y-intercept is $(0, -5)$. We find two other points.

When $x = 2$, $y = 2 - 5 = -3$.

When $x = 4$, $y = 4 - 5 = -1$.

x	y
0	−5
2	−3
4	−1

Plot these points, draw the line they determine, and label the graph $y = x - 5$.

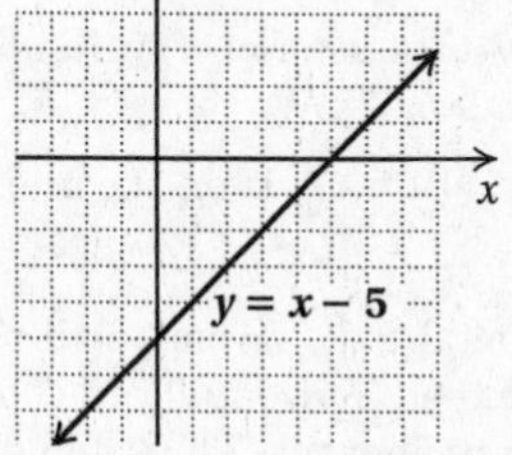

101.

$$3x - y = 3$$
$$3x - 3 = y$$
$$3x + (-3) = y$$

The y-intercept is $(0, -3)$. We find two other points.

When $x = 1$, $y = 3 \cdot 1 - 3 = 3 - 3 = 0$.

When $x = 2$, $y = 3 \cdot 2 - 3 = 6 - 3 = 3$.

x	y
0	−3
1	0
2	3

Plot these points, draw the line they determine, and label the graph $3x - y = 3$.

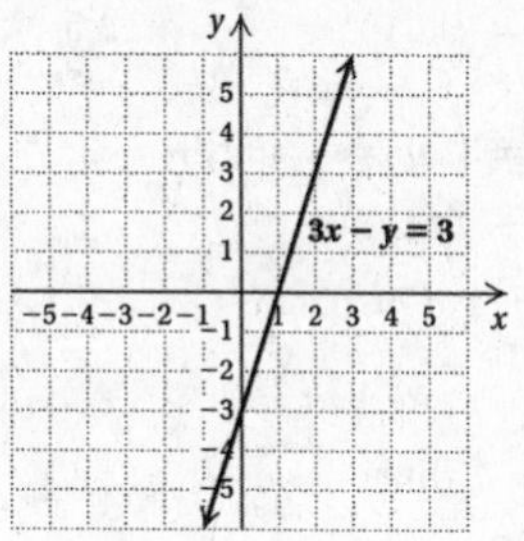

103. $2x = -10$

$x = -5$

Any ordered pair $(-5, y)$ is a solution. The variable x must be -5, but y can be any number we choose. A few solutions are listed below. Plot these points and draw the line.

x	y
−5	−3
−5	0
−5	4

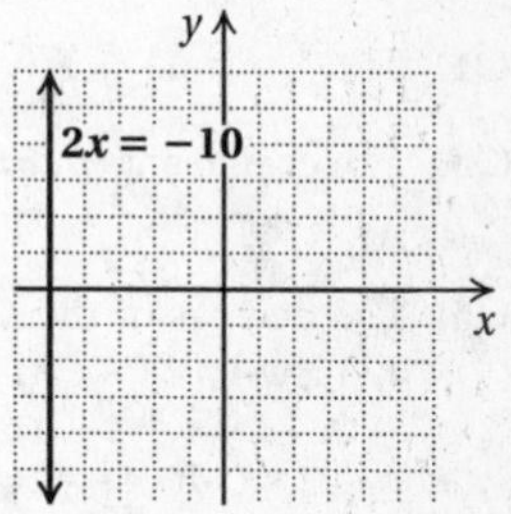

105. $8y - 16 = 0$

$8y = 16$

$y = 2$

Any ordered pair $(x, 2)$ is a solution. The variable y must be 2, but x can be any number we choose. A few solutions are listed below. Plot these points and draw the line.

x	y
−4	2
0	2
3	2

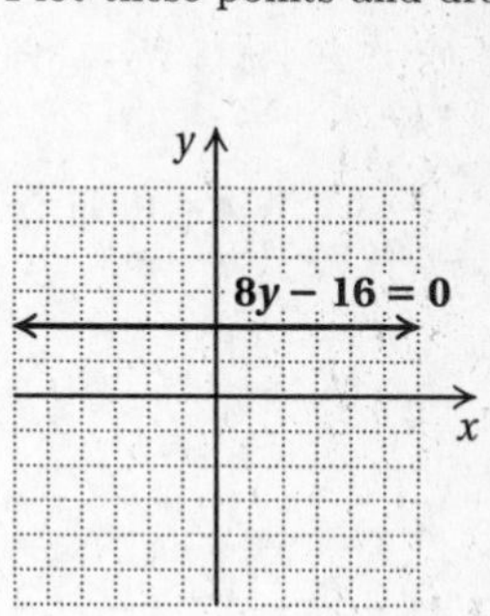

107. $\frac{(5.2\times10^6)(6.1\times10^{-11})}{1.28\times10^{-3}} = \frac{(5.2\cdot6.1)}{1.28}\times\frac{(10^6\cdot10^{-11})}{10^{-3}}$
$= 24.78125\times10^{-2}$
$= (2.478125\times10^1)\times10^{-2}$
$= 2.478125\times10^{-1}$

109. $\frac{(5^{12})^2}{5^{25}} = \frac{5^{24}}{5^{25}} = 5^{24-25} = 5^{-1} = \frac{1}{5}$

111. $\frac{(3^5)^4}{3^5\cdot3^4} = \frac{3^{5\cdot4}}{3^{5+4}} = \frac{3^{20}}{3^9} = 3^{20-9} = 3^{11}$

113. $\frac{49^{18}}{7^{35}} = \frac{(7^2)^{18}}{7^{35}} = \frac{7^{36}}{7^{35}} = 7$

115. $\frac{(0.4)^5}{[(0.4)^3]^2} = \frac{(0.4)^5}{(0.4)^6} = (0.4)^{-1} = \frac{1}{0.4}$, or 2.5

117. False; let $x = 2$, $y = 3$, $m = 4$, and $n = 2$:
$2^4\cdot3^2 = 16\cdot9 = 144$, but
$(2\cdot3)^{4\cdot2} = 6^8 = 1,679,616$

119. False; let $x = 5$, $y = 3$, and $m = 2$:
$(5-3)^2 = 2^2 = 4$, but
$5^2 - 3^2 = 25 - 9 = 16$

121. True; $(-x)^{2m} = (-1\cdot x)^{2m} = (-1)^{2m}\cdot x^{2m} = [(-1)^2]^m\cdot x^{2m} = 1^m\cdot x^{2m} = x^{2m}$

Exercise Set 4.3

RC1. $(-1)^2 - (-1) = 1 + 1 = 2$; answer (b) is correct.

RC3. The coefficients of $5x^4 - 3x + 7$ are 5, -3, and 7; answer (c) is correct.

RC5. The degrees of the terms of $5x^4 - 3x + 7$ are 4, 1, and 0; answer (a) is correct.

1. $-5x + 2 = -5\cdot4 + 2 = -20 + 2 = -18$;

$-5x + 2 = -5(-1) + 2 = 5 + 2 = 7$

3. $2x^2 - 5x + 7 = 2\cdot4^2 - 5\cdot4 + 7 = 2\cdot16 - 20 + 7 = 32 - 20 + 7 = 19$;

$2x^2 - 5x + 7 = 2(-1)^2 - 5(-1) + 7 = 2\cdot1 + 5 + 7 = 2 + 5 + 7 = 14$

5. $x^3 - 5x^2 + x = 4^3 - 5\cdot4^2 + 4 = 64 - 5\cdot16 + 4 = 64 - 80 + 4 = -12$;

$x^3 - 5x^2 + x = (-1)^3 - 5(-1)^2 + (-1) = -1 - 5\cdot1 - 1 = -1 - 5 - 1 = -7$

7. $\frac{1}{3}x + 5 = \frac{1}{3}(-2) + 5 = -\frac{2}{3} + 5 = -\frac{2}{3} + \frac{15}{3} = \frac{13}{3}$;
$\frac{1}{3}x + 5 = \frac{1}{3}\cdot0 + 5 = 0 + 5 = 5$

9. $x^2 - 2x + 1 = (-2)^2 - 2(-2) + 1 = 4 + 4 + 1 = 9$;

$x^2 - 2x + 1 = 0^2 - 2\cdot0 + 1 = 0 - 0 + 1 = 1$

11. $-3x^3 + 7x^2 - 3x - 2 = -3(-2)^3 + 7(-2)^2 - 3(-2) - 2 = -3(-8) + 7(4) - 3(-2) - 2 = 24 + 28 + 6 - 2 = 56$;

$-3x^3 + 7x^2 - 3x - 2 = -3\cdot0^3 + 7\cdot0^2 - 3\cdot0 - 2 = -3\cdot0 + 7\cdot0 - 0 - 2 = 0 + 0 - 0 - 2 = -2$

13. We evaluate the polynomial for $t = 10$:
$S = 11.12t^2 = 11.12(10)^2 = 11.12(100) = 1112$
The skydiver has fallen approximately 1112 ft.

15. $N = \frac{1}{3}x^3 + \frac{1}{2}x^2 + \frac{1}{6}x$
$= \frac{1}{3}\cdot5^3 + \frac{1}{2}\cdot5^2 + \frac{1}{6}\cdot5$
$= \frac{1}{3}\cdot125 + \frac{1}{2}\cdot25 + \frac{5}{6}$
$= \frac{125}{3} + \frac{25}{2} + \frac{5}{6}$
$= \frac{250}{6} + \frac{75}{6} + \frac{5}{6}$
$= \frac{330}{6}$
$= 55$

There are 55 oranges in a pyramid with 5 layers.

17. Locate -3 on the x-axis. Then move vertically to the graph and horizontally to the y-axis. It appears that the y-value that is paired with -3 is -4. Thus, the value of $y = 5 - x^2$ is -4 when $x = -3$.

Locate -1 on the x-axis. Then move vertically to the graph and horizontally to the y-axis. It appears that the y-value that is paired with -1 is 4. Thus, the value of $y = 5 - x^2$ is 4 when $x = -1$.

Locate 0 on the x-axis. Then move vertically to the graph. We arrive at a point on the y-axis with the y-value 5. Thus, the value of $5 - x^2$ is 5 when $x = 0$.

Locate 1.5 on the x-axis. Then move vertically to the graph and horizontally to the y-axis. It appears that the y-value that is paired with 1.5 is 2.75. Thus, the value of $y = 5 - x^2$ is 2.75 when $x = 1.5$.

Locate 2 on the x-axis. Then move vertically to the graph and horizontally to the y-axis. It appears that the y-value that is paired with 2 is 1. Thus, the value of $y = 5 - x^2$ is 1 when $x = 2$.

19. a) In 2010, $t = 2010 - 2000 = 10$.

$C = 0.27t + 2.97 = 0.27(10) + 2.97 = 2.7 + 2.97 = 5.67$

The average capacity of a solar-energy residential installation in 2010 was 5.67 kW.

b) We see that the point (10, 5.67) is on the graph, so the answer is correct.

21. Locate 10 on the horizontal axis. From there move vertically to the graph and then horizontally to the M-axis. This locates an M-value of about 9. Thus, about 9 words were memorized in 10 minutes.

23. Locate 8 on the horizontal axis. From there move vertically to the graph and then horizontally to the M-axis. This locates an M-value of about 6. Thus, the value of $-0.001t^3 + 0.1t^2$ for $t = 8$ is approximately 6.

25. Locate 13 on the horizontal axis. It is halfway between 12 and 14. From there move vertically to the graph and then horizontally to the M-axis. This locates an M-value of about 15. Thus, the value of $-0.001t^3 + 0.1t^2$ when t is 13 is approximately 15.

27. $2 - 3x + x^2 = 2 + (-3x) + x^2$

The terms are 2, $-3x$, and x^2.

29. $-2x^4 + \frac{1}{3}x^3 - x + 3 = -2x^4 + \frac{1}{3}x^3 + (-x) + 3$

The terms are $-2x^4$, $\frac{1}{3}x^3$, $-x$, and 3.

31. The polynomial $x^2 - 10x + 25$ is a *trinomial* because it has just three terms.

33. The polynomial $x^3 - 7x^2 + 2x - 4$ is *none of these* because it has more than three terms.

35. The polynomial $4x^2 - 25$ is a *binomial* because it has just two terms.

37. The polynomial $40x$ is a *monomial* because it has just one term.

39. $-3x + 6$

The coefficient of $-3x$, the first term, is -3.

The coefficient of 6, the second term, is 6.

41. $5x^2 + \frac{3}{4}x + 3$

The coefficient of $5x^2$, the first term, is 5.

The coefficient of $\frac{3}{4}x$, the second term, is $\frac{3}{4}$.

The coefficient of 3, the third term, is 3.

43. $-5x^4 + 6x^3 - 2.7x^2 + 8x - 2$

The coefficient of $-5x^4$, the first term, is -5.

The coefficient of $6x^3$, the second term, is 6.

The coefficient of $-2.7x^2$, the third term, is -2.7.

The coefficient of $8x$, the fourth term, is 8.

The coefficient of -2, the fifth term, is -2.

45. $2x - 4 = 2x^1 - 4x^0$

The degree of $2x$ is 1.

The degree of -4 is 0.

The degree of the polynomial is 1, the largest exponent.

47. $3x^2 - 5x + 2 = 3x^2 - 5x^1 + 2x^0$

The degree of $3x^2$ is 2.

The degree of $-5x$ is 1.

The degree of 2 is 0.

The degree of the polynomial is 2, the largest exponent.

49. $-7x^3 + 6x^2 + \frac{3}{5}x + 7 = -7x^3 + 6x^2 + \frac{3}{5}x^1 + 7x^0$

The degree of $-7x^3$ is 3.

The degree of $6x^2$ is 2.

The degree of $\frac{3}{5}x$ is 1.

The degree of 7 is 0.

The degree of the polynomial is 3, the largest exponent.

51. $x^2 - 3x + x^6 - 9x^4 = x^2 - 3x^1 + x^6 - 9x^4$

The degree of x^2 is 2.

The degree of $-3x$ is 1.

The degree of x^6 is 6.

The degree of $-9x^4$ is 4.

The degree of the polynomial is 6, the largest exponent.

53. See the answer section in the text.

55. $5x^3 + 6x^2 - 3x^2$

Like terms: $6x^2$ and $-3x^2$ — Same variable and exponent

57. $2x^4 + 5x - 7x - 3x^4$

Like terms: $2x^4$ and $-3x^4$ — Same variable and exponent

Like terms: $5x$ and $-7x$

59. $3x^5 - 7x + 8 + 14x^5 - 2x - 9$

Like terms: $3x^5$ and $14x^5$

Like terms: $-7x$ and $-2x$

Like terms: 8 and -9 — Constant terms are like terms.

61. $2x - 5x = (2 - 5)x = -3x$

63. $x - 9x = 1x - 9x = (1 - 9)x = -8x$

65. $5x^3 + 6x^3 + 4 = (5 + 6)x^3 + 4 = 11x^3 + 4$

67. $5x^3 + 6x - 4x^3 - 7x = (5 - 4)x^3 + (6 - 7)x =$

$1x^3 + (-1)x = x^3 - x$

69. $6b^5 + 3b^2 - 2b^5 - 3b^2 = (6 - 2)b^5 + (3 - 3)b^2 =$

$4b^5 + 0b^2 = 4b^5$

71. $\frac{1}{4}x^5 - 5 + \frac{1}{2}x^5 - 2x - 37 =$

$\left(\frac{1}{4} + \frac{1}{2}\right)x^5 - 2x + (-5 - 37) = \frac{3}{4}x^5 - 2x - 42$

73. $6x^2 + 2x^4 - 2x^2 - x^4 - 4x^2 =$

$6x^2 + 2x^4 - 2x^2 - 1x^4 - 4x^2 =$

$(6 - 2 - 4)x^2 + (2 - 1)x^4 = 0x^2 + 1x^4 =$

$0 + x^4 = x^4$

75. $\frac{1}{4}x^3 - x^2 - \frac{1}{6}x^2 + \frac{3}{8}x^3 + \frac{5}{16}x^3 =$

$\frac{1}{4}x^3 - 1x^2 - \frac{1}{6}x^2 + \frac{3}{8}x^3 + \frac{5}{16}x^3 =$

$\left(\frac{1}{4} + \frac{3}{8} + \frac{5}{16}\right)x^3 + \left(-1 - \frac{1}{6}\right)x^2 =$

$\left(\frac{4}{16} + \frac{6}{16} + \frac{5}{16}\right)x^3 + \left(-\frac{6}{6} - \frac{1}{6}\right)x^2 = \frac{15}{16}x^3 - \frac{7}{6}x^2$

77. $x^5 + x + 6x^3 + 1 + 2x^2 = x^5 + 6x^3 + 2x^2 + x + 1$

79. $5y^3 + 15y^9 + y - y^2 + 7y^8 =$
$15y^9 + 7y^8 + 5y^3 - y^2 + y$

81. $3x^4 - 5x^6 - 2x^4 + 6x^6 = x^4 + x^6 = x^6 + x^4$

83. $-2x + 4x^3 - 7x + 9x^3 + 8 = -9x + 13x^3 + 8 =$
$13x^3 - 9x + 8$

85. $3x + 3x + 3x - x^2 - 4x^2 = 9x - 5x^2 = -5x^2 + 9x$

87. $-x + \frac{3}{4} + 15x^4 - x - \frac{1}{2} - 3x^4 = -2x + \frac{1}{4} + 12x^4 =$
$12x^4 - 2x + \frac{1}{4}$

89. In the polynomial $x^3 - 27$, there are no x^2 or x terms. The x^2 term (or second-degree term) and the x term (or first-degree term) are missing.

91. In the polynomial $x^4 - x$, there are no x^3, x^2, or x^0 terms. The x^3 term (or third-degree term), the x^2 term (or second-degree term), and the x^0 term (or zero-degree term) are missing.

93. No terms are missing in the polynomial
$2x^3 - 5x^2 + x - 3$.

95.
$$x^3 - 27 = x^3 + 0x^2 + 0x - 27$$
$$x^3 - 27 = x^3 \qquad\qquad - 27$$

97.
$$x^4 - x = x^4 + 0x^3 + 0x^2 - x + 0x^0$$
$$x^4 - x = x^4 \qquad\qquad - x$$

99. There are no missing terms.

101. $1 + (-20)$

The difference in the absolute values is $20 - 1$, or 19. The negative number has the larger absolute value, so the answer is negative.

$1 + (-20) = -19$

103. $-4.2 + 1.95$

The difference in the absolute values is $4.2 - 1.95$, or 2.25. The negative number has the larger absolute value, so the answer is negative.

$-4.2 + 1.95 = -2.25$

105.
$$\begin{aligned}\frac{1}{8} - \frac{5}{6} &= \frac{1}{8} + \left(-\frac{5}{6}\right), \quad \text{LCM is 24}\\ &= \frac{1}{8}\cdot\frac{3}{3} + \left(-\frac{5}{6}\right)\left(\frac{4}{4}\right)\\ &= \frac{3}{24} + \left(-\frac{20}{24}\right)\\ &= -\frac{17}{24}\end{aligned}$$

107. $5.6 - 8.2 = 5.6 + (-8.2) = -2.6$

109. The product of two negative numbers is positive.
$(-6)(-3) = 18$

111. The product of a positive number and a negative number is negative.
$0.5(-1.2) = -0.6$

113. The quotient of two negative numbers is positive.
$-600 \div (-30) = \frac{600}{30} = \frac{20\cdot 30}{1\cdot 30} = \frac{20}{1}\cdot\frac{30}{30} = 20$

115. Zero divided by any nonzero number is 0.
$0 \div (-4) = 0$

117.
$$\begin{aligned}&6x^3\cdot 7x^2 - (4x^3)^2 + (-3x^3)^2 - (-4x^2)(5x^3)-\\ &10x^5 + 17x^6\\ &= 42x^5 - 16x^6 + 9x^6 - (-20x^5) - 10x^5 + 17x^6\\ &= 42x^5 - 16x^6 + 9x^6 + 20x^5 - 10x^5 + 17x^6\\ &= 10x^6 + 52x^5\end{aligned}$$

119. Answers may vary. Use any ax^5-term, where a is an integer, and 3 other terms with different degrees, each less than degree 5, and integer coefficients. Three answers are $-6x^5 + 14x^4 - x^2 + 11$, $x^5 - 8x^3 + 3x + 1$, and $23x^5 + 2x^4 - x^2 + 5x$.

121. Let $c =$ the coefficient of x^3. Solve:
$$\begin{aligned}c + (c-3) + 3(c-3) + (c+2) &= -4\\ c + c - 3 + 3c - 9 + c + 2 &= -4\\ 6c - 10 &= -4\\ 6c &= 6\\ c &= 1\end{aligned}$$

Coefficient of x^3, c: 1

Coefficient of x^2, $c - 3$: $1 - 3$, or -2

Coefficient of x, $3(c-3)$: $3(1-3)$, or -6

Coefficient remaining (constant term), $c + 2$:
$1 + 2$, or 3

The polynomial is $x^3 - 2x^2 - 6x + 3$.

123. Graph $y = 5 - x^2$. Then use VALUE from the CALC menu to find the y-values that correspond to $x = -3$, $x = -1$, $x = 0$, $x = 1.5$, and $x = 2$. As before, we find that these values are -4, 4, 5, 2.75, and 1, respectively.

125. Graph $y = -0.00006x^3 + 0.006x^2 - 0.1x + 1.9$. Then use VALUE from the CALC menu to find the y-values that correspond to $x = 20$ and $x = 40$. As before, we find that these values are 1.82 and 3.66, respectively. These results represent 1,820,000 and 3,660,000 hearing-impaired Americans.

Exercise Set 4.4

RC1. To find the opposite of a polynomial, we change the sign of every term. The given statement is false.

RC3. The statement is false. For example, $(-3x + 1) + (3x - 1) = 0$, a monomial.

1. $(3x + 2) + (-4x + 3) = (3 - 4)x + (2 + 3) = -x + 5$

3. $(-6x+2)+\left(x^2+\frac{1}{2}x-3\right)=x^2+\left(-6+\frac{1}{2}\right)x+(2-3)=$

$x^2+\left(-\frac{12}{2}+\frac{1}{2}\right)x+(2-3)=x^2-\frac{11}{2}x-1$

5. $(x^2-9)+(x^2+9)=(1+1)x^2+(-9+9)=2x^2$

7. $(3x^2-5x+10)+(2x^2+8x-40)=$

$(3+2)x^2+(-5+8)x+(10-40)=5x^2+3x-30$

9. $(1.2x^3+4.5x^2-3.8x)+(-3.4x^3-4.7x^2+23)=$

$(1.2-3.4)x^3+(4.5-4.7)x^2-3.8x+23=$

$-2.2x^3-0.2x^2-3.8x+23$

11. $(1+4x+6x^2+7x^3)+(5-4x+6x^2-7x^3)=$

$(1+5)+(4-4)x+(6+6)x^2+(7-7)x^3=$

$6+0x+12x^2+0x^3=6+12x^2$, or $12x^2+6$

13. $\left(\frac{1}{4}x^4+\frac{2}{3}x^3+\frac{5}{8}x^2+7\right)+\left(-\frac{3}{4}x^4+\frac{3}{8}x^2-7\right)=$

$\left(\frac{1}{4}-\frac{3}{4}\right)x^4+\frac{2}{3}x^3+\left(\frac{5}{8}+\frac{3}{8}\right)x^2+(7-7)=$

$-\frac{2}{4}x^4+\frac{2}{3}x^3+\frac{8}{8}x^2+0=$

$-\frac{1}{2}x^4+\frac{2}{3}x^3+x^2$

15. $(0.02x^5-0.2x^3+x+0.08)+(-0.01x^5+x^4-0.8x-0.02)=$

$(0.02-0.01)x^5+x^4-0.2x^3+(1-0.8)x+(0.08-0.02)=$

$0.01x^5+x^4-0.2x^3+0.2x+0.06$

17. $9x^8-7x^4+2x^2+5)+(8x^7+4x^4-2x)+$

$(-3x^4+6x^2+2x-1)=9x^8+8x^7+(-7+4-3)x^4+$

$(2+6)x^2+(-2+2)x+(5-1)=$

$9x^8+8x^7-6x^4+8x^2+4$

19. Rewrite the problem so the coefficients of like terms have the same number of decimal places.

$$\begin{array}{rrrrr}
0.15x^4 & +\ 0.10x^3 & -\ 0.90x^2 & & \\
 & -\ 0.01x^3 & +\ 0.01x^2 & +\ x & \\
1.25x^4 & & +\ 0.11x^2 & & +\ 0.01 \\
 & 0.27x^3 & & & +\ 0.99 \\
-0.35x^4 & & +\ 15.00x^2 & & -\ 0.03 \\
\hline
1.05x^4 & +\ 0.36x^3 & +\ 14.22x^2 & +\ x & +\ 0.97
\end{array}$$

21. We change the sign of the term inside the parentheses.

$-(-5x)=5x$

23. We change the sign of every term inside the parentheses.

$-(-x^2+\frac{3}{2}x-2)=x^2-\frac{3}{2}x+2$

25. We change the sign of every term inside the parentheses.

$-(12x^4-3x^3+3)=-12x^4+3x^3-3$

27. We change the sign of every term inside parentheses.

$-(3x-7)=-3x+7$

29. We change the sign of every term inside parentheses.

$-(4x^2-3x+2)=-4x^2+3x-2$

31. We change the sign of every term inside parentheses.

$-\left(-4x^4+6x^2+\frac{3}{4}x-8\right)=4x^4-6x^2-\frac{3}{4}x+8$

33. $(3x+2)-(-4x+3)=3x+2+4x-3$

Changing the sign of every term inside parentheses

$=7x-1$

35. $(-6x+2)-(x^2+x-3)=-6x+2-x^2-x+3$

$=-x^2-7x+5$

37. $(x^2-9)-(x^2+9)=x^2-9-x^2-9=-18$

39. $(6x^4+3x^3-1)-(4x^2-3x+3)$

$=6x^4+3x^3-1-4x^2+3x-3$

$=6x^4+3x^3-4x^2+3x-4$

41. $(1.2x^3+4.5x^2-3.8x)-(-3.4x^3-4.7x^2+23)$

$=1.2x^3+4.5x^2-3.8x+3.4x^3+4.7x^2-23$

$=4.6x^3+9.2x^2-3.8x-23$

43. $\frac{5}{8}x^3-\frac{1}{4}x-\frac{1}{3}-\left(-\frac{1}{8}x^3+\frac{1}{4}x-\frac{1}{3}\right)$

$=\frac{5}{8}x^3-\frac{1}{4}x-\frac{1}{3}+\frac{1}{8}x^3-\frac{1}{4}x+\frac{1}{3}$

$=\frac{6}{8}x^3-\frac{2}{4}x$

$=\frac{3}{4}x^3-\frac{1}{2}x$

45. $(0.08x^3-0.02x^2+0.01x)-(0.02x^3+0.03x^2-1)$

$=0.08x^3-0.02x^2+0.01x-0.02x^3-0.03x^2+1$

$=0.06x^3-0.05x^2+0.01x+1$

47.

$$\begin{array}{rrr}
x^2 & +\ 5x & +\ 6 \\
x^2 & +\ 2x & \\
\hline
\end{array}$$

$$\begin{array}{rrrl}
x^2 & +\ 5x & +\ 6 & \\
-x^2 & -\ 2x & & \text{Changing signs} \\
\hline
 & 3x & +\ 6 & \text{Adding}
\end{array}$$

49.

$$\begin{array}{rrrrr}
5x^4 & +\ 6x^3 & -\ 9x^2 & & \\
-6x^4 & -\ 6x^3 & & +\ 8x & +\ 9 \\
\hline
\end{array}$$

$$\begin{array}{rrrrrl}
5x^4 & +\ 6x^3 & -\ 9x^2 & & & \\
6x^4 & +\ 6x^3 & & -\ 8x & -\ 9 & \text{Changing signs} \\
\hline
11x^4 & +\ 12x^3 & -\ 9x^2 & -\ 8x & -\ 9 & \text{Adding}
\end{array}$$

51.

$$\begin{array}{rrrrrr}
x^5 & & & & & -\ 1 \\
x^5 & -\ x^4 & +\ x^3 & -\ x^2 & +\ x & -\ 1 \\
\hline
\end{array}$$

$$\begin{array}{rrrrrrl}
x^5 & & & & & -\ 1 & \\
-x^5 & +\ x^4 & -\ x^3 & +\ x^2 & -\ x & +\ 1 & \text{Changing signs} \\
\hline
 & x^4 & -\ x^3 & +\ x^2 & -\ x & & \text{Adding}
\end{array}$$

53. We add the lengths of the sides:

$$4a + 7 + a + \frac{1}{2}a + 3 + a + 2a + 3a$$

$$= \left(4 + 1 + \frac{1}{2} + 1 + 2 + 3\right)a + (7 + 3)$$

$$= 11\frac{1}{2}a + 10, \text{ or } \frac{23}{2}a + 10$$

55.

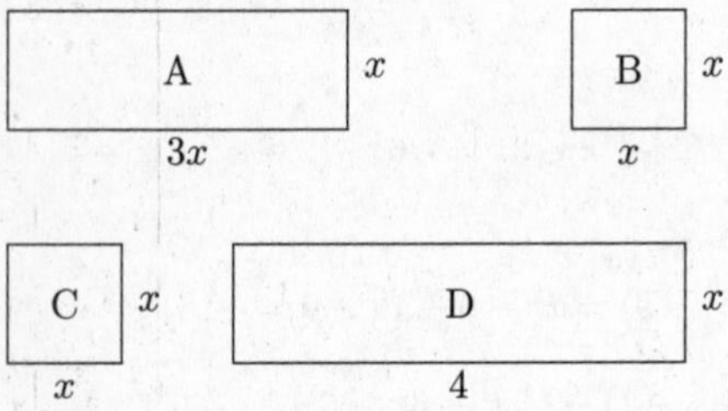

The area of a rectangle is the product of the length and width. The sum of the areas is found as follows:

$$\begin{array}{ccccccc} \text{Area of } A & + & \text{Area of } B & + & \text{Area of } C & + & \text{Area of } D \\ = 3x \cdot x & + & x \cdot x & + & x \cdot x & + & 4 \cdot x \\ = 3x^2 & + & x^2 & + & x^2 & + & 4x \\ = 5x^2 & + & 4x & & & & \end{array}$$

A polynomial for the sum of the areas is $5x^2 + 4x$.

57.

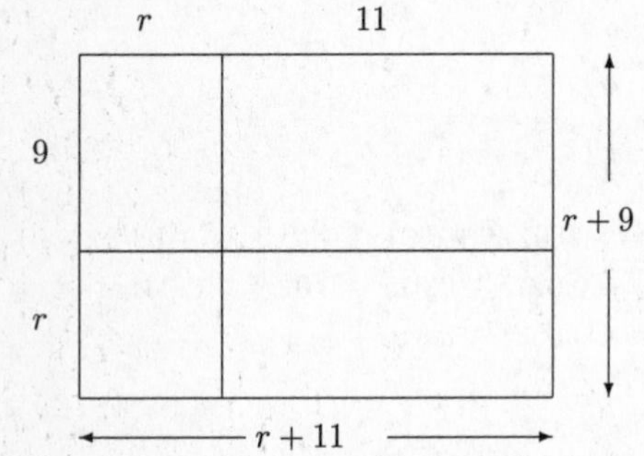

The length and width of the figure can be expressed as $r + 11$ and $r + 9$, respectively. The area of this figure (a rectangle) is the product of the length and width. An algebraic expression for the area is $(r + 11) \cdot (r + 9)$.

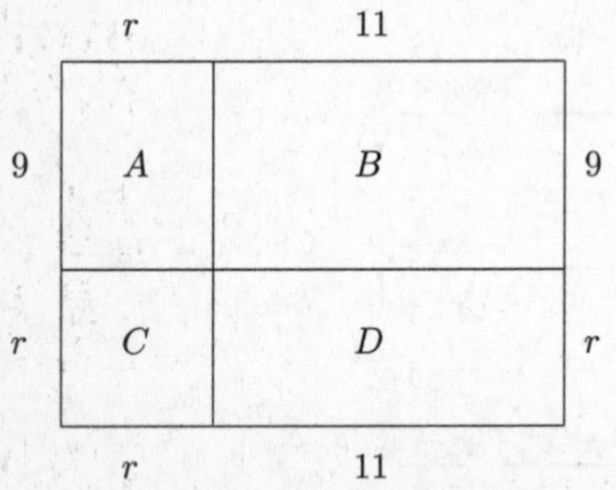

The area of the figure can also be found by adding the areas of the four rectangles A, B, C, and D. The area of a rectangle is the product of the length and the width.

$$\begin{array}{ccccccc} \text{Area of } A & + & \text{Area of } B & + & \text{Area of } C & + & \text{Area of } D \\ = 9 \cdot r & + & 11 \cdot 9 & + & r \cdot r & + & 11 \cdot r \\ = 9r & + & 99 & + & r^2 & + & 11r \end{array}$$

A second algebraic expression for the area of the figure is $9r + 99 + r^2 + 11r$, or $r^2 + 20r + 99$.

59.

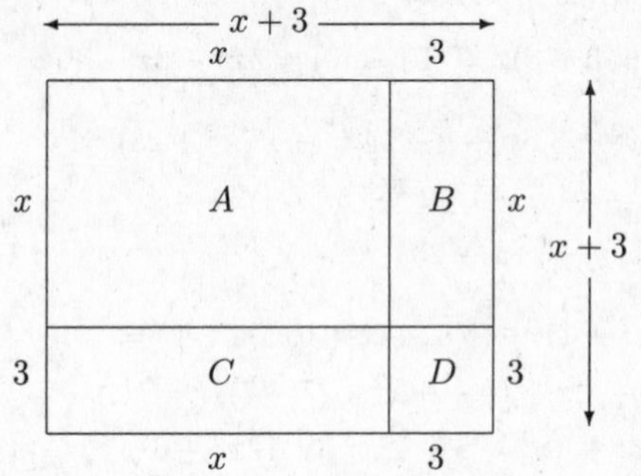

The length and width of the figure can each be expressed as $x + 3$. The area can be expressed as $(x + 3) \cdot (x + 3)$, or $(x + 3)^2$.

Another way to express the area is to find an expression for the sum of the areas of the four rectangles A, B, C, and D. The area of each rectangle is the product of its length and width.

$$\begin{array}{ccccccc} \text{Area of } A & + & \text{Area of } B & + & \text{Area of } C & + & \text{Area of } D \\ = x \cdot x & + & 3 \cdot x & + & 3 \cdot x & + & 3 \cdot 3 \\ = x^2 & + & 3x & + & 3x & + & 9 \end{array}$$

Then a second algebraic expression for the area of the figure is $x^2 + 3x + 3x + 9$, or $x^2 + 6x + 9$.

61.

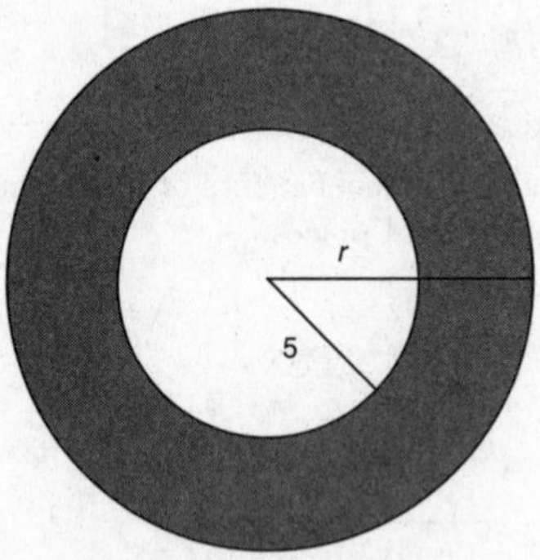

Familiarize. Recall that the area of a circle is the product of π and the square of the radius, r^2.

$$A = \pi r^2$$

Translate.

$$\begin{array}{ccccc} \text{Area of circle with radius } r & - & \text{Area of circle with radius 5} & = & \text{Shaded area} \\ \pi \cdot r^2 & - & \pi \cdot 5^2 & = & \text{Shaded area} \end{array}$$

Carry out. We simplify the expression.

$$\pi \cdot r^2 - \pi \cdot 5^2 = \pi r^2 - 25\pi$$

Check. We can go over our calculations. We can also assign some value to r, say 7, and carry out the computation in two ways.

Difference of areas: $\pi \cdot 7^2 - \pi \cdot 5^2 = 49\pi - 25\pi = 24\pi$

Substituting in the polynomial: $\pi \cdot 7^2 - 25\pi = 49\pi - 25\pi = 24\pi$

Since the results are the same, our solution is probably correct.

State. A polynomial for the shaded area is $\pi r^2 - 25\pi$.

63. ***Familiarize***. We label the figure with additional information.

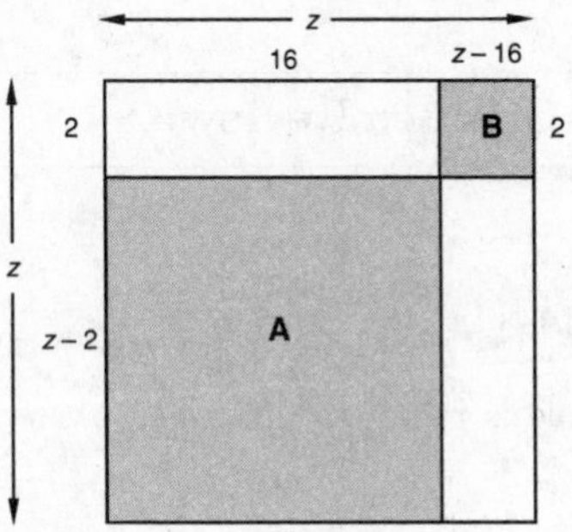

Translate.

Area of shaded sections = Area of A + Area of B

Area of shaded sections = $16(z-2) + 2(z-16)$

Carry out. We simplify the expression.

$16(z-2) + 2(z-16) = 16z - 32 + 2z - 32 = 18z - 64$

Check. We can go over the calculations. We can also assign some value to z, say 30, and carry out the computation in two ways.

Sum of areas:

$16 \cdot 28 + 2 \cdot 14 = 448 + 28 = 476$

Substituting in the polynomial:

$18 \cdot 30 - 64 = 540 - 64 = 476$

Since the results are the same, our solution is probably correct.

State. A polynomial for the shaded area is $18z - 64$.

65.
$$\begin{aligned} 8x + 3x &= 66 \\ 11x &= 66 && \text{Collecting like terms} \\ \frac{11x}{11} &= \frac{66}{11} && \text{Dividing by 11} \\ x &= 6 \end{aligned}$$

The solution is 6.

67.
$$\begin{aligned} \frac{3}{8}x + \frac{1}{4} - \frac{3}{4}x &= \frac{11}{16} + x, && \text{LCM is 16} \\ 16\left(\frac{3}{8}x + \frac{1}{4} - \frac{3}{4}x\right) &= 16\left(\frac{11}{16} + x\right) && \text{Clearing fractions} \\ 6x + 4 - 12x &= 11 + 16x \\ -6x + 4 &= 11 + 16x && \text{Collecting like terms} \\ -6x + 4 - 4 &= 11 + 16x - 4 && \text{Subtracting 4} \\ -6x &= 7 + 16x \\ -6x - 16x &= 7 + 16x - 16x && \text{Subtracting } 16x \\ -22x &= 7 \\ \frac{-22x}{-22} &= \frac{7}{-22} && \text{Dividing by } -22 \\ x &= -\frac{7}{22} \end{aligned}$$

The solution is $-\frac{7}{22}$.

69.
$$\begin{aligned} 1.5x - 2.7x &= 22 - 5.6x \\ 10(1.5x - 2.7x) &= 10(22 - 5.6x) && \text{Clearing decimals} \\ 15x - 27x &= 220 - 56x \\ -12x &= 220 - 56x && \text{Collecting like terms} \\ 44x &= 220 && \text{Adding } 56x \\ x &= \frac{220}{44} && \text{Dividing by 44} \\ x &= 5 && \text{Simplifying} \end{aligned}$$

The solution is 5.

71.
$$\begin{aligned} 6(y-3) - 8 &= 4(y+2) + 5 \\ 6y - 18 - 8 &= 4y + 8 + 5 && \text{Removing parentheses} \\ 6y - 26 &= 4y + 13 && \text{Collecting like terms} \\ 6y - 26 + 26 &= 4y + 13 + 26 && \text{Adding 26} \\ 6y &= 4y + 39 \\ 6y - 4y &= 4y + 39 - 4y && \text{Subtracting } 4y \\ 2y &= 39 \\ \frac{2y}{2} &= \frac{39}{2} && \text{Dividing by 2} \\ y &= \frac{39}{2} \end{aligned}$$

The solution is $\frac{39}{2}$.

73.
$$\begin{aligned} 3x - 7 &\le 5x + 13 \\ -2x - 7 &\le 13 && \text{Subtracting } 5x \\ -2x &\le 20 && \text{Adding 7} \\ x &\ge -10 && \text{Dividing by } -2 \text{ and reversing the inequality symbol} \end{aligned}$$

The solution set is $\{x|x \ge -10\}$.

75. ***Familiarize***. The surface area is $2lw + 2lh + 2wh$, where l = length, w = width, and h = height of the rectangular solid. Here we have $l = 3$, $w = w$, and $h = 7$.

Translate. We substitute in the formula above.

$2 \cdot 3 \cdot w + 2 \cdot 3 \cdot 7 + 2 \cdot w \cdot 7$

Carry out. We simplify the expression.

$$\begin{aligned}&2\cdot 3\cdot w+2\cdot 3\cdot 7+2\cdot w\cdot 7\\ &=6w+42+14w\\ &=20w+42\end{aligned}$$

Check. We can go over the calculations. We can also assign some value to w, say 6, and carry out the computation in two ways.

Using the formula: $2\cdot 3\cdot 6+2\cdot 3\cdot 7+2\cdot 6\cdot 7=$ $36+42+84=162$

Substituting in the polynomial: $20\cdot 6+42=$ $120+42=162$

Since the results are the same, our solution is probably correct.

State. A polynomial for the surface area is $20w+42$.

77. ***Familiarize.*** The surface area is $2lw+2lh+2wh$, where l = length, w = width, and h = height of the rectangular solid. Here we have $l=x$, $w=x$, and $h=5$.

Translate. We substitute in the formula above.

$$2\cdot x\cdot x+2\cdot x\cdot 5+2\cdot x\cdot 5$$

Carry out. We simplify the expression.

$$\begin{aligned}&2\cdot x\cdot x+2\cdot x\cdot 5+2\cdot x\cdot 5\\ &=2x^2+10x+10x\\ &=2x^2+20x\end{aligned}$$

Check. We can go over the calculations. We can also assign some value to x, say 3, and carry out the computation in two ways.

Using the formula: $2\cdot 3\cdot 3+2\cdot 3\cdot 5+2\cdot 3\cdot 5=$ $18+30+30=78$

Substituting in the polynomial: $2\cdot 3^2+20\cdot 3=$ $2\cdot 9+60=18+60=78$

Since the results are the same, our solution is probably correct.

State. A polynomial for the surface area is $2x^2+20x$.

79.

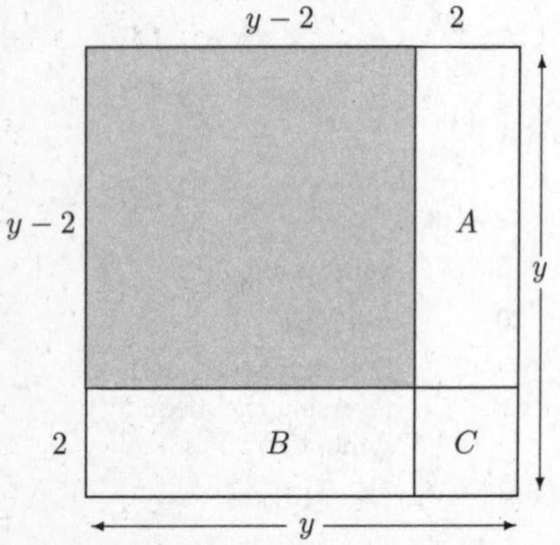

The shaded area is $(y-2)^2$. We find it as follows:

$$\begin{aligned}\text{Shaded area} &= \text{Area of square} - \text{Area of } A - \text{Area of } B - \text{Area of } C\\ (y-2)^2 &= y^2 - 2(y-2) - 2(y-2) - 2\cdot 2\\ (y-2)^2 &= y^2-2y+4-2y+4-4\\ (y-2)^2 &= y^2-4y+4\end{aligned}$$

81. $(7y^2-5y+6)-(3y^2+8y-12)+(8y^2-10y+3)$
$=7y^2-5y+6-3y^2-8y+12+8y^2-10y+3$
$=12y^2-23y+21$

83. $(-y^4-7y^3+y^2)+(-2y^4+5y-2)-(-6y^3+y^2)$
$=-y^4-7y^3+y^2-2y^4+5y-2+6y^3-y^2$
$=-3y^4-y^3+5y-2$

Chapter 4 Mid-Chapter Review

1. True; $a^n\cdot a^{-n}=a^0=1$.

2. False; $x^2\cdot x^3=x^{2+3}=x^5\neq x^6$.

3. False; the variable is raised to different powers, so these are not like terms.

4. True; $a^0=1$ for any nonzero number a.

5. $4w^3+6w-8w^3-3w=(4-8)w^3+(6-3)w$
$=-4w^3+3w$

6. $3y^4-y^2+11-(y^4-4y^2+5)$
$=3y^4-y^2+11-y^4+4y^2-5$
$=2y^4+3y^2+6$

7. $z^1=z$

8. $4.56^0=1$

9. $a^5=(-2)^5=-32$

10. $-x^3=-(-1)^3=-(-1)=1$

11. $5^3\cdot 5^4=5^{3+4}=5^7$

12. $(3a)^2(3a)^7=(3a)^{2+7}=(3a)^9$

13. $x^{-8}\cdot x^5=x^{-8+5}=x^{-3}=\dfrac{1}{x^3}$

14. $t^4\cdot t^{-4}=t^{4+(-4)}=t^0=1$

15. $\dfrac{7^8}{7^4}=7^{8-4}=7^4$

16. $\dfrac{x}{x^3}=x^{1-3}=x^{-2}=\dfrac{1}{x^2}$

17. $\dfrac{w^5}{w^{-3}}=w^{5-(-3)}=w^{5+3}=w^8$

18. $\dfrac{y^{-6}}{y^{-2}}=y^{-6-(-2)}=y^{-6+2}=y^{-4}=\dfrac{1}{y^4}$

19. $(3^5)^3=3^{5\cdot 3}=3^{15}$

20. $(x^{-3}y^2)^{-6}=x^{-3(-6)}y^{2(-6)}=x^{18}y^{-12}=\dfrac{x^{18}}{y^{12}}$

21. $\left(\frac{a^4}{5}\right)^6 = \frac{a^{4\cdot 6}}{5^6} = \frac{a^{24}}{5^6}$

22. $\left(\frac{2y^3}{xz^2}\right)^{-2} = \frac{2^{-2}y^{3(-2)}}{x^{-2}(z^2)^{-2}} = \frac{2^{-2}y^{-6}}{x^{-2}z^{-4}} = \frac{x^2z^4}{4y^6}$

23. 25,430,000

2.5,430,000.

↑________| 7 places

Large number, so exponent is positive.

$25,430,000 = 2.543 \times 10^7$

24. 0.00012

0.0001.2

|____↑ 4 places

Small number, so the exponent is negative.

$0.00012 = 1.2 \times 10^{-4}$

25. 3.6×10^{-5}

Negative exponent, so the answer is a small number.

0.00003.6

↑_____| 5 places

$3.6 \times 10^{-5} = 0.000036$

26. 1.44×10^8

Positive exponent, so the answer is a large number.

1.44000000.

|________↑ 8 places

$1.44 \times 10^8 = 144,000,000$

27. $(3 \times 10^6)(2 \times 10^{-3}) = (3 \times 2) \times (10^6 \times 10^{-3}) =$
6×10^3

28. $\frac{1.2 \times 10^{-4}}{2.4 \times 10^2} = \frac{1.2}{2.4} \times \frac{10^{-4}}{10^2} = 0.5 \times 10^{-6} =$
$(5 \times 10^{-1}) \times 10^{-6} = 5 \times 10^{-7}$

29. $-3x + 7 = -3(-3) + 7 = 9 + 7 = 16;$
$-3x + 7 = -3 \cdot 2 + 7 = -6 + 7 = 1$

30. $x^3 - 2x + 5 = (-3)^3 - 2(-3) + 5 = -27 + 6 + 5 = -16;$
$x^3 - 2x + 5 = 2^3 - 2 \cdot 2 + 5 = 8 - 4 + 5 = 9$

31. $3x - 2x^5 + x - 5x^2 + 2 = (3+1)x - 2x^5 - 5x^2 + 2 =$
$4x - 2x^5 - 5x^2 + 2$, or $-2x^5 - 5x^2 + 4x + 2$

32. $4x^3 - 9x^2 - 2x^3 + x^2 + 8x^6 = (4-2)x^3 + (-9+1)x^2 + 8x^6 =$
$2x^3 - 8x^2 + 8x^6$, or $8x^6 - 2x^3 - 8x^2$

33. The degree of $5x^3$ is 3.
The degree of $-x$, or $-x^1$, is 1.
The degree of 4, or 4^0, is 0.
The degree of the polynomial is 3.

34. The degree of $2x$, or $2x^1$, is 1.
The degree of $-x^4$ is 4.
The degree of $3x^6$ is 6.
The degree of the polynomial is 6.

35. $x - 9$ has two terms. It is a binomial.

36. $x^5 - 2x^3 + 6x^2$ has three terms. It is a trinomial.

37. $(3x^2 - 1) + (5x^2 + 6) = (3+5)x^2 + (-1+6) = 8x^2 + 5$

38. $(x^3 + 2x - 5) + (4x^3 - 2x^2 - 6) =$
$(1+4)x^3 - 2x^2 + 2x + (-5-6) =$
$5x^3 - 2x^2 + 2x - 11$

39. $(5x - 8) - (9x + 2) = 5x - 8 - 9x - 2 = -4x - 10$

40. $(0.1x^2 - 2.4x + 3.6) - (0.5x^2 + x - 5.4)$
$= 0.1x^2 - 2.4x + 3.6 - 0.5x^2 - x + 5.4$
$= -0.4x^2 - 3.4x + 9$

41. The areas of the rectangles are $3y$, y^2, and $2y^2$. The sum of the areas is $3y + y^2 + 2y^2 = 3y + 3y^2$.

42. Let s = the length of a side of the smaller square. Then $3s$ = the length of a side of the larger square. The area of the smaller square is s^2, and the area of the larger square is $(3s)^2$, or $9s^2$, so the area of the larger square is 9 times the area of the smaller square.

43. Let s = the width of the smaller cube. Then $2s$ = the width of the larger cube. The volume of the smaller cube is s^3, and the volume of the larger cube is $(2s)^3$, or $8s^3$, so the volume of the larger cube is 8 times the volume of the smaller cube.

44. Exponents are added when powers with like bases are multiplied. Exponents are multiplied when a power is raised to a power.

45. $3^{-29} = \frac{1}{3^{29}}$ and
$2^{-29} = \frac{1}{2^{29}}$. Since $3^{29} > 2^{29}$, we have $\frac{1}{3^{29}} < \frac{1}{2^{29}}$.

46. It is better to evaluate a polynomial after like terms have been collected, because there are fewer terms to evaluate.

47. Yes; consider the following.

$$(x^2 + 4) + (4x - 7) = x^2 + 4x - 3$$

Exercise Set 4.5

RC1. $8x \cdot 2x = (8 \cdot 2)(x \cdot x) = 16x^2$; answer (a) is correct.

RC3. $2x(8x - 1) = 2x \cdot 8x - 2x \cdot 1 = 16x^2 - 2x$; answer (d) is correct.

1. $(8x^2)(5) = (8 \cdot 5)x^2 = 40x^2$

3. $(-x^2)(-x) = (-1x^2)(-1x) = (-1)(-1)(x^2 \cdot x) = x^3$

5. $(8x^5)(4x^3) = (8 \cdot 4)(x^5 \cdot x^3) = 32x^8$

7. $(0.1x^6)(0.3x^5) = (0.1)(0.3)(x^6 \cdot x^5) = 0.03x^{11}$

9. $\left(-\frac{1}{5}x^3\right)\left(-\frac{1}{3}x\right) = \left(-\frac{1}{5}\right)\left(-\frac{1}{3}\right)(x^3 \cdot x) = \frac{1}{15}x^4$

11. $(-4x^2)(0) = 0$ Any number multiplied by 0 is 0.

13. $(3x^2)(-4x^3)(2x^6) = (3)(-4)(2)(x^2 \cdot x^3 \cdot x^6) = -24x^{11}$

15. $$\begin{aligned} 2x(-x+5) &= 2x(-x) + 2x(5) \\ &= -2x^2 + 10x \end{aligned}$$

17. $$\begin{aligned} -5x(x-1) &= -5x(x) - 5x(-1) \\ &= -5x^2 + 5x \end{aligned}$$

19. $$\begin{aligned} x^2(x^3+1) &= x^2(x^3) + x^2(1) \\ &= x^5 + x^2 \end{aligned}$$

21. $$\begin{aligned} 3x(2x^2 - 6x + 1) &= 3x(2x^2) + 3x(-6x) + 3x(1) \\ &= 6x^3 - 18x^2 + 3x \end{aligned}$$

23. $$\begin{aligned} -6x^2(x^2 + x) &= -6x^2(x^2) - 6x^2(x) \\ &= -6x^4 - 6x^3 \end{aligned}$$

25. $$\begin{aligned} 3y^2(6y^4 + 8y^3) &= 3y^2(6y^4) + 3y^2(8y^3) \\ &= 18y^6 + 24y^5 \end{aligned}$$

27. $$\begin{aligned} (x+6)(x+3) &= (x+6)x + (x+6)3 \\ &= x \cdot x + 6 \cdot x + x \cdot 3 + 6 \cdot 3 \\ &= x^2 + 6x + 3x + 18 \\ &= x^2 + 9x + 18 \end{aligned}$$

29. $$\begin{aligned} (x+5)(x-2) &= (x+5)x + (x+5)(-2) \\ &= x \cdot x + 5 \cdot x + x(-2) + 5(-2) \\ &= x^2 + 5x - 2x - 10 \\ &= x^2 + 3x - 10 \end{aligned}$$

31. $$\begin{aligned} (x-1)(x+4) &= (x-1)x + (x-1)4 \\ &= x \cdot x - 1 \cdot x + x \cdot 4 - 1 \cdot 4 \\ &= x^2 - x + 4x - 4 \\ &= x^2 + 3x - 4 \end{aligned}$$

33. $$\begin{aligned} (x-4)(x-3) &= (x-4)x + (x-4)(-3) \\ &= x \cdot x - 4 \cdot x + x(-3) - 4(-3) \\ &= x^2 - 4x - 3x + 12 \\ &= x^2 - 7x + 12 \end{aligned}$$

35. $$\begin{aligned} (x+3)(x-3) &= (x+3)x + (x+3)(-3) \\ &= x \cdot x + 3 \cdot x + x(-3) + 3(-3) \\ &= x^2 + 3x - 3x - 9 \\ &= x^2 - 9 \end{aligned}$$

37. $$\begin{aligned} (x-4)(x+4) &= (x-4)x + (x-4)4 \\ &= x \cdot x - 4 \cdot x + x \cdot 4 - 4 \cdot 4 \\ &= x^2 - 4x + 4x - 16 \\ &= x^2 - 16 \end{aligned}$$

39. $$\begin{aligned} (5+3x)(2+5x) &= (5+3x)2 + (5+3x)5x \\ &= 5 \cdot 2 + 3x \cdot 2 + 5 \cdot 5x + 3x \cdot 5x \\ &= 10 + 6x + 25x + 15x^2 \\ &= 10 + 31x + 15x^2 \end{aligned}$$

41. $$\begin{aligned} (5-x)(5-2x) &= (5-x)5 + (5-x)(-2x) \\ &= 5 \cdot 5 - x \cdot 5 + 5(-2x) - x(-2x) \\ &= 25 - 5x - 10x + 2x^2 \\ &= 25 - 15x + 2x^2 \end{aligned}$$

43. $$\begin{aligned} (2x+5)(2x+5) &= (2x+5)2x + (2x+5)5 \\ &= 2x \cdot 2x + 5 \cdot 2x + 2x \cdot 5 + 5 \cdot 5 \\ &= 4x^2 + 10x + 10x + 25 \\ &= 4x^2 + 20x + 25 \end{aligned}$$

45. $$\begin{aligned} (x-3)(x-3) &= (x-3)x + (x-3)(-3) \\ &= x \cdot x - 3 \cdot x + x(-3) - 3(-3) \\ &= x^2 - 3x - 3x + 9 \\ &= x^2 - 6x + 9 \end{aligned}$$

47. $$\begin{aligned} \left(x - \frac{5}{2}\right)\left(x + \frac{2}{5}\right) &= \left(x - \frac{5}{2}\right)x + \left(x - \frac{5}{2}\right)\frac{2}{5} \\ &= x \cdot x - \frac{5}{2} \cdot x + x \cdot \frac{2}{5} - \frac{5}{2} \cdot \frac{2}{5} \\ &= x^2 - \frac{5}{2}x + \frac{2}{5}x - 1 \\ &= x^2 - \frac{25}{10}x + \frac{4}{10}x - 1 \\ &= x^2 - \frac{21}{10}x - 1 \end{aligned}$$

49. $$\begin{aligned} (x-2.3)(x+4.7) &= (x-2.3)x + (x-2.3)4.7 \\ &= x \cdot x - 2.3 \cdot x + x \cdot 4.7 - 2.3(4.7) \\ &= x^2 - 2.3x + 4.7x - 10.81 \\ &= x^2 + 2.4x - 10.81 \end{aligned}$$

51. The length of the rectangle is $x+6$ and the width is $x+2$, so the area is $(x+6)(x+2)$. If we carry out the multiplication, we have $x^2 + 8x + 12$.

53. The length of the rectangle is $x+6$ and the width is $x+1$, so the area is $(x+6)(x+1)$. If we carry out the multiplication, we have $x^2 + 7x + 6$.

55. Illustrate $x(x+5)$ as the area of a rectangle with width x and length $x+5$.

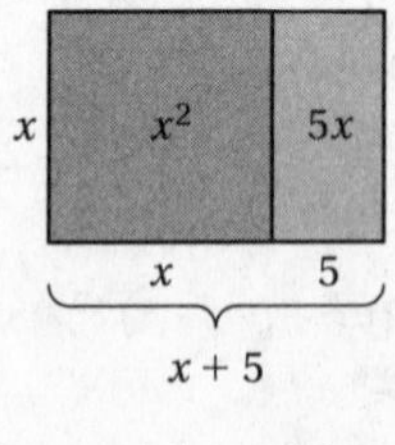

57. Illustrate $(x+1)(x+2)$ as the area of a rectangle with width $x+1$ and length $x+2$.

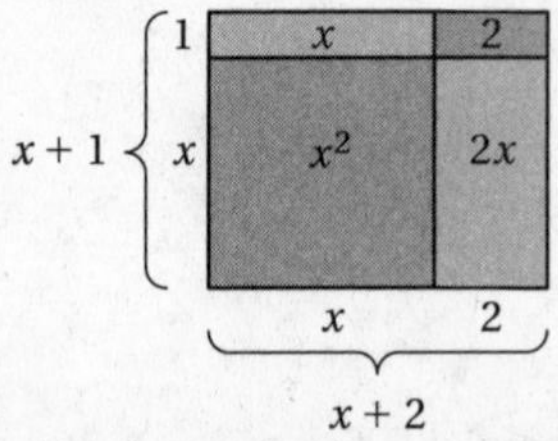

59. Illustrate $(x+5)(x+3)$ as the area of a rectangle with length $x+5$ and width $x+3$.

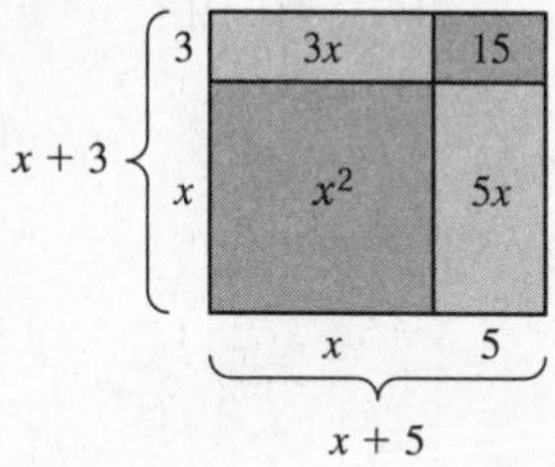

61.
$$\begin{aligned}&(x^2+x+1)(x-1)\\&=(x^2+x+1)x+(x^2+x+1)(-1)\\&=x^2\cdot x+x\cdot x+1\cdot x+x^2(-1)+x(-1)+1(-1)\\&=x^3+x^2+x-x^2-x-1\\&=x^3-1\end{aligned}$$

63.
$$\begin{aligned}&(2x+1)(2x^2+6x+1)\\&=2x(2x^2+6x+1)+1(2x^2+6x+1)\\&=2x\cdot 2x^2+2x\cdot 6x+2x\cdot 1+1\cdot 2x^2+1\cdot 6x+1\cdot 1\\&=4x^3+12x^2+2x+2x^2+6x+1\\&=4x^3+14x^2+8x+1\end{aligned}$$

65.
$$\begin{aligned}&(y^2-3)(3y^2-6y+2)\\&=y^2(3y^2-6y+2)-3(3y^2-6y+2)\\&=y^2\cdot 3y^2+y^2(-6y)+y^2\cdot 2-3\cdot 3y^2-3(-6y)-3\cdot 2\\&=3y^4-6y^3+2y^2-9y^2+18y-6\\&=3y^4-6y^3-7y^2+18y-6\end{aligned}$$

67.
$$\begin{aligned}&(x^3+x^2)(x^3+x^2-x)\\&=x^3(x^3+x^2-x)+x^2(x^3+x^2-x)\\&=x^3\cdot x^3+x^3\cdot x^2+x^3(-x)+x^2\cdot x^3+x^2\cdot x^2+x^2(-x)\\&=x^6+x^5-x^4+x^5+x^4-x^3\\&=x^6+2x^5-x^3\end{aligned}$$

69.
$$\begin{aligned}&(-5x^3-7x^2+1)(2x^2-x)\\&=(-5x^3-7x^2+1)2x^2+(-5x^3-7x^2+1)(-x)\\&=-5x^3\cdot 2x^2-7x^2\cdot 2x^2+1\cdot 2x^2-5x^3(-x)-7x^2(-x)+\\&\quad 1(-x)\\&=-10x^5-14x^4+2x^2+5x^4+7x^3-x\\&=-10x^5-9x^4+7x^3+2x^2-x\end{aligned}$$

71.
$$\begin{array}{rl} 1+x+x^2 & \text{Line up like terms}\\ \underline{-1-x+x^2} & \text{in columns}\\ x^2+x^3+x^4 & \text{Multiplying the top row by } x^2\\ -x-x^2-x^3 & \text{Multiplying by } -x\\ \underline{-1-x-x^2} & \text{Multiplying by } -1\\ -1-2x-x^2+x^4 & \end{array}$$

73.
$$\begin{array}{rl} 2t^2-t-4 & \\ \underline{3t^2+2t-1} & \\ -2t^2+t+4 & \text{Multiplying by } -1\\ 4t^3-2t^2-8t & \text{Multiplying by } 2t\\ \underline{6t^4-3t^3-12t^2} & \text{Multiplying by } 3t^2\\ 6t^4+t^3-16t^2-7t+4 & \end{array}$$

75.
$$\begin{array}{rl} x\quad -x^3\quad +x^5 & \\ \underline{-1+x^2\quad +x^4} & \text{Rewriting in ascending order}\\ x^5-x^7+x^9 & \text{Multiplying by } x^4\\ x^3-x^5+x^7 & \text{Multiplying by } x^2\\ \underline{-x+x^3-x^5} & \text{Multiplying by } -1\\ -x+2x^3-x^5\quad +x^9 & \end{array}$$

77. We will multiply horizontally while still aligning like terms.

$$\begin{array}{rl} (x+1)(x^3+7x^2+5x+4) & \\ =x^4+7x^3+5x^2+4x & \text{Multiplying by } x\\ \underline{+x^3+7x^2+5x+4} & \text{Multiplying by } 1\\ =x^4+8x^3+12x^2+9x+4 & \end{array}$$

79. We will multiply horizontally while still aligning like terms.

$$\begin{array}{r} \left(x-\frac{1}{2}\right)\left(2x^3-4x^2+3x-\frac{2}{5}\right)\\ =2x^4-4x^3+3x^2-\frac{2}{5}x\phantom{+\frac{1}{5}}\\ \underline{-x^3+2x^2-\frac{3}{2}x+\frac{1}{5}}\\ 2x^4-5x^3+5x^2-\frac{19}{10}x+\frac{1}{5} \end{array}$$

81. $x-(2x-3)=x-2x+3=-x+3$

83. $(10-2)(10+2)=8\cdot 12=96$

85. $15x-18y+12=3\cdot 5x-3\cdot 6y+3\cdot 4=$ $3(5x-6y+4)$

87. $-9x-45y+15=-3\cdot 3x-3\cdot 15y-3(-5)=$ $-3(3x+15y-5)$

89. The shaded area is the area of the large rectangle, $6y(14y-5)$ less the area of the unshaded rectangle, $3y(3y+5)$. We have:

$$\begin{aligned}&6y(14y-5)-3y(3y+5)\\&=84y^2-30y-9y^2-15y\\&=75y^2-45y\end{aligned}$$

91. Let $n=$ the missing number.

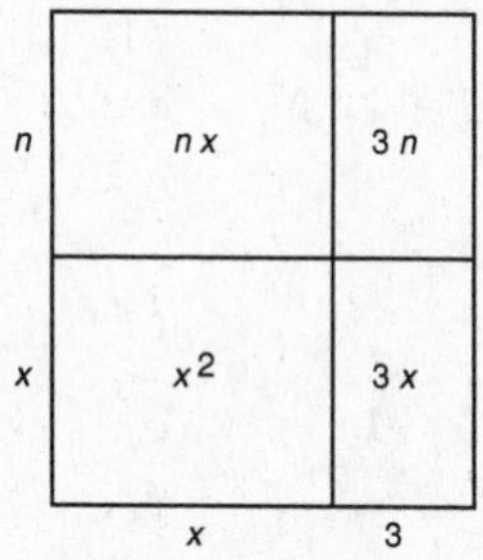

The area of the figure is $x^2+3x+nx+3n$. This is equivalent to $x^2+8x+15$, so we have $3x+nx=8x$ and $3n=15$. Solving either equation for n, we find that the missing number is 5.

93. We have a rectangular solid with dimensions x m by x m by $x+2$ m with a rectangular solid piece with dimensions 6 m by 5 m by 7 m cut out of it.

$$\text{Volume} = \begin{array}{c}\text{Volume of}\\ \text{large solid}\end{array} - \begin{array}{c}\text{Volume of}\\ \text{small solid}\end{array}$$
$$= (x \text{ m})(x \text{ m})(x+2 \text{ m}) - (6 \text{ m})(5 \text{ m})(7 \text{ m})$$
$$= x^2(x+2) \text{ m}^3 - 210 \text{ m}^3$$
$$= (x^3 + 2x^2 - 210) \text{ m}^3$$

95. $(x-2)(x-7)-(x-7)(x-2)$

First observe that, by the commutative law of multiplication, $(x-2)(x-7)$ and $(x-7)(x-2)$ are equivalent expressions. Then when we subtract $(x-7)(x-2)$ from $(x-2)(x-7)$, the result is 0.

97. $(x-a)(x-b)\cdots(x-x)(x-y)(x-z)$
$= (x-a)(x-b)\cdots 0\cdot(x-y)(x-z)$
$= 0$

Exercise Set 4.6

RC1. For the FOIL multiplication method, the initials F O I L represent the words first, <u>outside</u>, inside, and <u>last</u>.

RC3. The expression $(A+B)(A-B)$ is the product of the sum and the <u>difference</u> of the same two terms.

RC5. We can find the product of any two <u>binomials</u> using the FOIL method.

1. $(x+1)(x^2+3)$
F O I L
$= x\cdot x^2 + x\cdot 3 + 1\cdot x^2 + 1\cdot 3$
$= x^3 + 3x + x^2 + 3$

3. $(x^3+2)(x+1)$
F O I L
$= x^3\cdot x + x^3\cdot 1 + 2\cdot x + 2\cdot 1$
$= x^4 + x^3 + 2x + 2$

5. $(y+2)(y-3)$
F O I L
$= y\cdot y + y\cdot(-3) + 2\cdot y + 2\cdot(-3)$
$= y^2 - 3y + 2y - 6$
$= y^2 - y - 6$

7. $(3x+2)(3x+2)$
F O I L
$= 3x\cdot 3x + 3x\cdot 2 + 2\cdot 3x + 2\cdot 2$
$= 9x^2 + 6x + 6x + 4$
$= 9x^2 + 12x + 4$

9. $(5x-6)(x+2)$
F O I L
$= 5x\cdot x + 5x\cdot 2 + (-6)\cdot x + (-6)\cdot 2$
$= 5x^2 + 10x - 6x - 12$
$= 5x^2 + 4x - 12$

11. $(3t-1)(3t+1)$
F O I L
$= 3t\cdot 3t + 3t\cdot 1 + (-1)\cdot 3t + (-1)\cdot 1$
$= 9t^2 + 3t - 3t - 1$
$= 9t^2 - 1$

13. $(4x-2)(x-1)$
F O I L
$= 4x\cdot x + 4x\cdot(-1) + (-2)\cdot x + (-2)\cdot(-1)$
$= 4x^2 - 4x - 2x + 2$
$= 4x^2 - 6x + 2$

15. $\left(p-\frac{1}{4}\right)\left(p+\frac{1}{4}\right)$
F O I L
$= p\cdot p + p\cdot\frac{1}{4} + \left(-\frac{1}{4}\right)\cdot p + \left(-\frac{1}{4}\right)\cdot\frac{1}{4}$
$= p^2 + \frac{1}{4}p - \frac{1}{4}p - \frac{1}{16}$
$= p^2 - \frac{1}{16}$

17. $(x-0.1)(x+0.1)$
F O I L
$= x\cdot x + x\cdot(0.1) + (-0.1)\cdot x + (-0.1)(0.1)$
$= x^2 + 0.1x - 0.1x - 0.01$
$= x^2 - 0.01$

19. $(2x^2+6)(x+1)$
F O I L
$= 2x^3 + 2x^2 + 6x + 6$

21. $(-2x+1)(x+6)$
F O I L
$= -2x^2 - 12x + x + 6$
$= -2x^2 - 11x + 6$

23. $(a+7)(a+7)$
F O I L
$= a^2 + 7a + 7a + 49$
$= a^2 + 14a + 49$

25. $(1+2x)(1-3x)$
F O I L
$= 1 - 3x + 2x - 6x^2$
$= 1 - x - 6x^2$

27. $\left(\frac{3}{8}y-\frac{5}{6}\right)\left(\frac{3}{8}y-\frac{5}{6}\right)$
F O I L
$= \frac{9}{64}y^2 - \frac{15}{48}y - \frac{15}{48}y + \frac{25}{36}$
$= \frac{9}{64}y^2 - \frac{30}{48}y + \frac{25}{36}$
$= \frac{9}{64}y^2 - \frac{5}{8}y + \frac{25}{36}$

29. $(x^2+3)(x^3-1)$
$\quad\text{F}\quad\text{O}\quad\text{I}\quad\text{L}$
$=x^5-x^2+3x^3-3$

31. $(3x^2-2)(x^4-2)$
$\quad\text{F}\quad\text{O}\quad\text{I}\quad\text{L}$
$=3x^6-6x^2-2x^4+4$

33. $(2.8x-1.5)(4.7x+9.3)$
$\quad\text{F}\quad\text{O}\quad\text{I}\quad\text{L}$
$=2.8x(4.7x)+2.8x(9.3)-1.5(4.7x)-1.5(9.3)$
$=13.16x^2+26.04x-7.05x-13.95$
$=13.16x^2+18.99x-13.95$

35. $(3x^5+2)(2x^2+6)$
$\quad\text{F}\quad\text{O}\quad\text{I}\quad\text{L}$
$=6x^7+18x^5+4x^2+12$

37. $(4x^2+3)(x-3)$
$\quad\text{F}\quad\text{O}\quad\text{I}\quad\text{L}$
$=4x^3-12x^2+3x-9$

39. $(4y^4+y^2)(y^2+y)$
$\quad\text{F}\quad\text{O}\quad\text{I}\quad\text{L}$
$=4y^6+4y^5+y^4+y^3$

41. $(x+4)(x-4)$ Product of sum and difference of two terms
$=x^2-4^2$
$=x^2-16$

43. $(2x+1)(2x-1)$ Product of sum and difference of two terms
$=(2x)^2-1^2$
$=4x^2-1$

45. $(5m-2)(5m+2)$ Product of sum and difference of two terms
$=(5m)^2-2^2$
$=25m^2-4$

47. $(2x^2+3)(2x^2-3)$ Product of sum and difference of two terms
$=(2x^2)^2-3^2$
$=4x^4-9$

49. $(3x^4-4)(3x^4+4)$
$=(3x^4)^2-4^2$
$=9x^8-16$

51. $(x^6-x^2)(x^6+x^2)$
$=(x^6)^2-(x^2)^2$
$=x^{12}-x^4$

53. $(x^4+3x)(x^4-3x)$
$=(x^4)^2-(3x)^2$
$=x^8-9x^2$

55. $(x^{12}-3)(x^{12}+3)$
$=(x^{12})^2-3^2$
$=x^{24}-9$

57. $(2y^8+3)(2y^8-3)$
$=(2y^8)^2-3^2$
$=4y^{16}-9$

59. $\left(\frac{5}{8}x-4.3\right)\left(\frac{5}{8}x+4.3\right)$
$=\left(\frac{5}{8}x\right)^2-(4.3)^2$
$=\frac{25}{64}x^2-18.49$

61. $(x+2)^2=x^2+2\cdot x\cdot 2+2^2$ Square of a binomial sum
$=x^2+4x+4$

63. $(3x^2+1)$ Square of a binomial sum
$=(3x^2)^2+2\cdot 3x^2\cdot 1+1^2$
$=9x^4+6x^2+1$

65. $\left(a-\frac{1}{2}\right)^2$ Square of a binomial sum
$=a^2-2\cdot a\cdot\frac{1}{2}+\left(\frac{1}{2}\right)^2$
$=a^2-a+\frac{1}{4}$

67. $(3+x)^2=3^2+2\cdot 3\cdot x+x^2$
$=9+6x+x^2$

69. $(x^2+1)^2=(x^2)^2+2\cdot x^2\cdot 1+1^2$
$=x^4+2x^2+1$

71. $(2-3x^4)^2=2^2-2\cdot 2\cdot 3x^4+(3x^4)^2$
$=4-12x^4+9x^8$

73. $(5+6t^2)^2=5^2+2\cdot 5\cdot 6t^2+(6t^2)^2$
$=25+60t^2+36t^4$

75. $\left(x-\frac{5}{8}\right)^2=x^2-2\cdot x\cdot\frac{5}{8}+\left(\frac{5}{8}\right)^2$
$=x^2-\frac{5}{4}x+\frac{25}{64}$

77. $(3-2x^3)^2=3^2-2\cdot 3\cdot 2x^3+(2x^3)^2$
$=9-12x^3+4x^6$

79. $4x(x^2+6x-3)$ Product of a monomial and a trinomial
$=4x\cdot x^2+4x\cdot 6x+4x(-3)$
$=4x^3+24x^2-12x$

81. $\left(2x^2-\frac{1}{2}\right)\left(2x^2-\frac{1}{2}\right)$ Square of a binomial difference

$= (2x^2)^2 - 2\cdot 2x^2\cdot\frac{1}{2} + \left(\frac{1}{2}\right)^2$

$= 4x^4 - 2x^2 + \frac{1}{4}$

83. $(-1+3p)(1+3p)$

$= (3p-1)(3p+1)$ Product of the sum and difference of two terms

$= (3p)^2 - 1^2$

$= 9p^2 - 1$

85. $3t^2(5t^3 - t^2 + t)$ Product of a monomial and a trinomial

$= 3t^2\cdot 5t^3 + 3t^2(-t^2) + 3t^2\cdot t$

$= 15t^5 - 3t^4 + 3t^3$

87. $(6x^4+4)^2$ Square of a binomial sum

$= (6x^4)^2 + 2\cdot 6x^4\cdot 4 + 4^2$

$= 36x^8 + 48x^4 + 16$

89. $(3x+2)(4x^2+5)$ Product of two binomials; use FOIL

$= 3x\cdot 4x^2 + 3x\cdot 5 + 2\cdot 4x^2 + 2\cdot 5$

$= 12x^3 + 15x + 8x^2 + 10$

91. $(8-6x^4)^2$ Square of a binomial difference

$= 8^2 - 2\cdot 8\cdot 6x^4 + (6x^4)^2$

$= 64 - 96x^4 + 36x^8$

93.

$$\begin{array}{r} t^2+t+1 \\ t-1 \\ \hline -t^2-t-1 \\ t^3+t^2+t \quad\;\; \\ \hline t^3 \qquad\quad -1 \end{array}$$

95. $3^2 + 4^2 = 9 + 16 = 25$

$(3+4)^2 = 7^2 = 49$

97. $9^2 - 5^2 = 81 - 25 = 56$

$(9-5)^2 = 4^2 = 16$

99.

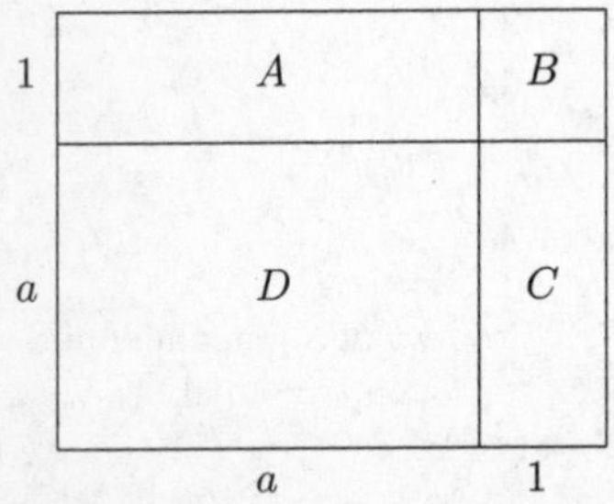

We can find the shaded area in two ways.

Method 1: The figure is a square with side $a+1$, so the area is $(a+1)^2 = a^2 + 2a + 1$.

Method 2: We add the areas of A, B, C, and D.

$1\cdot a + 1\cdot 1 + 1\cdot a + a\cdot a = a + 1 + a + a^2 = a^2 + 2a + 1$.

Either way we find that the total shaded area is $a^2 + 2a + 1$.

101.

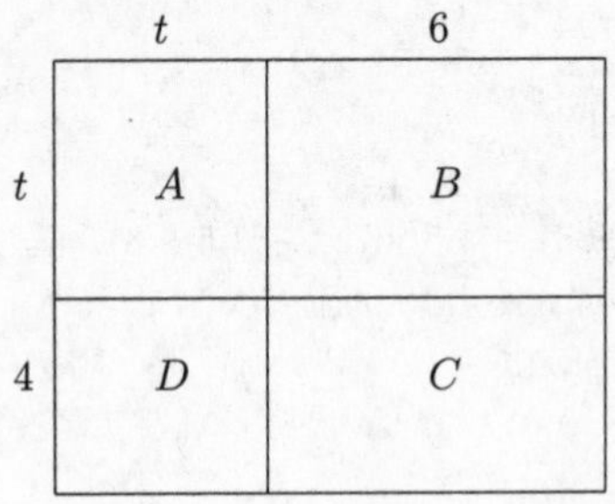

We can find the shaded area in two ways.

Method 1: The figure is a rectangle with dimensions $t+6$ by $t+4$, so the area is $(t+6)(t+4) = t^2 + 4t + 6t + 24 = t^2 + 10t + 24$.

Method 2: We add the areas of A, B, C, and D.

$t\cdot t + t\cdot 6 + 6\cdot 4 + 4\cdot t = t^2 + 6t + 24 + 4t = t^2 + 10t + 24$.

Either way, we find that the total shaded area is $t^2 + 10t + 24$.

103.

$$\begin{aligned} 3x - 8x &= 4(7-8x) \\ -5x &= 28 - 32x \\ 27x &= 28 \\ x &= \frac{28}{27} \end{aligned}$$

The solution is $\frac{28}{27}$.

105.

$$\begin{aligned} 5(2x-3) - 2(3x-4) &= 20 \\ 10x - 15 - 6x + 8 &= 20 \\ 4x - 7 &= 20 \\ 4x &= 27 \\ x &= \frac{27}{4} \end{aligned}$$

The solution is $\frac{27}{4}$.

107.

$$\begin{aligned} C &= ab - r \\ C + r &= ab \\ \frac{C+r}{a} &= b \end{aligned}$$

109. $5x(3x-1)(2x+3)$

$= 5x(6x^2 + 7x - 3)$ Using FOIL

$= 30x^3 + 35x^2 - 15x$

111. $[(a-5)(a+5)]^2$

$= (a^2 - 25)^2$ Finding the product of a sum and difference of same two terms

$= a^4 - 50a^2 + 625$ Squaring a binomial

113. $(3t^4-2)^2(3t^4+2)^2$

$= [(3t^4-2)(3t^4+2)]^2$

$= (9t^8 - 4)^2$

$= 81t^{16} - 72t^8 + 16$

115.
$$\begin{aligned}(x+2)(x-5) &= (x+1)(x-3)\\ x^2-5x+2x-10 &= x^2-3x+x-3\\ x^2-3x-10 &= x^2-2x-3\\ -3x-10 &= -2x-3 \quad \text{Adding } -x^2\\ -3x+2x &= 10-3 \quad \text{Adding } 2x \text{ and } 10\\ -x &= 7\\ x &= -7\end{aligned}$$
The solution is -7.

117. See the answer section in the text.

119. Enter $y_1 = (x-1)^2$ and $y_2 = x^2 - 2x + 1$. Then compare the graphs or the y_1-and y_2-values in a table. It appears that the graphs are the same and that the y_1-and y_2-values are the same, so $(x-1)^2 = x^2 - 2x + 1$ is correct.

121. Enter $y_1 = (x-3)(x+3)$ and $y_2 = x^2 - 6$. Then compare the graphs or the y_1-and y_2-values in a table. The graphs are not the same nor are the y_1-and y_2-values, so $(x-3)(x+3) = x^2 - 6$ is not correct.

Exercise Set 4.7

RC1. True

RC3. The variables do not have exactly the same exponents, so the statement is false.

1. We replace x by 3 and y by -2.

$x^2 - y^2 + xy = 3^2 - (-2)^2 + 3(-2) = 9 - 4 - 6 = -1$

3. We replace x by 3 and y by -2.

$x^2 - 3y^2 + 2xy = 3^2 - 3(-2)^2 + 2\cdot 3(-2) =$
$9 - 3\cdot 4 + 2\cdot 3(-2) = 9 - 12 - 12 = -15$

5. We replace x by 3, y by -2, and z by -5.

$8xyz = 8\cdot 3\cdot(-2)\cdot(-5) = 240$

7. We replace x by 3, y by -2, and z by -5.

$xyz^2 - z = 3(-2)(-5)^2 - (-5) = 3(-2)(25) - (-5) =$
$-150 + 5 = -145$

9. We replace h by 165 and A by 20.

$$\begin{aligned}C &= 0.041h - 0.018A - 2.69\\ &= 0.041(165) - 0.018(20) - 2.69\\ &= 6.765 - 0.36 - 2.69\\ &= 6.405 - 2.69\\ &= 3.715\end{aligned}$$

The lung capacity of a 20-year-old person who is 165 cm tall is 3.715 liters.

11. Replace w by 82, h by 185, and a by 67.

$$\begin{aligned}&19.18w + 7h - 9.52a + 92.4\\ &= 19.18(82) + 7(185) - 9.52(67) + 92.4\\ &= 1572.76 + 1295 - 637.84 + 92.4\\ &\approx 2322 \text{ calories}\end{aligned}$$

13. Replace h by 4.7, r by 1.2, and π by 3.14.

$$\begin{aligned}S &= 2\pi rh + 2\pi r^2\\ &\approx 2(3.14)(1.2)(4.7) + 2(3.14)(1.2)^2\\ &\approx 2(3.14)(1.2)(4.7) + 2(3.14)(1.44)\\ &\approx 35.4192 + 9.0432\\ &\approx 44.46\end{aligned}$$

The surface area of the can is about 44.46 in^2.

15. Evaluate the polynomial for $h = 7$, $r = 1\frac{1}{2}$, or $\frac{3}{2}$, and $\pi \approx 3.14$.

$$\begin{aligned}S &= 2\pi rh + \pi r^2\\ &\approx 2(3.14)\left(\frac{3}{2}\right)(7) + (3.14)\left(\frac{3}{2}\right)^2\\ &\approx 2(3.14)\left(\frac{3}{2}\right)(7) + (3.14)\left(\frac{9}{4}\right)\\ &\approx 65.94 + 7.065\\ &\approx 73.005\end{aligned}$$

The surface area is about 73.005 in^2.

17. $x^3y - 2xy + 3x^2 - 5$

Term	Coefficient	Degree	
x^3y	1	4	(Think: $x^3y = x^3y^1$)
$-2xy$	-2	2	(Think: $-2xy = -2x^1y^1$)
$3x^2$	3	2	
-5	-5	0	(Think: $-5 = -5x^0$)

The degree of the polynomial is the degree of the term of highest degree. The term of highest degree is x^3y. Its degree is 4. The degree of the polynomial is 4.

19. $17x^2y^3 - 3x^3yz - 7$

Term	Coefficient	Degree	
$17x^2y^3$	17	5	
$-3x^3yz$	-3	5	(Think: $-3x^3yz = -3x^3y^1z^1$)
-7	-7	0	(Think: $-7 = -7x^0$)

The terms of highest degree are $17x^2y^3$ and $-3x^3yz$. Each has degree 5. The degree of the polynomial is 5.

21. $a + b - 2a - 3b = (1-2)a + (1-3)b = -a - 2b$

23. $3x^2y - 2xy^2 + x^2$

There are *no* like terms, so none of the terms can be collected.

25.
$$\begin{aligned}&6au + 3av + 14au + 7av\\ &= (6+14)au + (3+7)av\\ &= 20au + 10av\end{aligned}$$

27.
$$\begin{aligned}&2u^2v - 3uv^2 + 6u^2v - 2uv^2\\ &= (2+6)u^2v + (-3-2)uv^2\\ &= 8u^2v - 5uv^2\end{aligned}$$

29.
$$\begin{aligned}&(2x^2 - xy + y^2) + (-x^2 - 3xy + 2y^2)\\ &= (2-1)x^2 + (-1-3)xy + (1+2)y^2\\ &= x^2 - 4xy + 3y^2\end{aligned}$$

31. $(r-2s+3)+(2r+s)+(s+4)$
$= (1+2)r+(-2+1+1)s+(3+4)$
$= 3r+0s+7$
$= 3r+7$

33. $(b^3a^2-2b^2a^3+3ba+4)+(b^2a^3-4b^3a^2+2ba-1)$
$= (1-4)b^3a^2+(-2+1)b^2a^3+(3+2)ba+(4-1)$
$= -3b^3a^2-b^2a^3+5ba+3$

35. $(a^3+b^3)-(a^2b-ab^2+b^3+a^3)$
$= a^3+b^3-a^2b+ab^2-b^3-a^3$
$= (1-1)a^3-a^2b+ab^2+(1-1)b^3$
$= -a^2b+ab^2$, or ab^2-a^2b

37. $(xy-ab-8)-(xy-3ab-6)$
$= xy-ab-8-xy+3ab+6$
$= (1-1)xy+(-1+3)ab+(-8+6)$
$= 2ab-2$

39. $(-2a+7b-c)-(-3b+4c-8d)$
$= -2a+7b-c+3b-4c+8d$
$= -2a+(7+3)b+(-1-4)c+8d$
$= -2a+10b-5c+8d$

41. $(3z-u)(2z+3u) = \overset{\text{F}}{6z^2} + \overset{\text{O}}{9zu} - \overset{\text{I}}{2uz} - \overset{\text{L}}{3u^2}$
$= 6z^2+7zu-3u^2$

43. $(a^2b-2)(a^2b-5) = \overset{\text{F}}{a^4b^2} - \overset{\text{O}}{5a^2b} - \overset{\text{I}}{2a^2b} + \overset{\text{L}}{10}$
$= a^4b^2-7a^2b+10$

45. $(a^3+bc)(a^3-bc) = (a^3)^2-(bc)^2$
$[(A+B)(A-B)=A^2-B^2]$
$= a^6-b^2c^2$

47.
$$\begin{array}{r} y^4x+y^2+1 \\ y^2+1 \\ \hline y^4x+y^2+1 \\ y^6x+y^4 \quad +y^2 \quad\;\; \\ \hline y^6x+y^4+y^4x+2y^2+1 \end{array}$$

49. $(3xy-1)(4xy+2)$
$= \overset{\text{F}}{12x^2y^2} + \overset{\text{O}}{6xy} - \overset{\text{I}}{4xy} - \overset{\text{L}}{2}$
$= 12x^2y^2+2xy-2$

51. $(3-c^2d^2)(4+c^2d^2)$
$= \overset{\text{F}}{12} + \overset{\text{O}}{3c^2d^2} - \overset{\text{I}}{4c^2d^2} - \overset{\text{L}}{c^4d^4}$
$= 12-c^2d^2-c^4d^4$

53. $(m^2-n^2)(m+n)$
$= \overset{\text{F}}{m^3} + \overset{\text{O}}{m^2n} - \overset{\text{I}}{mn^2} - \overset{\text{L}}{n^3}$

55. $(xy+x^5y^5)(x^4y^4-xy)$
$= \overset{\text{F}}{x^5y^5} - \overset{\text{O}}{x^2y^2} + \overset{\text{I}}{x^9y^9} - \overset{\text{L}}{x^6y^6}$
$= x^9y^9-x^6y^6+x^5y^5-x^2y^2$

57. $(x+h)^2$
$= x^2+2xh+h^2 \quad [(A+B)^2=A^2+2AB+B^2]$

59. $(3a+2b)^2$
$= (3a)^2+2\cdot 3a\cdot 2b+(2b)^2$
$[(A+B)^2=A^2+2AB+B^2]$
$= 9a^2+12ab+4b^2$

61. $(r^3t^2-4)^2$
$= (r^3t^2)^2-2\cdot r^3t^2\cdot 4+4^2$
$[(A-B)^2=A^2-2AB+B^2]$
$= r^6t^4-8r^3t^2+16$

63. $(p^4+m^2n^2)^2$
$= (p^4)^2+2\cdot p^4\cdot m^2n^2+(m^2n^2)^2$
$[(A+B)^2=A^2+2AB+B^2]$
$= p^8+2p^4m^2n^2+m^4n^4$

65. $3a(a-2b)^2 = 3a(a^2-4ab+4b^2)$
$= 3a^3-12a^2b+12ab^2$

67. $(m+n-3)^2$
$= [(m+n)-3]^2$
$= (m+n)^2-2(m+n)(3)+3^2$
$= m^2+2mn+n^2-6m-6n+9$

69. $(a+b)(a-b)=a^2-b^2$

71. $(2a-b)(2a+b)=(2a)^2-b^2=4a^2-b^2$

73. $(c^2-d)(c^2+d)=(c^2)^2-d^2$
$= c^4-d^2$

75. $(ab+cd^2)(ab-cd^2)=(ab)^2-(cd^2)^2$
$= a^2b^2-c^2d^4$

77. $(x+y-3)(x+y+3)$
$= [(x+y)-3][(x+y)+3]$
$= (x+y)^2-3^2$
$= x^2+2xy+y^2-9$

79. $[x+y+z][x-(y+z)]$
$= [x+(y+z)][x-(y+z)]$
$= x^2-(y+z)^2$
$= x^2-(y^2+2yz+z^2)$
$= x^2-y^2-2yz-z^2$

81. $(a+b+c)(a-b-c)$
$= [a+(b+c)][a-(b+c)]$
$= a^2-(b+c)^2$
$= a^2-(b^2+2bc+c^2)$
$= a^2-b^2-2bc-c^2$

83.
$$\begin{array}{llllll} x^2 & -\;4y & +\;2 & & & \\ 3x^2 & +\;5y & -\;3 & & & \\ \hline -3x^2 & +\;12y & -\;6 & & & \\ & +\;10y & & -\;20y^2 & +\;5x^2y & \\ 6x^2 & & & & -\;12x^2y & +\;3x^4 \\ \hline 3x^2 & +\;22y & -\;6 & -\;20y^2 & -\;7x^2y & +\;3x^4 \end{array}$$

We could also write the result as
$3x^4-7x^2y+3x^2-20y^2+22y-6$.

85. The first coordinate is positive and the second coordinate is negative, so $(2, -5)$ is in quadrant IV.

87. Both coordinates are positive, so $(16, 23)$ is in quadrant I.

89. -39 is 39 units from 0, so $|-39| = 39$.

91. -17 lies to the left of -5 on the number line, so $-17 < -5$.

93. It is helpful to add additional labels to the figure.

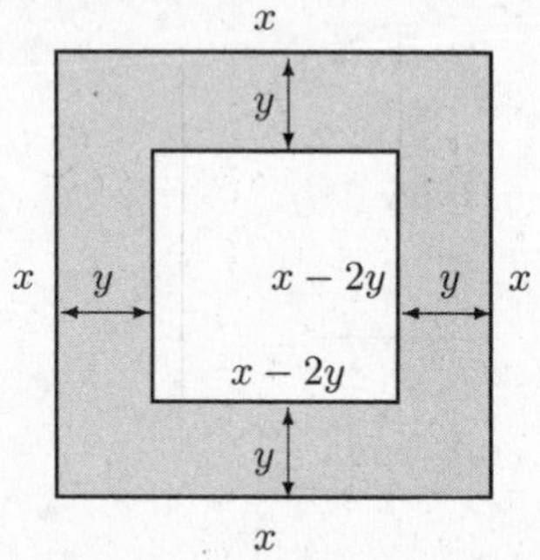

The area of the large square is $x \cdot x$, or x^2. The area of the small square is $(x-2y)(x-2y)$, or $(x-2y)^2$.

$$\text{Area of shaded region} = \text{Area of large square} - \text{Area of small square}$$

$$\text{Area of shaded region} = x^2 - (x-2y)^2$$

$$= x^2 - (x^2 - 4xy + 4y^2)$$
$$= x^2 - x^2 + 4xy - 4y^2$$
$$= 4xy - 4y^2$$

95. It is helpful to add additional labels to the figure.

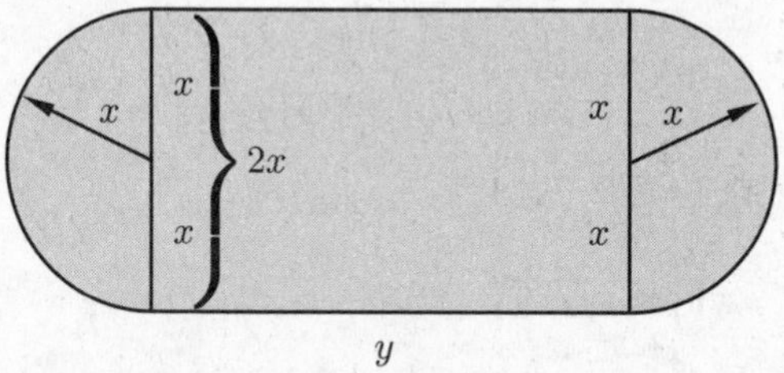

The two semicircles make a circle with radius x. The area of that circle is πx^2. The area of the rectangle is $2x \cdot y$. The sum of the two regions, $\pi x^2 + 2xy$, is the area of the shaded region.

97. The lateral surface area of the outer portion of the solid is the lateral surface area of a right circular cylinder with radius n and height h. The lateral surface area of the inner portion is the lateral surface area of a right circular cylinder with radius m and height h. Recall that the formula for the lateral surface area of a right circular cylinder with radius r and height h is $2\pi rh$.

The surface area of the top is the area of a circle with radius n less the area of a circle with radius m. The surface area of the bottom is the same as the surface area of the top.

Thus, the surface area of the solid is

$$2\pi nh + 2\pi mh + 2\pi n^2 - 2\pi m^2.$$

99. In the formula for the surface area of a silo, $S = 2\pi rh + \pi r^2$, the term πr^2 represents the area of the base. Since the base of the observatory rests on the ground, it will not need to be painted. Thus, we will subtract this term from the formula and find the remaining surface area, $2\pi rh$.

The height of the observatory is 40 ft and its radius is 30/2, or 15 ft, so the surface area is $2\pi rh \approx 2(3.14)(15)(40) \approx 3768\ \text{ft}^2$. Since $3768\ \text{ft}^2/250\ \text{ft}^2 = 15.072$, 16 gallons of paint should be purchased.

101. Substitute \$10,400 for P, 3.5% or 0.035 for r, and 5 for t.

$$P(1+r)^t$$
$$= \$10,400(1+0.035)^5$$
$$= \$10,400(1.035)^5$$
$$\approx \$12,351.94$$

Exercise Set 4.8

RC1. When dividing a monomial by a monomial, we <u>subtract</u> exponents and divide coefficients.

RC3. To carry out long division, we repeat the following process: divide, <u>multiply</u>, <u>subtract</u>, and bring down the next term.

1. $\dfrac{24x^4}{8} = \dfrac{24}{8} \cdot x^4 = 3x^4$

Check: We multiply.

$$3x^4 \cdot 8 = 24x^4$$

3. $\dfrac{25x^3}{5x^2} = \dfrac{25}{5} \cdot \dfrac{x^3}{x^2} = 5x^{3-2} = 5x$

Check: We multiply.

$$5x \cdot 5x^2 = 25x^3$$

5. $\dfrac{-54x^{11}}{-3x^8} = \dfrac{-54}{-3} \cdot \dfrac{x^{11}}{x^8} = 18x^{11-8} = 18x^3$

Check: We multiply.

$$18x^3(-3x^8) = -54x^{11}$$

7. $\dfrac{64a^5b^4}{16a^2b^3} = \dfrac{64}{16} \cdot \dfrac{a^5}{a^2} \cdot \dfrac{b^4}{b^3} = 4a^{5-2}b^{4-3} = 4a^3b$

Check: We multiply.

$$(4a^3b)(16a^2b^3) = 64a^5b^4$$

9. $\dfrac{24x^4 - 4x^3 + x^2 - 16}{8}$

$$= \frac{24x^4}{8} - \frac{4x^3}{8} + \frac{x^2}{8} - \frac{16}{8}$$
$$= 3x^4 - \frac{1}{2}x^3 + \frac{1}{8}x^2 - 2$$

Check: We multiply.

$$\begin{array}{r} 3x^4 - \frac{1}{2}x^3 + \frac{1}{8}x^2 - 2 \\ 8 \\ \hline 24x^4 - 4x^3 + x^2 - 16 \end{array}$$

11. $\dfrac{u-2u^2-u^5}{u}$

$= \dfrac{u}{u} - \dfrac{2u^2}{u} - \dfrac{u^5}{u}$

$= 1-2u-u^4$

Check: We multiply.

$$\begin{array}{r} 1 - 2u - u^4 \\ u \\ \hline u - 2u^2 - u^5 \end{array}$$

13. $(15t^3+24t^2-6t)\div(3t)$

$= \dfrac{15t^3+24t^2-6t}{3t}$

$= \dfrac{15t^3}{3t} + \dfrac{24t^2}{3t} - \dfrac{6t}{3t}$

$= 5t^2+8t-2$

Check: We multiply.

$$\begin{array}{r} 5t^2 + 8t - 2 \\ 3t \\ \hline 15t^3 + 24t^2 - 6t \end{array}$$

15. $(20x^6-20x^4-5x^2)\div(-5x^2)$

$= \dfrac{20x^6-20x^4-5x^2}{-5x^2}$

$= \dfrac{20x^6}{-5x^2} - \dfrac{20x^4}{-5x^2} - \dfrac{5x^2}{-5x^2}$

$= -4x^4-(-4x^2)-(-1)$

$= -4x^4+4x^2+1$

Check: We multiply.

$$\begin{array}{r} -4x^4 + 4x^2 + 1 \\ -5x^2 \\ \hline 20x^6 - 20x^4 - 5x^2 \end{array}$$

17. $(24x^5-40x^4+6x^3)\div(4x^3)$

$= \dfrac{24x^5-40x^4+6x^3}{4x^3}$

$= \dfrac{24x^5}{4x^3} - \dfrac{40x^4}{4x^3} + \dfrac{6x^3}{4x^3}$

$= 6x^2-10x+\dfrac{3}{2}$

Check: We multiply.

$$\begin{array}{r} 6x^2 - 10x + \frac{3}{2} \\ 4x^3 \\ \hline 24x^5 - 40x^4 + 6x^3 \end{array}$$

19. $\dfrac{18x^2-5x+2}{2}$

$= \dfrac{18x^2}{2} - \dfrac{5x}{2} + \dfrac{2}{2}$

$= 9x^2 - \dfrac{5}{2}x + 1$

Check: We multiply.

$$\begin{array}{r} 9x^2 - \frac{5}{2}x + 1 \\ 2 \\ \hline 18x^2 - 5x + 2 \end{array}$$

21. $\dfrac{12x^3+26x^2+8x}{2x}$

$= \dfrac{12x^3}{2x} + \dfrac{26x^2}{2x} + \dfrac{8x}{2x}$

$= 6x^2+13x+4$

Check: We multiply.

$$\begin{array}{r} 6x^2 + 13x + 4 \\ 2x \\ \hline 12x^3 + 26x^2 + 8x \end{array}$$

23. $\dfrac{9r^2s^2+3r^2s-6rs^2}{3rs}$

$= \dfrac{9r^2s^2}{3rs} + \dfrac{3r^2s}{3rs} - \dfrac{6rs^2}{3rs}$

$= 3rs+r-2s$

Check: We multiply.

$$\begin{array}{r} 3rs + r - 2s \\ 3rs \\ \hline 9r^2s^2 + 3r^2s - 6rs^2 \end{array}$$

25.
$$\begin{array}{rl} & x+2 \\ x+2 \,\big)\!\! & \overline{x^2+4x+4} \\ & \underline{x^2+2x} \\ & \quad 2x+4 \leftarrow (x^2+4x)-(x^2+2x) \\ & \quad \underline{2x+4} \\ & \qquad 0 \leftarrow (2x+4)-(2x+4) \end{array}$$

The answer is $x+2$.

27.
$$\begin{array}{rl} & x-5 \\ x-5 \,\big)\!\! & \overline{x^2-10x-25} \\ & \underline{x^2-5x} \\ & \quad -5x-25 \leftarrow (x^2-10x)-(x^2-5x) \\ & \quad \underline{-5x+25} \\ & \qquad -50 \leftarrow (-5x-25)-(-5x+25) \end{array}$$

The answer is $x-5+\dfrac{-50}{x-5}$.

29.
$$\begin{array}{rl} & x-2 \\ x+6 \,\big)\!\! & \overline{x^2+4x-14} \\ & \underline{x^2+6x} \\ & \quad -2x-14 \leftarrow (x^2+4x)-(x^2+6x) \\ & \quad \underline{-2x-12} \\ & \qquad -2 \leftarrow (-2x-14)-(-2x-12) \end{array}$$

The answer is $x-2+\dfrac{-2}{x+6}$.

31.
$$\begin{array}{rl} & x-3 \\ x+3 \,\big)\!\! & \overline{x^2+0x-9} \leftarrow \text{Filling in the missing term} \\ & \underline{x^2+3x} \\ & \quad -3x-9 \leftarrow x^2-(x^2+3x) \\ & \quad \underline{-3x-9} \\ & \qquad 0 \leftarrow (-3x-9)-(-3x-9) \end{array}$$

The answer is $x - 3$.

33.

$$x^4 - x^3 + x^2 - x + 1$$
$$x+1 \overline{\big)\, x^5+0x^4+0x^3+0x^2+0x+1} \leftarrow \text{Filling in missing terms}$$
$$\underline{x^5 + x^4}$$
$$-x^4 \quad \leftarrow x^5 - (x^5 + x^4)$$
$$\underline{-x^4 - x^3}$$
$$x^3 \quad \leftarrow -x^4 - (-x^4 - x^3)$$
$$\underline{x^3 + x^2}$$
$$-x^2 \quad \leftarrow x^3 - (x^3 + x^2)$$
$$\underline{-x^2 - x}$$
$$x + 1 \quad \leftarrow -x^2 - (-x^2 - x)$$
$$\underline{x + 1}$$
$$0 \leftarrow (x+1) - (x+1)$$

The answer is $x^4 - x^3 + x^2 - x + 1$.

35.

$$2x^2 - 7x + 4$$
$$4x+3 \overline{\big)\, 8x^3 - 22x^2 - 5x + 12}$$
$$\underline{8x^3 + 6x^2}$$
$$-28x^2 - 5x \quad \leftarrow (8x^3 - 22x^2) - (8x^3 + 6x^2)$$
$$\underline{-28x^2 - 21x}$$
$$16x + 12 \leftarrow (-28x^2 - 5x) - (-28x^2 - 21x)$$
$$\underline{16x + 12}$$
$$0 \leftarrow (16x + 12) - (16x + 12)$$

The answer is $2x^2 - 7x + 4$.

37.

$$x^3 - 6$$
$$x^3 - 7 \overline{\big)\, x^6 - 13x^3 + 42}$$
$$\underline{x^6 - 7x^3}$$
$$-6x^3 + 42 \leftarrow (x^6 - 13x^3) - (x^6 - 7x^3)$$
$$\underline{-6x^3 + 42}$$
$$0 \leftarrow (-6x^3 + 42) - (-6x^3 + 42)$$

The answer is $x^3 - 6$.

39.

$$t^2 + 1$$
$$t - 1 \overline{\big)\, t^3 - t^2 + t - 1}$$
$$\underline{t^3 - t^2} \quad \leftarrow (t^3 - t^2) - (t^3 - t^2)$$
$$0 + t - 1$$
$$\underline{t - 1} \leftarrow (t - 1) - (t - 1)$$
$$0$$

The answer is $t^2 + 1$.

41.

$$y^2 - 3y + 1$$
$$y + 2 \overline{\big)\, y^3 - y^2 - 5y - 3}$$
$$\underline{y^3 + 2y^2}$$
$$-3y^2 - 5y$$
$$\underline{-3y^2 - 6y}$$
$$y - 3$$
$$\underline{y + 2}$$
$$-5$$

The answer is $y^2 - 3y + 1 + \dfrac{-5}{y+2}$.

43.

$$3x^2 + x + 2$$
$$5x + 1 \overline{\big)\, 15x^3 + 8x^2 + 11x + 12}$$
$$\underline{15x^3 + 3x^2}$$
$$5x^2 + 11x$$
$$\underline{5x^2 + x}$$
$$10x + 12$$
$$\underline{10x + 2}$$
$$10$$

The answer is $3x^2 + x + 2 + \dfrac{10}{5x+1}$.

45.

$$6y^2 - 5$$
$$2y + 7 \overline{\big)\, 12y^3 + 42y^2 - 10y - 41}$$
$$\underline{12y^3 + 42y^2}$$
$$-10y - 41$$
$$\underline{-10y - 35}$$
$$-6$$

The answer is $6y^2 - 5 + \dfrac{-6}{2y+7}$.

47. $-13 = 8d - 5$

$-8 = 8d$ Adding 5

$-1 = d$ Dividing by 8

The solution is -1.

49. $4(x - 3) = 5(2 - 3x) + 1$

$4x - 12 = 10 - 15x + 1$ Removing parentheses

$4x - 12 = 11 - 15x$ Collecting like terms

$19x - 12 = 11$ Adding $15x$

$19x = 23$ Adding 12

$x = \dfrac{23}{19}$ Dividing by 19

The solution is $\dfrac{23}{19}$.

51. Amount of increase: $60 - 25 = 35$

Percent increase: $\dfrac{35}{25} = 1.4 = 140\%$

53. ***Familiarize.*** Let $w =$ the width. Then $w + 15 =$ the length. We draw a picture.

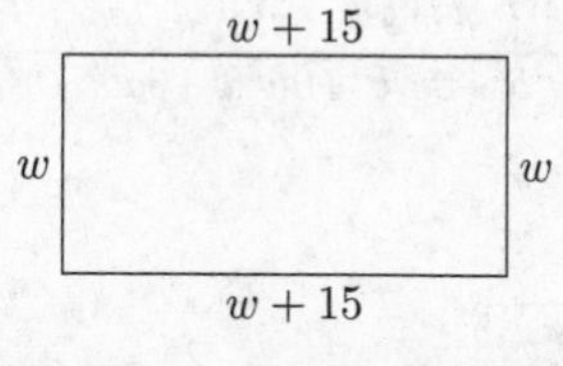

We will use the fact that the perimeter is 640 ft to find w (the width). Then we can find $w + 15$ (the length) and multiply the length and the width to find the area.

Translate.

Width + Width + Length + Length = Perimeter

$w + w + (w + 15) + (w + 15) = 640$

Solve.

$$\begin{aligned} w+w+(w+15)+(w+15) &= 640 \\ 4w+30 &= 640 \\ 4w &= 610 \\ w &= 152.5 \end{aligned}$$

If the width is 152.5, then the length is 152.5+15, or 167.5. The area is (167.5)(152.5), or 25, 543.75 ft^2.

Check. The length, 167.5 ft, is 15 ft greater than the width, 152.5 ft. The perimeter is $152.5+152.5+167.5+167.5$, or 640 ft. We should also recheck the computation we used to find the area. The answer checks.

State. The area is 25,543.75 ft^2.

55.

$$\begin{array}{r} x^2+5 \\ x^2+4\,\overline{)\,x^4+9x^2+20} \\ \underline{x^4+4x^2} \\ 5x^2+20 \\ \underline{5x^2+20} \\ 0 \end{array}$$

The answer is x^2+5.

57.

$$\begin{array}{r} a+3 \\ 5a^2-7a-2\,\overline{)\,5a^3+8a^2-23a-1} \\ \underline{5a^3-7a^2-2a} \\ 15a^2-21a-1 \\ \underline{15a^2-21a-6} \\ 5 \end{array}$$

The answer is $a+3+\dfrac{5}{5a^2-7a-2}$.

59. We rewrite the dividend in descending order.

$$\begin{array}{r} 2x^2+x-3 \\ 3x^3-2x-1\,\overline{)\,6x^5+3x^4-13x^3-4x^2+5x+3} \\ \underline{6x^5\qquad-4x^3-2x^2\qquad\qquad} \\ 3x^4-9x^3-2x^2+5x\quad \\ \underline{3x^4\qquad-2x^2-x\quad} \\ -9x^3\qquad+6x+3 \\ \underline{-9x^3\qquad+6x+3} \\ 0 \end{array}$$

The answer is $2x^2+x-3$.

61.

$$\begin{array}{r} a^5+a^4b+a^3b^2+a^2b^3+ab^4+b^5 \\ a-b\,\overline{)\,a^6+0a^5b+0a^4b^2+0a^3b^3+0a^2b^4+0ab^5-b^6} \\ \underline{a^6-a^5b} \\ a^5b \\ \underline{a^5b-a^4b^2} \\ a^4b^2 \\ \underline{a^4b^2-a^3b^3} \\ a^3b^3 \\ \underline{a^3b^3-a^2b^4} \\ a^2b^4 \\ \underline{a^2b^4-ab^5} \\ ab^5-b^6 \\ \underline{ab^5-b^6} \\ 0 \end{array}$$

The answer is $a^5+a^4b+a^3b^2+a^2b^3+ab^4+b^5$.

63.

$$\begin{array}{r} x+5 \\ x-1\,\overline{)\,x^2+4x+c} \\ \underline{x^2-x\quad} \\ 5x+c \\ \underline{5x-5} \\ c+5 \end{array}$$

We set the remainder equal to 0.

$$\begin{aligned} c+5 &= 0 \\ c &= -5 \end{aligned}$$

Thus, c must be -5.

65.

$$\begin{array}{r} c^2x+(-2c+c^2) \\ x-1\,\overline{)\,c^2x^2-2cx+1} \\ \underline{c^2x^2-c^2x} \\ (-2c+c^2)x+1 \\ \underline{(-2c+c^2)x-(-2c+c^2)} \\ 1+(-2c+c^2) \end{array}$$

We set the remainder equal to 0.

$$\begin{aligned} c^2-2c+1 &= 0 \\ (c-1)^2 &= 0 \\ c &= 1 \end{aligned}$$

Thus, c must be 1.

Chapter 4 Vocabulary Reinforcement

1. In the expression 7^5, the number 5 is the exponent.

2. The product rule asserts that when multiplying with exponential notation, if the bases are the same, keep the base and add the exponents.

3. An expression of the type ax^n, where a is a real-number constant and n is a nonnegative integer, is a monomial.

4. A trinomial is a polynomial with three terms, such as $5x^4-7x^2+4$.

5. The quotient rule asserts that when dividing with exponential notation, if the bases are the same, keep the base and subtract the exponent of the denominator from the exponent of the numerator.

6. If the exponents in a polynomial decrease from left to right, the polynomial is arranged in descending order.

7. The degree of a term is the sum of the exponents of the variables.

8. The number 2.3×10^{-5} is written in scientific notation.

Chapter 4 Concept Reinforcement

1. True; a trinomial is a polynomial with three terms.

2. False; $(x+y)^2=x^2+2xy+y^2\neq x^2+y^2$.

3. False; $(A-B)^2=A^2-2AB+B^2\neq A^2-B^2$.

4. True; $(A+B)(A-B) = A^2 - AB + AB - B^2 = A^2 - B^2$.

Chapter 4 Study Guide

1. $z^5 \cdot z^3 = z^{5+3} = z^8$

2. $\dfrac{a^4b^7}{a^2b} = \dfrac{a^4}{a^2} \cdot \dfrac{b^7}{b} = a^{4-2}b^{7-1} = a^2b^6$

3.
$$\begin{aligned}\left(\frac{x^{-4}y^2}{3z^3}\right)^3 &= \frac{(x^{-4}y^2)^3}{(3z^3)^3}\\ &= \frac{(x^{-4})^3(y^2)^3}{3^3(z^3)^3}\\ &= \frac{x^{-4\cdot 3}y^{2\cdot 3}}{27z^{3\cdot 3}}\\ &= \frac{x^{-12}y^6}{27z^9}\\ &= \frac{y^6}{27x^{12}z^9}\end{aligned}$$

4. 763,000

7.63,000.

↑ 5 places

Large number, so the exponent is positive.

$763,000 = 7.63 \times 10^5$

5. 3×10^{-4}

The exponent is negative so the number is small.

0.0003.

↑ 4 places

$3 \times 10^{-4} = 0.0003$

6.
$$\begin{aligned}\frac{3.6\times 10^3}{6.0\times 10^{-2}} &= \frac{3.6}{6.0} \times \frac{10^3}{10^{-2}} = 0.6 \times 10^5\\ &= (6\times 10^{-1}) \times 10^5 = 6\times 10^4\end{aligned}$$

7.
$$\begin{aligned}5x^4 - 6x^2 - 3x^4 + 2x^2 - 3 &= (5-3)x^4 + (-6+2)x^2 - 3\\ &= 2x^4 - 4x^2 - 3\end{aligned}$$

8.
$$\begin{aligned}&(3x^4 - 5x^2 - 4) + (x^3 + 3x^2 + 6)\\ &= 3x^4 + x^3 + (-5+3)x^2 + (-4+6)\\ &= 3x^4 + x^3 - 2x^2 + 2\end{aligned}$$

9. $(x^4 - 3x^2 + 2)(x^2 - 3)$
$$\begin{array}{r} x^4 - 3x^2 + 2 \\ x^2 - 3 \\ \hline -3x^4 + 9x^2 - 6 \\ x^6 - 3x^4 + 2x^2 \qquad \\ \hline x^6 - 6x^4 + 11x^2 - 6 \end{array}$$

10. We use FOIL.
$$\begin{aligned}(y+4)(2y+3) &= y\cdot 2y + y\cdot 3 + 4\cdot 2y + 4\cdot 3\\ &= 2y^2 + 3y + 8y + 12\\ &= 2y^2 + 11y + 12\end{aligned}$$

11. $(x+5)(x-5) = x^2 - 5^2 = x^2 - 25$

12.
$$\begin{aligned}(3w+4)^2 &= (3w)^2 + 2\cdot 3w\cdot 4 + 4^2\\ &= 9w^2 + 24w + 16\end{aligned}$$

13.
$$\begin{aligned}&(a^3b^2 - 5a^2b + 2ab) - (3a^3b^2 - ab^2 + 4ab)\\ &= a^3b^2 - 5a^2b + 2ab - 3a^3b^2 + ab^2 - 4ab\\ &= -2a^3b^2 - 5a^2b + ab^2 - 2ab\end{aligned}$$

14.
$$\begin{aligned}(5y^2 - 20y + 8) \div 5 &= \frac{5y^2 - 20y + 8}{5}\\ &= \frac{5y^2}{5} - \frac{20y}{5} + \frac{8}{5}\\ &= y^2 - 4y + \frac{8}{5}\end{aligned}$$

15.
$$\begin{array}{r|l} & x - 9 \\ \hline x+5 & x^2 - 4x + 3 \\ & x^2 + 5x \\ \hline & -9x + 3 \\ & -9x - 45 \\ \hline & 48 \end{array}$$

The answer is $x - 9 + \dfrac{48}{x+5}$.

Chapter 4 Review Exercises

1. $7^2 \cdot 7^{-4} = 7^{2+(-4)} = 7^{-2} = \dfrac{1}{7^2}$

2. $y^7 \cdot y^3 \cdot y = y^{7+3+1} = y^{11}$

3. $(3x)^5(3x)^9 = (3x)^{5+9} = (3x)^{14}$

4. $t^8 \cdot t^0 = t^8 \cdot 1 = t^8$, or

$t^8 \cdot t^0 = t^{8+0} = t^8$

5. $\dfrac{4^5}{4^2} = 4^{5-2} = 4^3$

6. $\dfrac{a^5}{a^8} = a^{5-8} = a^{-3} = \dfrac{1}{a^3}$

7. $\dfrac{(7x)^4}{(7x)^4} = 1$

8. $(3t^4)^2 = 3^2 \cdot (t^4)^2 = 9 \cdot t^{4\cdot 2} = 9t^8$

9. $(2x^3)^2(-3x)^2 = 2^2 \cdot (x^3)^2(-3)^2x^2 = 4\cdot x^6 \cdot 9 \cdot x^2 = 36x^8$

10. $\left(\dfrac{2x}{y}\right)^{-3} = \left(\dfrac{y}{2x}\right)^3 = \dfrac{y^3}{2^3\cdot x^3} = \dfrac{y^3}{8x^3}$

11. $\dfrac{1}{t^5} = t^{-5}$

12. $y^{-4} = \dfrac{1}{y^4}$

13. 0.00003. 28

↑ 5 places

Small number, so the exponent is negative.

$0.0000328 = 3.28 \times 10^{-5}$

14. 8.3×10^6

8.300000.

6 places

Positive exponent, so the answer is a large number.

$8.3 \times 10^6 = 8,300,000$

15. $$\begin{aligned}(3.8 \times 10^4)(5.5 \times 10^{-1}) &= (3.8 \cdot 5.5) \times (10^4 \cdot 10^{-1}) \\ &= 20.9 \times 10^3 \\ &= (2.09 \times 10) \times 10^3 \\ &= 2.09 \times 10^4\end{aligned}$$

16. $$\begin{aligned}\frac{1.28 \times 10^{-8}}{2.5 \times 10^{-4}} &= \frac{1.28}{2.5} \times \frac{10^{-8}}{10^{-4}} \\ &= 0.512 \times 10^{-4} \\ &= (5.12 \times 10^{-1}) \times 10^{-4} \\ &= 5.12 \times 10^{-5}\end{aligned}$$

17. $46 = 4.6 \times 10$

Also, $$\begin{aligned}340 \text{ million} &= 340 \times 10^6 \\ &= (3.4 \times 10^2) \times 10^6 \\ &= 3.4 \times 10^8.\end{aligned}$$

Then we have

$$\begin{aligned}(4.6 \times 10) \times (3.4 \times 10^8) &= (4.6 \times 3.4) \times (10 \times 10^8) \\ &= 15.64 \times 10^9 \\ &= (1.564 \times 10) \times 10^9 \\ &= 1.564 \times 10^{10}\end{aligned}$$

18. $x^2 - 3x + 6 = (-1)^2 - 3(-1) + 6 = 1 + 3 + 6 = 10$

19. $-4y^5 + 7y^2 - 3y - 2 = -4y^5 + 7y^2 + (-3y) + (-2)$

The terms are $-4y^5$, $7y^2$, $-3y$, and -2.

20. In the polynomial $x^3 + x$ there are no x^2 or x^0 terms. Thus, the x^2 term (or second-degree term) and the x^0 term (or zero-degree term) are missing.

21. $4x^3 + 6x^2 - 5x + \frac{5}{3} = 4x^3 + 6x^2 - 5x^1 + \frac{5}{3}x^0$

The degree of $4x^3$ is 3.

The degree of $6x^2$ is 2.

The degree of $-5x$ is 1.

The degree of $\frac{5}{3}$ is 0.

The degree of the polynomial is 3, the largest exponent.

22. The polynomial $4x^3 - 1$ is a binomial because it has just two terms.

23. The polynomial $4 - 9t^3 - 7t^4 + 10t^2$ is none of these because it has more than three terms.

24. The polynomial $7y^2$ is a monomial because it has just one term.

25. $$\begin{aligned}&3x^2 - 2x + 3 - 5x^2 - 1 - x \\ &= (3-5)x^2 + (-2-1)x + (3-1) \\ &= -2x^2 - 3x + 2\end{aligned}$$

26. $$\begin{aligned}&-x + \frac{1}{2} + 14x^4 - 7x^2 - 1 - 4x^4 \\ &= (14-4)x^4 - 7x^2 - x + \left(\frac{1}{2} - 1\right) \\ &= 10x^4 - 7x^2 - x - \frac{1}{2}\end{aligned}$$

27. $(3x^4 - x^3 + x - 4) + (x^5 + 7x^3 - 3x^2 - 5) + (-5x^4 + 6x^2 - x) = (3-5)x^4 + (-1+7)x^3 + (1-1)x + (-4-5) + x^5 + (-3+6)x^2 = -2x^4 + 6x^3 - 9 + x^5 + 3x^2$, or $x^5 - 2x^4 + 6x^3 + 3x^2 - 9$

28. $(3x^5 - 4x^4 + x^3 - 3) + (3x^4 - 5x^3 + 3x^2) + (-5x^5 - 5x^2) + (-5x^4 + 2x^3 + 5) = (3-5)x^5 + (-4+3-5)x^4 + (1-5+2)x^3 + (-3+5) + (3-5)x^2 = -2x^5 - 6x^4 - 2x^3 + 2 - 2x^2$, or $-2x^5 - 6x^4 - 2x^3 - 2x^2 + 2$

29. $$\begin{aligned}(5x^2 - 4x + 1) - (3x^2 + 1) &= 5x^2 - 4x + 1 - 3x^2 - 1 \\ &= 2x^2 - 4x\end{aligned}$$

30. $$\begin{aligned}&(3x^5 - 4x^4 + 3x^2 + 3) - (2x^5 - 4x^4 + 3x^3 + 4x^2 - 5) \\ &= 3x^5 - 4x^4 + 3x^2 + 3 - 2x^5 + 4x^4 - 3x^3 - 4x^2 + 5 \\ &= x^5 - 3x^3 - x^2 + 8\end{aligned}$$

31. $P = 2(w+3) + 2w = 2w + 6 + 2w = 4w + 6$

$A = w(w+3) = w^2 + 3w$

32. Regarding the figure as one large rectangle with length $t+4$ and width $t+3$, we have $(t+4)(t+3)$. We can also add the areas of the four smaller rectangles:

$3 \cdot t + 4 \cdot 3 + 4 \cdot t + t \cdot t$, or $3t + 12 + 4t + t^2$, or

$t^2 + 7t + 12$

33. $$\begin{aligned}\left(x + \frac{2}{3}\right)\left(x + \frac{1}{2}\right) &= x^2 + \frac{1}{2}x + \frac{2}{3}x + \frac{2}{6} \\ &= x^2 + \frac{3}{6}x + \frac{4}{6}x + \frac{1}{3} \\ &= x^2 + \frac{7}{6}x + \frac{1}{3}\end{aligned}$$

34. $$\begin{aligned}(7x+1)^2 &= (7x)^2 + 2 \cdot 7x \cdot 1 + 1^2 \\ &= 49x^2 + 14x + 1\end{aligned}$$

35.
$$\begin{array}{r} 4x^2 - 5x + 1 \\ 3x - 2 \\ \hline -8x^2 + 10x - 2 \\ 12x^3 - 15x^2 + 3x \\ \hline 12x^3 - 23x^2 + 13x - 2 \end{array}$$

36. $(3x^2 + 4)(3x^2 - 4) = (3x^2)^2 - 4^2 = 9x^4 - 16$

37. $$\begin{aligned}&5x^4(3x^3 - 8x^2 + 10x + 2) \\ &= 5x^4 \cdot 3x^3 - 5x^4 \cdot 8x^2 + 5x^4 \cdot 10x + 5x^4 \cdot 2 \\ &= 15x^7 - 40x^6 + 50x^5 + 10x^4\end{aligned}$$

38. $(x+4)(x-7) = x^2 - 7x + 4x - 28 = x^2 - 3x - 28$

39. $(3y^2 - 2y)^2 = (3y^2)^2 - 2 \cdot 3y^2 \cdot 2y + (2y)^2 = 9y^4 - 12y^3 + 4y^2$

40. $(2t^2 + 3)(t^2 - 7) = 2t^4 - 14t^2 + 3t^2 - 21 = 2t^4 - 11t^2 - 21$

41. $2 - 5xy + y^2 - 4xy^3 + x^6$
$= 2 - 5(-1)(2) + 2^2 - 4(-1)(2)^3 + (-1)^6$
$= 2 - 5(-1)(2) + 4 - 4(-1)(8) + 1$
$= 2 + 10 + 4 + 32 + 1$
$= 49$

42. $x^5y - 7xy + 9x^2 - 8$

Term	Coefficient	Degree	
x^5y	1	6	$(x^5y = 1 \cdot x^5y^1)$
$-7xy$	-7	2	$(-7xy = -7x^1y^1)$
$9x^2$	9	2	
-8	-8	0	$(-8 = -8x^0)$

The degree of the polynomial is the degree of the term of highest degree. The term of highest degree is x^5y. Its degree is 6, so the degree of the polynomial is 6.

43. $y + w - 2y + 8w - 5 = (1 - 2)y + (1 + 8)w - 5$
$= -y + 9w - 5$

44. $m^6 - 2m^2n + m^2n^2 + n^2m - 6m^3 + m^2n^2 + 7n^2m$
$= m^6 - 2m^2n + (1 + 1)m^2n^2 + (1 + 7)n^2m - 6m^3$
$= m^6 - 2m^2n + 2m^2n^2 + 8n^2m - 6m^3$

45. $(5x^2 - 7xy + y^2) + (-6x^2 - 3xy - y^2) + (x^2 + xy - 2y^2)$
$= (5 - 6 + 1)x^2 + (-7 - 3 + 1)xy + (1 - 1 - 2)y^2$
$= -9xy - 2y^2$

46. $(6x^3y^2 - 4x^2y - 6x) - (-5x^3y^2 + 4x^2y + 6x^2 - 6)$
$= 6x^3y^2 - 4x^2y - 6x + 5x^3y^2 - 4x^2y - 6x^2 + 6$
$= (6 + 5)x^3y^2 + (-4 - 4)x^2y - 6x - 6x^2 + 6$
$= 11x^3y^2 - 8x^2y - 6x - 6x^2 + 6$

47.
$$\begin{array}{r} p^2 + pq + q^2 \\ p - q \\ \hline -p^2q - pq^2 - q^3 \\ p^3 + p^2q + pq^2 \qquad \\ \hline p^3 \qquad\qquad\quad - q^3 \end{array}$$

48. $\left(3a^4 - \frac{1}{3}b^3\right)^2 = (3a^4)^2 - 2 \cdot 3a^4 \cdot \frac{1}{3}b^3 + \left(\frac{1}{3}b^3\right)^2$
$= 9a^8 - 2a^4b^3 + \frac{1}{9}b^6$

49. $\frac{10x^3 - x^2 + 6x}{2x} = \frac{10x^3}{2x} - \frac{x^2}{2x} + \frac{6x}{2x}$
$= 5x^2 - \frac{1}{2}x + 3$

50.
$$\begin{array}{r} 3x^2 - 7x + 4 \\ 2x+3 \overline{\smash{)}\, 6x^3 - 5x^2 - 13x + 13} \\ \underline{6x^3 + 9x^2} \qquad\qquad\quad \\ -14x^2 - 13x \qquad \\ \underline{-14x^2 - 21x} \qquad \\ 8x + 13 \\ \underline{8x + 12} \\ 1 \end{array}$$

The answer is $3x^2 - 7x + 4 + \frac{1}{2x + 3}$.

51. Locate -1 on the x-axis. Then move vertically to the graph and horizontally to the y-axis. It appears that the y-value that is paired with -1 is 0. Thus, the value of $y = 10x^3 - 10x$ is 0 when $x = -1$.

Locate -0.5 on the x-axis. Then move vertically to the graph and horizontally to the y-axis. It appears that the y-value that is paired with -0.5 is about 3.75. Thus, the value of $y = 10x^3 - 10x$ is 3.75 when $x = -0.5$.

Locate 0.5 on the x-axis. Then move vertically to the graph and horizontally to the y-axis. It appears that the y-value that is paired with 0.5 is about -3.75. Thus, the value of $y = 10x^3 - 10x$ is -3.75 when $x = 0.5$.

Locate 1 on the x-axis. Then move vertically to the graph and horizontally to the y-axis. It appears that the y-value that is paired with 1 is 0. Thus, the value of $y = 10x^3 - 10x$ is 0 when $x = 1$.

52. $(2x^2 - 3x + 4) - (x^2 + 2x) = 2x^2 - 3x + 4 - x^2 - 2x = x^2 - 5x + 4$

Answer B is correct.

53. $(x - 1)^2 = x^2 - 2 \cdot x \cdot 1 + 1^2 = x^2 - 2x + 1$

Answer D is correct.

54. $A = \frac{1}{2}bh$
$A = \frac{1}{2}(x + y)(x - y) = \frac{1}{2}(x^2 - y^2) = \frac{1}{2}x^2 - \frac{1}{2}y^2$

55. The shaded area is the area of a square with side 20 minus the area of 4 small squares, each with side a.

$A = 20^2 - 4 \cdot a^2 = 400 - 4a^2$

56. $-3x^5 \cdot 3x^3 - x^6(2x)^2 + (3x^4)^2 + (2x^2)^4 - 40x^2(x^3)^2$
$= -3x^5 \cdot 3x^3 - x^6(4x^2) + 9x^8 + 16x^8 - 40x^2(x^6)$
$= -9x^8 - 4x^8 + 9x^2 + 16x^8 - 40x^8$
$= -28x^8$

57.
$(x - 7)(x + 10) = (x - 4)(x - 6)$
$x^2 + 10x - 7x - 70 = x^2 - 6x - 4x + 24$
$x^2 + 3x - 70 = x^2 - 10x + 24$
$3x - 70 = -10x + 24$ Subtracting x^2
$13x - 70 = 24$ Adding $10x$
$13x = 94$ Adding 70
$x = \frac{94}{13}$

The solution is $\frac{94}{13}$.

58. Let P represent the other polynomial. Then we have $(x-1)P = x^5 - 1$, or $P = \dfrac{x^5-1}{x-1}$. We divide to find P.

$$\begin{array}{r|l} & x^4 + x^3 + x^2 + x + 1 \\ \hline x-1 & x^5 + 0x^4 + 0x^3 + 0x^2 + 0x - 1 \\ & \underline{x^5 - x^4} \\ & x^4 \\ & \underline{x^4 - x^3} \\ & x^3 \\ & \underline{x^3 - x^2} \\ & x^2 \\ & \underline{x^2 - x} \\ & x - 1 \\ & \underline{x - 1} \\ & 0 \end{array}$$

The other polynomial is $x^4 + x^3 + x^2 + x + 1$.

59. ***Familiarize***. Let w = the width of the garden, in feet. Then $2w$ = the length. From the drawing in the text we see that the width of the garden and the sidewalk together is $w + 4 + 4$, or $w + 8$, and the length of the garden and the sidewalk together is $2w + 4 + 4$, or $2w + 8$.

Translate. The area of the sidewalk is the area of the garden and sidewalk together minus the area of the garden. Recall that the formula for the area of a rectangle is $A = l \cdot w$. Thus, we have

$$1024 = (2w+8)(w+8) - 2w \cdot w.$$

Solve. We solve the equation.

$$\begin{aligned} 1024 &= (2w+8)(w+8) - 2w \cdot w \\ 1024 &= 2w^2 + 16w + 8w + 64 - 2w^2 \\ 1024 &= 24w + 64 \\ 960 &= 24w \\ 40 &= w \end{aligned}$$

If $w = 40$, then $2w = 2 \cdot 40 = 80$.

Check. The dimensions of the garden and the sidewalk together are $2 \cdot 40 + 8$ by $40 + 8$, or 88 by 48. Then the area of the garden and sidewalk together is 88 ft $\cdot$ 48 ft, or 4224 ft^2 and the area of the garden is 80 ft $\cdot$ 40 ft, or 3200 ft^2. Subtracting to find the area of the sidewalk, we get 4224 ft^2 – 3200 ft^2, or 1024 ft^2, so the answer checks.

State. The dimensions of the garden are 80 ft by 40 ft.

Chapter 4 Discussion and Writing Exercises

1. 578.6×10^{-7} is not in scientific notation because 578.6 is larger than 10.

2. When evaluating polynomials it is essential to know the order in which the operations are to be performed.

3. Label the figure as shown.

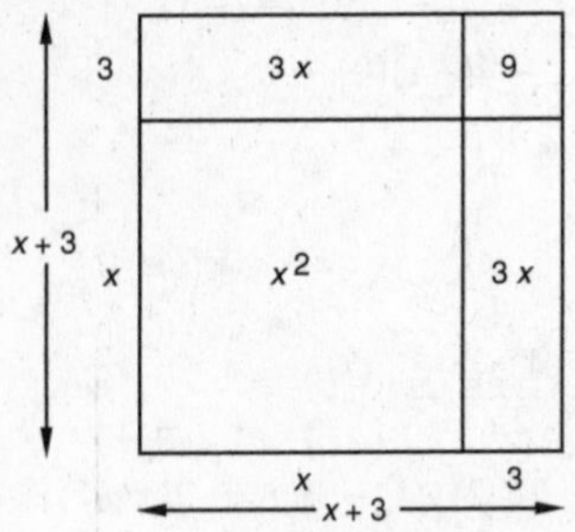

Then we see that the area of the figure is $(x+3)^2$, or $x^2 + 3x + 3x + 9 \neq x^2 + 9$.

4. Emma did not divide *each* term of the polynomial by the divisor. The first term was divided by $3x$, but the second was not. Multiplying Emma's "quotient" by the divisor $3x$ we get $12x^3 - 18x^2 \neq 12x^3 + 6x$. This should convince her that a mistake has been made.

5. Yes; for example, $(x^2+xy+1)+(3x-xy+2) = x^2+3x+3$.

6. Yes; consider $a + b + c + d$. This is a polynomial in 4 variables but it has degree 1.

Chapter 4 Test

1. $6^{-2} \cdot 6^{-3} = 6^{-2+(-3)} = 6^{-5} = \dfrac{1}{6^5}$

2. $x^6 \cdot x^2 \cdot x = x^{6+2+1} = x^9$

3. $(4a)^3 \cdot (4a)^8 = (4a)^{3+8} = (4a)^{11}$

4. $\dfrac{3^5}{3^2} = 3^{5-2} = 3^3$

5. $\dfrac{x^3}{x^8} = x^{3-8} = x^{-5} = \dfrac{1}{x^5}$

6. $\dfrac{(2x)^5}{(2x)^5} = 1$

7. $(x^3)^2 = x^{3 \cdot 2} = x^6$

8. $(-3y^2)^3 = (-3)^3(y^2)^3 = -27y^{2 \cdot 3} = -27y^6$

9. $(2a^3b)^4 = 2^4(a^3)^4 \cdot b^4 = 16a^{12}b^4$

10. $\left(\dfrac{ab}{c}\right)^3 = \dfrac{(ab)^3}{c^3} = \dfrac{a^3b^3}{c^3}$

11. $(3x^2)^3(-2x^5)^3 = 3^3(x^2)^3(-2)^3(x^5)^3 = 27x^6(-8)x^{15} = -216x^{21}$

12. $3(x^2)^3(-2x^5)^3 = 3x^6(-2)^3(x^5)^3 = 3x^6(-8)x^{15} = -24x^{21}$

13. $2x^2(-3x^2)^4 = 2x^2(-3)^4(x^2)^4 = 2x^2 \cdot 81x^8 = 162x^{10}$

14. $(2x)^2(-3x^2)^4 = 2^2x^2(-3)^4(x^2)^4 = 4x^2 \cdot 81x^8 = 324x^{10}$

15. $5^{-3} = \dfrac{1}{5^3}$

16. $\frac{1}{y^8} = y^{-8}$

17. 3,900,000,000

3.900,000,000.

↑ 9 places

Large number, so the exponent is positive.

$3,900,000,000 = 3.9 \times 10^9$

18. 5×10^{-8}

Negative exponent, so the answer is a small number.

0.00000005.

↑ 8 places

$5 \times 10^{-8} = 0.00000005$

19. $\frac{5.6 \times 10^6}{3.2 \times 10^{-11}} = \frac{5.6}{3.2} \times \frac{10^6}{10^{-11}} = 1.75 \times 10^{6-(-11)} = 1.75 \times 10^{17}$

20. $(2.4 \times 10^5)(5.4 \times 10^{16}) = (2.4 \cdot 5.4) \times (10^5 \cdot 10^{16}) =$
$12.96 \times 10^{21} = (1.296 \times 10) \times 10^{21} = 1.296 \times 10^{22}$

21. $\frac{5.7 \times 10^{23}}{6 \times 10^{21}} = 0.95 \times 10^2 = (9.5 \times 10^{-1}) \times 10^2 = 9.5 \times 10$

The mass of Saturn is 9.5×10 times the mass of Earth.

22. $x^5 + 5x - 1 = (-2)^5 + 5(-2) - 1 = -32 - 10 - 1 = -43$

23. $\frac{1}{3}x^5 - x + 7$

The coefficient of $\frac{1}{3}x^5$ is $\frac{1}{3}$.

The coefficient of $-x$, or $-1 \cdot x$, is -1.

The coefficient of 7 is 7.

24. $2x^3 - 4 + 5x + 3x^6$

The degree of $2x^3$ is 3.

The degree of -4, or $-4x^0$, is 0.

The degree of $5x$, or $5x^1$, is 1.

The degree of $3x^6$ is 6.

The degree of the polynomial is 6, the largest exponent.

25. $7 - x$ is a binomial because it has just 2 terms.

26. $4a^2 - 6 + a^2 = (4+1)a^2 - 6 = 5a^2 - 6$

27. $y^2 - 3y - y + \frac{3}{4}y^2 = \left(1 + \frac{3}{4}\right)y^2 + (-3-1)y =$
$\left(\frac{4}{4} + \frac{3}{4}\right)y^2 + (-3-1)y = \frac{7}{4}y^2 - 4y$

28.
$$\begin{aligned} &3 - x^2 + 2x^3 + 5x^2 - 6x - 2x + x^5 \\ &= 3 + (-1+5)x^2 + 2x^3 + (-6-2)x + x^5 \\ &= 3 + 4x^2 + 2x^3 - 8x + x^5 \\ &= x^5 + 2x^3 + 4x^2 - 8x + 3 \end{aligned}$$

29.
$$\begin{aligned} &(3x^5 + 5x^3 - 5x^2 - 3) + (x^5 + x^4 - 3x^3 - 3x^2 + 2x - 4) \\ &= (3+1)x^5 + x^4 + (5-3)x^3 + (-5-3)x^2 + 2x + (-3-4) \\ &= 4x^5 + x^4 + 2x^3 - 8x^2 + 2x - 7 \end{aligned}$$

30.
$$\begin{aligned} &\left(x^4 + \frac{2}{3}x + 5\right) + \left(4x^4 + 5x^2 + \frac{1}{3}x\right) \\ &= (1+4)x^4 + 5x^2 + \left(\frac{2}{3} + \frac{1}{3}\right)x + 5 \\ &= 5x^4 + 5x^2 + x + 5 \end{aligned}$$

31.
$$\begin{aligned} &(2x^4 + x^3 - 8x^2 - 6x - 3) - (6x^4 - 8x^2 + 2x) \\ &= 2x^4 + x^3 - 8x^2 - 6x - 3 - 6x^4 + 8x^2 - 2x \\ &= (2-6)x^4 + x^3 + (-8+8)x^2 + (-6-2)x - 3 \\ &= -4x^4 + x^3 - 8x - 3 \end{aligned}$$

32.
$$\begin{aligned} &(x^3 - 0.4x^2 - 12) - (x^5 + 0.3x^3 + 0.4x^2 + 9) \\ &= x^3 - 0.4x^2 - 12 - x^5 - 0.3x^3 - 0.4x^2 - 9 \\ &= -x^5 + (1-0.3)x^3 + (-0.4-0.4)x^2 + (-12-9) \\ &= -x^5 + 0.7x^3 - 0.8x^2 - 21 \end{aligned}$$

33. $-3x^2(4x^2 - 3x - 5) = -3x^2 \cdot 4x^2 - 3x^2(-3x) - 3x^2(-5) =$
$-12x^4 + 9x^3 + 15x^2$

34. $\left(x - \frac{1}{3}\right)^2 = x^2 - 2 \cdot x \cdot \frac{1}{3} + \left(\frac{1}{3}\right)^2 = x^2 - \frac{2}{3}x + \frac{1}{9}$

35. $(3x+10)(3x-10) = (3x)^2 - 10^2 = 9x^2 - 100$

36. $(3b+5)(b-3) = 3b^2 - 9b + 5b - 15 = 3b^2 - 4b - 15$

37. $(x^6 - 4)(x^8 + 4) = x^{14} + 4x^6 - 4x^8 - 16$, or
$x^{14} - 4x^8 + 4x^6 - 16$

38. $(8-y)(6+5y) = 48 + 40y - 6y - 5y^2 = 48 + 34y - 5y^2$

39.
$$\begin{array}{rrrr} & 3x^2 & -\ 5x & -\ 3 \\ & & 2x & +\ 1 \\ \hline & 3x^2 & -\ 5x & -\ 3 \\ 6x^3 & -\ 10x^2 & -\ 6x & \\ \hline 6x^3 & -\ 7x^2 & -\ 11x & -\ 3 \end{array}$$

40. $(5t+2)^2 = (5t)^2 + 2 \cdot 5t \cdot 2 + 2^2 = 25t^2 + 20t + 4$

41.
$$\begin{aligned} &x^3y - y^3 + xy^3 + 8 - 6x^3y - x^2y^2 + 11 \\ &= (1-6)x^3y - y^3 + xy^3 + (8+11) - x^2y^2 \\ &= -5x^3y - y^3 + xy^3 + 19 - x^2y^2 \end{aligned}$$

42.
$$\begin{aligned} &(8a^2b^2 - ab + b^3) - (-6ab^2 - 7ab - ab^3 + 5b^3) \\ &= 8a^2b^2 - ab + b^3 + 6ab^2 + 7ab + ab^3 - 5b^3 \\ &= 8a^2b^2 + (-1+7)ab + (1-5)b^3 + 6ab^2 + ab^3 \\ &= 8a^2b^2 + 6ab - 4b^3 + 6ab^2 + ab^3 \end{aligned}$$

43. $(3x^5 - 4y^5)(3x^5 + 4y^5) = (3x^5)^2 - (4y^5)^2 =$
$9x^{10} - 16y^{10}$

44.
$$\begin{aligned} &(12x^4 + 9x^3 - 15x^2) \div (3x^2) \\ &= \frac{12x^4 + 9x^3 - 15x^2}{3x^2} \\ &= \frac{12x^4}{3x^2} + \frac{9x^3}{3x^2} - \frac{15x^2}{3x^2} \\ &= 4x^2 + 3x - 5 \end{aligned}$$

45.

$$\begin{array}{r|l} & 2x^2 - 4x - 2 \\ 3x+2 & 6x^3 - 8x^2 - 14x + 13 \\ & \underline{6x^3 + 4x^2} \\ & \quad - 12x^2 - 14x \\ & \quad \underline{- 12x^2 - 8x} \\ & \qquad - 6x + 13 \\ & \qquad \underline{- 6x - 4} \\ & \qquad\qquad 17 \end{array}$$

The answer is $2x^2 - 4x - 2 + \frac{17}{3x+2}$.

46. Locate -1 on the x-axis. Then move vertically to the graph and horizontally to the y-axis. It appears that the y-value that is paired with -1 is 3. Thus, the value of $y = x^3 - 5x - 1$ is 3 when $x = -1$.

Locate -0.5 on the x-axis. Then move vertically to the graph and horizontally to the y-axis. It appears that the y-value that is paired with -0.5 is 1.5. Thus, the value of $y = x^3 - 5x - 1$ is 1.5 when $x = -0.5$.

Locate 0.5 on the x-axis. Then move vertically to the graph and horizontally to the y-axis. It appears that the y-value that is paired with 0.5 is -3.5. Thus, the value of $y = x^3 - 5x - 1$ is -3.5 when $x = 0.5$.

Locate 1 on the x-axis. Then move vertically to the graph and horizontally to the y-axis. It appears that the y-value that is paired with 1 is -5. Thus, the value of $y = x^3 - 5x - 1$ is -5 when $x = 1$.

Locate 1.1 on the x-axis. Then move vertically to the graph and horizontally to the y-axis. It appears that the y-value that is paired with 1.1 is -5.25. Thus, the value of $y = x^3 - 5x - 1$ is -5.25 when $x = 1.1$.

47. When we regard the figure as one large rectangle with dimensions $t + 2$ by $t + 2$, we can express the area as $(t + 2)(t + 2)$.

Next we will regard the figure as the sum of four smaller rectangles with dimensions t by t, 2 by t, 2 by t, and 2 by 2. The sum of the areas of these rectangles is $t \cdot t + 2 \cdot t + 2 \cdot t + 2 \cdot 2$, or $t^2 + 2t + 2t + 4$, or $t^2 + 4t + 4$.

48. Two sides have dimensions a by 5, two other sides have dimensions a by 9, and the two remaining sides have dimensions 9 by 5. Then the surface area is $2 \cdot a \cdot 5 + 2 \cdot a \cdot 9 + 2 \cdot 9 \cdot 5$, or $10a + 18a + 90$, or $28a + 90$. Answer B is correct.

49. Let l = the length of the box. Then the height is $l - 1$ and the width is $l - 2$. The volume of the box is length $\times$ width $\times$ height.

$$V = l(l-2)(l-1)$$
$$V = (l^2 - 2l)(l-1)$$
$$V = l^3 - l^2 - 2l^2 + 2l$$
$$V = l^3 - 3l^2 + 2l$$

50.
$$\begin{aligned} (x-5)(x+5) &= (x+6)^2 \\ x^2 - 25 &= x^2 + 12x + 36 \\ -25 &= 12x + 36 \qquad \text{Subtracting } x^2 \\ -61 &= 12x \\ -\frac{61}{12} &= x \end{aligned}$$

The solution is $-\frac{61}{12}$.

Cumulative Review Chapters 1 - 4

1. $\frac{x}{2y} = \frac{10}{2 \cdot 2} = \frac{10}{4} = \frac{5}{2}$

2. $2x^3 + x^2 - 3 = 2(-1)^3 + (-1)^2 - 3 = 2(-1) + 1 - 3 = -2 + 1 - 3 = -4$

3.
$$\begin{aligned} x^3y^2 + xy + 2xy^2 &= (-1)^3(2)^2 + (-1)(2) + 2(-1)(2)^2 \\ &= -1 \cdot 4 + (-1)(2) + 2(-1)(4) \\ &= -4 - 2 - 8 = -14 \end{aligned}$$

4. The distance of -4 from 0 is 4, so $|-4| = 4$.

5. The reciprocal of 5 is $\frac{1}{5}$, because $5 \cdot \frac{1}{5} = 1$.

6. $-\frac{3}{5} + \frac{5}{12} = -\frac{36}{60} + \frac{25}{60} = -\frac{11}{60}$

7. $3.4 - (-0.8) = 3.4 + 0.8 = 4.2$

8. $(-2)(-1.4)(2.6) = (2.8)(2.6) = 7.28$

9. $\frac{3}{8} \div \left(-\frac{9}{10}\right) = \frac{3}{8} \cdot \left(-\frac{10}{9}\right) = -\frac{3 \cdot 10}{8 \cdot 9} = -\frac{\cancel{3} \cdot \cancel{2} \cdot 5}{\cancel{2} \cdot 4 \cdot \cancel{3} \cdot 3} = -\frac{5}{12}$

10. $(1.1 \times 10^{10})(2 \times 10^{12}) = (1.1 \cdot 2) \times (10^{10} \cdot 10^{12}) = 2.2 \times 10^{22}$

11. $(3.2 \times 10^{-10}) \div (8 \times 10^{-6}) = \frac{3.2 \times 10^{-10}}{8 \times 10^{-6}} = 0.4 \times 10^{-4} = (4 \times 10^{-1}) \times 10^{-4} = 4 \times 10^{-5}$

12. $\frac{-9x}{3x} = \frac{-3 \cdot 3x}{1 \cdot 3x} = \frac{-3}{1} \cdot \frac{3x}{3x} = \frac{-3}{1} \cdot 1 = -3$

13. $y - (3y + 7) = y - 3y - 7 = -2y - 7$

14.
$$\begin{aligned} 3(x-1) - 2[x - (2x+7)] &= 3(x-1) - 2[x - 2x - 7] \\ &= 3(x-1) - 2[-x-7] \\ &= 3x - 3 + 2x + 14 \\ &= 5x + 11 \end{aligned}$$

15.
$$\begin{aligned} 2 - [32 \div (4 + 2^2)] &= 2 - [32 \div (4+4)] \\ &= 2 - [32 \div 8] \\ &= 2 - 4 \\ &= -2 \end{aligned}$$

16. $(x^4+3x^3-x+7)+(2x^5-3x^4+x-5)$
$=(1-3)x^4+3x^3+(-1+1)x+(7-5)+2x^5$
$=-2x^4+3x^3+2+2x^5$, or
$2x^5-2x^4+3x^3+2$

17. $(x^2+2xy)+(y^2-xy)+(2x^2-3y^2)$
$=(1+2)x^2+(2-1)xy+(1-3)y^2$
$=3x^2+xy-2y^2$

18. $(x^3+3x^2-4)-(-2x^2+x+3)=x^3+3x^2-4+2x^2-x-3$
$=x^3+(3+2)x^2-x+(-4-3)$
$=x^3+5x^2-x-7$

19. $\left(\frac{1}{3}x^2-\frac{1}{4}x-\frac{1}{5}\right)-\left(\frac{2}{3}x^2+\frac{1}{2}x-\frac{1}{5}\right)$
$=\frac{1}{3}x^2-\frac{1}{4}x-\frac{1}{5}-\frac{2}{3}x^2-\frac{1}{2}x+\frac{1}{5}$
$=\left(\frac{1}{3}-\frac{2}{3}\right)x^2+\left(-\frac{1}{4}-\frac{1}{2}\right)x+\left(-\frac{1}{5}+\frac{1}{5}\right)$
$=-\frac{1}{3}x^2-\frac{3}{4}x$

20. $3(4x-5y+7)=3\cdot 4x-3\cdot 5y+3\cdot 7=12x-15y+21$

21. $(-2x^3)(-3x^5)=(-2)(-3)x^3\cdot x^5=6x^{3+5}=6x^8$

22. $2x^2(x^3-2x^2+4x-5)$
$=2x^2\cdot x^3-2x^2\cdot 2x^2+2x^2\cdot 4x-2x^2\cdot 5$
$=2x^5-4x^4+8x^3-10x^2$

23.

$$\begin{array}{rrrrr}
 & & 3y^2 & +\ 5y & +\ 6 \\
 & & y^2 & & -\ 2 \\
\hline
 & & -\ 6y^2 & -\ 10y & -\ 12 \\
3y^4 & +\ 5y^3 & +\ 6y^2 & & \\
\hline
3y^4 & +\ 5y^3 & & -\ 10y & -\ 12
\end{array}$$

24.

$$\begin{array}{rrrrrr}
2p^3 & +\ p^2q & +\ pq^2 & & & \\
p & -\ pq & +\ q & & & \\
\hline
2p^3q & +\ p^2q^2 & +\ pq^3 & & & \\
 & & & -\ 2p^4q & -\ p^3q^2 & -\ p^2q^3 \\
2p^4\ +\ p^3q & +\ p^2q^2 & & & & \\
\hline
2p^4\ +\ 3p^3q & +\ 2p^2q^2 & +\ pq^3 & -\ 2p^4q & -\ p^3q^2 & -\ p^2q^3
\end{array}$$

25. $(2x+3)(3x+2)=6x^2+4x+9x+6=6x^2+13x+6$

26. $(3x^2+1)^2=(3x^2)^2+2\cdot 3x^2\cdot 1+1^2=9x^4+6x^2+1$

27. $\left(t+\frac{1}{2}\right)\left(t-\frac{1}{2}\right)=t^2-\left(\frac{1}{2}\right)^2=t^2-\frac{1}{4}$

28. $(2y^2+5)(2y^2-5)=(2y^2)^2-5^2=4y^4-25$

29. $(2x^4-3)(2x^2+3)=4x^6+6x^4-6x^2-9$

30. $(t-2t^2)^2=t^2-2\cdot t\cdot 2t^2+(2t^2)^2=t^2-4t^3+4t^4$

31. $(3p+q)(5p-2q)=15p^2-6pq+5pq-2q^2=$
$15p^2-pq-2q^2$

32. $(18x^3+6x^2-9x)\div 3x=\frac{18x^3+6x^2-9x}{3x}$
$=\frac{18x^3}{3x}+\frac{6x^2}{3x}-\frac{9x}{3x}$
$=6x^2+2x-3$

33.

$$\begin{array}{r|rrrr}
 & 3x^2 & -\ 2x & -\ 7 & \\
\hline
x+3 & 3x^3 & +\ 7x^2 & -\ 13x & -\ 21 \\
 & 3x^3 & +\ 9x^2 & & \\
\hline
 & & -\ 2x^2 & -\ 13x & \\
 & & -\ 2x^2 & -\ 6x & \\
\hline
 & & & -\ 7x & -\ 21 \\
 & & & -\ 7x & -\ 21 \\
\hline
 & & & & 0
\end{array}$$

The answer is $3x^2-2x-7$.

34. $1.5=2.7+x$
$1.5-2.7=2.7+x-2.7$
$-1.2=x$
The solution is -1.2.

35. $\frac{2}{7}x=-6$
$\frac{7}{2}\cdot\frac{2}{7}x=\frac{7}{2}(-6)$
$x=-\frac{7\cdot\cancel{2}\cdot 3}{\cancel{2}\cdot 1}$
$x=-21$
The solution is -21.

36. $5x-9=36$
$5x-9+9=36+9$
$5x=45$
$\frac{5x}{5}=\frac{45}{5}$
$x=9$
The solution is 9.

37. $\frac{2}{3}=\frac{-m}{10}$
$10\cdot\frac{2}{3}=10\left(\frac{-m}{10}\right)$
$\frac{20}{3}=-m$
$-1\cdot\frac{20}{3}=-1\cdot(-1\cdot m)$
$-\frac{20}{3}=m$
The solution is $-\frac{20}{3}$.

38. $5.4-1.9x=0.8x$
$5.4-1.9x+1.9x=0.8x+1.9x$
$5.4=2.7x$
$\frac{5.4}{2.7}=\frac{2.7x}{2.7}$
$2=x$
The solution is 2.

39.
$$\begin{aligned} x-\frac{7}{8} &= \frac{3}{4} \\ x-\frac{7}{8}+\frac{7}{8} &= \frac{3}{4}+\frac{7}{8} \\ x &= \frac{6}{8}+\frac{7}{8} \\ x &= \frac{13}{8} \end{aligned}$$
The solution is $\frac{13}{8}$.

40.
$$\begin{aligned} 2(2-3x) &= 3(5x+7) \\ 4-6x &= 15x+21 \\ 4-6x+6x &= 15x+21+6x \\ 4 &= 21x+21 \\ 4-21 &= 21x+21-21 \\ -17 &= 21x \\ \frac{-17}{21} &= \frac{21x}{21} \\ -\frac{17}{21} &= x \end{aligned}$$
The solution is $-\frac{17}{21}$.

41.
$$\begin{aligned} \frac{1}{4}x-\frac{2}{3} &= \frac{3}{4}+\frac{1}{3}x \\ \frac{1}{4}x-\frac{2}{3}-\frac{1}{4}x &= \frac{3}{4}+\frac{1}{3}x-\frac{1}{4}x \\ -\frac{2}{3} &= \frac{3}{4}+\frac{4}{12}x-\frac{3}{12}x \\ -\frac{2}{3} &= \frac{3}{4}+\frac{1}{12}x \\ -\frac{2}{3}-\frac{3}{4} &= \frac{3}{4}+\frac{1}{12}x-\frac{3}{4} \\ -\frac{8}{12}-\frac{9}{12} &= \frac{1}{12}x \\ -\frac{17}{12} &= \frac{1}{12}x \\ 12\left(-\frac{17}{12}\right) &= 12\left(\frac{1}{12}x\right) \\ -17 &= x \end{aligned}$$
The solution is -17.

42.
$$\begin{aligned} y+5-3y &= 5y-9 \\ -2y+5 &= 5y-9 \\ -2y+5-5y &= 5y-9-5y \\ -7y+5 &= -9 \\ -7y+5-5 &= -9-5 \\ -7y &= -14 \\ \frac{-7y}{-7} &= \frac{-14}{-7} \\ y &= 2 \end{aligned}$$
The solution is 2.

43.
$$\begin{aligned} \frac{1}{4}x-7 &< 5-\frac{1}{2}x \\ \frac{1}{4}x-7+\frac{1}{2}x &< 5-\frac{1}{2}x+\frac{1}{2}x \\ \frac{3}{4}x-7 &< 5 \\ \frac{3}{4}x-7+7 &< 5+7 \\ \frac{3}{4}x &< 12 \\ \frac{4}{3}\cdot\frac{3}{4}x &< \frac{4}{3}\cdot 12 \\ x &< \frac{4\cdot \cancel{3}\cdot 4}{\cancel{3}\cdot 1} \\ x &< 16 \end{aligned}$$
The solution set is $\{x|x<16\}$.

44.
$$\begin{aligned} 2(x+2) &\geq 5(2x+3) \\ 2x+4 &\geq 10x+15 \\ 2x+4-10x &\geq 10x+15-10x \\ -8x+4 &\geq 15 \\ -8x+4-4 &\geq 15-4 \\ -8x &\geq 11 \\ \frac{-8x}{-8} &\leq \frac{11}{-8} \quad \text{Reversing the inequality symbol} \\ x &\leq -\frac{11}{8} \end{aligned}$$
The solution set is $\left\{x\middle|x\leq -\frac{11}{8}\right\}$.

45.
$$\begin{aligned} A &= Qx-P \\ A+P &= Qx-P+P \\ A+P &= Qx \\ \frac{A+P}{Q} &= \frac{Qx}{Q} \\ \frac{A+P}{Q} &= x \end{aligned}$$

46. ***Familiarize***. Let $p =$ the price the store paid for the book. The store sells the book for a price of $p+80\%p$, or $p+0.8p$, or $1.8p$.

Translate.

$$\underbrace{\text{The selling price}}_{\downarrow\; 1.8p} \;\underset{\downarrow\; =}{\text{is}}\; \underset{\downarrow\; 6.30}{\$6.30.}$$

Solve.
$$\begin{aligned} 1.8p &= 6.30 \\ p &= \frac{6.30}{1.8} \\ p &= 3.5 \end{aligned}$$

Check. 80% of \$3.50 is $0.8\cdot\$3.50 = \$2.80$ and $\$3.50+\$2.80=\$6.30$. The answer checks.

State. The store paid \$3.50 for the book.

47. The area of the surface of the pool is πr^2 ft^2. The area of the raft is 6 ft $\cdot$ 3 ft = 18 ft^2. We subtract to find the area of the surface of the pool not covered by the raft. We have

πr^2 ft^2 − 18 ft^2, or $(\pi r^2 - 18)$ ft^2.

48. ***Familiarize.*** The page numbers are consecutive integers. Let n = the smaller number. Then $n+1$ = the larger number.

Translate.

$$\underbrace{\text{The sum of the page numbers}}_{\downarrow\; n+(n+1)} \;\; \underset{=}{\text{is}} \;\; \underset{37}{37.}$$

Solve.

$$\begin{aligned} n+(n+1) &= 37 \\ 2n+1 &= 37 \\ 2n &= 37-1 \\ 2n &= 36 \\ n &= 18 \end{aligned}$$

If $n = 18$, then $n+1 = 18+1 = 19$.

Check. 18 and 19 are consecutive numbers and $18+19 = 37$. The answer checks.

State. The page numbers are 18 and 19.

49. ***Familiarize.*** Let l = the length of the room, in feet. Then $l-4$ = the width. Recall that the formula for the perimeter of a rectangle is $P = 2l + 2w$.

Translate.

$$\underbrace{\text{Perimeter}}_{\downarrow\; 2l+2(l-4)} \;\; \underset{=}{\text{is}} \;\; \underset{88}{\underline{88\text{ ft}}}.$$

Solve.

$$\begin{aligned} 2l+2(l-4) &= 88 \\ 2l+2l-8 &= 88 \\ 4l-8 &= 88 \\ 4l &= 88+8 \\ 4l &= 96 \\ l &= \frac{96}{4} \\ l &= 24 \end{aligned}$$

If $l = 24$, then $l - 4 = 24 - 4 = 20$.

Check. The width, 20 ft, is 4 ft less than the length, 24 ft. The perimeter is $2\cdot 24$ ft $+ 2\cdot 20$ ft $= 48$ ft $+ 40$ ft $= 88$ ft. The answer checks.

State. The length is 24 ft, and the width is 20 ft.

50. ***Familiarize.*** Let x = the measure of the first angle. Then $5x$ = the measure of the second angle and $2(x+5x)$, or $2\cdot 6x$, or $12x$ = the measure of the third angle. Recall that the sum of the measures of the angles of a triangle is 180°.

Translate.

$$x + 5x + 12x = 180$$

Solve.

$$\begin{aligned} x+5x+12x &= 180 \\ 18x &= 180 \\ x &= \frac{180}{18} \\ x &= 10 \end{aligned}$$

We are asked to find only the measure of the first angle, but we will also find the measures of the other two angles so that we can check the answer.

If $x = 10$, then $5x = 5\cdot 10 = 50$ and $12x = 12\cdot 10 = 120$.

Check. The measure of the second angle, 50°, is five times 10°, the measure of the first angle; the measure of the third angle, 120°, is twice the sum of 10° and 50°, or 2(10°+50°), or 2·60°. Also, 10°+50°+120° = 180°. The answer checks.

State. The measure of the first angle is 10°.

51. $y^2\cdot y^{-6}\cdot y^8 = y^{2+(-6)+8} = y^4$

52. $\dfrac{x^6}{x^7} = x^{6-7} = x^{-1} = \dfrac{1}{x}$

53. $(-3x^3y^{-2})^3 = (-3)^3(x^3)^3(y^{-2})^3 = -27x^{3\cdot 3}y^{-2\cdot 3} = -27x^9y^{-6} = -\dfrac{27x^9}{y^6}$

54. $\dfrac{x^3x^{-4}}{x^{-5}x} = \dfrac{x^{3+(-4)}}{x^{-5+1}} = \dfrac{x^{-1}}{x^{-4}} = x^{-1-(-4)} = x^{-1+4} = x^3$

55. $4x - 5y = 20$

To find the x-intercept, let $y = 0$. Then solve for x.

$$\begin{aligned} 4x-5y &= 20 \\ 4x-5\cdot 0 &= 20 \\ 4x &= 20 \\ x &= 5 \end{aligned}$$

Thus, $(5,0)$ is the x-intercept.

To find the y-intercept, let $x = 0$. Then solve for y.

$$\begin{aligned} 4x-5y &= 20 \\ 4\cdot 0-5y &= 20 \\ -5y &= 20 \\ y &= -4 \end{aligned}$$

Thus, $(0,-4)$ is the y-intercept.

Plot these points and draw the graph.

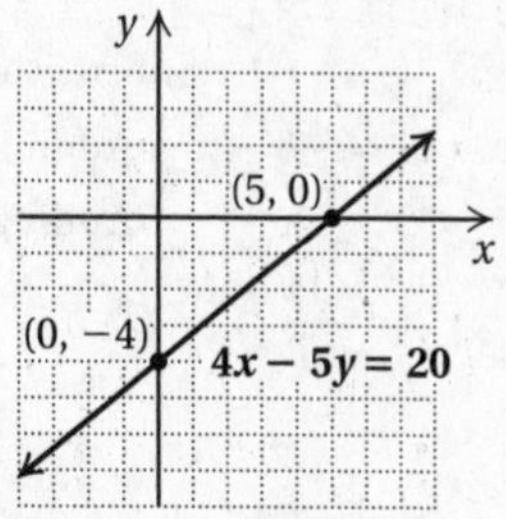

A third point should be used as a check. We substitute any value for x and solve for y.

We let $x = 4$. Then

$$\begin{aligned} 4x - 5y &= 20 \\ 4 \cdot 4 - 5y &= 20 \\ 16 - 5y &= 20 \\ -5y &= 4 \\ y &= -\frac{4}{5} \end{aligned}$$

The point $\left(4, -\frac{4}{5}\right)$ is on the graph, so the graph is probably correct.

56.

$$\begin{aligned} 3^2 &= 9 \\ 3^{-2} &= \frac{1}{3^2} = \frac{1}{9} \\ \left(\frac{1}{3}\right)^2 &= \frac{1}{9} \\ \left(\frac{1}{3}\right)^{-2} &= \left(\frac{3}{1}\right)^2 = 9 \\ -3^2 &= -9 \\ (-3)^2 &= 9 \\ \left(-\frac{1}{3}\right)^2 &= \frac{1}{9} \\ \left(-\frac{1}{3}\right)^{-2} &= \left(-\frac{3}{1}\right)^2 = 9 \end{aligned}$$

57. The area of the framed picture is x in.$\cdot x$ in., or x^2 in^2. The length of a side of the picture is $(x-2)$ in. Then the area of the picture is $(x-2)$ in.$\cdot(x-2)$ in., or (x^2-4x+4) in^2. We subtract to find the area of the frame.

$$x^2 - (x^2 - 4x + 4) = x^2 - x^2 + 4x - 4 = (4x - 4) \text{ in}^2$$

58.

$$\begin{aligned} &[(2x)^2 - (3x)^3 + 2x^2x^3 + (x^2)^2] + [5x^2(2x^3) - ((2x)^2)^2] \\ &= [4x^2 - 27x^3 + 2x^5 + x^4] + [10x^5 - (4x^2)^2] \\ &= [4x^2 - 27x^3 + 2x^5 + x^4] + [10x^5 - 16x^4] \\ &= 12x^5 - 15x^4 - 27x^3 + 4x^2 \end{aligned}$$

59. $(x-3)^2 + (2x+1)^2 = x^2 - 6x + 9 + 4x^2 + 4x + 1 = 5x^2 - 2x + 10$

60.

$$\begin{aligned} (x+3)(2x-5)+(x-1)^2 &= (3x+1)(x-3) \\ 2x^2-5x+6x-15+x^2-2x+1 &= 3x^2-9x+x-3 \\ 3x^2 - x - 14 &= 3x^2 - 8x - 3 \\ -x - 14 &= -8x - 3 \quad \text{Subtracting } 3x^2 \\ 7x - 14 &= -3 \quad \text{Adding } 8x \\ 7x &= 11 \quad \text{Adding 14} \\ x &= \frac{11}{7} \end{aligned}$$

The solution is $\frac{11}{7}$.

61. First we find $(2x^2 + x - 6) \div (2x - 3)$.

$$\begin{array}{r} x + 2 \\ 2x-3 \overline{)\, 2x^2 + x - 6} \\ \underline{2x^2 - 3x} \\ 4x - 6 \\ \underline{4x - 6} \\ 0 \end{array}$$

Next we find $(2x^2 - 9x - 5) \div (x - 5)$.

$$\begin{array}{r} 2x + 1 \\ x-5 \overline{)\, 2x^2 - 9x - 5} \\ \underline{2x^2 - 10x} \\ x - 5 \\ \underline{x - 5} \\ 0 \end{array}$$

Now we solve the resulting equation.

$$\begin{aligned} x + 2 &= 2x + 1 \\ -x + 2 &= 1 \\ -x &= -1 \\ x &= 1 \end{aligned}$$

The solution is 1.

62.

$$\begin{aligned} 20 - 3|x| &= 5 \\ -3|x| &= -15 \\ |x| &= 5 \end{aligned}$$

The solutions are the numbers whose distance from 0 is 5. They are -5 and 5.

63. First we find $(x^3 - 4x^2 - 17x + 60) \div (x - 5)$.

$$\begin{array}{r} x^2 + x - 12 \\ x-5 \overline{)\, x^3 - 4x^2 - 17x + 60} \\ \underline{x^3 - 5x^2} \\ x^2 - 17x \\ \underline{x^2 - 5x} \\ -12x + 60 \\ \underline{-12x + 60} \\ 0 \end{array}$$

Then we solve the resulting equation.

$$\begin{aligned} (x-3)(x+4) &= x^2 + x - 12 \\ x^2 + x - 12 &= x^2 + x - 12 \\ x - 12 &= x - 12 \quad \text{Subtracting } x^2 \\ -12 &= -12 \quad \text{TRUE} \end{aligned}$$

$-12 = -12$ is true for all real numbers. However, we divided by $x - 5$ on the right side of the original equation and $x - 5 = 0$ when $x = 5$. Thus, all real numbers except 5 are solutions of the equation.

Chapter 5

Polynomials: Factoring

Exercise Set 5.1

RC1. $(9x)(4x) = 36x^2$; answer (b) is correct.

RC3. Factoring completely, we have $36x^2 - 27x = 9x(4x - 3)$; answer (d) is correct.

1. $36 = 2 \cdot 2 \cdot 3 \cdot 3$

$42 = 2 \cdot 3 \cdot 7$

One factor of 2 and one factor of 3 are common to the prime factorizations. The GCF is $2 \cdot 3$, or 6.

3. $48 = 2 \cdot 2 \cdot 2 \cdot 2 \cdot 3$

$72 = 2 \cdot 2 \cdot 2 \cdot 3 \cdot 3$

$120 = 2 \cdot 2 \cdot 2 \cdot 3 \cdot 5$

Three factors of 2 and one factor of 3 are common to the prime factorizations. The GCF is $2 \cdot 2 \cdot 2 \cdot 3$, or 24.

5. $8 = 2 \cdot 2 \cdot 2$

$15 = 3 \cdot 5$

$40 = 2 \cdot 2 \cdot 2 \cdot 5$

There is no common prime factor, so the GCF is 1.

7. $x^2 = x^2$

$-6x = -1 \cdot 2 \cdot 3 \cdot x$

The coefficients have no common prime factor. The GCF of the powers of x is x because 1 is the smallest exponent of x. Thus the GCF is x.

9. $3x^4 = 3 \cdot x^4$

$x^2 = x^2$

The coefficients have no common prime factor. The GCF of the powers of x is x^2 because 2 is the smallest exponent of x. Thus the GCF is x^2.

11. $2x^2 = 2 \cdot x^2$

$2x = 2 \cdot x$

$-8 = -1 \cdot 2 \cdot 2 \cdot 2$

Each coefficient has a factor of 2. There are no other common prime factors. The GCF of the powers of x is 0 since -8 has no x-factor. Thus the GCF is 2.

13. $-17x^5y^3 = -1 \cdot 17 \cdot x^5 \cdot y^3$

$34x^3y^2 = 2 \cdot 17 \cdot x^3 \cdot y^2$

$51xy = 3 \cdot 17 \cdot x \cdot y$

Each coefficient has a factor of 17. There are no other common prime factors. The GCF of the powers of x is x because 1 is the smallest exponent of x. Similarly, the GCF of the powers of y is y because 1 is the smallest exponent of y. Thus the GCF is $17xy$.

15. $-x^2 = -1 \cdot x^2$

$-5x = -1 \cdot 5 \cdot x$

$-20x^3 = -1 \cdot 2 \cdot 2 \cdot 5 \cdot x^3$

The coefficients have no common prime factor. (Note that -1 is not a prime number.) The GCF of the powers of x is x because 1 is the smallest exponent of x. Thus the GCF is x.

17. $x^5y^5 = x^5 \cdot y^5$

$x^4y^3 = x^4 \cdot y^3$

$x^3y^3 = x^3 \cdot y^3$

$-x^2y^2 = -1 \cdot x^2 \cdot y^2$

There is no common prime factor. The GCF of the powers of x is x^2 because 2 is the smallest exponent of x. Similarly, the GCF of the powers of y is y^2 because 2 is the smallest exponent of y. Thus the GCF is x^2y^2.

19. $x^2 - 6x = x \cdot x - x \cdot 6$ Factoring each term

$= x(x - 6)$ Factoring out the common factor x

21. $2x^2 + 6x = 2x \cdot x + 2x \cdot 3$ Factoring each term

$= 2x(x + 3)$ Factoring out the common factor $2x$

23. $x^3 + 6x^2 = x^2 \cdot x + x^2 \cdot 6$ Factoring each term

$= x^2(x + 6)$ Factoring out x^2

25. $8x^4 - 24x^2 = 8x^2 \cdot x^2 - 8x^2 \cdot 3$

$= 8x^2(x^2 - 3)$ Factoring out $8x^2$

27. $2x^2 + 2x - 8 = 2 \cdot x^2 + 2 \cdot x - 2 \cdot 4$

$= 2(x^2 + x - 4)$ Factoring out 2

29. $17x^5y^3 + 34x^3y^2 + 51xy$

$= 17xy \cdot x^4y^2 + 17xy \cdot 2x^2y + 17xy \cdot 3$

$= 17xy(x^4y^2 + 2x^2y + 3)$

31. $6x^4 - 10x^3 + 3x^2 = x^2 \cdot 6x^2 - x^2 \cdot 10x + x^2 \cdot 3$

$= x^2(6x^2 - 10x + 3)$

33. $x^5y^5 + x^4y^3 + x^3y^3 - x^2y^2$

$= x^2y^2 \cdot x^3y^3 + x^2y^2 \cdot x^2y + x^2y^2 \cdot xy + x^2y^2(-1)$

$= x^2y^2(x^3y^3 + x^2y + xy - 1)$

35. $2x^7 - 2x^6 - 64x^5 + 4x^3$

$= 2x^3 \cdot x^4 - 2x^3 \cdot x^3 - 2x^3 \cdot 32x^2 + 2x^3 \cdot 2$

$= 2x^3(x^4 - x^3 - 32x^2 + 2)$

37. $1.6x^4 - 2.4x^3 + 3.2x^2 + 6.4x$
$= 0.8x(2x^3) - 0.8x(3x^2) + 0.8x(4x) + 0.8x(8)$
$= 0.8x(2x^3 - 3x^2 + 4x + 8)$

39. $\frac{5}{3}x^6 + \frac{4}{3}x^5 + \frac{1}{3}x^4 + \frac{1}{3}x^3$
$= \frac{1}{3}x^3(5x^3) + \frac{1}{3}x^3(4x^2) + \frac{1}{3}x^3(x) + \frac{1}{3}x^3(1)$
$= \frac{1}{3}x^3(5x^3 + 4x^2 + x + 1)$

41. $x^2(x+3) + 2(x+3)$

The binomial $x+3$ is common to both terms:

$x^2(x+3) + 2(x+3) = (x+3)(x^2+2)$

43. $4z^2(3z-1) + 7(3z-1)$

The binomial $3z-1$ is common to both terms.

$4z^2(3z-1) + 7(3z-1) = (3z-1)(4z^2+7)$

45. $2x^2(3x+2) + (3x+2) = 2x^2(3x+2) + 1\cdot(3x+2)$
$= (3x+2)(2x^2+1)$

47. $5a^3(2a-7) - (2a-7)$
$= 5a^3(2a-7) - 1(2a-7)$
$= (2a-7)(5a^3-1)$

49. $x^3 + 3x^2 + 2x + 6$
$= (x^3 + 3x^2) + (2x + 6)$
$= x^2(x+3) + 2(x+3)$ Factoring each binomial
$= (x+3)(x^2+2)$ Factoring out the common factor $x+3$

51. $2x^3 + 6x^2 + x + 3$
$= (2x^3 + 6x^2) + (x+3)$
$= 2x^2(x+3) + 1(x+3)$ Factoring each binomial
$= (x+3)(2x^2+1)$

53. $8x^3 - 12x^2 + 6x - 9 = 4x^2(2x-3) + 3(2x-3)$
$= (2x-3)(4x^2+3)$

55. $12p^3 - 16p^2 + 3p - 4$
$= 4p^2(3p-4) + 1(3p-4)$ Factoring 1 out of the second binomial
$= (3p-4)(4p^2+1)$

57. $5x^3 - 5x^2 - x + 1$
$= (5x^3 - 5x^2) + (-x + 1)$
$= 5x^2(x-1) - 1(x-1)$ Check: $-1(x-1) = -x+1$
$= (x-1)(5x^2-1)$

59. $x^3 + 8x^2 - 3x - 24 = x^2(x+8) - 3(x+8)$
$= (x+8)(x^2-3)$

61. $2x^3 - 8x^2 - 9x + 36 = 2x^2(x-4) - 9(x-4)$
$= (x-4)(2x^2-9)$

63. $(y+5)(y+7) = y^2 + 7y + 5y + 35$ Using FOIL
$= y^2 + 12y + 35$

65. $(y+7)(y-7) = y^2 - 7^2 = y^2 - 49$
$[(A+B))(A-B) = A^2 - B^2]$

67. $8x(2x^2 - 6x + 1) = 8x\cdot 2x^2 - 8x\cdot 6x + 8x\cdot 1 =$
$16x^3 - 48x^2 + 8x$

69. $(7w+6)^2 = (7w)^2 + 2\cdot 7w\cdot 6 + 6^2 = 49w^2 + 84w + 36$

71. $(4w-11)(4w+11) = (4w)^2 - (11)^2 = 16w^2 - 121$

73. $(3x-5y)(2x+7y) = 3x\cdot 2x + 3x\cdot 7y - 5y\cdot 2x - 5y\cdot 7y =$
$6x^2 + 21xy - 10xy - 35y^2 = 6x^2 + 11xy - 35y^2$

75. $4x^5 + 6x^3 + 6x^2 + 9 = 2x^3(2x^2+3) + 3(2x^2+3)$
$= (2x^3+3)(2x^2+3)$

77. $x^{12} + x^7 + x^5 + 1 = x^7(x^5+1) + (x^5+1)$
$= (x^7+1)(x^5+1)$

79. $p^3 + p^2 - 3p + 10 = p^2(p+1) - (3p-10)$

This polynomial is not factorable using factoring by grouping.

Exercise Set 5.2

RC1. The leading coefficient is the coefficient of the x^2-term; $x^2 = 1\cdot x^2$, so the leading coefficient is 1. The given statement is true.

RC3. The coefficient of the x-term is -3 so the given statement is true.

1. $x^2 + 8x + 15$

Since the constant term and coefficient of the middle term are both positive, we look for a factorization of 15 in which both factors are positive. Their sum must be 8.

Pairs of factors	Sums of factors
1, 15	16
3, 5	8

The numbers we want are 3 and 5.

$x^2 + 8x + 15 = (x+3)(x+5)$.

3. $x^2 + 7x + 12$

Since the constant term is positive and the coefficient of the middle term is positive, we look for a factorization of 12 in which both factors are positive. Their sum must be 7.

Pairs of factors	Sums of factors
1, 12	13
2, 6	8
3, 4	7

The numbers we want are 3 and 4.

$x^2 + 7x + 12 = (x+3)(x+4)$.

5. $x^2 - 6x + 9$

Since the constant term is positive and the coefficient of the middle term is negative, we look for a factorization of 9 in which both factors are negative. Their sum must be -6.

Pairs of factors	Sums of factors
$-1, -9$	-10
$-3, -3$	-6

The numbers we want are -3 and -3.

$x^2 - 6x + 9 = (x-3)(x-3)$, or $(x-3)^2$.

7. $x^2 - 5x - 14$

The constant term, -14, must be expressed as the product of a negative number and a positive number. Since the sum of these numbers must be the coefficient of the middle term, -5, the negative number must have the greater absolute value.

Pairs of factors	Sums of factors
$1, -14$	-13
$2, -7$	-5

The numbers we want are 2 and -7.

$x^2 - 5x - 14 = (x+2)(x-7)$.

9. $b^2 + 5b + 4$

Since the constant term is positive and the coefficient of the middle term is positive, we look for a factorization of 4 in which both factors are positive. Their sum must be 5.

Pairs of factors	Sums of factors
$1, 4$	5
$2, 2$	4

The numbers we want are 1 and 4.

$b^2 + 5b + 4 = (b+1)(b+4)$.

11. $t^2 + 3t - 18$

The constant term, -18, must be expressed as the product of a negative number and a positive number. Since the sum of these numbers must be the coefficient of the middle term, 3, the positive number must have the greater absolute value.

Pairs of factors	Sums of factors
$-1, 18$	17
$-2, 9$	7
$-3, 6$	3

The numbers we want are -3 and 6.

$t^2 + 3t - 18 = (t-3)(t+6)$

13. $d^2 - 7d + 10$

Since the constant term is positive and the coefficient of the middle term is negative, we look for a factorization of 10 in which both factors are negative. Their sum must be -7.

Pairs of factors	Sums of factors
$-1, -10$	-11
$-2, -5$	-7

The numbers we want are -2 and -5.

$d^2 - 7d + 10 = (d-2)(d-5)$.

15. $y^2 - 11y + 10$

Since the constant term is positive and the coefficient of the middle term is negative, we look for a factorization of 10 in which both factors are negative. Their sum must be -11.

Pairs of factors	Sums of factors
$-1, -10$	-11
$-2, -5$	-7

The numbers we want are -1 and -10.

$y^2 - 11y + 10 = (y-1)(y-10)$.

17. $x^2 + x + 1$

Since the constant term and the coefficient of the middle term are both positive, we look for a factorization of 1 in which both factors are positive. The sum must be 1. The only possible pair of factors is 1 and 1, but their sum is not 1. Thus, this polynomial is not factorable into binomials. It is prime.

19. $x^2 - 7x - 18$

The constant term, -18, must be expressed as the product of a negative number and a positive number. Since the sum of these numbers must be -7, the negative number must have the greater absolute value.

Pairs of factors	Sums of factors
$1, -18$	-17
$2, -9$	-7
$3, -6$	-3

The numbers we want are 2 and -9.

$x^2 - 7x - 18 = (x+2)(x-9)$.

21. $x^3 - 6x^2 - 16x = x(x^2 - 6x - 16)$

After factoring out the common factor, x, we consider $x^2-6x-16$. The constant term, -16, must be expressed as the product of a negative number and a positive number. Since the sum of these numbers must be -6, the negative number must have the larger absolute value.

Pairs of factors	Sums of factors
$1, -16$	-15
$2, -8$	-6

The numbers we want are 2 and -8.

Then $x^2 - 6x - 16 = (x+2)(x-8)$, so $x^3 - 6x^2 - 16x = x(x+2)(x-8)$.

23. $y^3 - 4y^2 - 45y = y(y^2 - 4y - 45)$

After factoring out the common factor, y, we consider $y^2-4y-45$. The constant term, -45, must be expressed as the product of a negative number and a positive number. Since the sum of these numbers must be -4, the negative number must have the larger absolute value.

Pairs of factors	Sums of factors
1, -45	-44
3, -15	-12
5, -9	-4

The numbers we want are 5 and -9.

Then $y^2 - 4y - 45 = (y+5)(y-9)$, so $y^3 - 4y^2 - 45y = y(y+5)(y-9)$.

25. $-2x - 99 + x^2 = x^2 - 2x - 99$

The constant term, -99, must be expressed as the product of a negative number and a positive number. Since the sum of these numbers must be -2, the negative number must have the larger absolute value.

Pairs of factors	Sums of factors
1, -99	-98
3, -33	-30
9, -11	-2

The numbers we want are 9 and -11.

$-2x - 99 + x^2 = (x+9)(x-11)$.

27. $c^4 + c^2 - 56$

Consider this trinomial as $(c^2)^2 + c^2 - 56$. We look for numbers p and q such that $c^4 + c^2 - 56 = (c^2+p)(c^2+q)$. The constant term, -56, must be expressed as the product of a negative number and a positive number. Since the sum of these numbers must be 1, the positive number must have the greater absolute value.

Pairs of factors	Sums of factors
-1, 56	55
-2, 28	26
-4, 14	10
-7, 8	1

The numbers we want are -7 and 8.

$c^4 + c^2 - 56 = (c^2 - 7)(c^2 + 8)$.

29. $a^4 + 2a^2 - 35$

Consider this trinomial as $(a^2)^2 + 2a^2 - 35$. We look for numbers p and q such that $a^4 + 2a^2 - 35 = (a^2+p)(a^2+q)$. The constant term, -35, must be expressed as the product of a negative number and a positive number. Since the sum of these numbers must be 2, the positive number must have the greater absolute value.

Pairs of factors	Sums of factors
-1, 35	34
-5, 7	2

The numbers we want are -5 and 7.

$a^4 + 2a^2 - 35 = (a^2 - 5)(a^2 + 7)$.

31. $x^2 + x - 42$

The constant term, -42, must be expressed as the product of a negative number and a positive number. Since the sum of these numbers must be 1, the positive number must have the greater absolute value.

Pairs of factors	Sums of factors
-1, 42	41
-2, 21	19
-3, 14	11
-6, 7	1

The numbers we want are -6 and 7.

$x^2 + x - 42 = (x-6)(x+7)$.

33. $7 - 2p + p^2 = p^2 - 2p + 7$

Since the constant term is positive and the coefficient of the middle term is negative, we look for a factorization of 7 in which both factors are negative. The sum must be -2. The only possible pair of factors is -1 and -7, but their sum is not -2. Thus, this polynomial is not factorable into binomials. It is prime.

35. $x^2 + 20x + 100$

We look for two factors, both positive, whose product is 100 and whose sum is 20.

They are 10 and 10. $10 \cdot 10 = 100$ and $10 + 10 = 20$.

$x^2 + 20x + 100 = (x+10)(x+10)$, or $(x+10)^2$.

37. $2z^3 - 2z^2 - 24z = 2z(z^2 - z - 12)$

After factoring out the common factor, we consider $z^2 - z - 12$. The constant term, -12, must be expressed as the product of a negative number and a positive number. Since the sum of these numbers must be -1, the negative number must have the larger absolute value.

Pairs of factors	Sums of factors
1, -12	-11
2, -6	-4
3, -4	-1

The numbers we want are 3 and -4.

Then $z^2 - z - 12 = (z+3)(z-4)$, so $2z^3 - 2z^2 - 24z = 2z(z+3)(z-4)$.

39. $3t^4 + 3t^3 + 3t^2 = 3t^2(t^2 + t + 1)$

After factoring out the common factor, we consider $t^2 + t + 1$. Since the constant term and the coefficient of the middle term are both positive, we look for a factorization of 1 in which both factors must be positive. Their sum must be 1. Since the only possible factorization of 1 is $1 \cdot 1$ and $1 + 1 \neq 1$, we see that $t^2 + t + 1$ cannot be factored as a product of polynomials. Thus, $3t^4 + 3t^3 + 3t^2 = 3t^2(t^2 + t + 1)$.

41. $x^4 - 21x^3 - 100x^2 = x^2(x^2 - 21x - 100)$

After factoring out the common factor, x^2, we consider $x^2 - 21x - 100$. We look for two factors whose product is -100 and whose sum is -21. They are 4 and -25. $4 \cdot (-25) = -100$ and $4 + (-25) = -21$.

Then $x^2-21x-100 = (x+4)(x-25)$, so $x^4-21x^3-100x^2 = x^2(x+4)(x-25)$.

43. $x^2 - 21x - 72$

We look for two factors whose product is -72 and whose sum is -21. They are 3 and -24.

$x^2 - 21x - 72 = (x+3)(x-24)$.

45. $x^2 - 25x + 144$

We look for two factors, both negative, whose product is 144 and whose sum is -25. They are -9 and -16.

$x^2 - 25x + 144 = (x-9)(x-16)$.

47. $a^2 + a - 132$

We look for two factors whose product is -132 and whose sum is 1. They are -11 and 12.

$a^2 + a - 132 = (a-11)(a+12)$.

49. $3t^2 + 6t + 3 = 3(t^2 + 2t + 1)$

After factoring out the common factor, we consider $t^2 + 2t + 1$. We look for two factors, both positive, whose product is 1 and whose sum is 2. They are 1 and 1.

Then $t^2+2t+1 = (t+1)(t+1)$, so $3t^2+6t+3 = 3(t+1)(t+1)$, or $3(t+1)^2$.

51. $w^4 - 8w^3 + 16w^2 = w^2(w^2 - 8w + 16)$

After factoring out the common factor, we consider $w^2 - 8w + 16$. We look for two factors, both negative, whose product is 16 and whose sum is -8. They are -4 and -4.

Then $w^2-8w+16 = (w-4)(w-4)$, so $w^4-8w^3+16w^2 = w^2(w-4)(w-4)$, or $w^2(w-4)^2$.

53. $30 + 7x - x^2 = -x^2 + 7x + 30 = -1(x^2 - 7x - 30)$

Now we factor $x^2-7x-30$. We look for two factors whose product is -30 and whose sum is -7.

Pairs of factors	Sums of factors
1, -30	-29
2, -15	-13
3, -10	-7
5, -6	-1

The numbers we want are 3 and -10. Then $x^2 - 7x - 30 = (x+3)(x-10)$, so we have:

$-x^2 + 7x + 30$

$= -1(x+3)(x-10)$

$= (-x-3)(x-10)$ Multiplying $x+3$ by -1

$= (x+3)(-x+10)$ Multiplying $x-10$ by -1

55. $24 - a^2 - 10a = -a^2 - 10a + 24 = -1(a^2 + 10a - 24)$

Now we factor $a^2+10a-24$. We look for two factors whose product is -24 and whose sum is 10.

Pairs of factors	Sums of factors
-1, 24	23
-2, 12	10
-3, 8	5
-4, 6	2

The numbers we want are -2 and 12. Then $a^2 + 10a - 24 = (a-2)(a+12)$, so we have:

$-a^2 - 10a + 24$

$= -1(a-2)(a+12)$

$= (-a+2)(a+12)$ Multiplying $a-2$ by -1

$= (a-2)(-a-12)$ Multiplying $a+12$ by -1

57. $120 - 23x + x^2 = x^2 - 23x + 120$

We look for two factors, both negative, whose product is 120 and whose sum is -23. They are -8 and -15.

$x^2 - 23x + 120 = (x-8)(x-15)$.

59. First write the polynomial in descending order and factor out -1.

$108 - 3x - x^2 = -x^2 - 3x + 108 = -1(x^2 + 3x - 108)$

Now we factor the polynomial $x^2 + 3x - 108$. We look for two factors whose product is -108 and whose sum is 3. They are -9 and 12.

$x^2 + 3x - 108 = (x-9)(x+12)$

The final answer must include -1 which was factored out above.

$-x^2 - 3x + 108$

$= -1(x-9)(x+12)$

$= (-x+9)(x+12)$ Multiplying $x-9$ by -1

$= (x-9)(-x-12)$ Multiplying $x+12$ by -1

61. $y^2 - 0.2y - 0.08$

We look for two factors whose product is -0.08 and whose sum is-0.2. They are -0.4 and 0.2.

$y^2 - 0.2y - 0.08 = (y-0.4)(y+0.2)$.

63. $p^2 + 3pq - 10q^2 = p^2 + 3qp - 10q^2$

Think of $3q$ as a "coefficient" of p. Then we look for factors of $-10q^2$ whose sum is $3q$. They are $5q$ and $-2q$.

$p^2 + 3pq - 10q^2 = (p+5q)(p-2q)$.

65. $84 - 8t - t^2 = -t^2 - 8t + 84 = -1(t^2 + 8t - 84)$

Now we factor $t^2+8t-84$. We look for two factors whose product is -84 and whose sum is 8. They are 14 and -6.

Then $t^2 + 8t - 84 = (t+14)(t-6)$, so we have:

$-t^2 - 8t + 84$

$= -1(t+14)(t-6)$

$= (-t-14)(t-6)$ Multiplying $t+14$ by -1

$= (t+14)(-t+6)$ Multiplying $t-6$ by -1

67. $m^2 + 5mn + 4n^2 = m^2 + 5nm + 4n^2$

We look for factors of $4n^2$ whose sum is $5n$. They are $4n$ and n.

$m^2 + 5mn + 4n^2 = (m+4n)(m+n)$

69. $s^2 - 2st - 15t^2 = s^2 - 2ts - 15t^2$

We look for factors of $-15t^2$ whose sum is $-2t$. They are $-5t$ and $3t$.

$s^2 - 2st - 15t^2 = (s-5t)(s+3t)$

71. $6a^{10} - 30a^9 - 84a^8 = 6a^8(a^2 - 5a - 14)$

After factoring out the common factor, $6a^8$, we consider $a^2 - 5a - 14$. We look for two factors whose product is -14 and whose sum is -5. They are 2 and -7.

$a^2 - 5a - 14 = (a+2)(a-7)$, so $6a^{10} - 30a^9 - 84a^8 = 6a^8(a+2)(a-7)$.

73.
$$\begin{aligned} 2x + 7 &= 0 \\ 2x &= -7 \\ x &= -\frac{7}{2} \end{aligned}$$

The solution is $-\frac{7}{2}$.

75.
$$\begin{aligned} 3x - 2 - x &= 5x - 9x \\ 2x - 2 &= -4x \\ 6x - 2 &= 0 \\ 6x &= 2 \\ x &= \frac{1}{3} \end{aligned}$$

The solution is $\frac{1}{3}$.

77.
$$\begin{aligned} 5(t-1) - 3 &= 4t - (7t - 2) \\ 5t - 5 - 3 &= 4t - 7t + 2 \\ 5t - 8 &= -3t + 2 \\ 8t - 8 &= 2 \\ 8t &= 10 \\ t &= \frac{10}{8} = \frac{5}{4} \end{aligned}$$

The solution is $\frac{5}{4}$.

79. $-2x < 48$

$x > -24$ Dividing by -2 and reversing the inequality symbol

The solution set is $\{x | x > -24\}$.

81.
$$\begin{aligned} -6(x-4) + 8(4-x) &\le 3(x-7) \\ -6x + 24 + 32 - 8x &\le 3x - 21 \\ -14x + 56 &\le 3x - 21 \\ 56 &\le 17x - 21 \\ 77 &\le 17x \\ \frac{77}{17} &\le x \end{aligned}$$

The solution set is $\left\{x \middle| x \ge \frac{77}{17}\right\}$.

83.
$$\begin{aligned} y &= mx + b \\ y - b &= mx \\ \frac{y-b}{m} &= x \end{aligned}$$

85. Let b = the attendance in 2011, in millions. Then the attendance in 2012 was $b + 2\% \cdot b$, or $100\%b + 2\%b$, or $102\%b$. Now we reword and translate.

$$\underbrace{74.97}_{\downarrow} \text{ is } 102\% \text{ of } \underbrace{\text{what number?}}$$
$$74.97 = 102\% \cdot b$$

We convert to decimal notation and then divide.

$$\begin{aligned} 74.97 &= 1.02 \cdot b \\ 73.5 &= b \end{aligned}$$

The attendance was 73.5 million in 2011.

87. $y^2 + my + 50$

We look for pairs of factors whose product is 50. The sum of each pair is represented by m.

Pairs of factors whose product is -50	Sums of factors
1, 50	51
-1, -50	-51
2, 25	27
-2, -25	-27
5, 10	15
-5, -10	-15

The polynomial $y^2 + my + 50$ can be factored if m is 51, -51, 27, -27, 15, or -15.

89. $x^2 - \frac{1}{2}x - \frac{3}{16}$

We look for two factors, one positive and one negative, whose product is $-\frac{3}{16}$ and whose sum is $-\frac{1}{2}$.

They are $-\frac{3}{4}$ and $\frac{1}{4}$.

$-\frac{3}{4} \cdot \frac{1}{4} = -\frac{3}{16}$ and $-\frac{3}{4} + \frac{1}{4} = -\frac{2}{4} = -\frac{1}{2}$.

$x^2 - \frac{1}{2}x - \frac{3}{16} = \left(x - \frac{3}{4}\right)\left(x + \frac{1}{4}\right)$

91. $x^2 + \frac{2}{3}x + \frac{1}{9}$

Since the constant term is positive and the coefficient of the middle term is positive, we look for a factorization of $\frac{1}{9}$ in which both factors are positive. Their sum must be $\frac{2}{3}$.

Pairs of factors	Sums of factors
$1, \frac{1}{9}$	$\frac{10}{9}$
$\frac{1}{3}, \frac{1}{3}$	$\frac{2}{3}$

The numbers we want are $\frac{1}{3}$ and $\frac{1}{3}$.

$x^2 + \frac{2}{3}x + \frac{1}{9} = \left(x + \frac{1}{3}\right)\left(x + \frac{1}{3}\right)$, or $\left(x + \frac{1}{3}\right)^2$.

93. $x^2 + \frac{30}{7}x - \frac{25}{7}$

We look for two factors, one positive and one negative, whose product is $-\frac{25}{7}$ and whose sum is $\frac{30}{7}$.

They are 5 and $-\frac{5}{7}$.

$5 \cdot \left(-\frac{5}{7}\right) = -\frac{25}{7}$ and $5 + \left(-\frac{5}{7}\right) = \frac{35}{7} + \left(-\frac{5}{7}\right) = \frac{30}{7}$.

$x^2 + \frac{30}{7}x - \frac{25}{7} = (x+5)\left(x - \frac{5}{7}\right)$

95. $b^{2n} + 7b^n + 10$

Consider this trinomial as $(b^n)^2 + 7b^n + 10$. We look for numbers p and q such that $b^{2n}+7b^n+10 = (b^n+p)(b^n+q)$. We find two factors, both positive, whose product is 10 and whose sum is 7. They are 5 and 2.

$b^{2n} + 7b^n + 10 = (b^n + 5)(b^n + 2)$

97. We first label the drawing with additional information.

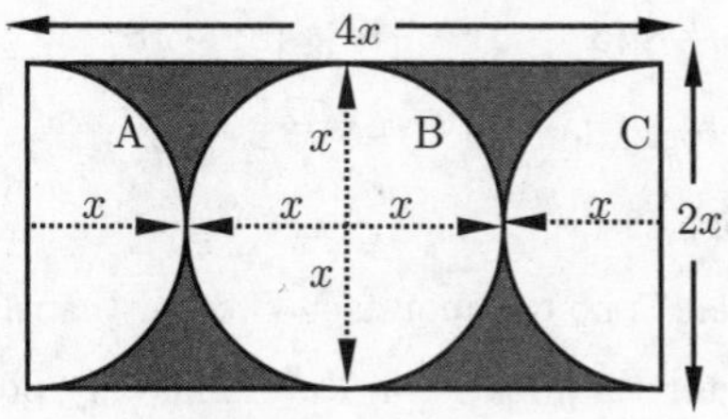

$4x$ represents the length of the rectangle and $2x$ the width. The area of the rectangle is $4x \cdot 2x$, or $8x^2$.

The area of semicircle A is $\frac{1}{2}\pi x^2$.

The area of circle B is πx^2.

The area of semicircle C is $\frac{1}{2}\pi x^2$.

$$\text{Area of shaded region} = \text{Area of rectangle} - \text{Area of } A - \text{Area of } B - \text{Area of } C$$

$$\text{Area of shaded region} = 8x^2 - \frac{1}{2}\pi x^2 - \pi x^2 - \frac{1}{2}\pi x^2$$

$$= 8x^2 - 2\pi x^2$$

$$= 2x^2(4-\pi)$$

The shaded area can be represented by $2x^2(4-\pi)$.

Exercise Set 5.3

RC1. True; see page 404 in the text.

RC3. False; another choice is x and $10x$.

1. $2x^2 - 7x - 4$

(1) Look for a common factor. There is none (other than 1 or -1).

(2) Factor the first term, $2x^2$. The only possibility is $2x$, x. The desired factorization is of the form:

$(2x+\quad)(x+\quad)$

(3) Factor the last term, -4, which is negative. The possibilities are -4, 1 and 4, -1 and 2, -2.

These factors can also be written as 1, -4 and -1, 4 and -2, 2.

(4) Look for combinations of factors from steps (2) and (3) such that the sum of their products is the middle term, $-7x$. We try some possibilities:

$(2x-4)(x+1) = 2x^2 - 2x - 4$

$(2x+4)(x-1) = 2x^2 + 2x - 4$

$(2x+2)(x-2) = 2x^2 - 2x - 4$

$(2x+1)(x-4) = 2x^2 - 7x - 4$

The factorization is $(2x+1)(x-4)$.

3. $5x^2 - x - 18$

(1) There is no common factor (other than 1 or -1).

(2) Factor the first term, $5x^2$. The only possibility is $5x$, x. The desired factorization is of the form:

$(5x+\quad)(x+\quad)$

(3) Factor the last term, -18. The possibilities are -18, 1 and 18, -1 and -9, 2 and 9, -2 and -6, 3 and 6, -3.

These factors can also be written as 1, -18 and -1, 18 and 2, -9 and -2, 9 and 3, -6 and -3, 6.

(4) Look for combinations of factors from steps (2) and (3) such that the sum of their products is the middle term, x. We try some possibilities:

$(5x-18)(x+1) = 5x^2 - 13x - 18$

$(5x+18)(x-1) = 5x^2 + 13x - 18$

$(5x+9)(x-2) = 5x^2 - x - 18$

The factorization is $(5x+9)(x-2)$.

5. $6x^2 + 23x + 7$

(1) There is no common factor (other than 1 or -1).

(2) Factor the first term, $6x^2$. The possibilities are $6x$, x and $3x$, $2x$. We have these as possibilities for factorizations:

$(6x+\quad)(x+\quad)$ and $(3x+\quad)(2x+\quad)$

(3) Factor the last term, 7. The possibilities are 7, 1 and -7, -1.

These factors can also be written as 1, 7 and -1, -7.

(4) Look for combinations of factors from steps (2) and (3) such that the sum of their products is the middle term, $23x$. Since all signs are positive, we need consider only plus signs. We try some possibilities:

$(6x+7)(x+1) = 6x^2 + 13x + 7$

$(3x+7)(2x+1) = 6x^2 + 17x + 7$

$(6x+1)(x+7) = 6x^2 + 43x + 7$

$(3x+1)(2x+7) = 6x^2 + 23x + 7$

The factorization is $(3x+1)(2x+7)$.

7. $3x^2 + 4x + 1$

(1) There is no common factor (other than 1 or -1).

(2) Factor the first term, $3x^2$. The only possibility is $3x$, x. The desired factorization is of the form:

$(3x+\quad)(x+\quad)$

(3) Factor the last term, 1. The possibilities are 1, 1 and −1, −1.

(4) Look for combinations of factors from steps (2) and (3) such that the sum of their products is the middle term, $4x$. Since all signs are positive, we need consider only plus signs. There is only one such possibility:

$(3x+1)(x+1) = 3x^2 + 4x + 1$

The factorization is $(3x+1)(x+1)$.

9. $4x^2 + 4x - 15$

(1) There is no common factor (other than 1 or −1).

(2) Factor the first term, $4x^2$. The possibilities are $4x$, x and $2x$, $2x$. We have these as possibilities for factorizations:

$(4x+\quad)(x+\quad)$ and $(2x+\quad)(2x+\quad)$

(3) Factor the last term, −15. The possibilities are 15, −1 and −15, 1 and 5, −3 and −5, 3.

These factors can also be written as −1, 15 and 1, −15 and −3, 5 and 3, −5.

(4) We try some possibilities:

$(4x+15)(x-1) = 4x^2 + 11x - 15$

$(2x+15)(2x-1) = 4x^2 + 28x - 15$

$(4x-15)(x+1) = 4x^2 - 11x - 15$

$(2x-15)(2x+1) = 4x^2 - 28x - 15$

$(4x+5)(x-3) = 4x^2 - 7x - 15$

$(2x+5)(2x-3) = 4x^2 + 4x - 15$

The factorization is $(2x+5)(2x-3)$.

11. $2x^2 - x - 1$

(1) There is no common factor (other than 1 or −1).

(2) Factor the first term, $2x^2$. The only possibility is $2x$, x. The desired factorization is of the form:

$(2x+\quad)(x+\quad)$

(3) Factor the last term, −1. The only possibility is −1, 1.

These factors can also be written as 1, −1.

(4) We try the possibilities:

$(2x-1)(x+1) = 2x^2 + x - 1$

$(2x+1)(x-1) = 2x^2 - x - 1$

The factorization is $(2x+1)(x-1)$.

13. $9x^2 + 18x - 16$

(1) There is no common factor (other than 1 or −1).

(2) Factor the first term, $9x^2$. The possibilities are $9x$, x and $3x$, $3x$. We have these as possibilities for factorizations:

$(9x+\quad)(x+\quad)$ and $(3x+\quad)(3x+\quad)$

(3) Factor the last term, −16. The possibilities are 16, −1 and −16, 1 and 8, −2 and −8, 2 and 4, −4.

These factors can also be written as −1, 16 and 1, −16 and −2, 8 and 2, −8 and −4, 4.

(4) We try some possibilities:

$(9x+16)(x-1) = 9x^2 + 7x - 16$

$(3x+16)(3x-1) = 9x^2 + 45x - 16$

$(9x-16)(x+1) = 9x^2 - 7x - 16$

$(3x-16)(3x+1) = 9x^2 - 45x - 16$

$(9x+8)(x-2) = 9x^2 - 10x - 16$

$(3x+8)(3x-2) = 9x^2 + 18x - 16$

The factorization is $(3x+8)(3x-2)$.

15. $3x^2 - 5x - 2$

(1) There is no common factor (other than 1 or −1).

(2) Factor the first term, $3x^2$. The only possibility is $3x$, x. The desired factorization is of the form:

$(3x+\quad)(x+\quad)$

(3) Factor the last term, −2. The possibilities are 2, −1 and −2 and 1.

These factors can also be written as −1, 2 and 1, −2.

(4) We try some possibilities:

$(3x+2)(x-1) = 3x^2 - x - 2$

$(3x-2)(x+1) = 3x^2 + x - 2$

$(3x-1)(x+2) = 3x^2 + 5x - 2$

$(3x+1)(x-2) = 3x^2 - 5x - 2$

The factorization is $(3x+1)(x-2)$.

17. $12x^2 + 31x + 20$

(1) There is no common factor (other than 1 or −1).

(2) Factor the first term, $12x^2$. The possibilities are $12x$, x and $6x$, $2x$ and $4x$, $3x$. We have these as possibilities for factorizations:

$(12x+\quad)(x+\quad)$ and $(6x+\quad)(2x+\quad)$ and

$(4x+\quad)(3x+\quad)$

(3) Factor the last term, 20. Since all signs are positive, we need consider only positive pairs of factors. Those factor pairs are 20, 1 and 10, 2 and 5, 4.

These factors can also be written as 1, 20 and 2, 10 and 4, 5.

(4) We can immediately reject all possibilities in which either factor has a common factor, such as $(12x+20)$ or $(6x+4)$, because we determined at the outset that there are no common factors. We try some of the remaining possibilities:

$(12x+1)(x+20) = 12x^2 + 241x + 20$

$(12x+5)(x+4) = 12x^2 + 53x + 20$

$(6x + 1)(2x + 20) = 12x^2 + 122x + 20$

$(4x + 5)(3x + 4) = 12x^2 + 31x + 20$

The factorization is $(4x + 5)(3x + 4)$.

19. $14x^2 + 19x - 3$

(1) There is no common factor (other than 1 or -1).

(2) Factor the first term, $14x^2$. The possibilities are $14x$, x and $7x$, $2x$. We have these as possibilities for factorizations:

$(14x+\quad)(x+\quad)$ and $(7x+\quad)(2x+\quad)$

(3) Factor the last term, -3. The possibilities are -1, 3 and -3, 1.

These factors can also be written as 3, -1 and 1, -3.

(4) We try some possibilities:

$(14x - 1)(x + 3) = 14x^2 + 41x - 3$

$(7x - 1)(2x + 3) = 7x^2 + 19x - 3$

The factorization is $(7x - 1)(2x + 3)$.

21. $9x^2 + 18x + 8$

(1) There is no common factor (other than 1 or -1).

(2) Factor the first term, $9x^2$. The possibilities are $9x$, x and $3x$, $3x$. We have these as possibilities for factorizations:

$(9x+\quad)(x+\quad)$ and $(3x+\quad)(3x+\quad)$

(3) Factor the last term, 8. Since all signs are positive, we need consider only positive pairs of factors. Those factor pairs are 8, 1 and 4, 2.

These factors can also be written as 1, 8 and 2, 4.

(4) We try some possibilities:

$(9x + 8)(x + 1) = 9x^2 + 17x + 8$

$(3x + 8)(3x + 1) = 9x^2 + 27x + 8$

$(9x + 4)(x + 2) = 9x^2 + 22x + 8$

$(3x + 4)(3x + 2) = 9x^2 + 18x + 8$

The factorization is $(3x + 4)(3x + 2)$.

23. $49 - 42x + 9x^2 = 9x^2 - 42x + 49$

(1) There is no common factor (other than 1 or -1).

(2) Factor the first term, $9x^2$. The possibilities are $9x$, x and $3x$, $3x$. We have these as possibilities for factorizations:

$(9x+\quad)(x+\quad)$ and $(3x+\quad)(3x+\quad)$

(3) Factor 49. Since 49 is positive and the middle term is negative, we need consider only negative pairs of factors. Those factor pairs are -49, -1 and -7, -7.

The first pair of factors can also be written as -1, -49.

(4) We try some possibilities:

$(9x - 49)(x - 1) = 9x^2 - 58x + 49$

$(3x - 49)(3x - 1) = 9x^2 - 150x + 49$

$(9x - 7)(x - 7) = 9x^2 - 70x + 49$

$(3x - 7)(3x - 7) = 9x^2 - 42x + 49$

The factorization is $(3x - 7)(3x - 7)$, or $(3x - 7)^2$. This can also be expressed as follows:

$(3x - 7)^2 = (-1)^2(3x - 7)^2 = [-1 \cdot (3x - 7)]^2 =$

$(-3x + 7)^2$, or $(7 - 3x)^2$

25. $24x^2 + 47x - 2$

(1) There is no common factor (other than 1 or -1).

(2) Factor the first term, $24x^2$. The possibilities are $24x$, x and $12x$, $2x$ and $6x$, $4x$ and $3x$, $8x$. We have these as possibilities for factorizations:

$(24x+\quad)(x+\quad)$ and $(12x+\quad)(2x+\quad)$ and

$(6x+\quad)(4x+\quad)$ and $(3x+\quad)(8x+\quad)$

(3) Factor the last term, -2. The possibilities are 2, -1 and -2, 1.

These factors can also be written as -1, 2 and 1, -2.

(4) We can immediately reject all possibilities in which either factor has a common factor, such as $(24x+2)$ or $(12x - 2)$, because we determined at the outset that there are no common factors. We try some of the remaining possibilities:

$(24x - 1)(x + 2) = 24x^2 + 47x - 2$

The factorization is $(24x - 1)(x + 2)$.

27. $35x^2 - 57x - 44$

(1) There is no common factor (other than 1 or -1).

(2) Factor the first term, $35x^2$. The possibilities are $35x$, x and $7x$, $5x$. We have these as possibilities for factorizations:

$(35x+\quad)(x+\quad)$ and $(7x+\quad)(5x+\quad)$

(3) Factor the last term, -44. The possibilities are 1, -44 and -1, 44 and 2, -22 and -2, 22 and 4, -11, and -4, 11.

These factors can also be written as -44, 1 and 44, -1 and -22, 2 and 22, -2 and -11, 4 and 11, -4.

(4) We try some possibilities:

$(35x + 1)(x - 44) = 35x^2 - 1539x - 44$

$(7x + 1)(5x - 44) = 35x^2 - 303x - 44$

$(35x + 2)(x - 22) = 35x^2 - 768x - 44$

$(7x + 2)(5x - 22) = 35x^2 - 144x - 44$

$(35x + 4)(x - 11) = 35x^2 - 381x - 44$

$(7x + 4)(5x - 11) = 35x^2 - 57x - 44$

The factorization is $(7x + 4)(5x - 11)$.

29. $20 + 6x - 2x^2 = -2x^2 + 6x + 20$

We factor out the common factor, -2. Factoring out -2 rather than 2 gives us a positive leading coefficient.

$-2(x^2 - 3x - 10)$

Then we factor the trinomial $x^2 - 3x - 10$. We look for a pair of factors whose product is -10 and whose sum is -3. The numbers are -5 and 2. The factorization of $x^2 - 3x - 10$ is $(x - 5)(x + 2)$. Then $20 + 6x - 2x^2 = -2(x-5)(x+2)$. If we think of -2 and $-1 \cdot 2$ then we can write other correct factorizations:

$20 + 6x - 2x^2$

$= 2(-x + 5)(x + 2)$ Multiplying $x - 5$ by -1

$= 2(x - 5)(-x - 2)$ Multplying $x + 2$ by -1

Note that we can also express $2(-x + 5)(x + 2)$ as $2(5 - x)(x + 2)$ since $-x + 5 = 5 - x$ by the commutative law of addition.

31. $12x^2 + 28x - 24$

(1) We factor out the common factor, 4: $4(3x^2 + 7x - 6)$

Then we factor the trinomial $3x^2 + 7x - 6$.

(2) Factor $3x^2$. The only possibility is $3x$, x. The desired factorization is of the form:

$$(3x+ \quad)(x+ \quad)$$

(3) Factor -6. The possibilities are 6, -1 and -6, 1 and 3, -2 and -3, 2.

These factors can also be written as -1, 6 and 1, -6 and -2, 3 and 2, -3.

(4) We can immediately reject all possibilities in which either factor has a common factor, such as $(3x + 6)$ or $(3x - 3)$, because we factored out the largest common factor at the outset. We try some of the remaining possibilities:

$$(3x - 1)(x + 6) = 3x^2 + 17x - 6$$

$$(3x - 2)(x + 3) = 3x^2 + 7x - 6$$

The factorization of $3x^2+7x-6$ is $(3x-2)(x+3)$. We must include the common factor in order to get a factorization of the original trinomial.

$12x^2 + 28x - 24 = 4(3x - 2)(x + 3)$

33. $30x^2 - 24x - 54$

(1) We factor out the common factor, 6: $6(5x^2 - 4x - 9)$

Then we factor the trinomial $5x^2 - 4x - 9$.

(2) Factor $5x^2$. The only possibility is $5x$, x. The desired factorization is of the form:

$$(5x+ \quad)(x+ \quad)$$

(3) Factor -9. The possibilities are 9, -1 and -9, 1 and -3, 3.

These factors can also be written as -1, 9 and 1, -9 and 3, -3.

(4) We try some possibilities:

$$(5x + 9)(x - 1) = 5x^2 + 4x - 9$$

$$(5x - 9)(x + 1) = 5x^2 - 4x - 9$$

The factorization of $5x^2-4x-9$ is $(5x-9)(x+1)$. We must include the common factor in order to get a factorization of the original trinomial.

$30x^2 - 24x - 54 = 6(5x - 9)(x + 1)$

35. $4y + 6y^2 - 10 = 6y^2 + 4y - 10$

(1) We factor out the common factor, 2: $2(3y^2 + 2y - 5)$

Then we factor the trinomial $3y^2 + 2y - 5$.

(2) Factor $3y^2$. The only possibility is $3y$, y. The desired factorization is of the form:

$$(3y+ \quad)(y+ \quad)$$

(3) Factor -5. The possibilities are 5, -1 and -5, 1.

These factors can also be written as -1, 5 and 1, -5.

(4) We try some possibilities:

$$(3y + 5)(y - 1) = 3y^2 + 2y - 5$$

Then $3y^2 + 2y - 5 = (3y + 5)(y - 1)$, so $6y^2 + 4y - 10 = 2(3y + 5)(y - 1)$.

37. $3x^2 - 4x + 1$

(1) There is no common factor (other than 1 or -1).

(2) Factor the first term, $3x^2$. The only possibility is $3x$, x. The desired factorization is of the form:

$$(3x+ \quad)(x+ \quad)$$

(3) Factor the last term, 1. Since 1 is positive and the middle term is negative, we need consider only negative factor pairs. The only such pair is -1, -1.

(4) There is only one possibility:

$$(3x - 1)(x - 1) = 3x^2 - 4x + 1$$

The factorization is $(3x - 1)(x - 1)$.

39. $12x^2 - 28x - 24$

(1) We factor out the common factor, 4: $4(3x^2 - 7x - 6)$

Then we factor the trinomial $3x^2 - 7x - 6$.

(2) Factor $3x^2$. The only possibility is $3x$, x. The desired factorization is of the form:

$$(3x+ \quad)(x+ \quad)$$

(3) Factor -6. The possibilities are 6, -1 and -6, 1 and 3, -2 and -3, 2.

These factors can also be written as -1, 6 and 1, -6 and -2, 3 and 2, -3.

(4) We can immediately reject all possibilities in which either factor has a common factor, such as $(3x - 6)$ or $(3x + 3)$, because we factored out the largest common factor at the outset. We try some of the remaining possibilities:

$$(3x - 1)(x + 6) = 3x^2 + 17x - 6$$
$$(3x - 2)(x + 3) = 3x^2 + 7x - 6$$
$$(3x + 2)(x - 3) = 3x^2 - 7x - 6$$

Then $3x^2 - 7x - 6 = (3x + 2)(x - 3)$, so $12x^2 - 28x - 24 = 4(3x + 2)(x - 3)$.

41. $-1 + 2x^2 - x = 2x^2 - x - 1$

(1) There is no common factor (other than 1 or -1).

(2) Factor the first term, $2x^2$. The only possibility is $2x$, x. The desired factorization is of the form:

$$(2x+ \quad)(x+ \quad)$$

(3) Factor -1. The only possibility is $1, -1$.

(4) We try some possibilities:

$$(2x + 1)(x - 1) = 2x^2 - x - 1$$

The factorization is $(2x + 1)(x - 1)$.

43. $9x^2 - 18x - 16$

(1) There is no common factor (other than 1 or -1).

(2) Factor the first term, $9x^2$. The possibilities are $9x$, x and $3x$, $3x$. We have these as possibilities for factorizations:

$$(9x+ \quad)(x+ \quad) \text{ and } (3x+ \quad)(3x+ \quad)$$

(3) Factor the last term, -16. The possibilities are $16, -1$ and $-16, 1$ and $8, -2$ and $-8, 2$ and $4, -4$.

These factors can also be written as $-1, 16$ and $1, -16$ and $-2, 8$ and and $2, -8$ and $-4, 4$.

(4) We try some possibilities:

$$(9x + 16)(x - 1) = 9x^2 + 7x - 16$$
$$(3x + 16)(3x - 1) = 9x^2 + 45x - 16$$
$$(9x + 8)(x - 2) = 9x^2 - 10x - 16$$
$$(3x + 8)(3x - 2) = 9x^2 + 18x - 16$$
$$(3x - 8)(3x + 2) = 9x^2 - 18x - 16$$

The factorization is $(3x - 8)(3x + 2)$.

45. $15x^2 - 25x - 10$

(1) Factor out the common factor, 5:

$5(3x^2 - 5x - 2)$

Then we factor the trinomial $3x^2 - 5x - 2$. This was done in Exercise 15. We know that $3x^2 - 5x - 2 = (3x + 1)(x - 2)$, so $15x^2 - 25x - 10 = 5(3x + 1)(x - 2)$.

47. $12p^3 + 31p^2 + 20p$

(1) We factor out the common factor, p:

$$p(12p^2 + 31p + 20)$$

Then we factor the trinomial $12p^2 + 31p + 20$. This was done in Exercise 17 although the variable is x in that exercise. We know that $12p^2 + 31p + 20 = (3p + 4)(4p + 5)$, so $12p^3 + 31p^2 + 20p = p(3p + 4)(4p + 5)$.

49. $16 + 18x - 9x^2 = -9x^2 + 18x + 16$
$= -1(9x^2 - 18x - 16)$
$= -1(3x - 8)(3x + 2)$ Using the result from Exercise 43

Other correct factorizations are:

$16 + 18x - 9x^2$
$= (-3x + 8)(3x + 2)$ Multiplying $3x - 8$ by -1
$= (3x - 8)(-3x - 2)$ Multiplying $3x + 2$ by -1

We can also express $(-3x + 8)(3x + 2)$ as $(8 - 3x)(3x + 2)$ since $-3x + 8 = 8 - 3x$ by the commutative law of addition.

51. $-15x^2 + 19x - 6 = -1(15x^2 - 19x + 6)$

Now we factor $15x^2 - 19x + 6$.

(1) There is no common factor (other than 1 or -1).

(2) Factor the first term, $15x^2$. The possibilities are $15x$, x and $5x$, $3x$. We have these as possibilities for factorizations:

$$(15x+ \quad)(x+ \quad) \text{ and } (5x+ \quad)(3x+ \quad)$$

(3) Factor the last term, 6. The possibilities are $6, 1$ and $-6, -1$ and $3, 2$ and $-3, -2$.

These factors can also be written as $1, 6$ and $-1, -6$ and $2, 3$ and $-2, -3$.

(4) We try some possibilities:

$$(15x + 1)(x + 6) = 15x^2 + 91x + 6$$
$$(5x + 3)(3x + 2) = 15x^2 - 19x + 6$$
$$(5x - 3)(3x - 2) = 15x^2 - 19x + 6$$

The factorization of $15x^2 - 19x + 6$ is $(5x - 3)(3x - 2)$. Then $-15x^2 + 19x - 6 = -1(5x - 3)(3x - 2)$. Other correct factorizations are:

$-15x^2 + 19x - 6$
$= (-5x + 3)(3x - 2)$ Multiplying $5x - 3$ by -1
$= (5x - 3)(-3x + 2)$ Multiplying $3x - 2$ by -1

Note that we can also express $(-5x + 3)(3x - 2)$ as $(3 - 5x)(3x - 2)$ since $-5x + 3 = 3 - 5x$ by the commutative law of addition. Similarly, we can express $(5x - 3)(-3x + 2)$ as $(5x - 3)(2 - 3x)$.

53. $14x^4 + 19x^3 - 3x^2$

(1) Factor out the common factor, x^2: $x^2(14x^2 + 19x - 3)$

Then we factor the trinomial $14x^2 + 19x - 3$. This was done in Exercise 19. We know that $14x^2 + 19x - 3 = (7x - 1)(2x + 3)$, so $14x^4 + 19x^3 - 3x^2 = x^2(7x - 1)(2x + 3)$.

55. $168x^3 - 45x^2 + 3x$

(1) Factor out the common factor, $3x$:

$3x(56x^2 - 15x + 1)$

Then we factor the trinomial $56x^2 - 15x + 1$.

(2) Factor $56x^2$. The possibilities are $56x$, x and $28x$, $2x$ and $14x$, $4x$ and $7x$, $8x$. We have these as possibilities for factorizations:

$(56x+\quad)(x+\quad)$ and $(28x+\quad)(2x+\quad)$ and

$(14x+\quad)(4x+\quad)$ and $(7x+\quad)(8x+\quad)$

(3) Factor 1. Since 1 is positive and the middle term is negative we need consider only the negative factor pair $-1, -1$.

(4) We try some possibilities:

$(56x - 1)(x - 1) = 56x^2 - 57x + 1$

$(28x - 1)(2x - 1) = 56x^2 - 30x + 1$

$(14x - 1)(4x - 1) = 56x^2 - 18x + 1$

$(7x - 1)(8x - 1) = 56x^2 - 15x + 1$

Then $56x^2 - 15x + 1 = (7x - 1)(8x - 1)$, so $168x^3 - 45x^2 + 3x = 3x(7x - 1)(8x - 1)$.

57. $15x^4 - 19x^2 + 6 = 15(x^2)^2 - 19x^2 + 6$

(1) There is no common factor (other than 1 or -1).

(2) Factor the first term, $15x^4$. The possibilities are $15x^2$, x^2 and $5x^2$, $3x^2$. We have these as possibilities for factorizations:

$(15x^2+\quad)(x^2+\quad)$ and $(5x^2+\quad)(3x^2+\quad)$

(3) Factor 6. Since 6 is positive and the middle term is negative, we need consider only negative factor pairs. Those pairs are -6, -1 and -3, -2.

These factors can also be written as -1, -6 and -2, -3.

(4) We can immediately reject all possibilities in which either factor has a common factor, such as $(15x^2-6)$ or $(3x^2 - 3)$, because we determined at the outset that there is no common factor. We try some of the remaining possibilities:

$(15x^2 - 1)(x^2 - 6) = 15x^4 - 91x^2 + 6$

$(15x^2 - 2)(x^2 - 3) = 15x^4 - 47x^2 + 6$

$(5x^2 - 6)(3x^2 - 1) = 15x^4 - 23x^2 + 6$

$(5x^2 - 3)(3x^2 - 2) = 15x^4 - 19x^2 + 6$

The factorization is $(5x^2 - 3)(3x^2 - 2)$.

59. $25t^2 + 80t + 64$

(1) There is no common factor (other than 1 or -1).

(2) Factor the first term, $25t^2$. The possibilities are $25t$, t and $5t$, $5t$. We have these as possibilities for factorizations:

$(25t+\quad)(t+\quad)$ and $(5t+\quad)(5t+\quad)$

(3) Factor the last term, 64. Since all signs are positive, we need consider only positive pairs of factors. Those factor pairs are 64, 1 and 32, 2 and 16, 4 and 8, 8.

These first three pairs can also be written as 1, 64 and 2, 32 and 4, 16.

(4) We try some possibilities:

$(25t + 64)(t + 1) = 25t^2 + 89t + 64$

$(5t + 32)(5t + 2) = 25t^2 + 170t + 64$

$(25t + 16)(t + 4) = 25t^2 + 116t + 64$

$(5t + 8)(5t + 8) = 25t^2 + 80t + 64$

The factorization is $(5t + 8)(5t + 8)$ or $(5t + 8)^2$.

61. $6x^3 + 4x^2 - 10x$

(1) Factor out the common factor, $2x$: $2x(3x^2 + 2x - 5)$

Then we factor the trinomial $3x^2 + 2x - 5$. We did this in Exercise 35 (after we factored 2 out of the original trinomial). We know that $3x^2 + 2x - 5 = (3x + 5)(x - 1)$, so $6x^3 + 4x^2 - 10x = 2x(3x + 5)(x - 1)$.

63. $25x^2 + 79x + 64$

We follow the same procedure as in Exercise 59. None of the possibilities works. Thus, $25x^2 + 79x + 64$ is not factorable. It is prime.

65. $6x^2 - 19x - 5$

(1) There is no common factor (other than 1 or -1).

(2) Factor the first term, $6x^2$. The possibilities are $6x$, x and $3x$, $2x$. We have these as possibilities for factorizations:

$(6x+\quad)(x+\quad)$ and $(3x+\quad)(2x+\quad)$

(3) Factor the last term, -5. The possibilities are -5, 1 and 5, -1.

These factors can also be written as 1, -5 and -1, 5.

(4) We try some possibilities:

$(6x - 5)(x + 1) = 6x^2 + x - 5$

$(6x + 5)(x - 1) = 6x^2 - x - 5$

$(6x + 1)(x - 5) = 6x^2 - 29x - 5$

$(6x - 1)(x + 5) = 6x^2 + 29x - 5$

$(3x - 5)(2x + 1) = 6x^2 - 7x - 5$

$(3x + 5)(2x - 1) = 6x^2 + 7x - 5$

$(3x + 1)(2x - 5) = 6x^2 - 13x - 5$

$(3x - 1)(2x + 5) = 6x^2 + 13x - 5$

None of the possibilities works. Thus, $6x^2 - 19x - 5$ is not factorable. It is prime.

67. $12m^2 - mn - 20n^2$

(1) There is no common factor (other than 1 or -1).

(2) Factor the first term, $12m^2$. The possibilities are $12m$, m and $6m$, $2m$ and $3m$, $4m$. We have these as possibilities for factorizations:

$(12m+\quad)(m+\quad)$ and $(6m+\quad)(2m+\quad)$

and $(3m+\quad)(4m+\quad)$

(3) Factor the last term, $-20n^2$. The possibilities are $20n$, $-n$ and $-20n$, n and $10n$, $-2n$ and $-10n$, $2n$ and $5n$, $-4n$ and $-5n$, $4n$.

These factors can also be written as $-n$, $20n$ and n, $-20n$ and $-2n$, $10n$ and $2n$, $-10n$ and $-4n$, $5n$ and $4n$, $-5n$.

(4) We can immediately reject all possibilities in which either factor has a common factor, such as $(12m + 20n)$ or $(4m - 2n)$, because we determined at the outset that there is no common factor. We try some of the remaining possibilities:

$$(12m - n)(m + 20n) = 12m^2 + 239mn - 20n^2$$
$$(12m + 5n)(m - 4n) = 12m^2 - 43mn - 20n^2$$
$$(3m - 20n)(4m + n) = 12m^2 - 77mn - 20n^2$$
$$(3m - 4n)(4m + 5n) = 12m^2 - mn - 20n^2$$

The factorization is $(3m - 4n)(4m + 5n)$.

69. $6a^2 - ab - 15b^2$

(1) There is no common factor (other than 1 or -1).

(2) Factor the first term, $6a^2$. The possibilities are $6a$, a and $3a$, $2a$. We have these as possibilities for factorizations:

$(6a+\quad)(a+\quad)$ and $(3a+\quad)(2a+\quad)$

(3) Factor the last term, $-15b^2$. The possibilities are $15b$, $-b$ and $-15b$, b and $5b$, $-3b$ and $-5b$, $3b$.

These factors can also be written as $-b$, $15b$ and b, $-15b$, and $-3b$, $5b$ and $3b$, $-5b$.

(4) We can immediately reject all possibilities in which either factor has a common factor, such as $(6a+15b)$ or $(3a - 3b)$, because we determined at the outset that there is no common factor. We try some of the remaining possibilities:

$$(6a - b)(a + 15b) = 6a^2 + 89ab - 15b^2$$
$$(3a - b)(2a + 15b) = 6a^2 + 43ab - 15b^2$$
$$(6a + 5b)(a - 3b) = 6a^2 - 13ab - 15b^2$$
$$(3a + 5b)(2a - 3b) = 6a^2 + ab - 15b^2$$
$$(3a - 5b)(2a + 3b) = 6a^2 - ab - 15b^2$$

The factorization is $(3a - 5b)(2a + 3b)$.

71. $9a^2 + 18ab + 8b^2$

(1) There is no common factor (other than 1 or -1).

(2) Factor the first term, $9a^2$. The possibilities are $9a$, a and $3a$, $3a$. We have these as possibilities for factorizations:

$(9a+\quad)(a+\quad)$ and $(3a+\quad)(3a+\quad)$

(3) Factor $8b^2$. Since all signs are positive, we need consider only pairs of factors with positive coefficients. Those factor pairs are $8b$, b and $4b$, $2b$.

These factors can also be written as b, $8b$ and $2b$, $4b$.

(4) We try some possibilities:

$$(9a + 8b)(a + b) = 9a^2 + 17ab + 8b^2$$
$$(3a + 8b)(3a + b) = 9a^2 + 27ab + 8b^2$$
$$(9a + 4b)(a + 2b) = 9a^2 + 22ab + 8b^2$$
$$(3a + 4b)(3a + 2b) = 9a^2 + 18ab + 8b^2$$

The factorization is $(3a + 4b)(3a + 2b)$.

73. $35p^2 + 34pt + 8t^2$

(1) There is no common factor (other than 1 or -1).

(2) Factor the first term, $35p^2$. The possibilities are $35p$, p and $7p$, $5p$. We have these as possibilities for factorizations:

$(35p+\quad)(p+\quad)$ and $(7p+\quad)(5p+\quad)$

(3) Factor $8t^2$. Since all signs are positive, we need consider only pairs of factors with positive coefficients. Those factor pairs are $8t$, t and $4t$, $2t$.

These factors can also be written as t, $8t$ and $2t$, $4t$.

(4) We try some possibilities:

$$(35p + 8t)(p + t) = 35p^2 + 43pt + 8t^2$$
$$(7p + 8t)(5p + t) = 35p^2 + 47pt + 8t^2$$
$$(35p + 4t)(p + 2t) = 35p^2 + 74pt + 8t^2$$
$$(7p + 4t)(5p + 2t) = 35p^2 + 34pt + 8t^2$$

The factorization is $(7p + 4t)(5p + 2t)$.

75. $18x^2 - 6xy - 24y^2$

(1) Factor out the common factor, 6:

$6(3x^2 - xy - 4y^2)$

Then we factor the trinomial $3x^2 - xy - 4y^2$.

(2) Factor $3x^2$. The only possibility is $3x$, x. The desired factorization is of the form:

$(3x+\quad)(x+\quad)$

(3) Factor $-4y^2$. The possibilities are $4y$, $-y$ and $-4y$, y and $2y$, $-2y$.

These factors can also be written as $-y$, $4y$ and y, $-4y$ and $-2y$, $2y$.

(4) We try some possibilities:

$$(3x + 4y)(x - y) = 3x^2 + xy - 4y^2$$
$$(3x - 4y)(x + y) = 3x^2 - xy - 4y^2$$

Then $3x^2 - xy - 4y^2 = (3x - 4y)(x + y)$, so
$18x^2 - 6xy - 24y^2 = 6(3x - 4y)(x + y)$.

77. Graph: $y = \frac{2}{5}x - 1$

Because the equation is in the form $y = mx + b$, we know the y-intercept is $(0, -1)$. We find two other points on the line, substituting multiples of 5 for x to avoid fractions.

When $x = -5$, $y = \frac{2}{5}(-5) - 1 = -2 - 1 = -3$.

When $x = 5$, $y = \frac{2}{5}(5) - 1 = 2 - 1 = 1$.

x	y
0	−1
−5	−3
5	1

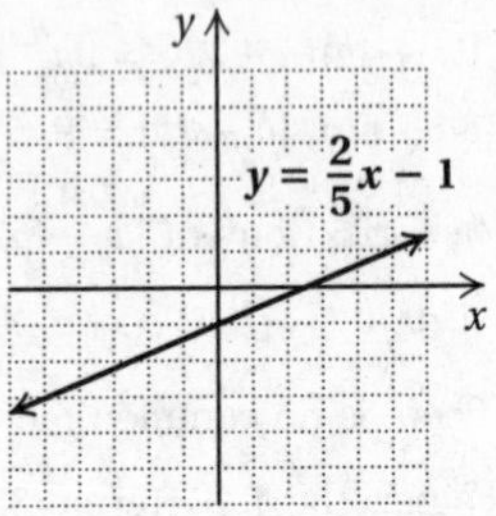

79. Graph: $x = 4 - 2y$

Find the x-intercept.

$x = 4 - 2 \cdot 0$

$x = 4$

The x-intercept is $(4, 0)$.

Find the y-intercept.

$0 = 4 - 2y$

$2y = 4$

$y = 2$

The y-intercept is $(0, 2)$.

Plot these points and draw the line through them. We could also find a third point as a check.

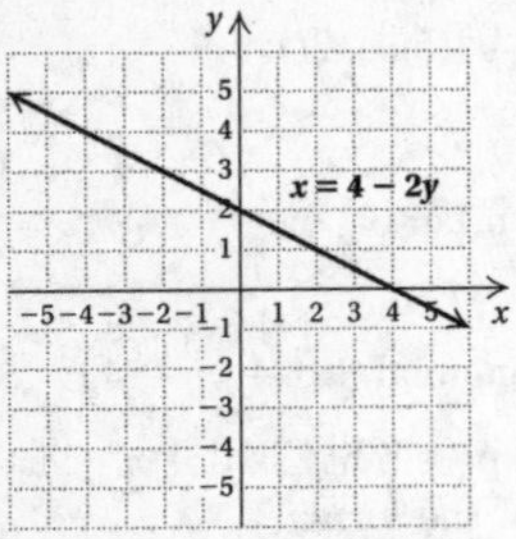

81. $x + y = 4$

To find the x-intercept, let $y = 0$. Then solve for x.

$x + y = 4$

$x + 0 = 4$

$x = 4$

The x-intercept is $(4, 0)$.

To find the y-intercept, let $x = 0$. Then solve for y.

$x + y = 4$

$0 + y = 4$

$y = 4$

The y-intercept is $(0, 4)$.

Plot these points and draw the line.

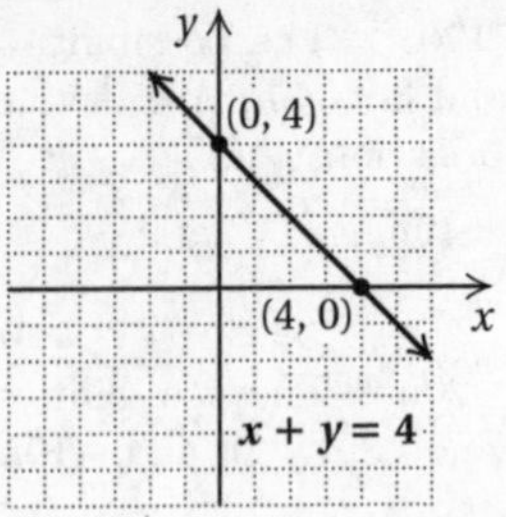

A third point should be used as a check. We substitute any value for x and solve for y. We let $x = 2$. Then

$x + y = 4$

$2 + y = 4$

$y = 2$

The point $(2, 2)$ is on the graph, so the graph is probably correct.

83. $5x - 3y = 15$

To find the x-intercept, let $y = 0$. Then solve for x.

$5x - 3y = 15$

$5x - 3 \cdot 0 = 15$

$5x = 15$

$x = 3$

The x-intercept is $(3, 0)$.

To find the y-intercept, let $x = 0$. Then solve for y.

$5x - 3y = 15$

$5 \cdot 0 - 3y = 15$

$-3y = 15$

$y = -5$

The y-intercept is $(0, -5)$.

Plot these points and draw the line.

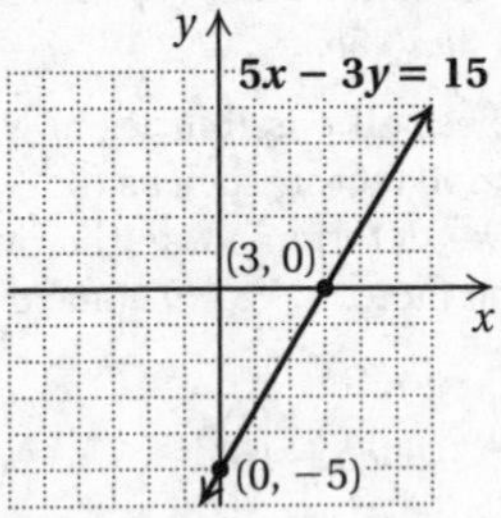

A third point should be used as a check. We substitute any value for x and solve for y. We let $x = 6$. Then

$5x - 3y = 15$

$5 \cdot 6 - 3y = 15$

$30 - 3y = 15$

$-3y = -15$

$y = 5$

The point $(6, 5)$ is on the graph, so the graph is probably correct.

85. $20x^{2n} + 16x^n + 3 = 20(x^n)^2 + 16x^n + 3$

(1) There is no common factor (other than 1 and −1).

(2) Factor the first term, $20x^{2n}$. The possibilities are $20x^n$, x^n and $10x^n$, $2x^n$ and $5x^n$, $4x^n$. We have these as possibilities for factorizations:

$(20x^n+\quad)(x^n+\quad)$ and $(10x^n+\quad)(2x^n+\quad)$ and $(5x^n+\quad)(4x^n+\quad)$

(3) Factor the last term, 3. Since all signs are positive, we need consider only the positive factor pair 3, 1.

(4) We try some possibilities:

$$(20x^n+3)(x^n+1) = 20x^{2n}+23x^n+3$$
$$(10x^n+3)(2x^n+1) = 20x^{2n}+16x^n+3$$

The factorization is $(10x^n+3)(2x^n+1)$.

87. $3x^{6a}-2x^{3a}-1 = 3(x^{3a})^2-2x^{3a}-1$

(1) There is no common factor (other than 1 or -1).

(2) Factor the first term, $3x^{6a}$. The only possibility is $3x^{3a}$, x^{3a}. The desired factorization is of the form:

$$(3x^{3a}+\quad)(x^{3a}+\quad)$$

(3) Factor the last term, -1. The only possibility is -1, 1.

(4) We try the possibilities:

$$(3x^{3a}-1)(x^{3a}+1) = 3x^{6a}+2x^{3a}-1$$
$$(3x^{3a}+1)(x^{3a}-1) = 3x^{6a}-2x^{3a}-1$$

The factorization is $(3x^{3a}+1)(x^{3a}-1)$.

89.-93. Left to the student

Exercise Set 5.4

RC1. Using the ac-method, multiply the leading coefficient 10 and the constant 2. The product is 20.

RC3. Split the middle term, $21x$, writing it as the sum of $20x$ and x.

1. $x^2+2x+7x+14 = (x^2+2x)+(7x+14)$
$= x(x+2)+7(x+2)$
$= (x+2)(x+7)$

3. $x^2-4x-x+4 = (x^2-4x)+(-x+4)$
$= x(x-4)-1(x-4)$
$= (x-4)(x-1)$

5. $6x^2+4x+9x+6 = (6x^2+4x)+(9x+6)$
$= 2x(3x+2)+3(3x+2)$
$= (3x+2)(2x+3)$

7. $3x^2-4x-12x+16 = (3x^2-4x)+(-12x+16)$
$= x(3x-4)-4(3x-4)$
$= (3x-4)(x-4)$

9. $35x^2-40x+21x-24 = (35x^2-40x)+(21x-24)$
$= 5x(7x-8)+3(7x-8)$
$= (7x-8)(5x+3)$

11. $4x^2+6x-6x-9 = (4x^2+6x)+(-6x-9)$
$= 2x(2x+3)-3(2x+3)$
$= (2x+3)(2x-3)$

13. $2x^4+6x^2+5x^2+15 = (2x^4+6x^2)+(5x^2+15)$
$= 2x^2(x^2+3)+5(x^2+3)$
$= (x^2+3)(2x^2+5)$

15. $2x^2+7x+6$

(1) First factor out a common factor, if any. There is none (other than 1 or -1).

(2) Multiply the leading coefficient, 2, and the constant, 6: $2\cdot 6=12$.

(3) Look for a factorization of 12 in which the sum of the factors is the coefficient of the middle term, 7.

Pairs of factors	Sums of factors
1, 12	13
2, 6	8
3, 4	7

(4) Split the middle term: $7x = 3x+4x$

(5) Factor by grouping:

$2x^2+7x+6 = 2x^2+3x+4x+6$
$= (2x^2+3x)+(4x+6)$
$= x(2x+3)+2(2x+3)$
$= (2x+3)(x+2)$

17. $3x^2-4x-15$

(1) First factor out a common factor, if any. There is none (other than 1 or -1).

(2) Multiply the leading coefficient, 3, and the constant, -15: $3(-15)=-45$.

(3) Look for a factorization of -45 in which the sum of the factors is the coefficient of the middle term, -4.

Pairs of factors	Sums of factors
−1, 45	44
1, −45	−44
−3, 15	12
3, −15	−12
−5, 9	4
5, −9	−4

(4) Split the middle term: $-4x = 5x-9x$

(5) Factor by grouping:

$3x^2-4x-15 = 3x^2+5x-9x-15$
$= (3x^2+5x)+(-9x-15)$
$= x(3x+5)-3(3x+5)$
$= (3x+5)(x-3)$

19. $5x^2 + 11x + 2$

(1) First factor out a common factor, if any. There is none (other than 1 or -1).

(2) Multiply the leading coefficient, 5, and the constant, 2: $5 \cdot 2 = 10$.

(3) Look for a factorization of 10 in which the sum of the factors is the coefficient of the middle term, 11. We see that the factors we need are 1 and 10.

(4) Split the middle term: $11x = x + 10x$

(5) Factor by grouping:

$$\begin{aligned}5x^2 + 11x + 2 &= 5x^2 + x + 10x + 2\\ &= (5x^2 + x) + (10x + 2)\\ &= x(5x + 1) + 2(5x + 1)\\ &= (5x + 1)(x + 2)\end{aligned}$$

21. $3x^2 - 4x + 1$

(1) First factor out a common factor, if any. There is none (other than 1 or -1).

(2) Multiply the leading coefficient, 3, and the constant, 1: $3 \cdot 1 = 3$.

(3) Look for a factorization of 3 in which the sum of the factors is the coefficient of the middle term, -4. The numbers we want are -1 and -3: $-1 \cdot (-3) = 3$ and $-1 + (-3) = -4$.

(4) Split the middle term: $-4x = -1x - 3x$

(5) Factor by grouping:

$$\begin{aligned}3x^2 - 4x + 1 &= 3x^2 - x - 3x + 1\\ &= (3x^2 - x) + (-3x + 1)\\ &= x(3x - 1) - 1(3x - 1)\\ &= (3x - 1)(x - 1)\end{aligned}$$

23. $6x^2 + 23x + 7$

(1) First factor out a common factor, if any. There is none (other than 1 or -1).

(2) Multiply the leading coefficient, 6, and the constant, 7: $6 \cdot 7 = 42$.

(3) Look for a factorization of 42 in which the sum of the factors is the coefficient of the middle term, 23. We only need to consider positive factors.

Pairs of factors	Sums of factors
1, 42	43
2, 21	23
3, 14	17
6, 7	13

(4) Split the middle term: $23x = 2x + 21x$

(5) Factor by grouping:

$$\begin{aligned}6x^2 + 23x + 7 &= 6x^2 + 2x + 21x + 7\\ &= (6x^2 + 2x) + (21x + 7)\\ &= 2x(3x + 1) + 7(3x + 1)\\ &= (3x + 1)(2x + 7)\end{aligned}$$

25. $4x^2 - 4x - 15$

(1) First factor out a common factor, if any. There is none (other than 1 or -1).

(2) Multiply the leading coefficient, 4, and the constant, -15: $4(-15) = -60$.

(3) Look for a factorization of -60 in which the sum of the factors is the coefficient of the middle term, -4.

Pairs of factors	Sums of factors
-1, 60	59
1, -60	-59
-2, 30	28
2, -30	-28
-3, 20	17
3, -20	-17
-4, 15	11
4, -15	-11
-5, 12	7
5, -12	-7
-6, 10	4
6, -10	-4

(4) Split the middle term: $-4x = 6x - 10x$

(5) Factor by grouping:

$$\begin{aligned}4x^2 - 4x - 15 &= 4x^2 + 6x - 10x - 15\\ &= (4x^2 + 6x) + (-10x - 15)\\ &= 2x(2x + 3) - 5(2x + 3)\\ &= (2x + 3)(2x - 5)\end{aligned}$$

27. $15x^2 + 19x - 10$

(1) First factor out a common factor, if any. There is none (other than 1 or -1).

(2) Multiply the leading coefficient, 15, and the constant, -10: $15(-10) = -150$.

(3) Look for a factorization of -150 in which the sum of the factors is the coefficient of the middle term, 19.

Pairs of factors	Sums of factors
-1, 150	149
-2, 75	73
-3, 50	47
-5, 30	25
-6, 25	19

(4) Split the middle term: $19x = -6x + 25x$

(5) Factor by grouping:

$$\begin{aligned}15x^2 + 19x - 10 &= 15x^2 - 6x + 25x - 10\\ &= 3x(5x - 2) + 5(5x - 2)\\ &= (5x - 2)(3x + 5)\end{aligned}$$

29. $9x^2 - 18x - 16$

(1) First factor out a common factor, if any. There is none (other than 1 or -1).

(2) Multiply the leading coefficient, 9, and the constant, -16: $9(-16) = -144$.

(3) Look for a factorization of -144, so the sum of the factors is the coefficient of the middle term, -18.

Pairs of factors	Sums of factors
$-1,\ 144$	143
$1, -144$	-143
$-2,\ 72$	70
$2, -72$	-70
$-3,\ 48$	45
$3, -48$	-45
$-4,\ 36$	32
$4, -36$	-32
$-6,\ 24$	18
$6, -24$	-18
$-8,\ 18$	10
$8, -18$	-10
$-9,\ 16$	7
$9, -16$	-7
$-12,\ 12$	0

(4) Split the middle term: $-18x = 6x - 24x$

(5) Factor by grouping:

$$\begin{aligned} 9x^2 - 18x - 16 &= 9x^2 + 6x - 24x - 16 \\ &= (9x^2 + 6x) + (-24x - 16) \\ &= 3x(3x + 2) - 8(3x + 2) \\ &= (3x + 2)(3x - 8) \end{aligned}$$

31. $3x^2 + 5x - 2$

(1) First factor out a common factor, if any. There is none (other than 1 or -1).

(2) Multiply the leading coefficient, 3, and the constant, -2: $3(-2) = -6$.

(3) Look for a factorization of -6 in which the sum of the factors is the coefficient of the middle term, 5. The numbers we want are -1 and 6: $-1(6) = -6$ and $-1 + 6 = 5$.

(4) Split the middle term: $5x = -1x + 6x$

(5) Factor by grouping:

$$\begin{aligned} 3x^2 + 5x - 2 &= 3x^2 - x + 6x - 2 \\ &= (3x^2 - x) + (6x - 2) \\ &= x(3x - 1) + 2(3x - 1) \\ &= (3x - 1)(x + 2) \end{aligned}$$

33. $12x^2 - 31x + 20$

(1) First factor out a common factor, if any. There is none (other than 1 or -1).

(2) Multiply the leading coefficient, 12, and the constant, 20: $12 \cdot 20 = 240$.

(3) Look for a factorization of 240 in which the sum of the factors is the coefficient of the middle term, -31. We only need to consider negative factors.

Pairs of factors	Sums of factors
$-1, -240$	-241
$-2, -120$	-122
$-3, -80$	-83
$-4, -60$	-64
$-5, -48$	-53
$-6, -40$	-46
$-8, -30$	-38
$-10, -24$	-34
$-12, -20$	-32
$-15, -16$	-31

(4) Split the middle term: $-31x = -15x - 16x$

(5) Factor by grouping:

$$\begin{aligned} 12x^2 - 31x + 20 &= 12x^2 - 15x - 16x + 20 \\ &= (12x^2 - 15x) + (-16x + 20) \\ &= 3x(4x - 5) - 4(4x - 5) \\ &= (4x - 5)(3x - 4) \end{aligned}$$

35. $14x^2 - 19x - 3$

(1) First factor out a common factor, if any. There is none (other than 1 or -1).

(2) Multiply the leading coefficient, 14, and the constant, -3: $14(-3) = -42$.

(3) Look for a factorization of -42 so that the sum of the factors is the coefficient of the middle term, -19.

Pairs of factors	Sums of factors
$-1,\ 42$	41
$1, -42$	-41
$-2,\ 21$	19
$2, -21$	-19
$-3,\ 14$	11
$3, -14$	-11
$-6,\ 7$	1
$6, -7$	-1

(4) Split the middle term: $-19x = 2x - 21x$

(5) Factor by grouping:

$$\begin{aligned} 14x^2 - 19x - 3 &= 14x^2 + 2x - 21x - 3 \\ &= (14x^2 + 2x) + (-21x - 3) \\ &= 2x(7x + 1) - 3(7x + 1) \\ &= (7x + 1)(2x - 3) \end{aligned}$$

37. $49 - 42x + 9x^2 = 9x^2 - 42x + 49$

(1) First factor out a common factor, if any. There is none (other than 1 or -1).

(2) Multiply the leading coefficient, 9, and the constant, 49: $9 \cdot 49 = 441$.

(3) Look for a factorization of 441 in which the sum of the factors is the coefficient of the middle term, -42. We only need to consider negative factors.

Pairs of factors	Sums of factors
$-1, -441$	-442
$-3, -147$	-150
$-7, -63$	-70
$-9, -49$	-58
$-21, -21$	-42

(4) Split the middle term: $-42x = -21x - 21x$

(5) Factor by grouping:

$$\begin{aligned} 9x^2 - 42x + 49 &= 9x^2 - 21x - 21x + 49 \\ &= (9x^2 - 21x) + (-21x + 49) \\ &= 3x(3x - 7) - 7(3x - 7) \\ &= (3x - 7)(3x - 7), \text{ or} \\ &\quad (3x - 7)^2 \end{aligned}$$

We could also write this as $(7 - 3x)^2$.

39. $9x^2 + 18x + 8$

(1) First factor out a common factor, if any. There is none (other than 1 or -1).

(2) Multiply the leading coefficient, 9, and the constant, 8: $9 \cdot 8 = 72$.

(3) Look for a factorization of 72 in which the sum of the factors is the coefficient of the middle term, 18. We only need to consider positive factors.

Pairs of factors	Sums of factors
1, 72	73
2, 36	38
3, 24	27
4, 18	22
6, 12	18
8, 9	17

(4) Split the middle term: $18x = 6x + 12x$

(5) Factor by grouping:

$$\begin{aligned} 9x^2 + 18x + 8 &= 9x^2 + 6x + 12x + 8 \\ &= (9x^2 + 6x) + (12x + 8) \\ &= 3x(3x + 2) + 4(3x + 2) \\ &= (3x + 2)(3x + 4) \end{aligned}$$

41. $5 - 9a^2 - 12a = -9a^2 - 12a + 5 = -1(9a^2 + 12a - 5)$

Now we factor $9a^2 + 12a - 5$.

(1) We have already factored out the common factor, -1, to make the leading coefficient positive.

(2) Multiply the leading coefficient, 9, and the constant, -5: $9(-5) = -45$.

(3) Look for a factorization of -45 in which the sum of the factors is the coefficient of the middle term, 12. The numbers we want are 15 and -3: $15(-3) = -45$ and $15 + (-3) = 12$.

(4) Split the middle term: $12a = 15a - 3a$

(5) Factor by grouping:

$$\begin{aligned} 9a^2 + 12a - 5 &= 9a^2 + 15a - 3a - 5 \\ &= (9a^2 + 15a) + (-3a - 5) \\ &= 3a(3a + 5) - (3a + 5) \\ &= (3a + 5)(3a - 1) \end{aligned}$$

Then we have

$$\begin{aligned} &5 - 9a^2 - 12a \\ &= -1(3a + 5)(3a - 1) \\ &= (3a + 5)(-3a + 1) \quad \text{Multiplying } 3a-1 \text{ by } -1 \\ &= (-3a - 5)(3a - 1) \quad \text{Multiplying } 3a+5 \text{ by } -1 \end{aligned}$$

Note that we can also express $(3a + 5)(-3a + 1)$ as $(3a + 5)(1 - 3a)$ since $-3a + 1 = 1 - 3a$ by the commutative law of addition.

43. $20 + 6x - 2x^2 = -2x^2 + 6x + 20$

(1) Factor out the common factor -2. We factor out -2 rather than 2 in order to make the leading coefficient of the trinomial factor positive.

$-2x^2 + 6x + 20 = -2(x^2 - 3x - 10)$

To factor $x^2 - 3x - 10$, we look for two factors of -10 whose sum is -3. The numbers we want are -5 and 2. Then $x^2 - 3x - 10 = (x - 5)(x + 2)$, so we have:

$$\begin{aligned} &20 + 6x - 2x^2 \\ &= -2(x - 5)(x + 2) \\ &= 2(-x + 5)(x + 2) \quad \text{Multiplying } x - 5 \text{ by } -1 \\ &= 2(x - 5)(-x - 2) \quad \text{Multiplying } x + 2 \text{ by } -1 \end{aligned}$$

Note that we can also express $2(-x + 5)(x + 2)$ as $2(5 - x)(x + 2)$ since $-x + 5 = 5 - x$ by the commutative law of addition.

45. $12x^2 + 28x - 24$

(1) Factor out the common factor, 4:

$12x^2 + 28x - 24 = 4(3x^2 + 7x - 6)$

(2) Now we factor the trinomial $3x^2 + 7x - 6$. Multiply the leading coefficient, 3, and the constant, -6: $3(-6) = -18$.

(3) Look for a factorization of -18 in which the sum of the factors is the coefficient of the middle term, 7. The numbers we want are 9 and -2: $9(-2) = -18$ and $9 + (-2) = 7$.

(4) Split the middle term: $7x = 9x - 2x$

(5) Factor by grouping:

$$\begin{aligned}3x^2+7x-6&=3x^2+9x-2x-6\\&=(3x^2+9x)+(-2x-6)\\&=3x(x+3)-2(x+3)\\&=(x+3)(3x-2)\end{aligned}$$

We must include the common factor to get a factorization of the original trinomial.

$12x^2+28x-24=4(x+3)(3x-2)$

47. $30x^2-24x-54$

(1) Factor out the common factor, 6.

$30x^2-24x-54=6(5x^2-4x-9)$

(2) Now we factor the trinomial $5x^2-4x-9$. Multiply the leading coefficient, 5, and the constant, -9: $5(-9)=-45$.

(3) Look for a factorization of -45 in which the sum of the factors is the coefficient of the middle term, -4. The numbers we want are -9 and 5: $-9\cdot 5=-45$ and $-9+5=-4$.

(4) Split the middle term: $-4x=-9x+5x$

(5) Factor by grouping:

$$\begin{aligned}5x^2-4x-9&=5x^2-9x+5x-9\\&=(5x^2-9x)+(5x-9)\\&=x(5x-9)+(5x-9)\\&=(5x-9)(x+1)\end{aligned}$$

We must include the common factor to get a factorization of the original trinomial.

$30x^2-24x-54=6(5x-9)(x+1)$

49. $4y+6y^2-10=6y^2+4y-10$

(1) Factor out the common factor, 2.

$6y^2+4y-10=2(3y^2+2y-5)$

(2) Now we factor the trinomial $3y^2+2y-5$. Multiply the leading coefficient, 3, and the constant, -5: $3(-5)=-15$.

(3) Look for a factorization of -15 in which the sum of the factors is the coefficient of the middle term, 2. The numbers we want are 5 and -3: $5(-3)=-15$ and $5+(-3)=2$.

(4) Split the middle term: $2y=5y-3y$

(5) Factor by grouping:

$$\begin{aligned}3y^2+2y-5&=3y^2+5y-3y-5\\&=(3y^2+5y)+(-3y-5)\\&=y(3y+5)-(3y+5)\\&=(3y+5)(y-1)\end{aligned}$$

We must include the common factor to get a factorization of the original trinomial.

$4y+6y^2-10=2(3y+5)(y-1)$

51. $3x^2-4x+1$

(1) There is no common factor (other than 1 or -1).

(2) Multiply the leading coefficient, 3, and the constant, 1: $3\cdot 1=3$.

(3) Look for a factorization of 3 in which the sum of the factors is the coefficient of the middle term, -4. The numbers we want are -1 and -3: $-1(-3)=3$ and $-1+(-3)=-4$.

(4) Split the middle term: $-4x=-1x-3x$

(5) Factor by grouping:

$$\begin{aligned}3x^2-4x+1&=3x^2-x-3x+1\\&=(3x^2-x)+(-3x+1)\\&=x(3x-1)-(3x-1)\\&=(3x-1)(x-1)\end{aligned}$$

53. $12x^2-28x-24$

(1) Factor out the common factor, 4:

$12x^2-28x-24=4(3x^2-7x-6)$

(2) Now we factor the trinomial $3x^2-7x-6$. Multiply the leading coefficient, 3, and the constant, -6: $3(-6)=-18$.

(3) Look for a factorization of -18 in which the sum of the factors is the coefficient of the middle term, -7. The numbers we want are -9 and 2: $-9\cdot 2=-18$ and $-9+2=-7$.

(4) Split the middle term: $-7x=-9x+2x$

(5) Factor by grouping:

$$\begin{aligned}3x^2-7x-6&=3x^2-9x+2x-6\\&=(3x^2-9x)+(2x-6)\\&=3x(x-3)+2(x-3)\\&=(x-3)(3x+2)\end{aligned}$$

We must include the common factor to get a factorization of the original trinomial.

$12x^2-28x-24=4(x-3)(3x+2)$

55. $-1+2x^2-x=2x^2-x-1$

(1) There is no common factor (other than 1 or -1).

(2) Multiply the leading coefficient, 2, and the constant, -1: $2(-1)=-2$.

(3) Look for a factorization of -2 in which the sum of the factors is the coefficient of the middle term, -1. The numbers we want are -2 and 1: $-2\cdot 1=-2$ and $-2+1=-1$.

(4) Split the middle term: $-x=-2x+1x$

(5) Factor by grouping:

$$\begin{aligned}2x^2-x-1&=2x^2-2x+x-1\\&=(2x^2-2x)+(x-1)\\&=2x(x-1)+(x-1)\\&=(x-1)(2x+1)\end{aligned}$$

57. $9x^2+18x-16$

(1) There is no common factor (other than 1 or -1).

(2) Multiply the leading coefficient, 9, and the constant, -16: $9(-16)=-144$.

(3) Look for a factorization of -144 in which the sum of the factors is the coefficient of the middle term, 18. The numbers we want are 24 and -6: $24(-6) = -144$ and $24 + (-6) = 18$.

(4) Split the middle term: $18x = 24x - 6x$

(5) Factor by grouping:

$$\begin{aligned} 9x^2 + 18x - 16 &= 9x^2 + 24x - 6x - 16 \\ &= (9x^2 + 24x) + (-6x - 16) \\ &= 3x(3x + 8) - 2(3x + 8) \\ &= (3x + 8)(3x - 2) \end{aligned}$$

59. $15x^2 - 25x - 10$

(1) Factor out the common factor, 5:

$15x^2 - 25x - 10 = 5(3x^2 - 5x - 2)$

(2) Now we factor the trinomial $3x^2 - 5x - 2$. Multiply the leading coefficient, 3, and the constant, -2: $3(-2) = -6$.

(3) Look for a factorization of -6 in which the sum of the factors is the coefficient of the middle term, -5. The numbers we want are -6 and 1: $-6 \cdot 1 = -6$ and $-6 + 1 = -5$.

(4) Split the middle term: $-5x = -6x + 1x$

(5) Factor by grouping:

$$\begin{aligned} 3x^2 - 5x - 2 &= 3x^2 - 6x + x - 2 \\ &= (3x^2 - 6x) + (x - 2) \\ &= 3x(x - 2) + (x - 2) \\ &= (x - 2)(3x + 1) \end{aligned}$$

We must include the common factor to get a factorization of the original trinomial.

$15x^2 - 25x - 10 = 5(x - 2)(3x + 1)$

61. $12p^3 + 31p^2 + 20p$

(1) Factor out the common factor, p:

$12p^3 + 31p^2 + 20p = p(12p^2 + 31p + 20)$

(2) Now we factor the trinomial $12p^2 + 31p + 20$. Multiply the leading coefficient, 12, and the constant, 20: $12 \cdot 20 = 240$.

(3) Look for a factorization of 240 in which the sum of the factors is the coefficient of the middle term, 31. The numbers we want are 15 and 16: $15 \cdot 16 = 240$ and $15 + 16 = 31$.

(4) Split the middle term: $31p = 15p + 16p$

(5) Factor by grouping:

$$\begin{aligned} 12p^2 + 31p + 20 &= 12p^2 + 15p + 16p + 20 \\ &= (12p^2 + 15p) + (16p + 20) \\ &= 3p(4p + 5) + 4(4p + 5) \\ &= (4p + 5)(3p + 4) \end{aligned}$$

We must include the common factor to get a factorization of the original trinomial.

$12p^3 + 31p^2 + 20p = p(4p + 5)(3p + 4)$

63. $4 - x - 5x^2 = -5x^2 - x + 4$

(1) Factor out -1 to make the leading coefficient positive:

$-5x^2 - x + 4 = -1(5x^2 + x - 4)$

(2) Now we factor the trinomial $5x^2 + x - 4$. Multiply the leading coefficient, 5, and the constant, -4: $5(-4) = -20$.

(3) Look for a factorization of -20 in which the sum of the factors is the coefficient of the middle term, 1. The numbers we want are 5 and -4: $5(-4) = -20$ and $5 + (-4) = 1$.

(4) Split the middle term: $x = 5x - 4x$

(5) Factor by grouping:

$$\begin{aligned} 5x^2 + x - 4 &= 5x^2 + 5x - 4x - 4 \\ &= (5x^2 + 5x) + (-4x - 4) \\ &= 5x(x + 1) - 4(x + 1) \\ &= (x + 1)(5x - 4) \end{aligned}$$

We must include the common factor to get a factorization of the original trinomial.

$$\begin{aligned} &4 - x - 5x^2 \\ &= -1(x + 1)(5x - 4) \\ &= (x + 1)(-5x + 4) \quad \text{Multiplying } 5x - 4 \text{ by } -1 \\ &= (-x - 1)(5x - 4) \quad \text{Multiplying } x + 1 \text{ by } -1 \end{aligned}$$

Note that we can also express $(x + 1)(-5x + 4)$ as $(x + 1)(4 - 5x)$ since $-5x + 4 = 4 - 5x$ by the commutative law of addition.

65. $33t - 15 - 6t^2 = -6t^2 + 33t - 15$

(1) Factor out the common factor, -3. We factor out -3 rather than 3 in order to make the leading coefficient of the trinomial factor positive.

$-6t^2 + 33t - 15 = -3(2t^2 - 11t + 5)$

(2) Now we factor the trinomial $2t^2 - 11t + 5$. Multiply the leading coefficient, 2, and the constant, 5: $2 \cdot 5 = 10$.

(3) Look for a factorization of 10 in which the sum of the factors is the coefficient of the middle term, -11. The numbers we want are -1 and -10: $-1(-10) = 10$ and $-1 + (-10) = -11$.

(4) Split the middle term: $-11t = -1t - 10t$

(5) Factor by grouping:

$$\begin{aligned} 2t^2 - 11t + 5 &= 2t^2 - t - 10t + 5 \\ &= (2t^2 - t) + (-10t + 5) \\ &= t(2t - 1) - 5(2t - 1) \\ &= (2t - 1)(t - 5) \end{aligned}$$

We must include the common factor to get a factorization of the original trinomial.

$$\begin{aligned} &33t - 15 - 6t^2 \\ &= -3(2t - 1)(t - 5) \\ &= 3(2t - 1)(-t + 5) \quad \text{Multiplying } t - 5 \text{ by } -1 \\ &= 3(-2t + 1)(t - 5) \quad \text{Multiplying } 2t - 1 \text{ by } -1 \end{aligned}$$

Note that we can also express $3(2t-1)(-t+5)$ as $3(2t-1)(5-t)$ since $-t+5=5-t$ by the commutative law of addition. Similarly, we can express $3(-2t+1)(t-5)$ as $3(1-2t)(t-5)$.

67. $14x^4+19x^3-3x^2$

(1) Factor out the common factor, x^2:

$14x^4+19x^3-3x^2 = x^2(14x^2+19x-3)$

(2) Now we factor the trinomial $14x^2+19x-3$. Multiply the leading coefficient, 14, and the constant, -3: $14(-3)=-42$.

(3) Look for a factorization of -42 in which the sum of the factors is the coefficient of the middle term, 19. The numbers we want are 21 and -2: $21(-2)=-42$ and $21+(-2)=19$.

(4) Split the middle term: $19x=21x-2x$

(5) Factor by grouping:

$$\begin{aligned}14x^2+19x-3 &= 14x^2+21x-2x-3\\ &= (14x^2+21x)+(-2x-3)\\ &= 7x(2x+3)-(2x+3)\\ &= (2x+3)(7x-1)\end{aligned}$$

We must include the common factor to get a factorization of the original trinomial.

$14x^4+19x^3-3x^2 = x^2(2x+3)(7x-1)$

69. $168x^3-45x^2+3x$

(1) Factor out the common factor, $3x$:

$168x^3-45x^2+3x = 3x(56x^2-15x+1)$

(2) Now we factor the trinomial $56x^2-15x+1$. Multiply the leading coefficient, 56, and the constant, 1: $56\cdot 1=56$.

(3) Look for a factorization of 56 in which the sum of the factors is the coefficient of the middle term, -15. The numbers we want are -7 and -8: $-7(-8)=56$ and $-7+(-8)=-15$.

(4) Split the middle term: $-15x=-7x-8x$

(5) Factor by grouping:

$$\begin{aligned}56x^2-15x+1 &= 56x^2-7x-8x+1\\ &= (56x^2-7x)+(-8x+1)\\ &= 7x(8x-1)-(8x-1)\\ &= (8x-1)(7x-1)\end{aligned}$$

We must include the common factor to get a factorization of the original trinomial.

$168x^3-45x^2+3x = 3x(8x-1)(7x-1)$

71. $15x^4-19x^2+6$

(1) There are no common factors (other than 1 or -1).

(2) Multiply the leading coefficient, 15, and the constant, 6: $15\cdot 6=90$.

(3) Look for a factorization of 90 in which the sum of the factors is the coefficient of the middle term, -19. The numbers we want are -9 and -10: $-9(-10)=90$ and $-9+(-10)=-19$.

(4) Split the middle term: $-19x^2=-9x^2-10x^2$

(5) Factor by grouping:

$$\begin{aligned}15x^4-19x^2+6 &= 15x^4-9x^2-10x^2+6\\ &= (15x^4-9x^2)+(-10x^2+6)\\ &= 3x^2(5x^2-3)-2(5x^2-3)\\ &= (5x^2-3)(3x^2-2)\end{aligned}$$

73. $25t^2+80t+64$

(1) There are no common factors (other than 1 or -1).

(2) Multiply the leading coefficient, 25, and the constant, 64: $25\cdot 64=1600$.

(3) Look for a factorization of 1600 in which the sum of the factors is the coefficient of the middle term, 80. The numbers we want are 40 and 40: $40\cdot 40=1600$ and $40+40=80$.

(4) Split the middle term: $80t=40t+40t$

(5) Factor by grouping:

$$\begin{aligned}25t^2+80t+64 &= 25t^2+40t+40t+64\\ &= (25t^2+40t)+(40t+64)\\ &= 5t(5t+8)+8(5t+8)\\ &= (5t+8)(5t+8),\text{ or}\\ &\quad (5t+8)^2\end{aligned}$$

75. $6x^3+4x^2-10x$

(1) Factor out the common factor, $2x$:

$6x^3+4x^2-10x = 2x(3x^2+2x-5)$

(2) - (5) Now we factor the trinomial $3x^2+2x-5$. We did this in Exercise 49, using the variable y rather than x. We found that $3x^2+2x-5=(3x+5)(x-1)$. We must include the common factor to get a factorization of the original trinomial.

$6x^3+4x^2-10x = 2x(3x+5)(x-1)$

77. $3x^2+9x+5$

(1) There are no common factors (other than 1 or -1).

(2) Multiply the leading coefficient, 3, and the constant, 5: $3\cdot 5=15$.

(3) Look for a factorization of 15 in which the sum of the factors is the coefficient of the middle term, 9. It is not possible to find such a pair of numbers. Thus, $3x^2+9x+5$ cannot be factored into a product of binomial factors. It is prime.

79. $6x^2-19x-5$

(1) There are no common factors (other than 1 or -1).

(2) Multiply the leading coefficient, 6, and the constant, -5: $6(-5)=-30$.

(3) Look for a factorization of -30 in which the sum of the factors is the coefficient of the middle term, -19. There is no such pair of numbers. Thus, $6x^2-19x-5$ cannot be factored into a product of binomial factors. It is prime.

81. $12m^2 - mn - 20n^2$

(1) There are no common factors (other than 1 or -1).

(2) Multiply the leading coefficient, 12, and the constant, -20: $12(-20) = -240$.

(3) Look for a factorization of -240 in which the sum of the factors is the coefficient of the middle term, -1. The numbers we want are 15 and -16: $15(-16) = -240$ and $15 + (-16) = -1$.

(4) Split the middle term: $-mn = 15mn - 16mn$

(5) Factor by grouping:

$$\begin{aligned} &12m^2 - mn - 20n^2 \\ &= 12m^2 + 15mn - 16mn - 20n^2 \\ &= (12m^2 + 15mn) + (-16mn - 20n^2) \\ &= 3m(4m + 5n) - 4n(4m + 5n) \\ &= (4m + 5n)(3m - 4n) \end{aligned}$$

83. $6a^2 - ab - 15b^2$

(1) There are no common factors (other than 1 or -1).

(2) Multiply the leading coefficient, 6, and the constant, -15: $6(-15) = -90$.

(3) Look for a factorization of -90 in which the sum of the factors is the coefficient of the middle term, -1. The numbers we want are -10 and 9: $-10 \cdot 9 = -90$ and $-10 + 9 = -1$.

(4) Split the middle term: $-ab = -10ab + 9ab$

(5) Factor by grouping:

$$\begin{aligned} 6a^2 - ab - 15b^2 &= 6a^2 - 10ab + 9ab - 15b^2 \\ &= (6a^2 - 10ab) + (9ab - 15b^2) \\ &= 2a(3a - 5b) + 3b(3a - 5b) \\ &= (3a - 5b)(2a + 3b) \end{aligned}$$

85. $9a^2 - 18ab + 8b^2$

(1) There are no common factors (other than 1 or -1).

(2) Multiply the leading coefficient, 9, and the constant, 8: $9 \cdot 8 = 72$.

(3) Look for a factorization of 72 in which the sum of the factors is the coefficient of the middle term, -18. The numbers we want are -6 and -12: $-6(-12) = 72$ and $-6 + (-12) = -18$.

(4) Split the middle term: $-18ab = -6ab - 12ab$

(5) Factor by grouping:

$$\begin{aligned} 9a^2 - 18ab + 8b^2 &= 9a^2 - 6ab - 12ab + 8b^2 \\ &= (9a^2 - 6ab) + (-12ab + 8b^2) \\ &= 3a(3a - 2b) - 4b(3a - 2b) \\ &= (3a - 2b)(3a - 4b) \end{aligned}$$

87. $35p^2 + 34pq + 8q^2$

(1) There are no common factors (other than 1 or -1).

(2) Multiply the leading coefficient, 35, and the constant, 8: $35 \cdot 8 = 280$.

(3) Look for a factorization of 280 in which the sum of the factors is the coefficient of the middle term, 34. The numbers we want are 14 and 20: $14 \cdot 20 = 280$ and $14 + 20 = 34$.

(4) Split the middle term: $34pq = 14pq + 20pq$

(5) Factor by grouping:

$$\begin{aligned} 35p^2 + 34pq + 8q^2 &= 35p^2 + 14pq + 20pq + 8q^2 \\ &= (35p^2 + 14pq) + (20pq + 8q^2) \\ &= 7p(5p + 2q) + 4q(5p + 2q) \\ &= (5p + 2q)(7p + 4q) \end{aligned}$$

89. $18x^2 - 6xy - 24y^2$

(1) Factor out the common factor, 6.

$18x^2 - 6xy - 24y^2 = 6(3x^2 - xy - 4y^2)$

(2) Now we factor the trinomial $3x^2 - xy - 4y^2$. Multiply the leading coefficient, 3, and the constant, -4: $3(-4) = -12$.

(3) Look for a factorization of -12 in which the sum of the factors is the coefficient of the middle term, -1. The numbers we want are -4 and 3: $-4 \cdot 3 = -12$ and $-4 + 3 = -1$.

(4) Split the middle term: $-xy = -4xy + 3xy$

(5) Factor by grouping:

$$\begin{aligned} 3x^2 - xy - 4y^2 &= 3x^2 - 4xy + 3xy - 4y^2 \\ &= (3x^2 - 4xy) + (3xy - 4y^2) \\ &= x(3x - 4y) + y(3x - 4y) \\ &= (3x - 4y)(x + y) \end{aligned}$$

We must include the common factor to get a factorization of the original trinomial.

$18x^2 - 6xy - 24y^2 = 6(3x - 4y)(x + y)$

91. $60x + 18x^2 - 6x^3 = -6x^3 + 18x^2 + 60x$

(1) Factor out the common factor, $-6x$. We factor out $-6x$ rather than $6x$ in order to have a positive leading coefficient in the trinomial factor.

$-6x^3 + 18x^2 + 60x = -6x(x^2 - 3x - 10)$

(2) - (5) We factor $x^2 - 3x - 10$ as we did in Exercise 43, getting the $(x - 5)(x + 2)$. Then we have:

$$\begin{aligned} &60x + 18x^2 - 6x^3 \\ &= -6x(x - 5)(x + 2) \\ &= 6x(-x + 5)(x + 2) \\ &\qquad \text{Multiplying } x - 5 \text{ by } -1 \\ &= 6x(x - 5)(-x - 2) \\ &\qquad \text{Multiplying } x + 2 \text{ by } -1 \end{aligned}$$

Note that we can express $6x(-x + 5)(x + 2)$ as $6x(5 - x)(x + 2)$ since $-x + 5 = 5 - x$ by the commutative law of addition.

93. $35x^5 - 57x^4 - 44x^3$

(1) We first factor out the common factor, x^3.

$x^3(35x^2 - 57x - 44)$

(2) Now we factor the trinomial $35x^2 - 57x - 44$. Multiply the leading coefficient, 35, and the constant, -44: $35(-44) = -1540$.

(3) Look for a factorization of -1540 in which the sum of the factors is the coefficient of the middle term, -57.

Pairs of factors	Sums of factors
7, −220	−213
10, −154	−144
11, −140	−129
14, −110	−96
20, −77	−57

(4) Split the middle term: $-57x = 20x - 77x$

(5) Factor by grouping:

$$\begin{aligned} 35x^2 - 57x - 44 &= 35x^2 + 20x - 77x - 44 \\ &= (35x^2 + 20x) + (-77x - 44) \\ &= 5x(7x+4) - 11(7x+4) \\ &= (7x+4)(5x-11) \end{aligned}$$

We must include the common factor to get a factorization of the original trinomial.

$$35x^5 - 57x^4 - 44x^3 = x^3(7x+4)(5x-11)$$

95. $(3x^4)^3 = 3^3(x^4)^3 = 27x^{4\cdot 3} = 27x^{12}$

97. $(x^2y)(x^3y^5) = (x^2x^3)(yy^5) = x^{2+3}y^{1+5} = x^5y^6$

99. 30,080,000,000

We move the decimal point 10 places to the left. This is a large number, so the exponent is positive.

$30,080,000,000 = 3.008 \times 10^{10}$

101. ***Familiarize***. We will use the formula $C = 2\pi r$, where C is circumference and r is radius, to find the radius in kilometers. Then we will multiply that number by 0.62 to find the radius in miles.

Translate.

$$\underbrace{\text{Circumference}}_{\downarrow\atop 40,000} \;\overset{=}{\underset{\approx}{\downarrow}}\; \underbrace{2\cdot\pi\cdot\text{radius}}_{\downarrow\atop 2(3.14)r}$$

Solve. First we solve the equation.

$40,000 \approx 2(3.14)r$

$40,000 \approx 6.28r$

$6369 \approx r$

Then we multiply to find the radius in miles:

$6369(0.62) \approx 3949$

Check. If $r = 6369$, then $2\pi r = 2(3.14)(6369) \approx 40,000$. We should also recheck the multiplication we did to find the radius in miles. Both values check.

State. The radius of the earth is about 6369 km or 3949 mi. (These values may differ slightly if a different approximation is used for π.)

103. $9x^{10} - 12x^5 + 4$

(a) First factor out a common factor, if any. There is none (other than 1 or -1).

(b) Multiply the leading coefficient, 9, and the constant, 4: $9 \cdot 4 = 36$.

(c) Look for a factorization of 36 in which the sum of the factors is the coefficient of the middle term, -12. The factors we want are -6 and -6.

(d) Split the middle term: $-12x^5 = -6x^5 - 6x^5$

(e) Factor by grouping:

$$\begin{aligned} 9x^{10} - 12x^5 + 4 &= 9x^{10} - 6x^5 - 6x^5 + 4 \\ &= (9x^{10} - 6x^5) + (-6x^5 + 4) \\ &= 3x^5(3x^5 - 2) - 2(3x^5 - 2) \\ &= (3x^5 - 2)(3x^5 - 2), \text{ or} \\ &= (3x^5 - 2)^2 \end{aligned}$$

105. $16x^{10} + 8x^5 + 1$

(a) First factor out a common factor, if any. There is none (other than 1 or -1).

(b) Multiply the leading coefficient, 16, and the constant, 1: $16 \cdot 1 = 16$.

(c) Look for a factorization of 16 in which the sum of the factors is the coefficient of the middle term, 8. The factors we want are 4 and 4.

(d) Split the middle term: $8x^5 = 4x^5 + 4x^5$

(e) Factor by grouping:

$$\begin{aligned} 16x^{10} + 8x^5 + 1 &= 16x^{10} + 4x^5 + 4x^5 + 1 \\ &= (16x^{10} + 4x^5) + (4x^5 + 1) \\ &= 4x^5(4x^5 + 1) + 1(4x^5 + 1) \\ &= (4x^5 + 1)(4x^5 + 1), \text{ or} \\ &= (4x^5 + 1)^2 \end{aligned}$$

107.-113. Left to the student

Chapter 5 Mid-Chapter Review

1. True; the smallest natural number is 1, and the smallest number in the set must be a factor of the GCF.

2. False; look for factors of c whose sum is b.

3. True; a prime polynomial cannot be factored, so it has no common factor other than 1 and -1.

4. False; because the constant term is positive and the middle term is negative, we need to consider only negative pairs of factors of 45.

5.
$$\begin{aligned} 10y^3 - 18y^2 + 12y &= 2y \cdot 5y^2 - 2y \cdot 9y + 2y \cdot 6 \\ &= 2y(5y^2 - 9y + 6) \end{aligned}$$

6. $2x^2 - x - 6$

$a \cdot c = 2 \cdot (-6) = -12$

$-x = -4x + 3x$

$$\begin{aligned}2x^2 - x - 6 &= 2x^2 - 4x + 3x - 6\\ &= 2x(x-2) + 3(x-2)\\ &= (x-2)(2x+3)\end{aligned}$$

7. $x^3 = x^3$

$3x = 3 \cdot x$

The coefficients have no common prime factor. The GCF of the powers of x is x because 1 is the smallest exponent of x. Thus the GCF is x.

8. $5x^4 = 5 \cdot x^4$

$x^2 = x^2$

The coefficients have no common prime factor. The GCF of the powers of x is x^2 because 2 is the smallest exponent of x. Thus the GCF is x^2.

9. $6x^5 = 2 \cdot 3 \cdot x^5$

$-12x^3 = -1 \cdot 2 \cdot 2 \cdot 3 \cdot x^3$

Each coefficient has factors of 2 and 3. The GCF of the powers of x is x^3 because 3 is the smallest exponent of x. Thus the GCF is $2 \cdot 3 \cdot x^3$, or $6x^3$.

10. $-8x = -1 \cdot 2 \cdot 2 \cdot 2 \cdot x$

$-12 = -1 \cdot 2 \cdot 2 \cdot 3$

$16x^2 = 2 \cdot 2 \cdot 2 \cdot 2 \cdot x^2$

Each coefficient has two factors of 2. The GCF of the powers of x is x^0 because -12 has no x-factor. Thus the GCF is $2 \cdot 2 \cdot x^0$, or 4.

11. $15x^3y^2 = 3 \cdot 5 \cdot x^3 \cdot y^2$

$5x^2y = 5 \cdot x^2 \cdot y$

$40x^4y^3 = 2 \cdot 2 \cdot 2 \cdot 5 \cdot x^4 \cdot y^3$

Each coefficient has a factor of 5. The GCF of the powers of x is x^2 because 2 is the smallest exponent of x. The GCF of the powers of y is y because 1 is the smallest exponent of y. Thus the GCF is $5x^2y$.

12. $x^2y^4 = x^2 \cdot y^4$

$-x^3y^3 = -1 \cdot x^3 \cdot y^3$

$x^3y^2 = x^3 \cdot y^2$

$x^5y^4 = x^5 \cdot y^4$

There is no common prime factor. The GCFs of the powers of x and y are x^2 and y^2, respectively, because 2 is the smallest exponent of each variable. Thus the GCF is x^2y^2.

13. $x^3 - 8x = x \cdot x^2 - x \cdot 8 = x(x^2 - 8)$

14. $3x^2 + 12x = 3x \cdot x + 3x \cdot 4 = 3x(x+4)$

15. $2y^2 + 8y - 4 = 2 \cdot y^2 + 2 \cdot 4y - 2 \cdot 2 = 2(y^2 + 4y - 2)$

The trinomial $y^2 + 4y - 2$ cannot be factored, so the factorization is complete.

16. $3t^6 - 5t^4 - 2t^3 = t^3 \cdot 3t^3 - t^3 \cdot 5t - t^3 \cdot 2 = t^3(3t^3 - 5t - 2)$

17. $x^2 + 4x + 3$

Look for a pair of factors of 3 whose sum is 4. The numbers we want are 1 and 3.

$x^2 + 4x + 3 = (x+1)(x+3)$

18. $z^2 - 4z + 4$

Look for a pair of factors of 4 whose sum is -4. The numbers we want are -2 and -2.

$z^2 - 4z + 4 = (z-2)(z-2)$, or $(z-2)^2$

19. $$\begin{aligned}x^2 + 4x^2 + 3x + 12 &= x^2(x+4) + 3(x+4)\\ &= (x+4)(x^2+3)\end{aligned}$$

20. $8y^5 - 48y^3 = 8y^3 \cdot y^2 - 8y^3 \cdot 6 = 8y^3(y^2 - 6)$

21. $6x^3y + 24x^2y^2 - 42xy^3 = 6xy \cdot x^2 + 6xy \cdot 4xy - 6xy \cdot 7y^2 = 6xy(x^2 + 4xy - 7y^2)$

The trinomial $x^2 + 4xy - 7y^2$ cannot be factored, so the factorization is complete.

22. $6 - 11t + 4t^2$, or $4t^2 - 11t + 6$

We will use the FOIL method.

(1) There is no common factor (other than 1 or -1).

(2) We can factor $4t^2$ as

$(4t+\quad)(t+\quad)$ or $(2t+\quad)(2t+\quad)$.

(3) The constant term is positive and the middle term is negative, so we look for pairs of negative factors of 6. The possibilities are -1, -6 and -2, -3.

The factors can also be written as -6, -1 and -3, -2.

(4) Look for combinations of factors from steps (2) and (3) such that the sum of their products is the middle term, $-11t$.

The factorization is $(4t-3)(t-2)$.

23. $z^2 + 4z - 5$

Look for a pair of factors of -5 whose sum is 4. The numbers we want are 5 and -1.

$z^2 + 4z - 5 = (z+5)(z-1)$

24. $$\begin{aligned}2z^3 + 8z^3 + 5z + 20 &= 2z^2(z+4) + 5(z+4)\\ &= (z+4)(2z^2+5)\end{aligned}$$

25. $$\begin{aligned}3p^3 - 2p^2 - 9p + 6 &= p^2(3p-2) - 3(3p-2)\\ &= (3p-2)(p^2-3)\end{aligned}$$

26. $10x^8 - 25x^6 - 15x^5 + 35x^3 = 5x^3(2x^5 - 5x^3 - 3x^2 + 7)$

27. $$\begin{aligned}2w^3 + 3w^2 - 6w - 9 &= w^2(2w+3) - 3(2w+3)\\ &= (2w+3)(w^2-3)\end{aligned}$$

28. $4x^4 - 5x^3 + 3x^2 = x^2(4x^2 - 5x + 3)$

The trinomial $4x^2 - 5x + 3$ cannot be factored, so the factorization is complete.

29. $6y^2 + 7y - 10$

We will use the *ac*-method.

(1) There is no common factor (other than 1 or -1).

(2) Multiply 6 and -10: $6(-10) = -60$.

(3) Look for a factorization of -60 in which the sum of the factors is 7. The numbers we want are 12 and -5.

(4) Split the middle term: $7y = 12y - 5y$.

(5) Factor by grouping.

$$\begin{aligned} 6y^2 + 7y - 10 &= 6y^2 + 12y - 5y - 10 \\ &= 6y(y+2) - 5(y+2) \\ &= (y+2)(6y-5) \end{aligned}$$

30. $3x^2 - 3x - 18 = 3(x^2 - x - 6)$

Consider $x^2 - x - 6$. Look for a pair of factors of -6 whose sum is -1. The numbers we want are -3 and 2.

$x^2 - x - 6 = (x-3)(x+2)$, so $3x^2 - 3x - 18 = 3(x-3)(x+2)$.

31. $$\begin{aligned} 6x^3 + 4x^2 + 3x + 2 &= 2x^2(3x+2) + (3x+2) \\ &= (3x+2)(2x^2+1) \end{aligned}$$

32. $15 - 8w + w^2$, or $w^2 - 8w + 15$

Look for a pair of factors of 15 whose sum is -8. The numbers we want are -3 and -5.

$15 - 8w + w^2 = (w-3)(w-5)$

33. $$\begin{aligned} 8x^3 + 20x^2 + 2x + 5 &= 4x^2(2x+5) + (2x+5) \\ &= (2x+5)(4x^2+1) \end{aligned}$$

34. $10z^2 - 21z - 10$

We will use the FOIL method.

(1) There is no common factor (other than 1 or -1).

(2) We can factor $10z^2$ as

$(z+\quad)(10z+\quad)$ or $(2z+\quad)(5z+\quad)$.

(3) Factor the last term, -10. The possibilities are -1, 10 and -10, 1 and -2, 5 and 2, -5.

The factors can also be written as 10, -1 and 1, -10 and 5, -2 and -5, 2.

(4) Look for combinations of factors from steps (2) and (3) such that the sum of their products is the middle term, $-21z$.

The factorization is $(2z-5)(5z+2)$.

35. $6x^2 + 7x + 2$

We will use the *ac*-method.

(1) There is no common factor (other than 1 or -1).

(2) Multiply 6 and 2: $6 \cdot 2 = 12$.

(3) Look for a factorization of 12 in which the sum of the factors is 7. The numbers we want are 3 and 4.

(4) Split the middle term: $7x = 3x + 4x$.

(5) Factor by grouping.

$$\begin{aligned} 6x^2 + 7x + 2 &= 6x^2 + 3x + 4x + 2 \\ &= 3x(2x+1) + 2(2x+1) \\ &= (2x+1)(3x+2) \end{aligned}$$

36. $x^2 - 10xy + 24y^2$

Look for a pair of factors of $24y^2$ whose sum is $-10y$. The factors we want are $-4y$ and $-6y$.

$x^2 - 10xy + 24y^2 = (x-4y)(x-6y)$

37. $$\begin{aligned} 6z^3 + 3z^2 + 2z + 1 &= 3z^2(2z+1) + (2z+1) \\ &= (2z+1)(3z^2+1) \end{aligned}$$

38. $a^3b^7 + a^4b^5 - a^2b^3 + a^5b^6 = a^2b^3(ab^4 + a^2b^2 - 1 + a^3b^3)$

39. $4y^2 - 7yz - 15z^2$

We will use the FOIL method.

(1) There is no common factor (other than 1 or -1).

(2) We can factor $4y^2$ as

$(4y+\quad)(y+\quad)$ or $(2y+\quad)(2y+\quad)$.

(3) Factor the last term, $-15z^2$. The possibilities are $-z$, $15z$ and z, $-15z$ and $-3z$, $5z$ and $3z$, $-5z$.

The factors can also be written as $15z$, $-z$ and $-15z$, z and $5z$, $-3z$ and $-5z$, $3z$.

(4) Look for combinations of factors from steps (2) and (3) such that the sum of their products is the middle term, $-7yz$.

The factorization is $(4y+5z)(y-3z)$.

40. $3x^3 + 21x^2 + 30x = 3x(x^2 + 7x + 10)$

Consider $x^2 + 7x + 10$. Look for a pair of factors of 10 whose sum is 7. The numbers we want are 2 and 5.

$x^2 + 7x + 10 = (x+2)(x+5)$, so $3x^3 + 21x^2 + 30x = 3x(x+2)(x+5)$.

41. $$\begin{aligned} x^3 - 3x^2 - 2x + 6 &= x^2(x-3) - 2(x-3) \\ &= (x-3)(x^2-2) \end{aligned}$$

42. $9y^2 + 6y + 1$

We will use the *ac*-method.

(1) There is no common factor (other than 1 or -1).

(2) Multiply 9 and 1: $9 \cdot 1 = 9$.

(3) Look for a factorization of 9 in which the sum of the factors is 6. The numbers we want are 3 and 3.

(4) Split the middle term: $6y = 3y + 3y$.

(5) Factor by grouping.

$$\begin{aligned} 9y^2 + 6y + 1 &= 9y^2 + 3y + 3y + 1 \\ &= 3y(3y+1) + (3y+1) \\ &= (3y+1)(3y+1), \text{ or } (3y+1)^2 \end{aligned}$$

43. $y^2 + 6y + 8$

Look for a pair of factors of 8 whose sum is 6. The numbers we want are 2 and 4.

$y^2 + 6y + 8 = (y+2)(y+4)$

44. $6y^2 + 33y + 45 = 3(2y^2 + 11y + 15)$

We will use the FOIL method to factor $2y^2 + 11y + 15$.

(1) There is no common factor (other than 1 or -1.)

(2) We can factor $2y^2$ as

$(2y+\quad)(y+\quad)$.

(3) The constant term and the middle term are both positive, so we look for pairs of positive factors of 15. The possibilities are

1, 15 and 3, 5.

We can also write these factors as

15, 1 and 5, 3.

(4) Look for combinations of factors from steps (2) and (3) such that the sum of their products is the middle term, $11y$.

The factorization of $2y^2+11y+15$ is $(2y+5)(y+3)$, so $6y^2 + 33y + 45 = 3(2y+5)(y+3)$.

45. $x^3 - 7x^2 + 4x - 28 = x^2(x-7) + 4(x-7)$
$= (x-7)(x^2+4)$

46. $4 + 3y - y^2 = -y^2 + 3y + 4 = -1(y^2 - 3y - 4)$

To factor $y^2 - 3y - 4$, look for a pair of factors of -4 whose sum is -3. The numbers we want are -4 and 1. Thus $y^2 - 3y - 4 = (y-4)(y+1)$. Then we have

$4 + 3y - y^2 = -1(y-4)(y+1)$, or
$(-y+4)(y+1)$, or
$(y-4)(-y-1)$

47. $16x^2 - 16x - 60 = 4(4x^2 - 4x - 15)$

We will factor $4x^2 - 4x - 15$ using the ac-method.

(1) There is no common factor (other than 1 or -1.)

(2) Multiply 4 and -15: $4(-15) = -60$.

(3) Look for a factorization of -60 in which the sum of the factors is -4. The numbers we want are 6 and -10.

(4) Split the middle term: $-4x = 6x - 10x$.

(5) Factor by grouping.

$4x^2 - 4x - 15 = 4x^2 + 6x - 10x - 15$
$= 2x(2x+3) - 5(2x+3)$
$= (2x+3)(2x-5)$

Then $16x^2 - 16x - 60 = 4(2x+3)(2x-5)$.

48. $10a^2 - 11ab + 3b^2$

We will use the FOIL method.

(1) There is no common factor (other than 1 or -1).

(2) We can factor $10a^2$ as

$(a+\quad)(10a+\quad)$ or $(2a+\quad)(5a+\quad)$.

(3) The constant term is positive and the middle term is negative, so we look for pairs of negative factors of $3b^2$. The possibilities are $-b$, $3b$ and b, $-3b$.

We can also write these factors as $3b$, $-b$ and $-3b$, b.

(4) Look for combinations of factors from steps (2) and (3) such that the sum of their products is the middle term, $-11ab$.

The factorization is $(2a-b)(5a-3b)$.

49. $6w^3 - 15w^2 - 10w + 25 = 3w^2(2w-5) - 5(2w-5)$
$= (2w-5)(3w^2-5)$

50. $y^3 + 9y^2 + 18y = y(y^2 + 9y + 18)$

To factor $y^2 + 9y + 18$, look for a pair of factors of 18 whose sum is 9. The numbers we want are 3 and 6. Thus $y^2 + 9y + 18 = (y+3)(y+6)$. Then $y^3 + 9y^2 + 18y = y(y+3)(y+6)$.

51. $4x^2 + 11xy + 6y^2$

We will use the ac-method.

(1) There is no common factor (other than 1 or -1).

(2) Multiply 4 and $6y^2$: $4 \cdot 6y^2 = 24y^2$

(3) Look for a factorization of $24y^2$ in which the sum of the factors is $11y$. The numbers we want are $3y$ and $8y$.

(4) Split the middle term: $11xy = 3xy + 8xy$.

(5) Factor by grouping:

$4x^2 + 11xy + 6y^2 = 4x^2 + 3xy + 8xy + 6y^2$
$= x(4x+3y) + 2y(4x+3y)$
$= (4x+3y)(x+2y)$

52. $6 - 5z - 6z^2 = -6z^2 - 5z + 6 = -1(6z^2 + 5z - 6)$

We will use the FOIL method to factor $6z^2 + 5z - 6$.

(1) There is no common factor (other than 1 or -1.)

(2) We can factor $6z^2$ as

$(z+\quad)(6z+\quad)$ or $(2z+\quad)(3z+\quad)$.

(3) Factor the last term, -6. The possibilities are -1, 6 and 1, -6 and -2, 3 and 2, -3.

The factors can also be written as 6, -1 and -6, 1 and 3, -2 and -3, 2.

(4) Look for combinations of factors from steps (2) and (3) such that the sum of their products is the middle term, $5z$.

The factorization of $6z^2 + 5z - 6$ is $(2z+3)(3z-2)$, so we have

$6 - 5z - 6z^2 = -1(2z+3)(3z-2)$, or
$(-2z-3)(3z-2)$, or
$(2z+3)(-3z+2)$

53. $12t^3 + 8t^2 - 9t - 6 = 4t^2(3t+2) - 3(3t+2)$
$= (3t+2)(4t^2-3)$

54. $y^2 + yz - 20z^2$

Look for a pair of factors of $-20z^2$ whose sum is z. The factors we want are $5z$ and $-4z$.

$y^2 + yz - 20z = (y+5z)(y-4z)$

55. $9x^2 - 6xy - 8y^2$

We will use the ac-method.

(1) There is no common factor (other than 1 or -1).

(2) Multiply 9 and $-8y^2$: $9(-8y^2) = -72y^2$.

(3) Look for a factorization of $-72y^2$ in which the sum of the factors is $-6y$. The factors we want and $-12y$ and $6y$.

(4) Split the middle term: $-6xy = -12xy + 6xy$.

(5) Factor by grouping.

$$\begin{aligned} 9x^2 - 6xy - 8y^2 &= 9x^2 - 12xy + 6xy - 8y^2 \\ &= 3x(3x - 4y) + 2y(3x - 4y) \\ &= (3x - 4y)(3x + 2y) \end{aligned}$$

56. $-3 + 8z + 3z^2 = 3z^2 + 8z - 3$

We will use the FOIL method.

(1) There is no common factor (other than 1 or -1).

(2) We can factor $3z^2$ as

$(3z+\quad)(z+\quad)$.

(3) We look for pairs of factors of -3. The possibilities are -1, 3 and 1, -3.

These factors can also be written as 3, -1 and -3, 1.

(4) Look for combinations of factors from steps (2) and (3) such that the sum of their products is the middle term, $8z$.

The factorization is $(3z - 1)(z + 3)$.

57. $m^2 - 6mn - 16n^2$

Look for a pair of factors of $-16n^2$ whose sum is $-6n$. The numbers we want are $-8n$ and $2n$.

$m^2 - 6mn - 16n^2 = (m - 8n)(m + 2n)$.

58. $2w^2 - 12w + 18 = 2(w^2 - 6w + 9)$

Consider $w^2 - 6w + 9$. Look for a pair of factors of 9 whose sum is -6. The numbers we want are -3 and -3.

$w^2 - 6w + 9 = (w - 3)(w - 3)$, so $2w^2 - 12w + 18 = 2(w - 3)(w - 3)$, or $2(w - 3)^2$.

59. $18t^3 - 18t^2 + 4t = 2t(9t^2 - 9t + 2)$

We will use the ac-method to factor $9t^2 - 9t + 2$.

(1) There is no common factor (other than 1 or -1.)

(2) Multiply 9 and 2: $9 \cdot 2 = 18$.

(3) Look for a factorization of 18 in which the sum of the factors is -9. The numbers we want are -6 and -3.

(4) Split the middle term: $-9t = -6t - 3t$.

(5) Factor by grouping.

$$\begin{aligned} 9t^2 - 9t + 2 &= 9t^2 - 6t - 3t + 2 \\ &= 3t(3t - 2) - (3t - 2) \\ &= (3t - 2)(3t - 1) \end{aligned}$$

Then $18t^3 - 18t^2 + 4t = 2t(3t - 2)(3t - 1)$.

60.
$$\begin{aligned} 5z^3 + 15z^2 + z + 3 &= 5z^2(z + 3) + (z + 3) \\ &= (z + 3)(5z^2 + 1) \end{aligned}$$

61. $-14 + 5t + t^2$, or $t^2 + 5t - 14$

Look for a pair of factors of -14 whose sum is 5. The numbers we want are 7 and -2.

$-14 + 5t + t^2 = (t + 7)(t - 2)$

62. $4t^2 - 20t + 25$

We will use the FOIL method.

(1) There is no common factor (other than 1 or -1).

(2) We can factor $4t^2$ as

$(4t+\quad)(t+\quad)$ or $(2t+\quad)(2t+\quad)$.

(3) Factor the last term, 25. The middle term is negative, so we consider only pairs of negative factors. The possibilities are -1, -25 and -5, -5.

The first pair of factors can also be written as -25, -1.

(4) Look for combinations of factors from steps (2) and (3) such that the sum of their products is the middle term, $-20t$.

The factorization is $(2t - 5)(2t - 5)$, or $(2t - 5)^2$.

63. $t^2 + 4t - 12$

Look for a pair of factors of -12 whose sum is 4. The numbers we want are 6 and -2.

$t^2 + 4t - 12 = (t + 6)(t - 2)$

64. $12 + 5z - 2z^2 = -1(2z^2 - 5z - 12)$

We use the FOIL method to factor $2z^2 - 5z - 12$.

(1) There is no common factor (other than 1 or -1).

(2) We can factor $2z^2$ as

$(2z+\quad)(z+\quad)$.

(3) Factor the last term, -12. The possibilities are -1, 12 and 1, -12 and -2, 6 and 2, -6 and 3, -4 and -3, 4.

These factors can also be written as 12, -1 and -12, 1 and 6, -2 and -6, 2 and -4, 3 and 4, -3.

(4) Look for combinations of factors from steps (2) and (3) such that the sum of their products is the middle term, $-5z$.

The factorization of $2z^2 - 5z - 12$ is $(2z + 3)(z - 4)$, so we have

$$\begin{aligned} 12 + 5z - 2z^2 &= -1(2z + 3)(z - 4), \text{ or} \\ &(-2z - 3)(z - 4), \text{ or} \\ &(2z + 3)(-z + 4) \end{aligned}$$

65. $12 + 4y - y^2 = -1(y^2 - 4y - 12)$

To factor $y^2 - 4y - 12$, look for a pair of factors of -12 whose sum is -4. The number we want are -6 and 2.

$y^2 - 4y - 12 = (y - 6)(y + 2)$, so we have

$$\begin{aligned} 12 + 4y - y^2 &= -1(y - 6)(y + 2), \text{ or} \\ &(-y + 6)(y + 2), \text{ or} \\ &(y - 6)(-y - 2) \end{aligned}$$

66. Find the product of two binomials. For example, $(ax^2+b)(cx+d)=acx^3+adx^2+bcx+bd$.

67. There is a finite number of pairs of numbers with the correct product, but there are infinitely many pairs with the correct sum.

68. Since both constants are negative, the middle term will be negative so $(x-17)(x-18)$ cannot be a factorization of $x^2+35x+306$.

69. No; both $2x+6$ and $2x+8$ contain a factor of 2, so $2\cdot 2$, or 4, must be factored out to reach the complete factorization. In other words, the largest common factor is 4 not 2.

Exercise Set 5.5

RC1. The statement is false; see page 418 in the text.

RC3. The statement is false; see page 420 in the text.

1. $x^2-14x+49$

(a) We know that x^2 and 49 are squares.

(b) There is no minus sign before either x^2 or 49.

(c) If we multiply the square roots, x and 7, and double the product, we get $2\cdot x\cdot 7=14x$. This is the opposite of the remaining term, $-14x$.

Thus, $x^2-14x+49$ is a trinomial square.

3. $x^2+16x-64$

Both x^2 and 64 are squares, but there is a minus sign before 64. Thus, $x^2+16x-64$ is not a trinomial square.

5. x^2-2x+4

(a) Both x^2 and 4 are squares.

(b) There is no minus sign before either x^2 or 4.

(c) If we multiply the square roots, x and 2, and double the product, we get $2\cdot x\cdot 2=4x$. This is neither the remaining term nor its opposite.

Thus, x^2-2x+4 is not a trinomial square.

7. $9x^2-24x+16$

(a) Both $9x^2$ and 16 are squares.

(b) There is no minus sign before either $9x^2$ or 16.

(c) If we multiply the square roots, $3x$ and 4, and double the product, we get $2\cdot 3x\cdot 4=24x$. This is the opposite of the remaining term, $-24x$.

Thus, $9x^2-24x+16$ is a trinomial square.

9.
$$\begin{aligned} x^2-14x+49 &= x^2 - 2\cdot x\cdot 7 + 7^2 = (x-7)^2 \\ & \quad\uparrow \quad\ \uparrow \quad \uparrow \quad \uparrow \qquad \uparrow \\ &= A^2 - 2\ \ A\ \ B + B^2 = (A-B)^2 \end{aligned}$$

11.
$$\begin{aligned} x^2+16x+64 &= x^2 + 2\cdot x\cdot 8 + 8^2 = (x+8)^2 \\ & \quad\uparrow \quad\ \uparrow \quad \uparrow \quad \uparrow \qquad \uparrow \\ &= A^2 + 2\ \ A\ \ B + B^2 = (A+B)^2 \end{aligned}$$

13. $x^2-2x+1=x^2-2\cdot x\cdot 1+1^2=(x-1)^2$

15.
$$\begin{aligned} 4+4x+x^2 &= x^2+4x+4 \qquad \text{Changing the order} \\ &= x^2+2\cdot x\cdot 2+2^2 \\ &= (x+2)^2 \end{aligned}$$

17. $y^2+12y+36=y^2+2\cdot y\cdot 6+6^2=(y+6)^2$

19.
$$\begin{aligned} 16+t^2-8t &= t^2-8t+16 \\ &= t^2-2\cdot t\cdot 4+4^2 \\ &= (t-4)^2 \end{aligned}$$

21. $q^4-6q^2+9=(q^2)^2-2\cdot q^2\cdot 3+3^2=(q^2-3)^2$

23.
$$\begin{aligned} 49+56y+16y^2 &= 16y^2+56y+49 \\ &= (4y)^2+2\cdot 4y\cdot 7+7^2 \\ &= (4y+7)^2 \end{aligned}$$

25.
$$\begin{aligned} 2x^2-4x+2 &= 2(x^2-2x+1) \\ &= 2(x^2-2\cdot x\cdot 1+1^2) \\ &= 2(x-1)^2 \end{aligned}$$

27.
$$\begin{aligned} x^3-18x^2+81x &= x(x^2-18x+81) \\ &= x(x^2-2\cdot x\cdot 9+9^2) \\ &= x(x-9)^2 \end{aligned}$$

29.
$$\begin{aligned} 12q^2-36q+27 &= 3(4q^2-12q+9) \\ &= 3[(2q)^2-2\cdot 2q\cdot 3+3^2] \\ &= 3(2q-3)^2 \end{aligned}$$

31.
$$\begin{aligned} 49-42x+9x^2 &= 7^2-2\cdot 7\cdot 3x+(3x)^2 \\ &= (7-3x)^2, \text{ or } (3x-7)^2 \end{aligned}$$

33.
$$\begin{aligned} 5y^4+10y^2+5 &= 5(y^4+2y^2+1) \\ &= 5[(y^2)^2+2\cdot y^2\cdot 1+1^2] \\ &= 5(y^2+1)^2 \end{aligned}$$

35.
$$\begin{aligned} 1+4x^4+4x^2 &= 1+4x^2+4x^4 \\ &= 1^2+2\cdot 1\cdot 2x^2+(2x^2)^2 \\ &= (1+2x^2)^2 \end{aligned}$$

37.
$$\begin{aligned} 4p^2+12pq+9q^2 &= (2p)^2+2\cdot 2p\cdot 3q+(3q)^2 \\ &= (2p+3q)^2 \end{aligned}$$

39.
$$\begin{aligned} a^2-6ab+9b^2 &= a^2-2\cdot a\cdot 3b+(3b)^2 \\ &= (a-3b)^2 \end{aligned}$$

41.
$$\begin{aligned} 81a^2-18ab+b^2 &= (9a)^2-2\cdot 9a\cdot b+b^2 \\ &= (9a-b)^2 \end{aligned}$$

43.
$$\begin{aligned} 36a^2+96ab+64b^2 &= 4(9a^2+24ab+16b^2) \\ &= 4[(3a)^2+2\cdot 3a\cdot 4b+(4b)^2] \\ &= 4(3a+4b)^2 \end{aligned}$$

45. x^2-4

(a) The first expression is a square: x^2

The second expression is a square: $4=2^2$

(b) The terms have different signs.

x^2-4 is a difference of squares.

47. $x^2 + 25$

The terms do not have different signs.

$x^2 + 25$ is not a difference of squares.

49. $x^2 - 45$

The number 45 is not a square.

$x^2 - 45$ is not a difference of squares.

51. $-25y^2 + 16x^2$

(a) The first expression is a square: $25y^2 = (5y)^2$

The second expression is a square: $16x^2 = (4x)^2$

(b) The terms have different signs.

$-25y^2 + 16x^2$ is a difference of squares.

53. $y^2 - 4 = y^2 - 2^2 = (y+2)(y-2)$

55. $p^2 - 1 = p^2 - 1^2 = (p+1)(p-1)$

57. $-49 + t^2 = t^2 - 49 = t^2 - 7^2 = (t+7)(t-7)$

59. $a^2 - b^2 = (a+b)(a-b)$

61. $25t^2 - m^2 = (5t)^2 - m^2 = (5t+m)(5t-m)$

63. $100 - k^2 = 10^2 - k^2 = (10+k)(10-k)$

65. $16a^2 - 9 = (4a)^2 - 3^2 = (4a+3)(4a-3)$

67. $4x^2 - 25y^2 = (2x)^2 - (5y)^2 = (2x+5y)(2x-5y)$

69. $8x^2 - 98 = 2(4x^2 - 49) = 2[(2x)^2 - 7^2] =$
$2(2x+7)(2x-7)$

71. $36x - 49x^3 = x(36 - 49x^2) = x[6^2 - (7x)^2] =$
$x(6+7x)(6-7x)$

73. $\frac{1}{16} - 49x^8 = \left(\frac{1}{4}\right)^2 - (7x^4)^2 = \left(\frac{1}{4} + 7x^4\right)\left(\frac{1}{4} - 7x^4\right)$

75. $0.09y^2 - 0.0004 = (0.3y)^2 - (0.02)^2 =$
$(0.3y + 0.02)(0.3y - 0.02)$

77. $49a^4 - 81 = (7a^2)^2 - 9^2 = (7a^2+9)(7a^2-9)$

79.
$$a^4 - 16$$
$$= (a^2)^2 - 4^2$$
$$= (a^2+4)(a^2-4) \quad \text{Factoring a difference of squares}$$
$$= (a^2+4)(a+2)(a-2) \quad \text{Factoring further: } a^2 - 4 \text{ is a difference of squares.}$$

81.
$$5x^4 - 405$$
$$5(x^4 - 81)$$
$$= 5[(x^2)^2 - 9^2]$$
$$= 5(x^2+9)(x^2-9)$$
$$= 5(x^2+9)(x+3)(x-3) \quad \text{Factoring } x^2 - 9$$

83.
$$1 - y^8$$
$$= 1^2 - (y^4)^2$$
$$= (1+y^4)(1-y^4)$$
$$= (1+y^4)(1+y^2)(1-y^2) \quad \text{Factoring } 1 - y^4$$
$$= (1+y^4)(1+y^2)(1+y)(1-y) \quad \text{Factoring } 1 - y^2$$

85.
$$x^{12} - 16$$
$$= (x^6)^2 - 4^2$$
$$= (x^6+4)(x^6-4)$$
$$= (x^6+4)(x^3+2)(x^3-2) \quad \text{Factoring } x^6 - 4$$

87. $y^2 - \frac{1}{16} = y^2 - \left(\frac{1}{4}\right)^2$
$= \left(y + \frac{1}{4}\right)\left(y - \frac{1}{4}\right)$

89. $25 - \frac{1}{49}x^2 = 5^2 - \left(\frac{1}{7}x\right)^2$
$= \left(5 + \frac{1}{7}x\right)\left(5 - \frac{1}{7}x\right)$

91.
$$16m^4 - t^4$$
$$= (4m^2)^2 - (t^2)^2$$
$$= (4m^2+t^2)(4m^2-t^2)$$
$$= (4m^2+t^2)(2m+t)(2m-t) \quad \text{Factoring } 4m^2 - t^2$$

93. $4x + 16y = 64$

To find the x-intercept, let $y = 0$ and solve for x.
$$4x + 16y = 64$$
$$4x + 16 \cdot 0 = 64$$
$$4x = 64$$
$$x = 16$$

The x-intercept is $(16, 0)$.

To find the y-intercept, let $x = 0$ and solve for y.
$$4x + 16y = 64$$
$$4 \cdot 0 + 16y = 64$$
$$16y = 64$$
$$y = 4$$

The y-intercept is $(0, 4)$.

95. $y = 2x - 5$

To find the x-intercept, let $y = 0$ and solve for x.
$$0 = 2x - 5$$
$$5 = 2x$$
$$\frac{5}{2} = x$$

The x-intercept is $\left(\frac{5}{2}, 0\right)$.

To find the y-intercept, let $x = 0$ and solve for y.
$$y = 2 \cdot 0 - 5$$
$$y = -5$$

The y-intercept is $(0, -5)$.

97. Graph $2x + 5y = 10$.

First we find the intercepts.

When $y = 0$, we have

$$2x + 5 \cdot 0 = 10$$
$$2x = 10$$
$$x = 5.$$

The x-intercept is $(5, 0)$.

When $x = 0$, we have

$$2 \cdot 0 + 5y = 10$$
$$5y = 10$$
$$y = 2.$$

The y-intercept is $(0, 2)$.

We draw the line through these points. We could find a third point as a check.

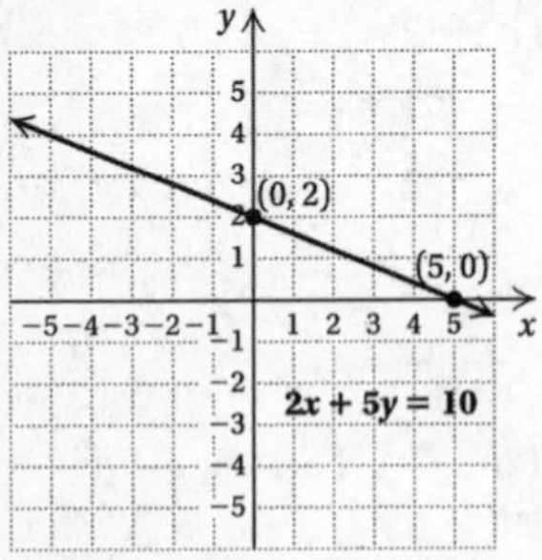

99. The shaded region is a square with sides of length $x-y-y$, or $x-2y$. Its area is $(x-2y)(x-2y)$, or $(x-2y)^2$. Multiplying, we get the polynomial $x^2 - 4xy + 4y^2$.

101. $49x^2 - 216$

There is no common factor. Also, $49x^2$ is a square, but 216 is not so this expression is not a difference of squares. It is not factorable. It is prime.

103. $x^2 + 22x + 121 = x^2 + 2 \cdot x \cdot 11 + 11^2$
$= (x + 11)^2$

105. $18x^3 + 12x^2 + 2x = 2x(9x^2 + 6x + 1)$
$= 2x[(3x)^2 + 2 \cdot 3x \cdot 1 + 1^2]$
$= 2x(3x + 1)^2$

107. $x^8 - 2^8$
$= (x^4 + 2^4)(x^4 - 2^4)$
$= (x^4 + 2^4)(x^2 + 2^2)(x^2 - 2^2)$
$= (x^4 + 2^4)(x^2 + 2^2)(x + 2)(x - 2)$, or
$= (x^4 + 16)(x^2 + 4)(x + 2)(x - 2)$

109. $3x^5 - 12x^3 = 3x^3(x^2 - 4) = 3x^3(x + 2)(x - 2)$

111. $18x^3 - \dfrac{8}{25}x = 2x\left(9x^2 - \dfrac{4}{25}\right) = 2x\left(3x + \dfrac{2}{5}\right)\left(3x - \dfrac{2}{5}\right)$

113. $0.49p - p^3 = p(0.49 - p^2) = p(0.7 + p)(0.7 - p)$

115. $0.64x^2 - 1.21 = (0.8x)^2 - (1.1)^2 = (0.8x + 1.1)(0.8x - 1.1)$

117. $(x+3)^2 - 9 = [(x+3)+3][(x+3)-3] = (x+6)x$, or $x(x+6)$

119. $x^2 - \left(\dfrac{1}{x}\right)^2 = \left(x + \dfrac{1}{x}\right)\left(x - \dfrac{1}{x}\right)$

121. $81 - b^{4k} = 9^2 - (b^{2k})^2$
$= (9 + b^{2k})(9 - b^{2k})$
$= (9 + b^{2k})[3^2 - (b^k)^2]$
$= (9 + b^{2k})(3 + b^k)(3 - b^k)$

123. $9b^{2n} + 12b^n + 4 = (3b^n)^2 + 2 \cdot 3b^n \cdot 2 + 2^2 =$
$(3b^n + 2)^2$

125. $(y + 3)^2 + 2(y + 3) + 1$
$= (y + 3)^2 + 2 \cdot (y + 3) \cdot 1 + 1^2$
$= [(y + 3) + 1]^2$
$= (y + 4)^2$

127. If $cy^2 + 6y + 1$ is the square of a binomial, then $2 \cdot a \cdot 1 = 6$ where $a^2 = c$. Then $a = 3$, so $c = a^2 = 3^2 = 9$. (The polynomial is $9y^2 + 6y + 1$.)

129. Enter $y_1 = x^2 + 9$ and $y_2 = (x + 3)(x + 3)$ and look at a table of values. The y_1-and y_2-values are not the same, so the factorization is not correct.

131. Enter $y_1 = x^2 + 9$ and $y_2 = (x + 3)^2$ and look at a table of values. The y_1-and y_2-values are not the same, so the factorization is not correct.

Exercise Set 5.6

RC1. Always look first for a <u>common</u> factor.

RC3. If there are three terms, determine whether the trinomial is a <u>square</u>.

RC5. Always factor <u>completely</u>.

1. $3x^2 - 192 = 3(x^2 - 64)$ — 3 is a common factor.
$= 3(x^2 - 8^2)$ — Difference of squares
$= 3(x + 8)(x - 8)$

3. $a^2 + 25 - 10a = a^2 - 10a + 25$
$= a^2 - 2 \cdot a \cdot 5 + 5^2$ — Trinomial square
$= (a - 5)^2$

5. $2x^2 - 11x + 12$

There is no common factor (other than 1). This polynomial has three terms, but it is not a trinomial square. Multiply the leading coefficient and the constant, 2 and 12: $2 \cdot 12 = 24$. Try to factor 24 so that the sum of the factors is -11. The numbers we want are -3 and -8: $-3(-8) = 24$ and $-3 + (-8) = -11$. Split the middle term and factor by grouping.

$2x^2 - 11x + 12 = 2x^2 - 3x - 8x + 12$
$= (2x^2 - 3x) + (-8x + 12)$
$= x(2x - 3) - 4(2x - 3)$
$= (2x - 3)(x - 4)$

7. $x^3 + 24x^2 + 144x$
$= x(x^2 + 24x + 144)$ x is a common factor.
$= x(x^2 + 2 \cdot x \cdot 12 + 12^2)$ Trinomial square
$= x(x + 12)^2$

9. $x^3 + 3x^2 - 4x - 12$
$= x^2(x + 3) - 4(x + 3)$ Factoring by grouping
$= (x + 3)(x^2 - 4)$
$= (x + 3)(x + 2)(x - 2)$ Factoring the difference of squares

11. $48x^2 - 3 = 3(16x^2 - 1)$ 3 is a common factor.
$= 3[(4x)^2 - 1^2]$ Difference of squares
$= 3(4x + 1)(4x - 1)$

13. $9x^3 + 12x^2 - 45x$
$= 3x(3x^2 + 4x - 15)$ $3x$ is a common factor.
$= 3x(3x - 5)(x + 3)$ Factoring the trinomial

15. $x^2 + 4$ is a *sum* of squares with no common factor. It cannot be factored. It is prime.

17. $x^4 + 7x^2 - 3x^3 - 21x = x(x^3 + 7x - 3x^2 - 21)$
$= x[x(x^2 + 7) - 3(x^2 + 7)]$
$= x[(x^2 + 7)(x - 3)]$
$= x(x^2 + 7)(x - 3)$

19. $x^5 - 14x^4 + 49x^3$
$= x^3(x^2 - 14x + 49)$ x^3 is a common factor.
$= x^3(x^2 - 2 \cdot x \cdot 7 + 7^2)$ Trinomial square
$= x^3(x - 7)^2$

21. $20 - 6x - 2x^2$
$= -2(-10 + 3x + x^2)$ -2 is a common factor.
$= -2(x^2 + 3x - 10)$ Writing in descending order
$= -2(x + 5)(x - 2),$ Using trial and error
or $2(-x - 5)(x - 2),$
or $2(x + 5)(-x + 2)$

23. $x^2 - 6x + 1$

There is no common factor (other than 1 or -1). This is not a trinomial square, because $-6x \neq 2 \cdot x \cdot 1$ and $-6x \neq -2 \cdot x \cdot 1$. We try factoring using the refined trial and error procedure. We look for two factors of 1 whose sum is -6. There are none. The polynomial cannot be factored. It is prime.

25. $4x^4 - 64$
$= 4(x^4 - 16)$ 4 is a common factor.
$= 4[(x^2)^2 - 4^2]$ Difference of squares
$= 4(x^2 + 4)(x^2 - 4)$ Difference of squares
$= 4(x^2 + 4)(x + 2)(x - 2)$

27. $1 - y^8$ Difference of squares
$= (1 + y^4)(1 - y^4)$ Difference of squares
$= (1 + y^4)(1 + y^2)(1 - y^2)$ Difference of squares
$= (1 + y^4)(1 + y^2)(1 + y)(1 - y)$

29. $x^5 - 4x^4 + 3x^3$
$= x^3(x^2 - 4x + 3)$ x^3 is a common factor.
$= x^3(x - 3)(x - 1)$ Factoring the trinomial using trial and error

31. $\frac{1}{81}x^6 - \frac{8}{27}x^3 + \frac{16}{9}$
$= \frac{1}{9}\left(\frac{1}{9}x^6 - \frac{8}{3}x^3 + 16\right)$ $\frac{1}{9}$ is a common factor.
$= \frac{1}{9}\left[\left(\frac{1}{3}x^3\right)^2 - 2 \cdot \frac{1}{3}x^3 \cdot 4 + 4^2\right]$ Trinomial square
$= \frac{1}{9}\left(\frac{1}{3}x^3 - 4\right)^2$

33. $mx^2 + my^2$
$= m(x^2 + y^2)$ m is a common factor.

The factor with more than one term cannot be factored further, so we have factored completely.

35. $9x^2y^2 - 36xy = 9xy(xy - 4)$

37. $2\pi rh + 2\pi r^2 = 2\pi r(h + r)$

39. $(a + b)(x - 3) + (a + b)(x + 4)$
$= (a + b)[(x - 3) + (x + 4)]$ $(a + b)$ is a common factor.
$= (a + b)(2x + 1)$

41. $(x - 1)(x + 1) - y(x + 1) = (x + 1)(x - 1 - y)$
$(x + 1)$ is a common factor.

43. $n^2 + 2n + np + 2p$
$= n(n + 2) + p(n + 2)$ Factoring by grouping
$= (n + 2)(n + p)$

45. $6q^2 - 3q + 2pq - p$
$= (6q^2 - 3q) + (2pq - p)$
$= 3q(2q - 1) + p(2q - 1)$ Factoring by grouping
$= (2q - 1)(3q + p)$

47. $4b^2 + a^2 - 4ab$
$= a^2 - 4ab + 4b^2$ Rearranging
$= a^2 - 2 \cdot a \cdot 2b + (2b)^2$ Trinomial square
$= (a - 2b)^2$

(Note that if we had rewritten the polynomial as $4b^2 - 4ab + a^2$, we might have written the result as $(2b - a)^2$. The two factorizations are equivalent.)

49. $16x^2 + 24xy + 9y^2$
$= (4x)^2 + 2 \cdot 4x \cdot 3y + (3y)^2$ Trinomial square
$= (4x + 3y)^2$

51. $49m^4 - 112m^2n + 64n^2$
$= (7m^2)^2 - 2 \cdot 7m^2 \cdot 8n + (8n)^2$ Trinomial square
$= (7m^2 - 8n)^2$

53. $y^4 + 10y^2z^2 + 25z^4$
$= (y^2)^2 + 2 \cdot y^2 \cdot 5z^2 + (5z^2)^2$ Trinomial square
$= (y^2 + 5z^2)^2$

55. $\frac{1}{4}a^2 + \frac{1}{3}ab + \frac{1}{9}b^2$
$= \left(\frac{1}{2}a\right)^2 + 2 \cdot \frac{1}{2}a \cdot \frac{1}{3}b + \left(\frac{1}{3}b\right)^2$
$= \left(\frac{1}{2}a + \frac{1}{3}b\right)^2$

57. $a^2 - ab - 2b^2 = (a - 2b)(a + b)$ Using trial and error

59. $2mn - 360n^2 + m^2$
$= m^2 + 2mn - 360n^2$ Rewriting
$= (m + 20n)(m - 18n)$ Using trial and error

61. $m^2n^2 - 4mn - 32 = (mn - 8)(mn + 4)$ Using trial and error

63. $r^5s^2 - 10r^4s + 16r^3$
$= r^3(r^2s^2 - 10rs + 16)$ r^3 is a common factor.
$= r^3(rs - 2)(rs - 8)$ Using trial and error

65. $a^5 + 4a^4b - 5a^3b^2$
$= a^3(a^2 + 4ab - 5b^2)$ a^3 is a common factor.
$= a^3(a + 5b)(a - b)$ Factoring the trinomial

67. $a^2 - \frac{1}{25}b^2$
$= a^2 - \left(\frac{1}{5}b\right)^2$ Difference of squares
$= \left(a + \frac{1}{5}b\right)\left(a - \frac{1}{5}b\right)$

69. $x^2 - y^2 = (x + y)(x - y)$ Difference of squares

71. $16 - p^4q^4$
$= 4^2 - (p^2q^2)^2$ Difference of squares
$= (4 + p^2q^2)(4 - p^2q^2)$ $4 - p^2q^2$ is a difference of squares.
$= (4 + p^2q^2)(2 + pq)(2 - pq)$

73. $1 - 16x^{12}y^{12}$
$= 1^2 - (4x^6y^6)^2$ Difference of squares
$= (1 + 4x^6y^6)(1 - 4x^6y^6)$ $1 - 4x^6y^6$ is a difference of squares.
$= (1 + 4x^6y^6)(1 + 2x^3y^3)(1 - 2x^3y^3)$

75. $q^3 + 8q^2 - q - 8$
$= q^2(q + 8) - (q + 8)$ Factoring by grouping
$= (q + 8)(q^2 - 1)$
$= (q + 8)(q + 1)(q - 1)$ Factoring the difference of squares

77. $6a^3b^3 - a^2b^2 - 2ab$
$= ab(6a^2b^2 - ab - 2)$ ab is a common factor
$= ab(2ab + 1)(3ab - 2)$ Factoring the trinomial

79. $m^4 - 5m^2 + 4$
$= (m^2 - 1)(m^2 - 4)$ Using trial and error
$= (m + 1)(m - 1)(m + 2)(m - 2)$ Factoring two differences of squares

81. $1.2 - 9.87 = 1.2 + (-9.87) = -8.67$

83. $\left(-\frac{1}{3}\right)\left(-\frac{3}{5}\right) = \frac{1 \cdot 3}{3 \cdot 5} = \frac{3}{3} \cdot \frac{1}{5} = \frac{1}{5}$

85. $-50 \div (-5)(-2) - 18 \div (-3)^2$
$= -50 \div (-5)(-2) - 18 \div 9$
$= 10(-2) - 2$
$= -20 - 2$
$= -22$

87. $-(-7) = 7$

89. $t^4 - 2t^2 + 1$ Trinomial square
$= (t^2 - 1)^2$ Difference of squares
$= [(t + 1)(t - 1)]^2$
$= (t + 1)^2(t - 1)^2$

91. $x^3 + 20 - (5x^2 + 4x) = x^3 + 20 - 5x^2 - 4x$
$= x^3 - 5x^2 - 4x + 20$
$= x^2(x - 5) - 4(x - 5)$
$= (x - 5)(x^2 - 4)$
$= (x - 5)(x + 2)(x - 2)$

93. $12.25x^2 - 7x + 1 = (3.5x)^2 - 2 \cdot (3.5x) \cdot 1 + 1^2$
$= (3.5x - 1)^2$

95. $18 + y^3 - 9y - 2y^2$
$= y^3 - 2y^2 - 9y + 18$
$= y^2(y - 2) - 9(y - 2)$
$= (y - 2)(y^2 - 9)$
$= (y - 2)(y + 3)(y - 3)$

97. $y^2(y - 1) - 2y(y - 1) + (y - 1)$
$= (y - 1)(y^2 - 2y + 1)$
$= (y - 1)(y - 1)^2$
$= (y - 1)^3$

99. $(y + 4)^2 + 2x(y + 4) + x^2$
$= (y + 4)^2 + 2 \cdot (y + 4) \cdot x + x^2$ Trinomial square
$= (y + 4 + x)^2$

Exercise Set 5.7

RC1. False; see page 436 in the text.

RC3. False; see Example 3 on page 437 in the text.

1. $(x+4)(x+9)=0$

$x+4=0$ *or* $x+9=0$ Using the principle of zero products

$x=-4$ *or* $x=-9$ Solving the two equations separately

Check:

For -4

$$\begin{array}{c|c} (x+4)(x+9) & 0 \\ \hline (-4+4)(-4+9) & ?\ 0 \\ 0\cdot 5 & \\ 0 & \text{TRUE} \end{array}$$

For -9

$$\begin{array}{c|c} (x+4)(x+9) & 0 \\ \hline (-9+4)(-9+9) & ?\ 0 \\ -5\cdot 0 & \\ 0 & \text{TRUE} \end{array}$$

The solutions are -4 and -9.

3. $(x+3)(x-8)=0$

$x+3=0$ *or* $x-8=0$ Using the principle of zero products

$x=-3$ *or* $x=8$

Check:

For -3

$$\begin{array}{c|c} (x+3)(x-8) & 0 \\ \hline (-3+3)(-3-8) & ?\ 0 \\ 0(-11) & \\ 0 & \text{TRUE} \end{array}$$

For 8

$$\begin{array}{c|c} (x+3)(x-8) & 0 \\ \hline (8+3)(8-8) & ?\ 0 \\ 11\cdot 0 & \\ 0 & \text{TRUE} \end{array}$$

The solutions are -3 and 8.

5. $(x+12)(x-11)=0$

$x+12=0$ *or* $x-11=0$

$x=-12$ *or* $x=11$

The solutions are -12 and 11.

7. $x(x+3)=0$

$x=0$ *or* $x+3=0$

$x=0$ *or* $x=-3$

The solutions are 0 and -3.

9. $0=y(y+18)$

$y=0$ *or* $y+18=0$

$y=0$ *or* $y=-18$

The solutions are 0 and -18.

11. $(2x+5)(x+4)=0$

$2x+5=0$ *or* $x+4=0$

$2x=-5$ *or* $x=-4$

$x=-\frac{5}{2}$ *or* $x=-4$

The solutions are $-\frac{5}{2}$ and -4.

13. $(5x+1)(4x-12)=0$

$5x+1=0$ *or* $4x-12=0$

$5x=-1$ *or* $4x=12$

$x=-\frac{1}{5}$ *or* $x=3$

The solutions are $-\frac{1}{5}$ and 3.

15. $(7x-28)(28x-7)=0$

$7x-28=0$ *or* $28x-7=0$

$7x=28$ *or* $28x=7$

$x=4$ *or* $x=\frac{7}{28}=\frac{1}{4}$

The solutions are 4 and $\frac{1}{4}$.

17. $2x(3x-2)=0$

$2x=0$ *or* $3x-2=0$

$x=0$ *or* $3x=2$

$x=0$ *or* $x=\frac{2}{3}$

The solutions are 0 and $\frac{2}{3}$.

19. $\left(\frac{1}{5}+2x\right)\left(\frac{1}{9}-3x\right)=0$

$\frac{1}{5}+2x=0$ *or* $\frac{1}{9}-3x=0$

$2x=-\frac{1}{5}$ *or* $-3x=-\frac{1}{9}$

$x=-\frac{1}{10}$ *or* $x=\frac{1}{27}$

The solutions are $-\frac{1}{10}$ and $\frac{1}{27}$.

21. $(0.3x-0.1)(0.05x+1)=0$

$0.3x-0.1=0$ *or* $0.05x+1=0$

$0.3x=0.1$ *or* $0.05x=-1$

$x=\frac{0.1}{0.3}$ *or* $x=-\frac{1}{0.05}$

$x=\frac{1}{3}$ *or* $x=-20$

The solutions are $\frac{1}{3}$ and -20.

23. $9x(3x-2)(2x-1)=0$

$9x=0 \quad or \quad 3x-2=0 \quad or \quad 2x-1=0$

$x=0 \quad or \quad 3x=2 \quad or \quad 2x=1$

$x=0 \quad or \quad x=\frac{2}{3} \quad or \quad x=\frac{1}{2}$

The solutions are 0, $\frac{2}{3}$, and $\frac{1}{2}$.

25. $x^2+6x+5=0$

$(x+5)(x+1)=0$ Factoring

$x+5=0 \quad or \quad x+1=0$ Using the principle of zero products

$x=-5 \quad or \quad x=-1$

The solutions are -5 and -1.

27. $x^2+7x-18=0$

$(x+9)(x-2)=0$ Factoring

$x+9=0 \quad or \quad x-2=0$ Using the principle of zero products

$x=-9 \quad or \quad x=2$

The solutions are -9 and 2.

29. $x^2-8x+15=0$

$(x-5)(x-3)=0$

$x-5=0 \quad or \quad x-3=0$

$x=5 \quad or \quad x=3$

The solutions are 5 and 3.

31. $x^2-8x=0$

$x(x-8)=0$

$x=0 \quad or \quad x-8=0$

$x=0 \quad or \quad x=8$

The solutions are 0 and 8.

33. $x^2+18x=0$

$x(x+18)=0$

$x=0 \quad or \quad x+18=0$

$x=0 \quad or \quad x=-18$

The solutions are 0 and -18.

35. $x^2=16$

$x^2-16=0$ Subtracting 16

$(x-4)(x+4)=0$

$x-4=0 \quad or \quad x+4=0$

$x=4 \quad or \quad x=-4$

The solutions are 4 and -4.

37. $9x^2-4=0$

$(3x-2)(3x+2)=0$

$3x-2=0 \quad or \quad 3x+2=0$

$3x=2 \quad or \quad 3x=-2$

$x=\frac{2}{3} \quad or \quad x=-\frac{2}{3}$

The solutions are $\frac{2}{3}$ and $-\frac{2}{3}$.

39. $0=6x+x^2+9$

$0=x^2+6x+9$ Writing in descending order

$0=(x+3)(x+3)$

$x+3=0 \quad or \quad x+3=0$

$x=-3 \quad or \quad x=-3$

There is only one solution, -3.

41. $x^2+16=8x$

$x^2-8x+16=0$ Subtracting $8x$

$(x-4)(x-4)=0$

$x-4=0 \quad or \quad x-4=0$

$x=4 \quad or \quad x=4$

There is only one solution, 4.

43. $5x^2=6x$

$5x^2-6x=0$

$x(5x-6)=0$

$x=0 \quad or \quad 5x-6=0$

$x=0 \quad or \quad 5x=6$

$x=0 \quad or \quad x=\frac{6}{5}$

The solutions are 0 and $\frac{6}{5}$.

45. $6x^2-4x=10$

$6x^2-4x-10=0$

$2(3x^2-2x-5)=0$

$2(3x-5)(x+1)=0$

$3x-5=0 \quad or \quad x+1=0$

$3x=5 \quad or \quad x=-1$

$x=\frac{5}{3} \quad or \quad x=-1$

The solutions are $\frac{5}{3}$ and -1.

47. $12y^2-5y=2$

$12y^2-5y-2=0$

$(4y+1)(3y-2)=0$

$4y+1=0 \quad or \quad 3y-2=0$

$4y=-1 \quad or \quad 3y=2$

$y=-\frac{1}{4} \quad or \quad y=\frac{2}{3}$

The solutions are $-\frac{1}{4}$ and $\frac{2}{3}$.

49.
$$t(3t+1)=2$$
$$3t^2+t=2 \quad \text{Multiplying on the left}$$
$$3t^2+t-2=0 \quad \text{Subtracting 2}$$
$$(3t-2)(t+1)=0$$
$$3t-2=0 \quad or \quad t+1=0$$
$$3t=2 \quad or \quad t=-1$$
$$t=\frac{2}{3} \quad or \quad t=-1$$
The solutions are $\frac{2}{3}$ and -1.

51.
$$100y^2=49$$
$$100y^2-49=0$$
$$(10y+7)(10y-7)=0$$
$$10y+7=0 \quad or \quad 10y-7=0$$
$$10y=-7 \quad or \quad 10y=7$$
$$y=-\frac{7}{10} \quad or \quad y=\frac{7}{10}$$
The solutions are $-\frac{7}{10}$ and $\frac{7}{10}$.

53.
$$x^2-5x=18+2x$$
$$x^2-5x-18-2x=0 \quad \text{Subtracting 18 and } 2x$$
$$x^2-7x-18=0$$
$$(x-9)(x+2)=0$$
$$x-9=0 \quad or \quad x+2=0$$
$$x=9 \quad or \quad x=-2$$
The solutions are 9 and -2.

55.
$$10x^2-23x+12=0$$
$$(5x-4)(2x-3)=0$$
$$5x-4=0 \quad or \quad 2x-3=0$$
$$5x=4 \quad or \quad 2x=3$$
$$x=\frac{4}{5} \quad or \quad x=\frac{3}{2}$$
The solutions are $\frac{4}{5}$ and $\frac{3}{2}$.

57. We let $y=0$ and solve for x.
$$0=x^2+3x-4$$
$$0=(x+4)(x-1)$$
$$x+4=0 \quad or \quad x-1=0$$
$$x=-4 \quad or \quad x=1$$
The x-intercepts are $(-4,0)$ and (1,0).

59. We let $y=0$ and solve for x
$$0=2x^2+x-10$$
$$0=(2x+5)(x-2)$$
$$2x+5=0 \quad or \quad x-2=0$$
$$2x=-5 \quad or \quad x=2$$
$$x=-\frac{5}{2} \quad or \quad x=2$$
The x-intercepts are $\left(-\frac{5}{2},0\right)$ and $(2,0)$.

61. We let $y=0$ and solve for x.
$$0=x^2-2x-15$$
$$0=(x-5)(x+3)$$
$$x-5=0 \quad or \quad x+3=0$$
$$x=5 \quad or \quad x=-3$$
The x-intercepts are $(5,0)$ and $(-3,0)$.

63. The solutions of the equation are the first coordinates of the x-intercepts of the graph. From the graph we see that the x-intercepts are $(-1,0)$ and $(4,0)$, so the solutions of the equation are -1 and 4.

65. The solutions of the equation are the first coordinates of the x-intercepts of the graph. From the graph we see that the x-intercepts are $(-1,0)$ and $(3,0)$, so the solutions of the equation are -1 and 3.

67. $(a+b)^2$

69.
$$-10x>1000$$
$$\frac{-10x}{-10}<\frac{1000}{-10} \quad \text{Dividing by } -10 \text{ and reversing the inequality symbol}$$
$$x<-100$$
The solution set is $\{x|x<-100\}$.

71.
$$3-2x-4x>-9$$
$$3-6x>-9$$
$$-6x>-12$$
$$x<2 \quad \text{Dividing by } -6 \text{ and reversing the inequality symbol}$$
The solution set is $\{x|x<2\}$.

73.
$$b(b+9)=4(5+2b)$$
$$b^2+9b=20+8b$$
$$b^2+9b-8b-20=0$$
$$b^2+b-20=0$$
$$(b+5)(b-4)=0$$
$$b+5=0 \quad or \quad b-4=0$$
$$b=-5 \quad or \quad b=4$$
The solutions are -5 and 4.

75.
$$(t-3)^2=36$$
$$t^2-6t+9=36$$
$$t^2-6t-27=0$$
$$(t-9)(t+3)=0$$
$$t-9=0 \quad or \quad t+3=0$$
$$t=9 \quad or \quad t=-3$$
The solutions are 9 and -3.

77.
$$x^2 - \frac{1}{64} = 0$$
$$\left(x - \frac{1}{8}\right)\left(x + \frac{1}{8}\right) = 0$$
$$x - \frac{1}{8} = 0 \quad or \quad x + \frac{1}{8} = 0$$
$$x = \frac{1}{8} \quad or \quad x = -\frac{1}{8}$$

The solutions are $\frac{1}{8}$ and $-\frac{1}{8}$.

79.
$$\frac{5}{16}x^2 = 5$$
$$\frac{5}{16}x^2 - 5 = 0$$
$$5\left(\frac{1}{16}x^2 - 1\right) = 0$$
$$5\left(\frac{1}{4}x - 1\right)\left(\frac{1}{4}x + 1\right) = 0$$
$$\frac{1}{4}x - 1 = 0 \quad or \quad \frac{1}{4}x + 1 = 0$$
$$\frac{1}{4}x = 1 \quad or \quad \frac{1}{4}x = -1$$
$$x = 4 \quad or \quad x = -4$$

The solutions are 4 and -4.

81. 2.33, 6.77

83. (a)
$$x = -3 \quad or \quad x = 4$$
$$x + 3 = 0 \quad or \quad x - 4 = 0$$
$$(x+3)(x-4) = 0 \quad \text{Principle of zero products}$$
$$x^2 - x - 12 = 0 \quad \text{Multiplying}$$

(b)
$$x = -3 \quad or \quad x = -4$$
$$x + 3 = 0 \quad or \quad x + 4 = 0$$
$$(x+3)(x+4) = 0$$
$$x^2 + 7x + 12 = 0$$

(c)
$$x = \frac{1}{2} \quad or \quad x = \frac{1}{2}$$
$$x - \frac{1}{2} = 0 \quad or \quad x - \frac{1}{2} = 0$$
$$\left(x - \frac{1}{2}\right)\left(x - \frac{1}{2}\right) = 0$$
$$x^2 - x + \frac{1}{4} = 0, \quad \text{or}$$
$$4x^2 - 4x + 1 = 0 \quad \text{Multiplying by 4}$$

(d)
$$(x-5)(x+5) = 0$$
$$x^2 - 25 = 0$$

(e)
$$(x-0)(x-0.1)\left(x - \frac{1}{4}\right) = 0$$
$$x\left(x - \frac{1}{10}\right)\left(x - \frac{1}{4}\right) = 0$$
$$x\left(x^2 - \frac{7}{20}x + \frac{1}{40}\right) = 0$$
$$x^3 - \frac{7}{20}x^2 + \frac{1}{40}x = 0, \quad \text{or}$$
$$40x^3 - 14x^2 + x = 0 \quad \text{Multiplying by 40}$$

Exercise Set 5.8

RC1. Let x and $x+1$ represent the integers. Then we have $x(x+1) = 20$. Answer (b) is correct.

RC3. Let x and $x+1$ represent the lengths of the legs. Then we have $x^2 + (x+1)^2 = 20^2$. Answer (d) is correct.

1. ***Familiarize.*** We make a drawing. Let w = the width, in inches. Then $3w$ = the length, in inches.

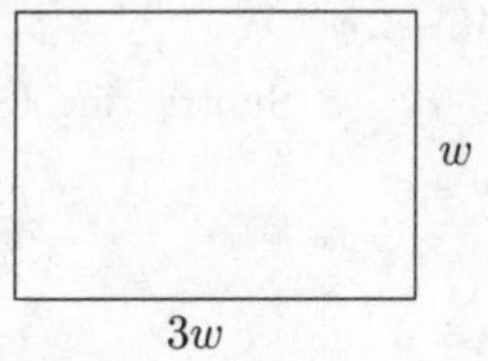

Recall that the area of a rectangle is length times width.

Translate. We reword the problem.

$$\underbrace{\text{Length}}\ \text{times}\ \text{width}\ \text{is}\ \underbrace{588\text{ in}^2}.$$
$$\downarrow \quad \downarrow \quad \downarrow \quad \downarrow \quad \downarrow$$
$$3w \quad \cdot \quad w \quad = \quad 588$$

Solve. We solve the equation.
$$3w \cdot w = 588$$
$$3w^2 = 588$$
$$3w^2 - 588 = 0$$
$$w^2 - 196 = 0$$
$$(w+14)(w-14) = 0$$
$$w + 14 = 0 \quad or \quad w - 14 = 0$$
$$w = -14 \quad or \quad w = 14$$

Check. Since the width must be positive, -14 cannot be a solution. If the width is 14 in., then the length is $3 \cdot 14$, or 42 in., and the area is $42 \cdot 14$, or 588 in^2. Thus, 14 checks.

State. The width is 14 in., and the length is 42 in.

3. ***Familiarize.*** Using the labels on the drawing in the text, we let w and $w+2$ represent the width and length of the screen, respectively, in cm.

Translate.

$$\text{Length}\ \text{times}\ \text{width}\ \text{is}\ \underbrace{24\text{ cm}^2}.$$
$$\downarrow \quad \downarrow \quad \downarrow \quad \downarrow \quad \downarrow$$
$$(w+2) \quad \cdot \quad w \quad = \quad 24$$

Solve.

$$(w+2)w = 24$$
$$w^2 + 2w = 24$$
$$w^2 + 2w - 24 = 0$$
$$(w+6)(w-4) = 0$$
$$w+6=0 \quad or \quad w-4=0$$
$$w=-6 \quad or \quad w=4$$

Check. Since the width must be positive, −6 cannot be a solution. If the width is 4 cm, then the length is (4+2) cm, or 6 cm, and the area is 6 cm · 4 cm = 24 cm^2. Thus, 4 checks.

State. The width is 4 cm, and the length is 6 cm.

5. ***Familiarize.*** Using the labels shown on the drawing in the text, we let h = the height, in cm, and $h+10$ = the base, in cm. Recall that the formula for the area of a triangle is $\frac{1}{2}\cdot$ (base) · (height).

Translate.

$$\underbrace{\tfrac{1}{2}}\ \text{times}\ \text{base}\ \text{times}\ \text{height}\ \text{is}\ \underbrace{28\text{ cm}^2}.$$
$$\frac{1}{2}\cdot(h+10)\cdot h = 28$$

Solve. We solve the equation.

$$\frac{1}{2}(h+10)h = 28$$
$$(h+10)h = 56 \quad \text{Multiplying by 2}$$
$$h^2+10h = 56$$
$$h^2+10h-56 = 0$$
$$(h+14)(h-4) = 0$$
$$h+14=0 \quad or \quad h-4=0$$
$$h=-14 \quad or \quad h=4$$

Check. Since the height of the triangle must be positive, −14 cannot be a solution. If the height is 4 cm, then the base is 4+10, or 14 cm, and the area is $\frac{1}{2}\cdot 14\cdot 4$, or 28 cm^2. Thus, 4 checks.

State. The height of the triangle is 4 cm, and the base is 14 cm.

7. ***Familiarize.*** Using the labels shown on the drawing in the text, we let h = the height of the triangle, in meters, and $\frac{1}{2}h$ = the length of the base, in meters. Recall that the formula for the area of a triangle is $\frac{1}{2}\cdot$ (base) · (height).

Translate.

$$\tfrac{1}{2}\ \text{times}\ \text{base}\ \text{times}\ \text{height}\ \text{is}\ \underbrace{64\text{ m}^2}.$$
$$\frac{1}{2}\cdot\frac{1}{2}h\cdot h = 64$$

Solve. We solve the equation.

$$\frac{1}{2}\cdot\frac{1}{2}h\cdot h = 64$$
$$\frac{1}{4}h^2 = 64$$
$$h^2 = 256 \quad \text{Multiplying by 4}$$
$$h^2 - 256 = 0$$
$$(h+16)(h-16) = 0$$
$$h+16=0 \quad or \quad h-16=0$$
$$h=-16 \quad or \quad h=16$$

Check. The height of the triangle cannot be negative, so −16 cannot be a solution. If the height is 16 m, then the length of the base is $\frac{1}{2}\cdot 16$ m, or 8 m, and the area is $\frac{1}{2}\cdot 8\text{ m}\cdot 16\text{ m} = 64\text{ m}^2$. These numbers check.

State. The length of the base is 8 m, and the height is 16 m.

9. ***Familiarize and Translate.*** Substitute 14 for n in the formula.

$$14^2 - 14 = N$$

Solve. We do the computation on the left.

$$14^2 - 14 = N$$
$$196 - 14 = N$$
$$182 = N$$

Check. We can redo the computation, or we can solve the equation $n^2 - n = 182$. The answer checks.

State. 182 games will be played.

11. ***Familiarize and Translate.*** Substitute 132 for N in the formula.

$$n^2 - n = 132$$

Solve.

$$n^2 - n = 132$$
$$n^2 - n - 132 = 0$$
$$(n-12)(n+11) = 0$$
$$n-12=0 \quad or \quad n+11=0$$
$$n=12 \quad or \quad n=-11$$

Check. The solutions of the equation are 12 and −11. Since the number of teams cannot be negative, −11 cannot be a solution. But 12 checks since $12^2 - 12 = 144 - 12 = 132$.

State. There are 12 teams in the league.

13. ***Familiarize.*** We will use the formula $N = \frac{1}{2}(n^2 - n)$.

Translate. Substitute 100 for n.

$$N = \frac{1}{2}(100^2 - 100)$$

Solve. We do the computation on the right.

$$N = \frac{1}{2}(10,000 - 100)$$
$$N = \frac{1}{2}(9900)$$
$$N = 4950$$

Check. We can redo the computation, or we can solve the equation $4950 = \frac{1}{2}(n^2 - n)$. The answer checks.

State. 4950 handshakes are possible.

15. ***Familiarize.*** We will use the formula $N = \frac{1}{2}(n^2 - n)$.

Translate. Substitute 300 for N.

$$300 = \frac{1}{2}(n^2 - n)$$

Solve. We solve the equation.

$$2 \cdot 300 = 2 \cdot \frac{1}{2}(n^2 - n) \quad \text{Multiplying by 2}$$
$$600 = n^2 - n$$
$$0 = n^2 - n - 600$$
$$0 = (n + 24)(n - 25)$$
$$n + 24 = 0 \quad \text{or} \quad n - 25 = 0$$
$$n = -24 \quad \text{or} \quad n = 25$$

Check. The number of people at a meeting cannot be negative, so -24 cannot be a solution. But 25 checks since $\frac{1}{2}(25^2 - 25) = \frac{1}{2}(625 - 25) = \frac{1}{2} \cdot 600 = 300$.

State. There were 25 people at the party.

17. ***Familiarize.*** The page numbers on facing pages are consecutive integers. Let x = the smaller integer. Then $x + 1$ = the larger integer.

Translate. We reword the problem.

$$\underbrace{\text{Smaller integer}}_{x} \ \underbrace{\text{times}}_{\cdot} \ \underbrace{\text{larger integer}}_{(x+1)} \ \underbrace{\text{is}}_{=} \ \underbrace{210.}_{210}$$

Solve. We solve the equation.

$$x(x + 1) = 210$$
$$x^2 + x = 210$$
$$x^2 + x - 210 = 0$$
$$(x + 15)(x - 14) = 0$$
$$x + 15 = 0 \quad \text{or} \quad x - 14 = 0$$
$$x = -15 \quad \text{or} \quad x = 14$$

Check. The solutions of the equation are -15 and 14. Since a page number cannot be negative, -15 cannot be a solution of the original problem. We only need to check 14. When $x = 14$, then $x + 1 = 15$, and $14 \cdot 15 = 210$. This checks.

State. The page numbers are 14 and 15.

19. ***Familiarize.*** Let x = the smaller even integer. Then $x + 2$ = the larger even integer.

Translate. We reword the problem.

$$\underbrace{\text{Smaller even integer}}_{x} \ \underbrace{\text{times}}_{\cdot} \ \underbrace{\text{larger even integer}}_{(x+2)} \ \underbrace{\text{is}}_{=} \ \underbrace{168.}_{168}$$

Solve.

$$x(x + 2) = 168$$
$$x^2 + 2x = 168$$
$$x^2 + 2x - 168 = 0$$
$$(x + 14)(x - 12) = 0$$
$$x + 14 = 0 \quad \text{or} \quad x - 12 = 0$$
$$x = -14 \quad \text{or} \quad x = 12$$

Check. The solutions of the equation are -14 and 12. When x is -14, then $x + 2$ is -12 and $-14(-12) = 168$. The numbers -14 and -12 are consecutive even integers which are solutions of the problem. When x is 12, then $x + 2$ is 14 and $12 \cdot 14 = 168$. The numbers 12 and 14 are also consecutive even integers which are solutions of the problem.

State. We have two solutions, each of which consists of a pair of numbers: -14 and -12, and 12 and 14.

21. ***Familiarize.*** Let x = the smaller odd integer. Then $x + 2$ = the larger odd integer.

Translate. We reword the problem.

$$\underbrace{\text{Smaller odd integer}}_{x} \ \underbrace{\text{times}}_{\cdot} \ \underbrace{\text{larger odd integer}}_{(x+2)} \ \underbrace{\text{is}}_{=} \ \underbrace{255.}_{255}$$

Solve.

$$x(x + 2) = 255$$
$$x^2 + 2x = 255$$
$$x^2 + 2x - 255 = 0$$
$$(x - 15)(x + 17) = 0$$
$$x - 15 = 0 \quad \text{or} \quad x + 17 = 0$$
$$x = 15 \quad \text{or} \quad x = -17$$

Check. The solutions of the equation are 15 and -17. When x is 15, then $x + 2$ is 17 and $15 \cdot 17 = 255$. The numbers 15 and 17 are consecutive odd integers which are solutions to the problem. When x is -17, then $x + 2$ is -15 and $-17(-15) = 255$. The numbers -17 and -15 are also consecutive odd integers which are solutions to the problem.

State. We have two solutions, each of which consists of a pair of numbers: 15 and 17, and -17 and -15.

23. ***Familiarize.*** Consider the drawing in the text. We let w = the width of Main Street, in feet.

Translate. Use the Pythagorean theorem.

$$a^2 + b^2 = c^2$$
$$24^2 + w^2 = 40^2$$

Solve. We solve the equation.

$$24^2 + w^2 = 40^2$$
$$576 + w^2 = 1600$$
$$w^2 - 1024 = 0$$
$$(w + 32)(w - 32) = 0$$
$$w + 32 = 0 \quad \text{or} \quad w - 32 = 0$$
$$w = -32 \quad \text{or} \quad w = 32$$

Check. The width of the street cannot be negative, so -32 cannot be a solution. If Main Street is 32 ft wide, we have $24^2 + 32^2 = 576 + 1024 = 1600$, which is 40^2. Thus, 32 ft checks.

State. Main Street is 32 ft wide.

25. ***Familiarize***. We make a drawing. Let x = the length of the unknown leg. Then $x + 2$ = the length of the hypotenuse.

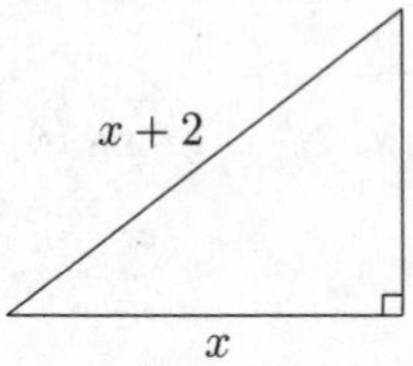

Translate. Use the Pythagorean theorem.

$$a^2 + b^2 = c^2$$
$$8^2 + x^2 = (x + 2)^2$$

Solve. We solve the equation.

$$8^2 + x^2 = (x + 2)^2$$
$$64 + x^2 = x^2 + 4x + 4$$
$$60 = 4x \quad \text{Subtracting } x^2 \text{ and } 4$$
$$15 = x$$

Check. When $x = 15$, then $x+2 = 17$ and $8^2 + 15^2 = 17^2$. Thus, 15 and 17 check.

State. The lengths of the hypotenuse and the other leg are 17 ft and 15 ft, respectively.

27. ***Familiarize***. Let x = the length of the unknown leg, in ft. Then $x + 200$ = the length of the hypotenuse.

Translate. We use the Pythagorean theorem.

$$a^2 + b^2 = c^2$$
$$400^2 + x^2 = (x + 200)^2$$

Solve.

$$400^2 + x^2 = (x + 200)^2$$
$$160,000 + x^2 = x^2 + 400x + 40,000$$
$$160,000 = 400x + 40,000 \quad \text{Subtracting } x^2$$
$$120,000 = 400x$$
$$300 = x$$

If $x = 300$, then $x + 200 = 300 + 200 = 500$.

Check. The hypotenuse is 200 ft longer than the 300 ft side. We have $400^2 + 300^2 = 160,00 + 90,000 = 250,000 = 500^2$, so the answer checks.

State. The dimensions of the garden were 300 ft by 400 ft by 500 ft.

29. ***Familiarize***. Let x = the length of the unknown leg, in meters. Then $x + 1$ = the length of the hypotenuse.

Translate.

$$a^2 + b^2 = c^2$$
$$7^2 + x^2 = (x + 1)^2$$

Solve.

$$7^2 + x^2 = (x + 1)^2$$
$$49 + x^2 = x^2 + 2x + 1$$
$$49 = 2x + 1 \quad \text{Subtracting } x^2$$
$$48 = 2x$$
$$24 = x$$

If $x = 24$, then $x + 1 = 24 + 1 = 25$.

Check. The lengths of the unknown leg and the hypotenuse are consecutive integers. We have $7^2 + 24^2 = 49 + 576 = 625 = 25^2$, so the answer checks.

State. The lengths of the other two sides are 24 m and 25 m.

31. ***Familiarize***. We label the drawing. Let x = the length of a side of the dining room, in ft. Then the dining room has dimensions x by x and the kitchen has dimensions x by 10. The entire rectangular space has dimension x by $x + 10$. Recall that we multiply these dimensions to find the area of the rectangle.

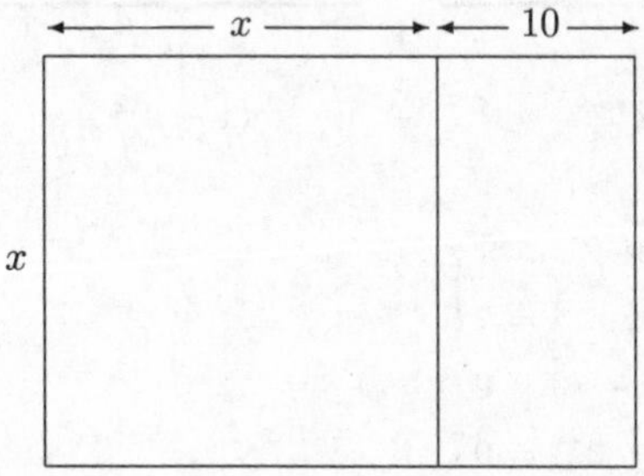

Translate.

$$\underbrace{\text{The area of the rectangular space}}_{x(x+10)} \quad \underset{=}{\text{is}} \quad \underbrace{264 \text{ ft}^2}_{264}.$$

Solve. We solve the equation.

$$x(x + 10) = 264$$
$$x^2 + 10x = 264$$
$$x^2 + 10x - 264 = 0$$
$$(x + 22)(x - 12) = 0$$
$$x + 22 = 0 \quad or \quad x - 12 = 0$$
$$x = -22 \quad or \quad x = 12$$

Check. Since the length of a side of the dining room must be positive, -22 cannot be a solution. If x is 12 ft, then $x+10$ is 22 ft, and the area of the space is $12 \cdot 22$, or 264 ft^2. The number 12 checks.

State. The dining room is 12 ft by 12 ft, and the kitchen is 12 ft by 10 ft.

33. ***Familiarize and Translate***. We substitute 32 for h in the formula.

$$32 = 48t - 16t^2$$

Solve.

$$32 = 48t - 16t^2$$
$$16t^2 - 48t + 32 = 0$$
$$t^2 - 3t + 2 = 0 \quad \text{Dividing by 16}$$
$$(t-1)(t-2) = 0$$
$$t - 1 = 0 \quad or \quad t - 2 = 0$$
$$t = 1 \quad or \quad t = 2$$

Check. When $t = 1$, $h = 48 \cdot 1 - 16 \cdot 1^2 = 48 - 16 = 32$. When $t = 2$, $h = 48 \cdot 2 - 16 \cdot 2^2 = 48 \cdot 2 - 16 \cdot 4 = 96 - 64 = 32$. Since the height is 32 ft for each value of t, the answer checks.

State. The rocket will be exactly 32 ft above the ground after 1 sec and again after 2 sec.

35. ***Familiarize.*** Let $x =$ the smaller odd positive integer. Then $x + 2 =$ the larger odd positive integer.

Translate.

$$\underbrace{\text{Square of the smaller odd positive integer}}_{x^2} \; + \; \underbrace{\text{Square of the larger odd positive integer}}_{(x+2)^2} \; \text{is} \; 74$$
$$x^2 + (x+2)^2 = 74$$

Solve.

$$x^2 + (x+2)^2 = 74$$
$$x^2 + x^2 + 4x + 4 = 74$$
$$2x^2 + 4x - 70 = 0$$
$$2(x^2 + 2x - 35) = 0$$
$$2(x+7)(x-5) = 0$$
$$x + 7 = 0 \quad or \quad x - 5 = 0$$
$$x = -7 \quad or \quad x = 5$$

Check. The solutions of the equation are -7 and 5. The problem asks for odd positive integers, so -7 cannot be a solution. When x is 5, $x + 2$ is 7. The numbers 5 and 7 are consecutive odd positive integers. The sum of their squares, $25 + 49$, is 74. The numbers check.

State. The integers are 5 and 7.

37. $-3.57 + 8.1$

The difference of the absolute values is $8.1 - 3.57$, or 4.53. The positive number has the larger absolute value, so the answer is positive.

$-3.57 + 8.1 = 4.53$

39. $(-2)(-4)(-5) = 8(-5) = -40$

41. $\dfrac{2 - |3 - 8|}{(-1-4)^2} = \dfrac{2 - |-5|}{(-5)^2} = \dfrac{2-5}{25} = \dfrac{-3}{25} = -\dfrac{3}{25}$

43. $2(y - 7) - (6y - 1) = 2y - 14 - 6y + 1 = -4y - 13$

45. ***Familiarize.*** Using the labels shown on the drawing in the text, we let $x =$ the width of the walk. Then the length and width of the rectangle formed by the pool and walk together are $40 + 2x$ and $20 + 2x$, respectively.

Translate.

$$\underbrace{\text{Area}}_{1500} \; \underbrace{\text{is}}_{=} \; \underbrace{\text{length}}_{(40+2x)} \; \underbrace{\text{times}}_{\cdot} \; \underbrace{\text{width.}}_{(20+2x)}$$
$$1500 = (40 + 2x) \cdot (20 + 2x)$$

Solve. We solve the equation.

$$1500 = (40 + 2x)(20 + 2x)$$
$$1500 = 2(20 + x) \cdot 2(10 + x) \quad \text{Factoring 2 out of each factor on the right}$$
$$1500 = 4 \cdot (20 + x)(10 + x)$$
$$375 = (20 + x)(10 + x) \quad \text{Dividing by 4}$$
$$375 = 200 + 30x + x^2$$
$$0 = x^2 + 30x - 175$$
$$0 = (x + 35)(x - 5)$$
$$x + 35 = 0 \quad or \quad x - 5 = 0$$
$$x = -35 \quad or \quad x = 5$$

Check. The solutions of the equation are -35 and 5. Since the width of the walk cannot be negative, -35 is not a solution. When $x = 5$, $40 + 2x = 40 + 2 \cdot 5$, or 50 and $20 + 2x = 20 + 2 \cdot 5$, or 30. The total area of the pool and walk is $50 \cdot 30$, or 1500 ft^2. This checks.

State. The width of the walk is 5 ft.

47. ***Familiarize.*** We make a drawing. Let $w =$ the width of the piece of cardboard. Then $2w =$ the length.

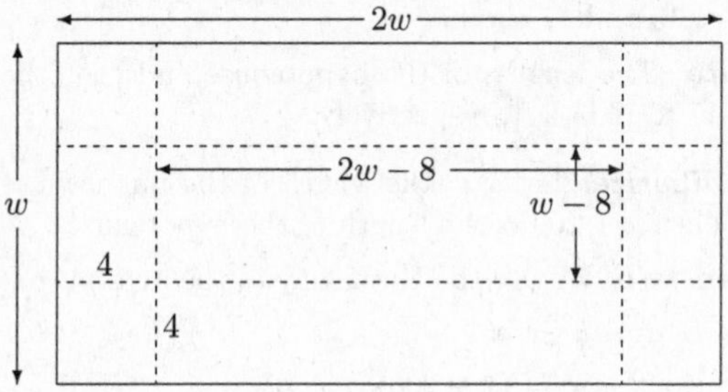

The box will have length $2w - 8$, width $w - 8$, and height 4. Recall that the formula for volume is $V =$ length $\times$ width $\times$ height.

Translate.

$$\underbrace{\text{The volume}}_{(2w-8)(w-8)(4)} \; \underbrace{\text{is}}_{=} \; \underbrace{616\text{cm}^3.}_{616}$$
$$(2w - 8)(w - 8)(4) = 616$$

Solve. We solve the equation.

$$(2w - 8)(w - 8)(4) = 616$$
$$(2w^2 - 24w + 64)(4) = 616$$
$$8w^2 - 96w + 256 = 616$$
$$8w^2 - 96w - 360 = 0$$
$$8(w^2 - 12w - 45) = 0$$
$$w^2 - 12w - 45 = 0 \quad \text{Dividing by 8}$$
$$(w - 15)(w + 3) = 0$$
$$w - 15 = 0 \quad or \quad w + 3 = 0$$
$$w = 15 \quad or \quad w = -3$$

Check. The width cannot be negative, so we only need to check 15. When $w = 15$, then $2w = 30$ and the dimensions of the box are $30 - 8$ by $15 - 8$ by 4, or 22 by 7 by 4. The volume is $22 \cdot 7 \cdot 4$, or 616.

State. The cardboard is 30 cm by 15 cm.

Chapter 5 Vocabulary Reinforcement

1. To factor a polynomial is to express it as a product.

2. A factor of a polynomial P is a polynomial that can be used to express P as a product.

3. A factorization of a polynomial is an expression that names that polynomial as a product.

4. When factoring, always look first for a common factor.

5. When factoring a polynomial with four terms, try factoring by grouping.

6. A trinomial square is the square of a binomial.

7. The principle of zero products states that if $ab = 0$, then $a = 0$ or $b = 0$.

8. The factorization of a difference of squares is the product of the sum and the difference of two terms.

Chapter 5 Concept Reinforcement

1. False; for example, $w + x + y + z$ cannot be factored by grouping.

2. This is true because the constant and the middle term are both positive.

3. False; $5 \cdot 0 = 0$, but $5 \neq 0$.

4. True; see page 438 in the text.

Chapter 5 Study Guide

1. $8x^3y^2 = 2 \cdot 2 \cdot 2 \cdot x^3 \cdot y^2$

$-20xy^3 = -1 \cdot 2 \cdot 2 \cdot 5 \cdot x \cdot y^3$

$32x^2y = 2 \cdot 2 \cdot 2 \cdot 2 \cdot 2 \cdot x^2 \cdot y$

Each coefficient has two factors of 2. The GCFs of the powers of x and of y are x and y, respectively, because 1 is the smallest exponent of both x and y. Thus the GCF is $2 \cdot 2 \cdot x \cdot y$, or $4xy$.

2. $27x^5 - 9x^3 + 18x^2 = 9x^2 \cdot 3x^3 - 9x^2 \cdot x + 9x^2 \cdot 2$

$= 9x^2(3x^3 - x + 2)$

3. $z^3 - 3z^2 + 4z - 12 = z^2(z - 3) + 4(z - 3)$

$= (z - 3)(z^2 + 4)$

4. $x^2 + 6x + 8$

Look for a pair of factors of 8 whose sum is 6. The numbers we want are 2 and 4.

$x^2 + 6x + 8 = (x + 2)(x + 4)$

5. $6z^2 - 21z - 12$

(1) Factor out the largest common factor.

$6z^2 - 21z - 12 = 3(2z^2 - 7z - 4)$

Now we factor $2z^2 - 7z - 4$.

(2) We can factor $2z^2$ as

$(2z+ \quad)(z+ \quad)$.

(3) We can factor -4 as

$-1, 4$ and $1, -4$ and $2, -2$.

These factors can also be written as

$4, -1$ and $-4, 1$ and $-2, 2$.

(4) Look for combinations of factors from steps (2) and (3) such that the sum of their products is the middle term, $-7z$.

The factorization of $2z^2 - 7z - 4$ is $(2z + 1)(z - 4)$, so $6z^2 - 21z - 12 = 3(2z + 1)(z - 4)$.

6. $6y^2 + 7y - 3$

(1) There is no common factor (other than 1 or -1).

(2) Multiply 6 and -3: $6(-3) = -18$.

(3) Look for a factorization of -18 in which the sum of the factors is 7. The numbers we want are 9 and -2.

(4) Split the middle term: $7y = 9y - 2y$.

(5) Factor by grouping.

$6y^2 + 7y - 3 = 6y^2 + 9y - 2y - 3$

$= 3y(2y + 3) - (2y + 3)$

$= (2y + 3)(3y - 1)$

7. $4x^2 + 4x + 1 = (2x)^2 + 2 \cdot 2x \cdot 1 + 1^2 = (2x + 1)^2$

8. $18x^2 - 8 = 2(9x^2 - 4)$

$= 2[(3x)^2 - 2^2]$

$= 2(3x + 2)(3x - 2)$

9. $x^2 + 4x = 5$

$x^2 + 4x - 5 = 0$

$(x + 5)(x - 1) = 0$

$x + 5 = 0 \quad \text{or} \quad x - 1 = 0$

$x = -5 \quad \text{or} \quad x = 1$

The solutions are -5 and 1.

Chapter 5 Review Exercises

1. $-15y^2 = -1 \cdot 3 \cdot 5 \cdot y^2$

$25y^6 = 5 \cdot 5 \cdot y^6$

Each coefficient has a factor of 5. There are no other common prime factors. The GCF of the powers of y is y^2 because 2 is the smallest exponent of y. Thus the GCF is $5y^2$.

2. $12x^3 = 2 \cdot 2 \cdot 3 \cdot x^3$

$-60x^2y = -1 \cdot 2 \cdot 2 \cdot 3 \cdot 5 \cdot x^2 \cdot y$

$36xy = 2 \cdot 2 \cdot 3 \cdot 3 \cdot x \cdot y$

Each coefficient has two factors of 2 and one factor of 3. There are no other common prime factors. The GCF of the powers of x is x because 1 is the smallest exponent of x. The GCF of the powers of y is 1 because $12x^3$ has no y-factor. Thus the GCF is $2 \cdot 2 \cdot 3 \cdot x \cdot 1$, or $12x$.

3. $5 - 20x^6$

$= 5(1 - 4x^6)$ 5 is a common factor.

$= 5(1 - 2x^3)(1 + 2x^3)$ Factoring the difference of squares

4. $x^2 - 3x = x(x - 3)$

5. $9x^2 - 4 = (3x + 2)(3x - 2)$ Factoring a difference of squares

6. $x^2 + 4x - 12$

We look for a pair of factors of -12 whose sum is 4. The numbers we need are 6 and -2.

$x^2 + 4x - 12 = (x + 6)(x - 2)$

7. $x^2 + 14x + 49 = x^2 + 2 \cdot x \cdot 7 + 7^2 = (x + 7)^2$

8. $6x^3 + 12x^2 + 3x = 3x(2x^2 + 4x + 1)$

The trinomial $2x^2 + 4x + 1$ cannot be factored, so the factorization is complete.

9. $x^3 + x^2 + 3x + 3$

$= (x^3 + x^2) + (3x + 3)$

$= x^2(x + 1) + 3(x + 1)$ Factoring by grouping

$= (x + 1)(x^2 + 3)$

10. $6x^2 - 5x + 1$

There is no common factor (other than 1). This polynomial has three terms, but it is not a trinomial square. Multiply the leading coefficient and the constant, 6 and 1: $6 \cdot 1 = 6$. Try to factor 6 so that the sum of the factors is -5. The numbers we want are -2 and -3: $-2(-3) = 6$ and $-2 + (-3) = -5$. Split the middle term and factor by grouping.

$$\begin{aligned} 6x^2 - 5x + 1 &= 6x^2 - 2x - 3x + 1 \\ &= (6x^2 - 2x) + (-3x + 1) \\ &= 2x(3x - 1) - 1(3x - 1) \\ &= (3x - 1)(2x - 1) \end{aligned}$$

11. $x^4 - 81 = (x^2 + 9)(x^2 - 9) = (x^2 + 9)(x + 3)(x - 3)$

12. $9x^3 + 12x^2 - 45x$

$= 3x(3x^2 + 4x - 15)$ $3x$ is a common factor.

$= 3x(3x - 5)(x + 3)$ Using trial and error

13. $2x^2 - 50 = 2(x^2 - 25) = 2(x + 5)(x - 5)$

14.
$$\begin{aligned} x^4 + 4x^3 - 2x - 8 &= (x^4 + 4x^3) + (-2x - 8) \\ &= x^3(x + 4) - 2(x + 4) \\ &= (x + 4)(x^3 - 2) \end{aligned}$$

15. $16x^4 - 1 = (4x^2 + 1)(4x^2 - 1) = (4x^2 + 1)(2x + 1)(2x - 1)$

16. $8x^6 - 32x^5 + 4x^4 = 4x^4(2x^2 - 8x + 1)$

The trinomial $2x^2 - 8x + 1$ cannot be factored, so the factorization is complete.

17.
$$\begin{aligned} 75 + 12x^2 + 60x &= 12x^2 + 60x + 75 \\ &= 3(4x^2 + 20x + 25) \\ &= 3(2x + 5)^2 \end{aligned}$$

18. $x^2 + 9$ is a sum of squares with no common factor, so it is prime.

19. $x^3 - x^2 - 30x = x(x^2 - x - 30) = x(x - 6)(x + 5)$

20. $4x^2 - 25 = (2x + 5)(2x - 5)$

21. $9x^2 + 25 - 30x = 9x^2 - 30x + 25 = (3x - 5)^2$

22.
$$\begin{aligned} 6x^2 - 28x - 48 &= 2(3x^2 - 14x - 24) \\ &= 2(3x + 4)(x - 6) \end{aligned}$$

23. $x^2 - 6x + 9 = (x - 3)^2$

24. $2x^2 - 7x - 4 = (2x + 1)(x - 4)$

25. $18x^2 - 12x + 2 = 2(9x^2 - 6x + 1) = 2(3x - 1)^2$

26. $3x^2 - 27 = 3(x^2 - 9) = 3(x + 3)(x - 3)$

27. $15 - 8x + x^2 = x^2 - 8x + 15 = (x - 3)(x - 5)$

28. $25x^2 - 20x + 4 = (5x - 2)^2$

29.
$$\begin{aligned} 49b^{10} + 4a^8 - 28a^4b^5 &= 49b^{10} - 28a^4b^5 + 4a^8 \\ &= (7b^5)^2 - 2 \cdot 7b^5 \cdot 2a^4 + (2a^4)^2 \\ &= (7b^5 - 2a^4)^2 \end{aligned}$$

30. $x^2y^2 + xy - 12 = (xy + 4)(xy - 3)$

31. $12a^2 + 84ab + 147b^2 = 3(4a^2 + 28ab + 49b^2) =$ $3(2a + 7b)^2$

32.
$$\begin{aligned} m^2 + 5m + mt + 5t &= (m^2 + 5m) + (mt + 5t) \\ &= m(m + 5) + t(m + 5) \\ &= (m + 5)(m + t) \end{aligned}$$

33. $32x^4 - 128y^4z^4 = 32(x^4 - 4y^4z^4) =$ $32(x^2 + 2y^2z^2)(x^2 - 2y^2z^2)$

34. $(x - 1)(x + 3) = 0$

$x - 1 = 0$ *or* $x + 3 = 0$

$x = 1$ *or* $x = -3$

The solutions are 1 and -3.

35. $x^2 + 2x - 35 = 0$

$(x + 7)(x - 5) = 0$

$x + 7 = 0$ *or* $x - 5 = 0$

$x = -7$ *or* $x = 5$

The solutions are -7 and 5.

36. $x^2 + 4x = 0$

$x(x+4) = 0$

$x = 0$ *or* $x + 4 = 0$

$x = 0$ *or* $x = -4$

The solutions are 0 and -4.

37. $3x^2 + 2 = 5x$

$3x^2 - 5x + 2 = 0$

$(3x-2)(x-1) = 0$

$3x - 2 = 0$ *or* $x - 1 = 0$

$3x = 2$ *or* $x = 1$

$x = \frac{2}{3}$ *or* $x = 1$

The solutions are $\frac{2}{3}$ and 1.

38. $x^2 = 64$

$x^2 - 64 = 0$

$(x+8)(x-8) = 0$

$x + 8 = 0$ *or* $x - 8 = 0$

$x = -8$ *or* $x = 8$

The solutions are -8 and 8.

39. $16 = x(x-6)$

$16 = x^2 - 6x$

$0 = x^2 - 6x - 16$

$0 = (x-8)(x+2)$

$x - 8 = 0$ *or* $x + 2 = 0$

$x = 8$ *or* $x = -2$

The solutions are 8 and -2.

40. Let $y = 0$ and solve for x.

$0 = x^2 + 9x + 20$

$0 = (x+5)(x+4)$

$x + 5 = 0$ *or* $x + 4 = 0$

$x = -5$ *or* $x = -4$

The x-intercepts are $(-5, 0)$ and $(-4, 0)$.

41. Let $y = 0$ and solve for x.

$0 = 2x^2 - 7x - 15$

$0 = (2x+3)(x-5)$

$2x + 3 = 0$ *or* $x - 5 = 0$

$2x = -3$ *or* $x = 5$

$x = -\frac{3}{2}$ *or* $x = 5$

The x-intercepts are $\left(-\frac{3}{2}, 0\right)$ and $(5, 0)$.

42. ***Familiarize***. Let $b =$ the length of the base, in cm. Then $b + 1 =$ the height.

Translate. We use the formula for the area of a triangle.

$A = \frac{1}{2}bh$

$15 = \frac{1}{2}b(b+1)$ Substituting

Solve.

$15 = \frac{1}{2}b(b+1)$

$15 = \frac{1}{2}b^2 + \frac{1}{2}b$

$0 = \frac{1}{2}b^2 + \frac{1}{2}b - 15$

$2 \cdot 0 = 2\left(\frac{1}{2}b^2 + \frac{1}{2}b - 15\right)$ Clearing fractions

$0 = b^2 + b - 30$

$0 = (b+6)(b-5)$

$b + 6 = 0$ *or* $b - 5 = 0$

$b = -6$ *or* $b = 5$

Check. The length of the base cannot be negative, so -6 cannot be a solution. If $b = 5$, then $b + 1 = 5 + 1 = 6$ and the area is $\frac{1}{2} \cdot 5 \cdot 6 = 15\text{ cm}^2$. The number 5 checks.

State. The base is 5 cm, and the height is 6 cm.

43. ***Familiarize***. Let $x =$ the smaller integer. Then $x + 2$ is the other integer.

Translate.

$$\underbrace{\text{Smaller even integer}}_{x}\ \underbrace{\text{times}}_{\cdot}\ \underbrace{\text{larger even integer}}_{(x+2)}\ \underbrace{\text{is}}_{=}\ \underbrace{288.}_{288}$$

Solve.

$x(x+2) = 288$

$x^2 + 2x = 288$

$x^2 + 2x - 288 = 0$

$(x+18)(x-16) = 0$

$x + 18 = 0$ *or* $x - 16 = 0$

$x = -18$ *or* $x = 16$

If $x = -18$, then $x + 2 = -18 + 2 = -16$.

If $x = 16$, then $x + 2 = 16 + 2 = 18$.

Check. -18 and -16 are consecutive even integers and $-18(-16) = 288$. Similarly, 16 and 18 are consecutive even integers and $16 \cdot 18 = 288$. Both pairs of integers check.

State. The integers are -18 and -16 or 16 and 18.

44. ***Familiarize***. Let $z =$ the length of the zipline, in ft. The zipline is the hypotenuse of the right triangle with legs of length 58 ft and 840 ft.

Translate. We use the Pythagorean theorem.

$a^2 + b^2 = c^2$

$58^2 + 840^2 = z^2$

Solve.

$$58^2 + 840^2 = z^2$$
$$708,964 = z^2$$
$$0 = z^2 - 708,964$$
$$0 = (z+842)(z-842)$$
$$z+842 = 0 \quad or \quad z-842 = 0$$
$$z = -842 \quad or \quad z = 842$$

Check. The length must be positive, so -842 cannot be a solution. We have $58^2 + 840^2 = 708,964 = 842^2$, so the answer checks.

State. The zipline is 842 ft long.

45. ***Familiarize.*** We make a drawing. Let d = the distance from the base of the tree to the point where each cable is attached to the tree. Then $d+1$ = the distance from the base of the tree to the point on the ground where each cable is anchored. The distances are in feet.

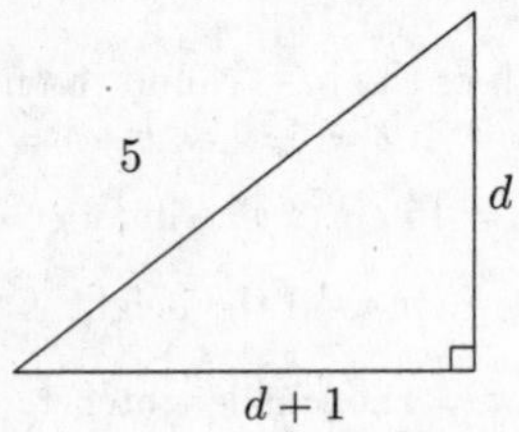

Translate. We use the Pythagorean theorem.

$$a^2 + b^2 = c^2$$
$$(d+1)^2 + d^2 = 5^2 \quad \text{Substituting}$$

Solve.

$$(d+1)^2 + d^2 = 5^2$$
$$d^2 + 2d + 1 + d^2 = 25$$
$$2d^2 + 2d + 1 = 25$$
$$2d^2 + 2d - 24 = 0$$
$$d^2 + d - 12 = 0 \quad \text{Dividing by 2}$$
$$(d+4)(d-3) = 0$$
$$d+4 = 0 \quad or \quad d-3 = 0$$
$$d = -4 \quad or \quad d = 3$$

Check. A distance cannot be negative, so we check only 3. If $d = 3$, then $d+1 = 3+1 = 4$ and $3^2 + 4^2 = 9 + 16 = 25 = 5^2$. The answer checks.

State. The distance from the base of the tree to the point where each cable is attached to the tree is 3 ft. The distance from the base of the tree to the point on the ground where each cable is anchored is 4 ft.

46. ***Familiarize.*** Let s = the length of a side of the original square, in km. Then $s+3$ = the length of a side of the enlarged square.

Translate.

$$\underbrace{\text{Area of enlarged square}}_{\downarrow} \quad \underset{\downarrow}{\text{is}} \quad \underbrace{81 \text{ km}^2}_{\downarrow}.$$
$$(s+3)^2 \quad = \quad 81$$

Solve.

$$(s+3)^2 = 81$$
$$s^2 + 6s + 9 = 81$$
$$s^2 + 6s - 72 = 0$$
$$(s+12)(s-6) = 0$$
$$s+12 = 0 \quad or \quad s-6 = 0$$
$$s = -12 \quad or \quad s = 6$$

Check. The length of a side of the square cannot be negative, so -12 cannot be a solution. If the length of a side of the original square is 6 km, then the length of a side of the enlarged square is $6+3$, or 9 km. The area of the enlarged square is $(9 \text{ km})^2$, or 81 km^2, so the answer checks.

State. The length of a side of the original square is 6 km.

47. $x^2 - 9x + 8 = (x-1)(x-8)$

Answer B is correct.

48. $15x^2 + 5x - 20 = 5(3x^2 + x - 4)$
$= 5(3x+4)(x-1)$

Answer A is correct.

49. ***Familiarize.*** Let w = the width of the margins, in cm. Then the printed area on each page has dimensions $20-2w$ by $15-2w$. The area of the margins constitutes one-half the area of each page, so the printed area also constitutes one-half of the area.

Translate.

$$\underbrace{\text{Printed area}}_{\downarrow} \quad \underset{\downarrow}{\text{is}} \quad \underbrace{\text{one-half}}_{\downarrow} \quad \underset{\downarrow}{\text{of}} \quad \underbrace{\text{total area.}}_{\downarrow}$$
$$(20-2w)(15-2w) \quad = \quad \frac{1}{2} \quad \cdot \quad 20 \cdot 15$$

Solve.

$$(20-2w)(15-2w) = \frac{1}{2} \cdot 20 \cdot 15$$
$$300 - 70w + 4w^2 = 150$$
$$150 - 70w + 4w^2 = 0$$
$$4w^2 - 70w + 150 = 0$$
$$2(2w^2 - 35w + 75) = 0$$
$$2w^2 - 35w + 75 = 0 \quad \text{Dividing by 2}$$
$$(2w-5)(w-15) = 0$$
$$2w-5 = 0 \quad or \quad w-15 = 0$$
$$2w = 5 \quad or \quad w = 15$$
$$w = 2.5 \quad or \quad w = 15$$

Check. If $w = 15$, then $20 - 2w$ and $15 - 2w$ are both negative. Since the dimensions of the printed area cannot be negative, 15 cannot be a solution. If $w = 2.5$, then $20 - 2w = 20 - 2(2.5) = 20 - 5 = 15$ and $15 - 2w = 15 - 2(2.5) = 15 - 5 = 10$. Thus the printed area is $15 \cdot 10$, or 150 cm^2. This is one-half of the total area of the page, $20 \cdot 15$, or 300 cm^2. The number 2.5 checks.

State. The width of the margins is 2.5 cm.

50. ***Familiarize***. Let $n =$ the number.

Translate.

$$\underbrace{\text{The cube of a number}}_{n^3} \quad \underbrace{\text{is}}_{=} \quad \underbrace{\text{twice}}_{2\cdot} \quad \underbrace{\text{the square of the number.}}_{n^2}$$

Solve.

$$n^3 = 2n^2$$
$$n^3 - 2n^2 = 0$$
$$n^2(n-2) = 0$$
$$n \cdot n(n-2) = 0$$
$$n = 0 \quad or \quad n = 0 \quad or \quad n - 2 = 0$$
$$n = 0 \quad or \quad n = 0 \quad or \quad n = 2$$

Check. If $n = 0$, then $n^3 = 0^3 = 0$, $n^2 = 0^2 = 0$, and $0 = 2 \cdot 0$. If $n = 2$, then $n^3 = 2^3 = 8$, $n^2 = 2^2 = 4$, and $8 = 2 \cdot 4$. Both numbers check.

State. The number is 0 or 2.

51. ***Familiarize***. Let $w =$ the width of the original rectangle, in inches. Then $2w =$ the length, in inches. The new length and width are $2w + 20$ and $w - 1$, respectively.

Translate.

$$\underbrace{\text{The new area}}_{(2w+20)(w-1)} \quad \underbrace{\text{is}}_{=} \quad \underbrace{160.}_{160}$$

Solve.

$$(2w + 20)(w - 1) = 160$$
$$2w^2 + 18w - 20 = 160$$
$$2w^2 + 18w - 180 = 0$$
$$2(w^2 + 9w - 90) = 0$$
$$w^2 + 9w - 90 = 0$$
$$(w + 15)(w - 6) = 0$$
$$w + 15 = 0 \quad or \quad w - 6 = 0$$
$$w = -15 \quad or \quad w = 6$$

Check. The dimensions of the rectangle cannot be negative, so -15 cannot be a solution. If $w = 6$, then $2w = 2 \cdot 6 = 12$, $2w + 20 = 12 + 20 = 32$ and $w - 1 = 6 - 1 = 5$. The area of a rectangle with dimensions 32 by 5 is $32 \cdot 5$, or 160, so the answer checks.

State. The length of the original rectangle is 12 in. and the width is 6 in.

52. ***Familiarize***. First we can use the Pythagorean theorem to find x, in ft. Then the height of the telephone pole is $x + 5$.

Translate. We use the Pythagorean theorem.

$$a^2 + b^2 = c^2$$
$$\left(\frac{1}{2}x + 1\right)^2 + x^2 = 34^2$$

Solve. We solve the equation.

$$\left(\frac{1}{2}x + 1\right)^2 + x^2 = 34^2$$
$$\frac{1}{4}x^2 + x + 1 + x^2 = 1156$$
$$x^2 + 4x + 4 + 4x^2 = 4624 \quad \text{Multiplying by 4}$$
$$5x^2 + 4x + 4 = 4624$$
$$5x^2 + 4x - 4620 = 0$$
$$(5x + 154)(x - 30) = 0$$
$$5x + 154 = 0 \quad or \quad x - 30 = 0$$
$$5x = -154 \quad or \quad x = 30$$
$$x = -30.8 \quad or \quad x = 30$$

Check. Since the length x must be positive, -30.8 cannot be a solution. If x is 30 ft, then $\frac{1}{2}x + 1$ is $\frac{1}{2} \cdot 30 + 1$, or 16 ft. Since $16^2 + 30^2 = 1156 = 34^2$, the number 30 checks. When x is 30 ft, then $x + 5$ is 35 ft.

State. The height of the telephone pole is 35 ft.

53. $x^2 + 25 = 0$

Since $x^2 + 25$ cannot be factored, the equation has no solution.

54. $(x - 2)(x + 3)(2x - 5) = 0$

$$x - 2 = 0 \quad or \quad x + 3 = 0 \quad or \quad 2x - 5 = 0$$
$$x = 2 \quad or \quad x = -3 \quad or \quad 2x = 5$$
$$x = 2 \quad or \quad x = -3 \quad or \quad x = \frac{5}{2}$$

The solutions are 2, -3, and $\frac{5}{2}$.

55. $(x-3)4x^2 + 3x(x-3) - (x-3)10 = 0$

$$(4x^2 + 3x - 10)(x - 3) = 0 \quad \text{Factoring out } x - 3$$
$$(4x - 5)(x + 2)(x - 3) = 0 \quad \text{Factoring } 4x^2 + 3x - 10$$
$$4x - 5 = 0 \quad or \quad x + 2 = 0 \quad or \quad x - 3 = 0$$
$$4x = 5 \quad or \quad x = -2 \quad or \quad x = 3$$
$$x = \frac{5}{4} \quad or \quad x = -2 \quad or \quad x = 3$$

The solutions are $\frac{5}{4}$, -2, and 3.

Chapter 5 Discussion and Writing Exercises

1. Although $x^3 - 8x^2 + 15x$ can be factored as $(x^2 - 5x)(x - 3)$, this is not a complete factorization of the polynomial since $x^2 - 5x = x(x - 5)$. The student should be advised *always* to look for a common factor first.

2. Josh is correct, because answers can easily be checked by multiplying.

3. For $x = -3$:

   $(x - 4)^2 = (-3 - 4)^2 = (-7)^2 = 49$

$(4-x)^2 = [4-(-3)]^2 = 7^2 = 49$

For $x = 1$:

$(x-4)^2 = (1-4)^2 = (-3)^2 = 9$

$(4-x)^2 = (4-1)^2 = 3^2 = 9$

In general, $(x-4)^2 = [-(-x+4)]^2 = [-(4-x)]^2 = (-1)^2(4-x)^2 = (4-x)^2$.

4. The equation is not in the form $ab = 0$. The correct procedure is:

$$\begin{aligned}(x-3)(x+4) &= 8\\ x^2+x-12 &= 8\\ x^2+x-20 &= 0\\ (x+5)(x-4) &= 0\\ x+5=0 \quad &or \quad x-4=0\\ x=-5 \quad &or \quad x=4\end{aligned}$$

The solutions are -5 and 4.

5. One solution of the equation is 0. Dividing both sides of the equation by x, leaving the solution $x = 3$, is equivalent to dividing by 0.

6. She could use the measuring sticks to draw a right angle as shown below. Then she could use the 3-ft and 4-ft sticks to extend one leg to 7 ft and the 4-ft and 5-ft sticks to extend the other leg to 9 ft.

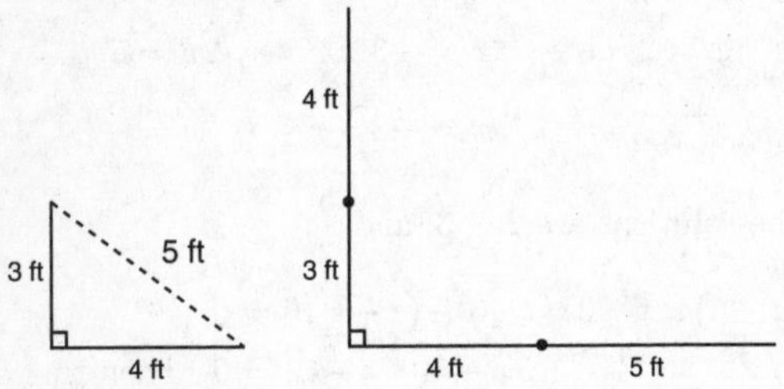

Next she could draw another right angle with either the 7-ft side or the 9-ft side as a side.

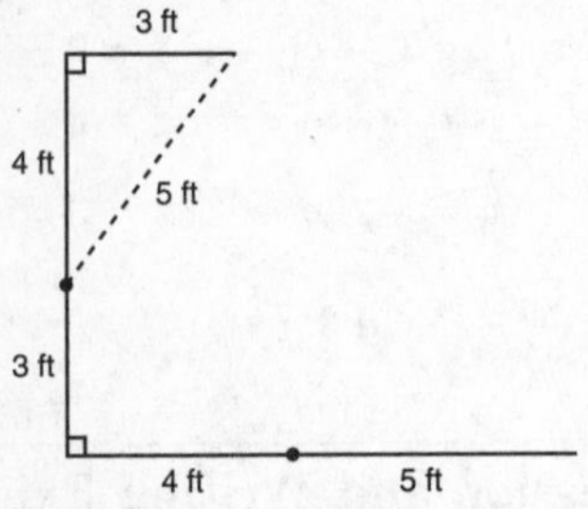

Then she could use the sticks to extend the other side to the appropriate length. Finally she would draw the remaining side of the rectangle.

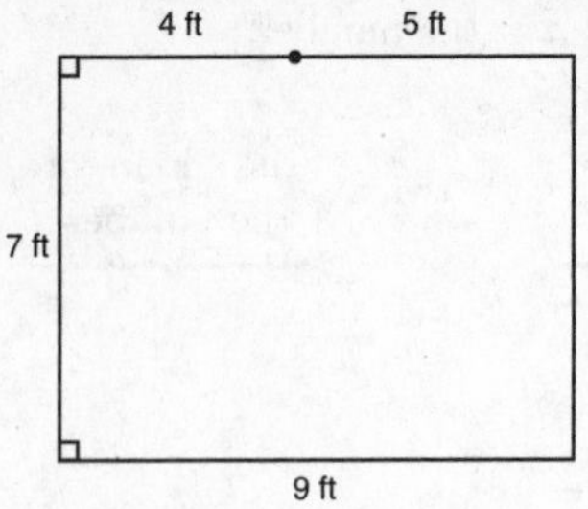

Chapter 5 Test

1. $28x^3 = 2 \cdot 2 \cdot 7 \cdot x^3$

$48x^7 = 2 \cdot 2 \cdot 2 \cdot 2 \cdot 3 \cdot x^7$

The coefficients each have two factors of 2. There are no other common prime factors. The GCF of the powers of x is x^3 because 3 is the smallest exponent of x. Thus the GCF is $2 \cdot 2 \cdot x^3$, or $4x^3$.

2. $x^2 - 7x + 10$

We look for a pair of factors of 10 whose sum is -7. The numbers we need are -2 and -5.

$$x^2 - 7x + 10 = (x-2)(x-5)$$

3. $$\begin{aligned}x^2 + 25 - 10x &= x^2 - 10x + 25\\ &= x^2 - 2 \cdot x \cdot 5 + 5^2\\ &= (x-5)^2\end{aligned}$$

4. $6y^2 - 8y^3 + 4y^4 = 4y^4 - 8y^3 + 6y^2 =$

$2y^2 \cdot 2y^2 - 2y^2 \cdot 4y + 2y^2 \cdot 3 = 2y^2(2y^2 - 4y + 3)$

Since $2y^2 - 4y + 3$ cannot be factored, the factorization is complete.

5. $$\begin{aligned}&x^3 + x^2 + 2x + 2\\ &= (x^3 + x^2) + (2x + 2)\\ &= x^2(x+1) + 2(x+1) \quad \text{Factoring by grouping}\\ &= (x+1)(x^2+2)\end{aligned}$$

6. $x^2 - 5x = x \cdot x - 5 \cdot x = x(x-5)$

7. $$\begin{aligned}&x^3 + 2x^2 - 3x\\ &= x(x^2 + 2x - 3) \quad x \text{ is a common factor.}\\ &= x(x+3)(x-1) \quad \text{Factoring the trinomial}\end{aligned}$$

8. $$\begin{aligned}&28x - 48 + 10x^2\\ &= 10x^2 + 28x - 48\\ &= 2(5x^2 + 14x - 24) \quad 2 \text{ is a common factor.}\\ &= 2(5x-6)(x+4) \quad \text{Factoring the trinomial}\end{aligned}$$

9. $$\begin{aligned}4x^2 - 9 &= (2x)^2 - 3^2 \quad \text{Difference of squares}\\ &= (2x+3)(2x-3)\end{aligned}$$

10. $x^2 - x - 12$

We look for a pair of factors of -12 whose sum is -1. The numbers we need are -4 and 3.

$$x^2 - x - 12 = (x-4)(x+3)$$

11. $6m^3 + 9m^2 + 3m$
$= 3m(2m^2 + 3m + 1)$ 3m is a common factor.
$= 3m(2m + 1)(m + 1)$ Factoring the trinomial

12. $3w^2 - 75 = 3(w^2 - 25)$ 3 is a common factor.
$= 3(w^2 - 5^2)$ Difference of squares
$= 3(w + 5)(w - 5)$

13. $60x + 45x^2 + 20$
$= 45x^2 + 60x + 20$
$= 5(9x^2 + 12x + 4)$ 5 is a common factor.
$= 5[(3x)^2 + 2 \cdot 3x \cdot 2 + 2^2]$ Trinomial square
$= 5(3x + 2)^2$

14. $3x^4 - 48$
$= 3(x^4 - 16)$ 3 is a common factor.
$= 3[(x^2)^2 - 4^2]$ Difference of squares
$= 3(x^2 + 4)(x^2 - 4)$
$= 3(x^2 + 4)(x^2 - 2^2)$ Difference of squares
$= 3(x^2 + 4)(x + 2)(x - 2)$

15. $49x^2 - 84x + 36$
$= (7x)^2 - 2 \cdot 7x \cdot 6 + 6^2$ Trinomial square
$= (7x - 6)^2$

16. $5x^2 - 26x + 5$

There is no common factor (other than 1). This polynomial has 3 terms, but it is not a trinomial square. Using the ac-method we first multiply the leading coefficient and the constant term: $5 \cdot 5 = 25$. Try to factor 25 so that the sum of the factors is -26. The numbers we want are -1 and -25: $-1(-25) = 25$ and $-1 + (-25) = -26$. Split the middle term and factor by grouping:

$$\begin{aligned} 5x^2 - 26x + 5 &= 5x^2 - x - 25x + 5 \\ &= (5x^2 - x) + (-25x + 5) \\ &= x(5x - 1) - 5(5x - 1) \\ &= (5x - 1)(x - 5) \end{aligned}$$

17. $x^4 + 2x^3 - 3x - 6$
$= (x^4 + 2x^3) + (-3x - 6)$
$= x^3(x + 2) - 3(x + 2)$ Factoring by grouping
$= (x + 2)(x^3 - 3)$

18. $80 - 5x^4$
$= 5(16 - x^4)$ 5 is a common factor.
$= 5[4^2 - (x^2)^2]$ Difference of squares
$= 5(4 + x^2)(4 - x^2)$
$= 5(4 + x^2)(2^2 - x^2)$ Difference of squares
$= 5(4 + x^2)(2 + x)(2 - x)$

19. $6t^3 + 9t^2 - 15t$
$= 3t(2t^2 + 3t - 5)$ 3t is a common factor.
$= 3t(2t + 5)(t - 1)$ Factoring the trinomial

20.
$$\begin{aligned} x^2 - 3x &= 0 \\ x(x - 3) &= 0 \\ x = 0 \quad &or \quad x - 3 = 0 \\ x = 0 \quad &or \quad x = 3 \end{aligned}$$
The solutions are 0 and 3.

21.
$$\begin{aligned} 2x^2 &= 32 \\ 2x^2 - 32 &= 0 \\ 2(x^2 - 16) &= 0 \\ 2(x + 4)(x - 4) &= 0 \\ x + 4 = 0 \quad &or \quad x - 4 = 0 \\ x = -4 \quad &or \quad x = 4 \end{aligned}$$
The solutions are -4 and 4.

22.
$$\begin{aligned} x^2 - x - 20 &= 0 \\ (x - 5)(x + 4) &= 0 \\ x - 5 = 0 \quad &or \quad x + 4 = 0 \\ x = 5 \quad &or \quad x = -4 \end{aligned}$$
The solutions are 5 and -4.

23.
$$\begin{aligned} 2x^2 + 7x &= 15 \\ 2x^2 + 7x - 15 &= 0 \\ (2x - 3)(x + 5) &= 0 \\ 2x - 3 = 0 \quad &or \quad x + 5 = 0 \\ 2x = 3 \quad &or \quad x = -5 \\ x = \frac{3}{2} \quad &or \quad x = -5 \end{aligned}$$
The solutions are $\frac{3}{2}$ and -5.

24.
$$\begin{aligned} x(x - 3) &= 28 \\ x^2 - 3x &= 28 \\ x^2 - 3x - 28 &= 0 \\ (x - 7)(x + 4) &= 0 \\ x - 7 = 0 \quad &or \quad x + 4 = 0 \\ x = 7 \quad &or \quad x = -4 \end{aligned}$$
The solutions are 7 and -4.

25. We let $y = 0$ and solve for x.
$$\begin{aligned} 0 &= x^2 - 2x - 35 \\ 0 &= (x + 5)(x - 7) \\ x + 5 = 0 \quad &or \quad x - 7 = 0 \\ x = -5 \quad &or \quad x = 7 \end{aligned}$$
The x-intercepts are $(-5, 0)$ and $(7, 0)$.

26. We let $y = 0$ and solve for x.
$$\begin{aligned} 0 &= 3x^2 - 5x + 2 \\ 0 &= (3x - 2)(x - 1) \\ 3x - 2 = 0 \quad &or \quad x - 1 = 0 \\ 3x = 2 \quad &or \quad x = 1 \\ x = \frac{2}{3} \quad &or \quad x = 1 \end{aligned}$$

The x-intercepts are $\left(\frac{2}{3}, 0\right)$ and $(1, 0)$.

27. ***Familiarize.*** Let w = the width, in meters. Then $w+2$ = the length. Recall that the area of a rectangle is (length) · (width).

Translate. We use the formula for the area of a rectangle.

$$48 = (w+2)w$$

Solve.

$$48 = (w+2)w$$
$$48 = w^2 + 2w$$
$$0 = w^2 + 2w - 48$$
$$0 = (w+8)(w-6)$$
$$w+8 = 0 \quad or \quad w-6 = 0$$
$$w = -8 \quad or \quad w = 6$$

Check. The width cannot be negative, so -8 cannot be a solution. If $w = 6$, then $w + 2 = 8$ and the area is $(8\text{ m}) \cdot (6\text{ m})$, or 48 m^2. The number 6 checks.

State. The length is 8 m and the width is 6 m.

28. ***Familiarize.*** Using the labels on the drawing in the text, we let h = the height of the triangle, in cm, and $2h+6$ = the base. Recall that the area of a triangle is $\frac{1}{2} \cdot$ (base) · (height).

Translate. We use the formula for the area of a triangle.

$$28 = \frac{1}{2} \cdot (2h+6) \cdot h$$

Solve.

$$28 = \frac{1}{2}(2h+6)h$$
$$28 = h^2 + 3h$$
$$0 = h^2 + 3h - 28$$
$$0 = (h+7)(h-4)$$
$$h+7 = 0 \quad or \quad h-4 = 0$$
$$h = -7 \quad or \quad h = 4$$

Check. The height cannot be negative, so -7 cannot be a solution. If $h = 4$, then $2h+6 = 2\cdot4+6 = 8+6 = 14$ and the area is $\frac{1}{2} \cdot (14\text{ cm}) \cdot (4\text{ cm})$, or 28 cm^2. The number 4 checks.

State. The height is 4 cm and the base is 14 cm.

29. ***Familiarize.*** Using the labels on the drawing in the text, we let x = the distance between the two marked points, in feet. If the corner is a right angle, the lengths 3 ft, 4 ft, and x will satisfy the Pythagorean theorem.

Translate.

$$a^2 + b^2 = c^2$$
$$3^2 + 4^2 = x^2 \quad \text{Substituting}$$

Solve.

$$3^2 + 4^2 = x^2$$
$$9 + 16 = x^2$$
$$25 = x^2$$
$$0 = x^2 - 25$$
$$0 = (x+5)(x-5)$$
$$x+5 = 0 \quad or \quad x-5 = 0$$
$$x = -5 \quad or \quad x = 5$$

Check. The distance cannot be negative, so -5 cannot be a solution. If $x = 5$, then we have $3^2 + 4^2 = 9 + 16 = 25 = 5^2$, so the answer checks.

State. The distance between the marked points should be 5 ft.

30. $2y^4 - 32 = 2(y^4 - 16)$

$$= 2(y^2+4)(y^2-4)$$
$$= 2(y^2+4)(y+2)(y-2)$$

Answer A is correct.

31. ***Familiarize.*** Let w = the width of the original rectangle, in meters. Then $5w$ = the length. The new width and length are $w+2$ and $5w-3$, respectively.

Translate. We will use the formula for the area of a rectangle, Area = (length) · (width).

$$60 = (5w-3)(w+2)$$

Solve.

$$60 = (5w-3)(w+2)$$
$$60 = 5w^2 + 7w - 6$$
$$0 = 5w^2 + 7w - 66$$
$$0 = (5w+22)(w-3)$$
$$5w+22 = 0 \quad or \quad w-3 = 0$$
$$5w = -22 \quad or \quad w = 3$$
$$w = -\frac{22}{5} \quad or \quad w = 3$$

Check. The width cannot be negative, so $-\frac{22}{5}$ cannot be a solution. If $w = 3$, then $5w = 5\cdot3 = 15$. The new dimensions are $w+2$, or $3+2$, or 5 and $5w-3$, or $15-3$, or 12, and the new area is $12\cdot5$, or 60. The number 3 checks.

State. The original length is 15 m and the original width is 3 m.

32. $(a+3)^2 - 2(a+3) - 35$

We can think of $a+3$ as the variable in this expression. Then we find a pair of factors of -35 whose sum is -2. The numbers we want are -7 and 5.

$(a+3)^2 - 2(a+3) - 35 = [(a+3)-7][(a+3)+5] = (a-4)(a+8)$

We could also do this exercise as follows:

$$(a+3)^2 - 2(a+3) - 35 = a^2 + 6a + 9 - 2a - 6 - 35$$
$$= a^2 + 4a - 32$$
$$= (a-4)(a+8)$$

33.

$$20x(x+2)(x-1) = 5x^3 - 24x - 14x^2$$
$$(20x^2+40x)(x-1) = 5x^3 - 24x - 14x^2$$
$$20x^3 + 20x^2 - 40x = 5x^3 - 24x - 14x^2$$
$$15x^3 + 34x^2 - 16x = 0$$
$$x(15x^2 + 34x - 16) = 0$$
$$x(3x+8)(5x-2) = 0$$

$$\begin{aligned} x = 0 \quad &or \quad 3x + 8 = 0 \quad &or \quad 5x - 2 = 0 \\ x = 0 \quad &or \quad 3x = -8 \quad &or \quad 5x = 2 \\ x = 0 \quad &or \quad x = -\frac{8}{3} \quad &or \quad x = \frac{2}{5} \end{aligned}$$

The solutions are 0, $-\frac{8}{3}$, and $\frac{2}{5}$.

34. $x^2 - y^2 = (x+y)(x-y) = 4 \cdot 6 = 24$, so choice D is correct.

Cumulative Review Chapters 1 - 5

1. $\frac{2}{3} = \frac{2}{3} \cdot \frac{7}{7} = \frac{14}{21}$

$\frac{5}{7} = \frac{5}{7} \cdot \frac{3}{3} = \frac{15}{21}$

Since $\frac{14}{21}$ is to the left of $\frac{15}{21}$ on the number line,

we have $\frac{14}{21} < \frac{15}{21}$, or $\frac{2}{3} < \frac{5}{7}$.

2. $-\frac{4}{7} = -\frac{4}{7} \cdot \frac{11}{11} = -\frac{44}{77}$

$-\frac{8}{11} = -\frac{8}{11} \cdot \frac{7}{7} = -\frac{56}{77}$

Since $-\frac{44}{77}$ is to the right of $-\frac{56}{77}$ on the number line, we have $-\frac{44}{77} > -\frac{56}{77}$, or $-\frac{4}{7} > -\frac{8}{11}$.

3. $$\begin{aligned} 2.06 + (-4.79) - (-3.08) &= 2.06 + (-4.79) + 3.08 \\ &= -2.73 + 3.08 \\ &= 0.35 \end{aligned}$$

4. We do the long division $5.652 \div 3.6$ and make the answer negative.

$$\begin{array}{r} 1.5\,7 \\ 3.6_{\wedge}\overline{)\,5.6_{\wedge}5\,2} \\ \underline{3\,6} \\ 2\,0\,5 \\ \underline{1\,8\,0} \\ 2\,5\,2 \\ \underline{2\,5\,2} \\ 0 \end{array}$$

We have $5.652 \div 3.6 = 1.57$, so $5.652 \div (-3.6) = -1.57$.

5. $$\begin{aligned} \left(\frac{2}{9}\right)\left(-\frac{3}{8}\right)\left(\frac{6}{7}\right) &= -\frac{2 \cdot 3 \cdot 6}{9 \cdot 8 \cdot 7} = -\frac{2 \cdot 3 \cdot 2 \cdot 3 \cdot 1}{3 \cdot 3 \cdot 2 \cdot 2 \cdot 2 \cdot 7} \\ &= -\frac{1}{2 \cdot 7} \cdot \frac{2 \cdot 3 \cdot 2 \cdot 3}{3 \cdot 3 \cdot 2 \cdot 2} \\ &= -\frac{1}{14} \end{aligned}$$

6. $$\begin{aligned} \frac{21}{5} \div \left(-\frac{7}{2}\right) &= \frac{21}{5} \cdot \left(-\frac{2}{7}\right) = -\frac{21 \cdot 2}{5 \cdot 7} \\ &= -\frac{3 \cdot 7 \cdot 2}{5 \cdot 7} = -\frac{3 \cdot 2}{5} \cdot \frac{7}{7} \\ &= -\frac{6}{5} \end{aligned}$$

7. $$\begin{aligned} &[3x + 2(x-1)] - [2x - (x+3)] \\ &= [3x + 2x - 2] - [2x - x - 3] \\ &= [5x - 2] - [x - 3] \\ &= 5x - 2 - x + 3 \\ &= 4x + 1 \end{aligned}$$

8. $$\begin{aligned} &1 - [14 + 28 \div 7 - (6 + 9 \div 3)] \\ &= 1 - [14 + 4 - (6 + 3)] \\ &= 1 - [14 + 4 - 9] \\ &= 1 - [9] \\ &= -8 \end{aligned}$$

9. $(2x^2y^{-1})^3 = 2^3(x^2)^3(y^{-1})^3 = 8x^6y^{-3} = \frac{8x^6}{y^3}$

10. $$\begin{aligned} \frac{3x^5}{4x^3} \cdot \frac{-2x^{-3}}{9x^2} &= -\frac{3x^5 \cdot 2x^{-3}}{4x^3 \cdot 9x^2} = -\frac{6x^2}{36x^5} \\ &= -\frac{6 \cdot x^2 \cdot 1}{6 \cdot 6 \cdot x^2 \cdot x^3} = -\frac{1}{6 \cdot x^3} \cdot \frac{6 \cdot x^2}{6 \cdot x^2} \\ &= -\frac{1}{6x^3} \end{aligned}$$

11. $$\begin{aligned} &(2x^2 - 3x^3 + x - 4) + (x^4 - x - 5x^2) \\ &= x^4 - 3x^3 + 2x^2 - 5x^2 + x - x - 4 \\ &= x^4 - 3x^3 - 3x^2 - 4 \end{aligned}$$

12. $$\begin{aligned} &(2x^2y^2 + xy - 2xy^2) - (2xy - 2xy^2 + x^2y) \\ &= 2x^2y^2 + xy - 2xy^2 - 2xy + 2xy^2 - x^2y \\ &= 2x^2y^2 + xy - 2xy - 2xy^2 + 2xy^2 - x^2y \\ &= 2x^2y^2 - xy - x^2y \end{aligned}$$

13. $$\begin{array}{r} x^2 + 3x + 2 \\ x-1\,\overline{)\,x^3 + 2x^2 - x + 1} \\ \underline{x^3 - x^2 } \\ 3x^2 - x \\ \underline{3x^2 - 3x } \\ 2x + 1 \\ \underline{2x - 2} \\ 3 \end{array}$$

The answer is $x^2 + 3x + 2 + \frac{3}{x-1}$.

14. $(2t-3)^2 = (2t)^2 - 2 \cdot 2t \cdot 3 + 3^2 = 4t^2 - 12t + 9$

15. $(x^2 - 3)(x^2 + 3) = (x^2)^2 - 3^2 = x^4 - 9$

16. We will use FOIL.

$$\begin{aligned} (2x+4)(3x-4) &= 2x \cdot 3x - 2x \cdot 4 + 4 \cdot 3x - 4 \cdot 4 \\ &= 6x^2 - 8x + 12x - 16 \\ &= 6x^2 + 4x - 16 \end{aligned}$$

17. $$\begin{aligned} 2x(x^3 + 3x^2 + 4x) &= 2x \cdot x^3 + 2x \cdot 3x^2 + 2x \cdot 4x \\ &= 2x^4 + 6x^3 + 8x^2 \end{aligned}$$

18. $$\begin{array}{r} 2y^2 + 3y + 4 \\ \underline{2y - 1} \\ -\,2y^2 - 3y - 4 \\ \underline{4y^3 + 6y^2 + 8y } \\ 4y^3 + 4y^2 + 5y - 4 \end{array}$$

19. $\left(x+\dfrac{2}{3}\right)\left(x-\dfrac{2}{3}\right)=x^2-\left(\dfrac{2}{3}\right)^2=x^2-\dfrac{4}{9}$

20. x^2+2x-8

Look for a pair of factors of -8 whose sum is 2. The numbers we want are 4 and -2.

$x^2+2x-8=(x+4)(x-2)$

21. $4x^2-25$

$=(2x)^2-5^2$ Difference of squares

$=(2x+5)(2x-5)$

22. $3x^3-4x^2+3x-4$

$=x^2(3x-4)+(3x-4)$

$=(3x-4)(x^2+1)$

23. $x^2-26x+169$

$=x^2-2\cdot x\cdot 13+13^2$ Trinomial square

$=(x-13)^2$

24. $75x^2-108y^2$

$=3(25x^2-36y^2)$ Factoring out the common factor

$=3[(5x)^2-(6y)^2]$ Difference of squares

$=3(5x+6y)(5x-6y)$

25. $6x^2-13x-63$

We will use the FOIL method.

(1) There is no common factor (other than 1 or -1).

(2) We can factor $6x^2$ as

$(6x+\quad)(x+\quad)$ or $(3x+\quad)(2x+\quad)$.

(3) Factor -63. The possibilities are -1, 63 and 1, -63 and -3, 21 and 3, -21 and -7, 9 and 7, -9.

These factors can also be written as 63, -1, and -63, 1 and 21, -3 and -21, 3 and 9, -7 and -9, 7.

(4) Look for combinations of factors from steps (2) and (3) such that the sum of their products is the middle term, $-13x$.

The factorization is $(3x+7)(2x-9)$.

26. x^4-2x^2-3

Look for a pair of factors of -3 whose sum is -2. The numbers we want are -3 and 1.

$x^4-2x^2-3=(x^2-3)(x^2+1)$

27. $4y^3-6y^2-4y+6=2(2y^3-3y^2-2y+3)$

$=2[y^2(2y-3)-(2y-3)]$

$=2(2y-3)(y^2-1)$

$=2(2y-3)(y+1)(y-1)$

28. $6p^2+pq-q^2$

We will use the ac-method.

(1) There is no common factor (other than 1 or -1).

(2) Multiply 6 and $-q^2$: $6(-q^2)=-6q^2$

(3) Find a factorization of $-6q^2$ in which the sum of the factors is q. The factors we want are $3q$ and $-2q$.

(4) Split the middle term: $pq=3pq-2pq$.

(5) Factor by grouping.

$$\begin{aligned}6p^2+pq-q^2&=6p^2+3pq-2pq-q^2\\&=3p(2p+q)-q(2p+q)\\&=(2p+q)(3p-q)\end{aligned}$$

29. $10x^3+52x^2+10x$

We will use the FOIL method.

(1) Factor out the common factor.

$10x^3+52x^2+10x=2x(5x^2+26x+5)$

Now factor $5x^2+26x+5$.

(2) We can factor $5x^2+26x+5$ as

$(5x+\quad)(x+\quad)$.

(3) The constant term and the middle term are both positive, so we find pairs of positive factors of 5. The only possibility is 1, 5. We can also write this as 5, 1.

(4) Look for combinations of factors from steps (2) and (3) such that the sum of their products is the middle term, $26x$.

The factorization of $5x^2+26x+5$ is $(5x+1)(x+5)$, so $10x^3+52x^5+10x=2x(5x+1)(x+5)$.

30. $49x^3-42x^2+9x$

$=x(49x^2-42x+9)$

$=x[(7x)^2-2\cdot 7x\cdot 3+3^2]$ Trinomial square

$=x(7x-3)^2$

31. $3x^2+5x-4$

We can try to factor this trinomial using the FOIL method or the ac-method. We find that it is prime.

32. $75x^3+27x=3x(25x^2+9)$

The binomial $25x^2+9$ cannot be factored, so the factorization is complete.

33. $3x^8-48y^8$

$=3(x^8-16y^8)$ Difference of squares

$=3(x^4+4y^4)(x^4-4y^4)$ Difference of squares

$=3(x^4+4y^4)(x^2+2y^2)(x^2-2y^2)$

34. $$\begin{aligned}14x^2+28+42x&=14x^2+42x+28\\&=14(x^2+3x+2)\end{aligned}$$

To factor x^2+3x+2, look for a pair of factors of 2 whose sum is 3. The numbers we want are 1 and 2.

Then $x^2+3x+2=(x+1)(x+2)$, so $14x^2+28+42x=14(x+1)(x+2)$.

35. $$\begin{aligned}2x^5-2x^3+x^2-1&=2x^3(x^2-1)+(x^2-1)\\&=(x^2-1)(2x^3+1)\\&=(x+1)(x-1)(2x^3+1)\end{aligned}$$

36. $3x - 5 = 2x + 10$

$x - 5 = 10$ Subtracting $2x$

$x = 15$ Adding 5

The solution is 15.

37. $3y + 4 > 5y - 8$

$-2y + 4 > -8$ Subtracting $5y$

$-2y > -12$ Subtracting 4

$y < 6$ Dividing by -2 and reversing the inequality symbol

The solution set is $\{y|y < 6\}$.

38. $(x - 15)\left(x + \frac{1}{4}\right) = 0$

$x - 15 = 0$ *or* $x + \frac{1}{4} = 0$

$x = 15$ *or* $x = -\frac{1}{4}$

The solutions are 15 and $-\frac{1}{4}$.

39. $-98x(x + 37) = 0$

$-98x = 0$ *or* $x + 37 = 0$

$x = 0$ *or* $x = -37$

The solutions are 0 and -37.

40. $x^3 + x^2 = 25x + 25$

$x^3 + x^2 - 25x - 25 = 0$

$x^2(x + 1) - 25(x + 1) = 0$

$(x + 1)(x^2 - 25) = 0$

$(x + 1)(x + 5)(x - 5) = 0$

$x + 1 = 0$ *or* $x + 5 = 0$ *or* $x - 5 = 0$

$x = -1$ *or* $x = -5$ *or* $x = 5$

The solutions are -1, -5 and 5.

41. $2x^2 = 72$

$2x^2 - 72 = 0$

$2(x^2 - 36) = 0$

$2(x + 6)(x - 6) = 0$

$x + 6 = 0$ *or* $x - 6 = 0$

$x = -6$ *or* $x = 6$

The solutions are -6 and 6.

42. $9x^2 + 1 = 6x$

$9x^2 - 6x + 1 = 0$

$(3x - 1)(3x - 1) = 0$

$3x - 1 = 0$ *or* $3x - 1 = 0$

$3x = 1$ *or* $3x = 1$

$x = \frac{1}{3}$ *or* $x = \frac{1}{3}$

The solution $\frac{1}{3}$.

43. $x^2 + 17x + 70 = 0$

$(x + 7)(x + 10) = 0$

$x + 7 = 0$ *or* $x + 10 = 0$

$x = -7$ *or* $x = -10$

The solutions are -7 and -10.

44. $14y^2 = 21y$

$14y^2 - 21y = 0$

$7y(2y - 3) = 0$

$7y = 0$ *or* $2y - 3 = 0$

$y = 0$ *or* $2y = 3$

$y = 0$ *or* $y = \frac{3}{2}$

The solutions are 0 and $\frac{3}{2}$.

45. $1.6 - 3.5x = 0.9$

$10(1.6 - 3.5x) = 10(0.9)$ Clearing decimals

$16 - 35x = 9$

$-35x = -7$

$x = \frac{1}{5}$, or 0.2

The solution is $\frac{1}{5}$, or 0.2.

46. $(x + 3)(x - 4) = 8$

$x^2 - x - 12 = 8$

$x^2 - x - 20 = 0$

$(x - 5)(x + 4) = 0$

$x - 5 = 0$ *or* $x + 4 = 0$

$x = 5$ *or* $x = -4$

The solutions are 5 and -4.

47. $1.5x - 3.6 \leq 1.3x + 0.4$

$10(1.5x - 3.6) \leq 10(1.3x + 0.4)$ Clearing decimals

$15x - 36 \leq 13x + 4$

$2x \leq 40$ Subtracting $13x$ and adding 36

$x \leq 20$

The solution set is $\{x|x \leq 20\}$.

48. $2x - [3x - (2x + 3)] = 3x + [4 - (2x + 1)]$

$2x - [3x - 2x - 3] = 3x + [4 - 2x - 1]$

$2x - [x - 3] = 3x + [3 - 2x]$

$2x - x + 3 = 3x + 3 - 2x$

$x + 3 = x + 3$

$3 = 3$ Subtracting x

We get an equation that is true for all real numbers. Thus, all real numbers are solutions of the original equation.

49. $y = mx + b$

$y - b = mx$ Subtracting b

$\frac{y - b}{m} = x$ Dividing by m

50. ***Familiarize.*** Let x and $x+2$ represent the integers.

Translate.

$$\underbrace{\text{The sum of the integers}}_{\downarrow} \underset{\downarrow}{\text{is}} \underset{\downarrow}{102.}$$
$$x + (x+2) = 102$$

Solve.

$$x + (x+2) = 102$$
$$2x + 2 = 102$$
$$2x = 100$$
$$x = 50$$

If $x = 50$, then $x + 2 = 50 + 2 = 52$.

Check. 50 and 52 are consecutive even integers and their sum is $50 + 52$, or 102. The answer checks.

State. The integers are 50 and 52.

51. ***Familiarize.*** Let x and $x+2$ represent the integers.

Translate.

$$\underbrace{\text{The product of the integers}}_{\downarrow} \underset{\downarrow}{\text{is}} \underset{\downarrow}{360.}$$
$$x(x+2) = 360$$

Solve.

$$x(x+2) = 360$$
$$x^2 + 2x = 360$$
$$x^2 + 2x - 360 = 0$$
$$(x+20)(x-18) = 0$$
$$x + 20 = 0 \quad or \quad x - 18 = 0$$
$$x = -20 \quad or \quad x = 18$$

If $x = -20$, then $x + 2 = -20 + 2 = -18$.

If $x = 18$, then $x + 2 = 18 + 2 = 20$.

Check. -20 and -18 are consecutive even integers and $-20(-18) = 360$. Also, 18 and 20 are consecutive even integers and $18 \cdot 20 = 360$. Both pairs of integers check.

State. The integers are -20 and -18 or 18 and 20.

52. ***Familiarize.*** Let h = the height of the window, in feet. Then $h + 3$ = the length.

Translate. We multiply the length and the height (or width) to find the area.

$$(h+3)h = 18$$

Solve.

$$(h+3)h = 18$$
$$h^2 + 3h = 18$$
$$h^3 + 3h - 18 = 0$$
$$(h+6)(h-3) = 0$$
$$h + 6 = 0 \quad or \quad h - 3 = 0$$
$$h = -6 \quad or \quad h = 3$$

Check. Since the height of the window cannot be negative, we check only 3. If $h = 3$, then the length is $3 + 3$, or 6. We see that the length is 3 ft more than the height. The area is 6 ft $\cdot$ 3 ft, or 18 ft^2. The answer checks.

State. The length of the window is 6 ft, and the height is 3 ft.

53. ***Familiarize.*** Let w = the width of the lot, in meters. Then $w + 200$ = the length.

Translate. We use the formula $P = 2l + 2w$.

$$1000 = 2(w + 200) + 2w$$

Solve.

$$1000 = 2(w + 200) + 2w$$
$$1000 = 2w + 400 + 2w$$
$$1000 = 4w + 400$$
$$600 = 4w$$
$$150 = w$$

If $w = 150$, then $w + 200 = 150 + 200 = 350$.

Check. 350 m is 200 m longer than 150 m, and $2 \cdot 350 + 2 \cdot 150 = 700 + 300 = 1000$ m. The answer checks.

State. The lot is 350 m by 150 m.

54. ***Familiarize.*** Let b = the amount that was borrowed. The simple interest after 1 year is $12\% \cdot b$, or $0.12b$.

Translate.

$$\underbrace{\text{Amount borrowed}}_{\downarrow} \underset{\downarrow}{\text{plus}} \underset{\downarrow}{\text{interest}} \underset{\downarrow}{\text{is}} \underbrace{\text{amount to pay off loan.}}_{\downarrow}$$
$$b \quad + \quad 0.12b \quad = \quad 7280$$

Solve.

$$b + 0.12b = 7280$$
$$1.12b = 7280 \quad (b = 1 \cdot b)$$
$$b = 6500 \quad \text{Dividing by 1.12}$$

Check. 12% of \$6500 is \$780, and \$6500 + \$780 = \$7280. The answer checks.

State. \$6500 was borrowed.

55. ***Familiarize.*** Let h = the length of the hypotenuse, in meters. Then the length of one leg of the triangle is $h - 9$.

Translate. We use the Pythagorean theorem, $a^2 + b^2 = c^2$, substituting 15 for a, $h - 9$ for b, and h for c.

$$15^2 + (h-9)^2 = h^2$$

Solve.

$$15^2 + (h-9)^2 = h^2$$
$$225 + h^2 - 18h + 81 = h^2$$
$$225 - 18h + 81 = 0 \quad \text{Subtracting } h^2$$
$$306 - 18h = 0$$
$$306 = 18h$$
$$17 = h$$

Check. If $h = 17$, then $h - 9 = 17 - 9 = 8$. We have $15^2 + 8^2 = 225 + 64 = 289 = 17^2$. The answer checks.

State. The length of the hypotenuse is 17 m.

56. ***Familiarize.*** Let x = the length of the first piece of wire, in meters. Then $2x$ = the length of the second piece and $\frac{1}{3}x$ = the length of the third piece.

Translate.

First piece	plus	Second piece	plus	Third piece	is	Total length.
↓	↓	↓	↓	↓	↓	↓
x	$+$	$2x$	$+$	$\frac{1}{3}x$	$=$	100

Solve.

$$x + 2x + \frac{1}{3}x = 100$$
$$\frac{10}{3}x = 100$$
$$x = \frac{3}{10} \cdot 100 = \frac{300}{10} = 30$$

If $x = 30$, then $2x = 2 \cdot 30 = 60$ and $\frac{1}{3}x = \frac{1}{3} \cdot 30 = 10$.

Check. The length of the second piece of wire, 60 m, is twice 30 m, the length of the first piece. Also, the length of the third piece, 10 m, is $\frac{1}{3}$ of 30 m. The sum of the length is $30 + 60 + 10 = 100$ m. The answer checks.

State. The lengths of the first, second, and third pieces of wire are 30 m, 60 m, and 10 m, respectively.

57. ***Familiarize.*** Let p = the price of the shoes before the reduction. Then the price after the reduction is $p - 25\%$ of p, or $p - 0.25p$, or $0.75p$.

Translate.

Price after reduction	is	\$21.75.
↓	↓	↓
$0.75p$	$=$	21.75

Solve.

$$0.75p = 21.75$$
$$p = \frac{21.75}{0.75} = 29$$

Check. 25% of \$29 is 0.25(\$29), or \$7.25, and \$29 − \$7.25 = \$21.75, the reduced price. The answer checks.

State. Before the reduction, the price of the shoes was \$29.

58. ***Familiarize.*** Let x = the width of the base, in ft. Then the height is also x.

Translate.

$$A = \frac{1}{2}bh$$
$$98 = \frac{1}{2} \cdot x \cdot x$$

Solve.

$$98 = \frac{1}{2} \cdot x \cdot x$$
$$196 = x^2$$
$$0 = x^2 - 196$$
$$0 = (x + 14)(x - 14)$$
$$x + 14 = 0 \quad or \quad x - 14 = 0$$
$$x = -14 \quad or \quad x = 14$$

Check. The lengths must be positive, so −14 cannot be a solution. When the base and height are each 14, we have $\frac{1}{2} \cdot 14 \cdot 14 = 98$, the area. The answer checks.

State. The height and the base are each 14 ft.

59. $3x + 4y = -12$

To find the x-intercept, substitute 0 for y and solve for x.

$$3x + 4 \cdot 0 = -12$$
$$3x = -12$$
$$x = -4$$

The x-intercept is $(-4, 0)$.

To find the y-intercept, substitute 0 for x and solve for y.

$$3 \cdot 0 + 4y = -12$$
$$4y = -12$$
$$y = -3$$

The y-intercept is $(0, -3)$.

Find a third point as a check. Let $x = 4$.

$$3 \cdot 4 + 4y = -12$$
$$12 + 4y = -12$$
$$4y = -24$$
$$y = -6$$

A third point is $(4, -6)$. Plot this point and the intercepts and draw the graph.

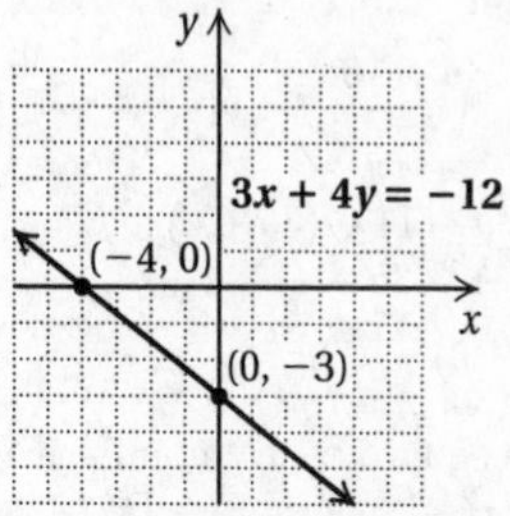

60. $(x + 3)(x - 5) \leq (x + 2)(x - 1)$

$$x^2 - 2x - 15 \leq x^2 + x - 2$$
$$-3x - 15 \leq -2 \quad \text{Subtracting } x^2 \text{ and } x$$
$$-3x \leq 13$$
$$x \geq -\frac{13}{3} \quad \text{Dividing by } -3 \text{ and reversing the inequality symbol}$$

The solution set is $\left\{x \middle| x \geq -\frac{13}{3}\right\}$.

61.
$$\frac{x-3}{2}-\frac{2x+5}{26}=\frac{4x+11}{13},\ \text{LCD is 26}$$
$$26\left(\frac{x-3}{2}-\frac{2x+5}{26}\right)=26\cdot\frac{4x+11}{13}$$
$$26\cdot\frac{x-3}{2}-26\cdot\frac{2x+5}{26}=2(4x+11)$$
$$13(x-3)-(2x+5)=8x+22$$
$$13x-39-2x-5=8x+22$$
$$11x-44=8x+22$$
$$3x=66$$
$$x=22$$

The solution is 22.

62.
$$(x+1)^2=25$$
$$x^2+2x+1=25$$
$$x^2+2x-24=0$$
$$(x+6)(x-4)=0$$
$$x+6=0\quad or\quad x-4=0$$
$$x=-6\quad or\quad x=4$$

The solutions are −6 and 4.

63.
$$x^2(x-3)-x(x-3)-2(x-3)$$
$$=(x-3)(x^2-x-2)$$
$$=(x-3)(x-2)(x+1)$$

64.
$$4a^2-4a+1-9b^2-24b-16$$
$$=4a^2-4a+1-(9b^2+24b+16)$$
$$=(2a-1)^2-(3b+4)^2$$
$$=[(2a-1)+(3b+4)][(2a-1)-(3b+4)]$$
$$=(2a-1+3b+4)(2a-1-3b-4)$$
$$=(2a+3b+3)(2a-3b-5)$$

65. $cx^2-40x+16$ will be the square of a binomial if it is of the form $(Ax-B)^2$, or $A^2x^2-2\cdot Ax\cdot B+B^2$. Then
$$-40x=-2\cdot Ax\cdot 4$$
$$-40x=-8Ax$$
$$5=A$$

Then we have $(5x-4)^2=25x^2-40x+16$, so $c=25$.

66. ***Familiarize.*** Let $r=$ the radius of the original circle, in cm. Then $r+2=$ the radius of the new circle.

Translate.

$$\underbrace{\text{Area of new circle}}_{\downarrow\ \pi(r+2)^2}\ \ \underset{=\ 4\ \ \cdot}{\text{is 4 times}}\ \ \underbrace{\text{Area of original circle.}}_{\downarrow\ \pi r^2}$$

Solve.
$$\pi(r+2)^2=4\cdot\pi r^2$$
$$\pi(r^2+4r+4)=4\pi r^2$$
$$r^2+4r+4=4r^2\quad\text{Dividing by }\pi$$
$$0=3r^2-4r-4$$
$$0=(3r+2)(r-2)$$
$$3r+2=0\quad or\quad r-2=0$$
$$3r=-2\quad or\quad r=2$$
$$r=-\frac{2}{3}\quad or\quad r=2$$

Check. The length of the radius cannot be negative, so we check only 2. The area of the original circle is $\pi\cdot 2^2=4\pi$. The radius of the new circle is $2+2$, or 4, and the area is $\pi\cdot 4^2=16\pi$. Since $16\pi=4\cdot 4\pi$, the answer checks.

State. The length of the radius of the original circle is 2 cm.

Chapter 6

Rational Expressions and Equations

Exercise Set 6.1

RC1. Expressions that have the same value for all allowable replacements are called equivalent expressions.

RC3. A rational expression can be written as a quotient of two polynomials.

1. $\frac{-3}{2x}$

To determine the numbers for which the rational expression is not defined, we set the denominator equal to 0 and solve:

$$2x = 0$$
$$x = 0$$

The expression is not defined for the replacement number 0.

3. $\frac{5}{x-8}$

To determine the numbers for which the rational expression is not defined, we set the denominator equal to 0 and solve:

$$x - 8 = 0$$
$$x = 8$$

The expression is not defined for the replacement number 8.

5. $\frac{3}{2y+5}$

Set the denominator equal to 0 and solve:

$$2y + 5 = 0$$
$$2y = -5$$
$$y = -\frac{5}{2}$$

The expression is not defined for the replacement number $-\frac{5}{2}$.

7. $\frac{x^2+11}{x^2-3x-28}$

Set the denominator equal to 0 and solve:

$$x^2 - 3x - 28 = 0$$
$$(x-7)(x+4) = 0$$
$$x - 7 = 0 \quad \text{or} \quad x + 4 = 0$$
$$x = 7 \quad \text{or} \quad x = -4$$

The expression is not defined for the replacement numbers 7 and −4.

9. $\frac{m^3-2m}{m^2-25}$

Set the denominator equal to 0 and solve:

$$m^2 - 25 = 0$$
$$(m+5)(m-5) = 0$$
$$m + 5 = 0 \quad \text{or} \quad m - 5 = 0$$
$$m = -5 \quad \text{or} \quad m = 5$$

The expression is not defined for the replacement numbers −5 and 5.

11. $\frac{x-4}{3}$

Since the denominator is the constant 3, there are no replacement numbers for which the expression is not defined.

13. $\frac{4x}{4x} \cdot \frac{3x^2}{5y} = \frac{(4x)(3x^2)}{(4x)(5y)}$ Multiplying the numerators and the denominators

15. $\frac{2x}{2x} \cdot \frac{x-1}{x+4} = \frac{2x(x-1)}{2x(x+4)}$ Multiplying the numerators and the denominators

17. $\frac{3-x}{4-x} \cdot \frac{-1}{-1} = \frac{(3-x)(-1)}{(4-x)(-1)}$

19. $\frac{y+6}{y+6} \cdot \frac{y-7}{y+2} = \frac{(y+6)(y-7)}{(y+6)(y+2)}$

21.
$$\frac{8x^3}{32x} = \frac{8 \cdot x \cdot x^2}{8 \cdot 4 \cdot x} \quad \text{Factoring numerator and denominator}$$
$$= \frac{8x}{8x} \cdot \frac{x^2}{4} \quad \text{Factoring the rational expression}$$
$$= 1 \cdot \frac{x^2}{4} \quad \left(\frac{8x}{8x} = 1\right)$$
$$= \frac{x^2}{4} \quad \text{We removed a factor of 1.}$$

23.
$$\frac{48p^7q^5}{18p^5q^4} = \frac{8 \cdot 6 \cdot p^5 \cdot p^2 \cdot q^4 \cdot q}{6 \cdot 3 \cdot p^5 \cdot q^4} \quad \text{Factoring numerator and denominator}$$
$$= \frac{6p^5q^4}{6p^5q^4} \cdot \frac{8p^2q}{3} \quad \text{Factoring the rational expression}$$
$$= 1 \cdot \frac{8p^2q}{3} \quad \left(\frac{6p^5q^4}{6p^5q^4} = 1\right)$$
$$= \frac{8p^2q}{3} \quad \text{Removing a factor of 1}$$

25.
$$\frac{4x-12}{4x} = \frac{4(x-3)}{4 \cdot x}$$
$$= \frac{4}{4} \cdot \frac{x-3}{x}$$
$$= 1 \cdot \frac{x-3}{x}$$
$$= \frac{x-3}{x}$$

27. $$\frac{3m^2+3m}{6m^2+9m}=\frac{3m(m+1)}{3m(2m+3)}=\frac{3m}{3m}\cdot\frac{m+1}{2m+3}=1\cdot\frac{m+1}{2m+3}=\frac{m+1}{2m+3}$$

29. $$\frac{a^2-9}{a^2+5a+6}=\frac{(a-3)(a+3)}{(a+2)(a+3)}=\frac{a-3}{a+2}\cdot\frac{a+3}{a+3}=\frac{a-3}{a+2}\cdot 1=\frac{a-3}{a+2}$$

31. $$\frac{a^2-10a+21}{a^2-11a+28}=\frac{(a-7)(a-3)}{(a-7)(a-4)}=\frac{a-7}{a-7}\cdot\frac{a-3}{a-4}=1\cdot\frac{a-3}{a-4}=\frac{a-3}{a-4}$$

33. $$\frac{x^2-25}{x^2-10x+25}=\frac{(x-5)(x+5)}{(x-5)(x-5)}=\frac{x-5}{x-5}\cdot\frac{x+5}{x-5}=1\cdot\frac{x+5}{x-5}=\frac{x+5}{x-5}$$

35. $$\frac{a^2-1}{a-1}=\frac{(a-1)(a+1)}{a-1}=\frac{a-1}{a-1}\cdot\frac{a+1}{1}=1\cdot\frac{a+1}{1}=a+1$$

37. $\frac{x^2+1}{x+1}$ cannot be simplified.

Neither the numerator nor the denominator can be factored.

39. $$\frac{6x^2-54}{4x^2-36}=\frac{2\cdot 3(x^2-9)}{2\cdot 2(x^2-9)}=\frac{2(x^2-9)}{2(x^2-9)}\cdot\frac{3}{2}=1\cdot\frac{3}{2}=\frac{3}{2}$$

41. $$\frac{6t+12}{t^2-t-6}=\frac{6(t+2)}{(t-3)(t+2)}=\frac{6}{t-3}\cdot\frac{t+2}{t+2}=\frac{6}{t-3}\cdot 1=\frac{6}{t-3}$$

43. $$\frac{2t^2+6t+4}{4t^2-12t-16}=\frac{2(t^2+3t+2)}{4(t^2-3t-4)}=\frac{2(t+2)(t+1)}{2\cdot 2(t-4)(t+1)}=\frac{2(t+1)}{2(t+1)}\cdot\frac{t+2}{2(t-4)}=1\cdot\frac{t+2}{2(t-4)}=\frac{t+2}{2(t-4)}$$

45. $$\frac{t^2-4}{(t+2)^2}=\frac{(t-2)(t+2)}{(t+2)(t+2)}=\frac{t-2}{t+2}\cdot\frac{t+2}{t+2}=\frac{t-2}{t+2}\cdot 1=\frac{t-2}{t+2}$$

47. $$\frac{6-x}{x-6}=\frac{-(-6+x)}{x-6}=\frac{-1(x-6)}{x-6}=-1\cdot\frac{x-6}{x-6}=-1\cdot 1=-1$$

49. $$\frac{a-b}{b-a}=\frac{-(-a+b)}{b-a}=\frac{-1(b-a)}{b-a}=-1\cdot\frac{b-a}{b-a}=-1\cdot 1=-1$$

51. $$\frac{6t-12}{2-t}=\frac{-6(-t+2)}{2-t}=\frac{-6(2-t)}{2-t}=\frac{-6(2-t)}{2-t}=-6$$

53. $$\frac{x^2-1}{1-x} = \frac{(x+1)(x-1)}{-1(-1+x)}$$
$$= \frac{(x+1)(x-1)}{-1(x-1)}$$
$$= \frac{(x+1)\cancel{(x-1)}}{-1\cancel{(x-1)}}$$
$$= -(x+1)$$
$$= -x-1$$

55. $$\frac{6qt-3t^4}{t^3-2q} = \frac{3t(2q-t^3)}{t^3-2q}$$
$$= \frac{3t(2q-t^3)}{-1(-t^3+2q)}$$
$$= \frac{3t(2q-t^3)}{-1(2q-t^3)}$$
$$= \frac{3t}{-1}\cdot\frac{2q-t^3}{2q-t^3}$$
$$= -3t$$

57. $\frac{4x^3}{3x}\cdot\frac{14}{x} = \frac{4x^3\cdot 14}{3x\cdot x}$ Multiplying the numerators and the denominators

$= \frac{4\cdot x\cdot x\cdot x\cdot 14}{3\cdot x\cdot x}$ Factoring the numerator and the denominator

$= \frac{4\cdot \cancel{x}\cdot \cancel{x}\cdot x\cdot 14}{3\cdot \cancel{x}\cdot \cancel{x}}$ Removing a factor of 1

$= \frac{56x}{3}$ Simplifying

59. $\frac{3c}{d^2}\cdot\frac{4d}{6c^3} = \frac{3c\cdot 4d}{d^2\cdot 6c^3}$ Multiplying the numerators and the denominators

$= \frac{3\cdot c\cdot 2\cdot 2\cdot d}{d\cdot d\cdot 3\cdot 2\cdot c\cdot c\cdot c}$ Factoring the numerator and the denominator

$$= \frac{\cancel{3}\cdot \cancel{c}\cdot \cancel{2}\cdot 2\cdot \cancel{d}}{\cancel{d}\cdot d\cdot \cancel{3}\cdot \cancel{2}\cdot \cancel{c}\cdot c\cdot c}$$
$$= \frac{2}{dc^2}$$

61. $$\frac{x+4}{x}\cdot\frac{x^2-3x}{x^2+x-12} = \frac{(x+4)(x^2-3x)}{x(x^2+x-12)}$$
$$= \frac{(x+4)(x)(x-3)}{x(x+4)(x-3)}$$
$$= 1$$

63. $$\frac{a^2-9}{a^2}\cdot\frac{a^2-3a}{a^2+a-12} = \frac{(a-3)(a+3)(a)(a-3)}{a\cdot a(a+4)(a-3)}$$
$$= \frac{\cancel{(a-3)}(a+3)(\cancel{a})(a-3)}{\cancel{a}\cdot a(a+4)\cancel{(a-3)}}$$
$$= \frac{(a-3)(a+3)}{a(a+4)}$$

65. $$\frac{4a^2}{3a^2-12a+12}\cdot\frac{3a-6}{2a} = \frac{4a^2(3a-6)}{(3a^2-12a+12)2a}$$
$$= \frac{2\cdot 2\cdot a\cdot a\cdot 3\cdot (a-2)}{3\cdot (a-2)\cdot (a-2)\cdot 2\cdot a}$$
$$= \frac{\cancel{2}\cdot 2\cdot \cancel{a}\cdot a\cdot \cancel{3}\cdot \cancel{(a-2)}}{\cancel{3}\cdot \cancel{(a-2)}\cdot (a-2)\cdot \cancel{2}\cdot \cancel{a}}$$
$$= \frac{2a}{a-2}$$

67. $$\frac{t^4-16}{t^4-1}\cdot\frac{t^2+1}{t^2+4}$$
$$= \frac{(t^4-16)(t^2+1)}{(t^4-1)(t^2+4)}$$
$$= \frac{(t^2+4)(t+2)(t-2)(t^2+1)}{(t^2+1)(t+1)(t-1)(t^2+4)}$$
$$= \frac{\cancel{(t^2+4)}(t+2)(t-2)\cancel{(t^2+1)}}{\cancel{(t^2+1)}(t+1)(t-1)\cancel{(t^2+4)}}$$
$$= \frac{(t+2)(t-2)}{(t+1)(t-1)}$$

69. $$\frac{(x+4)^3}{(x+2)^3}\cdot\frac{x^2+4x+4}{x^2+8x+16}$$
$$= \frac{(x+4)^3(x^2+4x+4)}{(x+2)^3(x^2+8x+16)}$$
$$= \frac{(x+4)(x+4)(x+4)(x+2)(x+2)}{(x+2)(x+2)(x+2)(x+4)(x+4)}$$
$$= \frac{\cancel{(x+4)}\cancel{(x+4)}(x+4)\cancel{(x+2)}\cancel{(x+2)}}{\cancel{(x+2)}\cancel{(x+2)}(x+2)\cancel{(x+4)}\cancel{(x+4)}}$$
$$= \frac{x+4}{x+2}$$

71. $$\frac{5a^2-180}{10a^2-10}\cdot\frac{20a+20}{2a-12} = \frac{(5a^2-180)(20a+20)}{(10a^2-10)(2a-12)}$$
$$= \frac{5(a+6)(a-6)(2)(10)(a+1)}{10(a+1)(a-1)(2)(a-6)}$$
$$= \frac{5(a+6)\cancel{(a-6)}\cancel{(2)}\cancel{(10)}\cancel{(a+1)}}{\cancel{10}\cancel{(a+1)}(a-1)\cancel{(2)}\cancel{(a-6)}}$$
$$= \frac{5(a+6)}{a-1}$$

73. Graph: $x+y=-1$

Find the x-intercept.

$$x+0=-1$$
$$x=-1$$

The x-intercept is $(-1,0)$.

Find the y-intercept.

$$0+y=-1$$
$$y=-1$$

The y-intercept is $(0,-1)$.

Find a third point as a check. Let $x=2$.

$$2+y=-1$$
$$y=-3$$

A third point is $(2, -3)$. Plot this point and the intercepts and draw the graph.

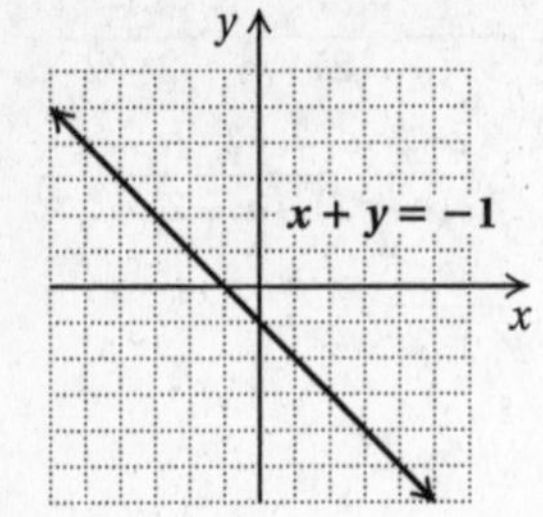

75. $x^5 - 2x^4 - 35x^3 = x^3(x^2 - 2x - 35) = x^3(x-7)(x+5)$

77. $16 - t^4 = 4^2 - (t^2)^2$ Difference of squares
$= (4+t^2)(4-t^2)$
$= (4+t^2)(2^2-t^2)$ Difference of squares
$= (4+t^2)(2+t)(2-t)$

79. $\dfrac{x^2-y^2}{(x-y)^2} \cdot \dfrac{x^2-2xy+y^2}{x^2-4xy-5y^2}$

$= \dfrac{(x+y)(x-y)(x-y)(x-y)}{(x-y)(x-y)(x-5y)(x+y)}$

$= \dfrac{\cancel{(x+y)}\cancel{(x-y)}\cancel{(x-y)}(x-y)}{\cancel{(x-y)}\cancel{(x-y)}(x-5y)\cancel{(x+y)}}$

$= \dfrac{x-y}{x-5y}$

81. $\dfrac{(x-1)}{x^2+1} \cdot \dfrac{x^4-1}{(x-1)^2} \cdot \dfrac{x^2-1}{x^4-2x^2+1} =$

$\dfrac{(x-1)(x^4-1)(x^2-1)}{(x^2+1)(x-1)^2(x^4-2x^2+1)} =$

$\dfrac{(x-1)(x^4-1)(x^2-1)}{(x^2+1)(x-1)^2(x^2-1)^2} =$

$\dfrac{(x-1)(x^2+1)(x^2-1)(x+1)(x-1)}{(x^2+1)(x-1)(x-1)(x^2-1)(x^2-1)} =$

$\dfrac{\cancel{(x-1)}\cancel{(x^2+1)}\cancel{(x^2-1)}\cancel{(x+1)}\cancel{(x-1)} \cdot 1}{\cancel{(x^2+1)}\cancel{(x-1)}\cancel{(x-1)}\cancel{(x^2-1)}\cancel{(x+1)}(x-1)} =$

$\dfrac{1}{x-1}$

Exercise Set 6.2

RC1. The reciprocal of $\dfrac{2}{x-2}$ is $\dfrac{x-2}{2}$ because $\dfrac{2}{x-2} \cdot \dfrac{x-2}{2} = 1$. Answer (d) is correct.

RC3. $\dfrac{2}{x} \cdot \dfrac{x}{2} = \dfrac{2 \cdot x}{x \cdot 2} = 1$; answer (a) is correct.

RC5. $\dfrac{1}{x} \div \dfrac{1}{2} = \dfrac{1}{x} \cdot \dfrac{2}{1} = \dfrac{2}{x}$; answer (b) is correct.

1. The reciprocal of $\dfrac{4}{x}$ is $\dfrac{x}{4}$ because $\dfrac{4}{x} \cdot \dfrac{x}{4} = 1$.

3. The reciprocal of $x^2 - y^2$ is $\dfrac{1}{x^2-y^2}$ because $\dfrac{x^2-y^2}{1} \cdot \dfrac{1}{x^2-y^2} = 1$.

5. The reciprocal of $\dfrac{1}{a+b}$ is $a+b$ because $\dfrac{1}{a+b} \cdot (a+b) = 1$.

7. The reciprocal of $\dfrac{x^2+2x-5}{x^2-4x+7}$ is $\dfrac{x^2-4x+7}{x^2+2x-5}$ because $\dfrac{x^2+2x-5}{x^2-4x+7} \cdot \dfrac{x^2-4x+7}{x^2+2x-5} = 1$.

9. $\dfrac{2}{5} \div \dfrac{4}{3} = \dfrac{2}{5} \cdot \dfrac{3}{4}$ Multiplying by the reciprocal of the divisor

$= \dfrac{2 \cdot 3}{5 \cdot 4}$

$= \dfrac{2 \cdot 3}{5 \cdot 2 \cdot 2}$ Factoring the denominator

$= \dfrac{\cancel{2} \cdot 3}{5 \cdot \cancel{2} \cdot 2}$ Removing a factor of 1

$= \dfrac{3}{10}$ Simplifying

11. $\dfrac{2}{x} \div \dfrac{8}{x} = \dfrac{2}{x} \cdot \dfrac{x}{8}$ Multiplying by the reciprocal of the divisor

$= \dfrac{2 \cdot x}{x \cdot 8}$

$= \dfrac{2 \cdot x \cdot 1}{x \cdot 2 \cdot 4}$ Factoring the numerator and the denominator

$= \dfrac{\cancel{2} \cdot \cancel{x} \cdot 1}{\cancel{x} \cdot \cancel{2} \cdot 4}$ Removing a factor of 1

$= \dfrac{1}{4}$ Simplifying

13. $\dfrac{a}{b^2} \div \dfrac{a^2}{b^3} = \dfrac{a}{b^2} \cdot \dfrac{b^3}{a^2}$ Multiplying by the reciprocal of the divisor

$= \dfrac{a \cdot b^3}{b^2 \cdot a^2}$

$= \dfrac{a \cdot b^2 \cdot b}{b^2 \cdot a \cdot a}$

$= \dfrac{\cancel{a} \cdot \cancel{b^2} \cdot b}{\cancel{b^2} \cdot \cancel{a} \cdot a}$

$= \dfrac{b}{a}$

15. $\dfrac{a+2}{a-3} \div \dfrac{a-1}{a+3} = \dfrac{a+2}{a-3} \cdot \dfrac{a+3}{a-1}$

$= \dfrac{(a+2)(a+3)}{(a-3)(a-1)}$

17. $$\frac{x^2-1}{x} \div \frac{x+1}{x-1} = \frac{x^2-1}{x} \cdot \frac{x-1}{x+1}$$
$$= \frac{(x^2-1)(x-1)}{x(x+1)}$$
$$= \frac{(x-1)(x+1)(x-1)}{x(x+1)}$$
$$= \frac{(x-1)\cancel{(x+1)}(x-1)}{x\cancel{(x+1)}}$$
$$= \frac{(x-1)^2}{x}$$

19. $$\frac{x+1}{6} \div \frac{x+1}{3} = \frac{x+1}{6} \cdot \frac{3}{x+1}$$
$$= \frac{(x+1)\cdot 3}{6(x+1)}$$
$$= \frac{3(x+1)}{2\cdot 3(x+1)}$$
$$= \frac{1\cdot \cancel{3}\cancel{(x+1)}}{2\cdot \cancel{3}\cancel{(x+1)}}$$
$$= \frac{1}{2}$$

21. $$\frac{5x-5}{16} \div \frac{x-1}{6} = \frac{5x-5}{16} \cdot \frac{6}{x-1}$$
$$= \frac{(5x-5)\cdot 6}{16(x-1)}$$
$$= \frac{5(x-1)\cdot 2\cdot 3}{2\cdot 8(x-1)}$$
$$= \frac{5\cancel{(x-1)}\cdot \cancel{2}\cdot 3}{\cancel{2}\cdot 8\cancel{(x-1)}}$$
$$= \frac{15}{8}$$

23. $$\frac{-6+3x}{5} \div \frac{4x-8}{25} = \frac{-6+3x}{5} \cdot \frac{25}{4x-8}$$
$$= \frac{(-6+3x)\cdot 25}{5(4x-8)}$$
$$= \frac{3(x-2)\cdot 5\cdot 5}{5\cdot 4(x-2)}$$
$$= \frac{3\cancel{(x-2)}\cdot \cancel{5}\cdot 5}{\cancel{5}\cdot 4\cancel{(x-2)}}$$
$$= \frac{15}{4}$$

25. $$\frac{a+2}{a-1} \div \frac{3a+6}{a-5} = \frac{a+2}{a-1} \cdot \frac{a-5}{3a+6}$$
$$= \frac{(a+2)(a-5)}{(a-1)(3a+6)}$$
$$= \frac{(a+2)(a-5)}{(a-1)\cdot 3\cdot (a+2)}$$
$$= \frac{\cancel{(a+2)}(a-5)}{(a-1)\cdot 3\cdot \cancel{(a+2)}}$$
$$= \frac{a-5}{3(a-1)}$$

27. $$\frac{x^2-4}{x} \div \frac{x-2}{x+2} = \frac{x^2-4}{x} \cdot \frac{x+2}{x-2}$$
$$= \frac{(x^2-4)(x+2)}{x(x-2)}$$
$$= \frac{(x-2)(x+2)(x+2)}{x(x-2)}$$
$$= \frac{\cancel{(x-2)}(x+2)(x+2)}{x\cancel{(x-2)}}$$
$$= \frac{(x+2)^2}{x}$$

29. $$\frac{x^2-9}{4x+12} \div \frac{x-3}{6} = \frac{x^2-9}{4x+12} \cdot \frac{6}{x-3}$$
$$= \frac{(x^2-9)\cdot 6}{(4x+12)(x-3)}$$
$$= \frac{(x-3)(x+3)\cdot 3\cdot 2}{2\cdot 2(x+3)(x-3)}$$
$$= \frac{\cancel{(x-3)}\cancel{(x+3)}\cdot 3\cdot \cancel{2}}{\cancel{2}\cdot 2\cancel{(x+3)}\cancel{(x-3)}}$$
$$= \frac{3}{2}$$

31. $$\frac{c^2+3c}{c^2+2c-3} \div \frac{c}{c+1} = \frac{c^2+3c}{c^2+2c-3} \cdot \frac{c+1}{c}$$
$$= \frac{(c^2+3c)(c+1)}{(c^2+2c-3)c}$$
$$= \frac{c(c+3)(c+1)}{(c+3)(c-1)c}$$
$$= \frac{\cancel{c}\cancel{(c+3)}(c+1)}{\cancel{(c+3)}(c-1)\cancel{c}}$$
$$= \frac{c+1}{c-1}$$

33. $$\frac{2y^2-7y+3}{2y^2+3y-2} \div \frac{6y^2-5y+1}{3y^2+5y-2}$$
$$= \frac{2y^2-7y+3}{2y^2+3y-2} \cdot \frac{3y^2+5y-2}{6y^2-5y+1}$$
$$= \frac{(2y^2-7y+3)(3y^2+5y-2)}{(2y^2+3y-2)(6y^2-5y+1)}$$
$$= \frac{(2y-1)(y-3)(3y-1)(y+2)}{(2y-1)(y+2)(3y-1)(2y-1)}$$
$$= \frac{\cancel{(2y-1)}(y-3)\cancel{(3y-1)}\cancel{(y+2)}}{\cancel{(2y-1)}\cancel{(y+2)}\cancel{(3y-1)}(2y-1)}$$
$$= \frac{y-3}{2y-1}$$

35. $$\frac{x^2-1}{4x+4} \div \frac{2x^2-4x+2}{8x+8} = \frac{x^2-1}{4x+4}\cdot\frac{8x+8}{2x^2-4x+2}$$
$$= \frac{(x^2-1)(8x+8)}{(4x+4)(2x^2-4x+2)}$$
$$= \frac{(x+1)(x-1)(2)(4)(x+1)}{4(x+1)(2)(x-1)(x-1)}$$
$$= \frac{\cancel{(x+1)}\cancel{(x-1)}\cancel{(2)}\cancel{(4)}(x+1)}{\cancel{4}\cancel{(x+1)}\cancel{(2)}(x-1)\cancel{(x-1)}}$$
$$= \frac{x+1}{x-1}$$

37. ***Familiarize.*** Let $s =$ Camila's score on the last test.

Translate. The average of the four scores must be at least 90. This means it must be greater than or equal to 90. We translate.

$$\frac{96+98+89+s}{4} \geq 90$$

Solve. We solve the inequality. First we multiply by 4 to clear the fraction.

$$4\left(\frac{96+98+89+s}{4}\right) \geq 4\cdot 90$$
$$96+98+89+s \geq 360$$
$$283+s \geq 360$$
$$s \geq 77 \quad \text{Subtracting 283}$$

Check. We can do a partial check by substituting a value for s less than 77 and a value for s greater than 77.

For $s = 76$: $\dfrac{96+98+89+76}{4} = 89.75 < 90$

For $s = 78$: $\dfrac{96+98+89+78}{4} = 90.25 \leq 90$

Since the average is less than 90 for a value of s less than 77 and greater than or equal to 90 for a value greater than or equal to 77, the answer is probably correct.

State. The scores on the last test that will earn Camila an A are $\{s|s \geq 77\}$.

39. $(8x^3-3x^2+7)-(8x^2+3x-5) =$
$8x^3-3x^2+7-8x^2-3x+5 =$
$8x^3-11x^2-3x+12$

41. $$(2x^{-3}y^4)^2 = 2^2(x^{-3})^2(y^4)^2$$
$$= 2^2x^{-6}y^8 \quad \text{Multiplying exponents}$$
$$= 4x^{-6}y^8 \quad (2^2 = 4)$$
$$= \frac{4y^8}{x^6} \quad \left(x^{-6} = \frac{1}{x^6}\right)$$

43. $$\left(\frac{2x^3}{y^5}\right)^2 = \frac{2^2(x^3)^2}{(y^5)^2}$$
$$= \frac{2^2x^6}{y^{10}} \quad \text{Multiplying exponents}$$
$$= \frac{4x^6}{y^{10}} \quad (2^2 = 4)$$

45. $$\frac{3a^2-5ab-12b^2}{3ab+4b^2} \div (3b^2-ab)$$
$$= \frac{3a^2-5ab-12b^2}{3ab+4b^2}\cdot\frac{1}{3b^2-ab}$$
$$= \frac{(3a+4b)(a-3b)}{b(3a+4b)\cdot b(3b-a)}$$
$$= \frac{(3a+4b)(-1)(3b-a)}{b(3a+4b)\cdot b(3b-a)}$$
$$= \frac{\cancel{(3a+4b)}(-1)\cancel{(3b-a)}}{b\cancel{(3a+4b)}\cdot b\cancel{(3b-a)}}$$
$$= -\frac{1}{b^2}$$

47. $$\frac{a^2b^2+3ab^2+2b^2}{a^2b^4+4b^4} \div (5a^2+10a)$$
$$= \frac{a^2b^2+3ab^2+2b^2}{a^2b^4+4b^4}\cdot\frac{1}{5a^2+10a}$$
$$= \frac{a^2b^2+3ab^2+2b^2}{(a^2b^4+4b^4)(5a^2+10a)}$$
$$= \frac{b^2(a^2+3a+2)}{b^4(a^2+4)(5a)(a+2)}$$
$$= \frac{b^2(a+1)(a+2)}{b^2\cdot b^2(a^2+4)(5a)(a+2)}$$
$$= \frac{\cancel{b^2}(a+1)\cancel{(a+2)}}{\cancel{b^2}\cdot b^2(a^2+4)(5a)\cancel{(a+2)}}$$
$$= \frac{a+1}{5ab^2(a^2+4)}$$

Exercise Set 6.3

RC1. To add $\frac{5}{16}+\frac{7}{24}$, we begin by finding a common denominator.

RC3. That number becomes the least common denominator of the two fractions.

RC5. The LCM is $2\cdot 2\cdot 2\cdot 2\cdot 3$.

1. $12 = 2\cdot 2\cdot 3$
$27 = 3\cdot 3\cdot 3$
$\text{LCM} = 2\cdot 2\cdot 3\cdot 3\cdot 3$, or 108

3. $8 = 2\cdot 2\cdot 2$
$9 = 3\cdot 3$
$\text{LCM} = 2\cdot 2\cdot 2\cdot 3\cdot 3$, or 72

5. $6 = 2\cdot 3$
$9 = 3\cdot 3$
$21 = 3\cdot 7$
$\text{LCM} = 2\cdot 3\cdot 3\cdot 7$, or 126

7. $24 = 2 \cdot 2 \cdot 2 \cdot 3$

$36 = 2 \cdot 2 \cdot 3 \cdot 3$

$40 = 2 \cdot 2 \cdot 2 \cdot 5$

$\text{LCM} = 2 \cdot 2 \cdot 2 \cdot 3 \cdot 3 \cdot 5$, or 360

9. $10 = 2 \cdot 5$

$100 = 2 \cdot 2 \cdot 5 \cdot 5$

$500 = 2 \cdot 2 \cdot 5 \cdot 5 \cdot 5$

$\text{LCM} = 2 \cdot 2 \cdot 5 \cdot 5 \cdot 5$, or 500

(We might have observed at the outset that both 10 and 100 are factors of 500, so the LCM is 500.)

11. $24 = 2 \cdot 2 \cdot 2 \cdot 3$

$18 = 2 \cdot 3 \cdot 3$

$\text{LCD} = 2 \cdot 2 \cdot 2 \cdot 3 \cdot 3$, or 72

$$\frac{7}{24} + \frac{11}{18} = \frac{7}{2 \cdot 2 \cdot 2 \cdot 3} \cdot \frac{3}{3} + \frac{11}{2 \cdot 3 \cdot 3} \cdot \frac{2 \cdot 2}{2 \cdot 2}$$
$$= \frac{21}{2 \cdot 2 \cdot 2 \cdot 3 \cdot 3} + \frac{44}{2 \cdot 2 \cdot 2 \cdot 3 \cdot 3}$$
$$= \frac{65}{72}$$

13. $$\frac{1}{6} + \frac{3}{40}$$
$$= \frac{1}{2 \cdot 3} + \frac{3}{2 \cdot 2 \cdot 2 \cdot 5}$$

LCD is $2 \cdot 2 \cdot 2 \cdot 3 \cdot 5$, or 120

$$= \frac{1}{2 \cdot 3} \cdot \frac{2 \cdot 2 \cdot 5}{2 \cdot 2 \cdot 5} + \frac{3}{2 \cdot 2 \cdot 2 \cdot 5} \cdot \frac{3}{3}$$
$$= \frac{20 + 9}{2 \cdot 2 \cdot 2 \cdot 3 \cdot 5}$$
$$= \frac{29}{120}$$

15. $$\frac{1}{20} + \frac{1}{30} + \frac{2}{45}$$
$$= \frac{1}{2 \cdot 2 \cdot 5} + \frac{1}{2 \cdot 3 \cdot 5} + \frac{2}{3 \cdot 3 \cdot 5}$$

LCD is $2 \cdot 2 \cdot 3 \cdot 3 \cdot 5$, or 180

$$= \frac{1}{2 \cdot 2 \cdot 5} \cdot \frac{3 \cdot 3}{3 \cdot 3} + \frac{1}{2 \cdot 3 \cdot 5} \cdot \frac{2 \cdot 3}{2 \cdot 3} + \frac{2}{3 \cdot 3 \cdot 5} \cdot \frac{2 \cdot 2}{2 \cdot 2}$$
$$= \frac{9 + 6 + 8}{2 \cdot 2 \cdot 3 \cdot 3 \cdot 5}$$
$$= \frac{23}{180}$$

17. $6x^2 = 2 \cdot 3 \cdot x \cdot x$

$12x^3 = 2 \cdot 2 \cdot 3 \cdot x \cdot x \cdot x$

$\text{LCM} = 2 \cdot 2 \cdot 3 \cdot x \cdot x \cdot x$, or $12x^3$

19. $2x^2 = 2 \cdot x \cdot x$

$6xy = 2 \cdot 3 \cdot x \cdot y$

$18y^2 = 2 \cdot 3 \cdot 3 \cdot y \cdot y$

$\text{LCM} = 2 \cdot 3 \cdot 3 \cdot x \cdot x \cdot y \cdot y$, or $18x^2y^2$

21. $2(y-3) = 2 \cdot (y-3)$

$6(y-3) = 2 \cdot 3 \cdot (y-3)$

$\text{LCM} = 2 \cdot 3 \cdot (y-3)$, or $6(y-3)$

23. t, $t+2$, $t-2$

The expressions are not factorable, so the LCM is their product:

$\text{LCM} = t(t+2)(t-2)$

25. $x^2 - 4 = (x+2)(x-2)$

$x^2 + 5x + 6 = (x+3)(x+2)$

$\text{LCM} = (x+2)(x-2)(x+3)$

27. $t^3 + 4t^2 + 4t = t(t^2 + 4t + 4) = t(t+2)(t+2)$

$t^2 - 4t = t(t-4)$

$\text{LCM} = t(t+2)(t+2)(t-4) = t(t+2)^2(t-4)$

29. $a + 1 = a + 1$

$(a-1)^2 = (a-1)(a-1)$

$a^2 - 1 = (a+1)(a-1)$

$\text{LCM} = (a+1)(a-1)(a-1) = (a+1)(a-1)^2$

31. $m^2 - 5m + 6 = (m-3)(m-2)$

$m^2 - 4m + 4 = (m-2)(m-2)$

$\text{LCM} = (m-3)(m-2)(m-2) = (m-3)(m-2)^2$

33. $2 + 3x = 2 + 3x$

$4 - 9x^2 = (2+3x)(2-3x)$

$2 - 3x = 2 - 3x$

$\text{LCM} = (2+3x)(2-3x)$

35. $10v^2 + 30v = 10v(v+3) = 2 \cdot 5 \cdot v(v+3)$

$5v^2 + 35v + 60 = 5(v^2 + 7v + 12)$

$= 5(v+4)(v+3)$

$\text{LCM} = 2 \cdot 5 \cdot v(v+3)(v+4) = 10v(v+3)(v+4)$

37. $9x^3 - 9x^2 - 18x = 9x(x^2 - x - 2)$

$= 3 \cdot 3 \cdot x(x-2)(x+1)$

$6x^5 - 24x^4 + 24x^3 = 6x^3(x^2 - 4x + 4)$

$= 2 \cdot 3 \cdot x \cdot x \cdot x(x-2)(x-2)$

$\text{LCM} = 2 \cdot 3 \cdot 3 \cdot x \cdot x \cdot x(x-2)(x-2)(x+1) =$
$18x^3(x-2)^2(x+1)$

39. $x^5 + 4x^4 + 4x^3 = x^3(x^2 + 4x + 4)$

$= x \cdot x \cdot x(x+2)(x+2)$

$3x^2 - 12 = 3(x^2 - 4) = 3(x+2)(x-2)$

$2x + 4 = 2(x+2)$

$\text{LCM} = 2 \cdot 3 \cdot x \cdot x \cdot x(x+2)(x+2)(x-2)$

$= 6x^3(x+2)^2(x-2)$

41. $24w^4 = 2 \cdot 2 \cdot 2 \cdot 3 \cdot w \cdot w \cdot w \cdot w$

$w^2 = w \cdot w$

$10w^3 = 2 \cdot 5 \cdot w \cdot w \cdot w$

$w^6 = w \cdot w \cdot w \cdot w \cdot w \cdot w$

$\text{LCM} = 2 \cdot 2 \cdot 2 \cdot 3 \cdot 5 \cdot w \cdot w \cdot w \cdot w \cdot w \cdot w = 120w^6$

43. $40x^3 = 2 \cdot 2 \cdot 2 \cdot 5 \cdot x \cdot x \cdot x$

$24x^4 = 2 \cdot 2 \cdot 2 \cdot 3 \cdot x \cdot x \cdot x \cdot x$

$\text{LCM} = 2 \cdot 2 \cdot 2 \cdot 3 \cdot 5 \cdot x \cdot x \cdot x \cdot x = 120x^4$

$\text{GCF} = 2 \cdot 2 \cdot 2 \cdot x \cdot x \cdot x = 8x^3$

$120x^4(8x^3) = 960x^7$

45. $16x^5 = 2 \cdot 2 \cdot 2 \cdot 2 \cdot x \cdot x \cdot x \cdot x \cdot x$

$48x^6 = 2 \cdot 2 \cdot 2 \cdot 2 \cdot 3 \cdot x \cdot x \cdot x \cdot x \cdot x \cdot x$

$\text{LCM} = 2 \cdot 2 \cdot 2 \cdot 2 \cdot 3 \cdot x \cdot x \cdot x \cdot x \cdot x \cdot x = 48x^6$

$\text{GCF} = 2 \cdot 2 \cdot 2 \cdot 2 \cdot x \cdot x \cdot x \cdot x \cdot x = 16x^5$

$48x^6 \cdot 16x^5 = 768x^{11}$

47. $20x^2 = 2 \cdot 2 \cdot 5 \cdot x \cdot x$

$10x = 2 \cdot 5 \cdot x$

$\text{LCM} = 2 \cdot 2 \cdot 5 \cdot x \cdot x = 20x^2$

$\text{GCF} = 2 \cdot 5 \cdot x = 10x$

$20x^2(10x) = 200x^3$

49. The time it takes Pedro and Maria to meet again at the starting place is the LCM of the times it takes them to complete one round of the course.

$6 = 2 \cdot 3$

$8 = 2 \cdot 2 \cdot 2$

$\text{LCM} = 2 \cdot 2 \cdot 2 \cdot 3$, or 24

It takes 24 min.

Exercise Set 6.4

RC1. $15 = 3 \cdot 5$

$10 = 2 \cdot 5$

The LCD is $2 \cdot 3 \cdot 5$, or 30.

$$\frac{2x^2}{15} + \frac{3x}{10} = \frac{2x^2}{15} \cdot \left(\frac{2}{2}\right) + \frac{3x}{10} \cdot \left(\frac{3}{3}\right) = \frac{4x^2}{30} + \frac{9x}{30}$$

RC3. The LCD is $(x+2)(x-3)$.

$$\frac{3x}{x+2} + \frac{2}{x-3}$$

$$= \frac{3x}{x+2} \cdot \left(\frac{x-3}{x-3}\right) + \frac{2}{x-3} \cdot \left(\frac{x+2}{x+2}\right)$$

$$= \frac{3x(x-3)}{(x+2)(x-3)} + \frac{2(x+2)}{(x-3)(x+2)}$$

1. $\frac{5}{8} + \frac{3}{8} = \frac{5+3}{8} = \frac{8}{8} = 1$

3. $\frac{1}{3+x} + \frac{5}{3+x} = \frac{1+5}{3+x} = \frac{6}{3+x}$

5. $\frac{4x+6}{2x-1} + \frac{5-8x}{-1+2x} = \frac{4x+6}{2x-1} + \frac{5-8x}{2x-1} \quad (-1+2x = 2x-1)$

$$= \frac{4x+6+5-8x}{2x-1}$$

$$= \frac{-4x+11}{2x-1}$$

7. $\frac{2}{x} + \frac{5}{x^2} = \frac{2}{x} + \frac{5}{x \cdot x} \qquad \text{LCD} = x \cdot x, \text{ or } x^2$

$$= \frac{2}{x} \cdot \frac{x}{x} + \frac{5}{x \cdot x}$$

$$= \frac{2x+5}{x^2}$$

9. $\left.\begin{array}{l} 6r = 2 \cdot 3 \cdot r \\ 8r = 2 \cdot 2 \cdot 2 \cdot r \end{array}\right\} \text{LCD} = 2 \cdot 2 \cdot 2 \cdot 3 \cdot r, \text{ or } 24r$

$$\frac{5}{6r} + \frac{7}{8r} = \frac{5}{6r} \cdot \frac{4}{4} + \frac{7}{8r} \cdot \frac{3}{3}$$

$$= \frac{20+21}{24r}$$

$$= \frac{41}{24r}$$

11. $\left.\begin{array}{l} xy^2 = x \cdot y \cdot y \\ x^2y = x \cdot x \cdot y \end{array}\right\} \text{LCD} = x \cdot x \cdot y \cdot y, \text{ or } x^2y^2$

$$\frac{4}{xy^2} + \frac{6}{x^2y} = \frac{4}{xy^2} \cdot \frac{x}{x} + \frac{6}{x^2y} \cdot \frac{y}{y}$$

$$= \frac{4x+6y}{x^2y^2} = \frac{2(2x+3y)}{x^2y^2}$$

13. $\left.\begin{array}{l} 9t^3 = 3 \cdot 3 \cdot t \cdot t \cdot t \\ 6t^2 = 2 \cdot 3 \cdot t \cdot t \end{array}\right\} \text{LCD} = 2 \cdot 3 \cdot 3 \cdot t \cdot t \cdot t, \text{ or } 18t^3$

$$\frac{2}{9t^3} + \frac{1}{6t^2} = \frac{2}{9t^3} \cdot \frac{2}{2} + \frac{1}{6t^2} \cdot \frac{3t}{3t}$$

$$= \frac{4+3t}{18t^3}$$

15. $\text{LCD} = x^2y^2$ (See Exercise 11.)

$$\frac{x+y}{xy^2} + \frac{3x+y}{x^2y} = \frac{x+y}{xy^2} \cdot \frac{x}{x} + \frac{3x+y}{x^2y} \cdot \frac{y}{y}$$

$$= \frac{x(x+y)+y(3x+y)}{x^2y^2}$$

$$= \frac{x^2+xy+3xy+y^2}{x^2y^2}$$

$$= \frac{x^2+4xy+y^2}{x^2y^2}$$

17. The denominators do not factor, so the LCD is their product, $(x-2)(x+2)$.

$$\frac{3}{x-2} + \frac{3}{x+2} = \frac{3}{x-2} \cdot \frac{x+2}{x+2} + \frac{3}{x+2} \cdot \frac{x-2}{x-2}$$

$$= \frac{3(x+2)+3(x-2)}{(x-2)(x+2)}$$

$$= \frac{3x+6+3x-6}{(x-2)(x+2)}$$

$$= \frac{6x}{(x-2)(x+2)}$$

19. $\left.\begin{array}{l}3x = 3\cdot x\\ x+1 = x+1\end{array}\right\}\text{LCD} = 3x(x+1)$

$$\frac{3}{x+1}+\frac{2}{3x} = \frac{3}{x+1}\cdot\frac{3x}{3x}+\frac{2}{3x}\cdot\frac{x+1}{x+1}$$
$$= \frac{9x+2(x+1)}{3x(x+1)}$$
$$= \frac{9x+2x+2}{3x(x+1)}$$
$$= \frac{11x+2}{3x(x+1)}$$

21. $\left.\begin{array}{l}x^2-16 = (x+4)(x-4)\\ x-4 = x-4\end{array}\right\}\text{LCD} = (x+4)(x-4)$

$$\frac{2x}{x^2-16}+\frac{x}{x-4} = \frac{2x}{(x+4)(x-4)}+\frac{x}{x-4}\cdot\frac{x+4}{x+4}$$
$$= \frac{2x+x(x+4)}{(x+4)(x-4)}$$
$$= \frac{2x+x^2+4x}{(x+4)(x-4)}$$
$$= \frac{x^2+6x}{(x+4)(x-4)}$$
$$= \frac{x(x+6)}{(x+4)(x-4)}$$

23. $\frac{5}{z+4}+\frac{3}{3z+12} = \frac{5}{z+4}+\frac{3}{3(z+4)}$ LCD $= 3(z+4)$

$$= \frac{5}{z+4}\cdot\frac{3}{3}+\frac{3}{3(z+4)}$$
$$= \frac{15+3}{3(z+4)} = \frac{18}{3(z+4)}$$
$$= \frac{3\cdot 6}{3(z+4)} = \frac{\not{3}\cdot 6}{\not{3}(z+4)}$$
$$= \frac{6}{z+4}$$

25. $\frac{3}{x-1}+\frac{2}{(x-1)^2}$ LCD $= (x-1)^2$

$$= \frac{3}{x-1}\cdot\frac{x-1}{x-1}+\frac{2}{(x-1)^2}$$
$$= \frac{3(x-1)+2}{(x-1)^2}$$
$$= \frac{3x-3+2}{(x-1)^2}$$
$$= \frac{3x-1}{(x-1)^2}$$

27. $\frac{4a}{5a-10}+\frac{3a}{10a-20} = \frac{4a}{5(a-2)}+\frac{3a}{2\cdot 5(a-2)}$

LCD $= 2\cdot 5(a-2)$

$$= \frac{4a}{5(a-2)}\cdot\frac{2}{2}+\frac{3a}{2\cdot 5(a-2)}$$
$$= \frac{8a+3a}{10(a-2)}$$
$$= \frac{11a}{10(a-2)}$$

29. $\frac{x+4}{x}+\frac{x}{x+4}$ LCD $= x(x+4)$

$$= \frac{x+4}{x}\cdot\frac{x+4}{x+4}+\frac{x}{x+4}\cdot\frac{x}{x}$$
$$= \frac{(x+4)^2+x^2}{x(x+4)}$$
$$= \frac{x^2+8x+16+x^2}{x(x+4)}$$
$$= \frac{2x^2+8x+16}{x(x+4)} = \frac{2(x^2+4x+8)}{x(x+4)}$$

31. $\frac{4}{a^2-a-2}+\frac{3}{a^2+4a+3}$

$$= \frac{4}{(a-2)(a+1)}+\frac{3}{(a+3)(a+1)}$$

LCD $= (a-2)(a+1)(a+3)$

$$= \frac{4}{(a-2)(a+1)}\cdot\frac{a+3}{a+3}+\frac{3}{(a+3)(a+1)}\cdot\frac{a-2}{a-2}$$
$$= \frac{4(a+3)+3(a-2)}{(a-2)(a+1)(a+3)}$$
$$= \frac{4a+12+3a-6}{(a-2)(a+1)(a+3)}$$
$$= \frac{7a+6}{(a-2)(a+1)(a+3)}$$

33. $\frac{x+3}{x-5}+\frac{x-5}{x+3}$ LCD $= (x-5)(x+3)$

$$= \frac{x+3}{x-5}\cdot\frac{x+3}{x+3}+\frac{x-5}{x+3}\cdot\frac{x-5}{x-5}$$
$$= \frac{(x+3)^2+(x-5)^2}{(x-5)(x+3)}$$
$$= \frac{x^2+6x+9+x^2-10x+25}{(x-5)(x+3)}$$
$$= \frac{2x^2-4x+34}{(x-5)(x+3)} = \frac{2(x^2-2x+17)}{(x-5)(x+3)}$$

35. $\frac{a}{a^2-1}+\frac{2a}{a^2-a}$

$$= \frac{a}{(a+1)(a-1)}+\frac{2a}{a(a-1)}$$

LCD $= a(a+1)(a-1)$

$$= \frac{a}{(a+1)(a-1)}\cdot\frac{a}{a}+\frac{2a}{a(a-1)}\cdot\frac{a+1}{a+1}$$
$$= \frac{a^2+2a(a+1)}{a(a+1)(a-1)} = \frac{a^2+2a^2+2a}{a(a+1)(a-1)}$$
$$= \frac{3a^2+2a}{a(a+1)(a-1)} = \frac{a(3a+2)}{a(a+1)(a-1)}$$
$$= \frac{\not{a}(3a+2)}{\not{a}(a+1)(a-1)} = \frac{3a+2}{(a+1)(a-1)}$$

37. $\frac{7}{8}+\frac{5}{-8}=\frac{7}{8}+\frac{5}{-8}\cdot\frac{-1}{-1}$
$=\frac{7}{8}+\frac{-5}{8}$
$=\frac{7+(-5)}{8}$
$=\frac{2}{8}=\frac{\cancel{2}\cdot 1}{4\cdot\cancel{2}}$
$=\frac{1}{4}$

39. $\frac{3}{t}+\frac{4}{-t}=\frac{3}{t}+\frac{4}{-t}\cdot\frac{-1}{-1}$
$=\frac{3}{t}+\frac{-4}{t}$
$=\frac{3+(-4)}{t}$
$=\frac{-1}{t}$
$=-\frac{1}{t}$

41. $\frac{2x+7}{x-6}+\frac{3x}{6-x}=\frac{2x+7}{x-6}+\frac{3x}{6-x}\cdot\frac{-1}{-1}$
$=\frac{2x+7}{x-6}+\frac{-3x}{x-6}$
$=\frac{(2x+7)+(-3x)}{x-6}$
$=\frac{-x+7}{x-6}$, or $\frac{7-x}{x-6}$, or $\frac{x-7}{6-x}$

43. $\frac{y^2}{y-3}+\frac{9}{3-y}=\frac{y^2}{y-3}+\frac{9}{3-y}\cdot\frac{-1}{-1}$
$=\frac{y^2}{y-3}+\frac{-9}{y-3}$
$=\frac{y^2+(-9)}{y-3}$
$=\frac{y^2-9}{y-3}$
$=\frac{(y+3)(y-3)}{y-3}$
$=\frac{(y+3)\cancel{(y-3)}}{1\cancel{(y-3)}}$
$=y+3$

45. $\frac{b-7}{b^2-16}+\frac{7-b}{16-b^2}=\frac{b-7}{b^2-16}+\frac{7-b}{16-b^2}\cdot\frac{-1}{-1}$
$=\frac{b-7}{b^2-16}+\frac{b-7}{b^2-16}$
$=\frac{(b-7)+(b-7)}{b^2-16}$
$=\frac{2b-14}{b^2-16}=\frac{2(b-7)}{(b+4)(b-4)}$

47. $\frac{a^2}{a-b}+\frac{b^2}{b-a}=\frac{a^2}{a-b}+\frac{b^2}{b-a}\cdot\frac{-1}{-1}$
$=\frac{a^2}{a-b}+\frac{-b^2}{a-b}$
$=\frac{a^2+(-b^2)}{a-b}$
$=\frac{a^2-b^2}{a-b}$
$=\frac{(a+b)(a-b)}{a-b}$
$=\frac{(a+b)\cancel{(a-b)}}{1\cancel{(a-b)}}$
$=a+b$

49. $\frac{x+3}{x-5}+\frac{2x-1}{5-x}+\frac{2(3x-1)}{x-5}$
$=\frac{x+3}{x-5}+\frac{2x-1}{5-x}\cdot\frac{-1}{-1}+\frac{2(3x-1)}{x-5}$
$=\frac{x+3}{x-5}+\frac{1-2x}{x-5}+\frac{2(3x-1)}{x-5}$
$=\frac{(x+3)+(1-2x)+(6x-2)}{x-5}$
$=\frac{5x+2}{x-5}$

51. $\frac{2(4x+1)}{5x-7}+\frac{3(x-2)}{7-5x}+\frac{-10x-1}{5x-7}$
$=\frac{2(4x+1)}{5x-7}+\frac{3(x-2)}{7-5x}\cdot\frac{-1}{-1}+\frac{-10x-1}{5x-7}$
$=\frac{2(4x+1)}{5x-7}+\frac{-3(x-2)}{5x-7}+\frac{-10x-1}{5x-7}$
$=\frac{(8x+2)+(-3x+6)+(-10x-1)}{5x-7}$
$=\frac{-5x+7}{5x-7}$
$=\frac{-1(5x-7)}{5x-7}$
$=\frac{-1\cancel{(5x-7)}}{\cancel{5x-7}}$
$=-1$

53. $\frac{x+1}{(x+3)(x-3)}+\frac{4(x-3)}{(x-3)(x+3)}+\frac{(x-1)(x-3)}{(3-x)(x+3)}$
$=\frac{x+1}{(x+3)(x-3)}+\frac{4(x-3)}{(x-3)(x+3)}+\frac{(x-1)(x-3)}{(3-x)(x+3)}\cdot\frac{-1}{-1}$
$=\frac{x+1}{(x+3)(x-3)}+\frac{4(x-3)}{(x-3)(x+3)}+\frac{-1(x^2-4x+3)}{(x-3)(x+3)}$
$=\frac{(x+1)+(4x-12)+(-x^2+4x-3)}{(x+3)(x-3)}$
$=\frac{-x^2+9x-14}{(x+3)(x-3)}$

55. $\dfrac{6}{x-y}+\dfrac{4x}{y^2-x^2}$

$=\dfrac{6}{x-y}+\dfrac{4x}{(y-x)(y+x)}$

$=\dfrac{6}{x-y}+\dfrac{4x}{(y-x)(y+x)}\cdot\dfrac{-1}{-1}$

$=\dfrac{6}{x-y}+\dfrac{-4x}{(x-y)(x+y)}$

$[-1(y-x)=x-y; y+x=x+y]$

$\text{LCD}=(x-y)(x+y)$

$=\dfrac{6}{x-y}\cdot\dfrac{x+y}{x+y}+\dfrac{-4x}{(x-y)(x+y)}$

$=\dfrac{6(x+y)-4x}{(x-y)(x+y)}$

$=\dfrac{6x+6y-4x}{(x-y)(x+y)}$

$=\dfrac{2x+6y}{(x-y)(x+y)}$

$=\dfrac{2(x+3y)}{(x-y)(x+y)}$

57. $\dfrac{4-a}{25-a^2}+\dfrac{a+1}{a-5}$

$=\dfrac{4-a}{25-a^2}\cdot\dfrac{-1}{-1}+\dfrac{a+1}{a-5}$

$=\dfrac{a-4}{a^2-25}+\dfrac{a+1}{a-5}$

$=\dfrac{a-4}{(a+5)(a-5)}+\dfrac{a+1}{a-5}$

$\text{LCD}=(a+5)(a-5)$

$=\dfrac{a-4}{(a+5)(a-5)}+\dfrac{a+1}{a-5}\cdot\dfrac{a+5}{a+5}$

$=\dfrac{a-4}{(a+5)(a-5)}+\dfrac{(a+1)(a+5)}{(a+5)(a-5)}$

$=\dfrac{(a-4)+(a+1)(a+5)}{(a+5)(a-5)}$

$=\dfrac{a-4+a^2+6a+5}{(a+5)(a-5)}$

$=\dfrac{a^2+7a+1}{(a+5)(a-5)}$

59. $\dfrac{2}{t^2+t-6}+\dfrac{3}{t^2-9}$

$=\dfrac{2}{(t+3)(t-2)}+\dfrac{3}{(t+3)(t-3)}$

$\text{LCD}=(t+3)(t-2)(t-3)$

$=\dfrac{2}{(t+3)(t-2)}\cdot\dfrac{t-3}{t-3}+\dfrac{3}{(t+3)(t-3)}\cdot\dfrac{t-2}{t-2}$

$=\dfrac{2(t-3)+3(t-2)}{(t+3)(t-2)(t-3)}$

$=\dfrac{2t-6+3t-6}{(t+3)(t-2)(t-3)}$

$=\dfrac{5t-12}{(t+3)(t-2)(t-3)}$

61. $(x^2+x)-(x+1)=x^2+x-x-1=x^2-1$

63. $\left(\dfrac{x^{-4}}{y^7}\right)^3=\dfrac{(x^{-4})^3}{(y^7)^3}=\dfrac{x^{-12}}{y^{21}}=\dfrac{1}{x^{12}y^{21}}$

65.
$3x-7=5x+9$
$-2x-7=9$ Subtracting $5x$
$-2x=16$ Adding 7
$x=-8$ Dividing by -2

The solution is -8.

67. $y=\dfrac{1}{2}x-5=\dfrac{1}{2}x+(-5)$

The y-intercept is $(0,-5)$. We find two other pairs.

When $x=2$, $y=\dfrac{1}{2}\cdot 2-5=1-5=-4$.

When $x=4$, $y=\dfrac{1}{2}\cdot 4-5=2-5=-3$.

x	y
0	-5
2	-4
4	-3

Plot these points, draw the line they determine, and label the graph $y=\dfrac{1}{2}x-5$.

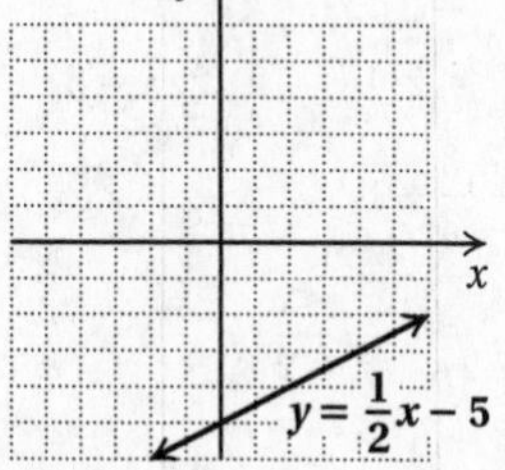

69. $y=3$

Any ordered pair $(x,3)$ is a solution. The variable y must be 3, but x can be any number we choose. A few solutions are listed below. Plot these points and draw the line.

x	y
-4	3
0	3
3	3

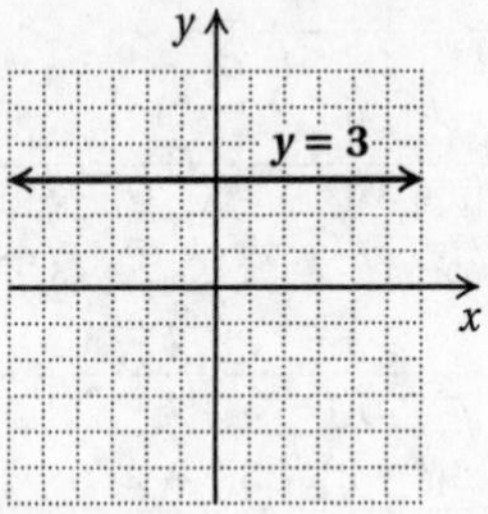

71. To find the perimeter we add the lengths of the sides:

$$\frac{y+4}{3}+\frac{y+4}{3}+\frac{y-2}{5}+\frac{y-2}{5} \quad \text{LCD} = 3\cdot 5$$

$$= \frac{y+4}{3}\cdot\frac{5}{5}+\frac{y+4}{3}\cdot\frac{5}{5}+\frac{y-2}{5}\cdot\frac{3}{3}+\frac{y-2}{5}\cdot\frac{3}{3}$$

$$= \frac{5y+20+5y+20+3y-6+3y-6}{3\cdot 5}$$

$$= \frac{16y+28}{15}$$

To find the area we multiply the length and the width:

$$\left(\frac{y+4}{3}\right)\left(\frac{y-2}{5}\right) = \frac{(y+4)(y-2)}{3\cdot 5} = \frac{y^2+2y-8}{15}$$

73. $$\frac{5}{z+2}+\frac{4z}{z^2-4}+2$$

$$= \frac{5}{z+2}+\frac{4z}{(z+2)(z-2)}+\frac{2}{1} \quad \text{LCD} = (z+2)(z-2)$$

$$= \frac{5}{z+2}\cdot\frac{z-2}{z-2}+\frac{4z}{(z+2)(z-2)}+\frac{2}{1}\cdot\frac{(z+2)(z-2)}{(z+2)(z-2)}$$

$$= \frac{5z-10+4z+2(z^2-4)}{(z+2)(z-2)}$$

$$= \frac{5z-10+4z+2z^2-8}{(z+2)(z-2)} = \frac{2z^2+9z-18}{(z+2)(z-2)}$$

$$= \frac{(2z-3)(z+6)}{(z+2)(z-2)}$$

75. $$\frac{3z^2}{z^4-4}+\frac{5z^2-3}{2z^4+z^2-6}$$

$$= \frac{3z^2}{(z^2+2)(z^2-2)}+\frac{5z^2-3}{(2z^2-3)(z^2+2)}$$

$$\text{LCD} = (z^2+2)(z^2-2)(2z^2-3)$$

$$= \frac{3z^2}{(z^2+2)(z^2-2)}\cdot\frac{2z^2-3}{2z^2-3}+ \frac{5z^2-3}{(2z^2-3)(z^2+2)}\cdot\frac{z^2-2}{z^2-2}$$

$$= \frac{6z^4-9z^2+5z^4-13z^2+6}{(z^2+2)(z^2-2)(2z^2-3)}$$

$$= \frac{11z^4-22z^2+6}{(z^2+2)(z^2-2)(2z^2-3)}$$

Exercise Set 6.5

RC1. $$\frac{10x}{x-7}-\frac{3x+5}{x-7} = \frac{10x-(3x+5)}{x-7} = \frac{10x-3x-5}{x-7} = \frac{7x-5}{x-7}$$

RC3. $$\frac{9y-2}{y^2-10}-\frac{y+1}{y^2-10} = \frac{9y-2-(y+1)}{y^2-10} = \frac{9y-2-y-1}{y^2-10} = \frac{8y-3}{y^2-10}$$

1. $$\frac{7}{x}-\frac{3}{x} = \frac{7-3}{x} = \frac{4}{x}$$

3. $$\frac{y}{y-4}-\frac{4}{y-4} = \frac{y-4}{y-4} = 1$$

5. $$\frac{2x-3}{x^2+3x-4}-\frac{x-7}{x^2+3x-4}$$

$$= \frac{2x-3-(x-7)}{x^2+3x-4}$$

$$= \frac{2x-3-x+7}{x^2+3x-4}$$

$$= \frac{x+4}{x^2+3x-4}$$

$$= \frac{x+4}{(x+4)(x-1)}$$

$$= \frac{\cancel{(x+4)}\cdot 1}{\cancel{(x+4)}(x-1)}$$

$$= \frac{1}{x-1}$$

7. $$\frac{a-2}{10}-\frac{a+1}{5} = \frac{a-2}{10}-\frac{a+1}{5}\cdot\frac{2}{2} \quad \text{LCD} = 10$$

$$= \frac{a-2}{10}-\frac{2(a+1)}{10}$$

$$= \frac{(a-2)-2(a+1)}{10}$$

$$= \frac{a-2-2a-2}{10}$$

$$= \frac{-a-4}{10}$$

9. $$\frac{4z-9}{3z}-\frac{3z-8}{4z} = \frac{4z-9}{3z}\cdot\frac{4}{4}-\frac{3z-8}{4z}\cdot\frac{3}{3}$$

$$\text{LCD} = 3\cdot 4\cdot z, \text{ or } 12z$$

$$= \frac{16z-36}{12z}-\frac{9z-24}{12z}$$

$$= \frac{16z-36-(9z-24)}{12z}$$

$$= \frac{16z-36-9z+24}{12z}$$

$$= \frac{7z-12}{12z}$$

11. $\dfrac{4x+2t}{3xt^2}-\dfrac{5x-3t}{x^2t}$ LCD $=3x^2t^2$

$$=\frac{4x+2t}{3xt^2}\cdot\frac{x}{x}-\frac{5x-3t}{x^2t}\cdot\frac{3t}{3t}$$
$$=\frac{4x^2+2tx}{3x^2t^2}-\frac{15xt-9t^2}{3x^2t^2}$$
$$=\frac{4x^2+2tx-(15xt-9t^2)}{3x^2t^2}$$
$$=\frac{4x^2+2tx-15xt+9t^2}{3x^2t^2}$$
$$=\frac{4x^2-13xt+9t^2}{3x^2t^2}$$

13. $\dfrac{5}{x+5}-\dfrac{3}{x-5}$ LCD $=(x+5)(x-5)$

$$=\frac{5}{x+5}\cdot\frac{x-5}{x-5}-\frac{3}{x-5}\cdot\frac{x+5}{x+5}$$
$$=\frac{5x-25}{(x+5)(x-5)}-\frac{3x+15}{(x+5)(x-5)}$$
$$=\frac{5x-25-(3x+15)}{(x+5)(x-5)}$$
$$=\frac{5x-25-3x-15}{(x+5)(x-5)}$$
$$=\frac{2x-40}{(x+5)(x-5)}$$
$$=\frac{2(x-20)}{(x+5)(x-5)}$$

15. $\dfrac{3}{2t^2-2t}-\dfrac{5}{2t-2}$

$$=\frac{3}{2t(t-1)}-\frac{5}{2(t-1)}\quad \text{LCD} = 2t(t-1)$$
$$=\frac{3}{2t(t-1)}-\frac{5}{2(t-1)}\cdot\frac{t}{t}$$
$$=\frac{3}{2t(t-1)}-\frac{5t}{2t(t-1)}$$
$$=\frac{3-5t}{2t(t-1)}$$

17. $\dfrac{2s}{t^2-s^2}-\dfrac{s}{t-s}$ LCD $=(t-s)(t+s)$

$$=\frac{2s}{(t-s)(t+s)}-\frac{s}{t-s}\cdot\frac{t+s}{t+s}$$
$$=\frac{2s}{(t-s)(t+s)}-\frac{st+s^2}{(t-s)(t+s)}$$
$$=\frac{2s-(st+s^2)}{(t-s)(t+s)}$$
$$=\frac{2s-st-s^2}{(t-s)(t+s)}$$
$$=\frac{s(2-t-s)}{(t-s)(t+s)}$$

19. $\dfrac{y-5}{y}-\dfrac{3y-1}{4y}=\dfrac{y-5}{y}\cdot\dfrac{4}{4}-\dfrac{3y-1}{4y}$ LCD $=4y$

$$=\frac{4y-20}{4y}-\frac{3y-1}{4y}$$
$$=\frac{4y-20-(3y-1)}{4y}$$
$$=\frac{4y-20-3y+1}{4y}$$
$$=\frac{y-19}{4y}$$

21. $\dfrac{a}{x+a}-\dfrac{a}{x-a}$ LCD $=(x+a)(x-a)$

$$=\frac{a}{x+a}\cdot\frac{x-a}{x-a}-\frac{a}{x-a}\cdot\frac{x+a}{x+a}$$
$$=\frac{ax-a^2}{(x+a)(x-a)}-\frac{ax+a^2}{(x+a)(x-a)}$$
$$=\frac{ax-a^2-(ax+a^2)}{(x+a)(x-a)}$$
$$=\frac{ax-a^2-ax-a^2}{(x+a)(x-a)}$$
$$=\frac{-2a^2}{(x+a)(x-a)}$$

23. $\dfrac{11}{6}-\dfrac{5}{-6}=\dfrac{11}{6}-\dfrac{5}{-6}\cdot\dfrac{-1}{-1}$

$$=\frac{11}{6}-\frac{-5}{6}$$
$$=\frac{11-(-5)}{6}$$
$$=\frac{11+5}{6}$$
$$=\frac{16}{6}$$
$$=\frac{8}{3}$$

25. $\dfrac{5}{a}-\dfrac{8}{-a}=\dfrac{5}{a}-\dfrac{8}{-a}\cdot\dfrac{-1}{-1}$

$$=\frac{5}{a}-\frac{-8}{a}$$
$$=\frac{5-(-8)}{a}$$
$$=\frac{5+8}{a}$$
$$=\frac{13}{a}$$

27. $\dfrac{4}{y-1}-\dfrac{4}{1-y}=\dfrac{4}{y-1}-\dfrac{4}{1-y}\cdot\dfrac{-1}{-1}$
$=\dfrac{4}{y-1}-\dfrac{4(-1)}{(1-y)(-1)}$
$=\dfrac{4}{y-1}-\dfrac{-4}{y-1}$
$=\dfrac{4-(-4)}{y-1}$
$=\dfrac{4+4}{y-1}$
$=\dfrac{8}{y-1}$

29. $\dfrac{3-x}{x-7}-\dfrac{2x-5}{7-x}=\dfrac{3-x}{x-7}-\dfrac{2x-5}{7-x}\cdot\dfrac{-1}{-1}$
$=\dfrac{3-x}{x-7}-\dfrac{(2x-5)(-1)}{(7-x)(-1)}$
$=\dfrac{3-x}{x-7}-\dfrac{5-2x}{x-7}$
$=\dfrac{(3-x)-(5-2x)}{x-7}$
$=\dfrac{3-x-5+2x}{x-7}$
$=\dfrac{x-2}{x-7}$

31. $\dfrac{a-2}{a^2-25}-\dfrac{6-a}{25-a^2}=\dfrac{a-2}{a^2-25}-\dfrac{6-a}{25-a^2}\cdot\dfrac{-1}{-1}$
$=\dfrac{a-2}{a^2-25}-\dfrac{(6-a)(-1)}{(25-a^2)(-1)}$
$=\dfrac{a-2}{a^2-25}-\dfrac{a-6}{a^2-25}$
$=\dfrac{(a-2)-(a-6)}{a^2-25}$
$=\dfrac{a-2-a+6}{a^2-25}$
$=\dfrac{4}{a^2-25}=\dfrac{4}{(a+5)(a-5)}$

33. $\dfrac{4-x}{x-9}-\dfrac{3x-8}{9-x}=\dfrac{4-x}{x-9}-\dfrac{3x-8}{9-x}\cdot\dfrac{-1}{-1}$
$=\dfrac{4-x}{x-9}-\dfrac{8-3x}{x-9}$
$=\dfrac{(4-x)-(8-3x)}{x-9}$
$=\dfrac{4-x-8+3x}{x-9}$
$=\dfrac{2x-4}{x-9}=\dfrac{2(x-2)}{x-9}$

35. $\dfrac{5x}{x^2-9}-\dfrac{4}{3-x}$
$=\dfrac{5x}{(x+3)(x-3)}-\dfrac{4}{3-x}$ $x-3$ and $3-x$ are opposites
$=\dfrac{5x}{(x+3)(x-3)}-\dfrac{4}{3-x}\cdot\dfrac{-1}{-1}$
$=\dfrac{5x}{(x+3)(x-3)}-\dfrac{-4}{x-3}$ LCD $=(x+3)(x-3)$
$=\dfrac{5x}{(x+3)(x-3)}-\dfrac{-4}{x-3}\cdot\dfrac{x+3}{x+3}$
$=\dfrac{5x}{(x+3)(x-3)}-\dfrac{-4x-12}{(x+3)(x-3)}$
$=\dfrac{5x-(-4x-12)}{(x+3)(x-3)}$
$=\dfrac{5x+4x+12}{(x+3)(x-3)}$
$=\dfrac{9x+12}{(x+3)(x-3)}$
$=\dfrac{3(3x+4)}{(x+3)(x-3)}$

37. $\dfrac{t^2}{2t^2-2t}-\dfrac{1}{2t-2}$
$=\dfrac{t^2}{2t(t-1)}-\dfrac{1}{2(t-1)}$ LCD $=2t(t-1)$
$=\dfrac{t^2}{2t(t-1)}-\dfrac{1}{2(t-1)}\cdot\dfrac{t}{t}$
$=\dfrac{t^2}{2t(t-1)}-\dfrac{t}{2t(t-1)}$
$=\dfrac{t^2-t}{2t(t-1)}$
$=\dfrac{t(t-1)}{2t(t-1)}$
$=\dfrac{\not{t}(\not{t}-\not{1})(1)}{2\not{t}(\not{t}-\not{1})}$
$=\dfrac{1}{2}$

39. $\dfrac{x}{x^2+5x+6}-\dfrac{2}{x^2+3x+2}$

$=\dfrac{x}{(x+3)(x+2)}-\dfrac{2}{(x+2)(x+1)}$

LCD $=(x+3)(x+2)(x+1)$

$=\dfrac{x}{(x+3)(x+2)}\cdot\dfrac{x+1}{x+1}-\dfrac{2}{(x+2)(x+1)}\cdot\dfrac{x+3}{x+3}$

$=\dfrac{x^2+x}{(x+3)(x+2)(x+1)}-\dfrac{2x+6}{(x+3)(x+2)(x+1)}$

$=\dfrac{x^2+x-(2x+6)}{(x+3)(x+2)(x+1)}$

$=\dfrac{x^2+x-2x-6}{(x+3)(x+2)(x+1)}$

$=\dfrac{x^2-x-6}{(x+3)(x+2)(x+1)}$

$=\dfrac{(x-3)(x+2)}{(x+3)(x+2)(x+1)}$

$=\dfrac{(x-3)\cancel{(x+2)}}{(x+3)\cancel{(x+2)}(x+1)}$

$=\dfrac{x-3}{(x+3)(x+1)}$

41. $\dfrac{3(2x+5)}{x-1}-\dfrac{3(2x-3)}{1-x}+\dfrac{6x+1}{x-1}$

$=\dfrac{3(2x+5)}{x-1}-\dfrac{3(2x-3)}{1-x}\cdot\dfrac{-1}{-1}+\dfrac{6x-1}{x-1}$

$=\dfrac{3(2x+5)}{x-1}-\dfrac{-3(2x-3)}{x-1}+\dfrac{6x-1}{x-1}$

$=\dfrac{(6x+15)-(-6x+9)+(6x-1)}{x-1}$

$=\dfrac{6x+15+6x-9+6x-1}{x-1}$

$=\dfrac{18x+5}{x-1}$

43. $\dfrac{x-y}{x^2-y^2}+\dfrac{x+y}{x^2-y^2}-\dfrac{2x}{x^2-y^2}$

$=\dfrac{x-y+x+y-2x}{x^2-y^2}$

$=\dfrac{0}{x^2-y^2}$

$=0$

45. $\dfrac{2(x-1)}{2x-3}-\dfrac{3(x+2)}{2x-3}-\dfrac{x-1}{3-2x}$

$=\dfrac{2(x-1)}{2x-3}-\dfrac{3(x+2)}{2x-3}-\dfrac{x-1}{3-2x}\cdot\dfrac{-1}{-1}$

$=\dfrac{2(x-1)}{2x-3}-\dfrac{3(x+2)}{2x-3}-\dfrac{1-x}{2x-3}$

$=\dfrac{(2x-2)-(3x+6)-(1-x)}{2x-3}$

$=\dfrac{2x-2-3x-6-1+x}{2x-3}$

$=\dfrac{-9}{2x-3}$

47. $\dfrac{10}{2y-1}-\dfrac{6}{1-2y}+\dfrac{y}{2y-1}+\dfrac{y-4}{1-2y}$

$=\dfrac{10}{2y-1}-\dfrac{6}{1-2y}\cdot\dfrac{-1}{-1}+\dfrac{y}{2y-1}+\dfrac{y-4}{1-2y}\cdot\dfrac{-1}{-1}$

$=\dfrac{10}{2y-1}-\dfrac{-6}{2y-1}+\dfrac{y}{2y-1}+\dfrac{4-y}{2y-1}$

$=\dfrac{10-(-6)+y+4-y}{2y-1}$

$=\dfrac{10+6+y+4-y}{2y-1}$

$=\dfrac{20}{2y-1}$

49. $\dfrac{a+6}{4-a^2}-\dfrac{a+3}{a+2}+\dfrac{a-3}{2-a}$

$=\dfrac{a+6}{(2+a)(2-a)}-\dfrac{a+3}{2+a}+\dfrac{a-3}{2-a}$

$a+2=2+a$; LCD $=(2+a)(2-a)$

$=\dfrac{a+6}{(2+a)(2-a)}-\dfrac{a+3}{2+a}\cdot\dfrac{2-a}{2-a}+\dfrac{a-3}{2-a}\cdot\dfrac{2+a}{2+a}$

$=\dfrac{(a+6)-(a+3)(2-a)+(a-3)(2+a)}{(2+a)(2-a)}$

$=\dfrac{a+6-(-a^2-a+6)+(a^2-a-6)}{(2+a)(2-a)}$

$=\dfrac{a+6+a^2+a-6+a^2-a-6}{(2+a)(2-a)}$

$=\dfrac{2a^2+a-6}{(2+a)(2-a)}$

$=\dfrac{(2a-3)(a+2)}{(2+a)(2-a)}$

$=\dfrac{(2a-3)\cancel{(2+a)}}{\cancel{(2+a)}(2-a)}$

$=\dfrac{2a-3}{2-a}$

51. $$\frac{2z}{1-2z}+\frac{3z}{2z+1}-\frac{3}{4z^2-1}$$
$$=\frac{2z}{1-2z}\cdot\frac{-1}{-1}+\frac{3z}{2z+1}-\frac{3}{4z^2-1}$$
$$=\frac{-2z}{2z-1}+\frac{3z}{2z+1}-\frac{3}{(2z-1)(2z+1)}$$
$$\text{LCD} = (2z-1)(2z+1)$$
$$=\frac{-2z}{2z-1}\cdot\frac{2z+1}{2z+1}+\frac{3z}{2z+1}\cdot\frac{2z-1}{2z-1}-\frac{3}{(2z-1)(2z+1)}$$
$$=\frac{(-4z^2-2z)+(6z^2-3z)-3}{(2z-1)(2z+1)}$$
$$=\frac{2z^2-5z-3}{(2z-1)(2z+1)}$$
$$=\frac{(z-3)(2z+1)}{(2z-1)(2z+1)}$$
$$=\frac{(z-3)\cancel{(2z+1)}}{(2z-1)\cancel{(2z+1)}}$$
$$=\frac{z-3}{2z-1}$$

53. $$\frac{1}{x+y}-\frac{1}{x-y}+\frac{2x}{x^2-y^2}$$
$$=\frac{1}{x+y}-\frac{1}{x-y}+\frac{2x}{(x+y)(x-y)}$$
$$\text{LCD} = (x+y)(x-y)$$
$$=\frac{1}{x+y}\cdot\frac{x-y}{x-y}-\frac{1}{x-y}\cdot\frac{x+y}{x+y}+\frac{2x}{(x+y)(x-y)}$$
$$=\frac{x-y-(x+y)+2x}{(x+y)(x-y)}$$
$$=\frac{x-y-x-y+2x}{(x+y)(x-y)}$$
$$=\frac{2x-2y}{(x+y)(x-y)}$$
$$=\frac{2(x-y)}{(x+y)(x-y)}$$
$$=\frac{2\cancel{(x-y)}}{(x+y)\cancel{(x-y)}}$$
$$=\frac{2}{x+y}$$

55. $(a^2b^{-5})^{-4}=a^{2(-4)}b^{-5(-4)}=a^{-8}b^{20}=\dfrac{b^{20}}{a^8}$

57. $3x^4\cdot 10x^8=3\cdot 10\cdot x^4\cdot x^8=30x^{4+8}=30x^{12}$

59. $$\frac{4}{7}+3x=\frac{1}{2}x-\frac{3}{14},\quad \text{LCD is 14}$$
$$14\left(\frac{4}{7}+3x\right)=14\left(\frac{1}{2}x-\frac{3}{14}\right)$$
$$14\cdot\frac{4}{7}+14\cdot 3x=14\cdot\frac{1}{2}x-14\cdot\frac{3}{14}$$
$$8+42x=7x-3$$
$$8+35x=-3$$
$$35x=-11$$
$$x=-\frac{11}{35}$$

The solution is $-\dfrac{11}{35}$.

61. The shaded area has dimensions $x-6$ by $x-3$. Then the area is $(x-6)(x-3)$, or $x^2-9x+18$.

63. Let l = the length of the missing side.
$$\frac{a^2-5a-9}{a-6}+\frac{a^2-6}{a-6}+l=2a+5$$
$$\frac{2a^2-5a-15}{a-6}+l=2a+5$$
$$l=2a+5-\frac{2a^2-5a-15}{a-6}$$
$$l=(2a+5)\cdot\frac{a-6}{a-6}-\frac{2a^2-5a-15}{a-6}$$
$$l=\frac{2a^2-7a-30}{a-6}-\frac{2a^2-5a-15}{a-6}$$
$$l=\frac{2a^2-7a-30-(2a^2-5a-15)}{a-6}$$
$$l=\frac{2a^2-7a-30-2a^2+5a+15}{a-6}$$
$$l=\frac{-2a-15}{a-6}$$

The length of the missing side is $\dfrac{-2a-15}{a-6}$.

Now find the area.
$$A=\frac{1}{2}\cdot b\cdot h$$
$$A=\frac{1}{2}\left(\frac{-2a-15}{a-6}\right)\left(\frac{a^2-6}{a-6}\right)$$
$$A=\frac{(-2a-15)(a^2-6)}{2(a-6)^2},\text{ or}$$
$$A=\frac{-2a^3-15a^2+12a+90}{2a^2-24a+72}$$

Chapter 6 Mid-Chapter Review

1. False; the reciprocal of $\dfrac{3-w}{w+2}$ is $\dfrac{w+2}{3-w}$, not $\dfrac{w-3}{w+2}$.

2. True; we must avoid *denominators* of 0. The numerator has no bearing on whether a rational expression is defined.

3. True; see pages 485 and 493 in the text.

4. False; x is a factor of the numerator, but it is a *term* of the denominator.

5. True; see page 481 in the text.

6. $\frac{x-1}{x-2}-\frac{x+1}{x+2}-\frac{x-6}{4-x^2}$

$=\frac{x-1}{x-2}-\frac{x+1}{x+2}-\frac{x-6}{4-x^2}\cdot\frac{-1}{-1}$

$=\frac{x-1}{x-2}-\frac{x+1}{x+2}-\frac{6-x}{x^2-4}$

$=\frac{x-1}{x-2}-\frac{x+1}{x+2}-\frac{6-x}{(x-2)(x+2)}$

$=\frac{x-1}{x-2}\cdot\frac{x+2}{x+2}-\frac{x+1}{x+2}\cdot\frac{x-2}{x-2}-\frac{6-x}{(x-2)(x+2)}$

$=\frac{x^2+x-2}{(x-2)(x+2)}-\frac{x^2-x-2}{(x-2)(x+2)}-\frac{6-x}{(x-2)(x+2)}$

$=\frac{x^2+x-2-(x^2-x-2)-(6-x)}{(x-2)(x+2)}$

$=\frac{x^2+x-2-x^2+x+2-6+x}{(x-2)(x+2)}$

$=\frac{3x-6}{(x-2)(x+2)}$

$=\frac{3(x-2)}{(x-2)(x+2)}=\frac{x-2}{x-2}\cdot\frac{3}{x+2}=\frac{3}{x+2}$

7. $\frac{t^2-16}{3}$

Since the denominator is the constant 3, there are no numbers for which the expression is not defined.

8. $\frac{x-8}{x^2-11x+24}$

Set the denominator equal to 0 and solve for x.

$x^2-11x+24=0$

$(x-3)(x-8)=0$

$x-3=0 \quad or \quad x-8=0$

$x=3 \quad or \quad x=8$

The expression is not defined for the numbers 3 and 8.

9. $\frac{7}{2w-7}$

Set the denominator equal to 0 and solve for x.

$2w-7=0$

$2w=7$

$w=\frac{7}{2}$

The expression is not defined for the number $\frac{7}{2}$.

10. $\frac{x^2+2x+3}{x^2-9}=\frac{(x+1)(x+3)}{(x+3)(x-3)}$

$=\frac{x+3}{x+3}\cdot\frac{x+1}{x-3}$

$=1\cdot\frac{x+1}{x-3}$

$=\frac{x+1}{x-3}$

11. $\frac{6y^2+12y-48}{3y^2-9y+6}=\frac{6(y^2+2y-8)}{3(y^2-3y+2)}$

$=\frac{2\cdot 3(y+4)(y-2)}{3(y-2)(y-1)}$

$=\frac{3(y-2)}{3(y-2)}\cdot\frac{2(y+4)}{y-1}$

$=1\cdot\frac{2(y+4)}{y-1}$

$=\frac{2(y+4)}{y-1}$

12. $\frac{r-s}{s-r}=\frac{-1(-r+s)}{s-r}=\frac{-1(s-r)}{s-r}$

$=-1\cdot\frac{s-r}{s-r}$

$=-1\cdot 1=-1$

13. The reciprocal of $-x+3$ is $\frac{1}{-x+3}$, because $-x+3\cdot\frac{1}{-x+3}=1$. Since $-x+3$ can be written equivalently as $3-x$, we can also write the reciprocal as $\frac{1}{3-x}$.

14. $x^2-100=(x+10)(x-10)$

$10x^3=2\cdot 5\cdot x\cdot x\cdot x$

$x^2-20x+100=(x-10)(x-10)$

$\text{LCM}=2\cdot 5\cdot x\cdot x\cdot x(x+10)(x-10)(x-10)=$
$10x^3(x+10)(x-10)^2$

15. $\frac{a^2-a-2}{a^2-a-6}\div\frac{a^2-2a}{2a+a^2}=\frac{a^2-a-2}{a^2-a-6}\cdot\frac{2a+a^2}{a^2-2a}$

$=\frac{(a-2)(a+1)(a)(2+a)}{(a-3)(a+2)(a)(a-2)}$

$=\frac{\cancel{(a-2)}(a+1)\cancel{(a)}\cancel{(2+a)}}{(a-3)\cancel{(a+2)}\cancel{(a)}\cancel{(a-2)}}$

$=\frac{a+1}{a-3}$

16. $\frac{3y}{y^2-7y+10}-\frac{2y}{y^2-8y+15}$

$=\frac{3y}{(y-2)(y-5)}-\frac{2y}{(y-3)(y-5)}$,

LCD is $(y-2)(y-5)(y-3)$

$=\frac{3y}{(y-2)(y-5)}\cdot\frac{y-3}{y-3}-\frac{2y}{(y-3)(y-5)}\cdot\frac{y-2}{y-2}$

$=\frac{3y(y-3)-2y(y-2)}{(y-2)(y-5)(y-3)}$

$=\frac{3y^2-9y-2y^2+4y}{(y-2)(y-5)(y-3)}$

$=\frac{y^2-5y}{(y-2)(y-5)(y-3)}$

$=\frac{y(y-5)}{(y-2)(y-5)(y-3)}=\frac{y-5}{y-5}\cdot\frac{y}{(y-2)(y-3)}$

$=\frac{y}{(y-2)(y-3)}$

17. $\dfrac{x^2}{x-11}+\dfrac{121}{11-x}=\dfrac{x^2}{x-11}+\dfrac{121}{11-x}\cdot\dfrac{-1}{-1}$

$=\dfrac{x^2}{x-11}+\dfrac{-121}{x-11}$

$=\dfrac{x^2-121}{x-11}$

$=\dfrac{(x+11)(x-11)}{x-11}=\dfrac{x+11}{1}\cdot\dfrac{x-11}{x-11}$

$=x+11$

18. $\dfrac{x^2-y^2}{(x-y)^2}\cdot\dfrac{1}{x+y}=\dfrac{(x+y)(x-y)\cdot 1}{(x-y)^2(x+y)}$

$=\dfrac{\cancel{(x+y)}\cancel{(x-y)}\cdot 1}{\cancel{(x-y)}(x-y)\cancel{(x+y)}}$

$=\dfrac{1}{x-y}$

19. $\dfrac{3a-b}{a^2b}+\dfrac{a+2b}{ab^2}$, LCD is a^2b^2

$=\dfrac{3a-b}{a^2b}\cdot\dfrac{b}{b}+\dfrac{a+2b}{ab^2}\cdot\dfrac{a}{a}$

$=\dfrac{(3a-b)b+(a+2b)a}{a^2b^2}$

$=\dfrac{3ab-b^2+a^2+2ab}{a^2b^2}$

$=\dfrac{a^2+5ab-b^2}{a^2b^2}$

20. $\dfrac{5x}{x^2-4}-\dfrac{3}{x}+\dfrac{4}{x+2}$

$=\dfrac{5x}{(x+2)(x-2)}-\dfrac{3}{x}+\dfrac{4}{x+2}$, LCD is $x(x+2)(x-2)$

$=\dfrac{5x}{(x+2)(x-2)}\cdot\dfrac{x}{x}-\dfrac{3}{x}\cdot\dfrac{(x+2)(x-2)}{(x+2)(x-2)}+\dfrac{4}{x+2}\cdot\dfrac{x(x-2)}{x(x-2)}$

$=\dfrac{5x^2-3(x^2-4)+4x(x-2)}{x(x+2)(x-2)}$

$=\dfrac{5x^2-3x^2+12+4x^2-8x}{x(x+2)(x-2)}$

$=\dfrac{6x^2-8x+12}{x(x+2)(x-2)}$

$=\dfrac{2(3x^2-4x+6)}{x(x+2)(x-2)}$

21. $\dfrac{2}{x-2}\div\dfrac{1}{x+3}=\dfrac{2}{x-2}\cdot\dfrac{x+3}{1}=\dfrac{2(x+3)}{x-2}$

Answer E is correct.

22. $\dfrac{1}{x+3}-\dfrac{2}{x-2}=\dfrac{1}{x+3}\cdot\dfrac{x-2}{x-2}-\dfrac{2}{x-2}\cdot\dfrac{x+3}{x+3}$

$=\dfrac{x-2-2(x+3)}{(x+3)(x-2)}$

$=\dfrac{x-2-2x-6}{(x+3)(x-2)}$

$=\dfrac{-x-8}{(x+3)(x-2)}$

Answer A is correct.

23. $\dfrac{2}{x-2}-\dfrac{1}{x+3}=\dfrac{2}{x-2}\cdot\dfrac{x+3}{x+3}-\dfrac{1}{x+3}\cdot\dfrac{x-2}{x-2}$

$=\dfrac{2(x+3)-(x-2)}{(x-2)(x+3)}$

$=\dfrac{2x+6-x+2}{(x-2)(x+3)}$

$=\dfrac{x+8}{(x-2)(x+3)}$

Answer D is correct.

24. $\dfrac{1}{x+3}\div\dfrac{2}{x-2}=\dfrac{1}{x+3}\cdot\dfrac{x-2}{2}=\dfrac{x-2}{2(x+3)}$

Answer B is correct.

25. $\dfrac{2}{x-2}+\dfrac{1}{x+3}=\dfrac{2}{x-2}\cdot\dfrac{x+3}{x+3}+\dfrac{1}{x+3}\cdot\dfrac{x-2}{x-2}$

$=\dfrac{2(x+3)+x-2}{(x-2)(x+3)}$

$=\dfrac{2x+6+x-2}{(x-2)(x+3)}$

$=\dfrac{3x+4}{(x-2)(x+3)}$

Answer F is correct.

26. $\dfrac{2}{x-2}\cdot\dfrac{1}{x+3}=\dfrac{2}{(x-2)(x+3)}$

Answer C is correct.

27. If the numbers have a common factor, their product contains that factor more than the greatest number of times it occurs in any one factorization. In this case, their product is not their least common multiple.

28. Yes; consider the product $\dfrac{a}{b}\cdot\dfrac{c}{d}=\dfrac{ac}{bd}$. The reciprocal of the product is $\dfrac{bd}{ac}$. This is equal to the product of the reciprocals of the two original factors: $\dfrac{b}{a}\cdot\dfrac{d}{c}=\dfrac{bd}{ac}$.

29. Although multiplying the denominators of the expressions being added results in a common denominator, it is often not the *least* common denominator. Using a common denominator other than the LCD makes the expressions more complicated, requires additional simplification after the addition has been performed, and leaves more room for error.

30. Their sum is zero. Another explanation is that $-\left(\dfrac{1}{3-x}\right)=\dfrac{1}{-(3-x)}=\dfrac{1}{x-3}$.

31. $\dfrac{x+3}{x-5}$ is undefined for $x=5$, $\dfrac{x-7}{x+1}$ is undefined for $x=-1$, and $\dfrac{x+1}{x-7}$ $\left(\text{the reciprocal of } \dfrac{x-7}{x+1}\right)$ is undefined for $x=7$.

32. The binomial is a factor of the trinomial.

Exercise Set 6.6

RC1. The expression given in the text is a complex rational expression.

RC3. The least common denominator of the rational expressions $\frac{8}{x}$, $\frac{5}{9}$, and $\frac{2}{x}$ is $9x$.

1. $\dfrac{1+\frac{9}{16}}{1-\frac{3}{4}}$ LCM of the denominators is 16.

$= \dfrac{1+\frac{9}{16}}{1-\frac{3}{4}} \cdot \dfrac{16}{16}$ Multiplying by 1 using $\frac{16}{16}$

$= \dfrac{\left(1+\frac{9}{16}\right)16}{\left(1-\frac{3}{4}\right)16}$ Multiplying numerator and denominator by 16

$= \dfrac{1(16)+\frac{9}{16}(16)}{1(16)-\frac{3}{4}(16)}$

$= \dfrac{16+9}{16-12}$

$= \dfrac{25}{4}$

3. $\dfrac{1-\frac{3}{5}}{1+\frac{1}{5}}$

$= \dfrac{1\cdot\frac{5}{5}-\frac{3}{5}}{1\cdot\frac{5}{5}+\frac{1}{5}}$ Getting a common denominator in numerator and in denominator

$= \dfrac{\frac{5}{5}-\frac{3}{5}}{\frac{5}{5}+\frac{1}{5}}$

$= \dfrac{\frac{2}{5}}{\frac{6}{5}}$ Subtracting in numerator; adding in denominator

$= \dfrac{2}{5}\cdot\dfrac{5}{6}$ Multiplying by the reciprocal of the divisor

$= \dfrac{2\cdot 5}{5\cdot 2\cdot 3}$

$= \dfrac{\cancel{2}\cdot\cancel{5}\cdot 1}{\cancel{5}\cdot\cancel{2}\cdot 3}$

$= \dfrac{1}{3}$

5. $\dfrac{\frac{1}{2}+\frac{3}{4}}{\frac{5}{8}-\frac{5}{6}} = \dfrac{\frac{1}{2}\cdot\frac{2}{2}+\frac{3}{4}}{\frac{5}{8}\cdot\frac{3}{3}-\frac{5}{6}\cdot\frac{4}{4}}$ Getting a common denominator in numerator and in denominator

$= \dfrac{\frac{2}{4}+\frac{3}{4}}{\frac{15}{24}-\frac{20}{24}}$

$= \dfrac{\frac{5}{4}}{\frac{-5}{24}}$ Adding in numerator; subtracting in denominator

$= \dfrac{5}{4}\cdot\dfrac{24}{-5}$ Multiplying by the reciprocal of the divisor

$= \dfrac{5\cdot 4\cdot 6}{4\cdot(-1)\cdot 5}$

$= \dfrac{\cancel{5}\cdot\cancel{4}\cdot 6}{\cancel{4}\cdot(-1)\cdot\cancel{5}}$

$= -6$

7. $\dfrac{\frac{1}{x}+3}{\frac{1}{x}-5}$ LCM of the denominators is x.

$= \dfrac{\frac{1}{x}+3}{\frac{1}{x}-5}\cdot\dfrac{x}{x}$ Multiplying by 1 using $\frac{x}{x}$

$= \dfrac{\left(\frac{1}{x}+3\right)x}{\left(\frac{1}{x}-5\right)x}$

$= \dfrac{\frac{1}{x}\cdot x+3\cdot x}{\frac{1}{x}\cdot x-5\cdot x}$

$= \dfrac{1+3x}{1-5x}$

9. $\dfrac{4-\frac{1}{x^2}}{2-\frac{1}{x}}$ LCM of the denominators is x^2.

$= \dfrac{4-\frac{1}{x^2}}{2-\frac{1}{x}}\cdot\dfrac{x^2}{x^2}$

$= \dfrac{\left(4-\frac{1}{x^2}\right)x^2}{\left(2-\frac{1}{x}\right)x^2}$

$= \dfrac{4\cdot x^2-\frac{1}{x^2}\cdot x^2}{2\cdot x^2-\frac{1}{x}\cdot x^2}$

$= \dfrac{4x^2-1}{2x^2-x}$

$= \dfrac{(2x+1)(2x-1)}{x(2x-1)}$ Factoring numerator and denominator

$= \dfrac{(2x+1)\cancel{(2x-1)}}{x\cancel{(2x-1)}}$

$= \dfrac{2x+1}{x}$

11. $\dfrac{8+\frac{8}{d}}{1+\frac{1}{d}} = \dfrac{8\cdot\frac{d}{d}+\frac{8}{d}}{1\cdot\frac{d}{d}+\frac{1}{d}}$

$= \dfrac{\frac{8d+8}{d}}{\frac{d+1}{d}}$

$= \dfrac{8d+8}{d}\cdot\dfrac{d}{d+1}$

$= \dfrac{8(d+1)(d)}{d(d+1)}$

$= \dfrac{8\cancel{(d+1)}\cancel{(d)}}{\cancel{d}\cancel{(d+1)}(1)}$

$= 8$

13. $\dfrac{\frac{x}{8}-\frac{8}{x}}{\frac{1}{8}+\frac{1}{x}}$ LCM of the denominators is $8x$.

$= \dfrac{\frac{x}{8}-\frac{8}{x}}{\frac{1}{8}+\frac{1}{x}}\cdot\dfrac{8x}{8x}$

$= \dfrac{\left(\frac{x}{8}-\frac{8}{x}\right)8x}{\left(\frac{1}{8}+\frac{1}{x}\right)8x}$

$= \dfrac{\frac{x}{8}(8x)-\frac{8}{x}(8x)}{\frac{1}{8}(8x)+\frac{1}{x}(8x)}$

$= \dfrac{x^2-64}{x+8}$

$= \dfrac{(x+8)(x-8)}{x+8}$

$= \dfrac{\cancel{(x+8)}(x-8)}{1\cancel{(x+8)}}$

$= x-8$

15. $\dfrac{1+\frac{1}{y}}{1-\frac{1}{y^2}} = \dfrac{1\cdot\frac{y}{y}+\frac{1}{y}}{1\cdot\frac{y^2}{y^2}-\frac{1}{y^2}}$

$= \dfrac{\frac{y+1}{y}}{\frac{y^2-1}{y^2}}$

$= \dfrac{y+1}{y}\cdot\dfrac{y^2}{y^2-1}$

$= \dfrac{(y+1)y\cdot y}{y(y+1)(y-1)}$

$= \dfrac{\cancel{(y+1)}\cancel{y}\cdot y}{\cancel{y}\cancel{(y+1)}(y-1)}$

$= \dfrac{y}{y-1}$

17. $\dfrac{\dfrac{1}{5}-\dfrac{1}{a}}{\dfrac{5-a}{5}}$ LCM of the denominators is $5a$.

$$= \frac{\dfrac{1}{5}-\dfrac{1}{a}}{\dfrac{5-a}{5}} \cdot \frac{5a}{5a}$$

$$= \frac{\left(\dfrac{1}{5}-\dfrac{1}{a}\right)5a}{\left(\dfrac{5-a}{5}\right)5a}$$

$$= \frac{\dfrac{1}{5}(5a)-\dfrac{1}{a}(5a)}{a(5-a)}$$

$$= \frac{a-5}{5a-a^2}$$

$$= \frac{a-5}{-a(-5+a)}$$

$$= \frac{1(\cancel{a-5})}{-a(\cancel{a-5})}$$

$$= -\frac{1}{a}$$

19. $\dfrac{\dfrac{1}{a}+\dfrac{1}{b}}{\dfrac{1}{a^2}-\dfrac{1}{b^2}}$ LCM of the denominators is a^2b^2.

$$= \frac{\dfrac{1}{a}+\dfrac{1}{b}}{\dfrac{1}{a^2}-\dfrac{1}{b^2}} \cdot \frac{a^2b^2}{a^2b^2}$$

$$= \frac{\left(\dfrac{1}{a}+\dfrac{1}{b}\right)\cdot a^2b^2}{\left(\dfrac{1}{a^2}-\dfrac{1}{b^2}\right)\cdot a^2b^2}$$

$$= \frac{\dfrac{1}{a}\cdot a^2b^2+\dfrac{1}{b}\cdot a^2b^2}{\dfrac{1}{a^2}\cdot a^2b^2-\dfrac{1}{b^2}\cdot a^2b^2}$$

$$= \frac{ab^2+a^2b}{b^2-a^2}$$

$$= \frac{ab(b+a)}{(b+a)(b-a)}$$

$$= \frac{ab(\cancel{b+a})}{(\cancel{b+a})(b-a)}$$

$$= \frac{ab}{b-a}$$

21. $\dfrac{\dfrac{p}{q}+\dfrac{q}{p}}{\dfrac{1}{p}+\dfrac{1}{q}}$ LCM of the denominators is pq.

$$= \frac{\left(\dfrac{p}{q}+\dfrac{q}{p}\right)\cdot pq}{\left(\dfrac{1}{p}+\dfrac{1}{q}\right)\cdot pq}$$

$$= \frac{\dfrac{p}{q}\cdot pq+\dfrac{q}{p}\cdot pq}{\dfrac{1}{p}\cdot pq+\dfrac{1}{q}\cdot pq}$$

$$= \frac{p^2+q^2}{q+p}$$

23. $\dfrac{\dfrac{2}{a}+\dfrac{4}{a^2}}{\dfrac{5}{a^3}-\dfrac{3}{a}}$ LCD is a^3

$$= \frac{\dfrac{2}{a}+\dfrac{4}{a^2}}{\dfrac{5}{a^3}-\dfrac{3}{a}} \cdot \frac{a^3}{a^3}$$

$$= \frac{\dfrac{2}{a}\cdot a^3+\dfrac{4}{a^2}\cdot a^3}{\dfrac{5}{a^3}\cdot a^3-\dfrac{3}{a}\cdot a^3}$$

$$= \frac{2a^2+4a}{5-3a^2}$$

$$= \frac{2a(a+2)}{5-3a^2}$$

25. $\dfrac{\dfrac{2}{7a^4}-\dfrac{1}{14a}}{\dfrac{3}{5a^2}+\dfrac{2}{15a}} = \dfrac{\dfrac{2}{7a^4}\cdot\dfrac{2}{2}-\dfrac{1}{14a}\cdot\dfrac{a^3}{a^3}}{\dfrac{3}{5a^2}\cdot\dfrac{3}{3}+\dfrac{2}{15a}\cdot\dfrac{a}{a}}$

$$= \frac{\dfrac{4-a^3}{14a^4}}{\dfrac{9+2a}{15a^2}}$$

$$= \frac{4-a^3}{14a^4}\cdot\frac{15a^2}{9+2a}$$

$$= \frac{15\cdot\cancel{a^2}(4-a^3)}{14a^2\cdot\cancel{a^2}(9+2a)}$$

$$= \frac{15(4-a^3)}{14a^2(9+2a)}$$

27. $$\frac{\frac{a}{b}+\frac{c}{d}}{\frac{b}{a}+\frac{d}{c}} = \frac{\frac{a}{b}\cdot\frac{d}{d}+\frac{c}{d}\cdot\frac{b}{b}}{\frac{b}{a}\cdot\frac{c}{c}+\frac{d}{c}\cdot\frac{a}{a}}$$
$$= \frac{\frac{ad+bc}{bd}}{\frac{bc+ad}{ac}}$$
$$= \frac{ad+bc}{bd}\cdot\frac{ac}{bc+ad}$$
$$= \frac{ac(ad+bc)}{bd(bc+ad)}$$
$$= \frac{ac}{bd}\cdot\frac{ad+bc}{bc+ad}$$
$$= \frac{ac}{bd}\cdot 1$$
$$= \frac{ac}{bd}$$

29. $$\frac{\frac{x}{5y^3}+\frac{3}{10y}}{\frac{3}{10y}+\frac{x}{5y^3}}$$

Observe that, by the commutative law of addition, the numerator and denominator are equivalent, so the result is 1. We could also simplify this expression as follows:

$$\frac{\frac{x}{5y^3}+\frac{3}{10y}}{\frac{3}{10y}+\frac{x}{5y^3}} = \frac{\frac{x}{5y^3}+\frac{3}{10y}}{\frac{3}{10y}+\frac{x}{5y^3}}\cdot\frac{10y^3}{10y^3}$$
$$= \frac{\frac{x}{5y^3}\cdot 10y^3+\frac{3}{10y}\cdot 10y^3}{\frac{3}{10y}\cdot 10y^3+\frac{x}{5y^3}\cdot 10y^3}$$
$$= \frac{2x+3y^2}{3y^2+2x}$$
$$= 1$$

31. $$\frac{\frac{3}{x+1}+\frac{1}{x}}{\frac{2}{x+1}+\frac{3}{x}} = \frac{\frac{3}{x+1}+\frac{1}{x}}{\frac{2}{x+1}+\frac{3}{x}}\cdot\frac{x(x+1)}{x(x+1)}$$
$$= \frac{\frac{3}{x+1}\cdot x(x+1)+\frac{1}{x}\cdot x(x+1)}{\frac{2}{x+1}\cdot x(x+1)+\frac{3}{x}\cdot x(x+1)}$$
$$= \frac{3x+x+1}{2x+3(x+1)}$$
$$= \frac{4x+1}{2x+3x+3}$$
$$= \frac{4x+1}{5x+3}$$

33. $$4-\frac{1}{6}x \geq -12$$
$$-\frac{1}{6}x \geq -16$$
$$-6\left(-\frac{1}{6}x\right) \leq -6(-16)$$
$$x \leq 96$$

The solution set is $\{x|x \leq 96\}$.

35. $$1.5x+19.2 < 4.2-3.5x$$
$$5x+19.2 < 4.2$$
$$5x < -15$$
$$x < -3$$

The solution set is $\{x|x < -3\}$.

37. ***Familiarize***. Let w = the width of the rectangle. Then $w+3$ = the length. Recall that the formula for the area of a rectangle is $A = lw$ and the formula for the perimeter of a rectangle is $P = 2l+2w$.

Translate. We substitute in the formula for area.

$$10 = lw$$
$$10 = (w+3)w$$

Solve.

$$10 = (w+3)w$$
$$10 = w^2+3w$$
$$0 = w^2+3w-10$$
$$0 = (w+5)(w-2)$$
$$w+5=0 \quad\text{or}\quad w-2=0$$
$$w=-5 \quad\text{or}\quad w=2$$

Check. Since the width cannot be negative, we only check 2. If $w=2$, then $w+3=2+3$, or 5. Since $2\cdot 5=10$, the given area, the answer checks. Now we find the perimeter:

$$P = 2l+2w$$
$$P = 2\cdot 5+2\cdot 2$$
$$P = 10+4$$
$$P = 14$$

We can check this by repeating the calculation.

State. The perimeter is 14 yd.

39.
$$1+\cfrac{1}{1+\cfrac{1}{1+\cfrac{1}{1+\cfrac{1}{x}}}}=1+\cfrac{1}{1+\cfrac{1}{1+\cfrac{1}{\frac{x+1}{x}}}}$$
$$=1+\cfrac{1}{1+\cfrac{1}{1+\cfrac{x}{x+1}}}$$
$$=1+\cfrac{1}{1+\cfrac{1}{\frac{x+1+x}{x+1}}}$$
$$=1+\cfrac{1}{1+\cfrac{1}{\frac{2x+1}{x+1}}}$$
$$=1+\cfrac{1}{1+\cfrac{x+1}{2x+1}}$$
$$=1+\cfrac{1}{\frac{2x+1+x+1}{2x+1}}$$
$$=1+\cfrac{1}{\frac{3x+2}{2x+1}}$$
$$=1+\frac{2x+1}{3x+2}$$
$$=\frac{3x+2+2x+1}{3x+2}$$
$$=\frac{5x+3}{3x+2}$$

Exercise Set 6.7

RC1. We are combining (adding) rational expressions, so the result will be a rational expression.

RC3. We are combining (subtracting) rational expressions, so the result will be a rational expression.

RC5. We are solving an equation, so the result will be a solution or solutions.

RC7. We are combining (multiplying) rational expressions, so the result will be a rational expression.

1.
$$\frac{4}{5}-\frac{2}{3}=\frac{x}{9},\ \text{LCM}=45$$
$$45\left(\frac{4}{5}-\frac{2}{3}\right)=45\cdot\frac{x}{9}$$
$$45\cdot\frac{4}{5}-45\cdot\frac{2}{3}=45\cdot\frac{x}{9}$$
$$36-30=5x$$
$$6=5x$$
$$\frac{6}{5}=x$$

Check:

$\frac{4}{5}-\frac{2}{3}=\frac{x}{9}$	
$\frac{4}{5}-\frac{2}{3}$?	$\frac{\frac{6}{5}}{9}$
$\frac{12}{15}-\frac{10}{15}$	$\frac{6}{5}\cdot\frac{1}{9}$
$\frac{2}{15}$	$\frac{2}{15}$ TRUE

This checks, so the solution is $\frac{6}{5}$.

3.
$$\frac{3}{5}+\frac{1}{8}=\frac{1}{x},\ \text{LCM}=40x$$
$$40x\left(\frac{3}{5}+\frac{1}{8}\right)=40x\cdot\frac{1}{x}$$
$$40x\cdot\frac{3}{5}+40x\cdot\frac{1}{8}=40x\cdot\frac{1}{x}$$
$$24x+5x=40$$
$$29x=40$$
$$x=\frac{40}{29}$$

Check:

$\frac{3}{5}+\frac{1}{8}=\frac{1}{x}$	
$\frac{3}{5}+\frac{1}{8}$?	$\frac{1}{\frac{40}{29}}$
$\frac{24}{40}+\frac{5}{40}$	$1\cdot\frac{29}{40}$
$\frac{29}{40}$	$\frac{29}{40}$ TRUE

This checks, so the solution is $\frac{40}{29}$.

5.
$$\frac{3}{8}+\frac{4}{5}=\frac{x}{20},\ \text{LCM}=40$$
$$40\left(\frac{3}{8}+\frac{4}{5}\right)=40\cdot\frac{x}{20}$$
$$40\cdot\frac{3}{8}+40\cdot\frac{4}{5}=40\cdot\frac{x}{20}$$
$$15+32=2x$$
$$47=2x$$
$$\frac{47}{2}=x$$

Check:

$\frac{3}{8}+\frac{4}{5}=\frac{x}{20}$	
$\frac{3}{8}+\frac{4}{5}$?	$\frac{\frac{47}{2}}{20}$
$\frac{15}{40}+\frac{32}{40}$	$\frac{47}{2}\cdot\frac{1}{20}$
$\frac{47}{40}$	$\frac{47}{40}$ TRUE

This checks, so the solution is $\frac{47}{2}$.

7.
$$\frac{1}{x}=\frac{2}{3}-\frac{5}{6},\ \text{LCM}=6x$$
$$6x\cdot\frac{1}{x}=6x\left(\frac{2}{3}-\frac{5}{6}\right)$$
$$6x\cdot\frac{1}{x}=6x\cdot\frac{2}{3}-6x\cdot\frac{5}{6}$$
$$6=4x-5x$$
$$6=-x$$
$$-6=x$$

Check:

$\frac{1}{x}=\frac{2}{3}-\frac{5}{6}$	
$\frac{1}{-6}$?	$\frac{2}{3}-\frac{5}{6}$
$-\frac{1}{6}$	$\frac{4}{6}-\frac{5}{6}$
	$-\frac{1}{6}$ TRUE

This checks, so the solution is -6.

9.
$$\frac{1}{6}+\frac{1}{8}=\frac{1}{t},\ \text{LCM}=24t$$
$$24t\left(\frac{1}{6}+\frac{1}{8}\right)=24t\cdot\frac{1}{t}$$
$$24t\cdot\frac{1}{6}+24t\cdot\frac{1}{8}=24t\cdot\frac{1}{t}$$
$$4t+3t=24$$
$$7t=24$$
$$t=\frac{24}{7}$$

Check:

$\frac{1}{6}+\frac{1}{8}=\frac{1}{t}$	
$\frac{1}{6}+\frac{1}{8}$?	$\frac{1}{24/7}$
$\frac{4}{24}+\frac{3}{24}$	$1\cdot\frac{7}{24}$
$\frac{7}{24}$	$\frac{7}{24}$ TRUE

This checks, so the solution is $\frac{24}{7}$.

11.
$$x+\frac{4}{x}=-5,\ \text{LCM}=x$$
$$x\left(x+\frac{4}{x}\right)=x(-5)$$
$$x\cdot x+x\cdot\frac{4}{x}=x(-5)$$
$$x^2+4=-5x$$
$$x^2+5x+4=0$$
$$(x+4)(x+1)=0$$
$$x+4=0\quad\text{or}\quad x+1=0$$
$$x=-4\quad\text{or}\quad x=-1$$

Check:

$x+\frac{4}{x}=-5$	
$-4+\frac{4}{-4}$?	-5
$-4-1$	
-5	TRUE

$x+\frac{4}{x}=-5$	
$-1+\frac{4}{-1}$?	-5
$-1-4$	
-5	TRUE

Both of these check, so the two solutions are -4 and -1.

13.
$$\frac{x}{4}-\frac{4}{x}=0,\ \text{LCM}=4x$$
$$4x\left(\frac{x}{4}-\frac{4}{x}\right)=4x\cdot 0$$
$$4x\cdot\frac{x}{4}-4x\cdot\frac{4}{x}=4x\cdot 0$$
$$x^2-16=0$$
$$(x+4)(x-4)=0$$
$$x+4=0\quad\text{or}\quad x-4=0$$
$$x=-4\quad\text{or}\quad x=4$$

Check:

$\frac{x}{4}-\frac{4}{x}=0$	
$\frac{-4}{4}-\frac{4}{-4}$?	0
$-1-(-1)$	
$-1+1$	
0	TRUE

$\frac{x}{4}-\frac{4}{x}=0$	
$\frac{4}{4}-\frac{4}{4}$?	0
$1-1$	
0	TRUE

Both of these check, so the two solutions are -4 and 4.

15.
$$\frac{5}{x}=\frac{6}{x}-\frac{1}{3},\ \text{LCM}=3x$$
$$3x\cdot\frac{5}{x}=3x\left(\frac{6}{x}-\frac{1}{3}\right)$$
$$3x\cdot\frac{5}{x}=3x\cdot\frac{6}{x}-3x\cdot\frac{1}{3}$$
$$15=18-x$$
$$-3=-x$$
$$3=x$$

Check:

$$\begin{array}{c|c} \frac{5}{x} = \frac{6}{x} - \frac{1}{3} & \\ \hline \frac{5}{3} \ ? & \frac{6}{3} - \frac{1}{3} \\ & \frac{5}{3} \quad \text{TRUE} \end{array}$$

This checks, so the solution is 3.

17.
$$\frac{5}{3x} + \frac{3}{x} = 1, \text{ LCM} = 3x$$
$$3x\left(\frac{5}{3x} + \frac{3}{x}\right) = 3x \cdot 1$$
$$3x \cdot \frac{5}{3x} + 3x \cdot \frac{3}{x} = 3x \cdot 1$$
$$5 + 9 = 3x$$
$$14 = 3x$$
$$\frac{14}{3} = x$$

Check:

$$\begin{array}{c|c} \frac{5}{3x} + \frac{3}{x} = 1 & \\ \hline \frac{5}{3 \cdot (14/3)} + \frac{3}{(14/3)} & ? \ 1 \\ \frac{5}{14} + \frac{9}{14} & \\ \frac{14}{14} & \\ 1 & \text{TRUE} \end{array}$$

This checks, so the solution is $\frac{14}{3}$.

19.
$$\frac{t-2}{t+3} = \frac{3}{8}, \text{ LCM} = 8(t+3)$$
$$8(t+3)\left(\frac{t-2}{t+3}\right) = 8(t+3)\left(\frac{3}{8}\right)$$
$$8(t-2) = 3(t+3)$$
$$8t - 16 = 3t + 9$$
$$5t = 25$$
$$t = 5$$

Check:

$$\begin{array}{c|c} \frac{t-2}{t+3} = \frac{3}{8} & \\ \hline \frac{5-2}{5+3} \ ? & \frac{3}{8} \\ \frac{3}{8} & \text{TRUE} \end{array}$$

This checks, so the solution is 5.

21.
$$\frac{2}{x+1} = \frac{1}{x-2}, \text{ LCM} = (x+1)(x-2)$$
$$(x+1)(x-2) \cdot \frac{2}{x+1} = (x+1)(x-2) \cdot \frac{1}{x-2}$$
$$2(x-2) = x+1$$
$$2x - 4 = x + 1$$
$$x = 5$$

This checks, so the solution is 5.

23.
$$\frac{x}{6} - \frac{x}{10} = \frac{1}{6}, \text{ LCM} = 30$$
$$30\left(\frac{x}{6} - \frac{x}{10}\right) = 30 \cdot \frac{1}{6}$$
$$30 \cdot \frac{x}{6} - 30 \cdot \frac{x}{10} = 30 \cdot \frac{1}{6}$$
$$5x - 3x = 5$$
$$2x = 5$$
$$x = \frac{5}{2}$$

This checks, so the solution is $\frac{5}{2}$.

25.
$$\frac{t+2}{5} - \frac{t-2}{4} = 1, \text{ LCM} = 20$$
$$20\left(\frac{t+2}{5} - \frac{t-2}{4}\right) = 20 \cdot 1$$
$$20\left(\frac{t+2}{5}\right) - 20\left(\frac{t-2}{4}\right) = 20 \cdot 1$$
$$4(t+2) - 5(t-2) = 20$$
$$4t + 8 - 5t + 10 = 20$$
$$-t + 18 = 20$$
$$-t = 2$$
$$t = -2$$

This checks, so the solution is -2.

27.
$$\frac{5}{x-1} = \frac{3}{x+2},$$
$$\text{LCD} = (x-1)(x+2)$$
$$(x-1)(x+2) \cdot \frac{5}{x-1} = (x-1)(x+2) \cdot \frac{3}{x+2}$$
$$5(x+2) = 3(x-1)$$
$$5x + 10 = 3x - 3$$
$$2x = -13$$
$$x = -\frac{13}{2}$$

This checks, so the solution is $-\frac{13}{2}$.

29.

$$\frac{a-3}{3a+2} = \frac{1}{5}, \text{ LCM} = 5(3a+2)$$
$$5(3a+2)\cdot\frac{a-3}{3a+2} = 5(3a+2)\cdot\frac{1}{5}$$
$$5(a-3) = 3a+2$$
$$5a-15 = 3a+2$$
$$2a = 17$$
$$a = \frac{17}{2}$$

This checks, so the solution is $\frac{17}{2}$.

31.

$$\frac{x-1}{x-5} = \frac{4}{x-5}, \text{ LCM} = x-5$$
$$(x-5)\cdot\frac{x-1}{x-5} = (x-5)\cdot\frac{4}{x-5}$$
$$x-1 = 4$$
$$x = 5$$

The number 5 is not a solution because it makes a denominator zero. Thus, there is no solution.

33.

$$\frac{2}{x+3} = \frac{5}{x}, \text{ LCM} = x(x+3)$$
$$x(x+3)\cdot\frac{2}{x+3} = x(x+3)\cdot\frac{5}{x}$$
$$2x = 5(x+3)$$
$$2x = 5x+15$$
$$-15 = 3x$$
$$-5 = x$$

This checks, so the solution is -5.

35.

$$\frac{x-2}{x-3} = \frac{x-1}{x+1}, \text{ LCM} = (x-3)(x+1)$$
$$(x-3)(x+1)\cdot\frac{x-2}{x-3} = (x-3)(x+1)\cdot\frac{x-1}{x+1}$$
$$(x+1)(x-2) = (x-3)(x-1)$$
$$x^2-x-2 = x^2-4x+3$$
$$-x-2 = -4x+3$$
$$3x = 5$$
$$x = \frac{5}{3}$$

This checks, so the solution is $\frac{5}{3}$.

37.

$$\frac{1}{x+3}+\frac{1}{x-3} = \frac{1}{x^2-9},$$
$$\text{LCM} = (x+3)(x-3)$$
$$(x+3)(x-3)\left(\frac{1}{x+3}+\frac{1}{x-3}\right) = (x+3)(x-3)\cdot\frac{1}{(x+3)(x-3)}$$
$$(x-3)+(x+3) = 1$$
$$2x = 1$$
$$x = \frac{1}{2}$$

This checks, so the solution is $\frac{1}{2}$.

39.

$$\frac{x}{x+4}-\frac{4}{x-4} = \frac{x^2+16}{x^2-16},$$
$$\text{LCM} = (x+4)(x-4)$$
$$(x+4)(x-4)\left(\frac{x}{x+4}-\frac{4}{x-4}\right) = (x+4)(x-4)\cdot\frac{x^2+16}{(x+4)(x-4)}$$
$$x(x-4)-4(x+4) = x^2+16$$
$$x^2-4x-4x-16 = x^2+16$$
$$x^2-8x-16 = x^2+16$$
$$-8x-16 = 16$$
$$-8x = 32$$
$$x = -4$$

The number -4 is not a solution because it makes a denominator zero. Thus, there is no solution.

41.

$$\frac{4-a}{8-a} = \frac{4}{a-8} \quad 8-a \text{ and } a-8 \text{ are opposites}$$
$$\frac{4-a}{8-a}\cdot\frac{-1}{-1} = \frac{4}{a-8}$$
$$\frac{a-4}{a-8} = \frac{4}{a-8}, \text{ LCM} = a-8$$
$$(a-8)\left(\frac{a-4}{a-8}\right) = (a-8)\left(\frac{4}{a-8}\right)$$
$$a-4 = 4$$
$$a = 8$$

The number 8 is not a solution because it makes a denominator zero. Thus, there is no solution.

43.

$$2-\frac{a-2}{a+3} = \frac{a^2-4}{a+3}, \text{ LCM} = a+3$$
$$(a+3)\left(2-\frac{a-2}{a+3}\right) = (a+3)\cdot\frac{a^2-4}{a+3}$$
$$2(a+3)-(a-2) = a^2-4$$
$$2a+6-a+2 = a^2-4$$
$$0 = a^2-a-12$$
$$0 = (a-4)(a+3)$$
$$a-4 = 0 \quad \text{or} \quad a+3 = 0$$
$$a = 4 \quad \text{or} \quad a = -3$$

Only 4 checks, so the solution is 4.

45.

$$\frac{x+1}{x+2} = \frac{x+3}{x+4},$$
$$\text{LCM} = (x+2)(x+4)$$
$$(x+2)(x+4)\left(\frac{x+1}{x+2}\right) = (x+2)(x+4)\left(\frac{x+3}{x+4}\right)$$
$$(x+4)(x+1) = (x+2)(x+3)$$
$$x^2+5x+4 = x^2+5x+6$$
$$4 = 6 \quad \text{Subtracting } x^2 \text{ and } 5x$$

We get a false equation, so the original equation has no solution.

47. $4a - 3 = \frac{a+13}{a+1}$, LCM $= a+1$

$$(a+1)(4a-3) = (a+1)\cdot\frac{a+13}{a+1}$$
$$4a^2 + a - 3 = a + 13$$
$$4a^2 - 16 = 0$$
$$4(a+2)(a-2) = 0$$
$$a+2 = 0 \quad \text{or} \quad a-2 = 0$$
$$a = -2 \quad \text{or} \quad a = 2$$

Both of these check, so the two solutions are -2 and 2.

49. $\frac{4}{y-2} - \frac{2y-3}{y^2-4} = \frac{5}{y+2}$, LCM $= (y+2)(y-2)$

$$(y+2)(y-2)\left(\frac{4}{y-2} - \frac{2y-3}{(y+2)(y-2)}\right) = (y+2)(y-2)\cdot\frac{5}{y+2}$$
$$4(y+2) - (2y-3) = 5(y-2)$$
$$4y + 8 - 2y + 3 = 5y - 10$$
$$2y + 11 = 5y - 10$$
$$21 = 3y$$
$$7 = y$$

This checks, so the solution is 7.

51. $(2x^3 - 4x^2 + x - 7) + (4x^4 + x^3 + 4x^2 + x)$
$= 4x^4 + 3x^3 + 2x - 7$

53. $50p^2 - 100 = 50(p^2 - 2)$ Factoring out the common factor

Since $p^2 - 2$ cannot be factored, we have factored completely.

55. ***Familiarize***. Let x = the smaller even integer. Then $x + 2$ = the larger even integer.

Translate. We reword the problem.

Smaller even integer times larger even integer is 360.

$$x \cdot (x+2) = 360$$

Solve.

$$x(x+2) = 360$$
$$x^2 + 2x = 360$$
$$x^2 + 2x - 360 = 0$$
$$(x+20)(x-18) = 0$$
$$x + 20 = 0 \quad \text{or} \quad x - 18 = 0$$
$$x = -20 \quad \text{or} \quad x = 18$$

Check. The solutions of the equation are -20 and 18. When $x = -20$, then $x + 2 = -18$ and $-20(-18) = 360$. The numbers -20 and -18 are consecutive even integers which are solutions to the problem. When $x = 18$, then $x + 2 = 20$ and $18 \cdot 20 = 360$. The numbers 18 and 20 are also consecutive even integers which are solutions to the problem.

State. We have two solutions, each of which consists of a pair of numbers: -20 and -18, and 18 and 20.

57. $$\frac{x}{x^2+3x-4} + \frac{x+1}{x^2+6x+8} = \frac{2x}{x^2+x-2}$$
$$\frac{x}{(x+4)(x-1)} + \frac{x+1}{(x+4)(x+2)} = \frac{2x}{(x+2)(x-1)}$$
$$x(x+2) + (x+1)(x-1) = 2x(x+4)$$

Multiplying by the LCM, $(x+4)(x-1)(x+2)$

$$x^2 + 2x + x^2 - 1 = 2x^2 + 8x$$
$$2x^2 + 2x - 1 = 2x^2 + 8x$$
$$2x - 1 = 8x$$
$$-1 = 6x$$
$$-\frac{1}{6} = x$$

This checks, so the solution is $-\frac{1}{6}$.

Exercise Set 6.8

RC1. If two triangles are similar, then their corresponding angles have the same measures and their corresponding sides are proportional.

RC3. An equality of ratios, $\frac{A}{B} = \frac{C}{D}$, is called a proportion.

RC5. Rate equals distance divided by time.

1. ***Familiarize***. We complete the table shown in the text.

$d = r \cdot t$

	Distance	Speed	Time	
Car	150	r	t	$\rightarrow 150 = r(t)$
Truck	350	$r + 40$	t	$\rightarrow 350 = (r+40)t$

Translate. We apply the formula $d = rt$ along the rows of the table to obtain two equations:

$$150 = rt,$$
$$350 = (r+40)t$$

Then we solve each equation for t and set the results equal.

Solving $150 = rt$ for t: $t = \frac{150}{r}$

Solving $350 = (r+40)t$ for t: $t = \frac{350}{r+40}$

Thus, we have

$$\frac{150}{r} = \frac{350}{r+40}.$$

Solve. We multiply by the LCM, $r(r+40)$.

$$r(r+40)\cdot\frac{150}{r} = r(r+40)\cdot\frac{350}{r+40}$$
$$150(r+40) = 350r$$
$$150r + 6000 = 350r$$
$$6000 = 200r$$
$$30 = r$$

Check. If r is 30 km/h, then $r + 40$ is 70 km/h. The time for the car is 150/30, or 5 hr. The time for the truck is 350/70, or 5 hr. The times are the same. The values check.

State. The speed of Sarah's car is 30 km/h, and the speed of Rick's truck is 70 km/h.

3. ***Familiarize.*** Let r represent the giraffe's speed. Then $r+8$ represents the ostrich's speed. Let t represent time. We organize the information in a table.

$$d = r \cdot t$$

	Distance	Speed	Time
Giraffe	4	r	t
Ostrich	5	$r+8$	t

Translate. The rows of the table give us two equations.

$$4 = rt,$$
$$5 = (r+8)t$$

We can solve each equation for t and set the results equal.

Solving $4 = rt$ for t: $t = \frac{4}{r}$

Solving $5 = (r+8)t$ for t: $t = \frac{5}{r+8}$

Then $\frac{4}{r} = \frac{5}{r+8}$.

Solve. We multiply on both sides by the LCM, $r(r+8)$.

$$r(r+8)\cdot\frac{4}{r} = r(r+8)\cdot\frac{5}{r+8}$$
$$4(r+8) = 5r$$
$$4r+32 = 5r$$
$$32 = r$$

If $r = 32$, then $r + 8 = 32 + 8 = 40$.

Check. At 32 mph, the giraffe can run 4 mi in $\frac{4}{32}$, or $\frac{1}{8}$ hr. At 40 mph, the ostrich can run 5 mi in $\frac{5}{40}$, or $\frac{1}{8}$ hr. The times are the same so the answer checks.

State. A giraffe can run at a speed of 32 mph, and an ostrich can run 40 mph.

5. ***Familiarize.*** Let r = Kelly's speed, in km/h, and t = the time the bicyclists travel, in hours. Organize the information in a table.

	Distance	Speed	Time
Hank	42	$r-5$	t
Kelly	57	r	t

Translate. We can replace the t's in the table above using the formula $t = d/r$.

	Distance	Speed	Time
Hank	42	$r-5$	$\frac{42}{r-5}$
Kelly	57	r	$\frac{57}{r}$

Since the times are the same for both bicyclists, we have the equation

$$\frac{42}{r-5} = \frac{57}{r}.$$

Solve. We first multiply by the LCD, $r(r-5)$.

$$r(r-5)\cdot\frac{42}{r-5} = r(r-5)\cdot\frac{57}{r}$$
$$42r = 57(r-5)$$
$$42r = 57r - 285$$
$$-15r = -285$$
$$r = 19$$

If $r = 19$, then $r - 5 = 14$.

Check. If Hank's speed is 14 km/h and Kelly's speed is 19 km/h, then Hank bicycles 5 km/h slower than Kelly. Hank's time is 42/14, or 3 hr. Kelly's time is 57/19, or 3 hr. Since the times are the same, the answer checks.

State. Hank travels at 14 km/h, and Kelly travels at 19 km/h.

7. ***Familiarize.*** We let r represent the speed going. Then $2r$ is the speed returning. We let t represent the time going. Then $t-3$ represents the time returning. We organize the information in a table.

$$d = r \cdot t$$

	Distance	Speed	Time
Going	120	r	t
Returning	120	$2r$	$t-3$

Translate. The rows of the table give us two equations:

$$120 = rt,$$
$$120 = 2r(t-3)$$

We can solve each equation for r and set the results equal.

Solving $120 = rt$ for r: $r = \frac{120}{t}$

Solving $120 = 2r(t-3)$ for r: $r = \frac{120}{2(t-3)}$, or

$$r = \frac{60}{t-3}$$

Then $\frac{120}{t} = \frac{60}{t-3}$.

Solve. We multiply on both sides by the LCM, $t(t-3)$.

$$t(t-3)\cdot\frac{120}{t} = t(t-3)\cdot\frac{60}{t-3}$$
$$120(t-3) = 60t$$
$$120t - 360 = 60t$$
$$-360 = -60t$$
$$6 = t$$

Then substitute 6 for t in either equation to find r, the speed going:

$$r = \frac{120}{t}$$

$$r = \frac{120}{6} \quad \text{Substituting 6 for } t$$

$$r = 20$$

Check. If $r = 20$ and $t = 6$, then $2r = 2 \cdot 20$, or 40 mph and $t - 3 = 6 - 3$, or 3 hr. The distance going is $6 \cdot 20$, or 120 mi. The distance returning is $40 \cdot 3$, or 120 mi. The numbers check.

State. The speed going is 20 mph.

9. ***Familiarize.*** Let r = Ralph's speed, in km/h. Then Bonnie's speed is $r + 3$. Also set t = the time, in hours, that Ralph and Bonnie walk. We organize the information in a table.

	Distance	Speed	Time
Ralph	7.5	r	t
Bonnie	12	$r+3$	t

Translate. We can replace the t's in the table shown above using the formula $t = d/r$.

	Distance	Speed	Time
Ralph	7.5	r	$\frac{7.5}{r}$
Bonnie	12	$r+3$	$\frac{12}{r+3}$

Since the times are the same for both walkers, we have the equation

$$\frac{7.5}{r} = \frac{12}{r+3}.$$

Solve. We first multiply by the LCD, $r(r+3)$.

$$r(r+3) \cdot \frac{7.5}{r} = r(r+3) \cdot \frac{12}{r+3}$$

$$7.5(r+3) = 12r$$

$$7.5r + 22.5 = 12r$$

$$22.5 = 4.5r$$

$$5 = r$$

If $r = 5$, then $r + 3 = 8$.

Check. If Ralph's speed is 5 km/h and Bonnie's speed is 8 km/h, then Bonnie walks 3 km/h faster than Ralph. Ralph's time is 7.5/5, or 1.5 hr. Bonnie's time is 12/8, or 1.5 hr. Since the times are the same, the answer checks.

State. Ralph's speed is 5 km/h, and Bonnie's speed is 8 km/h.

11. ***Familiarize.*** Let t = the time Tory travels. Then $t+2$ = Emilio's time. We organize the information in a table.

	Distance	Speed	Time
Tory	40	r	t
Emilio	100	r	$t+2$

Translate. From the rows of the table we have two equations.

$$40 = rt$$

$$100 = r(t+2)$$

We solve each equation for r and set the results equal.

Solving $40 = rt$ for r: $r = \frac{40}{t}$

Solving $100 = r(t+2)$ for r: $r = \frac{100}{t+2}$

Thus, we have

$$\frac{40}{t} = \frac{100}{t+2}.$$

Solve. We multiply by the LCM, $t(t+2)$.

$$t(t+2) \cdot \frac{40}{t} = t(t+2) \cdot \frac{100}{t+2}$$

$$40(t+2) = 100t$$

$$40t + 80 = 100t$$

$$80 = 60t$$

$$\frac{4}{3} = t$$

Check. If Tory travels 40 km in $\frac{4}{3}$ hr, then her speed is $\frac{40}{4/3}$, or $40 \cdot \frac{3}{4}$, or $\frac{120}{4}$, or 30 km/h. If Tory travels for $\frac{4}{3}$ hr, then Emilio travels for $\frac{4}{3} + 2$, or $\frac{10}{3}$ hr. If he travels 100 km in this time, then his speed is $\frac{100}{10/3}$, or $100 \cdot \frac{3}{10}$, or $\frac{300}{10}$, or 30 km/h. Since the speeds are the same, the answer checks.

State. It took Tory $\frac{4}{3}$ hr, or $1\frac{1}{3}$ hr, to navigate the 40 km.

13. ***Familiarize.*** Let t = the number of minutes it takes Nicole and Glen to weed the garden, working together.

Translate. We use the work principle.

$$t\left(\frac{1}{50}\right) + t\left(\frac{1}{40}\right) = 1, \text{ or } \frac{t}{50} + \frac{t}{40} = 1$$

Solve. We solve the equation.

$$\frac{t}{50} + \frac{t}{40} = 1, \text{ LCM} = 200$$

$$200\left(\frac{t}{50} + \frac{t}{40}\right) = 200 \cdot 1$$

$$200 \cdot \frac{t}{50} + 200 \cdot \frac{t}{40} = 200$$

$$4t + 5t = 200$$

$$9t = 200$$

$$t = \frac{200}{9}, \text{ or } 22\frac{2}{9}$$

Check. In $\frac{200}{9}$ min, the portion of the job done is $\frac{1}{50} \cdot \frac{200}{9} + \frac{1}{40} \cdot \frac{200}{9} = \frac{4}{9} + \frac{5}{9} = 1$. The answer checks.

State. It would take $22\frac{2}{9}$ min to weed the garden if Nicole and Glen worked together.

15. ***Familiarize.*** The job takes Vern 45 min working alone and Nina 60 min working alone. Then in 1 minute Vern does $\frac{1}{45}$ of the job and Nina does $\frac{1}{60}$ of the job. Working together, they can do $\frac{1}{45}+\frac{1}{60}$, or $\frac{7}{180}$ of the job in 1 minute. In 20 minutes, Vern does $\frac{20}{45}$ of the job and Nina does $\frac{20}{60}$ of the job. Working together, they can do $\frac{20}{45}+\frac{20}{60}$, or $\frac{7}{9}$ of the job. In 30 minutes, they can do $\frac{30}{45}+\frac{30}{60}$, or $\frac{7}{6}$ of the job which is more of the job than needs to be done. The answer is somewhere between 20 minutes and 30 minutes.

Translate. If they work together t minutes, then Vern does $t\left(\frac{1}{45}\right)$ of the job and Nina does $t\left(\frac{1}{60}\right)$ of the job. We want some number t such that

$$t\left(\frac{1}{45}\right)+t\left(\frac{1}{60}\right)=1, \text{ or } \frac{t}{45}+\frac{t}{60}=1.$$

Solve. We solve the equation.

$$\begin{aligned}
\frac{t}{45}+\frac{t}{60}&=1, \text{ LCM} = 180\\
180\left(\frac{t}{45}+\frac{t}{60}\right)&=180\cdot 1\\
180\cdot\frac{t}{45}+180\cdot\frac{t}{60}&=180\\
4t+3t&=180\\
7t&=180\\
t&=\frac{180}{7}, \text{ or } 25\frac{5}{7}
\end{aligned}$$

Check. The check can be done by repeating the computations. We also have a partial check in that we expected from our familiarization step that the answer would be between 20 minutes and 30 minutes.

State. It would take them $25\frac{5}{7}$ minutes to complete the job working together.

17. ***Familiarize.*** Let t = the time it would take Henry and Carly to prepare the trays, working together.

Translate. We use the work principle.

$$\frac{t}{a}+\frac{t}{b}=1$$

$$\frac{t}{8.5}+\frac{t}{10.4}=1$$

Solve.

$$\begin{aligned}
8.5(10.4)\left(\frac{t}{8.5}+\frac{t}{10.4}\right)&=8.5(10.4)\cdot 1\\
10.4t+8.5t&=88.4\\
18.9t&=88.4\\
t&\approx 4.7
\end{aligned}$$

Check. In 4.7 hr Henry does $\frac{4.7}{8.5}$, or about 0.55, of the job and Carly does $\frac{4.7}{10.4}$, or about 0.45, of the job. Together they do 0.55+0.45, or 1 complete job, so the answer checks.

State. Working together, Henry and Carly can prepare the trays in about 4.7 hr.

19. ***Familiarize.*** The job takes Peggyann 9 hours working alone and Matthew 7 hours working alone. Then in 1 hour Peggyann does $\frac{1}{9}$ of the job and Matthew does $\frac{1}{7}$ of the job. Working together they can do $\frac{1}{9}+\frac{1}{7}$, or $\frac{16}{63}$ of the job in 1 hour. In two hours, Peggyann does $2\left(\frac{1}{9}\right)$ of the job and Matthew does $2\left(\frac{1}{7}\right)$ of the job. Working together they can do $2\left(\frac{1}{9}\right)+2\left(\frac{1}{7}\right)$, or $\frac{32}{63}$ of the job in two hours. In five hours they can do $5\left(\frac{1}{9}\right)+5\left(\frac{1}{7}\right)$, or $\frac{80}{63}$, or $1\frac{17}{63}$ of the job which is more of the job than needs to be done. The answer is somewhere between 2 hr and 5 hr.

Translate. If they work together t hours, Peggyann does $t\left(\frac{1}{9}\right)$ of the job and Matthew does $t\left(\frac{1}{7}\right)$ of the job. We want some number t such that

$$t\left(\frac{1}{9}\right)+t\left(\frac{1}{7}\right)=1, \text{ or } \frac{t}{9}+\frac{t}{7}=1.$$

Solve. We solve the equation.

$$\begin{aligned}
\frac{t}{9}+\frac{t}{7}&=1, \text{ LCM} = 63\\
63\left(\frac{t}{9}+\frac{t}{7}\right)&=63\cdot 1\\
63\cdot\frac{t}{9}+63\cdot\frac{t}{7}&=63\\
7t+9t&=63\\
16t&=63\\
t&=\frac{63}{16}, \text{ or } 3\frac{15}{16}
\end{aligned}$$

Check. The check can be done by repeating the computations. We also have a partial check in that we expected from our familiarization step that the answer would be between 2 hr and 5 hr.

State. Working together, it takes them $3\frac{15}{16}$ hr to complete the job.

21. ***Familiarize.*** Let t = the number of minutes it would take the two machines to make one copy of the report, working together.

Translate. We use the work principle.

$$t\left(\frac{1}{10}\right)+t\left(\frac{1}{6}\right)=1, \text{ or } \frac{t}{10}+\frac{t}{6}=1$$

Solve. We solve the equation.

$$\frac{t}{10}+\frac{t}{6}=1,\ \text{LCM}=30$$
$$30\left(\frac{t}{10}+\frac{t}{6}\right)=30\cdot 1$$
$$30\cdot\frac{t}{10}+30\cdot\frac{t}{6}=30$$
$$3t+5t=30$$
$$8t=30$$
$$t=\frac{15}{4},\ \text{or } 3\frac{3}{4}$$

Check. In $\frac{15}{4}$ min, the portion of the job done is $\frac{1}{10}\cdot\frac{15}{4}+\frac{1}{6}\cdot\frac{15}{4}=\frac{3}{8}+\frac{5}{8}=1$. The answer checks.

State. It would take the two machines $3\frac{3}{4}$ min to make one copy of the report, working together.

23. $\frac{60 \text{ students}}{18 \text{ teachers}}=\frac{60}{18}$ students/teacher = $\frac{10}{3}$ students/teacher

25. $\frac{4.6 \text{ km}}{2 \text{ hr}}=2.3$ km/h

27. ***Familiarize.*** A 120-lb person should eat at least 44 g of protein each day, and we wish to find the minimum protein required for a 180-lb person. We can set up ratios. We let p = the minimum number of grams of protein a 180-lb person should eat each day.

Translate. If we assume the rates of protein intake are the same, the ratios are the same and we have an equation.

$$\begin{array}{r}\text{Protein}\rightarrow\\ \text{Weight}\rightarrow\end{array}\frac{44}{120}=\frac{p}{180}\begin{array}{l}\leftarrow\text{Protein}\\ \leftarrow\text{Weight}\end{array}$$

Solve. We solve the proportion.

$$360\cdot\frac{44}{120}=360\cdot\frac{p}{180}\quad\text{Multiplying by the LCM, 360}$$
$$3\cdot 44=2\cdot p$$
$$132=2p$$
$$66=p$$

Check. $\frac{44}{120}=\frac{4\cdot 11}{4\cdot 30}=\frac{\not{4}\cdot 11}{\not{4}\cdot 30}=\frac{11}{30}$ and $\frac{66}{180}=\frac{6\cdot 11}{6\cdot 30}=\frac{\not{6}\cdot 11}{\not{6}\cdot 30}=\frac{11}{30}$. The ratios are the same.

State. A 180-lb person should eat a minimum of 66 g of protein each day.

29. ***Familiarize.*** 10 cc of human blood contains 1.2 grams of hemoglobin, and we wish to find how many grams of hemoglobin are contained in 16 cc of the same blood. We can set up ratios. Let H = the amount of hemoglobin in 16 cc of the same blood.

Translate. Assuming the two ratios are the same, we can translate to a proportion.

$$\begin{array}{r}\text{Grams}\rightarrow\\ \text{cm}^3\rightarrow\end{array}\frac{H}{16}=\frac{1.2}{10}\begin{array}{l}\leftarrow\text{Grams}\\ \leftarrow\text{cm}^3\end{array}$$

Solve. We solve the proportion.

We multiply by 16 to get H alone.

$$16\cdot\frac{H}{16}=16\cdot\frac{1.2}{10}$$
$$H=\frac{19.2}{10}$$
$$H=1.92$$

Check.

$\frac{1.92}{16}=0.12$ and $\frac{1.2}{10}=0.12$. The ratios are the same.

State. 16 cc of the same blood would contain 1.92 grams of hemoglobin.

31. ***Familiarize.*** Let g = the number of gallons of gas required for 176 mi of city driving.

Translate. We translate to a proportion.

$$\begin{array}{r}\text{Gas}\rightarrow\\ \text{Miles}\rightarrow\end{array}\frac{11}{242}=\frac{g}{176}\begin{array}{l}\leftarrow\text{Gas}\\ \leftarrow\text{Miles}\end{array}$$

Solve. We multiply by 176 to get g alone.

$$176\cdot\frac{11}{242}=176\cdot\frac{g}{176}$$
$$\frac{1936}{242}=g$$
$$8=g$$

Check. $\frac{11}{242}=0.0\overline{45}$ and $\frac{8}{176}=0.0\overline{45}$. The ratios are the same, so the answer checks.

State. Traveling 176 mi takes 8 gal of gas.

33. ***Familiarize.*** The ratio of trout tagged to the total trout population, P, is $\frac{112}{P}$. Of the 82 trout checked later, 32 were tagged. The ratio of trout tagged to trout checked is $\frac{32}{82}$.

Translate. Assuming the two ratios are the same, we can translate to a proportion.

$$\begin{array}{c}\text{Trout tagged}\\ \text{originally}\end{array}\longrightarrow\frac{112}{P}=\frac{32}{82}\longleftarrow\begin{array}{c}\text{Tagged trout}\\ \text{caught later}\end{array}$$
$$\begin{array}{c}\text{Trout}\\ \text{population}\end{array}\longrightarrow\qquad\qquad\longleftarrow\begin{array}{c}\text{Trout caught}\\ \text{later}\end{array}$$

Solve. We solve the equation.

$$82P\cdot\frac{112}{P}=82P\cdot\frac{32}{82}\quad\text{Multiplying by the LCM, } 82P$$
$$82\cdot 112=P\cdot 32$$
$$9184=32P$$
$$287=P$$

Check. $\frac{112}{287}\approx 0.390$ and $\frac{32}{82}\approx 0.390$. The ratios are the same.

State. The trout population is 287.

35. ***Familiarize.*** A sample of 144 firecrackers contained 9 duds, and we wish to find how many duds could be expected in a sample of 3200 firecrackers. We can set up ratios, letting $d =$ the number of duds expected in a sample of 3200 firecrackers.

Translate. Assuming the rates of occurrence of duds are the same, we can translate to a proportion.

$$\begin{array}{r} \text{Duds} \rightarrow \\ \text{Sample size} \rightarrow \end{array} \frac{9}{144} = \frac{d}{3200} \begin{array}{l} \leftarrow \text{Duds} \\ \leftarrow \text{Sample size} \end{array}$$

Solve. We solve the equation. We multiply by 3200 to get d alone.

$$3200 \cdot \frac{9}{144} = 3200 \cdot \frac{d}{3200}$$
$$\frac{28,800}{144} = d$$
$$200 = d$$

Check. $\frac{9}{144} = 0.0625$ and $\frac{200}{3200} = 0.0625$. The ratios are the same.

State. You would expect 200 duds in a sample of 3200 firecrackers.

37. ***Familiarize.*** Let $b =$ the number of bags that would be mishandled for 1450 passengers.

Translate. We translate to a proportion.

$$\begin{array}{r} \text{Bags} \rightarrow \\ \text{Passengers} \rightarrow \end{array} \frac{3.39}{1000} = \frac{b}{1450} \begin{array}{l} \leftarrow \text{Bags} \\ \leftarrow \text{Passengers} \end{array}$$

Solve.

$$1450 \cdot \frac{3.39}{1000} = 1450 \cdot \frac{b}{1450}$$
$$\frac{4915.5}{1000} = b$$
$$4.92 \approx b$$

Check. $\frac{3.39}{1000} = 0.00339$ and $\frac{4.92}{1450} \approx 0.00339$. The ratios are the same, so the answer checks.

State. About 4.92 bags would be mishandled for 1450 passengers.

39. ***Familiarize.*** Let $h =$ the amount of honey, in pounds, that 35,000 trips to flowers would produce.

Translate. We translate to a proportion.

$$\begin{array}{r} \text{Honey} \rightarrow \\ \text{Trips} \rightarrow \end{array} \frac{1}{20,000} = \frac{h}{35,000} \begin{array}{l} \leftarrow \text{Honey} \\ \leftarrow \text{Trips} \end{array}$$

Solve. We solve the proportion.

$$35,000 \cdot \frac{1}{20,000} = 35,000 \cdot \frac{h}{35,000}$$
$$1.75 = h$$

Check. $\frac{1}{20,000} = 0.00005$ and $\frac{1.75}{35,000} = 0.00005$. The ratios are the same.

State. 35,000 trips to gather nectar will produce 1.75 lb of honey.

41. We write a proportion and then solve it.

$$\frac{b}{6} = \frac{7}{4}$$
$$b = \frac{7}{4} \cdot 6 \qquad \text{Multiplying by 6}$$
$$b = \frac{42}{4}$$
$$b = \frac{21}{2}, \text{ or } 10.5$$

(Note that the proportions $\frac{6}{b} = \frac{4}{7}$, $\frac{b}{7} = \frac{6}{4}$, or $\frac{7}{b} = \frac{4}{6}$ could also be used.)

43. We write a proportion and then solve it.

$$\frac{4}{f} = \frac{6}{4}$$
$$4f \cdot \frac{4}{f} = 4f \cdot \frac{6}{4}$$
$$16 = 6f$$
$$\frac{8}{3} = f \qquad \text{Simplifying}$$

(One of the following proportions could also be used: $\frac{f}{4} = \frac{4}{6}$, $\frac{4}{f} = \frac{9}{6}$, $\frac{f}{4} = \frac{6}{9}$, $\frac{4}{9} = \frac{f}{6}$, $\frac{9}{4} = \frac{6}{f}$)

45. We write a proportion and then solve it.

$$\frac{h}{7} = \frac{10}{6}$$
$$h = \frac{10}{6} \cdot 7 \quad \text{Multiplying by 7}$$
$$h = \frac{70}{6}$$
$$h = \frac{35}{3} \qquad \text{Simplifying}$$

(Note that the proportions $\frac{7}{h} = \frac{6}{10}$, $\frac{h}{10} = \frac{7}{6}$, or $\frac{10}{h} = \frac{6}{7}$ could also be used.)

47. ***Familiarize.*** Let $B =$ the estimated largemouth bass population.

Translate. We translate to a proportion.

$$\begin{array}{c} \text{Bass tagged} \\ \text{originally} \\ \text{Bass} \\ \text{population} \end{array} \begin{array}{c} \longrightarrow \\ \longrightarrow \end{array} \frac{300}{P} = \frac{15}{85} \begin{array}{c} \longleftarrow \\ \longleftarrow \end{array} \begin{array}{c} \text{Tagged bass} \\ \text{caught later} \\ \text{Bass caught} \\ \text{later} \end{array}$$

Solve.

$$85P \cdot \frac{300}{P} = 85P \cdot \frac{15}{85}$$
$$25,500 = 15P$$
$$1700 = P$$

Check. $\frac{300}{1700} \approx 0.176$ and $\frac{15}{85} \approx 0.176$. The ratios are the same, so the answer checks.

State. The largemouth bass population is estimated to be 1700.

49. $m = \dfrac{-6-(-6)}{0-7} = \dfrac{0}{-7} = 0$

51. $x^5 \cdot x^6 = x^{5+6} = x^{11}$

53. $x^{-5} \cdot x^{-6} = x^{-5+(-6)} = x^{-11} = \dfrac{1}{x^{11}}$

55. Graph: $y = -\dfrac{3}{4}x + 2$

We select some x-values and compute y-values. We use multiples of 4 to avoid fractions.

If $x = -4$, then $y = -\dfrac{3}{4}(-4) + 2 = 5$.

If $x = 0$, then $y = -\dfrac{3}{4} \cdot 0 + 2 = 2$.

If $x = 4$, then $y = -\dfrac{3}{4} \cdot 4 + 2 = -1$.

x	y	(x, y)
-4	5	$(-4, 5)$
0	2	$(0, 2)$
4	-1	$(4, -1)$

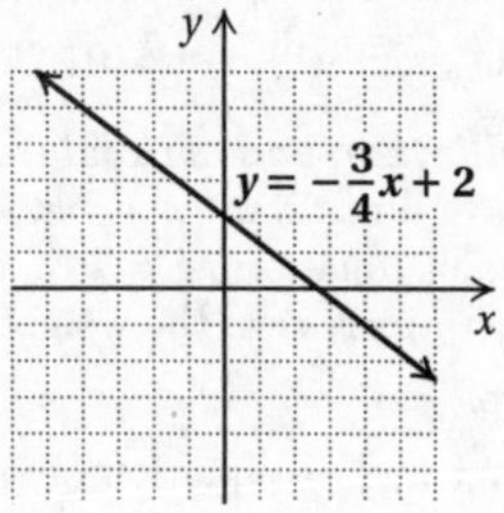

57. Graph: $x = -3$

The graph is a vertical line with x-intercept $(-3, 0)$.

59. ***Familiarize***. Let $t =$ the number of minutes after 5:00 at which the hands of the clock will first be together. While the minute hand moves through t minutes, the hour hand moves through $t/12$ minutes. At 5:00 the hour hand is on the 25-minute mark. We wish to find when a move of the minute hand through t minutes is equal to $25 + t/12$ minutes.

Translate. We use the last sentence of the familiarization step to write an equation.

$$t = 25 + \frac{t}{12}$$

Solve. We solve the equation.

$$t = 25 + \frac{t}{12}$$

$$12 \cdot t = 12\left(25 + \frac{t}{12}\right)$$

$$12t = 300 + t \qquad \text{Multiplying by 12}$$

$$11t = 300$$

$$t = \frac{300}{11} \text{ or } 27\frac{3}{11}$$

Check. At $27\dfrac{3}{11}$ minutes after 5:00, the minute hand is at the $27\dfrac{3}{11}$-minutes mark and the hour hand is at the $25 + \dfrac{27\frac{3}{11}}{12}$-minute mark. Simplifying $25 + \dfrac{27\frac{3}{11}}{12}$, we get

$$25 + \frac{\frac{300}{11}}{12} = 25 + \frac{300}{11} \cdot \frac{1}{12} = 25 + \frac{25}{11} = 25 + 2\frac{3}{11} = 27\frac{3}{11}.$$

Thus, the hands are together.

State. The hands are first together $27\dfrac{3}{11}$ minutes after 5:00.

Exercise Set 6.9

RC1. (b)

RC3. (f)

RC5. (c)

1. We substitute to find k.

$y = kx$

$36 = k \cdot 9$ Substituting 36 for y and 9 for x

$\dfrac{36}{9} = k$

$4 = k$ k is the variation constant.

The equation of the variation is $y = 4x$.

To find the value of y when $x = 20$ we substitute 20 for x in the equation of variation.

$y = 4x$

$y = 4 \cdot 20$

$y = 80$

The value of y is 80 when $x = 20$.

3. We substitute to find k.

$y = kx$

$0.8 = k \cdot 0.5$ Substituting 0.8 for y and 0.5 for x

$\dfrac{0.8}{0.5} = k$

$1.6 = k$ k is the variation constant.

The equation of the variation is $y = 1.6x$.

To find the value of y when $x = 20$ we substitute 20 for x in the equation of variation.

$y = 1.6x$

$y = 1.6(20)$

$y = 32$

The value of y is 32 when $x = 20$.

5. We substitute to find k.

$y = kx$

$630 = k \cdot 175$ Substituting 630 for y and 175 for x

$\dfrac{630}{175} = k$

$3.6 = k$ k is the variation constant.

The equation of the variation is $y = 3.6x$.

To find the value of y when $x = 20$ we substitute 20 for x in the equation of variation.

$y = 3.6x$

$y = 3.6(20)$

$y = 72$

The value of y is 72 when $x = 20$.

7. We substitute to find k.

$y = kx$

$500 = k \cdot 60$ Substituting 500 for y and 60 for x

$\frac{500}{60} = k$

$\frac{25}{3} = k$ k is the variation constant.

The equation of the variation is $y = \frac{25}{3}x$.

To find the value of y when $x = 20$ we substitute 20 for x in the equation of variation.

$y = \frac{25}{3}x$

$y = \frac{25}{3} \cdot 20$

$y = \frac{500}{3}$

The value of y is $\frac{500}{3}$ when $x = 20$.

9. ***Familiarize and Translate***. The problem states that we have direct variation between the variables P and H. Thus, an equation $P = kH$, $k > 0$, applies. As the number of hours increases, the paycheck increases.

Solve.

a) First find an equation of variation.

$P = kH$

$180 = k \cdot 15$ Substituting 180 for P and 15 for H

$\frac{180}{15} = k$

$12 = k$

The equation of variation is $P = 12H$.

b) Use the equation to find the pay for 35 hours work.

$P = 12H$

$P = 12(35)$ Substituting 35 for H

$P = 420$

Check. This check might be done by repeating the computations. We might also do some reasoning about the answer. The paycheck increased from \$180 to \$420. Similarly, the hours increased from 15 to 35.

State. a) The equation of variation is $P = 12H$.

b) For 35 hours work, the paycheck is \$420.

11. ***Familiarize and Translate***. The problem states that we have direct variation between the variables C and S. Thus, an equation $C = kS$, $k > 0$, applies. As the depth increases, the cost increases.

Solve.

a) First find an equation of variation.

$C = kS$

$67.5 = k \cdot 6$ Substituting 75 for C and 6 for S

$\frac{67.5}{6} = k$

$11.25 = k$

The equation of variation is $C = 11.25S$.

b) Use the equation to find the cost of filling the sandbox to a depth of 9 inches.

$C = 11.25S$

$C = 11.25(9)$ Substituting 9 for S

$C = 101.25$

Check. In addition to repeating the computations, we can also do some reasoning. The depth increased from 6 inches to 9 inches. Similarly, the cost increased from \$67.50 to \$101.25.

State. a) The equation of variation is $C = 11.25S$.

b) The sand will cost \$101.25.

13. ***Familiarize and Translate***. The problem states that we have direct variation between the variables M and E. Thus, an equation $M = kE$, $k > 0$, applies. As the weight on earth increases, the weight on the moon increases.

Solve.

a) First find an equation of variation.

$M = kE$

$32 = k \cdot 192$ Substituting 32 for M and 192 for E

$\frac{32}{192} = k$

$\frac{1}{6} = k$

The equation of variation is $M = \frac{1}{6}E$.

b) Use the equation to find how much a 110-lb person would weigh on the moon.

$M = \frac{1}{6}E$

$M = \frac{1}{6} \cdot 110$ Substituting 110 for E

$M = \frac{110}{6}$, or $18.\overline{3}$

c) Use the equation to find how much a person who weighs 5 lb on the moon would weigh on Earth.

$M = \frac{1}{6}E$

$5 = \frac{1}{6}E$

$30 = E$ Multiplying by 6

Check. In addition to repeating the computations we can do some reasoning. When the weight on Earth decreased from 192 lb to 110 lb, the weight on the moon decreased from 32 lb to $18.\overline{3}$ lb. Similarly, when the weight on the moon decreased from 32 lb to 5 lb, the weight on Earth decreased from 192 lb to 30 lb.

State. a) The equation of variation is $M = \frac{1}{6}E$.

b) Elizabeth, who weighs 110 lb on Earth, would weigh $18.\overline{3}$ lb on the moon.

c) Jasmine, who weighs 5 lb on the moon, would weigh 30 lb on Earth.

15. ***Familiarize and Translate.*** The problem states that we have direct variation between the variables N and S. Thus, an equation $N = kS$, $k > 0$, applies. As the speed of the internal processor increases, the number of instructions increases.

Solve.

a) First find an equation of variation.

$$N = kS$$

$$2,000,000 = k \cdot 25 \quad \text{Substituting 2,000,000 for } N \text{ and 25 for } S$$

$$\frac{2,000,000}{25} = k$$

$$80,000 = k$$

The equation of variation is $N = 80,000S$.

b) Use the equation to find how many instructions the processor will perform at a speed of 200 megahertz.

$$N = 80,000S$$

$$N = 80,000 \cdot 200 \quad \text{Substituting 200 for } S$$

$$N = 16,000,000$$

Check. In addition to repeating the computations we can do some reasoning. The speed of the processor increased from 25 to 200 megahertz. Similarly, the number of instructions performed per second increased from 2,000,000 to 16,000,000.

State. a) The equation of variation is $N = 80,000S$.

b) The processor will perform 16,000,000 instructions per second running at a speed of 200 megahertz.

17. ***Familiarize and Translate.*** This problem states that we have direct variation between the variables S and W. Thus, an equation $S = kW$, $k > 0$, applies. As the weight increases, the number of servings increases.

Solve. First find an equation of variation.

$$S = kW$$

$$70 = k \cdot 9 \quad \text{Substituting 70 for } S \text{ and 9 for } W$$

$$\frac{70}{9} = k$$

The equation of variation is $S = \frac{70}{9}W$.

Use the equation to find the number of servings from 12 kg of round steak.

$$S = \frac{70}{9}W$$

$$S = \frac{70}{9} \cdot 12 \quad \text{Substituting 12 for } W$$

$$S = \frac{840}{9}$$

$$S = \frac{280}{3}, \text{ or } 93\frac{1}{3}$$

Check. A check can always be done by repeating the computations. We can also do some reasoning about the answer. When the weight increased from 9 kg to 12 kg, the number of servings increased from 70 to $93\frac{1}{3}$.

State. $93\frac{1}{3}$ servings can be obtained from 12 kg of round steak.

19. We substitute to find k.

$$y = \frac{k}{x}$$

$$3 = \frac{k}{25} \quad \text{Substituting 3 for } y \text{ and 25 for } x$$

$$25 \cdot 3 = k$$

$$75 = k$$

The equation of variation is $y = \frac{75}{x}$.

To find the value of y when $x = 10$ we substitute 10 for x in the equation of variation.

$$y = \frac{75}{x}$$

$$y = \frac{75}{10}$$

$$y = \frac{15}{2}, \text{ or } 7.5$$

The value of y is $\frac{15}{2}$, or 7.5, when $x = 10$.

21. We substitute to find k.

$$y = \frac{k}{x}$$

$$10 = \frac{k}{8} \quad \text{Substituting 10 for } y \text{ and 8 for } x$$

$$8 \cdot 10 = k$$

$$80 = k$$

The equation of variation is $y = \frac{80}{x}$.

To find the value of y when $x = 10$ we substitute 10 for x in the equation of variation.

$$y = \frac{80}{x}$$

$$y = \frac{80}{10}$$

$$y = 8$$

The value of y is 8 when $x = 10$.

23. We substitute to find k.

$$y = \frac{k}{x}$$

$$6.25 = \frac{k}{0.16} \quad \text{Substituting 6.25 for } y \text{ and 0.16 for } x$$

$$0.16(6.25) = k$$

$$1 = k$$

The equation of variation is $y = \frac{1}{x}$.

To find the value of y when $x = 10$ we substitute 10 for x in the equation of variation.

$$y = \frac{1}{x}$$

$$y = \frac{1}{10}$$

The value of y is $\frac{1}{10}$ when $x = 10$.

25. We substitute to find k.

$$y = \frac{k}{x}$$

$$50 = \frac{k}{42} \quad \text{Substituting 50 for } y \text{ and 42 for } x$$

$$42 \cdot 50 = k$$

$$2100 = k$$

The equation of variation is $y = \frac{2100}{x}$.

To find the value of y when $x = 10$ we substitute 10 for x in the equation of variation.

$$y = \frac{2100}{x}$$

$$y = \frac{2100}{10}$$

$$y = 210$$

The value of y is 210 when $x = 10$.

27. We substitute to find k.

$$y = \frac{k}{x}$$

$$0.2 = \frac{k}{0.3} \quad \text{Substituting 0.2 for } y \text{ and 0.3 for } x$$

$$0.06 = k$$

The equation of variation is $y = \frac{0.06}{x}$.

To find the value of y when $x = 10$ we substitute 10 for x in the equation of variation.

$$y = \frac{0.06}{x}$$

$$y = \frac{0.06}{10}$$

$$y = \frac{6}{1000} \quad \text{Multiplying } \frac{0.06}{10} \text{ by } \frac{100}{100}$$

$$y = \frac{3}{500}, \text{ or } 0.006$$

The value of y is $\frac{3}{500}$, or 0.006, when $x = 10$.

29. a) It seems reasonable that, as the number of hours of production increases, the number of compact-disc players produced will increase, so direct variation might apply.

b) ***Familiarize.*** Let H = the number of hours the production line is working, and let P = the number of compact-disc players produced. An equation $P = kH$, $k > 0$, applies. (See part (a)).

Translate. We write an equation of variation.

Number of players produced varies directly as hours of production. This translates to $P = kH$.

Solve.

a) First we find an equation of variation.

$$P = kH$$

$$15 = k \cdot 8 \quad \text{Substituting 8 for } H \text{ and 15 for } P$$

$$\frac{15}{8} = k$$

The equation of variation is $P = \frac{15}{8}H$.

b) Use the equation to find the number of players produced in 37 hr.

$$P = \frac{15}{8}H$$

$$P = \frac{15}{8} \cdot 37 \quad \text{Substituting 37 for } H$$

$$P = \frac{555}{8} = 69\frac{3}{8}$$

Check. In addition to repeating the computations, we can do some reasoning. The number of hours increased from 8 to 37. Similarly, the number of compact disc players produced increased from 15 to $69\frac{3}{8}$.

State. About $69\frac{3}{8}$ compact-disc players can be produced in 37 hr.

31. a) It seems reasonable that, as the number of workers increases, the number of hours required to do the job decreases, so inverse variation might apply.

b) ***Familiarize.*** Let T = the time required to cook the meal and N = the number of cooks. An equation $T = k/N$, $k > 0$, applies. (See part (a)).

Translate. We write an equation of variation. Time varies inversely as the number of cooks. This translates to $T = \frac{k}{N}$.

Solve.

a) First find the equation of variation.

$$T = \frac{k}{N}$$

$$4 = \frac{k}{9} \quad \text{Substituting 4 for } T \text{ and 9 for } N$$

$$36 = k$$

The equation of variation is $T = \dfrac{36}{N}$.

b) Use the equation to find the amount of time it takes 8 cooks to prepare the dinner.

$$T = \frac{36}{N}$$

$$T = \frac{36}{8} \quad \text{Substituting 8 for } N$$

$$T = 4.5$$

Check. The check might be done by repeating the computation. We might also analyze the results. The number of cooks decreased from 9 to 8, and the time increased from 4 hr to 4.5 hr. This is what we would expect with inverse variation.

State. It will take 8 cooks 4.5 hr to prepare the dinner.

33. ***Familiarize.*** The problem states that we have inverse variation between the variables N and P. Thus, an equation $N = k/P$, $k > 0$, applies. As the miles per gallon rating increases, the number of gallons required to travel the fixed distance decreases.

Translate. We write an equation of variation. Number of gallons varies inversely as miles per gallon rating. This translates to $N = \dfrac{k}{P}$.

Solve.

a) First find an equation of variation.

$$N = \frac{k}{P}$$

$$20 = \frac{k}{14} \quad \text{Substituting 20 for } N \text{ and 14 for } P$$

$$280 = k$$

The equation is $N = \dfrac{280}{P}$.

b) Use the equation to find the number of gallons of gasoline needed for a car that gets 28 mpg.

$$N = \frac{k}{P}$$

$$N = \frac{280}{28} \quad \text{Substituting 28 for } P$$

$$N = 10$$

Check. In addition to repeating the computations, we can analyze the results. The number of miles per gallon increased from 14 to 28, and the number of gallons required decreased from 20 to 10. This is what we would expect with inverse variation.

State. a) The equation of variation is $N = \dfrac{280}{P}$.

b) A car that gets 28 mpg will need 10 gallons of gasoline to travel the fixed distance.

35. ***Familiarize.*** The problem states that we have inverse variation between the variables I and R. Thus, an equation $I = k/R$, $k > 0$, applies. As the resistance increases, the current decreases.

Translate. We write an equation of variation. Current varies inversely as resistance. This translates to $I = \dfrac{k}{R}$.

Solve.

a) First find an equation of variation.

$$I = \frac{k}{R}$$

$$96 = \frac{k}{20} \quad \text{Substituting 96 for } I \text{ and 20 for } R$$

$$1920 = k$$

The equation of variation is $I = \dfrac{1920}{R}$.

b) Use the equation to find the current when the resistance is 60 ohms.

$$I = \frac{1920}{R}$$

$$I = \frac{1920}{60} \quad \text{Substituting 60 for } R$$

$$I = 32$$

Check. The check might be done by repeating the computations. We might also analyze the results. The resistance increased from 20 ohms to 60 ohms, and the current decreased from 96 amperes to 32 amperes. This is what we would expect with inverse variation.

State. a) The equation of variation is $I = \dfrac{1920}{R}$.

b) The current is 32 amperes when the resistance is 60 ohms.

37. ***Familiarize.*** The problem states that we have inverse variation between the variables m and n. Thus, an equation $m = k/n$, $k > 0$, applies. As the number of questions increases, the number of minutes allowed for each question decreases.

Translate. We write an equation of variation. Time allowed per question varies inversely as the number of questions.

Solve.

a) First find an equation of variation.

$$m = \frac{k}{n}$$

$$2.5 = \frac{k}{16} \quad \text{Substituting 2.5 for } m \text{ and 16 for } n$$

$$40 = k$$

The equation of variation is $m = \dfrac{40}{n}$.

b) Use the equation to find the number of questions on a quiz when students have 4 min per question.

$$m = \frac{40}{n}$$

$$4 = \frac{40}{n} \quad \text{Substituting 4 for } m$$

$$4n = 40 \quad \text{Multiplying by } n$$

$$n = 10 \quad \text{Dividing by 4}$$

Check. The check might be done by repeating the computations. We might also analyze the results. The time allowed for each question increased from 2.5 min to 4 min, and the number of questions decreased from 16 to 10. This is what we would expect with inverse variation.

State. a) The equation of variation is $m = \frac{40}{n}$.

b) There would be 10 questions on a quiz for which students have 4 min per question.

39. ***Familiarize***. The problem states that we have inverse variation between the variables A and d. Thus, an equation $A = k/d$, $k > 0$, applies. As the distance increases, the apparent size decreases.

Translate. We write an equation of variation. Apparent size varies inversely as the distance. This translates to $A = \frac{k}{d}$.

Solve.

a) First find an equation of variation.

$$A = \frac{k}{d}$$

$$27.5 = \frac{k}{30} \quad \text{Substituting 27.5 for } A \text{ and 30 for } d$$

$$825 = k$$

The equation of variation is $A = \frac{825}{d}$.

b) Use the equation to find the apparent size when the distance is 100 ft.

$$A = \frac{825}{d}$$

$$A = \frac{825}{100} \quad \text{Substituting 100 for } d$$

$$A = 8.25$$

Check. The check might be done by repeating the computations. We might also analyze the results. The distance increased from 30 ft to 100 ft, and the apparent size decreased from 27.5 ft to 8.25 ft. This is what we would expect with inverse variation.

State. The flagpole will appear to be 8.25 ft tall when it is 100 ft from the observer.

41. $\frac{x+2}{x+5} = \frac{x-4}{x-6}$, LCM is $(x+5)(x-6)$

$$(x+5)(x-6) \cdot \frac{x+2}{x+5} = (x+5)(x-6) \cdot \frac{x-4}{x-6}$$

$$(x-6)(x+2) = (x+5)(x-4)$$

$$x^2 - 4x - 12 = x^2 + x - 20$$

$$-4x - 12 = x - 20 \quad \text{Subtracting } x^2$$

$$-5x = -8 \quad \text{Subtracting } x \text{ and adding 12}$$

$$x = \frac{8}{5}$$

The number $\frac{8}{5}$ checks and is the solution.

43. We do the divisions in order from left to right.

$$3^7 \div 3^4 \div 3^3 \div 3 = 3^3 \div 3^3 \div 3$$

$$= 1 \div 3$$

$$= \frac{1}{3}$$

45. $-5^2 + 4 \cdot 6 = -25 + 4 \cdot 6$

$$= -25 + 24$$

$$= -1$$

47. $P^2 = kt$

49. $P = kV^3$

Chapter 6 Vocabulary Reinforcement

1. A complex rational expression is a rational expression that has one or more rational expressions within its numerator or its denominator.

2. An equality of ratios, $\frac{A}{B} = \frac{C}{D}$, is called a proportion.

3. Two expressions are reciprocals of each other if their product is 1.

4. Expressions that have the same value for all allowable replacements are called equivalent expressions.

5. Expressions of the form $a - b$ and $b - a$ are opposites of each other.

6. In similar triangles, corresponding angles have the same measure and the lengths of corresponding sides are proportional.

7. When a situation translates to an equation described by $y = \frac{k}{x}$, with k a positive constant, we say that y varies inversely as x. The equation $y = \frac{k}{x}$ is called an equation of inverse variation.

8. When a situation translates to an equation described by $y = kx$, with k a positive constant, we say that y varies directly as x. The equation $y = kx$ is called an equation of direct variation.

Chapter 6 Concept Reinforcement

1. True; see page 467 in the text.

2. False; $-(y+5) = -y-5 \neq y-5$.

3. True; $-(2-x) = -2+x = x-2$.

Chapter 6 Study Guide

1. $c^2 - 11c + 30 = 0$

$(c-5)(c-6) = 0$

$c-5=0$ *or* $c-6=0$

$c=5$ *or* $c=6$

The rational expression is not defined for 5 and 6.

2. $$\frac{2x^2-2}{4x^2+24x+20} = \frac{2(x^2-1)}{4(x^2+6x+5)} = \frac{2(x+1)(x-1)}{2\cdot 2(x+1)(x+5)} = \frac{2(x+1)}{2(x+1)}\cdot\frac{x-1}{2(x+5)} = \frac{x-1}{2(x+5)}$$

3. $$\frac{2y^2+7y-15}{5y^2-45}\cdot\frac{y-3}{2y-3} = \frac{2y^2+7y-15)(y-3)}{(5y^2-45)(2y-3)} = \frac{(2y-3)(y+5)(y-3)}{5(y^2-9)(2y-3)} = \frac{\cancel{(2y-3)}(y+5)\cancel{(y-3)}}{5(y+3)\cancel{(y-3)}\cancel{(2y-3)}} = \frac{y+5}{5(y+3)}$$

4. $$\frac{b^2+3b-28}{b^2+5b-24}\div\frac{b-4}{b-3} = \frac{b^2+3b-28}{b^2+5b-24}\cdot\frac{b-3}{b-4} = \frac{(b^2+3b-28)(b-3)}{(b^2+5b-24)(b-4)} = \frac{(b+7)\cancel{(b-4)}\cancel{(b-3)}}{(b+8)\cancel{(b-3)}\cancel{(b-4)}} = \frac{b+7}{b+8}$$

5. $\frac{5}{18}+\frac{7}{60}$

$18 = 2\cdot 3\cdot 3$

$60 = 2\cdot 2\cdot 3\cdot 5$

The LCD is $2\cdot 2\cdot 3\cdot 3\cdot 5$, or 180.

$$\frac{5}{18}+\frac{7}{60} = \frac{5}{18}\cdot\frac{10}{10}+\frac{7}{60}\cdot\frac{3}{3} = \frac{50}{180}+\frac{21}{180} = \frac{71}{180}$$

6. $x^2-7x-18 = (x-9)(x+2)$

$x^2-18 = (x+9)(x-9)$

$\text{LCM} = (x-9)(x+2)(x+9)$

7. $$\frac{x}{x-4}+\frac{2x-4}{4-x} = \frac{x}{x-4}+\frac{2x-4}{4-x}\cdot\frac{-1}{-1} = \frac{x}{x-4}+\frac{-2x+4}{x-4} = \frac{x-2x+4}{x-4} = \frac{-x+4}{x-4} = \frac{-1(x-4)}{x-4} = -1\cdot\frac{x-4}{x-4} = -1$$

8. $$\frac{x}{x^2+x-2}-\frac{5}{x^2-1}$$

$$= \frac{x}{(x+2)(x-1)}-\frac{5}{(x+1)(x-1)},$$

LCM is $(x+2)(x-1)(x+1)$

$$= \frac{x}{(x+2)(x-1)}\cdot\frac{x+1}{x+1}-\frac{5}{(x+1)(x-1)}\cdot\frac{x+2}{x+2} = \frac{x(x+1)-5(x+2)}{(x+2)(x-1)(x+1)} = \frac{x^2+x-5x-10}{(x+2)(x-1)(x+1)} = \frac{x^2-4x-10}{(x+2)(x-1)(x+1)}$$

9. $$\frac{\frac{2}{5}-\frac{1}{y}}{\frac{3}{y}-\frac{1}{3}}$$ (The LCM of 5, y, and 3 is $15y$).

$$= \frac{\frac{2}{5}-\frac{1}{y}}{\frac{3}{y}-\frac{1}{3}}\cdot\frac{15y}{15y} = \frac{\frac{2}{5}\cdot 15y-\frac{1}{y}\cdot 15y}{\frac{3}{y}\cdot 15y-\frac{1}{3}\cdot 15y} = \frac{6y-15}{45-5y} = \frac{3(2y-5)}{5(9-y)}$$

10. $\frac{1}{x} = \frac{2}{3-x}$, LCM is $x(3-x)$

$x(3-x)\cdot\frac{1}{x} = x(3-x)\cdot\frac{2}{3-x}$

$3-x = 2x$

$3 = 3x$

$1 = x$

The number 1 checks. It is the solution.

11. $y = kx$

$60 = k(0.4)$ Substituting

$150 = k$ Variation constant

The equation of variation is $y = 150x$.

Substitute 2 for x and find y.

$$y = 150x = 150 \cdot 2 = 300$$

The value of y is 300 when $x = 2$.

12. $y = \frac{k}{x}$

$150 = \frac{k}{1.5}$ Substituting

$225 = k$ Variation constant

The equation of variation is $y = \frac{225}{x}$.

Substitute 10 for x and find y.

$$y = \frac{225}{x} = \frac{225}{10} = 22.5$$

The value of y is 22.5 when $x = 10$.

Chapter 6 Review Exercises

1. $\frac{3}{x}$

The denominator is 0 when $x = 0$, so the expression is not defined for the replacement number 0.

2. $\frac{4}{x-6}$

To determine the numbers for which the rational expression is not defined, we set the denominator equal to 0 and solve:

$$x - 6 = 0$$
$$x = 6$$

The expression is not defined for the replacement number 6.

3. $\frac{x+5}{x^2-36}$

To determine the numbers for which the rational expression is not defined, we set the denominator equal to 0 and solve:

$$x^2 - 36 = 0$$
$$(x+6)(x-6) = 0$$
$$x+6 = 0 \quad \text{or} \quad x-6 = 0$$
$$x = -6 \quad \text{or} \quad x = 6$$

The expression is not defined for the replacement numbers -6 and 6.

4. $\frac{x^2-3x+2}{x^2+x-30}$

To determine the numbers for which the rational expression is not defined, we set the denominator equal to 0 and solve:

$$x^2 + x - 30 = 0$$
$$(x+6)(x-5) = 0$$
$$x+6 = 0 \quad \text{or} \quad x-5 = 0$$
$$x = -6 \quad \text{or} \quad x = 5$$

The expression is not defined for the replacement numbers -6 and 5.

5. $\frac{-4}{(x+2)^2}$

To determine the numbers for which the rational expression is not defined, we set the denominator equal to 0 and solve:

$$(x+2)^2 = 0$$
$$(x+2)(x+2) = 0$$
$$x+2 = 0 \quad \text{or} \quad x+2 = 0$$
$$x = -2 \quad \text{or} \quad x = -2$$

The expression is not defined for the replacement number -2.

6. $\frac{x-5}{5}$

Since the denominator is the constant 5, there are no replacement numbers for which the expression is not defined.

7. $$\frac{4x^2-8x}{4x^2+4x} = \frac{4x(x-2)}{4x(x+1)} = \frac{4x}{4x} \cdot \frac{x-2}{x+1} = 1 \cdot \frac{x-2}{x+1} = \frac{x-2}{x+1}$$

8. $$\frac{14x^2-x-3}{2x^2-7x+3} = \frac{(2x-1)(7x+3)}{(2x-1)(x-3)} = \frac{2x-1}{2x-1} \cdot \frac{7x+3}{x-3} = 1 \cdot \frac{7x+3}{x-3} = \frac{7x+3}{x-3}$$

9. $$\frac{(y-5)^2}{y^2-25} = \frac{(y-5)(y-5)}{(y+5)(y-5)} = \frac{y-5}{y+5} \cdot \frac{y-5}{y-5} = \frac{y-5}{y+5} \cdot 1 = \frac{y-5}{y+5}$$

10. $$\frac{a^2-36}{10a} \cdot \frac{2a}{a+6} = \frac{(a^2-36)(2a)}{10a(a+6)} = \frac{(a+6)(a-6) \cdot 2 \cdot a}{2 \cdot 5 \cdot a \cdot (a+6)} = \frac{\cancel{(a+6)}(a-6) \cdot \cancel{2} \cdot \cancel{a}}{\cancel{2} \cdot 5 \cdot \cancel{a} \cdot \cancel{(a+6)}} = \frac{a-6}{5}$$

11. $\dfrac{6t-6}{2t^2+t-1}\cdot\dfrac{t^2-1}{t^2-2t+1}$

$=\dfrac{(6t-6)(t^2-1)}{(2t^2+t-1)(t^2-2t+1)}$

$=\dfrac{6(t-1)(t+1)(t-1)}{(2t-1)(t+1)(t-1)(t-1)}$

$=\dfrac{6\cancel{(t-1)}\cancel{(t+1)}\cancel{(t-1)}}{(2t-1)\cancel{(t+1)}\cancel{(t-1)}\cancel{(t-1)}}$

$=\dfrac{6}{2t-1}$

12. $\dfrac{10-5t}{3}\div\dfrac{t-2}{12t}=\dfrac{10-5t}{3}\cdot\dfrac{12t}{t-2}$

$=\dfrac{(10-5t)(12t)}{3(t-2)}$

$=\dfrac{5(2-t)\cdot 3\cdot 4t}{3(t-2)}$

$=\dfrac{5(-1)(t-2)\cdot 3\cdot 4t}{3(t-2)}$ $\quad 2-t=-1(t-2)$

$=\dfrac{5(-1)\cancel{(t-2)}\cdot\cancel{3}\cdot 4t}{\cancel{3}\cancel{(t-2)}\cdot 1}$

$=-20t$

13. $\dfrac{4x^4}{x^2-1}\div\dfrac{2x^3}{x^2-2x+1}=\dfrac{4x^4}{x^2-1}\cdot\dfrac{x^2-2x+1}{2x^3}$

$=\dfrac{4x^4(x^2-2x+1)}{(x^2-1)(2x^3)}$

$=\dfrac{2\cdot 2\cdot x\cdot x^3(x-1)(x-1)}{(x+1)(x-1)\cdot 2\cdot x^3}$

$=\dfrac{\cancel{2}\cdot 2\cdot x\cdot\cancel{x^3}\cancel{(x-1)}(x-1)}{(x+1)\cancel{(x-1)}\cdot\cancel{2}\cdot\cancel{x^3}}$

$=\dfrac{2x(x-1)}{x+1}$

14. $3x^2=3\cdot x\cdot x$

$10xy=2\cdot 5\cdot x\cdot y$

$15y^2=3\cdot 5\cdot y\cdot y$

LCM $=2\cdot 3\cdot 5\cdot x\cdot x\cdot y\cdot y$, or $30x^2y^2$

15. $a-2=a-2$

$4a-8=4(a-2)$

LCM $=4(a-2)$

16. $y^2-y-2=(y-2)(y+1)$

$y^2-4=(y+2)(y-2)$

LCM $=(y-2)(y+1)(y+2)$

17. $\dfrac{x+8}{x+7}+\dfrac{10-4x}{x+7}=\dfrac{x+8+10-4x}{x+7}=\dfrac{-3x+18}{x+7}=$

$\dfrac{-3(x-6)}{x+7}$

18. $\dfrac{3}{3x-9}+\dfrac{x-2}{3-x}=\dfrac{3}{3(x-3)}+\dfrac{x-2}{3-x}$

$=\dfrac{3}{3(x-3)}+\dfrac{x-2}{3-x}\cdot\dfrac{-1}{-1}$

$=\dfrac{3}{3(x-3)}+\dfrac{-1(x-2)}{-1(3-x)}$

$=\dfrac{3}{3(x-3)}+\dfrac{-x+2}{x-3}$

$=\dfrac{3}{3(x-3)}+\dfrac{-x+2}{x-3}\cdot\dfrac{3}{3}$

$=\dfrac{3}{3(x-3)}+\dfrac{-3x+6}{3(x-3)}$

$=\dfrac{3-3x+6}{3(x-3)}$

$=\dfrac{-3x+9}{3(x-3)}$

$=\dfrac{-3(x-3)}{3(x-3)}$

$=\dfrac{-1\cdot\cancel{3}\cancel{(x-3)}}{1\cdot\cancel{3}\cancel{(x-3)}}$

$=-1$

19. $\dfrac{2a}{a+1}+\dfrac{4a}{a^2-1}$

$=\dfrac{2a}{a+1}+\dfrac{4a}{(a+1)(a-1)}$, LCM is $(a+1)(a-1)$

$=\dfrac{2a}{a+1}\cdot\dfrac{a-1}{a-1}+\dfrac{4a}{(a+1)(a-1)}$

$=\dfrac{2a(a-1)+4a}{(a+1)(a-1)}$

$=\dfrac{2a^2-2a+4a}{(a+1)(a-1)}$

$=\dfrac{2a^2+2a}{(a+1)(a-1)}$

$=\dfrac{2a(a+1)}{(a+1)(a-1)}$

$=\dfrac{2a\cancel{(a+1)}}{\cancel{(a+1)}(a-1)}$

$=\dfrac{2a}{a-1}$

20. $\dfrac{d^2}{d-c}+\dfrac{c^2}{c-d}=\dfrac{d^2}{d-c}+\dfrac{c^2}{c-d}\cdot\dfrac{-1}{-1}$

$=\dfrac{d^2}{d-c}+\dfrac{-c^2}{d-c}$

$=\dfrac{d^2-c^2}{d-c}$

$=\dfrac{(d+c)(d-c)}{d-c}$

$=\dfrac{(d+c)\cancel{(d-c)}}{\cancel{(d-c)}\cdot 1}$

$=d+c$

21. $\dfrac{6x-3}{x^2-x-12}-\dfrac{2x-15}{x^2-x-12}=\dfrac{6x-3-(2x-15)}{x^2-x-12}$

$=\dfrac{6x-3-2x+15}{x^2-x-12}$

$=\dfrac{4x+12}{x^2-x-12}$

$=\dfrac{4(x+3)}{(x-4)(x+3)}$

$=\dfrac{4(\cancel{x+3})}{(x-4)(\cancel{x+3})}$

$=\dfrac{4}{x-4}$

22. $\dfrac{3x-1}{2x}-\dfrac{x-3}{x}$, LCM is $2x$

$=\dfrac{3x-1}{2x}-\dfrac{x-3}{x}\cdot\dfrac{2}{2}$

$=\dfrac{3x-1}{2x}-\dfrac{2(x-3)}{2x}$

$=\dfrac{3x-1-2(x-3)}{2x}$

$=\dfrac{3x-1-2x+6}{2x}$

$=\dfrac{x+5}{2x}$

23. $\dfrac{x+3}{x-2}-\dfrac{x}{2-x}=\dfrac{x+3}{x-2}-\dfrac{x}{2-x}\cdot\dfrac{-1}{-1}$

$=\dfrac{x+3}{x-2}-\dfrac{-x}{x-2}$

$=\dfrac{x+3-(-x)}{x-2}$

$=\dfrac{x+3+x}{x-2}$

$=\dfrac{2x+3}{x-2}$

24. $\dfrac{1}{x^2-25}-\dfrac{x-5}{x^2-4x-5}$

$=\dfrac{1}{(x+5)(x-5)}-\dfrac{x-5}{(x-5)(x+1)}$,
LCM is $(x+5)(x-5)(x+1)$

$=\dfrac{1}{(x+5)(x-5)}\cdot\dfrac{x+1}{x+1}-\dfrac{x-5}{(x-5)(x+1)}\cdot\dfrac{x+5}{x+5}$

$=\dfrac{x+1}{(x+5)(x-5)(x+1)}-\dfrac{(x-5)(x+5)}{(x+5)(x-5)(x+1)}$

$=\dfrac{x+1-(x^2-25)}{(x+5)(x-5)(x+1)}$

$=\dfrac{x+1-x^2+25}{(x+5)(x-5)(x+1)}$

$=\dfrac{-x^2+x+26}{(x+5)(x-5)(x+1)}$

25. $\dfrac{3x}{x+2}-\dfrac{x}{x-2}+\dfrac{8}{x^2-4}$

$=\dfrac{3x}{x+2}-\dfrac{x}{x-2}+\dfrac{8}{(x+2)(x-2)}$, LCM is $(x+2)(x-2)$

$=\dfrac{3x}{x+2}\cdot\dfrac{x-2}{x-2}-\dfrac{x}{x-2}\cdot\dfrac{x+2}{x+2}+\dfrac{8}{(x+2)(x-2)}$

$=\dfrac{3x(x-2)}{(x+2)(x-2)}-\dfrac{x(x+2)}{(x+2)(x-2)}+\dfrac{8}{(x+2)(x-2)}$

$=\dfrac{3x(x-2)-x(x+2)+8}{(x+2)(x-2)}$

$=\dfrac{3x^2-6x-x^2-2x+8}{(x+2)(x-2)}$

$=\dfrac{2x^2-8x+8}{(x+2)(x-2)}$

$=\dfrac{2(x^2-4x+4)}{(x+2)(x-2)}$

$=\dfrac{2(x-2)(x-2)}{(x+2)(x-2)}$

$=\dfrac{2(\cancel{x-2})(x-2)}{(x+2)(\cancel{x-2})}$

$=\dfrac{2(x-2)}{x+2}$

26. $\dfrac{\frac{1}{z}+1}{\frac{1}{z^2}-1}$ LCM of the denominators is z^2.

$=\dfrac{\frac{1}{z}+1}{\frac{1}{z^2}-1}\cdot\dfrac{z^2}{z^2}$

$=\dfrac{\left(\frac{1}{z}+1\right)z^2}{\left(\frac{1}{z^2}-1\right)z^2}$

$=\dfrac{\frac{1}{z}\cdot z^2+1\cdot z^2}{\frac{1}{z^2}\cdot z^2-1\cdot z^2}$

$=\dfrac{z+z^2}{1-z^2}$

$=\dfrac{z(1+z)}{(1+z)(1-z)}$

$=\dfrac{z(\cancel{1+z})}{(\cancel{1+z})(1-z)}$

$=\dfrac{z}{1-z}$

27. $\dfrac{\dfrac{c}{d}-\dfrac{d}{c}}{\dfrac{1}{c}+\dfrac{1}{d}}$

$= \dfrac{\dfrac{c}{d}\cdot\dfrac{c}{c}-\dfrac{d}{c}\cdot\dfrac{d}{d}}{\dfrac{1}{c}\cdot\dfrac{d}{d}+\dfrac{1}{d}\cdot\dfrac{c}{c}}$ Getting a common denominator in numerator and in denominator

$= \dfrac{\dfrac{c^2}{cd}-\dfrac{d^2}{cd}}{\dfrac{d}{cd}+\dfrac{c}{cd}}$

$= \dfrac{\dfrac{c^2-d^2}{cd}}{\dfrac{d+c}{cd}}$

$= \dfrac{c^2-d^2}{cd}\cdot\dfrac{cd}{d+c}$

$= \dfrac{(c^2-d^2)cd}{cd(d+c)}$

$= \dfrac{(c+d)(c-d)cd}{cd(d+c)}$

$= \dfrac{\cancel{(c+d)}(c-d)\cancel{cd}}{\cancel{cd}\,\cancel{(d+c)}\cdot 1}$

$= c-d$

28. $\dfrac{3}{y}-\dfrac{1}{4}=\dfrac{1}{y}$, LCM $= 4y$

$$4y\left(\frac{3}{y}-\frac{1}{4}\right)=4y\cdot\frac{1}{y}$$
$$4y\cdot\frac{3}{y}-4y\cdot\frac{1}{4}=4y\cdot\frac{1}{y}$$
$$12-y=4$$
$$-y=-8$$
$$y=8$$

This checks, so the solution is 8.

29. $\dfrac{15}{x}-\dfrac{15}{x+2}=2$, LCM $= x(x+2)$

$$x(x+2)\left(\frac{15}{x}-\frac{15}{x+2}\right)=x(x+2)\cdot 2$$
$$x(x+2)\cdot\frac{15}{x}-x(x+2)\cdot\frac{15}{x+2}=2x(x+2)$$
$$15(x+2)-15x=2x(x+2)$$
$$15x+30-15x=2x^2+4x$$
$$30=2x^2+4x$$
$$0=2x^2+4x-30$$
$$0=2(x^2+2x-15)$$
$$0=x^2+2x-15 \quad \text{Dividing by 2}$$
$$0=(x+5)(x-3)$$

$x+5=0$ *or* $x-3=0$

$x=-5$ *or* $x=3$

Both numbers check. The solutions are -5 and 3.

30. ***Familiarize.*** Let $t =$ the time the job would take if the crews worked together.

Translate. We use the work principle, substituting 9 for a and 12 for b.

$$\frac{t}{a}+\frac{t}{b}=1$$
$$\frac{t}{9}+\frac{t}{12}=1 \quad \text{Substituting}$$

Solve.

$$\frac{t}{9}+\frac{t}{12}=1,\ \text{LCM}=36$$
$$36\left(\frac{t}{9}+\frac{t}{12}\right)=36\cdot 1$$
$$36\cdot\frac{t}{9}+36\cdot\frac{t}{12}=36$$
$$4t+3t=36$$
$$7t=36$$
$$t=\frac{36}{7},\ \text{or}\ 5\frac{1}{7}$$

Check. In $\dfrac{36}{7}$ hr, the portion of the job done is $\dfrac{36}{7}\cdot\dfrac{1}{9}+\dfrac{36}{7}\cdot\dfrac{1}{12}=\dfrac{4}{7}+\dfrac{3}{7}=1.$ The answer checks.

State. The job would take $5\frac{1}{7}$ hr if the crews worked together.

31. ***Familiarize.*** Let $r =$ the speed of the slower plane, in mph. Then $r+80 =$ the speed of the faster plane. We organize the information in a table.

	Distance	Speed	Time
Slower plane	950	r	t
Faster plane	1750	$r+80$	t

Translate. We use the formula $d = rt$ in each row of the table to obtain two equations.

$$950=rt,$$
$$1750=(r+80)t$$

Since the times are the same, we solve each equation for t and set the results equal to each other.

$$950=rt,\ \text{so}\ t=\frac{950}{r}.$$
$$1750=(r+80)t,\ \text{so}\ t=\frac{1750}{r+80}.$$

Then we have

$$\frac{950}{r}=\frac{1750}{r+80}.$$

Solve.

$$\frac{950}{r} = \frac{1750}{r+80}, \text{ LCM} = r(r+80)$$

$$r(r+80)\cdot\frac{950}{r} = r(r+80)\cdot\frac{1750}{r+80}$$

$$950(r+80) = 1750r$$

$$950r + 76{,}000 = 1750r$$

$$76{,}000 = 800r$$

$$95 = r$$

If $95 = r$, then $r + 80 = 95 + 80 = 175$.

Check. If the speeds are 95 mph and 175 mph, then the speed of the faster plane is 80 mph faster than the speed of the slower plane. At 95 mph, the slower plane travels 950 mi in 950/95, or 10 hr. At 175 mph, the faster plane travels 1750 mi in 1750/175, or 10 hr. The times are the same, so the answer checks.

State. The speed of the slower plane is 95 mph; the speed of the faster plane is 175 mph.

32. ***Familiarize.*** Let r = the speed of the slower train, in km/h. Then $r + 40$ = the speed of the faster train. We organize the information in a table.

	Distance	Speed	Time
Slower train	60	r	t
Faster train	70	$r+40$	t

Translate. We use the formula $d = rt$ in each row of the table to obtain two equations.

$$60 = rt,$$

$$70 = (r+40)t$$

Since the times are the same, we solve each equation for t and set the results equal to each other.

$$60 = rt, \text{ so } t = \frac{60}{r}.$$

$$70 = (r+40)t, \text{ so } t = \frac{70}{r+40}.$$

Then we have

$$\frac{60}{r} = \frac{70}{r+40}.$$

Solve.

$$\frac{60}{r} = \frac{70}{r+40}, \text{ LCM} = r(r+40)$$

$$r(r+40)\cdot\frac{60}{r} = r(r+40)\cdot\frac{70}{r+40}$$

$$60(r+40) = 70r$$

$$60r + 2400 = 70r$$

$$2400 = 10r$$

$$240 = r$$

If $r = 240$, then $r + 40 = 240 + 40 = 280$.

Check. If the speeds are 240 km/h and 280 km/h, then the speed of the faster train is 40 km/h faster than the speed of the slower train. At 240 km/h, the slower train travels 60 km in 60/240, or 1/4 hr. At 280 km/h, the faster train travels 70 km in 70/280, or 1/4 hr. Since the times are the same, the answer checks.

State. The speed of the slower train is 240 km/h; the speed of the faster train is 280 km/h.

33. ***Familiarize.*** We can translate to a proportion, letting d = the number of defective calculators that can be expected in a sample of 5000.

Translate.

$$\begin{array}{r} \text{Number defective} \rightarrow \\ \text{Sample size} \rightarrow \end{array} \frac{8}{250} = \frac{d}{5000} \begin{array}{l} \leftarrow \text{Number defective} \\ \leftarrow \text{Sample size} \end{array}$$

Solve. We solve the proportion.

$$\frac{8}{250} = \frac{d}{5000}$$

$$5000\cdot\frac{8}{250} = 5000\cdot\frac{d}{5000}$$

$$160 = d$$

Check.

$$\frac{8}{250} = 0.032 \text{ and } \frac{160}{5000} = 0.032.$$

The ratios are the same, so the answer checks.

State. You would expect to find 160 defective calculators in a sample of 5000.

34. a) Let x = the number of cups of onion that would be used. Then we can write and solve a proportion.

$$\frac{6}{13} = \frac{x}{2}$$

$$2\cdot\frac{6}{13} = 2\cdot\frac{x}{2}$$

$$\frac{12}{13} = x$$

Thus, $\frac{12}{13}$ cup of onion would be used.

b) Let c = the number of cups of cheese that would be used. We can write and solve a proportion.

$$\frac{5}{7} = \frac{3}{c}$$

$$7c\cdot\frac{5}{7} = 7c\cdot\frac{3}{c}$$

$$5c = 21$$

$$c = \frac{21}{5}, \text{ or } 4\frac{1}{5}$$

Thus, $4\frac{1}{5}$ cups of cheese would be used.

c) Let c = the number of cups of cheese that would be used. We can write and solve a proportion.

$$\frac{9}{14} = \frac{6}{c}$$

$$14c\cdot\frac{9}{14} = 14c\cdot\frac{6}{c}$$

$$9c = 84$$

$$c = \frac{84}{9} = \frac{28}{3}, \text{ or } 9\frac{1}{3}$$

Thus, $9\frac{1}{3}$ cups of cheese would be used.

35. ***Familiarize***. The ratio of blue whales tagged to the total blue whale population, P, is $\frac{500}{P}$. Of the 400 blue whales checked later, 20 were tagged. The ratio of blue whales tagged to blue whales checked is $\frac{20}{400}$.

Translate. Assuming the two ratios are the same, we can translate to a proportion.

$$\begin{array}{rcl} \text{Whales tagged originally} \longrightarrow & \dfrac{500}{P} = \dfrac{20}{400} & \longleftarrow \text{Tagged whales caught later} \\ \text{Whale population} \longrightarrow & & \longleftarrow \text{Whales caught later} \end{array}$$

Solve. We solve the proportion.

$$400P \cdot \frac{500}{P} = 400P \cdot \frac{20}{400} \quad \text{Multiplying by the LCM, } 400P$$
$$400 \cdot 500 = P \cdot 20$$
$$200,000 = 20P$$
$$10,000 = P$$

Check.

$$\frac{500}{10,000} = \frac{1}{20} \text{ and } \frac{20}{400} = \frac{1}{20}.$$

The ratios are the same.

State. The blue whale population is about 10,000.

36. We write a proportion and solve it.

$$\frac{3.4}{8.5} = \frac{2.4}{x}$$
$$8.5x \cdot \frac{3.4}{8.5} = 8.5x \cdot \frac{2.4}{x}$$
$$3.4x = 20.4$$
$$x = 6$$

(Note that the proportions $\frac{8.5}{3.4} = \frac{x}{2.4}$, $\frac{8.5}{x} = \frac{3.4}{2.4}$, and $\frac{x}{8.5} = \frac{2.4}{3.4}$ could also be used.)

37. We substitute to find k.

$$y = kx$$
$$12 = k \cdot 4$$
$$3 = k$$

The equation of variation is $y = 3x$.

To find the value of y when $x = 20$ we substitute 20 for x in the equation of variation.

$$y = 3x$$
$$y = 3 \cdot 20$$
$$y = 60$$

The value of y is 60 when $x = 20$.

38. We substitute to find k.

$$y = kx$$
$$0.4 = k \cdot 0.5$$
$$\frac{0.4}{0.5} = k$$
$$\frac{4}{5} = k \quad \left(\frac{0.4}{0.5} = \frac{0.4}{0.5} \cdot \frac{10}{10} = \frac{4}{5}\right)$$

The equation of variation is $y = \frac{4}{5}x$.

To find the value of y when $x = 20$ we substitute 20 for x in the equation of variation.

$$y = \frac{4}{5}x$$
$$y = \frac{4}{5} \cdot 20$$
$$y = 16$$

The value of y is 16 when $x = 20$.

39. We substitute to find k.

$$y = \frac{k}{x}$$
$$5 = \frac{k}{6}$$
$$30 = k$$

The equation of variation is $y = \frac{30}{x}$.

To find the value of y when $x = 5$ we substitute 5 for x in the equation of variation.

$$y = \frac{30}{x}$$
$$y = \frac{30}{5}$$
$$y = 6$$

The value of y is 6 when $x = 5$.

40. We substitute to find k.

$$y = \frac{k}{x}$$
$$0.5 = \frac{k}{2}$$
$$1 = k$$

The equation of variation is $y = \frac{1}{x}$.

To find the value of y when $x = 5$ we substitute 5 for x in the equation of variation.

$$y = \frac{1}{x}$$
$$y = \frac{1}{5}$$

The value of y is $\frac{1}{5}$ when $x = 5$.

41. We substitute to find k.

$$y = \frac{k}{x}$$
$$1.3 = \frac{k}{0.5}$$
$$0.65 = k$$

The equation of variation is $y = \dfrac{0.65}{x}$.

To find the value of y when $x = 5$ we substitute 5 for x in the equation of variation.

$$y = \frac{0.65}{x}$$
$$y = \frac{0.65}{5}$$
$$y = 0.13$$

The value of y is 0.13 when $x = 5$.

42. ***Familiarize and Translate.*** The problem states that we have direct variation between the variables P and H. Thus, an equation $P = kH$, $k > 0$, applies.

Solve.

a) First we find an equation of variation.

$$P = kH$$
$$165 = k \cdot 20$$
$$8.25 = k$$

The equation of variation is $P = 8.25H$.

b) Use the equation to find the pay for 35 hr of work.

$$P = 8.25H$$
$$P = 8.25(35)$$
$$P = 288.75$$

Check. We can repeat the computations. Also note that when the number of hours worked increased from 20 to 35, the pay increased from \$165.00 to \$288.75 so the answer seems reasonable.

State. The pay for 35 hr of work is \$288.75.

43. ***Familiarize and Translate.*** Let M = the number of washing machines used and T = the time required to do the laundry. The problem states that we have inverse variation between T and M, so an equation $T = k/M$, $k > 0$, applies.

Solve.

a) First we find an equation of variation.

$$T = \frac{k}{M}$$
$$5 = \frac{k}{2}$$
$$10 = k$$

The equation of variation is $T = \dfrac{10}{M}$.

b) Use the equation to find the time required to do the laundry if 10 washing machines are used.

$$T = \frac{10}{M}$$
$$T = \frac{10}{10}$$
$$T = 1$$

Check. We can repeat the computations. Also note that as the number of washing machines increased from 5 to 10, the time required to do the laundry decreased from 2 hr to 1 hr, so the answer seems reasonable.

State. It would take 1 hr for 10 washing machines to do the laundry.

44. $\dfrac{3x^2 - 2x - 1}{3x^2 + x}$

Set the denominator equal to 0 and solve for x.

$$3x^2 + x = 0$$
$$x(3x + 1) = 0$$
$$x = 0 \quad or \quad 3x + 1 = 0$$
$$x = 0 \quad or \quad 3x = -1$$
$$x = 0 \quad or \quad x = -\frac{1}{3}$$

The expression is not defined for 0 and $-\dfrac{1}{3}$. Answer C is correct.

45. $\dfrac{1}{x-5} - \dfrac{1}{x+5}$, LCM is $(x-5)(x+5)$

$$= \frac{1}{x-5} \cdot \frac{x+5}{x+5} - \frac{1}{x+5} \cdot \frac{x-5}{x-5}$$
$$= \frac{x+5-(x-5)}{(x-5)(x+5)}$$
$$= \frac{x+5-x+5}{(x-5)(x+5)}$$
$$= \frac{10}{(x-5)(x+5)}$$

Answer A is correct.

46. $\dfrac{2a^2 + 5a - 3}{a^2} \cdot \dfrac{5a^3 + 30a^2}{2a^2 + 7a - 4} \div \dfrac{a^2 + 6a}{a^2 + 7a + 12}$

$$= \frac{2a^2 + 5a - 3}{a^2} \cdot \frac{5a^3 + 30a^2}{2a^2 + 7a - 4} \cdot \frac{a^2 + 7a + 12}{a^2 + 6a}$$
$$= \frac{(2a-1)(a+3)}{a^2} \cdot \frac{5a^2(a+6)}{(2a-1)(a+4)} \cdot \frac{(a+3)(a+4)}{a(a+6)}$$
$$= \frac{(2a-1)(a+3)(5a^2)(a+6)(a+3)(a+4)}{a^2(2a-1)(a+4)(a)(a+6)}$$
$$= \frac{\cancel{(2a-1)}(a+3) \cdot 5 \cdot \cancel{a^2}\,\cancel{(a+6)}(a+3)\cancel{(a+4)}}{\cancel{a^2}\,\cancel{(2a-1)}\cancel{(a+4)}(a)\cancel{(a+6)}}$$
$$= \frac{5(a+3)^2}{a}$$

47. $\dfrac{A+B}{B} = \dfrac{C+D}{D}$

$$\frac{A}{B} + \frac{B}{B} = \frac{C}{D} + \frac{D}{D}$$
$$\frac{A}{B} + 1 = \frac{C}{D} + 1$$
$$\frac{A}{B} = \frac{C}{D}$$

The two given proportions are equivalent.

Chapter 6 Discussion and Writing Exercises

1. No; when adding, no sign changes are required so the result is the same regardless of parentheses. When subtracting, however, the sign of each term of the expression being subtracted must be changed and parentheses are needed to make sure this is done.

2. Graph each side of the equation and determine the number of points of intersection of the graphs.

3. Canceling removes a factor of 1, allowing us to rewrite $a \cdot 1$ as a.

4. The larger the average gain per play, the smaller the number of plays required to go 80 yd. Thus, inverse variation applies.

5. Form a rational expression that has factors of $x+3$ and $x-4$ in the denominator.

6. If we multiply both sides of a rational equation by a variable expression in order to clear fractions, it is possible that the variable expression is equal to 0. Thus, an equivalent equation might not be produced.

Chapter 6 Test

1. $\dfrac{8}{2x}$

To determine the numbers for which the rational expression is not defined, we set the denominator equal to 0 and solve:

$$2x = 0$$
$$x = 0$$

The expression is not defined for the replacement number 0.

2. $\dfrac{5}{x+8}$

To determine the numbers for which the rational expression is not defined, we set the denominator equal to 0 and solve:

$$x+8 = 0$$
$$x = -8$$

The expression is not defined for the replacement number -8.

3. $\dfrac{x-7}{x^2-49}$

To determine the numbers for which the rational expression is not defined, we set the denominator equal to 0 and solve:

$$x^2 - 49 = 0$$
$$(x+7)(x-7) = 0$$
$$x+7 = 0 \quad or \quad x-7 = 0$$
$$x = -7 \quad or \quad x = 7$$

The expression is not defined for the replacement numbers -7 and 7.

4. $\dfrac{x^2+x-30}{x^2-3x+2}$

To determine the numbers for which the rational expression is not defined, we set the denominator equal to 0 and solve:

$$x^2 - 3x + 2 = 0$$
$$(x-1)(x-2) = 0$$
$$x-1 = 0 \quad or \quad x-2 = 0$$
$$x = 1 \quad or \quad x = 2$$

The expression is not defined for the replacement numbers 1 and 2.

5. $\dfrac{11}{(x-1)^2}$

To determine the numbers for which the rational expression is not defined, we set the denominator equal to 0 and solve:

$$(x-1)^2 = 0$$
$$(x-1)(x-1) = 0$$
$$x-1 = 0 \quad or \quad x-1 = 0$$
$$x = 1 \quad or \quad x = 1$$

The expression is not defined for the replacement number 1.

6. $\dfrac{x+2}{2}$

Since the denominator is the constant 2, there are no replacement numbers for which the expression is not defined.

7. $$\frac{6x^2+17x+7}{2x^2+7x+3} = \frac{(2x+1)(3x+7)}{(2x+1)(x+3)}$$
$$= \frac{2x+1}{2x+1} \cdot \frac{3x+7}{x+3}$$
$$= 1 \cdot \frac{3x+7}{x+3}$$
$$= \frac{3x+7}{x+3}$$

8. $$\frac{a^2-25}{6a} \cdot \frac{3a}{a-5} = \frac{(a^2-25)(3a)}{6a(a-5)}$$
$$= \frac{(a+5)(a-5) \cdot 3 \cdot a}{2 \cdot 3 \cdot a \cdot (a-5)}$$
$$= \frac{(a+5)\cancel{(a-5)} \cdot \cancel{3} \cdot \cancel{a}}{2 \cdot \cancel{3} \cdot \cancel{a} \cdot \cancel{(a-5)}}$$
$$= \frac{a+5}{2}$$

9. $\dfrac{25x^2-1}{9x^2-6x} \div \dfrac{5x^2+9x-2}{3x^2+x-2}$

$= \dfrac{25x^2-1}{9x^2-6x} \cdot \dfrac{3x^2+x-2}{5x^2+9x-2}$

$= \dfrac{(25x^2-1)(3x^2+x-2)}{(9x^2-6x)(5x^2+9x-2)}$

$= \dfrac{(5x+1)(5x-1)(3x-2)(x+1)}{3x(3x-2)(5x-1)(x+2)}$

$= \dfrac{(5x+1)\cancel{(5x-1)}\cancel{(3x-2)}(x+1)}{3x\cancel{(3x-2)}\cancel{(5x-1)}(x+2)}$

$= \dfrac{(5x+1)(x+1)}{3x(x+2)}$

10. $y^2-9=(y+3)(y-3)$

$y^2+10y+21=(y+3)(y+7)$

$y^2+4y-21=(y-3)(y+7)$

$\text{LCM} = (y+3)(y-3)(y+7)$

11. $\dfrac{16+x}{x^3}+\dfrac{7-4x}{x^3}=\dfrac{16+x+7-4x}{x^3}=\dfrac{23-3x}{x^3}$

12. $\dfrac{5-t}{t^2+1}-\dfrac{t-3}{t^2+1}=\dfrac{5-t-(t-3)}{t^2+1}$

$=\dfrac{5-t-t+3}{t^2+1}$

$=\dfrac{8-2t}{t^2+1}$

13. $\dfrac{x-4}{x-3}+\dfrac{x-1}{3-x}=\dfrac{x-4}{x-3}+\dfrac{x-1}{3-x}\cdot\dfrac{-1}{-1}$

$=\dfrac{x-4}{x-3}+\dfrac{-x+1}{x-3}$

$=\dfrac{x-4-x+1}{x-3}$

$=\dfrac{-3}{x-3}$

14. $\dfrac{x-4}{x-3}-\dfrac{x-1}{3-x}=\dfrac{x-4}{x-3}-\dfrac{x-1}{3-x}\cdot\dfrac{-1}{-1}$

$=\dfrac{x-4}{x-3}-\dfrac{-x+1}{x-3}$

$=\dfrac{x-4-(-x+1)}{x-3}$

$=\dfrac{x-4+x-1}{x-3}$

$=\dfrac{2x-5}{x-3}$

15. $\dfrac{5}{t-1}+\dfrac{3}{t}$, LCD is $t(t-1)$.

$=\dfrac{5}{t-1}\cdot\dfrac{t}{t}+\dfrac{3}{t}\cdot\dfrac{t-1}{t-1}$

$=\dfrac{5t}{t(t-1)}+\dfrac{3(t-1)}{t(t-1)}$

$=\dfrac{5t}{t(t-1)}+\dfrac{3t-3}{t(t-1)}$

$=\dfrac{5t+3t-3}{t(t-1)}$

$=\dfrac{8t-3}{t(t-1)}$

16. $\dfrac{1}{x^2-16}-\dfrac{x+4}{x^2-3x-4}$

$=\dfrac{1}{(x+4)(x-4)}-\dfrac{x+4}{(x-4)(x+1)}$, LCD is $(x+4)(x-4)(x+1)$.

$=\dfrac{1}{(x+4)(x-4)}\cdot\dfrac{x+1}{x+1}-\dfrac{x+4}{(x-4)(x+1)}\cdot\dfrac{x+4}{x+4}$

$=\dfrac{x+1-(x+4)(x+4)}{(x+4)(x-4)(x+1)}$

$=\dfrac{x+1-(x^2+8x+16)}{(x+4)(x-4)(x+1)}$

$=\dfrac{x+1-x^2-8x-16}{(x+4)(x-4)(x+1)}$

$=\dfrac{-x^2-7x-15}{(x+4)(x-4)(x+1)}$

17. $\dfrac{1}{x-1}+\dfrac{4}{x^2-1}-\dfrac{2}{x^2-2x+1}$

$=\dfrac{1}{x-1}+\dfrac{4}{(x+1)(x-1)}-\dfrac{2}{(x-1)(x-1)}$,

LCD is $(x+1)(x-1)(x-1)$

$=\dfrac{1}{x-1}\cdot\dfrac{(x+1)(x-1)}{(x+1)(x-1)}+\dfrac{4}{(x+1)(x-1)}\cdot\dfrac{x-1}{x-1}-$

$\dfrac{2}{(x-1)(x-1)}\cdot\dfrac{x+1}{x+1}$

$=\dfrac{(x+1)(x-1)+4(x-1)-2(x+1)}{(x+1)(x-1)(x-1)}$

$=\dfrac{x^2-1+4x-4-2x-2}{(x+1)(x-1)(x-1)}$

$=\dfrac{x^2+2x-7}{(x+1)(x-1)^2}$

18. We multiply the numerator and the denominator by the LCM of the denominators, y^2.

$$\frac{9-\dfrac{1}{y^2}}{3-\dfrac{1}{y}} = \frac{9-\dfrac{1}{y^2}}{3-\dfrac{1}{y}} \cdot \frac{y^2}{y^2}$$

$$= \frac{\left(9-\dfrac{1}{y^2}\right)\cdot y^2}{\left(3-\dfrac{1}{y}\right)\cdot y^2}$$

$$= \frac{9\cdot y^2 - \dfrac{1}{y^2}\cdot y^2}{3\cdot y^2 - \dfrac{1}{y}\cdot y^2}$$

$$= \frac{9y^2-1}{3y^2-y}$$

$$= \frac{(3y+1)(3y-1)}{y(3y-1)}$$

$$= \frac{(3y+1)\cancel{(3y-1)}}{y\cancel{(3y-1)}}$$

$$= \frac{3y+1}{y}$$

19. $\dfrac{7}{y} - \dfrac{1}{3} = \dfrac{1}{4}$, LCM is $12y$

$$12y\left(\frac{7}{y}-\frac{1}{3}\right) = 12y\cdot\frac{1}{4}$$

$$12y\cdot\frac{7}{y} - 12y\cdot\frac{1}{3} = 12y\cdot\frac{1}{4}$$

$$84-4y = 3y$$

$$84 = 7y$$

$$12 = y$$

The number 12 checks, so it is the solution.

20. $\dfrac{15}{x} - \dfrac{15}{x-2} = -2$, LCM is $x(x-2)$

$$x(x-2)\left(\frac{15}{x}-\frac{15}{x-2}\right) = x(x-2)(-2)$$

$$x(x-2)\cdot\frac{15}{x} - x(x-2)\cdot\frac{15}{x-2} = -2x(x-2)$$

$$15(x-2)-15x = -2x(x-2)$$

$$15x-30-15x = -2x^2+4x$$

$$-30 = -2x^2+4x$$

$$2x^2-4x-30 = 0$$

$$2(x^2-2x-15) = 0$$

$$x^2-2x-15 = 0 \quad \text{Dividing by 2}$$

$$(x-5)(x+3) = 0$$

$$x-5=0 \quad or \quad x+3=0$$

$$x=5 \quad or \quad x=-3$$

Both numbers check. The solutions are 5 and -3.

21. We substitute to find k.

$$y = kx$$

$$6 = k\cdot 3$$

$$2 = k$$

The equation of variation is $y = 2x$.

To find the value of y when $x = 25$ we substitute 25 for x in the equation of variation.

$$y = 2x$$

$$y = 2\cdot 25$$

$$y = 50$$

The value of y is 50 when $x = 25$.

22. We substitute to find k.

$$y = kx$$

$$1.5 = k\cdot 3$$

$$0.5 = k$$

The equation of variation is $y = 0.5x$.

To find the value of y when $x = 25$ we substitute 25 for x in the equation of variation.

$$y = 0.5x$$

$$y = 0.5(25)$$

$$y = 12.5$$

The value of y is 12.5 when $x = 25$.

23. We substitute to find k.

$$y = \frac{k}{x}$$

$$6 = \frac{k}{3}$$

$$18 = k$$

The equation of variation is $y = \dfrac{18}{x}$.

To find the value of y when $x = 100$ we substitute 100 for x in the equation of variation.

$$y = \frac{18}{x}$$

$$y = \frac{18}{100}$$

$$y = \frac{9}{50}$$

The value of y is $\dfrac{9}{50}$ when $x = 100$.

24. We substitute to find k.

$$y = \frac{k}{x}$$

$$11 = \frac{k}{2}$$

$$22 = k$$

The equation of variation is $y = \dfrac{22}{x}$.

To find the value of y when $x = 100$ we substitute 100 for x in the equation of variation.

$$y = \frac{22}{x}$$
$$y = \frac{22}{100}$$
$$y = \frac{11}{50}$$

The value of y is $\frac{11}{50}$ when $x = 100$.

25. ***Familiarize and Translate***. The problem states that we have direct variation between the variables d and t. Thus, an equation $d = kt$, $k > 0$, applies.

Solve.

a) First find an equation of variation.

$$d = kt$$
$$60 = k \cdot \frac{1}{2}$$
$$120 = k$$

The equation of variation is $d = 120t$.

b) Use the equation to find the distance the train will travel in 2 hr.

$$d = 120t$$
$$d = 120 \cdot 2$$
$$d = 240$$

Check. We can repeat the computations. Also note that when the time increases from $\frac{1}{2}$ hr to 2 hr, the distance traveled increases from 60 km to 240 km so the answer seems reasonable.

State. The train will travel 240 km in 2 hr.

26. ***Familiarize and Translate***. Let T = the time required to do the job and let M = the number of concrete mixers used. We have inverse variation between T and M so an equation $T = \frac{k}{M}$, $k > 0$, applies.

Solve.

a) First we find an equation of variation.

$$T = \frac{k}{M}$$
$$3 = \frac{k}{2}$$
$$6 = k$$

The equation of variation is $T = \frac{6}{M}$.

b) Use the equation to find the time required to do the job when 5 mixers are used.

$$T = \frac{6}{M}$$
$$T = \frac{6}{5}$$
$$T = 1\frac{1}{5}$$

Check. We can repeat the computations. Also note that as the number of mixers increased from 2 to 5, the time required to mix the concrete decreased from 3 hr to $1\frac{1}{5}$ hr, so the answer seems reasonable.

State. It would take $1\frac{1}{5}$ hr to do the job if 5 concrete mixers are used.

27. ***Familiarize***. We can translate to a proportion, letting d = the number of defective spark plugs that would be expected in a sample of 500.

Translate.

$$\begin{array}{r} \text{Number defective} \rightarrow \\ \text{Sample size} \rightarrow \end{array} \frac{4}{125} = \frac{d}{500} \begin{array}{l} \leftarrow \text{Number defective} \\ \leftarrow \text{Sample size} \end{array}$$

Solve. We solve the proportion.

$$\frac{4}{125} = \frac{d}{500}$$
$$500 \cdot \frac{4}{125} = 500 \cdot \frac{d}{500}$$
$$16 = d$$

Check.

$$\frac{4}{125} = 0.032 \text{ and } \frac{16}{500} = 0.032.$$

The ratios are the same, so the answer checks.

State. You would expect to find 16 defective spark plugs in a sample of 500.

28. ***Familiarize***. The ratio of zebras tagged to the total zebra population, P, is $\frac{15}{P}$. Of the 20 zebras checked later, 6 were tagged. The ratio of zebras tagged to zebras checked is $\frac{6}{20}$.

Translate. Assuming the two ratios are the same, we can translate to a proportion.

$$\begin{array}{c} \text{Zebras tagged} \\ \text{originally} \\ \text{Zebra} \\ \text{population} \end{array} \begin{array}{c} \longrightarrow \\ \longrightarrow \end{array} \frac{15}{P} = \frac{6}{20} \begin{array}{c} \longleftarrow \\ \longleftarrow \end{array} \begin{array}{c} \text{Tagged zebras} \\ \text{caught later} \\ \text{Zebras caught} \\ \text{later} \end{array}$$

Solve. We solve the proportion.

$$\frac{15}{P} = \frac{6}{20}$$
$$20P \cdot \frac{15}{P} = 20P \cdot \frac{6}{20}$$
$$300 = 6P$$
$$50 = P$$

Check.

$$\frac{15}{50} = 0.3 \text{ and } \frac{6}{20} = 0.3.$$

The ratios are the same, so the answer checks.

State. The zebra population is 50.

29. ***Familiarize***. Let t = the time, in minutes, required to copy the report using both copy machines working together.

Translate. We use the work principle, substituting 20 for a and 30 for b.

$$\frac{t}{a} + \frac{t}{b} = 1$$
$$\frac{t}{20} + \frac{t}{30} = 1$$

Solve.

$$\frac{t}{20}+\frac{t}{30}=1,\text{ LCM is }60$$
$$60\left(\frac{t}{20}+\frac{t}{30}\right)=60\cdot 1$$
$$60\cdot\frac{t}{20}+60\cdot\frac{t}{30}=60$$
$$3t+2t=60$$
$$5t=60$$
$$t=12$$

Check. In 12 min, the portion of the job done is

$$12\cdot\frac{1}{20}+12\cdot\frac{1}{30}=\frac{3}{5}+\frac{2}{5}=1.$$

The answer checks.

State. It would take 12 min to copy the report using both machines working together.

30. *Familiarize*. Let r = Marilyn's speed, in km/h. Then $r+20$ = Craig's speed. We organize the information in a table.

	Distance	Speed	Time
Marilyn	225	r	t
Craig	325	$r+20$	t

Translate. We use the formula $d=rt$ in each row of the table to obtain two equations.

$$225=rt,$$
$$325=(r+20)t$$

Since the times are the same, we solve each equation for t and set the results equal to each other.

$$225=rt,\text{ so } t=\frac{225}{r}.$$
$$325=(r+20)t,\text{ so } t=\frac{325}{r+20}.$$

Then we have

$$\frac{225}{r}=\frac{325}{r+20}.$$

Solve.

$$\frac{225}{r}=\frac{325}{r+20},\text{ LCM is }r(r+20)$$
$$r(r+20)\cdot\frac{225}{r}=r(r+20)\cdot\frac{325}{r+20}$$
$$225(r+20)=325r$$
$$225r+4500=325r$$
$$4500=100r$$
$$45=r$$

If $r=45$, then $r+20=45+20=65$.

Check. If Marilyn's speed is 45 km/h and Craig's speed is 65 km/h, then Craig's speed is 20 km/h faster than Marilyn's. At 45 km/h, Marilyn travels 225 km in 225/45, or 5 hr. At 65 km/h, Craig travels 325 km in 325/65, or 5 hr. Since the times are the same, the answer checks.

State. The speed of Marilyn's car is 45 km/h and the speed of Craig's car is 65 km/h.

31. We write a proportion and solve it.

$$\frac{12}{9}=\frac{20}{x}$$
$$9x\cdot\frac{12}{9}=9x\cdot\frac{20}{x}$$
$$12x=180$$
$$x=15$$

(Note that the proportions $\frac{9}{12}=\frac{x}{20}$, $\frac{12}{20}=\frac{9}{x}$, and $\frac{20}{12}=\frac{x}{9}$ could also be used.)

32.

$$\frac{2}{x-4}+\frac{2x}{x^2-16}=\frac{1}{x+4},$$
$$\text{LCM}=(x-4)(x+4)$$
$$(x-4)(x+4)\left(\frac{2}{x-4}+\frac{2x}{(x+4)(x-4)}\right)=(x-4)(x+4)\cdot\frac{1}{x+4}$$
$$2(x+4)+2x=x-4$$
$$2x+8+2x=x-4$$
$$4x+8=x-4$$
$$3x=-12$$
$$x=-4$$

The number -4 is not a solution because it makes a denominator zero. Thus, there is no solution. The correct answer is D.

33. *Familiarize*. Let r = the number of hours it would take Rema to do the job working alone. Then $r+6$ = the number of hours it would take Reggie to do the job working alone.

Translate. We use the work principle, substituting $2\frac{6}{7}$, or $\frac{20}{7}$, for t, r for a, and $r+6$ for b.

$$\frac{t}{a}+\frac{t}{b}=1$$
$$\frac{\frac{20}{7}}{r}+\frac{\frac{20}{7}}{r+6}=1$$

Solve.

$$\frac{\frac{20}{7}}{r}+\frac{\frac{20}{7}}{r+6}=1$$
$$\frac{20}{7}\cdot\frac{1}{r}+\frac{20}{7}\cdot\frac{1}{r+6}=1$$
$$\frac{20}{7r}+\frac{20}{7(r+6)}=1,\text{ LCM is }7r(r+6)$$
$$7r(r+6)\left(\frac{20}{7r}+\frac{20}{7(r+6)}\right)=7r(r+6)\cdot 1$$
$$7r(r+6)\cdot\frac{20}{7r}+7r(r+6)\cdot\frac{20}{7(r+6)}=7r(r+6)$$
$$20(r+6)+20r=7r^2+42r$$
$$20r+120+20r=7r^2+42r$$
$$40r+120=7r^2+42r$$
$$0=7r^2+2r-120$$
$$0=(7r+30)(r-4)$$

$$7r + 30 = 0 \quad or \quad r - 4 = 0$$
$$7r = -30 \quad or \quad r = 4$$
$$r = -\frac{30}{7} \quad or \quad r = 4$$

Check. Since the time cannot be negative, $-\frac{30}{7}$ cannot be a solution. If $r = 4$, then $r + 6 = 4 + 6 = 10$. In $2\frac{6}{7}$ hr, or $\frac{20}{7}$ hr, then the portion of the job done is

$$\frac{\frac{20}{7}}{4} + \frac{\frac{20}{7}}{10} = \frac{20}{7} \cdot \frac{1}{4} + \frac{20}{7} \cdot \frac{1}{10} = \frac{5}{7} + \frac{2}{7} = 1.$$

The answer checks.

State. Working alone it would take Rema 4 hr to do the job and it would take Reggie 10 hr.

34.
$$1 + \frac{1}{1 + \frac{1}{1 + \frac{1}{a}}} = 1 + \frac{1}{1 + \frac{1}{\frac{a}{a} + \frac{1}{a}}}$$
$$= 1 + \frac{1}{1 + \frac{1}{\frac{a+1}{a}}}$$
$$= 1 + \frac{1}{1 + 1 \cdot \frac{a}{a+1}}$$
$$= 1 + \frac{1}{1 + \frac{a}{a+1}}$$
$$= 1 + \frac{1}{\frac{a+1}{a+1} + \frac{a}{a+1}}$$
$$= 1 + \frac{1}{\frac{a+1+a}{a+1}}$$
$$= 1 + \frac{1}{\frac{2a+1}{a+1}}$$
$$= 1 + 1 \cdot \frac{a+1}{2a+1}$$
$$= 1 + \frac{a+1}{2a+1}$$
$$= \frac{2a+1}{2a+1} + \frac{a+1}{2a+1}$$
$$= \frac{2a+1+a+1}{2a+1}$$
$$= \frac{3a+2}{2a+1}$$

Cumulative Review Chapters 1 - 6

1. The distance of 3.5 from 0 is 3.5, so $|3.5| = 3.5$.

2. $x^3 - 2x^2 + x - 1$

The degree of x^3 is 3.

The degree of $-2x^2$ is 2.

The degree of x, or x^1, is 1.

The degree of -1, or $-1 \cdot x^0$, is 0.

The degree of the polynomial is 3, the largest exponent.

3. If a free-range egg has 35% less cholesterol than a factory farm egg, then it has 65% of the cholesterol of a factory farm egg. We reword and translate. Let $c =$ the number of mg of cholesterol in a free-range egg.

$$\text{What is } 65\% \text{ of } \underbrace{423 \text{ mg}}?$$
$$c = 65\% \cdot 423$$

We convert 65% to decimal notation and multiply.

$$c = 65\% \cdot 423 = 0.65 \cdot 423 = 274.95 \approx 275$$

A free-range egg contains about 275 mg of cholesterol.

4. First we find the amount of decrease.

$$2629 - 2438 = 191$$

Now we reword and translate. Let $p =$ the percent of decrease.

$$\underbrace{191 \text{ ft}^2} \text{ is } \underbrace{\text{what percent}} \text{ of } \underbrace{2629 \text{ ft}^2}$$
$$191 = p \cdot 2629$$

We divide by 2629 on both sides of the equation.

$$191 = p \cdot 2629$$
$$\frac{191}{2629} = p$$
$$0.073 \approx p$$
$$7.3\% \approx p$$

The percent of decrease was about 7.3%.

5. Let $a =$ the amount originally borrowed. At a 6% rate, the amount of simple interest owed after 1 yr is $6\%a$, or $0.06a$. We reword the problem and translate to an equation.

$$\underbrace{\text{Amount borrowed}} \text{ plus interest is } \$2650.$$
$$a + 0.06a = 2650$$

Now we solve the equation.

$$a + 0.06a = 2650$$
$$1.06a = 2650$$
$$a = \frac{2650}{1.06}$$
$$a = 2500$$

The amount originally borrowed was $2500.

6. ***Familiarize***. Let $r =$ the speed of the faster car, in mph. Then $r - 10 =$ the speed of the slower car. We organize the information in a table.

	Distance	Speed	Time
Faster car	105	r	t
Slower car	75	$r - 10$	t

Translate. We use the formula $d = rt$ in each row of the table to obtain two equations.

$$105 = rt,$$
$$75 = (r - 10)t$$

Since the times are the same, we solve each equation for t and set the results equal to each other.

$$105 = rt, \text{ so } t = \frac{105}{r}.$$
$$75 = (r - 10)t, \text{ so } t = \frac{75}{r - 10}.$$

Then we have

$$\frac{105}{r} = \frac{75}{r - 10}.$$

Solve.

$$\frac{105}{r} = \frac{75}{r - 10}, \text{ LCM is } r(r - 10)$$
$$r(r - 10) \cdot \frac{105}{r} = r(r - 10) \cdot \frac{75}{r - 10}$$
$$105(r - 10) = 75r$$
$$105r - 1050 = 75r$$
$$-1050 = -30r$$
$$35 = r$$

If $r = 35$, then $r - 10 = 35 - 10 = 25$.

Check. If the speeds are 35 mph and 25 mph, then one car travels 10 mph slower than the other. At 35 mph, the faster car travels 105 mi in 105/35, or 3 hr. At 25 mph, the slower car travels 75 mi in 75/25, or 3 hr. The times are the same, so the answer checks.

State. The speed of the faster car is 35 mph, and the slower car's speed is 25 mph.

7. ***Familiarize***. Let $s =$ the length of a side of the original square, in ft. Then $s + 2 =$ the length of a side of the enlarged square. Recall that the area of a square with side s is given by $s \cdot s$, or s^2.

Translate.

$$\underbrace{\text{Area of original square}}_{\downarrow\; s^2} \quad \underbrace{\text{plus}}_{\downarrow\; +} \quad \underbrace{\text{Area of enlarged square}}_{\downarrow\; (s+2)^2} \quad \underbrace{\text{is}}_{\downarrow\; =} \quad \underbrace{452 \text{ ft}^2}_{\downarrow\; 452}$$

Solve.

$$s^2 + (s + 2)^2 = 452$$
$$s^2 + s^2 + 4s + 4 = 452$$
$$2s^2 + 4s + 4 = 452$$
$$2s^2 + 4s - 448 = 0$$
$$2(s^2 + 2s - 224) = 0$$
$$s^2 + 2s - 224 = 0$$
$$(s - 14)(s + 16) = 0$$
$$s - 14 = 0 \quad or \quad s + 16 = 0$$
$$s = 14 \quad or \quad s = -16$$

Check. Since the length of a side cannot be negative, -16 cannot be a solution. If $s = 14$, then $s + 2 = 14 + 2 = 16$. The area of square with a side of 14 ft is $(14 \text{ ft})^2$, or 196 ft^2, and the area of a square with a side of 16 ft is $(16 \text{ ft})^2$, or 256 ft^2. The sum of the area is $196 \text{ ft}^2 + 256 \text{ ft}^2$, or 452 ft^2, so the answer checks.

State. The length of a side of the original square is 14 ft.

8. a) $M = kB$

$$70 = k \cdot 175$$
$$0.4 = k$$

The equation of variation is $M = 0.4B$.

b) We use the equation to find the value of M when $B = 192$ lb.

$$M = 0.4B$$
$$M = 0.4(192)$$
$$M = 76.8$$

Mike's muscle weight is 76.8 lb.

9. $x^2 - 3x^3 - 4x^2 + 5x^3 - 2$

$$= -3x^3 + 5x^3 + x^2 - 4x^2 - 2$$
$$= (-3 + 5)x^3 + (1 - 4)x^2 - 2$$
$$= 2x^3 - 3x^2 - 2$$

10. $\frac{1}{2}x - \left[\frac{3}{8}x - \left(\frac{2}{3} + \frac{1}{4}x\right) - \frac{1}{3}\right]$

$$= \frac{1}{2}x - \left[\frac{3}{8}x - \frac{2}{3} - \frac{1}{4}x - \frac{1}{3}\right]$$
$$= \frac{1}{2}x - \left[\frac{1}{8}x - 1\right]$$
$$= \frac{1}{2}x - \frac{1}{8}x + 1$$
$$= \frac{3}{8}x + 1$$

11. $\left(\frac{2x^3}{3x^{-1}}\right)^{-2} = \left(\frac{2x^{3-(-1)}}{3}\right)^{-2} = \left(\frac{2x^4}{3}\right)^{-2} =$

$$\left(\frac{3}{2x^4}\right)^2 = \frac{3^2}{(2x^4)^2} = \frac{9}{2^2(x^4)^2} = \frac{9}{4x^8}$$

12. We begin by getting a common denominator in the numerator and in the denominator.

$$\frac{\frac{4}{x}-\frac{6}{x^2}}{\frac{5}{x}+\frac{7}{2x}} = \frac{\frac{4}{x}\cdot\frac{x}{x}-\frac{6}{x^2}}{\frac{5}{x}\cdot\frac{2}{2}+\frac{7}{2x}}$$

$$= \frac{\frac{4x}{x^2}-\frac{6}{x^2}}{\frac{10}{2x}+\frac{7}{2x}}$$

$$= \frac{\frac{4x-6}{x^2}}{\frac{17}{2x}}$$

$$= \frac{4x-6}{x^2}\cdot\frac{2x}{17}$$

$$= \frac{(4x-6)(2x)}{17x^2}$$

$$= \frac{(4x-6)\cdot 2\cdot \cancel{x}}{17\cdot \cancel{x}\cdot x}$$

$$= \frac{8x-12}{17x} = \frac{4(2x-3)}{17x}$$

13. $(5xy^2 - 6x^2y^2 - 3xy^3) - (-4xy^3 + 7xy^2 - 2x^2y^2)$

$= 5xy^2 - 6x^2y^2 - 3xy^3 + 4xy^3 - 7xy^2 + 2x^2y^2$

$= (5-7)xy^2 + (-6+2)x^2y^2 + (-3+4)xy^3$

$= -2xy^2 - 4x^2y^2 + xy^3$

14. $(4x^4+6x^3-6x^2-4)+(2x^5+2x^4-4x^3-4x^2+3x-5)$

$= 2x^5+(4+2)x^4+(6-4)x^3+(-6-4)x^2+3x+(-4-5)$

$= 2x^5 + 6x^4 + 2x^3 - 10x^2 + 3x - 9$

15. $$\frac{2y+4}{21}\cdot\frac{7}{y^2+4y+4} = \frac{(2y+4)(7)}{21(y^2+4y+4)}$$

$$= \frac{2(y+2)(7)}{3\cdot 7(y+2)(y+2)}$$

$$= \frac{2\cancel{(y+2)}\cancel{(7)}}{3\cdot \cancel{7}\cancel{(y+2)}(y+2)}$$

$$= \frac{2}{3(y+2)}$$

16. $$\frac{x^2-9}{x^2+8x+15} \div \frac{x-3}{2x+10} = \frac{x^2-9}{x^2+8x+15}\cdot\frac{2x+10}{x-3}$$

$$= \frac{(x^2-9)(2x+10)}{(x^2+8x+15)(x-3)}$$

$$= \frac{(x+3)(x-3)(2)(x+5)}{(x+3)(x+5)(x-3)}$$

$$= \frac{\cancel{(x+3)}\cancel{(x-3)}(2)\cancel{(x+5)}}{\cancel{(x+3)}\cancel{(x+5)}\cancel{(x-3)}\cdot 1}$$

$= 2$

17. $$\frac{x^2}{x-4}+\frac{16}{4-x} = \frac{x^2}{x-4}+\frac{16}{4-x}\cdot\frac{-1}{-1}$$

$$= \frac{x^2}{x-4}+\frac{-16}{x-4}$$

$$= \frac{x^2-16}{x-4}$$

$$= \frac{(x+4)(x-4)}{x-4}$$

$$= \frac{(x+4)\cancel{(x-4)}}{\cancel{(x-4)}\cdot 1}$$

$= x+4$

18. $$\frac{5x}{x^2-4}-\frac{-3}{2-x} = \frac{5x}{(x+2)(x-2)}-\frac{-3}{2-x}$$

$$= \frac{5x}{(x+2)(x-2)}-\frac{-3}{2-x}\cdot\frac{-1}{-1}$$

$$= \frac{5x}{(x+2)(x-2)}-\frac{3}{x-2}$$

$$= \frac{5x}{(x+2)(x-2)}-\frac{3}{x-2}\cdot\frac{x+2}{x+2}$$

$$= \frac{5x}{(x+2)(x-2)}-\frac{3(x+2)}{(x+2)(x-2)}$$

$$= \frac{5x-3(x+2)}{(x+2)(x-2)}$$

$$= \frac{5x-3x-6}{(x+2)(x-2)}$$

$$= \frac{2x-6}{(x+2)(x-2)} = \frac{2(x-3)}{(x+2)(x-2)}$$

19. We use FOIL.

$(2.5a + 7.5)(0.4a - 1.2) = a^2 - 3a + 3a - 9 = a^2 - 9$

20. $(6x-5)^2 = (6x)^2 - 2\cdot 6x\cdot 5 + 5^2 = 36x^2 - 60x + 25$

21. $(2x^3+1)(2x^3-1) = (2x^3)^2 - 1^2 = 4x^6 - 1$

22. $9a^2 + 52a - 12$

There is no common factor (other than 1). This polynomial has three terms but it is not a trinomial square. Using the *ac*-method, we multiply the leading coefficient and the constant term: $9(-12) = -108$. Now try to factor -108 so that the sum of the factors is 52. The numbers we want are 54 and -2: $54(-2) = -108$ and $54 + (-2) = 52$. Use these factors to split the middle term and then factor by grouping.

$9a^2 + 52a - 12 = 9a^2 + 54a - 2a - 12$

$= (9a^2 + 54a) + (-2a - 12)$

$= 9a(a+6) - 2(a+6)$

$= (9a-2)(a+6)$

23. $9x^2 - 30xy + 25y^2 = (3x)^2 - 2\cdot 3x\cdot 5y + (5y)^2 = (3x-5y)^2$

24. $49x^2 - 1 = (7x)^2 - 1^2 = (7x+1)(7x-1)$

25. $x - [x - (x - 1)] = 2$
$x - [x - x + 1] = 2$
$x - [1] = 2$
$x - 1 = 2$
$x = 3$

The solution is 3.

26. $2x^2 + 7x = 4$
$2x^2 + 7x - 4 = 0$
$(2x - 1)(x + 4) = 0$
$2x - 1 = 0 \quad or \quad x + 4 = 0$
$2x = 1 \quad or \quad x = -4$
$x = \frac{1}{2} \quad or \quad x = -4$

The solutions are $\frac{1}{2}$ and -4.

27. $x^2 = 10x$
$x^2 - 10x = 0$
$x(x - 10) = 0$
$x = 0 \quad or \quad x - 10 = 0$
$x = 0 \quad or \quad x = 10$

The solutions are 0 and 10.

28. $3(x - 2) \le 4(x + 5)$
$3x - 6 \le 4x + 20$
$-6 \le x + 20$
$-26 \le x$

The solution set is $\{x | x \ge -26\}$.

29. $\frac{5x - 2}{4} - \frac{4x - 5}{3} = 1$, LCM is 12
$12\left(\frac{5x - 2}{4} - \frac{4x - 5}{3}\right) = 12 \cdot 1$
$12 \cdot \frac{5x - 2}{4} - 12 \cdot \frac{4x - 5}{3} = 12$
$3(5x - 2) - 4(4x - 5) = 12$
$15x - 6 - 16x + 20 = 12$
$-x + 14 = 12$
$-x = -2$
$x = 2$

The solution is 2.

30. $t = ax + ay$
$t = a(x + y)$
$\frac{t}{x + y} = a$

31. $m = \frac{y_2 - y_1}{x_2 - x_1} = \frac{-1 - 6}{-2 - (-2)} = \frac{-7}{0}$

The slope is not defined.

32. $m = \frac{y_2 - y_1}{x_2 - x_1} = \frac{1 - (-2)}{-4 - 3} = \frac{3}{-7} = -\frac{3}{7}$

33. $m = \frac{y_2 - y_1}{x_2 - x_1} = \frac{-5 - 4}{3\frac{1}{2} - \left(-\frac{1}{2}\right)} = \frac{-9}{3\frac{1}{2} + \frac{1}{2}} = \frac{-9}{4} = -\frac{9}{4}$

34. $m = \frac{y_2 - y_1}{x_2 - x_1} = \frac{\frac{3}{4} - \frac{3}{4}}{-4 - (-7)} = \frac{0}{-4 + 7} = \frac{0}{3} = 0$

35. y-intercept:
$-8 \cdot 0 - 24y = 48$
$-24y = 48$
$y = -2$

The y-intercept is $(0, -2)$.

x-intercept:
$-8x - 24 \cdot 0 = 48$
$-8x = 48$
$x = -6$

The x-intercept is $(-6, 0)$.

36. y-intercept:
$15 - 40 \cdot 0 = -120y$
$15 = -120y$
$-\frac{1}{8} = y$

The y-intercept is $\left(0, -\frac{1}{8}\right)$.

x-intercept:
$15 - 40x = -120 \cdot 0$
$15 - 40x = 0$
$15 = 40x$
$\frac{3}{8} = x$

The x-intercept is $\left(\frac{3}{8}, 0\right)$.

37. The graph of $y = 25$ is a horizontal line with y-intercept $(0, 25)$. There is no x-intercept.

38. The graph of $x = -\frac{1}{4}$ is a vertical line with x-intercept $\left(-\frac{1}{4}, 0\right)$. There is no y-intercept.

39. $x = -3$

We can think of this equation as $x + 0 \cdot y = -3$. No matter what number we choose for y, the value of x will be -3. Some solutions are shown below. The graph is a vertical line.

x	y
-3	3
-3	0
-3	4

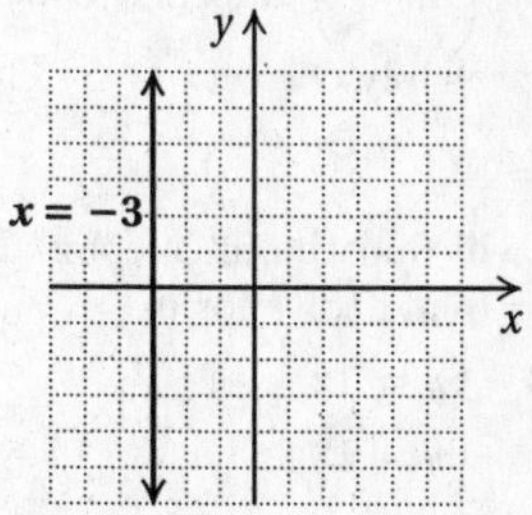

40. $y = -3$

We can think of this equation as $0 \cdot x + y = -3$. No matter what number we choose for x, the value of y will be -3. The graph is a horizontal line with y-intercept $(0, -3)$.

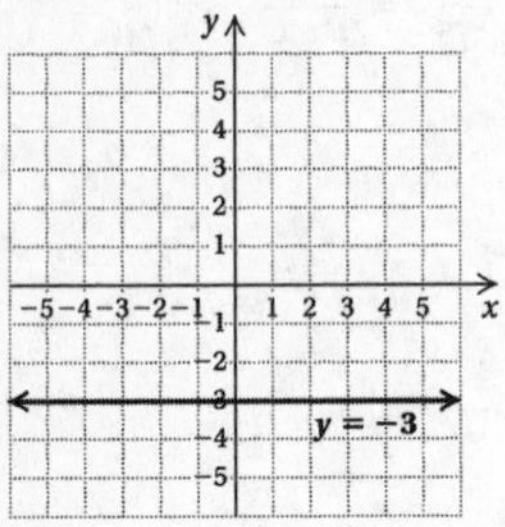

41. $3x - 5y = 15$

First we will find the intercepts. To find the y-intercept we let $x = 0$ and solve for y.

$$3 \cdot 0 - 5y = 15$$
$$-5y = 15$$
$$y = -3$$

The y-intercept is $(0, -3)$.

To find the x-intercept we let $y = 0$ and solve for x.

$$3x - 5 \cdot 0 = 15$$
$$3x = 15$$
$$x = 5$$

The x-intercept is $(5, 0)$.

Plot these points and draw the line containing them.

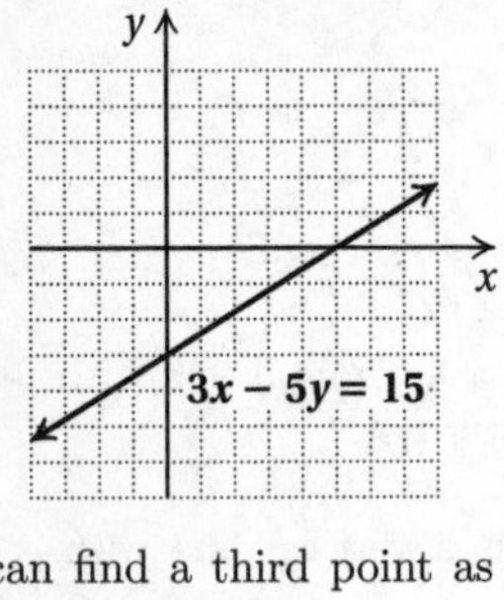

We can find a third point as a check. When $x = 3$, we have

$$3 \cdot 3 - 5y = 15$$
$$9 - 5y = 15$$
$$-5y = 6$$
$$y = -\frac{6}{5}$$

The point $\left(3, -\frac{6}{5}\right)$ appears to be on the graph, so the graph is probably correct.

42. $2x - 6y = 12$

First we will find the intercepts. To find the y-intercept we let $x = 0$ and solve for y.

$$2 \cdot 0 - 6y = 12$$
$$-6y = 12$$
$$y = -2$$

The y-intercept is $(0, -2)$.

To find the x-intercept we let $y = 0$ and solve for x.

$$2x - 6 \cdot 0 = 12$$
$$2x = 12$$
$$x = 6$$

The x-intercept is $(6, 0)$.

Plot these points and draw the line containing them.

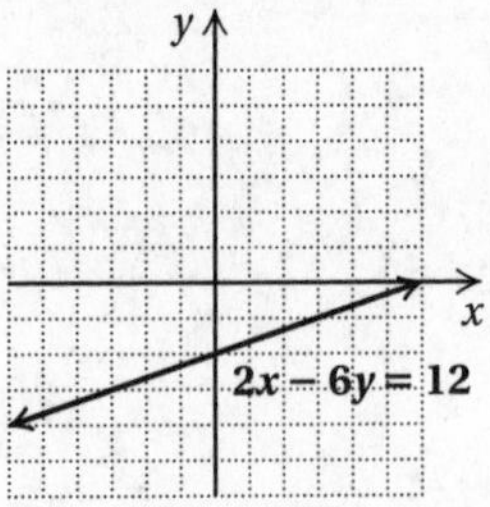

We can find a third point as a check. When $x = -3$, we have

$$2(-3) - 6y = 12$$
$$-6 - 6y = 12$$
$$-6y = 18$$
$$y = -3.$$

The point $(-3, -3)$ appears to be on the graph, so the graph is probably correct.

43. $y = -\frac{1}{3}x - 2$

We find some ordered pairs that are on the graph.

When $x = -3$, $y = -\frac{1}{3}(-3) - 2 = 1 - 2 = -1$.

When $x = 0$, $y = -\frac{1}{3} \cdot 0 - 2 = -2$.

When $x = 3$, $y = -\frac{1}{3} \cdot 3 - 2 = -1 - 2 = -3$.

We plot these points and draw the line.

x	y
-3	-1
0	-2
3	-3

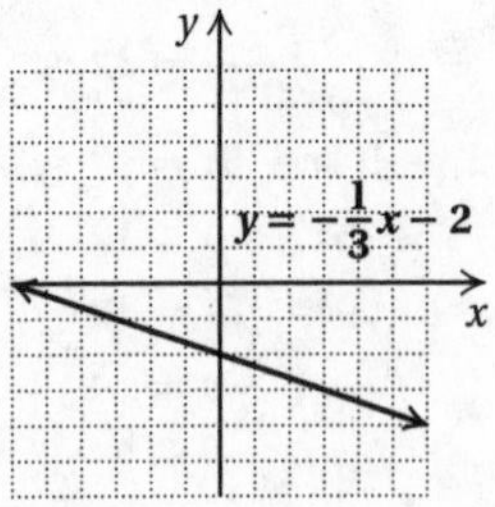

44. $x - y = -5$

First we will find the intercepts. To find the y-intercept we let $x = 0$ and solve for y.

$$0 - y = -5$$
$$-y = -5$$
$$y = 5$$

The y-intercept is $(0, 5)$.

To find the x-intercept we let $y = 0$ and solve for x.

$$x - 0 = -5$$
$$x = -5$$

The x-intercept is $(-5, 0)$.

Plot these points and draw the line containing them.

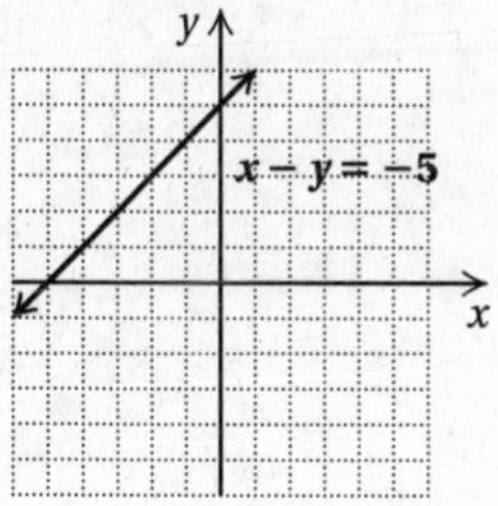

We can find a third point as a check. When $x = -3$, we have

$$-3 - y = -5$$
$$-y = -2$$
$$y = 2.$$

The point $(-3, 2)$ appears to be on the graph, so the graph is probably correct.

45. $\dfrac{\dfrac{1}{x} + x}{2 + \dfrac{1}{x-3}}$

$\dfrac{1}{x}$ is not defined when $x = 0$; $x - 3$ is not defined when $x - 3 = 0$, or $x = 3$. In addition, the expression is not defined when $2 + \dfrac{1}{x-3}$ is 0. We find the value(s) of x for which this is the case.

$$2 + \frac{1}{x-3} = 0$$
$$(x-3)\left(2 + \frac{1}{x-3}\right) = (x-3) \cdot 0$$
$$2(x-3) + (x-3) \cdot \frac{1}{x-3} = 0$$
$$2x - 6 + 1 = 0$$
$$2x - 5 = 0$$
$$2x = 5$$
$$x = \frac{5}{2}$$

Thus, the expression is not defined for 0, 3, and $\dfrac{5}{2}$.

Chapter 7

Systems of Equations

Exercise Set 7.1

RC1. True; see page 552 in the text.

RC3. True; see page 555 in the text.

1. We check by substituting alphabetically 1 for x and 5 for y.

$5x - 2y = -5$	
$5 \cdot 1 - 2 \cdot 5 \ ? \ -5$	
$5 - 10$	
-5	TRUE

$3x - 7y = -32$	
$3 \cdot 1 - 7 \cdot 5 \ ? \ -32$	
$3 - 35$	
-32	TRUE

The ordered pair $(1, 5)$ is a solution of both equations, so it is a solution of the system of equations.

3. We check by substituting alphabetically 4 for a and 2 for b.

$3b - 2a = -2$	
$3 \cdot 2 - 2 \cdot 4 \ ? \ -2$	
$6 - 8$	
-2	TRUE

$b + 2a = 8$	
$2 + 2 \cdot 4 \ ? \ 8$	
$2 + 8$	
10	FALSE

The ordered pair $(4, 2)$ is not a solution of $b + 2a = 8$, so it is not a solution of the system of equations.

5. We check by substituting alphabetically 15 for x and 20 for y.

$3x - 2y = 5$	
$3 \cdot 15 - 2 \cdot 20 \ ? \ 5$	
$45 - 40$	
5	TRUE

$6x - 5y = -10$	
$6 \cdot 15 - 5 \cdot 20 \ ? \ -10$	
$90 - 100$	
-10	TRUE

The ordered pair $(15, 20)$ is a solution of both equations, so it is a solution of the system of equations.

7. We check by substituting alphabetically -1 for x and 1 for y.

$x = -1$	
$-1 \ ? \ -1$	TRUE

$x - y = -2$	
$-1 - 1 \ ? \ -2$	
-2	TRUE

The ordered pair $(-1, 1)$ is a solution of both equations, so it is a solution of the system of equations.

9. We check by substituting alphabetically 18 for x and 3 for y.

$y = \frac{1}{6}x$	
$3 \ ? \ \frac{1}{6} \cdot 18$	
	3 TRUE

$2x - y = 33$	
$2 \cdot 18 - 3 \ ? \ 33$	
$36 - 3$	
33	TRUE

The ordered pair $(18, 3)$ is a solution of both equations, so it is a solution of the system of equations.

11. We graph the equations.

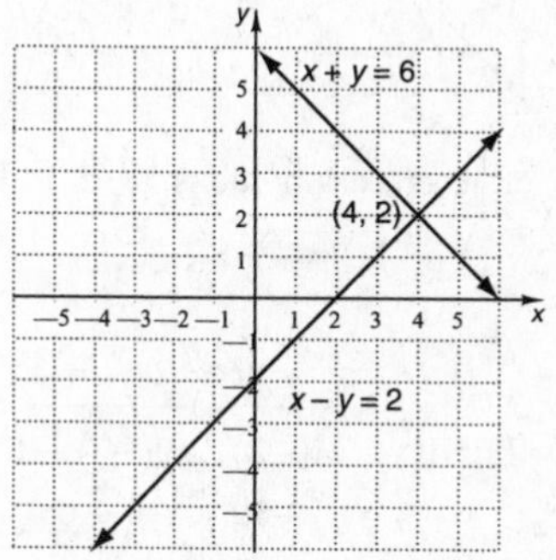

The point of intersection looks as if it has coordinates $(4, 2)$.

Check:

$x - y = 2$	
$4 - 2 \ ? \ 2$	
2	TRUE

$x + y = 6$	
$4 + 2 \ ? \ 6$	
6	TRUE

The solution is $(4, 2)$.

13. We graph the equations.

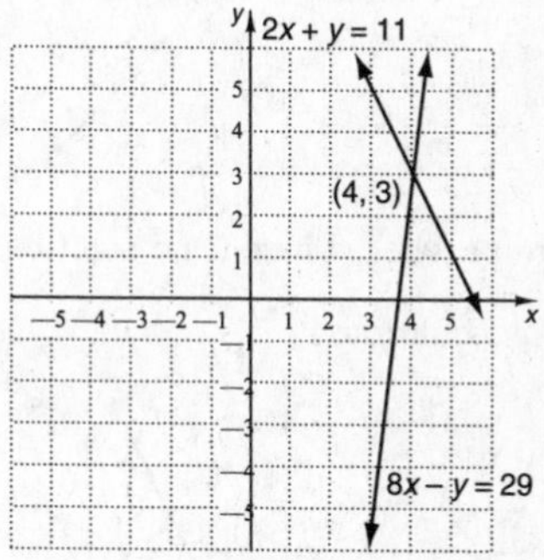

The point of intersection looks as if it has coordinates $(4, 3)$.

Check:

$8x - y = 29$	
$8 \cdot 4 - 3 \ ? \ 29$	
$32 - 3$	
29	TRUE

$2x + y = 11$	
$2 \cdot 4 + 3 \ ? \ 11$	
$8 + 3$	
11	TRUE

The solution is $(4, 3)$.

15. We graph the equations.

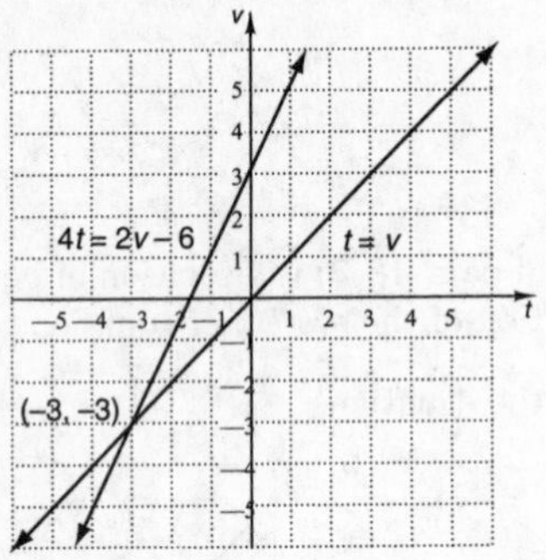

The point of intersection looks as if it has coordinates $(-3,-3)$.

Check:

$$\begin{array}{c|c} t & v \\ \hline -3 & -3 \end{array}\quad \text{TRUE}$$

$t = v$: $-3 \;?\; -3$ TRUE

$$\begin{array}{r|l} 4t & = 2v-6 \\ \hline 4(-3) & ?\; 2(-3)-6 \\ -12 & -6-6 \\ & -12 \end{array}\quad \text{TRUE}$$

The solution is $(-3,-3)$.

17. We graph the equations.

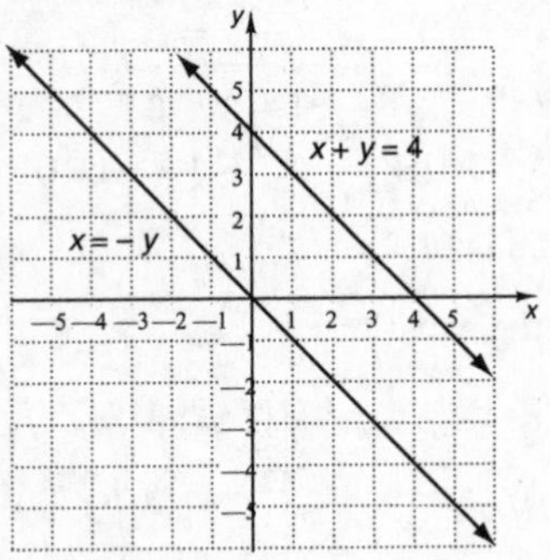

The lines are parallel. There is no solution.

19. We graph the equations.

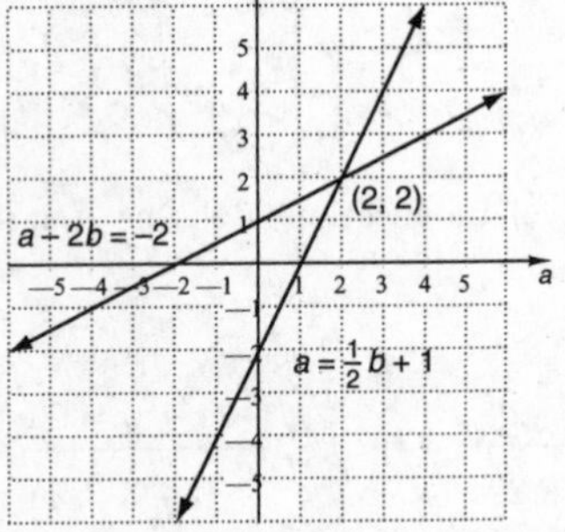

The point of intersection looks as if it has coordinates $(2,2)$.

Check:

$$\begin{array}{r|l} a & = \frac{1}{2}b+1 \\ \hline 2 & ?\; \frac{1}{2}\cdot 2+1 \\ & 1+1 \\ & 2 \end{array}\quad \text{TRUE}$$

$$\begin{array}{r|l} a-2b & = -2 \\ \hline 2-2\cdot 2 & ?\; -2 \\ 2-4 & \\ -2 & \end{array}\quad \text{TRUE}$$

The solution is $(2,2)$.

21. We graph the equations.

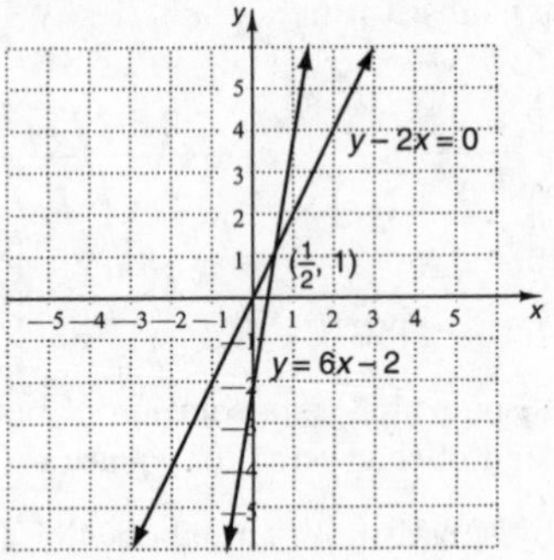

The point of intersection looks as if it has coordinates $\left(\frac{1}{2},1\right)$.

Check:

$$\begin{array}{r|l} y-2x & = 0 \\ \hline 1-2\cdot\frac{1}{2} & ?\; 0 \\ 1-1 & \\ 0 & \end{array}\quad \text{TRUE}$$

$$\begin{array}{r|l} y & = 6x-2 \\ \hline 1 & ?\; 6\cdot\frac{1}{2}-2 \\ & 3-2 \\ & 1 \end{array}\quad \text{TRUE}$$

The solution is $\left(\frac{1}{2},1\right)$.

23. We graph the equations.

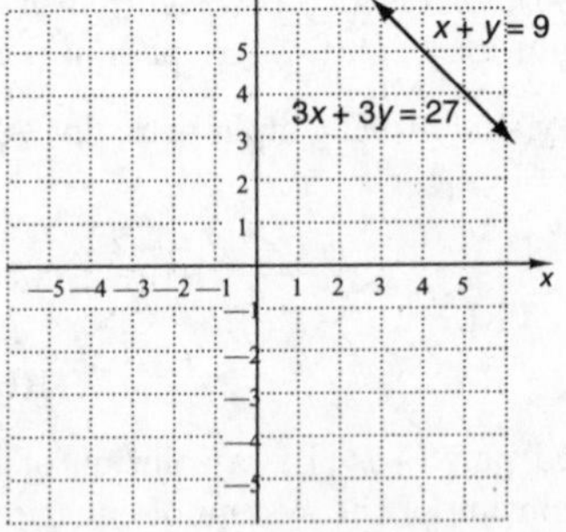

The lines coincide. The system has an infinite number of solutions.

25. We graph the equations.

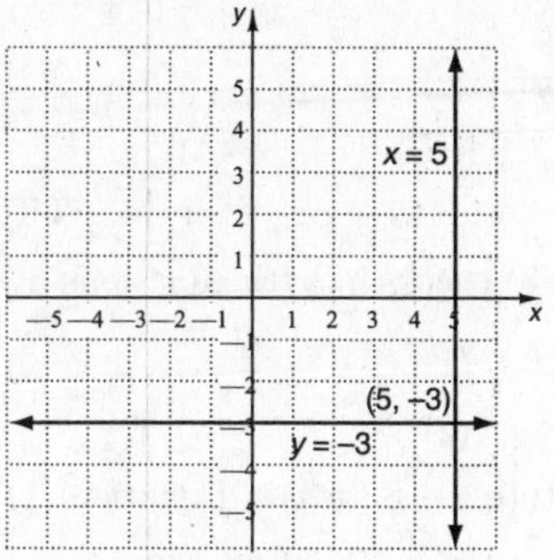

The point of intersection looks as if it has coordinates $(5, -3)$.

Check:

$x = 5$		$y = -3$	
$5 \; ? \; 5$	TRUE	$-3 \; ? \; -3$	TRUE

The solution is $(5, -3)$.

27. $$\frac{x^2-25}{x^2-10x+25} = \frac{(x+5)(x-5)}{(x-5)(x-5)} = \frac{x-5}{x-5}\cdot\frac{x+5}{x-5} = \frac{x+5}{x-5}$$

29. $$\frac{3y-12}{4-y} = \frac{3(y-4)}{-1(-4+y)} = \frac{3(\cancel{y-4})}{-1(\cancel{y-4})} = -3$$

31. The polynomial has exactly three terms, so it is a trinomial.

33. The polynomial has exactly one term, so it is a monomial.

35. $(2, -3)$ is a solution of $Ax - 3y = 13$. Substitute 2 for x and -3 for y and solve for A.

$$Ax - 3y = 13$$
$$A \cdot 2 - 3(-3) = 13$$
$$2A + 9 = 13$$
$$2A = 4$$
$$A = 2$$

$(2, -3)$ is a solution of $x - By = 8$. Substitute 2 for x and -3 for y and solve for B.

$$x - By = 8$$
$$2 - B(-3) = 8$$
$$2 + 3B = 8$$
$$3B = 6$$
$$B = 2$$

37. Answers may vary. Any two equations with a solution of $(6, -2)$ will do. One possibility is

$$x + y = 4,$$
$$x - y = 8.$$

39.-41. Left to the student

Exercise Set 7.2

RC1. True; see page 559 in the text.

RC3. False; we write the ordered pair in alphabetical order of the variables.

1. $x = -2y, \quad (1)$
$x + 4y = 2 \quad (2)$

We substitute $-2y$ for x in Equation (2) and solve for y.

$x + 4y = 2$	Equation (2)
$-2y + 4y = 2$	Substituting
$2y = 2$	Collecting like terms
$y = 1$	Dividing by 2

Next we substitute 1 for y in either equation of the original system and solve for x.

$x = -2y$	Equation (1)
$x = -2 \cdot 1$	
$x = -2$	

We check the ordered pair $(-2, 1)$.

$x = -2y$		$x + 4y = 2$	
$-2 \; ? \; -2 \cdot 1$		$-2 + 4 \cdot 1 \; ? \; 2$	
	-2 TRUE	$-2 + 4$	
		2	TRUE

Since $(-2, 1)$ checks in both equations, it is the solution.

3. $y = x - 6, \quad (1)$
$x + y = -2 \quad (2)$

We substitute $x - 6$ for y in Equation (2) and solve for x.

$x + y = -2$	Equation (2)
$x + (x - 6) = -2$	Substituting
$2x - 6 = -2$	Collecting like terms
$2x = 4$	Adding 6
$x = 2$	Dividing by 2

Next we substitute 2 for x in either equation of the original system and solve for y. We choose Equation (1) since it has y alone on one side.

$y = x - 6$	Equation (1)
$y = 2 - 6$	Substituting
$y = -4$	

We check the ordered pair $(2, -4)$.

$y = x - 6$		$x + y = -2$	
$-4 \; ? \; 2 - 6$		$2 + (-4) \; ? \; -2$	
-4	TRUE	-2	TRUE

Since $(2, -4)$ checks in both equations, it is the solution.

5. $y = 2x - 5$, (1)

$3y - x = 5$ (2)

We substitute $2x - 5$ for y in Equation (2) and solve for x.

$$\begin{aligned} 3y - x &= 5 && \text{Equation (2)} \\ 3(2x - 5) - x &= 5 && \text{Substituting} \\ 6x - 15 - x &= 5 && \text{Removing parentheses} \\ 5x - 15 &= 5 && \text{Collecting like terms} \\ 5x &= 20 && \text{Adding 15} \\ x &= 4 && \text{Dividing by 5} \end{aligned}$$

Next we substitute 4 for x in either equation of the original system and solve for y.

$$\begin{aligned} y &= 2x - 5 && \text{Equation (1)} \\ y &= 2 \cdot 4 - 5 && \text{Substituting} \\ y &= 8 - 5 \\ y &= 3 \end{aligned}$$

We check the ordered pair $(4, 3)$.

$y = 2x - 5$		$3y - x = 5$	
$3 \; ? \; 2 \cdot 4 - 5$		$3 \cdot 3 - 4 \; ? \; 5$	
$8 - 5$		$9 - 4$	
3	TRUE	5	TRUE

Since $(4, 3)$ checks in both equations, it is the solution.

7. $x = y + 5$, (1)

$2x + y = 1$ (2)

We substitute $y + 5$ for x in Equation (2) and solve for y.

$$\begin{aligned} 2x + y &= 1 && \text{Equation (2)} \\ 2(y + 5) + y &= 1 && \text{Substituting} \\ 2y + 10 + y &= 1 && \text{Removing parentheses} \\ 3y + 10 &= 1 && \text{Collecting like terms} \\ 3y &= -9 && \text{Subtracting 10} \\ y &= -3 && \text{Dividing by 3} \end{aligned}$$

Next we substitute -3 for y in either equation of the original system and solve for x. We choose Equation (1) since it has x alone on one side.

$$\begin{aligned} x &= y + 5 && \text{Equation (1)} \\ x &= -3 + 5 && \text{Substituting} \\ x &= 2 \end{aligned}$$

We check the ordered pair $(2, -3)$.

$x = y + 5$		$2x + y = 1$	
$2 \; ? \; 3 + 5$		$2 \cdot 2 - 3 \; ? \; 1$	
2	TRUE	$4 - 3$	
		1	TRUE

Since $(2, -3)$ checks in both equations, it is the solution.

9. $x + y = 10$, (1)

$y = x + 8$ (2)

We substitute $x + 8$ for y in Equation (1) and solve for x.

$$\begin{aligned} x + y &= 10 && \text{Equation (1)} \\ x + (x + 8) &= 10 && \text{Substituting} \\ 2x + 8 &= 10 && \text{Collecting like terms} \\ 2x &= 2 && \text{Subtracting 8} \\ x &= 1 && \text{Dividing by 2} \end{aligned}$$

Next we substitute 1 for x in either equation of the original system and solve for y. We choose Equation (2) since it has y alone on one side.

$$\begin{aligned} y &= x + 8 && \text{Equation (2)} \\ y &= 1 + 8 && \text{Substituting} \\ y &= 9 \end{aligned}$$

We check the ordered pair $(1, 9)$.

$x + y = 10$		$y = x + 8$	
$1 + 9 \; ? \; 10$		$9 \; ? \; 1 + 8$	
10	TRUE	9	TRUE

Since $(1, 9)$ checks in both equations, it is the solution.

11. $2x + y = 5$, (1)

$x = y + 7$ (2)

We substitute $y + 7$ for x in Equation (1) and solve for y.

$$\begin{aligned} 2x + y &= 5 && \text{Equation (1)} \\ 2(y + 7) + y &= 5 && \text{Substituting} \\ 2y + 14 + y &= 5 && \text{Removing parentheses} \\ 3y + 14 &= 5 && \text{Collecting like terms} \\ 3y &= -9 && \text{Subtracting 14} \\ y &= -3 && \text{Dividing by 3} \end{aligned}$$

Now we substitute -3 for y in Equation (2) and find x.

$$x = y + 7 = -3 + 7 = 4$$

The ordered pair $(4, -3)$ checks in both equations. It is the solution.

13. $x - y = 6$, (1)

$x + y = -2$ (2)

We solve Equation (1) for x.

$$\begin{aligned} x - y &= 6 && \text{Equation (1)} \\ x &= y + 6 && \text{Adding } y && (3) \end{aligned}$$

We substitute $y+6$ for x in Equation (2) and solve for y.

$x+y=-2$ Equation (2)
$(y+6)+y=-2$ Substituting
$2y+6=-2$ Collecting like terms
$2y=-8$ Subtracting 6
$y=-4$ Dividing by 2

Now we substitute -4 for y in Equation (3) and compute x.

$x=y+6=-4+6=2$

The ordered pair $(2,-4)$ checks in both equations. It is the solution.

15. $y-2x=-6$, (1)
$2y-x=5$ (2)

We solve Equation (1) for y.

$y-2x=-6$ Equation (1)
$y=2x-6$ (3)

We substitute $2x-6$ for y in Equation (2) and solve for x.

$2y-x=5$ Equation (2)
$2(2x-6)-x=5$ Substituting
$4x-12-x=5$ Removing parentheses
$3x-12=5$ Collecting like terms
$3x=17$ Adding 12
$x=\frac{17}{3}$ Dividing by 3

We substitute $\frac{17}{3}$ for x in Equation (3) and compute y.

$$y=2x-6=2\left(\frac{17}{3}\right)-6=\frac{34}{3}-\frac{18}{3}=\frac{16}{3}$$

The ordered pair $\left(\frac{17}{3},\frac{16}{3}\right)$ checks in both equations. It is the solution.

17. $r-2s=0$, (1)
$4r-3s=15$ (2)

We solve Equation (1) for r.

$r-2s=0$ Equation (1)
$r=2s$ (3)

We substitute $2s$ for r in Equation (2) and solve for s.

$4r-3s=15$ Equation (2)
$4(2s)-3s=15$ Substituting
$8s-3s=15$ Removing parentheses
$5s=15$ Collecting like terms
$s=3$ Dividing by 5

Now we substitute 3 for s in Equation (3) and compute r.

$r=2s=2\cdot 3=6$

The ordered pair $(6,3)$ checks in both equations. It is the solution.

19. $2x+3y=-2$, (1)
$2x-y=9$ (2)

We solve Equation (2) for y.

$2x-y=9$ Equation (2)
$2x=9+y$ Adding y
$2x-9=y$ Subtracting 9 (3)

We substitute $2x-9$ for y in Equation (1) and solve for x.

$2x+3y=-2$ Equation (1)
$2x+3(2x-9)=-2$ Substituting
$2x+6x-27=-2$ Removing parentheses
$8x-27=-2$ Collecting like terms
$8x=25$ Adding 27
$x=\frac{25}{8}$ Dividing by 8

Now we substitute $\frac{25}{8}$ for x in Equation (3) and compute y.

$$y=2x-9=2\left(\frac{25}{8}\right)-9=\frac{25}{4}-\frac{36}{4}=-\frac{11}{4}$$

The ordered pair $\left(\frac{25}{8},-\frac{11}{4}\right)$ checks in both equations. It is the solution.

21. $x+3y=5$, (1)
$3x+5y=3$ (2)

We solve Equation (1) for x.

$x+3y=5$
$x=-3y+5$ Subtracting $3y$ (3)

We substitute $-3y+5$ for x in Equation (2) and solve for y.

$3x+5y=3$ Equation (2)
$3(-3y+5)+5y=3$ Substituting
$-9y+15+5y=3$ Removing parentheses
$-4y+15=3$ Collecting like terms
$-4y=-12$ Subtracting 15
$y=3$ Dividing by -4

Now we substitute 3 for y in Equation (3) and compute x.

$x=-3y+5=-3\cdot 3+5=-9+5=-4$

The ordered pair $(-4,3)$ checks in both equations. It is the solution.

23. $x-y=-3$, (1)
$2x+3y=-6$ (2)

We solve Equation (1) for x.

$x-y=-3$ Equation (1)
$x=y-3$ (3)

We substitute $y-3$ for x in Equation (2) and solve for y.

$$\begin{aligned} 2x+3y &= -6 && \text{Equation (2)}\\ 2(y-3)+3y &= -6 && \text{Substituting}\\ 2y-6+3y &= -6 && \text{Removing parentheses}\\ 5y-6 &= -6 && \text{Collecting like terms}\\ 5y &= 0 && \text{Adding 6}\\ y &= 0 && \text{Dividing by 5} \end{aligned}$$

Now we substitute 0 for y in Equation (3) and compute x.

$$x = y-3 = 0-3 = -3$$

The ordered pair $(-3, 0)$ checks in both equations. It is the solution.

25. ***Familiarize.*** We make a drawing. We let l = the length and w = the width, in inches.

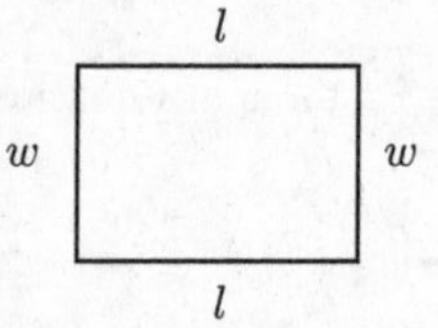

Translate. The perimeter is $2l+2w$. We translate the first statement.

$$\underbrace{\text{The perimeter}}_{\downarrow \atop 2l+2w} \ \underset{\downarrow \atop =}{\text{is}} \ \underbrace{\text{10 in.}}_{\downarrow \atop 10}$$

We translate the second statement.

$$\underbrace{\text{The length}}_{\downarrow \atop l} \ \underset{\downarrow \atop =}{\text{is}} \ \underbrace{\text{2 in. more than the width.}}_{\downarrow \atop w+2}$$

The resulting system is

$$2l+2w = 10, \quad (1)$$
$$l = w+2. \quad (2)$$

Solve. We solve the system. We substitute $w+2$ for l in Equation (1) and solve for w.

$$\begin{aligned} 2l+2w &= 10 && \text{Equation (1)}\\ 2(w+2)+2w &= 10 && \text{Substituting}\\ 2w+4+2w &= 10 && \text{Removing parentheses}\\ 4w+4 &= 10 &&\\ 4w &= 6 && \text{Collecting like terms}\\ w &= \frac{3}{2}, \text{ or } 1\frac{1}{2} && \end{aligned}$$

Now we substitute $\frac{3}{2}$ for w in Equation (2) and solve for l.

$$\begin{aligned} l &= w+2 && \text{Equation (2)}\\ l &= \frac{3}{2}+2 && \text{Substituting}\\ l &= \frac{3}{2}+\frac{4}{2} &&\\ l &= \frac{7}{2}, \text{ or } 3\frac{1}{2} && \end{aligned}$$

Check. A possible solution is a length of $\frac{7}{2}$, or $3\frac{1}{2}$ in. and a width of $\frac{3}{2}$, or $1\frac{1}{2}$ in. The perimeter would be $2\cdot\frac{7}{2}+2\cdot\frac{3}{2}$, or $7+3$, or 10 in. Also, the width plus 2 is $\frac{3}{2}+2$, or $\frac{7}{2}$, which is the length. These numbers check.

State. The length is $3\frac{1}{2}$ in., and the width is $1\frac{1}{2}$ in.

27. ***Familiarize.*** Let l = the length, in mi, and w = the width, in mi. Recall that the perimeter of a rectangle with length l and width w is given by $2l+2w$.

Translate.

$$\underbrace{\text{The perimeter}}_{\downarrow \atop 2l+2w} \ \underset{\downarrow \atop =}{\text{is}} \ \underbrace{\text{1280 mi.}}_{\downarrow \atop 1280}$$

$$\underbrace{\text{The width}}_{\downarrow \atop w} \ \underset{\downarrow \atop =}{\text{is}} \ \underbrace{\text{the length}}_{\downarrow \atop l} \ \underset{\downarrow \atop -}{\text{less}} \ \underbrace{\text{90 mi.}}_{\downarrow \atop 90}$$

The resulting system is

$$2l+2w = 1280, \quad (1)$$
$$w = l-90. \quad (2)$$

Solve. We solve the system. Substitute $l-90$ for w in the first equation and solve for l.

$$\begin{aligned} 2l+2(l-90) &= 1280\\ 2l+2l-180 &= 1280\\ 4l-180 &= 1280\\ 4l &= 1460\\ l &= 365 \end{aligned}$$

Now substitute 365 for l in Equation (2).

$$w = l-90 = 365-90 = 275$$

Check. If the length is 365 mi and the width is 275 mi, then the width is 90 mi less than the length and the perimeter is $2\cdot 365+2\cdot 275$, or $730+550$, or 1280. The answer checks.

State. The length is 365 mi, and the width is 275 mi.

29. ***Familiarize.*** Let l = the length and w = the width, in yards. The perimeter is $l+l+w+w$, or $2l+2w$.

Translate.

$$\underbrace{\text{The perimeter}}_{\downarrow \atop 2l+2w} \ \underset{\downarrow \atop =}{\text{is}} \ \underbrace{\text{340 yd.}}_{\downarrow \atop 340}$$

$$\underbrace{\text{The length}}_{\downarrow \atop l} \ \underset{\downarrow \atop =}{\text{is}} \ \underbrace{\text{10 yd less than twice the width.}}_{\downarrow \atop 2w-10}$$

The resulting system is

$$2l+2w = 340, \quad (1)$$
$$l = 2w-10. \quad (2)$$

Solve. We solve the system. We substitute $2w-10$ for l in Equation (1) and solve for w.

$$2l + 2w = 340 \quad (1)$$
$$2(2w - 10) + 2w = 340$$
$$4w - 20 + 2w = 340$$
$$6w - 20 = 340$$
$$6w = 360$$
$$w = 60$$

Next we substitute 60 for w in Equation (2) and solve for l.

$$l = 2w - 10 = 2 \cdot 60 - 10 = 120 - 10 = 110$$

Check. The perimeter is $2 \cdot 110 + 2 \cdot 60$, or 340 yd. Also 10 yd less than twice the width is $2 \cdot 60 - 10 = 120 - 10 = 110$. The answer checks.

State. The length is 110 yd, and the width is 60 yd.

31. ***Familiarize***. We let $x =$ the larger number and $y =$ the smaller number.

Translate. We translate the first statement.

$$\underbrace{\text{The sum of two numbers}}_{x + y} \ \underbrace{\text{is}}_{=} \ \underbrace{37.}_{37}$$

Now we translate the second statement.

$$\underbrace{\text{One number}}_{x} \ \underbrace{\text{is}}_{=} \ \underbrace{5}_{5} \ \underbrace{\text{more than}}_{+} \ \underbrace{\text{the other.}}_{y}$$

The resulting system is

$$x + y = 37, \quad (1)$$
$$x = 5 + y. \quad (2)$$

Solve. We solve the system of equations. We substitute $5 + y$ for x in Equation (1) and solve for y.

$$x + y = 37 \quad \text{Equation (1)}$$
$$(5 + y) + y = 37 \quad \text{Substituting}$$
$$5 + 2y = 37 \quad \text{Collecting like terms}$$
$$2y = 32 \quad \text{Subtracting 5}$$
$$y = 16 \quad \text{Dividing by 2}$$

We go back to the original equations and substitute 16 for y. We use Equation (2).

$$x = 5 + y \quad \text{Equation (2)}$$
$$x = 5 + 16 \quad \text{Substituting}$$
$$x = 21$$

Check. The sum of 21 and 16 is 37. The number 21 is 5 more than the number 16. These numbers check.

State. The numbers are 21 and 16.

33. ***Familiarize***. Let $x =$ one number and $y =$ the other.

Translate. We reword and translate.

$$\underbrace{\text{The sum of two numbers}}_{x + y} \ \underbrace{\text{is}}_{=} \ \underbrace{52.}_{52}$$

$$\underbrace{\text{The difference of two numbers}}_{x - y} \ \underbrace{\text{is}}_{=} \ \underbrace{28.}_{28}$$

(The second statement could also be translated as $y - x = 28$.)

The resulting system is

$$x + y = 52, \quad (1)$$
$$x - y = 28. \quad (2)$$

Solve. We solve the system. First we solve Equation (2) for x.

$$x - y = 28 \quad \text{Equation (2)}$$
$$x = y + 28 \quad \text{Adding } y \quad (3)$$

We substitute $y + 28$ for x in Equation (1) and solve for y.

$$x + y = 52 \quad \text{Equation (1)}$$
$$(y + 28) + y = 52 \quad \text{Substituting}$$
$$2y + 28 = 52 \quad \text{Collecting like terms}$$
$$2y = 24 \quad \text{Subtracting 28}$$
$$y = 12 \quad \text{Dividing by 2}$$

Now we substitute 12 for y in Equation (3) and compute x.

$$x = y + 28 = 12 + 28 = 40$$

Check. The sum of 40 and 12 is 52, and their difference is 28. These numbers check.

State. The numbers are 40 and 12.

35. ***Familiarize***. We let $x =$ the larger number and $y =$ the smaller number.

Translate. We translate the first statement.

$$\underbrace{\text{The difference between two numbers}}_{x - y} \ \underbrace{\text{is}}_{=} \ \underbrace{12.}_{12}$$

Now we translate the second statement.

$$\underbrace{\text{Two times the larger number}}_{2x} \ \underbrace{\text{is}}_{=} \ \underbrace{\text{five times the smaller.}}_{5y}$$

The resulting system is

$$x - y = 12, \quad (1)$$
$$2x = 5y. \quad (2)$$

Solve. We solve the system. First we solve Equation (1) for x.

$$x - y = 12 \quad \text{Equation (1)}$$
$$x = y + 12 \quad \text{Adding } y \quad (3)$$

We substitute $y + 12$ for x in Equation (2) and solve for y.

$$\begin{aligned} 2x &= 5y && \text{Equation (2)} \\ 2(y+12) &= 5y && \text{Substituting} \\ 2y+24 &= 5y && \text{Removing parentheses} \\ 24 &= 3y && \text{Subtracting 2y} \\ 8 &= y && \text{Dividing by 3} \end{aligned}$$

Now we substitute 8 for y in Equation (3) and compute x.

$$x = y + 12 = 8 + 12 = 20$$

Check. The difference between 20 and 8 is 12. Two times 20, or 40, is five times 8. These numbers check.

State. The numbers are 20 and 8.

37. ***Familiarize.*** Let n and h represent the number of hours Americans spend, per month on average, watching videos on Netflix and Hulu, respectively.

Translate.

$$\underbrace{\text{Netflix hours}}_{\downarrow \; n} \;\; \underset{\downarrow \; +}{\text{plus}} \;\; \underbrace{\text{Hulu hours}}_{\downarrow \; h} \;\; \underset{\downarrow \; =}{\text{is}} \;\; \underbrace{\text{Total hours}}_{\downarrow \; 12.5}$$

$$\underbrace{\text{Netflix hours}}_{\downarrow \; n} \;\; \underset{\downarrow \; =}{\text{are}} \;\; \underbrace{\text{Hulu hours}}_{\downarrow \; h} \;\; \underset{\downarrow \; +}{\text{plus}} \;\; \underbrace{\text{7.5 hours}}_{\downarrow \; 7.5}$$

The resulting system is

$$n + h = 12.5, \quad (1)$$
$$n = h + 7.5. \quad (2)$$

Solve. We first substitute $h + 7.5$ for n in Equation (1) and solve for h.

$$\begin{aligned} n + h &= 12.5 \\ (h + 7.5) + h &= 12.5 \\ 2h + 7.5 &= 12.5 \\ 2h &= 5 \\ h &= 2.5 \end{aligned}$$

Now substitute 2.5 for h in Equation (2) and calculate n.

$$\begin{aligned} n &= h + 7.5 \\ n &= 2.5 + 7.5 \\ n &= 10 \end{aligned}$$

Check. The total number of hours is $10 + 2.5$, or 12.5. The Netflix hours, 10, are 7.5 more than the number of Hulu hours, 2.5. The answer checks.

State. On average, Americans spend 10 hours per month watching videos on Netflix and 2.5 hours per month watching videos on Hulu.

39. Graph: $2x - 3y = 6$

To find the x-intercept, let $y = 0$. Then solve for x.

$$\begin{aligned} 2x - 3 \cdot 0 &= 6 \\ 2x &= 6 \\ x &= 3 \end{aligned}$$

The x-intercept is $(3, 0)$.

To find the y-intercept, let $x = 0$. Then solve for y.

$$\begin{aligned} 2 \cdot 0 - 3y &= 6 \\ -3y &= 6 \\ y &= -2 \end{aligned}$$

The y-intercept is $(0, -2)$.

We plot these points and draw the line.

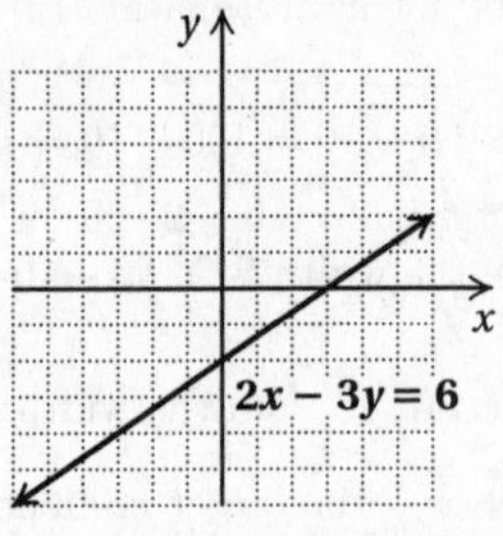

A third point should be used as a check. We let $x = -3$:

$$\begin{aligned} 2(-3) - 3y &= 6 \\ -6 - 3y &= 6 \\ -3y &= 12 \\ y &= -4 \end{aligned}$$

The point $(-3, -4)$ is on the graph, so our graph is probably correct.

41. Graph: $y = 2x - 5$

We select several values for x and compute the corresponding y-values.

When $x = 0$, $y = 2 \cdot 0 - 5 = 0 - 5 = -5$.

When $x = 2$, $y = 2 \cdot 2 - 5 = 4 - 5 = -1$.

When $x = 4$, $y = 2 \cdot 4 - 5 = 8 - 5 = 3$.

x	y	(x, y)
0	-5	$(0, -5)$
2	-1	$(2, -1)$
4	3	$(4, 3)$

We plot these points and draw the line connecting them.

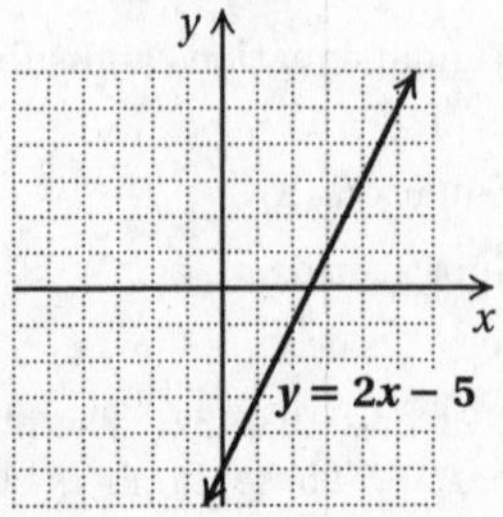

43. $6x^2 - 13x + 6$

The possibilities are $(x+\quad)(6x+\quad)$ and $(2x+\quad)(3x+\quad)$. We look for a pair of factors of the last term, 6, which produces the correct middle term. Since the last term is positive and the middle term is negative, we need only consider negative pairs. The factorization is $(2x - 3)(3x - 2)$.

45. $4x^2 + 3x + 2$

The possibilities are $(x+\ \ )(4x+\ \ )$ and $(2x+\ \ )(2x+\ \ )$. We look for a pair of factors of the last term, 2, which produce the correct middle term. Since the last term and the middle term are both positive, we need only consider positive pairs. We find that there is no possibility that works. The trinomial cannot be factored.

47. $\dfrac{x^{-2}}{x^{-5}} = x^{-2-(-5)} = x^3$

49. $x^{-2} \cdot x^{-5} = x^{-2+(-5)} = x^{-7} = \dfrac{1}{x^7}$

51. First put the equations in "$y =$" form by solving for y. We get

$$y_1 = x - 5,$$
$$y_2 = (-1/2)x + 7/2.$$

Then graph these equations in the standard window and use the INTERSECT feature from the CALC menu to find the coordinates of the point of intersection of the graphs. The solution is $(5.\overline{6}, 0.\overline{6})$.

53. First put the equations in "$y =$" form by solving for y. We get

$$y_1 = 2.35x - 5.97,$$
$$y_2 = (1/2.14)x + (4.88/2.14).$$

Then graph these equations in the standard window and use the INTERSECT feature from the CALC menu to find the coordinates of the point of intersection of the graphs. The solution is approximately $(4.38, 4.33)$.

55. ***Familiarize***. Let $s =$ the perimeter of a softball diamond, in yards, and $b =$ the perimeter of a baseball diamond, in yards.

Translate.

Perimeter of a softball diamond is $\frac{2}{3}$ of perimeter of a baseball diamond.

$$s = \frac{2}{3} \cdot b$$

The sum of the perimeters is 200 yd.

$$s + b = 200$$

The resulting system is

$$s = \frac{2}{3}b, \quad (1)$$
$$s + b = 200. \quad (2)$$

Solve. We solve the system of equations. We substitute $\frac{2}{3}b$ for s in Equation (2) and solve for b.

$$s + b = 200$$
$$\frac{2}{3}b + b = 200$$
$$\frac{5}{3}b = 200$$
$$\frac{3}{5} \cdot \frac{5}{3}b = \frac{3}{5} \cdot 200$$
$$b = 120$$

Next we substitute 120 for b in Equation (1) and solve for s.

$$s = \frac{2}{3}b = \frac{2}{3} \cdot 120 = 80$$

Each diamond has four sides of equal length, so we divide each perimeter by 4 to find the distance between bases in each sport. For the softball diamond the distance is 80/4, or 20 yd. For the baseball diamond it is 120/4, or 30 yd.

Check. The perimeter of the softball diamond, 80 yd, is $\frac{2}{3}$ of 120 yd, the perimeter of the baseball diamond. The sum of the perimeters is $80 + 120$, or 200 yd. We can also recheck the calculations of the distances between the bases. The answer checks.

State. The distance between bases on a softball diamond is 20 yd and the distance between bases on a baseball diamond is 30 yd.

Exercise Set 7.3

RC1. False; see Example 1 on page 566 in the text.

RC3. True; see Example 2 on page 567 in the text.

RC5. False; see Example 7 on page 570 in the text.

1.
$$\begin{aligned} x - y &= 7 \quad &(1)\\ x + y &= 5 \quad &(2)\\ \hline 2x &= 12 \quad &\text{Adding}\\ x &= 6 \quad &\text{Dividing by 2} \end{aligned}$$

Substitute 6 for x in either of the original equations and solve for y.

$$\begin{aligned} x + y &= 5 \quad &\text{Equation (2)}\\ 6 + y &= 5 \quad &\text{Substituting}\\ y &= -1 \quad &\text{Subtracting 6} \end{aligned}$$

Check:

$x - y = 7$	
$6 - (-1)$	? 7
$6 + 1$	
7	TRUE

$x + y = 5$	
$6 + (-1)$	? 5
5	TRUE

Since $(6, -1)$ checks, it is the solution.

3.
$$\begin{aligned} x + y &= 8 \quad &(1)\\ -x + 2y &= 7 \quad &(2)\\ \hline 3y &= 15 \quad &\text{Adding}\\ y &= 5 \quad &\text{Dividing by 3} \end{aligned}$$

Substitute 5 for y in either of the original equations and solve for x.

$$\begin{aligned} x + y &= 8 \quad &\text{Equation (1)}\\ x + 5 &= 8 \quad &\text{Substituting}\\ x &= 3 \end{aligned}$$

Check:

$$\begin{array}{c|c} x+y=8 & \\ \hline 3+5 \ ? \ 8 & \\ 8 & \text{TRUE} \end{array} \qquad \begin{array}{c|c} -x+2y=7 & \\ \hline -3+2\cdot 5 \ ? \ 7 & \\ -3+10 & \\ 7 & \text{TRUE} \end{array}$$

Since $(3,5)$ checks, it is the solution.

5.
$$\begin{array}{rll} 5x - y = 5 & (1) \\ 3x + y = 11 & (2) \\ \hline 8x \quad = 16 & \text{Adding} \\ x = 2 & \text{Dividing by 8} \end{array}$$

Substitute 2 for x in either of the original equations and solve for y.

$$\begin{aligned} 3x+y &= 11 \quad \text{Equation (2)} \\ 3\cdot 2+y &= 11 \quad \text{Substituting} \\ 6+y &= 11 \\ y &= 5 \end{aligned}$$

Check:

$$\begin{array}{c|c} 5x-y=5 & \\ \hline 5\cdot 2-5 \ ? \ 5 & \\ 10-5 & \\ 5 & \text{TRUE} \end{array} \qquad \begin{array}{c|c} 3x+y=11 & \\ \hline 3\cdot 2+5 \ ? \ 11 & \\ 6+5 & \\ 11 & \text{TRUE} \end{array}$$

Since $(2,5)$ checks, it is the solution.

7.
$$\begin{array}{rll} 4a + 3b = 7 & (1) \\ -4a + b = 5 & (2) \\ \hline 4b = 12 & \text{Adding} \\ b = 3 & \end{array}$$

Substitute 3 for b in either of the original equations and solve for a.

$$\begin{aligned} 4a+3b &= 7 \quad \text{Equation (1)} \\ 4a+3\cdot 3 &= 7 \quad \text{Substituting} \\ 4a+9 &= 7 \\ 4a &= -2 \\ a &= -\frac{1}{2} \end{aligned}$$

Check:

$$\begin{array}{c|c} 4a+3b=7 & \\ \hline 4\left(-\frac{1}{2}\right)+3\cdot 3 \ ? \ 7 & \\ -2+9 & \\ 7 & \text{TRUE} \end{array} \qquad \begin{array}{c|c} -4a+b=5 & \\ \hline -4\left(-\frac{1}{2}\right)+3 \ ? \ 5 & \\ 2+3 & \\ 5 & \text{TRUE} \end{array}$$

Since $\left(-\frac{1}{2},3\right)$ checks, it is the solution.

9.
$$\begin{array}{rll} 8x - 5y = -9 & (1) \\ 3x + 5y = -2 & (2) \\ \hline 11x \quad = -11 & \text{Adding} \\ x = -1 & \end{array}$$

Substitute -1 for x in either of the original equations and solve for y.

$$\begin{aligned} 3x+5y &= -2 \quad \text{Equation (2)} \\ 3(-1)+5y &= -2 \quad \text{Substituting} \\ -3+5y &= -2 \\ 5y &= 1 \\ y &= \frac{1}{5} \end{aligned}$$

Check:

$$\begin{array}{c|c} 8x-5y=-9 & \\ \hline 8(-1)-5\left(\frac{1}{5}\right) \ ? \ -9 & \\ -8-1 & \\ -9 & \text{TRUE} \end{array} \qquad \begin{array}{c|c} 3x+5y=-2 & \\ \hline 3(-1)+5\left(\frac{1}{5}\right) \ ? \ -2 & \\ -3+1 & \\ -2 & \text{TRUE} \end{array}$$

Since $\left(-1,\frac{1}{5}\right)$ checks, it is the solution.

11.
$$\begin{array}{rl} 4x - 5y = 7 & \\ -4x + 5y = 7 & \\ \hline 0 = 14 & \text{Adding} \end{array}$$

We obtain a false equation, $0 = 14$, so there is no solution.

13.
$$\begin{array}{rl} x + y = -7, & (1) \\ 3x + y = -9 & (2) \end{array}$$

We multiply on both sides of Equation (1) by -1 and then add.

$$\begin{array}{rll} -x - y = 7 & \text{Multiplying by } -1 \\ 3x + y = -9 & \text{Equation (2)} \\ \hline 2x \quad = -2 & \text{Adding} \\ x = -1 & \end{array}$$

Substitute -1 for x in one of the original equations and solve for y.

$$\begin{aligned} x+y &= -7 \quad \text{Equation (1)} \\ -1+y &= -7 \quad \text{Substituting} \\ y &= -6 \end{aligned}$$

Check:

$$\begin{array}{c|c} x+y=-7 & \\ \hline -1+(-6) \ ? \ -7 & \\ -7 & \text{TRUE} \end{array} \qquad \begin{array}{c|c} 3x+y=-9 & \\ \hline 3(-1)+(-6) \ ? \ -9 & \\ -3-6 & \\ -9 & \text{TRUE} \end{array}$$

Since $(-1,-6)$ checks, it is the solution.

15.
$$\begin{array}{rl} 3x - y = 8, & (1) \\ x + 2y = 5 & (2) \end{array}$$

We multiply on both sides of Equation (1) by 2 and then add.

$$\begin{array}{rll} 6x - 2y = 16 & \text{Multiplying by 2} \\ x + 2y = 5 & \text{Equation (2)} \\ \hline 7x \quad = 21 & \text{Adding} \\ x = 3 & \end{array}$$

Substitute 3 for x in one of the original equations and solve for y.

$x + 2y = 5$ Equation (2)

$3 + 2y = 5$ Substituting

$2y = 2$

$y = 1$

Check:

$3x - y = 8$	
$3 \cdot 3 - 1 \ ? \ 8$	
$9 - 1$	
8	TRUE

$x + 2y = 5$	
$3 + 2 \cdot 1 \ ? \ 5$	
$3 + 2$	
5	TRUE

Since $(3, 1)$ checks, it is the solution.

17. $x - y = 5$, (1)

$4x - 5y = 17$ (2)

We multiply on both sides of Equation (1) by -4 and then add.

$-4x + 4y = -20$ Multiplying by -4

$4x - 5y = 17$ Equation (2)

$-y = -3$ Adding

$y = 3$

Substitute 3 for y in one of the original equations and solve for x.

$x - y = 5$ Equation (1)

$x - 3 = 5$ Substituting

$x = 8$

Check:

$x - y = 5$	
$8 - 3 \ ? \ 5$	
5	TRUE

$4x - 5y = 17$	
$4 \cdot 8 - 5 \cdot 3 \ ? \ 17$	
$32 - 15$	
17	TRUE

Since $(8, 3)$ checks, it is the solution.

19. $2w - 3z = -1$, (1)

$3w + 4z = 24$ (2)

We use the multiplication principle with both equations and then add.

$8w - 12z = -4$ Multiplying (1) by 4

$9w + 12z = 72$ Multiplying (2) by 3

$17w = 68$ Adding

$w = 4$

Substitute 4 for w in one of the original equations and solve for z.

$3w + 4z = 24$ Equation (2)

$3 \cdot 4 + 4z = 24$ Substituting

$12 + 4z = 24$

$4z = 12$

$z = 3$

Check:

$2w - 3z = -1$	
$2 \cdot 4 - 3 \cdot 3 \ ? \ -1$	
$8 - 9$	
-1	TRUE

$3w + 4z = 24$	
$3 \cdot 4 + 4 \cdot 3 \ ? \ 24$	
$12 + 12$	
24	TRUE

Since $(4, 3)$ checks, it is the solution.

21. $2a + 3b = -1$, (1)

$3a + 5b = -2$ (2)

We use the multiplication principle with both equations and then add.

$-10a - 15b = 5$ Multiplying (1) by -5

$9a + 15b = -6$ Multiplying (2) by 3

$-a = -1$ Adding

$a = 1$

Substitute 1 for a in one of the original equations and solve for b.

$2a + 3b = -1$ Equation (1)

$2 \cdot 1 + 3b = -1$ Substituting

$2 + 3b = -1$

$3b = -3$

$b = -1$

Check:

$2a + 3b = -1$	
$2 \cdot 1 + 3(-1) \ ? \ -1$	
$2 - 3$	
-1	TRUE

$3a + 5b = -2$	
$3 \cdot 1 + 5(-1) \ ? \ -2$	
$3 - 5$	
-2	TRUE

Since $(1, -1)$ checks, it is the solution.

23. $x = 3y$, (1)

$5x + 14 = y$ (2)

We first get each equation in the form $Ax + By = C$.

$x - 3y = 0$, (1a) Adding $-3y$

$5x - y = -14$ (2a) Adding $-y - 14$

We multiply by -5 on both sides of Equation (1a) and add.

$-5x + 15y = 0$ Multiplying by -5

$5x - y = -14$

$14y = -14$ Adding

$y = -1$

Substitute -1 for y in Equation (1) and solve for x.

$x - 3y = 0$

$x - 3(-1) = 0$ Substituting

$x + 3 = 0$

$x = -3$

Check:

$x - 3y = 0$	
$-3 - 3(-1)$? 0	
$-3 + 3$	
0	TRUE

$5x - y = -14$	
$5(-3) - (-1)$? -14	
$-15 + 1$	
-14	TRUE

Since $(-3, -1)$ checks, it is the solution.

25. $2x + 5y = 16,$ (1)

$3x - 2y = 5$ (2)

We use the multiplication principle with both equations and then add.

$4x + 10y = 32$ Multiplying (1) by 2

$15x - 10y = 25$ Multiplying (2) by 5

$19x = 57$

$x = 3$

Substitute 3 for x in one of the original equations and solve for y.

$2x + 5y = 16$ Equation (1)

$2 \cdot 3 + 5y = 16$ Substituting

$6 + 5y = 16$

$5y = 10$

$y = 2$

Check:

$2x + 5y = 16$	
$2 \cdot 3 + 5 \cdot 2$? 16	
$6 + 10$	
16	TRUE

$3x - 2y = 5$	
$3 \cdot 3 - 2 \cdot 2$? 5	
$9 - 4$	
5	TRUE

Since $(3, 2)$ checks, it is the solution.

27. $p = 32 + q,$ (1)

$3p = 8q + 6$ (2)

First we write each equation in the form $Ap + Bq = C$.

$p - q = 32,$ (1a) Subtracting q

$3p - 8q = 6$ (2a) Subtracting $8q$

Now we multiply both sides of Equation (1a) by -3 and then add.

$-3p + 3q = -96$ Multiplying by -3

$3p - 8q = 6$ Equation (2a)

$-5q = -90$ Adding

$q = 18$

Substitute 18 for q in Equation (1) and solve for p.

$p = 32 + q$

$p = 32 + 18$ Substituting

$p = 50$

Check:

$p - q = 32$	
$50 - 18$? 32	
32	TRUE

$3p - 8q = 6$	
$3 \cdot 50 - 8 \cdot 18$? 6	
$150 - 144$	
6	TRUE

Since $(50, 18)$ checks, it is the solution.

29. $3x - 2y = 10,$ (1)

$-6x + 4y = -20$ (2)

We multiply by 2 on both sides of Equation (1) and add.

$6x - 4y = 20$

$-6x + 4y = -20$

$0 = 0$

We get an obviously true equation, so the system has an infinite number of solutions.

31. $0.06x + 0.05y = 0.07,$

$0.04x - 0.03y = 0.11$

We first multiply each equation by 100 to clear the decimals.

$6x + 5y = 7,$ (1)

$4x - 3y = 11$ (2)

We use the multiplication principle with both equations of the resulting system.

$18x + 15y = 21$ Multiplying (1) by 3

$20x - 15y = 55$ Multiplying (2) by 5

$38x = 76$ Adding

$x = 2$

Substitute 2 for x in Equation (1) and solve for y.

$6x + 5y = 7$

$6 \cdot 2 + 5y = 7$

$12 + 5y = 7$

$5y = -5$

$y = -1$

Check:

$0.06x + 0.05y = 0.07$	
$0.06(2) + 0.05(-1)$? 0.07	
$0.12 - 0.05$	
0.07	TRUE

$0.04x - 0.03y = 0.11$	
$0.04(2) - 0.03(-1)$? 0.11	
$0.08 + 0.03$	
0.11	TRUE

Since $(2, -1)$ checks, it is the solution.

33. $\frac{1}{3}x + \frac{3}{2}y = \frac{5}{4}$,

$\frac{3}{4}x - \frac{5}{6}y = \frac{3}{8}$

First we clear the fractions. We multiply on both sides of the first equation by 12 and on both sides of the second equation by 24.

$$12\left(\frac{1}{3}x + \frac{3}{2}y\right) = 12 \cdot \frac{5}{4}$$
$$12 \cdot \frac{1}{3}x + 12 \cdot \frac{3}{2}y = 15$$
$$4x + 18y = 15$$
$$24\left(\frac{3}{4}x - \frac{5}{6}y\right) = 24 \cdot \frac{3}{8}$$
$$24 \cdot \frac{3}{4}x - 24 \cdot \frac{5}{6}y = 9$$
$$18x - 20y = 9$$

The resulting system is

$4x + 18y = 15$, (1)

$18x - 20y = 9$. (2)

We use the multiplication principle with both equations.

$$\begin{array}{rll} 72x + 324y = & 270 & \text{Multiplying (1) by 18} \\ -72x + 80y = & -36 & \text{Multiplying (2) by } -4 \\ \hline 404y = & 234 & \end{array}$$
$$y = \frac{234}{404}, \text{ or } \frac{117}{202}$$

Substitute $\frac{117}{202}$ for y in Equation (1) and solve for x.

$$4x + 18\left(\frac{117}{202}\right) = 15$$
$$4x + \frac{1053}{101} = 15$$
$$4x = \frac{462}{101}$$
$$x = \frac{1}{4} \cdot \frac{462}{101}$$
$$x = \frac{231}{202}$$

The ordered pair $\left(\frac{231}{202}, \frac{117}{202}\right)$ checks in both equations. It is the solution.

35. $-4.5x + 7.5y = 6$,

$-x + 1.5y = 5$

First we clear the decimals by multiplying by 10 on both sides of each equation.

$$10(-4.5x + 7.5y) = 10 \cdot 6$$
$$-45x + 75y = 60$$
$$10(-x + 1.5y) = 10 \cdot 5$$
$$-10x + 15y = 50$$

The resulting system is

$-45x + 75y = 60$, (1)

$-10x + 15y = 50$. (2)

We multiply both sides of Equation (2) by -5 and then add.

$$\begin{array}{rll} -45x + 75y = & 60 & \text{Equation (1)} \\ 50x - 75y = & -250 & \text{Multiplying by } -5 \\ \hline 5x \quad = & -190 & \text{Adding} \end{array}$$
$$x = -38$$

Substitute -38 for x in Equation (2) and solve for y.

$$-10x + 15y = 50$$
$$-10(-38) + 15y = 50$$
$$380 + 15y = 50$$
$$15y = -330$$
$$y = -22$$

The ordered pair $(-38, -22)$ checks in both equations. It is the solution.

37. $2t - 13 - t = 5t + 7$

$$t - 13 = 5t + 7$$
$$-4t - 13 = 7$$
$$-4t = 20$$
$$t = -5$$

The solution is -5.

39. $m - (5 - m) = 2(m + 1)$

$$m - 5 + m = 2m + 2$$
$$2m - 5 = 2m + 2$$
$$-5 = 2$$

We get a false equation. The equation has no solution.

41. $2x^2 = x$

$$2x^2 - x = 0$$
$$x(2x - 1) = 0$$
$$x = 0 \quad or \quad 2x - 1 = 0$$
$$x = 0 \quad or \quad 2x = 1$$
$$x = 0 \quad or \quad x = \frac{1}{2}$$

The solutions are 0 and $\frac{1}{2}$.

43. $\frac{a-3}{a-1} = \frac{2}{5}$ LCD $= 5(a-1)$

$$5(a-1) \cdot \frac{a-3}{a-1} = 5(a-1) \cdot \frac{2}{5}$$
$$5(a-3) = 2(a-1)$$
$$5a - 15 = 2a - 2$$
$$3a - 15 = -2$$
$$3a = 13$$
$$a = \frac{13}{3}$$

The solution is $\frac{13}{3}$.

45.-47. Left to the student

49.-53. Left to the student

55. $3(x-y)=9,$
$x+y=7$

First we remove parentheses in the first equation.

$$\begin{aligned} 3x - 3y &= 9, \quad (1)\\ x + y &= 7 \quad (2) \end{aligned}$$

Then we multiply Equation (2) by 3 and add.

$$\begin{aligned} 3x - 3y &= 9\\ 3x + 3y &= 21\\ \hline 6x \quad &= 30\\ x &= 5 \end{aligned}$$

Now we substitute 5 for x in Equation (2) and solve for y.

$$\begin{aligned} x+y &= 7\\ 5+y &= 7\\ y &= 2 \end{aligned}$$

The ordered pair $(5,2)$ checks and is the solution.

57. $2(5a-5b)=10,$
$-5(6a+2b)=10$

First we remove parentheses.

$$\begin{aligned} 10a - 10b &= 10, \quad (1)\\ -30a - 10b &= 10 \quad (2) \end{aligned}$$

Then we multiply Equation (2) by -1 and add.

$$\begin{aligned} 10a - 10b &= 10\\ 30a + 10b &= -10\\ \hline 40a \quad &= 0\\ a &= 0 \end{aligned}$$

Substitute 0 for a in Equation (1) and solve for b.

$$\begin{aligned} 10\cdot 0 - 10b &= 10\\ -10b &= 10\\ b &= -1 \end{aligned}$$

The ordered pair $(0,-1)$ checks and is the solution.

59. $y=-\frac{2}{7}x+3, \quad (1)$
$y=\frac{4}{5}x+3 \qquad (2)$

Observe that these equations represent lines with different slopes and the same y-intercept. Thus, their point of intersection is the y-intercept, $(0,3)$ and this is the solution of the system of equations.

We could also solve this system of equations algebraically. First substitute $\frac{4}{5}x+3$ for y in Equation (1) and solve for x.

$$\begin{aligned} \frac{4}{5}x+3 &= -\frac{2}{7}x+3\\ 35\left(\frac{4}{5}x+3\right) &= 35\left(-\frac{2}{7}x+3\right) \quad \text{Clearing fractions}\\ 35\cdot\frac{4}{5}x+35\cdot 3 &= 35\left(-\frac{2}{7}x\right)+35\cdot 3\\ 28x+105 &= -10x+105\\ 28x &= -10x\\ 38x &= 0\\ x &= 0 \end{aligned}$$

Now substitute 0 for x in one of the original equations and find y. We will use Equation (1).

$$y=-\frac{2}{7}x+3=-\frac{2}{7}\cdot 0+3=0+3=3$$

The ordered pair $(0,3)$ checks and is the solution.

61. $y=ax+b, \quad (1)$
$y=x+c \qquad (2)$

Substitute $x+c$ for y in Equation (1) and solve for x.

$$\begin{aligned} y &= ax+b\\ x+c &= ax+b \quad \text{Substituting}\\ x-ax &= b-c\\ (1-a)x &= b-c\\ x &= \frac{b-c}{1-a} \end{aligned}$$

Substitute $\frac{b-c}{1-a}$ for x in Equation (2) and simplify to find y.

$$\begin{aligned} y &= x+c\\ y &= \frac{b-c}{1-a}+c\\ y &= \frac{b-c}{1-a}+c\cdot\frac{1-a}{1-a}\\ y &= \frac{b-c+c-ac}{1-a}\\ y &= \frac{b-ac}{1-a} \end{aligned}$$

The ordered pair $\left(\frac{b-c}{1-a},\frac{b-ac}{1-a}\right)$ checks and is the solution. This ordered pair could also be expressed as $\left(\frac{c-b}{a-1},\frac{ac-b}{a-1}\right)$.

Chapter 7 Mid-Chapter Review

1. False; a solution of a system of equations must make *both* equations true.

2. False; see page 555 in the text.

3. True; the graphs of the equations are parallel lines, so there is no point of intersection.

4. True; the vertical line $x = a$ and the horizontal line $y = b$ intersect at the point (a, b).

5. $x + y = -1, \quad (1)$

$y = x - 3 \quad (2)$

$$x + (x - 3) = -1$$
$$2x - 3 = -1$$
$$2x = -1 + 3$$
$$2x = 2$$
$$x = 1$$

$$y = 1 - 3$$
$$y = -2$$

The solution is $(1, -2)$.

6.
$$\begin{array}{r} 2x - 3y = 7 \\ x + 3y = -10 \\ \hline 3x + 0y = -3 \end{array}$$
$$3x = -3$$
$$x = -1$$

$$-1 + 3y = -10$$
$$3y = -9$$
$$y = -3$$

The solution is $(-1, -3)$.

7.
$$\begin{array}{c|l} \multicolumn{2}{c}{x + y = 1} \\ \hline -4 + 5 \ ? \ 1 & \\ 1 & \text{TRUE} \end{array} \qquad \begin{array}{c|l} \multicolumn{2}{c}{x = y - 9} \\ \hline -4 \ ? \ 5 - 9 & \\ & -4 \quad \text{TRUE} \end{array}$$

The ordered pair $(-4, 5)$ is a solution of both equations, so it is a solution of the system of equations.

8.
$$\begin{array}{c|l} \multicolumn{2}{c}{x = y + 10} \\ \hline 6 \ ? \ -4 + 10 & \\ & 6 \quad \text{TRUE} \end{array} \qquad \begin{array}{c|l} \multicolumn{2}{c}{x - y = 2} \\ \hline 6 - (-4) \ ? \ 2 & \\ 10 & \text{FALSE} \end{array}$$

The ordered pair $(6, -4)$ is not a solution of $x - y = 2$, so it is not a solution of the system of equations.

9.
$$\begin{array}{c|l} \multicolumn{2}{c}{3x + 5y = 2} \\ \hline 3(-1) + 5 \cdot 1 \ ? \ 2 & \\ -3 + 5 & \\ 2 & \text{TRUE} \end{array} \qquad \begin{array}{c|l} \multicolumn{2}{c}{2x - y = -1} \\ \hline 2(-1) - 1 \ ? \ -1 & \\ -2 - 1 & \\ -3 & \text{FALSE} \end{array}$$

The ordered pair $(-1, 1)$ is not a solution of $2x - y = -1$, so it is not a solution of the system of equations.

10.
$$\begin{array}{c|l} \multicolumn{2}{c}{2x + y = 1} \\ \hline 2 \cdot 2 - 3 \ ? \ 1 & \\ 4 - 3 & \\ 1 & \text{TRUE} \end{array} \qquad \begin{array}{c|l} \multicolumn{2}{c}{3x - 2y = 12} \\ \hline 3 \cdot 2 - 2(-3) \ ? \ 12 & \\ 6 + 6 & \\ 12 & \text{TRUE} \end{array}$$

The ordered pair $(2, -3)$ is a solution of both equations, so it is a solution of the system of equations.

11. We graph the equations.

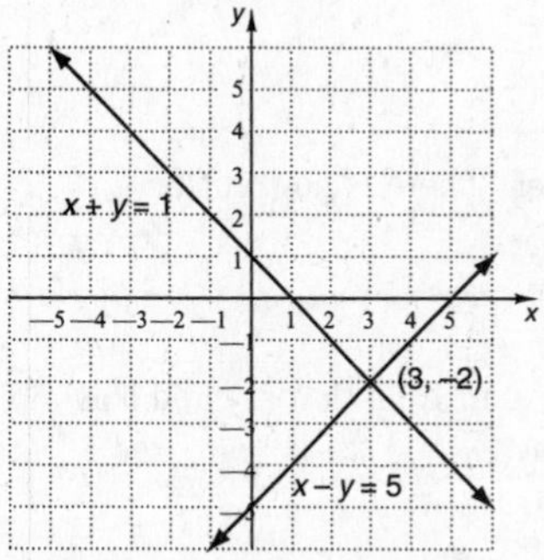

The point of intersection looks as if it has coordinates $(3, -2)$. The ordered pair $(3, -2)$ checks in both equations. It is the solution.

12. We graph the equations.

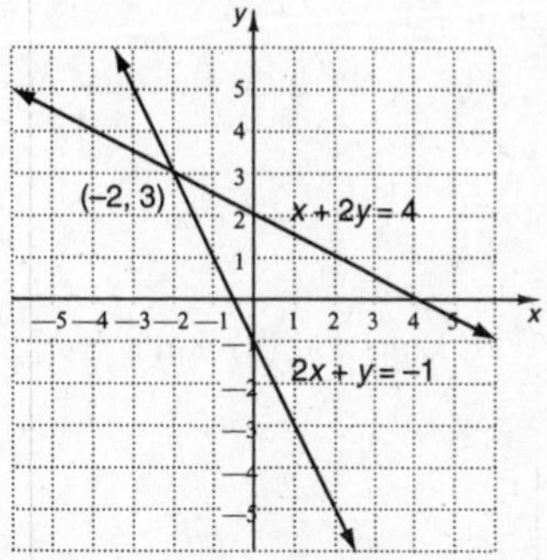

The point of intersection looks as if it has coordinates $(-2, 3)$. The ordered pair $(-2, 3)$ checks in both equations. It is the solution.

13. We graph the equations.

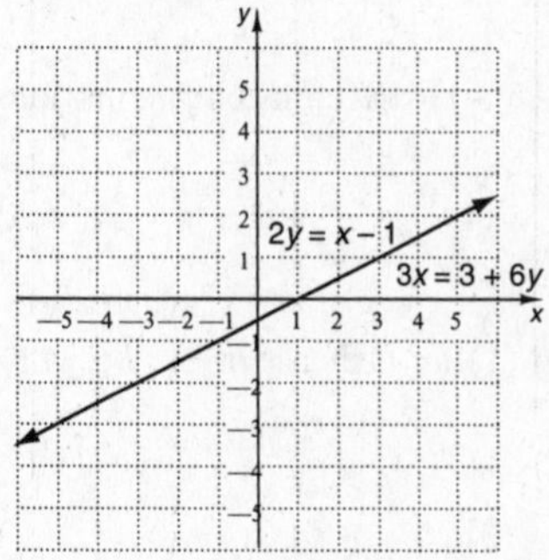

The lines coincide. The system has an infinite number of solutions.

14. We graph the equations.

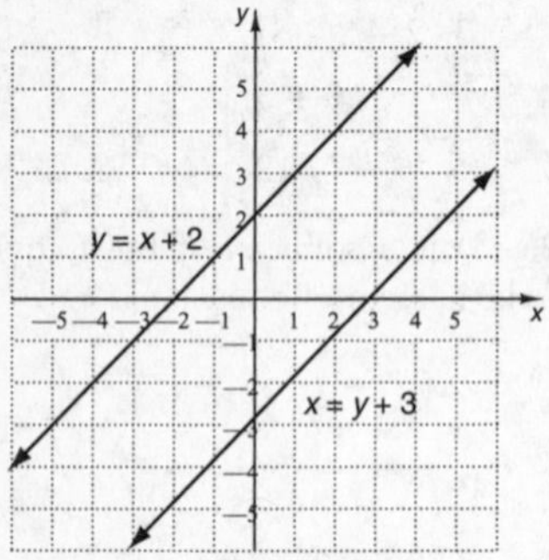

The lines are parallel. There is no solution.

15. $x + y = 2,\quad (1)$
$y = x - 8\quad (2)$

Substitute $x - 8$ for y in Equation (1) and solve for x.

$$\begin{aligned} x + y &= 2 \\ x + (x - 8) &= 2 \\ 2x - 8 &= 2 \\ 2x &= 10 \\ x &= 5 \end{aligned}$$

Now substitute 5 for x in Equation (2) and find y.

$$y = x - 8 = 5 - 8 = -3$$

The ordered pair $(5, -3)$ checks in both equations. It is the solution.

16. $x = y - 1,\quad (1)$
$2x - 5y = 1\quad (2)$

Substitute $y - 1$ for x in Equation (2) and solve for y.

$$\begin{aligned} 2x - 5y &= 1 \\ 2(y - 1) - 5y &= 1 \\ 2y - 2 - 5y &= 1 \\ -3y - 2 &= 1 \\ -3y &= 3 \\ y &= -1 \end{aligned}$$

Now substitute -1 for y in Equation (1) and find x.

$$x = y - 1 = -1 - 1 = -2$$

The ordered pair $(-2, -1)$ checks in both equations. It is the solution.

17. $x + y = 1,\quad (1)$
$3x + 6y = 1\quad (2)$

First solve Equation (1) for either x or y. We will choose to solve for y.

$$\begin{aligned} x + y &= 1 \\ y &= 1 - x \quad (3) \end{aligned}$$

Substitute $1 - x$ for y in Equation (2) and solve for x.

$$\begin{aligned} 3x + 6y &= 1 \\ 3x + 6(1 - x) &= 1 \\ 3x + 6 - 6x &= 1 \\ -3x + 6 &= 1 \\ -3x &= -5 \\ x &= \frac{5}{3} \end{aligned}$$

Substitute $\frac{5}{3}$ for x in Equation (3) and find y.

$$y = 1 - x = 1 - \frac{5}{3} = -\frac{2}{3}$$

The ordered pair $\left(\frac{5}{3}, -\frac{2}{3}\right)$ checks in both equations. It is the solution.

18. $2x + y = 2,\quad (1)$
$2x - y = -1\quad (2)$

First solve Equation (1) for y.

$$\begin{aligned} 2x + y &= 2 \\ y &= 2 - 2x \quad (3) \end{aligned}$$

Now substitute $2 - 2x$ for y in Equation (2) and solve for x.

$$\begin{aligned} 2x - y &= -1 \\ 2x - (2 - 2x) &= -1 \\ 2x - 2 + 2x &= -1 \\ 4x - 2 &= -1 \\ 4x &= 1 \\ x &= \frac{1}{4} \end{aligned}$$

Substitute $\frac{1}{4}$ for x in Equation (3) and find y.

$$y = 2 - 2x = 2 - 2 \cdot \frac{1}{4} = 2 - \frac{1}{2} = \frac{3}{2}$$

The ordered pair $\left(\frac{1}{4}, \frac{3}{2}\right)$ checks in both equations. It is the solution.

19.
$$\begin{aligned} x + y &= 3 \\ -x - y &= 5 \\ \hline 0 &= 8 \end{aligned}$$

We get a false equation. The system of equations has no solution.

20.
$$\begin{aligned} 3x - 2y &= 2, \quad (1) \\ 5x + 2y &= -2 \quad (2) \\ \hline 8x \quad &= 0 \quad \text{Adding} \\ x &= 0 \end{aligned}$$

Substitute 0 for x in one of the original equations and solve for y.

$$\begin{aligned} 5x + 2y &= -2 \quad (2) \\ 5 \cdot 0 + 2y &= -2 \\ 2y &= -2 \\ y &= -1 \end{aligned}$$

The ordered pair $(0,-1)$ checks in both equations. It is the solution.

21. $2x+3y=1,\quad (1)$

$3x+2y=-6\quad (2)$

Multiply the first equation by 2 and the second by -3 and then add.

$$\begin{aligned} 4x+6y&=2\\ -9x-6y&=18\\ \hline -5x\qquad&=20\\ x&=-4 \end{aligned}$$

Substitute -4 for x in one of the original equations and solve for y.

$$\begin{aligned} 2x+3y&=1\quad (1)\\ 2(-4)+3y&=1\\ -8+3y&=1\\ 3y&=9\\ y&=3 \end{aligned}$$

The ordered pair $(-4,3)$ checks in both equations. It is the solution.

22. $2x-3y=6\quad (1)$

$-4x+6y=-12\quad (2)$

Multiply Equation (1) by 2 and then add.

$$\begin{aligned} 4x-6y&=12\\ -4x+6y&=-12\\ \hline 0&=0 \end{aligned}$$

We get an equation that is true for all values of x and y. The system of equations has an infinite number of solutions.

23. ***Familiarize.*** Let $l=$ the length of the rug, in feet, and $w=$ the width.

Translate. The formula for the perimeter of a rectangle gives us one equation.

$2l+2w=18$

The width is one foot shorter than the length, so we have a second equation.

$w=l-1$

The resulting system of equations is

$2l+2w=18,\quad (1)$

$w=l-1.\quad (2)$

Solve. First we substitute $l-1$ for w in Equation (1) and solve for l.

$$\begin{aligned} 2l+2w&=18\\ 2l+2(l-1)&=18\\ 2l+2l-2&=18\\ 4l-2&=18\\ 4l&=20\\ l&=5 \end{aligned}$$

Now substitute 5 for l in Equation (2) and find w.

$w=l-1=5-1=4.$

Check. If the length is 5 ft and the width is 4 ft, the perimeter is $2\cdot5+2\cdot4=10+8=18$ ft. The width, 4 ft, is 1 ft shorter than the length, 5 ft. The answer checks.

State. The length is 5 ft, and the width is 4 ft.

24. ***Familiarize.*** Let x and y represent the numbers.

Translate.

The sum of the numbers is 18.

$$x+y=18$$

The difference of the numbers is 86.

$$x-y=86$$

The resulting system of equations is

$x+y=18,\quad (1)$

$x-y=86.\quad (2)$

Solve. We add the equations.

$$\begin{aligned} x+y&=18\\ x-y&=86\\ \hline 2x\qquad&=104\\ x&=52 \end{aligned}$$

Now substitute 52 for x in one of the original equations and solve for y.

$$\begin{aligned} x+y&=18\quad (1)\\ 52+y&=18\\ y&=-34 \end{aligned}$$

Check. $52+(-34)=18$ and $52-(-34)=52+34=86$. The answer checks.

State. The numbers are 52 and -34.

25. ***Familiarize.*** Let $x=$ the larger number and $y=$ the smaller number.

Translate.

The difference of the numbers is 4.

$$x-y=4$$

Two times the larger number is three times the smaller number.

$$2\cdot x=3\cdot y$$

The resulting system of equations is

$x-y=4,\quad (1)$

$2x=3y.\quad (2)$

Solve. First solve Equation (1) for x.

$$\begin{aligned} x-y&=4\\ x&=y+4\quad (3) \end{aligned}$$

Now substitute $y+4$ for x in Equation (2) and solve for y.

$$\begin{aligned} 2x &= 3y \\ 2(y+4) &= 3y \\ 2y+8 &= 3y \\ 8 &= y \end{aligned}$$

Substitute 8 for y in Equation (3) and find x.

$$x = y+4 = 8+4 = 12$$

Check. $12-8=4$ and $2\cdot 12 = 24 = 3\cdot 8$. The answer checks.

State. The numbers are 12 and 8.

26. We know that the first coordinate of the point of intersection is 2. Substitute 2 for x in either $y = 3x-1$ or $y = 9-2x$ and find y, the second coordinate of the point of intersection.

27. The coordinates of the point of intersection of the graphs are not integers, so it is difficult to determine the solution from the graph.

28. The equations have the same coefficients of x and y but different constant terms. This means their graphs have the same slope but different y-intercepts. Thus, they have no points in common and the system of equations has no solution.

29. This is not the best approach, in general. If the first equation has x alone on one side, for instance, or if the second equation has a variable alone on one side, solving for y in the first equation is inefficient. This procedure could also introduce fractions in the computations unnecessarily.

Exercise Set 7.4

RC1. The mixture will have a total of <u>10</u> gal.

RC3. Berry Choice is 15% fruit juice, so the amount of fruit juice in y gal of Berry Choice is <u>$0.15y$</u>.

1. ***Familiarize.*** Let p = the number of points a discussion post is worth and r = the number of points a reply is worth. Then Ana received $3p$ points for her posts and $10r$ points for her replies. Jae received $8p$ points for her posts and $1\cdot r$, or r, points for her replies.

Translate. Ana received 95 points, so we have one equation: $3p+10r = 95$. Jae received 125 points, so we have a second equation: $8p+r = 125$.

The resulting system is

$$\begin{aligned} 3p + 10r &= 95, \quad (1) \\ 8p + r &= 125. \quad (2) \end{aligned}$$

Solve. Multiply Equation (2) by -10 and add.

$$\begin{aligned} 3p + 10r &= 95 \\ -80p - 10r &= -1250 \\ \hline -77p &= -1155 \\ p &= 15 \end{aligned}$$

Substitute 15 for p in one of the original equations and solve for r.

$$\begin{aligned} 8p+r &= 125 \quad (2) \\ 8\cdot 15 + r &= 125 \\ 120 + r &= 125 \\ r &= 5 \end{aligned}$$

Check. For 3 posts and 10 replies Ana would earn $3\cdot 15 + 10\cdot 5$ or $45+50$, or 95 points. For 8 posts and 1 reply Jae would receive $8\cdot 15+5$, or $120+5$, or 125 points. The answer checks.

State. A discussion post is worth 15 points, and a reply is worth 5 points.

3. ***Familiarize.*** Let a = the number of adults and c = the number of children. We organize the information in a table.

	Adults	Children	Total
Number of visitors	a	c	320
Price	\$4	\$2	
Money taken in	$4a$	$2c$	\$820

Translate. The first and last rows of the table give us two equations. The total number of visitors was 320.

$$a + c = 320$$

Total receipts were \$820.

$$4a + 2c = 820$$

The resulting system of equations is

$$\begin{aligned} a + c &= 320, \quad (1) \\ 4a + 2c &= 820. \quad (2) \end{aligned}$$

Solve. Multiply Equation (1) by -2 and then add.

$$\begin{aligned} -2a - 2c &= -640 \\ 4a + 2c &= 820 \\ \hline 2a &= 180 \\ a &= 90 \end{aligned}$$

Now substitute 90 for a in one of the original equations and solve for c.

$$\begin{aligned} a + c &= 320 \quad (1) \\ 90 + c &= 320 \\ c &= 230 \end{aligned}$$

Check. The total number of visitors was $90+230$, or 320. The receipts from 90 adult tickets were $\$4\cdot 90$, or \$360, and the receipts from 230 children's tickets were $\$2\cdot 230$, or \$460. Then total receipts were $\$360+\$460$, or \$820. The answer checks.

State. 90 adults and 230 children visited the exhibit.

5. ***Familiarize.*** Let x = the number of 4×6 prints and y = the number of 5×7 prints. We organize the information in a table.

	4×6	5×7	Total
Number ordered	x	y	36
Price	\$0.19	\$1.80	
Amount paid	$0.19x$	$1.8y$	\$16.50

Translate. The first and last rows of the table give us two equations. Lucy ordered 36 prints, so we have

$$x + y = 36.$$

The total cost of the prints was \$16.50, so we have

$$0.19x + 1.8y = 16.50.$$

We can multiply the second equation by 100 to clear the decimals. The resulting system of equations is

$$x + y = 36, \quad (1)$$
$$19x + 180y = 1650. \quad (2)$$

Solve. Multiply Equation (1) by -19 and then add.

$$\begin{aligned} -19x - 19y &= -684 \\ 19x + 180y &= 1650 \\ \hline 161y &= 966 \\ y &= 6 \end{aligned}$$

Substitute 6 for y in one of the original equations and solve for x.

$$\begin{aligned} x + y &= 36 \quad (1) \\ x + 6 &= 36 \\ x &= 30 \end{aligned}$$

Check. A total of $30 + 6$, or 36, prints were ordered. The 30 4×6 prints cost $\$0.19 \times 30$, or \$5.70, and the 6 5×7 prints cost $\$1.80 \times 6$, or \$10.80. The total cost of the prints was $\$5.70 + \$10.80$, or \$16.50. The answer checks.

State. Lucy ordered 30 4×6 prints and 6 5×7 prints.

7. ***Familiarize.*** Let x = the number of two-point baskets made and y = the number of three-point baskets made. Then $2x$ points were scored from two-point baskets and $3y$ points were scored from three-point baskets.

Translate. Since 40 baskets were made we have one equation: $x + y = 40$. Since 85 points were scored, we have a second equation: $2x + 3y = 85$.

The resulting system is

$$x + y = 40, \quad (1)$$
$$2x + 3y = 85. \quad (2)$$

Solve. We use the elimination method. First we multiply both sides of Equation (1) by -2 and add.

$$\begin{aligned} -2x - 2y &= -80 \\ 2x + 3y &= 85 \\ \hline y &= 5 \end{aligned}$$

Now we substitute 5 for y in Equation (1) and solve for x.

$$\begin{aligned} x + y &= 40 \\ x + 5 &= 40 \\ x &= 35 \end{aligned}$$

Check. If 35 two-point baskets and 5 three-point baskets are made, then a total of $35 + 5$, or 40, baskets are made. The points scored are $2 \cdot 35 + 3 \cdot 5$, or $70 + 15$, or 85. The answer checks.

State. The Spurs made 35 two-point shots and 5 three-point shots.

9. ***Familiarize.*** Let x = the number of \$50 bonds and y = the number of \$100 bonds. Then the total value of the \$50 bonds is $50x$ and the total value of the \$100 bonds is $100y$.

Translate.

$$\underbrace{\text{Total value of bonds}}_{50x + 100y} \ \underbrace{\text{is}}_{=} \ \underbrace{\$1250.}_{1250}$$

$$\underbrace{\text{Number of \$50 bonds}}_{x} \ \underbrace{\text{is}}_{=} \ \underbrace{7}_{7} \ \underbrace{\text{more than}}_{+} \ \underbrace{\text{number of \$100 bonds.}}_{y}$$

The resulting system is

$$50x + 100y = 1250, \quad (1)$$
$$x = 7 + y. \quad (2)$$

Solve. We use the substitution method, substituting $7+y$ for x in Equation (1).

$$\begin{aligned} 50x + 100y &= 1250 \quad (1) \\ 50(7 + y) + 100y &= 1250 \\ 350 + 50y + 100y &= 1250 \\ 350 + 150y &= 1250 \\ 150y &= 900 \\ y &= 6 \end{aligned}$$

Now we substitute 6 for y in Equation (2) to find x.

$$x = 7 + y \quad (2)$$
$$x = 7 + 6 = 13$$

Check. If there are 13 \$50 bonds and 6 \$100 bonds, there are 7 more \$50 bonds than \$100 bonds. The total value of the bonds is $\$50 \cdot 13 + \$100 \cdot 6$, or $\$650 + \$600$, or \$1250. The answer checks.

State. Cassandra has 13 \$50 bonds and 6 \$100 bonds.

11. ***Familiarize.*** We complete the table in the text. Note that x represents the number of liters of solution A to be used and y represents the number of liters of solution B.

Type of solution	A	B	Mixture
Amount of solution	x	y	100 L
Percent of acid	50%	80%	68%
Amount of acid in solution	$50\%x$	$80\%y$	$68\% \times 100$, or 68 L

Equation from first row: $x + y = 100$

Equation from third row: $50\%x + 80\%y = 68$

Translate. The first and third rows of the table give us two equations. Since the total amount of solution is 100 liters, we have

$$x + y = 100.$$

The amount of acid in the mixture is to be 68% of 100, or 68 liters. The amounts of acid from the two solutions are $50\%x$ and $80\%y$. Thus

$$50\%x + 80\%y = 68,$$
or $$0.5x + 0.8y = 68,$$
or $$5x + 8y = 680 \quad \text{Clearing decimals}$$

The resulting system is

$$x + y = 100, \quad (1)$$
$$5x + 8y = 680. \quad (2)$$

Solve. We use the elimination method. We multiply on both sides of Equation (1) by -5 and then add.

$$\begin{aligned} -5x - 5y &= -500 \quad \text{Multiplying by } -5 \\ 5x + 8y &= 680 \\ \hline 3y &= 180 \\ y &= 60 \end{aligned}$$

We go back to Equation (1) and substitute 60 for y.

$$\begin{aligned} x + y &= 100 \\ x + 60 &= 100 \\ x &= 40 \end{aligned}$$

Check. We consider $x = 40$ and $y = 60$. The sum is 100. Now 50% of 40 is 20 and 80% of 60 is 48. These add up to 68. The numbers check.

State. 40 liters of solution A and 60 liters of solution B should be used.

13. ***Familiarize***. We let $x =$ the number of pounds of hay and $y =$ the number of pounds of grain that should be fed to the horse each day. We arrange the information in a table.

Type of feed	Hay	Grain	Mixture
Amount of feed	x	y	15
Percent of protein	6%	12%	8%
Amount of protein in mixture	$6\%x$	$12\%y$	$8\% \times 15$, or 1.2 lb

Translate. The first and last rows of the table give us two equations. The total amount of feed is 15 lb, so we have

$$x + y = 15.$$

The amount of protein in the mixture is to be 8% of 15 lb, or 1.2 lb. The amounts of protein from the two feeds are $6\%x$ and $12\%y$. Thus

$$6\%x + 12\%y = 1.2, \text{ or}$$
$$0.06x + 0.12y = 1.2, \text{ or}$$
$$6x + 12y = 120 \quad \text{Clearing decimals}$$

The resulting system is

$$x + y = 15, \quad (1)$$
$$6x + 12y = 120. \quad (2)$$

Solve. We use the elimination method. Multiply on both sides of Equation (1) by -6 and then add.

$$\begin{aligned} -6x - 6y &= -90 \\ 6x + 12y &= 120 \\ \hline 6y &= 30 \\ y &= 5 \end{aligned}$$

We go back to Equation (1) and substitute 5 for y.

$$\begin{aligned} x + y &= 15 \\ x + 5 &= 15 \\ x &= 10 \end{aligned}$$

Check. The sum of 10 and 5 is 15. Also, 6% of 10 is 0.6 and 12% of 5 is 0.6, and $0.6 + 0.6 = 1.2$. These numbers check.

State. Brianna should feed her horse 10 lb of hay and 5 lb of grain each day.

15. ***Familiarize***. Let d represent the number of dimes and q the number of quarters. Then, $10d$ represents the value of the dimes in cents, and $25q$ represents the value of the quarters in cents. The total value is \$15.25, or 1525¢. The total number of coins is 103.

Translate.

$$\underbrace{\text{Number of dimes}}_{d} \;\; \underset{+}{\text{plus}} \;\; \underbrace{\text{number of quarters}}_{q} \;\; \underset{=}{\text{is}} \;\; \underbrace{103.}_{103}$$

$$\underbrace{\text{Value of dimes}}_{10d} \;\; \underset{+}{\text{plus}} \;\; \underbrace{\text{value of quarters}}_{25q} \;\; \underset{=}{\text{is}} \;\; \underbrace{\$15.25.}_{1525}$$

The resulting system is

$$d + q = 103, \quad (1)$$
$$10d + 25q = 1525. \quad (2)$$

Solve. We use the addition method. We multiply Equation (1) by -10 and then add.

$$\begin{aligned} -10d - 10q &= -1030 \quad \text{Multiplying by } -10 \\ 10d + 25q &= 1525 \\ \hline 15q &= 495 \quad \text{Adding} \\ q &= 33 \end{aligned}$$

Now we substitute 33 for q in one of the original equations and solve for d.

$$\begin{aligned} d + q &= 103 \quad (1) \\ d + 33 &= 103 \quad \text{Substituting} \\ d &= 70 \end{aligned}$$

Check. The number of dimes plus the number of quarters is 70+33, or 103. The total value in cents is $10 \cdot 70 + 25 \cdot 33$, or 700+825, or 1525. This is equal to \$15.25. This checks.

State. There are 70 dimes and 33 quarters.

17. ***Familiarize.*** We complete the table in the text. Note that x represents the number of pounds of Brazilian coffee to be used and y represents the number of pounds of Turkish coffee.

Type of coffee	Brazilian	Turkish	Mixture
Cost of coffee	\$19	\$22	\$20
Amount (in pounds)	x	y	300
Mixture	$19x$	$22y$	\$20(300), or \$6000

Equation from second row: $x + y = 300$

Equation from third row: $19x + 22y = 6000$

Translate. The second and third rows of the table give us two equations. Since the total amount of the mixture is 300 lb, we have

$$x + y = 300.$$

The value of the Brazilian coffee is $19x$ (x lb at \$19 per pound), the value of the Turkish coffee is $22y$ (y lb at \$22 per pound), and the value of the mixture is \$20(300) or \$6000. Thus we have

$$19x + 22y = 6000.$$

The resulting system is

$$x + y = 300, \quad (1)$$
$$19x + 22y = 6000. \quad (2)$$

Solve. We use the elimination method. We multiply on both sides of Equation (1) by -19 and then add.

$$-19x - 19y = -5700 \quad \text{Multiplying by } -19$$
$$19x + 22y = 6000$$
$$3y = 300$$
$$y = 100$$

We go back to Equation (1) and substitute 100 for y.

$$x + y = 300$$
$$x + 100 = 300$$
$$x = 200$$

Check. The sum of 200 and 100 is 300. The value of the mixture is \$19(200)+\$22(100), or \$3800+\$2200, or \$6000. These values check.

State. 200 lb of Brazilian coffee and 100 lb of Turkish coffee should be used.

19. ***Familiarize.*** Let x = the number of pounds of the 70% mixture and y = the number of pounds of the 45% mixture to be used. We organize the information in the table.

Percent of cashews	70%	45%	
Amount	x	y	60
Mixture	$70\%x$, or $0.7x$	$45\%y$, or $0.45y$	60(60%) or 60(0.6) or 36

Translate. The last two rows of the table give us two equations. The total weight of the mixture is 60 lb, so we have

$$x + y = 60.$$

The amount of cashews in the mixture is 36 lb, so we have

$$0.7x + 0.45y = 36, \text{ or}$$
$$70x + 45y = 3600 \quad \text{Clearing decimals}$$

The resulting system is

$$x + y = 60, \quad (1)$$
$$70x + 45y = 3600. \quad (2)$$

Solve. We use the elimination method. We multiply Equation (1) by -45 and then add.

$$-45x - 45y = -2700$$
$$70x + 45y = 3600$$
$$25x = 900$$
$$x = 36$$

Next we substitute 36 for x in one of the original equations and solve for y.

$$x + y = 60 \quad (1)$$
$$36 + y = 60$$
$$y = 24$$

Check. The total weight of the mixture is 36 lb+24 lb, or 60 lb. The amount of cashews in the mixture is 0.7(36 lb)+ 0.45(24 lb), or 36 lb. Since 36 lb is 60% of 60 lb, the answer checks.

State. The new mixture should contain 36 lb of the 70% cashew mixture and 24 lb of the 45% cashew mixture.

21. ***Familiarize.*** Let x = the number of ounces of Dr. Zeke's cough syrup and y = the number of ounces of Vitabrite cough syrup that should be used. We organize the information in a table.

	Dr. Zeke's	Vitabrite	Mixture
Percent of alcohol	2%	5%	3%
Amount	x	y	80
Mixture	$2\%x$, or $0.02x$	$5\%x$, or $0.05y$	$3\% \cdot 80$, or $0.03 \cdot 80$, or 2.4

Translate. The last two rows of the table give us a system of equations.

$$x + y = 80,$$
$$0.02x + 0.05y = 2.4.$$

Clearing decimals, we have

$$x + y = 80, \quad (1)$$
$$2x + 5y = 240. \quad (2)$$

Solve. We use the elimination method. First we multiply Equation (1) by -2 and then add.

$$\begin{aligned} -2x - 2y &= -160 \\ 2x + 5y &= 240 \\ \hline 3y &= 80 \\ y &= 26\frac{2}{3} \end{aligned}$$

Substitute $26\frac{2}{3}$ for y in one of the original equations and solve for x.

$$\begin{aligned} x + y &= 80 \\ x + 26\frac{2}{3} &= 80 \\ x &= 53\frac{1}{3} \end{aligned}$$

Check. The number of ounces in the mixture is $53\frac{1}{3} + 26\frac{2}{3}$, or 80. The amount of alcohol in the mixture is $0.02\left(53\frac{1}{3}\right) + 0.05\left(26\frac{2}{3}\right)$, or 2.4 oz. Since 2.4 oz is 3% of 80 oz, the answer checks.

State. The mixture should contain $53\frac{1}{3}$ oz of Dr. Zeke's cough syrup and $26\frac{2}{3}$ oz of Vitabrite cough syrup.

23. ***Familiarize***. We arrange the information in a table. Let a = the number of type A questions and b = the number of type B questions.

Type of question	A	B	Mixture (Test)
Number	a	b	16
Time	3 min	6 min	
Value	10 points	15 points	
Mixture (Test)	$3a$ min, $10a$ points	$6b$ min, $15b$ points	60 min, 180 points

Translate. The table actually gives us three equations. Since the total number of questions is 16, we have

$$a + b = 16.$$

The total time is 60 min, so we have

$$3a + 6b = 60.$$

The total number of points is 180, so we have

$$10a + 15b = 180.$$

The resulting system is

$$\begin{aligned} a + b &= 16, &(1) \\ 3a + 6b &= 60, &(2) \\ 10a + 15b &= 180. &(3) \end{aligned}$$

Solve. We will solve the system composed of Equations (1) and (2) and then check to see that this solution also satisfies Equation (3). We multiply equation (1) by -3 and add.

$$\begin{aligned} -3a - 3b &= -48 \\ 3a + 6b &= 60 \\ \hline 3b &= 12 \\ b &= 4 \end{aligned}$$

Now we substitute 4 for b in Equation (1) and solve for a.

$$\begin{aligned} a + b &= 16 \\ a + 4 &= 16 \\ a &= 12 \end{aligned}$$

Check. We consider $a = 12$ questions and $b = 4$ questions. The total number of questions is 16. The time required is $3 \cdot 12 + 6 \cdot 4$, or $36 + 24$, or 60 min. The total points are $10 \cdot 12 + 15 \cdot 4$, or $120 + 60$, or 180. These values check.

State. 12 questions of type A and 4 questions of type B were answered. Assuming all the answers were correct, the score was 180 points.

25. ***Familiarize***. Let k = the age of the Kuyatt's house now and m = the age of the Marconi's house now. Eight years ago the houses' ages were $k - 8$ and $m - 8$.

Translate. We reword and translate.

The age of the Kuyatt's house is twice the age of the Marconi's house.

$$k = 2 \cdot m$$

The age of the Kuyatt's house 8 years ago was three times the age of the Marconi's house 8 years ago.

$$k - 8 = 3 \cdot (m - 8)$$

The resulting system is

$$\begin{aligned} k &= 2m, &(1) \\ k - 8 &= 3(m - 8). &(2) \end{aligned}$$

Solve. We use the substitution method. We substitute $2m$ for k in Equation (2) and solve for m.

$$\begin{aligned} k - 8 &= 3(m - 8) \\ 2m - 8 &= 3(m - 8) \\ 2m - 8 &= 3m - 24 \\ -8 &= m - 24 \\ 16 &= m \end{aligned}$$

We find k by substituting 16 for m in Equation (1).

$$\begin{aligned} k &= 2m \\ k &= 2 \cdot 16 \\ k &= 32 \end{aligned}$$

Check. The age of the Kuyatt's house, 32 years, is twice the age of the Marconi's house, 16 years. Eight years ago, when the Kuyatt's house was 24 years old and the Marconi's house was 8 years old, the Kuyatt's house was three times as old as the Marconi's house. These numbers check.

State. The Kuyatt's house is 32 years old, and the Marconi's house is 16 years old.

27. ***Familiarize***. Let R = Randy's age now and M = Marie's age now. In twelve years their ages will be $R + 12$ and $M + 12$.

Translate. We reword and translate.

Randy's age is four times Marie's age.

$$R = 4 \cdot M$$

Marie's age in 12 years will be half of Randy's age in 12 years.

$$M + 12 = \frac{1}{2} \cdot (R + 12)$$

The resulting system is

$$R = 4M, \quad (1)$$
$$M + 12 = \frac{1}{2}(R + 12). \quad (2)$$

Solve. We use the substitution method. We substitute $4M$ for R in Equation (2) and solve for M.

$$M + 12 = \frac{1}{2}(R + 12)$$
$$M + 12 = \frac{1}{2}(4M + 12)$$
$$M + 12 = 2M + 6$$
$$12 = M + 6$$
$$6 = M$$

We find R by substituting 6 for M in Equation (1).

$$R = 4M$$
$$R = 4 \cdot 6$$
$$R = 24$$

Check. Randy's age now, 24, is 4 times 6, Marie's age. In 12 yr, when Randy will be 36 and Marie 18, Marie's age will be half of Randy's age. These numbers check.

State. Randy is 24 years old now, and Marie is 6.

29. ***Familiarize***. Let x = the smaller angle and y = the larger angle.

Translate. We reword the problem.

The smaller angle plus the larger angle is 180°.

$$x + y = 180$$

The larger angle is 30° more than 2 times the smaller angle.

$$y = 30 + 2 \cdot x$$

The resulting system is

$$x + y = 180,$$
$$y = 30 + 2x.$$

Solve. We solve the system. We will use the elimination method although we could also easily use the substitution method. First we get the second equation in the form $Ax + By = C$.

$$x + y = 180 \quad (1)$$
$$-2x + y = 30 \quad (2) \text{ Adding } -2x$$

Now we multiply Equation (2) by -1 and add.

$$\begin{aligned} x + y &= 180 \\ 2x - y &= -30 \\ \hline 3x \quad &= 150 \\ x &= 50 \end{aligned}$$

Then we substitute 50 for x in Equation (1) and solve for y.

$$x + y = 180 \quad \text{Equation (1)}$$
$$50 + y = 180 \quad \text{Substituting}$$
$$y = 130$$

Check. The sum of the angles is $50° + 130°$, or 180°, so the angles are supplementary. Also, 30° more than two times the 50° angle is $30° + 2 \cdot 50°$, or $30° + 100°$, or 130°, the other angle. These numbers check.

State. The angles are 50° and 130°.

31. ***Familiarize***. We let x = the larger angle and y = the smaller angle.

Translate. We reword and translate the first statement.

The sum of two angles is 90°.

$$x + y = 90$$

We reword and translate the second statement.

The difference of two angles is 34°.

$$x - y = 34$$

The resulting system is

$$x + y = 90,$$
$$x - y = 34.$$

Solve. We solve the system.

$$\begin{aligned} x + y &= 90, \quad (1) \\ x - y &= 34 \quad (2) \\ \hline 2x \quad &= 124 \quad \text{Adding} \\ x &= 62 \end{aligned}$$

Now we substitute 62 for x in Equation (1) and solve for y.

$$x + y = 90 \quad \text{Equation (1)}$$
$$62 + y = 90 \quad \text{Substituting}$$
$$y = 28$$

Check. The sum of the angles is $62° + 28°$, or 90°, so the angles are complementary. The difference of the angles is $62° - 28°$, or 34°. These numbers check.

State. The angles are 62° and 28°.

33. ***Familiarize***. Let $x =$ the number of gallons of 87-octane gas and $y =$ the number of gallons of 93-octane gas that should be used. We arrange the information in a table.

Type of gas	87-octane	93-octane	Mixture
Amount of gas	x	y	18
Octane rating	87	93	89
Mixture	$87x$	$93y$	$18 \cdot 89$, or 1602

Translate. The first and last rows of the table give us a system of equations.

$$x + y = 18, \quad (1)$$
$$87x + 93y = 1602 \quad (2)$$

Solve. We multiply Equation (1) by -87 and then add.

$$\begin{aligned} -87x - 87y &= -1566 \\ 87x + 93y &= 1602 \\ \hline 6y &= 36 \\ y &= 6 \end{aligned}$$

Then substitute 6 for y in Equation (1) and solve for x.

$$x + y = 18$$
$$x + 6 = 18$$
$$x = 12$$

Check. The total amount of gas is 12 gal + 6 gal, or 18 gal. Also $87(12) + 93(6) = 1044 + 558 = 1602$. The answer checks.

State. 12 gal of 87-octane gas and 6 gal of 93-octane gas should be blended.

35. ***Familiarize***. We let $x =$ the number of pages in large type and $y =$ the number of pages in small type. We arrange the information in a table.

Size of type	Large	Small	Mixture (Book)
Words per page	830	1050	
Number of pages	x	y	12
Number of words	$830x$	$1050y$	11,720

Translate. The last two rows of the table give us two equations. The total number of pages in the document is 12, so we have

$$x + y = 12.$$

The number of words on the pages with large type is $830x$ (x pages with 830 words per page), and the number of words on the pages with small type is $1050y$ (y pages with 1050 words per page). The total number of words is 11,720, so we have

$$830x + 1050y = 11,720.$$

The resulting system is

$$x + y = 12, \quad (1)$$
$$830x + 1050y = 11,720. \quad (2)$$

Solve. We use the elimination method. We multiply on both sides of Equation (1) by -830 and then add.

$$\begin{aligned} -830x - 830y &= -9,960 \quad \text{Multiplying by } -830 \\ 830x + 1050y &= 11,720 \\ \hline 220y &= 1760 \\ y &= 8 \end{aligned}$$

We go back to Equation (1) and substitute 8 for y.

$$x + y = 12$$
$$x + 8 = 12$$
$$x = 4$$

Check. The sum of 4 and 8 is 12. The number of words in large type is $830 \cdot 4$, or 3320, and the number of words in small type is $1050 \cdot 8$, or 8400. Then the total number of words is $3320 + 8400$, or 11,720. These numbers check.

State. There were 4 pages in large type and 8 pages in small type.

37. $(2x^2 - 3) - (x^2 - x - 3) = 2x^2 - 3 - x^2 + x + 3 = x^2 + x$

39. $(t^2 + 1.2)^2 = (t^2)^2 + 2 \cdot t^2 \cdot 1.2 + (1.2)^2 = t^4 + 2.4t^2 + 1.44$

41. $(x - 1)(x^2 + x + 1) = x^3 + x^2 + x - x^2 - x - 1 = x^3 - 1$

43.
$$\begin{aligned} \frac{x+7}{x^2-1} - \frac{3}{x+1} &= \frac{x+7}{(x+1)(x-1)} - \frac{3}{x+1} \\ &= \frac{x+7}{(x+1)(x-1)} - \frac{3}{x+1} \cdot \frac{x-1}{x-1} \\ &= \frac{x+7-3(x-1)}{(x+1)(x-1)} \\ &= \frac{x+7-3x+3}{(x+1)(x-1)} \\ &= \frac{-2x+10}{(x+1)(x-1)} \\ &= \frac{-2(x-5)}{(x+1)(x-1)} \end{aligned}$$

45.
$$\begin{aligned} &\frac{1}{x} - \frac{1}{x^2} + \frac{1}{x+1} \\ &= \frac{1}{x} \cdot \frac{x(x+1)}{x(x+1)} - \frac{1}{x^2} \cdot \frac{x+1}{x+1} + \frac{1}{x+1} \cdot \frac{x^2}{x^2} \\ &= \frac{x(x+1) - (x+1) + x^2}{x^2(x+1)} \\ &= \frac{x^2 + x - x - 1 + x^2}{x^2(x+1)} \\ &= \frac{2x^2 - 1}{x^2(x+1)} \end{aligned}$$

47. $\dfrac{10c^2d}{3x^2} \div \dfrac{30cd}{x^3} = \dfrac{10c^2d}{3x^2} \cdot \dfrac{x^3}{30cd}$

$= \dfrac{10c^2dx^3}{3x^2 \cdot 30cd}$

$= \dfrac{10 \cdot c \cdot c \cdot d \cdot x \cdot x^2}{3 \cdot x^2 \cdot 3 \cdot 10 \cdot c \cdot d}$

$= \dfrac{10 \cdot c \cdot d \cdot x^2}{10 \cdot c \cdot d \cdot x^2} \cdot \dfrac{c \cdot x}{3 \cdot 3}$

$= \dfrac{cx}{9}$

49. $y = -2x - 3$

The equation is in the form $y = mx + b$, so the y-intercept is $(0, -3)$.

To find the x-intercept, we let $y = 0$ and solve for x.

$0 = -2x - 3$

$2x = -3$

$x = -\dfrac{3}{2}$

The x-intercept is $\left(-\dfrac{3}{2}, 0\right)$.

We plot the intercepts and draw the line.

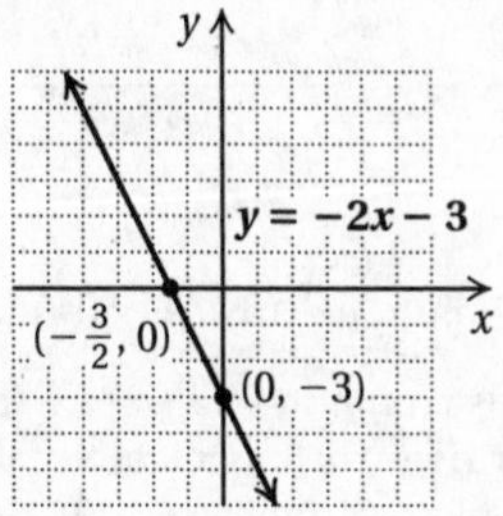

A third point should be used as a check.

For example, let $x = -3$. Then

$y = -2(-3) - 3 = 6 - 3 = 3.$

It appears that the point $(-3, 3)$ is on the graph, so the graph is probably correct.

51. $5x - 2y = -10$

To find the y-intercept, let $x = 0$ and solve for y.

$5 \cdot 0 - 2y = -10$

$-2y = -10$

$y = 5$

The y-intercept is $(0, 5)$.

To find the x-intercept, let $y = 0$ and solve for x.

$5x - 2 \cdot 0 = -10$

$5x = -10$

$x = -2$

The x-intercept is $(-2, 0)$.

We plot the intercepts and draw the line.

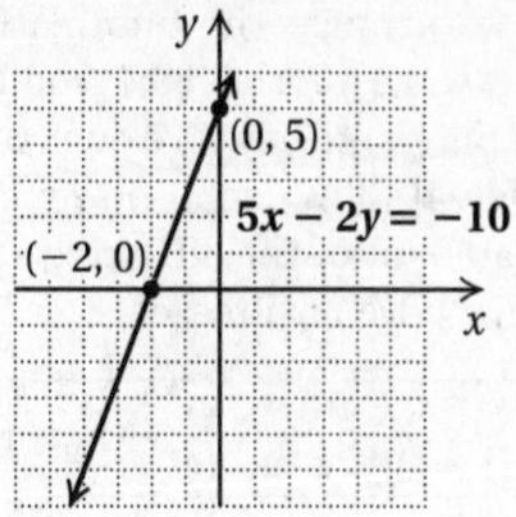

A third point should be used as a check. For example, let $x = -4$. then

$5(-4) - 2y = -10$

$-20 - 2y = -10$

$-2y = 10$

$y = -5$

It appears that the point $(-4, -5)$ is on the graph, so the graph is probably correct.

53. ***Familiarize***. We arrange the information in a table. Let x = the number of liters of skim milk and y = the number of liters of 3.2% milk.

Type of milk	4.6%	Skim	3.2% (Mixture)
Amount of milk	100 L	x	y
Percent of butterfat	4.6%	0%	3.2%
Amount of butterfat in milk	4.6% × 100, or 4.6 L	0% · x, or 0 L	3.2%y

Translate. The first and third rows of the table give us two equations.

Amount of milk: $100 + x = y$

Amount of butterfat: $4.6 + 0 = 3.2\%y$, or $4.6 = 0.032y$.

The resulting system is

$100 + x = y,$

$4.6 = 0.032y.$

Solve. We solve the second equation for y.

$4.6 = 0.032y$

$\dfrac{4.6}{0.032} = y$

$143.75 = y$

We substitute 143.75 for y in the first equation and solve for x.

$100 + x = y$

$100 + x = 143.75$

$x = 43.75$

Check. We consider $x = 43.75$ L and $y = 143.75$ L. The difference between 143.75 L and 43.75 L is 100 L. There is no butterfat in the skim milk. There are 4.6 liters of butterfat in the 100 liters of the 4.6% milk. Thus there are 4.6 liters of butterfat in the mixture. This checks because 3.2% of 143.75 is 4.6.

State. 43.75 L of skim milk should be used.

55. ***Familiarize***. In a table we arrange the information regarding the solution after some of the 30% solution is drained and replaced with pure antifreeze. We let x represent the amount of the original (30%) solution remaining, and we let y represent the amount of the 30% mixture that is drained and replaced with pure antifreeze.

Type of solution	Original (30%)	Pure antifreeze	Mixture
Amount of solution	x	y	16
Percent of antifreeze	30%	100%	50%
Amount of antifreeze in solution	$0.3x$	$1 \cdot y$, or y	$0.5(16)$, or 8

Translate. The table gives us two equations.

Amount of solution: $x + y = 16$

Amount of antifreeze in solution: $0.3x + y = 8$, or $3x + 10y = 80$

The resulting system is

$$x + y = 16, \quad (1)$$
$$3x + 10y = 80. \quad (2)$$

Solve. We multiply Equation (1) by -3 and then add.

$$-3x - 3y = -48$$
$$3x + 10y = 80$$
$$7y = 32$$
$$y = \frac{32}{7}, \text{ or } 4\frac{4}{7}$$

Then we substitute $4\frac{4}{7}$ for y in Equation (1) and solve for x.

$$x + y = 16$$
$$x + 4\frac{4}{7} = 16$$
$$x = 11\frac{3}{7}$$

Check. When $x = 11\frac{3}{7}$ L and $y = 4\frac{4}{7}$ L, the total is 16 L. The amount of antifreeze in the mixture is $0.3\left(11\frac{3}{7}\right) + 4\frac{4}{7}$, or $\frac{3}{10} \cdot \frac{80}{7} + \frac{32}{7}$, or $\frac{24}{7} + \frac{32}{7} = \frac{56}{7}$, or 8 L. This is 50% of 16 L, so the numbers check.

State. $4\frac{4}{7}$ L of the original mixture should be drained and replaced with pure antifreeze.

57. ***Familiarize***. Let x = the tens digit and y = the ones digit. Then the number is $10x + y$.

Translate. The number is six times the sum of its digits, so we have

$$10x + y = 6(x + y)$$
$$10x + y = 6x + 6y$$
$$4x - 5y = 0.$$

The tens digit is 1 more than the ones digit so we have

$$x = y + 1.$$

The resulting system is

$$4x - 5y = 0, \quad (1)$$
$$x = y + 1. \quad (2)$$

Solve. First substitute $y + 1$ for x in Equation (1) and solve for y.

$$4x - 5y = 0 \quad (1)$$
$$4(y + 1) - 5y = 0$$
$$4y + 4 - 5y = 0$$
$$-y + 4 = 0$$
$$4 = y$$

Now substitute 4 for y in one of the original equations and solve for x.

$$x = y + 1 \quad (2)$$
$$x = 4 + 1$$
$$x = 5$$

Check. If the number is 54, then the sum of the digits is $5 + 4$, or 9, and $54 = 6 \cdot 9$. Also, the tens digit, 5, is one more than the ones digit, 4. The answer checks.

State. The number is 54.

Exercise Set 7.5

RC1. $d = rt$

$d = 60t$

If Troy drove t hr at 60 mph, he traveled $\underline{60t}$ mi.

RC3. Rosa's motorboat travels r mph in still water. If she is motoring upstream in a river with a current of 4 mph, she is moving at a rate of $\underline{r - 4}$ mph.

1. ***Familiarize***. We first make a drawing.

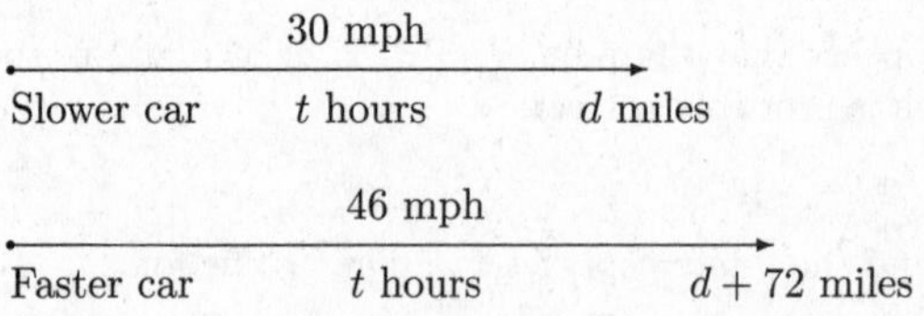

We let d = the distance the slower car travels. Then $d + 72$ = the distance the faster car travels. We call the time t. We complete the table in the text, filling in the distances as well as the other information.

$$d = r \cdot t$$

	Distance	Speed	Time
Slower car	d	30	t
Faster car	$d + 72$	46	t

Translate. We get an equation $d = rt$ from each row of the table. Thus we have

$$d = 30t, \quad (1)$$
$$d + 72 = 46t. \quad (2)$$

Solve. We use the substitution method. We substitute $30t$ for d in Equation (2).

$$\begin{aligned} d + 72 &= 46t \\ 30t + 72 &= 46t \quad \text{Substituting} \\ 72 &= 16t \quad \text{Subtracting } 30t \\ 4.5 &= t \quad \text{Dividing by 16} \end{aligned}$$

Check. In 4.5 hr the slower car travels 30(4.5), or 135 mi, and the faster car travels 46(4.5), or 207 mi. Since 207 is 72 more than 135, our result checks.

State. The cars will be 72 mi apart in 4.5 hr.

3. ***Familiarize.*** First make a drawing.

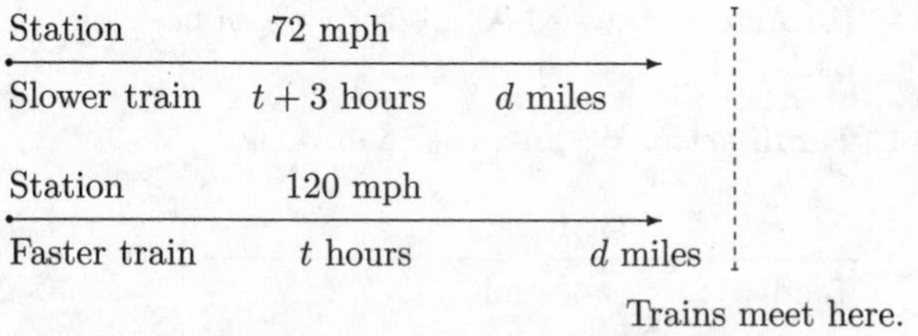

From the drawing we see that the distances are the same. Let's call the distance d. Let t represent the time for the faster train and $t+3$ represent the time for the slower train. We complete the table in the text.

$$d \quad = \quad r \quad \cdot \quad t$$

	Distance	Speed	Time
Slower train	d	72	$t+3$
Faster train	d	120	t

Equation from first row: $d = 72(t+3)$

Equation from second row: $d = 120t$

Translate. Using $d = rt$ in each row of the table, we get the following system of equations:

$$\begin{aligned} d &= 72(t+3), \quad (1) \\ d &= 120t. \qquad\quad (2) \end{aligned}$$

Solve. Substitute $120t$ for d in Equation (1) and solve for t.

$$\begin{aligned} d &= 72(t+3) \\ 120t &= 72(t+3) \quad \text{Substituting} \\ 120t &= 72t + 216 \\ 48t &= 216 \\ t &= \frac{216}{48} \\ t &= 4.5 \end{aligned}$$

Check. When $t = 4.5$ hours, the faster train will travel 120(4.5), or 540 mi, and the slower train will travel 72(7.5), or 540 mi. In both cases we get the distance 540 mi.

State. In 4.5 hours after the second train leaves, the second train will overtake the first train. We can also state the answer as 7.5 hours after the first train leaves.

5. ***Familiarize.*** We first make a drawing.

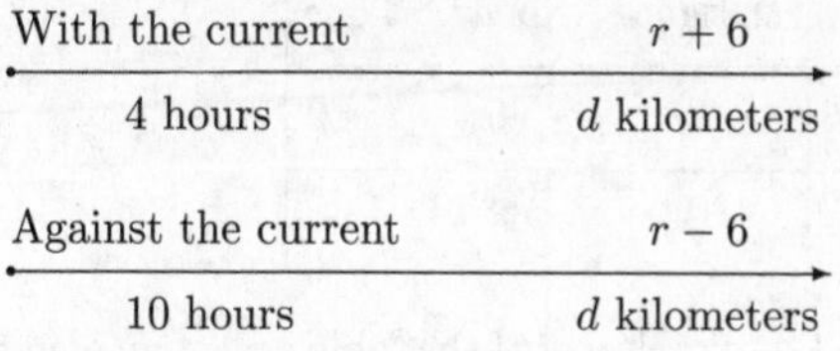

From the drawing we see that the distances are the same. Let d represent the distance. Let r represent the speed of the canoe in still water. Then, when the canoe is traveling with the current, its speed is $r + 6$. When it is traveling against the current, its speed is $r - 6$. We complete the table in the text.

$$d \quad = \quad r \quad \cdot \quad t$$

	Distance	Speed	Time
With current	d	$r+6$	4
Against current	d	$r-6$	10

Equation from first row: $d = (r+6)4$

Equation from second row: $d = (r-6)10$

Translate. Using $d = rt$ in each row of the table, we get the following system of equations:

$$\begin{aligned} d &= (r+6)4, \quad (1) \\ d &= (r-6)10 \quad (2) \end{aligned}$$

Solve. Substitute $(r+6)4$ for d in Equation (2) and solve for r.

$$\begin{aligned} d &= (r-6)10 \\ (r+6)4 &= (r-6)10 \quad \text{Substituting} \\ 4r + 24 &= 10r - 60 \\ 84 &= 6r \\ 14 &= r \end{aligned}$$

Check. When $r = 14$, $r + 6 = 20$ and $20 \cdot 4 = 80$, the distance. When $r = 14$, $r - 6 = 8$ and $8 \cdot 10 = 80$. In both cases, we get the same distance.

State. The speed of the canoe in still water is 14 km/h.

7. ***Familiarize.*** First make a drawing.

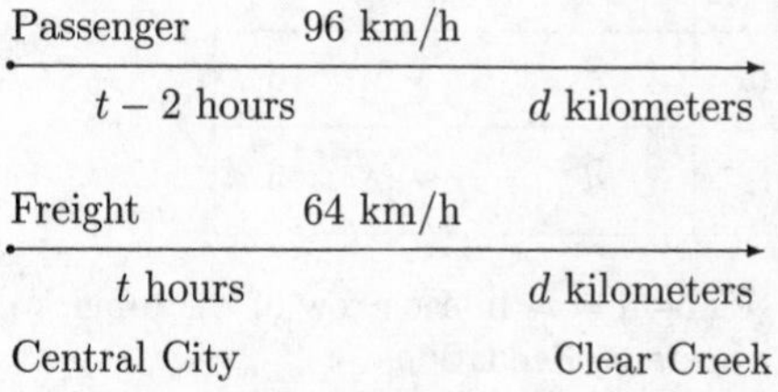

From the drawing we see that the distances are the same. Let d represent the distance. Let t represent the time for the freight train. Then the time for the passenger train is $t - 2$. We organize the information in a table.

$d = r \cdot t$

	Distance	Speed	Time
Passenger	d	96	$t-2$
Freight	d	64	t

Translate. From each row of the table we get an equation.

$d = 96(t-2)$, (1)

$d = 64t$ (2)

Solve. Substitute $64t$ for d in Equation (1) and solve for t.

$$\begin{aligned} d &= 96(t-2) \\ 64t &= 96(t-2) \quad \text{Substituting} \\ 64t &= 96t - 192 \\ 192 &= 32t \\ 6 &= t \end{aligned}$$

Next we substitute 6 for t in one of the original equations and solve for d.

$$\begin{aligned} d &= 64t \quad \text{Equation (2)} \\ d &= 64 \cdot 6 \quad \text{Substituting} \\ d &= 384 \end{aligned}$$

Check. If the time is 6 hr, then the distance the passenger train travels is $96(6-2)$, or 384 km. The freight train travels $64(6)$, or 384 km. The distances are the same.

State. It is 384 km from Central City to Clear Creek.

9. ***Familiarize***. We first make a drawing.

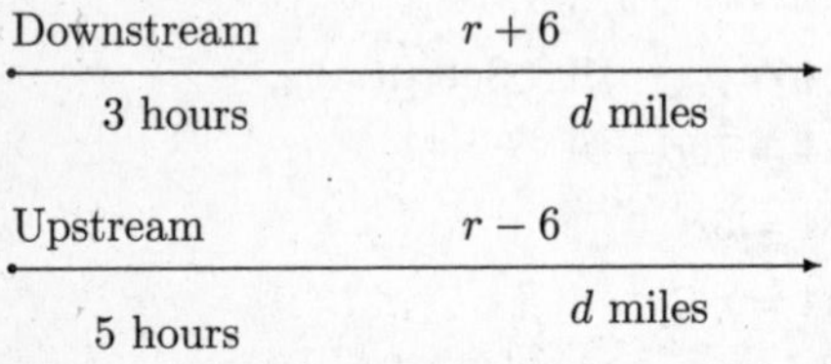

We let r represent the speed of the boat in still water and d represent the distance Antoine traveled downstream before he turned back. We organize the information in a table.

$d = r \cdot t$

	Distance	Speed	Time
Downstream	d	$r+6$	3
Upstream	d	$r-6$	5

Translate. Using $d = rt$ in each row of the table, we get the following system of equations:

$d = (r+6)3$, (1)

$d = (r-6)5$ (2)

Solve. Substitute $(r+6)3$ for d in Equation (2) and solve for r.

$$\begin{aligned} d &= (r-6)5 \\ (r+6)3 &= (r-6)5 \quad \text{Substituting} \\ 3r + 18 &= 5r - 30 \\ 48 &= 2r \\ 24 &= r \end{aligned}$$

If $r = 24$, then $d = (r+6)3 = (24+6)3 = 30 \cdot 3 = 90$.

Check. If $r = 24$, then $r + 6 = 24 + 6 = 30$ and $r - 6 = 24 - 6 = 18$. If Antoine travels for 3 hr at 30 mph, then he travels $3 \cdot 30$, or 90 mi, downstream. If he travels for 5 hr at 18 mph, then he also travels $5 \cdot 18$, or 90 mi, upstream. Since the distances are the same, the answer checks.

State. (a) Antoine must travel at a speed of 24 mph.

(b) Antoine traveled 90 mi downstream before he turned back.

11. ***Familiarize***. We first make a drawing.

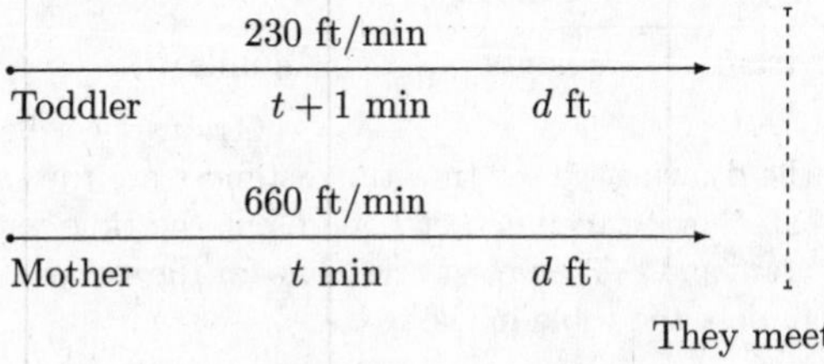

From the drawing we see that the distances are the same. Let's call the distance d. Let t = the time the mother runs. Then $t + 1$ = the time the toddler runs. We arrange the information in a table.

$d = r \cdot t$

	Distance	Speed	Time
Toddler	d	230	$t+1$
Mother	d	660	t

Translate. Using $d = rt$ in each row of the table we get two equations.

$d = 230(t+1)$, (1)

$d = 660t$ (2)

Solve. Substitute $660t$ for d in Equation (1) and solve for t.

$$\begin{aligned} d &= 230(t+1) \\ 660t &= 230(t+1) \quad \text{Substituting} \\ 660t &= 230t + 230 \\ 430t &= 230 \\ t &= \frac{230}{430}, \text{ or } \frac{23}{43} \end{aligned}$$

Check. When $t = \dfrac{23}{43}$ the toddler will travel $230\left(1\dfrac{23}{43}\right)$, or $230 \cdot \dfrac{66}{43}$, or $\dfrac{15,180}{43}$ ft and the mother will travel $660 \cdot \dfrac{23}{43}$, or $\dfrac{15,180}{43}$ ft. Since the distances are the same, our result checks.

State. The mother will overtake the toddler $\frac{23}{43}$ min after she starts running. We can also state the answer as $1\frac{23}{43}$ min after the toddler starts running.

13. ***Familiarize.*** First make a drawing.

Home t hr 45 mph | $(2 - t)$ hr 6 mph Work

Motorcycle distance | Walking distance

← 25 miles →

Let t represent the time the motorcycle was driven. Then $2 - t$ represents the time the rider walked. We organize the information in a table.

$d \quad = \quad r \quad \cdot \quad t$

	Distance	Speed	Time
Motorcycling	Motorcycle distance	45	t
Walking	Walking distance	6	$2 - t$
Total	25		

Translate. From the drawing we see that

Motorcycle distance + Walking distance = 25

Then using $d = rt$ in each row of the table we get

$$45t + 6(2 - t) = 25$$

Solve. We solve this equation for t.

$$45t + 12 - 6t = 25$$
$$39t + 12 = 25$$
$$39t = 13$$
$$t = \frac{13}{39}$$
$$t = \frac{1}{3}$$

Check. The problem asks us to find how far the motorcycle went before it broke down. If $t = \frac{1}{3}$, then $45t$ (the distance the motorcycle traveled) $= 45 \cdot \frac{1}{3}$, or 15 and $6(2 - t)$ (the distance walked) $= 6\left(2 - \frac{1}{3}\right) = 6 \cdot \frac{5}{3}$, or 10. The total of these distances is 25, so $\frac{1}{3}$ checks.

State. The motorcycle went 15 miles before it broke down.

15. $25x^2 - 81 = (5x)^2 - 9^2 = (5x + 9)(5x - 9)$

17.
$$9y^3 - 12y^2 + 4y = y(9y^2 - 12y + 4)$$
$$= y[(3y)^2 - 2 \cdot 3y \cdot 2 + 2^2]$$
$$= y(3y - 2)^2$$

19. ***Familiarize.*** We arrange the information in a table. Let d = the length of the route and t = Lindbergh's time. Note that 16 hr and 57 min $= 16\frac{57}{60}$ hr $= 16.95$ hr.

$d \quad = \quad r \quad \cdot \quad t$

	Distance	Speed	Time
Lindbergh	d	107.4	t
Hughes	d	217.1	$t - 16.95$

Translate. From the rows of the table we get two equations.

$$d = 107.4t, \qquad (1)$$
$$d = 217.1(t - 16.95) \quad (2)$$

Solve. We substitute $107.4t$ for d in Equation (2) and solve for t.

$$d = 217.1(t - 16.95)$$
$$107.4t = 217.1(t - 16.95)$$
$$107.4t = 217.1t - 3679.845$$
$$-109.7t = -3679.845$$
$$t \approx 33.54$$

Now we go back to Equation (1) and substitute 33.54 for t.

$$d = 107.4t$$
$$d = 107.4(33.54)$$
$$d \approx 3602$$

Check. When $t \approx 33.54$, Lindbergh traveled $107.4(33.54) \approx 3602$ mi, and Hughes traveled $217.1(16.59) \approx 3602$ mi. Since the distances are the same, our result checks.

State. The route was 3602 mi long. (Answers may vary slightly due to rounding differences.)

Chapter 7 Vocabulary Reinforcement

1. A solution of a system of two equations in two variables is an ordered pair that makes both equations true.
2. To solve a system of equations graphically, we graph both equations and find the coordinates of any points of intersection.
3. The substitution method is an algebraic method for solving systems of equations.
4. If, when solving a system algebraically, we obtain a false equation, then the system has no solution.
5. If the graphs of the equations in a system of two equations are parallel, then the system has no solution.
6. If the graphs of the equations in a system of two equations are the same line, then the system has infinitely many solutions.

Chapter 7 Concept Reinforcement

1. False; there are only three possibilities: no solution, exactly one solution, or an infinite number of solutions. See page 555 in the text.
2. True; the point(s) of intersection is/are solutions of both equations.
3. True; this indicates that there are no ordered pairs that are solutions of both equations.
4. False; the ordered pairs must be solutions of both equations.

Chapter 7 Study Guide

1.

$x+3y=1$		$y=x+3$	
$-2+3\cdot 1\ ?\ 1$		$1\ ?\ -2+3$	
$-2+3$			1 TRUE
1	TRUE		

The ordered pair $(-2,1)$ is a solution of both equations, so it is a solution of the system of equations.

2. We graph the equations.

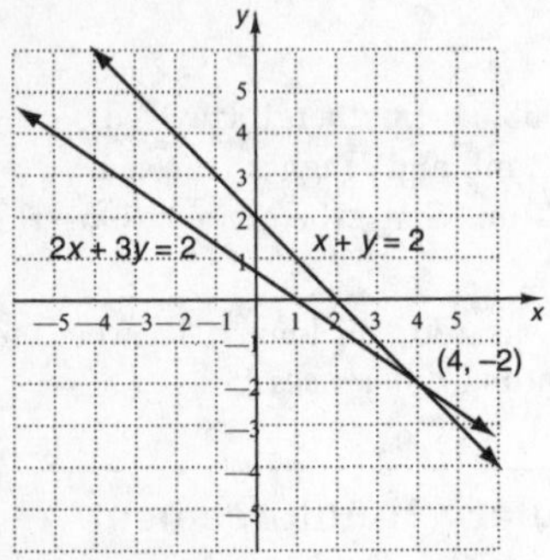

The point of intersection appears to be $(4,-2)$. This checks in both equations, so it is the solution.

3. $x + y = -1$, (1)

$2x + 5y = 1$ (2)

To solve this system of equations by the substitution method, we first solve one of the equations for one of the variables.

$x+y=-1$ (1)

$y=-x-1$ (3)

Substitute $-x-1$ for y in Equation (2) and solve for x.

$$2x+5y=1$$
$$2x+5(-x-1)=1$$
$$2x-5x-5=1$$
$$-3x-5=1$$
$$-3x=6$$
$$x=-2$$

Now substitute -2 for x in Equation (3) and find y.

$$y=-x-1=-(-2)-1=2-1=1$$

The ordered pair $(-2,1)$ checks in both equations. It is the solution.

4. $3x + 2y = 6$, (1)

$x - y = 7$ (2)

Multiply Equation (2) by 2 and then add.

$$3x + 2y = 6$$
$$2x - 2y = 14$$
$$5x = 20$$
$$x = 4$$

Now substitute 4 for x in one of the original equations and solve for y.

$x-y=7$ (2)

$$4-y=7$$
$$-y=3$$
$$y=-3$$

The ordered pair $(4,-3)$ checks in both equations. It is the solution.

Chapter 7 Review Exercises

1. We check by substituting alphabetically 6 for x and -1 for y.

$x-y=3$	
$6-(-1)\ ?\ 3$	
$6+1$	
7	FALSE

Since $(6,-1)$ is not a solution of the first equation, it is not a solution of the system of equations.

2. We check by substituting alphabetically 2 for x and -3 for y.

$2x+y=1$		$x-y=5$	
$2\cdot 2+(-3)\ ?\ 1$		$2-(-3)\ ?\ 5$	
$4-3$		$2+3$	
1	TRUE	5	TRUE

The ordered pair $(2,-3)$ is a solution of both equations, so it is a solution of the system of equations.

3. We check by substituting alphabetically -2 for x and 1 for y.

$x+3y=1$		$2x-y=-5$	
$-2+3\cdot 1\ ?\ 1$		$2(-2)-1\ ?\ -5$	
$-2+3$		$-4-1$	
1	TRUE	-5	TRUE

The ordered pair $(-2,1)$ is a solution of both equations, so it is a solution of the system of equations.

4. We check by substituting alphabetically -4 for x and -1 for y.

$$\begin{array}{c|c} \multicolumn{2}{c}{x - y = 3} \\ \hline -4-(-1) & ?\ 3 \\ -4+1 & \\ -3 & \text{FALSE} \end{array}$$

Since $(-4, -1)$ is not a solution of the first equation, it is not a solution of the system of equations.

5. We graph the equations.

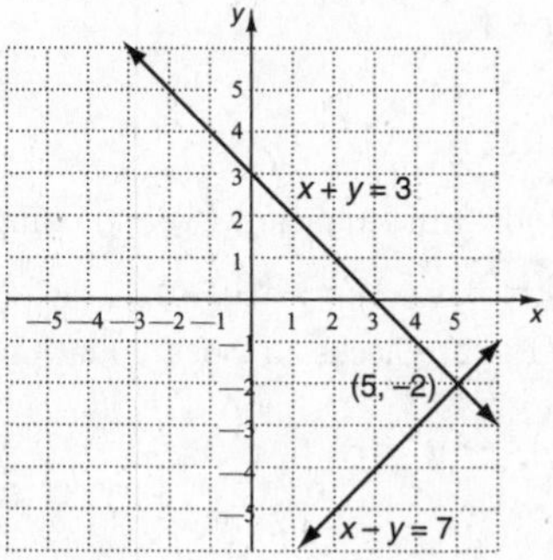

The point of intersection looks as if it has coordinates $(5, -2)$.

Check:

$$\begin{array}{c|c} \multicolumn{2}{c}{x + y = 3} \\ \hline 5+(-2) & ?\ 3 \\ 3 & \text{TRUE} \end{array} \qquad \begin{array}{c|c} \multicolumn{2}{c}{x - y = 7} \\ \hline 5-(-2) & ?\ 7 \\ 5+2 & \\ 7 & \text{TRUE} \end{array}$$

The solution is $(5, -2)$.

6. We graph the equations.

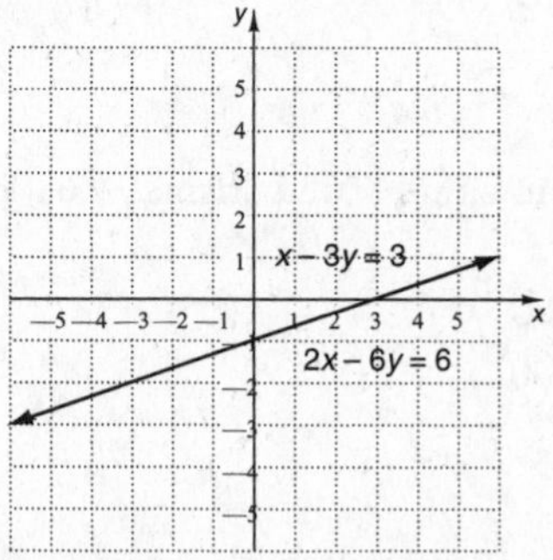

The graphs are the same. The system of equations has an infinite number of solutions.

7. We graph the equations.

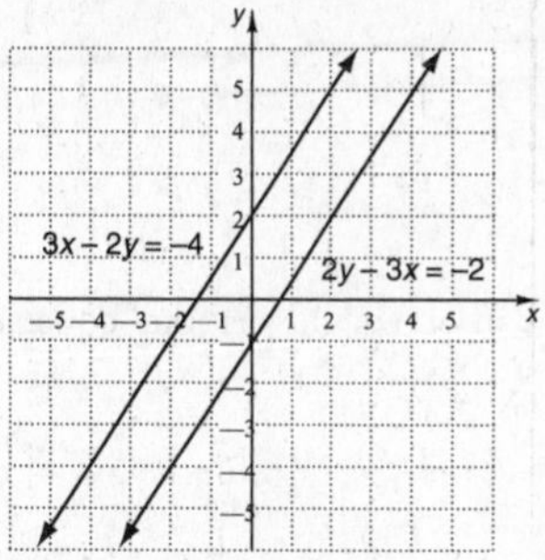

The lines are parallel. There is no solution.

8. $y = 5 - x, \quad (1)$

$3x - 4y = -20 \quad (2)$

We substitute $5 - x$ for y in Equation (2) and solve for x.

$$\begin{aligned} 3x - 4y &= -20 \quad (2) \\ 3x - 4(5-x) &= -20 \\ 3x - 20 + 4x &= -20 \\ 7x - 20 &= -20 \\ 7x &= 0 \\ x &= 0 \end{aligned}$$

Next we substitute 0 for x in one of the original equations and solve for y.

$$\begin{aligned} y &= 5 - x \quad (1) \\ y &= 5 - 0 \\ y &= 5 \end{aligned}$$

The ordered pair $(0, 5)$ checks in both equations. It is the solution.

9. $x + y = 6, \quad (1)$

$y = 3 - 2x \quad (2)$

We substitute $3 - 2x$ for y in Equation (1) and solve for x.

$$\begin{aligned} x + y &= 6 \quad (1) \\ x + (3 - 2x) &= 6 \\ -x + 3 &= 6 \\ -x &= 3 \\ x &= -3 \end{aligned}$$

Now we substitute -3 for x in one of the original equations and solve for y.

$$\begin{aligned} y &= 3 - 2x \quad (2) \\ y &= 3 - 2(-3) \\ y &= 3 + 6 \\ y &= 9 \end{aligned}$$

The ordered pair $(-3, 9)$ checks in both equations. It is the solution.

10. $x - y = 4, \quad (1)$

$y = 2 - x \quad (2)$

We substitute $2 - x$ for y in Equation (1) and solve for x.

$$\begin{aligned} x - y &= 4 \quad (1) \\ x - (2 - x) &= 4 \\ x - 2 + x &= 4 \\ 2x - 2 &= 4 \\ 2x &= 6 \\ x &= 3 \end{aligned}$$

Now substitute 3 for x in one of the original equations and solve for y.

$$\begin{aligned} y &= 2 - x \quad (2) \\ y &= 2 - 3 \\ y &= -1 \end{aligned}$$

The ordered pair $(3, -1)$ checks in both equations. It is the solution.

11. $s + t = 5, \quad (1)$
$s = 13 - 3t \quad (2)$

We substitute $13 - 3t$ for s in Equation (1) and solve for t.

$$\begin{aligned} s + t &= 5 \quad (1) \\ (13 - 3t) + t &= 5 \\ 13 - 2t &= 5 \\ -2t &= -8 \\ t &= 4 \end{aligned}$$

Now substitute 4 for t in one of the original equations and solve for s.

$$\begin{aligned} s &= 13 - 3t \quad (2) \\ s &= 13 - 3 \cdot 4 \\ s &= 13 - 12 \\ s &= 1 \end{aligned}$$

The ordered pair $(1, 4)$ checks in both equations. It is the solution.

12. $x + 2y = 6, \quad (1)$
$2x + 3y = 8 \quad (2)$

We solve Equation (1) for x.

$$\begin{aligned} x + 2y &= 6 \quad (1) \\ x &= -2y + 6 \quad (3) \end{aligned}$$

We substitute $-2y + 6$ for x in Equation (2) and solve for y.

$$\begin{aligned} 2x + 3y &= 8 \quad (2) \\ 2(-2y + 6) + 3y &= 8 \\ -4y + 12 + 3y &= 8 \\ -y + 12 &= 8 \\ -y &= -4 \\ y &= 4 \end{aligned}$$

Now substitute 4 for y in Equation (3) and compute x.

$$x = -2y + 6 = -2 \cdot 4 + 6 = -8 + 6 = -2$$

The ordered pair $(-2, 4)$ checks in both equations. It is the solution.

13. $3x + y = 1, \quad (1)$
$x - 2y = 5 \quad (2)$

We solve Equation (2) for x.

$$\begin{aligned} x - 2y &= 5 \quad (1) \\ x &= 2y + 5 \quad (3) \end{aligned}$$

We substitute $2y + 5$ for x in Equation (1) and solve for y.

$$\begin{aligned} 3x + y &= 1 \quad (1) \\ 3(2y + 5) + y &= 1 \\ 6y + 15 + y &= 1 \\ 7y + 15 &= 1 \\ 7y &= -14 \\ y &= -2 \end{aligned}$$

Now substitute -2 for y in Equation (3) and compute x.

$$x = 2y + 5 = 2(-2) + 5 = -4 + 5 = 1$$

The ordered pair $(1, -2)$ checks in both equations. It is the solution.

14.
$$\begin{aligned} x + y &= 4, \quad (1) \\ 2x - y &= 5 \quad (2) \\ \hline 3x \quad &= 9 \\ x &= 3 \end{aligned}$$

Substitute 3 for x in either of the original equations and solve for y.

$$\begin{aligned} x + y &= 4 \quad (1) \\ 3 + y &= 4 \\ y &= 1 \end{aligned}$$

The ordered pair $(3, 1)$ checks in both equations. It is the solution.

15.
$$\begin{aligned} x + 2y &= 9, \quad (1) \\ 3x - 2y &= -5 \quad (2) \\ \hline 4x \quad &= 4 \\ x &= 1 \end{aligned}$$

Substitute 1 for x in either of the original equations and solve for y.

$$\begin{aligned} x + 2y &= 9 \quad (1) \\ 1 + 2y &= 9 \quad (2) \\ 2y &= 8 \\ y &= 4 \end{aligned}$$

The ordered pair $(1, 4)$ checks in both equations. It is the solution.

16. $x - y = 8, \quad (1)$
$2x - 2y = 7 \quad (2)$

Multiply Equation (1) by -2 and then add.

$$\begin{aligned} -2x + 2y &= -16, \quad (1) \\ 2x - 2y &= 7 \quad (2) \\ \hline 0 &= -9 \end{aligned}$$

We get a false equation. The system of equations has no solution.

17. $2x + 3y = 8,\quad (1)$

$5x + 2y = -2\quad (2)$

We use the multiplication principle with both equations and then add.

$$\begin{array}{rll} 4x + 6y &= 16 & \text{Multiplying (1) by 2} \\ -15x - 6y &= 6 & \text{Multiplying (2) by } -3 \\ \hline -11x \quad &= 22 & \\ x &= -2 & \end{array}$$

Substitute -2 for x in one of the original equations and solve for y.

$$\begin{aligned} 2x + 3y &= 8 \quad (1) \\ 2(-2) + 3y &= 8 \\ -4 + 3y &= 8 \\ 3y &= 12 \\ y &= 4 \end{aligned}$$

The ordered pair $(-2, 4)$ checks in both equations. It is the solution.

18. $5x - 2y = 2,\quad (1)$

$3x - 7y = 36\quad (2)$

We use the multiplication principle with both equations and then add.

$$\begin{array}{rll} 35x - 14y &= 14 & \text{Multiplying (1) by 7} \\ -6x + 14y &= -72 & \text{Multiplying (2) by } -2 \\ \hline 29x \quad &= -58 & \\ x &= -2 & \end{array}$$

Substitute -2 for x in one of the original equations and solve for y.

$$\begin{aligned} 5x - 2y &= 2 \quad (1) \\ 5(-2) - 2y &= 2 \\ -10 - 2y &= 2 \\ -2y &= 12 \\ y &= -6 \end{aligned}$$

The ordered pair $(-2, -6)$ checks in both equations. It is the solution.

19. $-x - y = -5,\quad (1)$

$2x - y = 4\quad (2)$

We multiply Equation (1) by -1 and then add.

$$\begin{array}{rl} x + y &= 5 \\ 2x - y &= 4 \\ \hline 3x \quad &= 9 \\ x &= 3 \end{array}$$

Substitute 3 for x in one of the original equations and solve for y.

$$\begin{aligned} -x - y &= -5 \quad (1) \\ -3 - y &= -5 \\ -y &= -2 \\ y &= 2 \end{aligned}$$

The ordered pair $(3, 2)$ checks in both equations. It is the solution.

20. $6x + 2y = 4,\quad (1)$

$10x + 7y = -8\quad (2)$

We use the multiplication principle with both equations and then add.

$$\begin{array}{rll} 42x + 14y &= 28 & \text{Multiplying (1) by 7} \\ -20x - 14y &= 16 & \text{Multiplying (2) by } -2 \\ \hline 22x \quad &= 44 & \\ x &= 2 & \end{array}$$

Substitute 2 for x in one of the original equations and solve for y.

$$\begin{aligned} 6x + 2y &= 4 \quad (1) \\ 6 \cdot 2 + 2y &= 4 \\ 12 + 2y &= 4 \\ 2y &= -8 \\ y &= -4 \end{aligned}$$

The ordered pair $(2, -4)$ checks in both equations. It is the solution.

21. $-6x - 2y = 5,\quad (1)$

$12x + 4y = -10\quad (2)$

We multiply Equation (1) by 2 and then add.

$$\begin{array}{rl} -12x - 4y &= 10 \\ 12x + 4y &= -10 \\ \hline 0 &= 0 \end{array}$$

We get an obviously true equation, so the system has an infinite number of solutions.

22. $\frac{2}{3}x + y = -\frac{5}{3}$

$x - \frac{1}{3}y = -\frac{13}{3}$

First we multiply both sides of each equation to clear the fractions.

$$\begin{aligned} 3\left(\frac{2}{3}x + y\right) &= 3\left(-\frac{5}{3}\right) \\ 3 \cdot \frac{2}{3}x + 3y &= -5 \\ 2x + 3y &= -5 \end{aligned}$$

$$\begin{aligned} 3\left(x - \frac{1}{3}y\right) &= 3\left(-\frac{13}{3}\right) \\ 3x - 3 \cdot \frac{1}{3}y &= -13 \\ 3x - y &= -13 \end{aligned}$$

The resulting system is

$2x + 3y = -5,\quad (1)$

$3x - y = -13.\quad (2)$

Now we multiply Equation (2) by 3 and then add.

$$\begin{array}{rl} 2x + 3y &= -5 \\ 9x - 3y &= -39 \\ \hline 11x \quad &= -44 \\ x &= -4 \end{array}$$

Substitute -4 for x in Equation (1) and solve for y.

$$\begin{aligned} 2x+3y&=-5\\ 2(-4)+3y&=-5\\ -8+3y&=-5\\ 3y&=3\\ y&=1 \end{aligned}$$

The ordered pair $(-4,1)$ checks in both equations. It is the solution.

23. ***Familiarize***. We make a drawing. We let $l =$ the length and $w =$ the width, in cm.

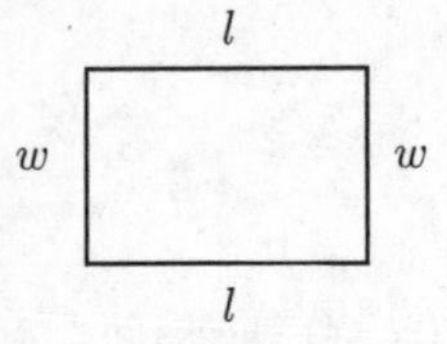

Translate. The perimeter is $2l+2w$. We translate the first statement.

$$\underbrace{\text{The perimeter}}_{2l+2w}\ \underbrace{\text{is}}_{=}\ \underbrace{\text{96 cm.}}_{96}$$

We translate the second statement.

$$\underbrace{\text{The length}}_{l}\ \underbrace{\text{is}}_{=}\ \underbrace{\text{27 cm}}_{27}\ \underbrace{\text{more than}}_{+}\ \underbrace{\text{the width.}}_{w}$$

The resulting system is

$$\begin{aligned} 2l+2w&=96, \quad (1)\\ l&=27+w. \quad (2) \end{aligned}$$

Solve. First we substitute $27+w$ for l in Equation (1) and solve for w.

$$\begin{aligned} 2l+2w&=96 \quad (1)\\ 2(27+w)+2w&=96\\ 54+2w+2w&=96\\ 54+4w&=96\\ 4w&=42\\ w&=10.5 \end{aligned}$$

Now we substitute 10.5 for w in Equation (2) and find l.

$$l=27+w=27+10.5=37.5$$

Check. If the length is 37.5 cm and the width is 10.5 cm, then the perimeter is $2(37.5)+2(10.5)$, or $75+21$, or 96 cm. Also, the length is 27 cm more than the width. The answer checks.

State. The length of the rectangle is 37.5 cm, and the width is 10.5 cm.

24. ***Familiarize***. Let $x =$ the number of orchestra seats sold and $y =$ the number of balcony seats sold. We organize the information in a table.

	Orchestra	Balcony	Total
Price	\$25	\$18	
Number bought	x	y	508
Receipts	$25x$	$18y$	\$11,223

Translate. The last two rows of the table give us a system of equations.

$$\begin{aligned} x+y&=508, \quad (1)\\ 25x+18y&=11,223. \quad (2) \end{aligned}$$

Solve. First we multiply Equation (1) by -18 and then add.

$$\begin{aligned} -18x-18y&=-9144\\ 25x+18y&=11,223\\ \hline 7x\qquad&=2079\\ x&=297 \end{aligned}$$

Now we substitute 297 for x in Equation (1) and solve for y.

$$\begin{aligned} x+y&=508\\ 297+y&=508\\ y&=211 \end{aligned}$$

Check. The total number of tickets sold was $297+211$, or 508. The total receipts were $\$25\cdot 297+\$18\cdot 211$, or $\$7425+\$3798$, or \$11,223. The answer checks.

State. 297 orchestra seats and 211 balcony seats were sold.

25. ***Familiarize***. Let $c =$ the number of liters of Clear Shine and $s =$ the number of liters of Sunstream window cleaner to be used in the mixture. We organize the information in a table.

Type of cleaner	Clear Shine	Sunstream	Mixture
Amount used	c	s	80
Percent of alcohol	30%	60%	45%
Amount of alcohol in solution	$0.3c$	$0.6s$	$45\% \times 80$, or 36 L

Translate. The first and third rows of the table give us a system of equations.

$$\begin{aligned} c+s&=80,\\ 0.3c+0.6s&=36 \end{aligned}$$

After we clear decimals we have

$$\begin{aligned} c+s&=80, \quad (1)\\ 3c+6s&=360. \quad (2) \end{aligned}$$

Solve. First we multiply Equation (1) by -3 and then add.

$$\begin{aligned} -3c - 3s &= -240 \\ 3c + 6s &= 360 \\ \hline 3s &= 120 \\ s &= 40 \end{aligned}$$

Now we substitute 40 for s in Equation (1) and solve for c.

$$\begin{aligned} c + s &= 80 \\ c + 40 &= 80 \\ c &= 40 \end{aligned}$$

Check. If 40 L of Clear Shine and 40 L of Sunstream are used, then there is 40 L + 40 L, or 80 L, of solution. The amount of alcohol in the solution is $0.3(40) + 0.6(40)$, or $12 + 24$, or 36 L. The answer checks.

State. 40 L of each window cleaner should be used.

26. ***Familiarize***. Let $x =$ the weight of the Asian elephant and $y =$ the weight of the African elephant, in kg.

Translate.

$$\underbrace{\text{African elephant's weight}}_{\displaystyle y} \;\underset{=}{\text{is}}\; \underbrace{2400\text{ kg}}_{\displaystyle 2400} \;\underbrace{\text{more than}}_{\displaystyle +}\; \underbrace{\text{Asian elephant's weight.}}_{\displaystyle x}$$

$$\underbrace{\text{Asian elephant's weight}}_{\displaystyle x} \;\underset{+}{\text{plus}}\; \underbrace{\text{African elephant's weight}}_{\displaystyle y} \;\underset{=}{\text{is}}\; \underbrace{12{,}000\text{ kg}}_{\displaystyle 12{,}000}$$

The resulting system is

$$\begin{aligned} y &= 2400 + x, &(1) \\ x + y &= 12{,}000. &(2) \end{aligned}$$

Solve. First we substitute $2400 + x$ for y in Equation (2) and solve for x.

$$\begin{aligned} x + y &= 12{,}000 \\ x + (2400 + x) &= 12{,}000 \\ 2x + 2400 &= 12{,}000 \\ 2x &= 9600 \\ x &= 4800 \end{aligned}$$

Now substitute 4800 for x in Equation (1) and find y.

$$y = 2400 + x = 2400 + 4800 = 7200$$

Check. If the Asian elephant weighs 7200 kg and the African elephant weighs 4800 kg, then the African elephant weighs 2400 kg more than the Asian elephant and their total weight is 4800 + 7200, or 12,000 kg. The answer checks.

State. The Asian elephant weighs 4800 kg, and the African elephant weighs 7200 kg.

27. ***Familiarize***. Let $x =$ the number of pounds of peanuts and $y =$ the number of pounds of fancy nuts to be used. We organize the information in a table.

Type of nuts	Peanuts	Fancy	Mixture
Cost per pound	\$4.50	\$7.00	
Amount	x	y	13
Mixture	$4.5x$	$7y$	\$71

Translate. The last two rows of the table give us a system of equations.

$$\begin{aligned} x + y &= 13, \\ 4.5x + 7y &= 71 \end{aligned}$$

After clearing decimals, we have

$$\begin{aligned} x + y &= 13, &(1) \\ 45x + 70y &= 710. &(2) \end{aligned}$$

Solve. First we multiply Equation (1) by -45 and then add.

$$\begin{aligned} -45x - 45y &= -585 \\ 45x + 70y &= 710 \\ \hline 25y &= 125 \\ y &= 5 \end{aligned}$$

Now substitute 5 for y in one of the original equations and solve for x.

$$\begin{aligned} x + y &= 13 \quad (1) \\ x + 5 &= 13 \\ x &= 8 \end{aligned}$$

Check. If 8 lb of peanuts and 5 lb of fancy nuts are used, the mixture weighs 13 lb. The value of the mixture is $\$4.50(8) + \$7.00(5) = \$36 + \$35 = \$71$. The answer checks.

State. 8 lb of peanuts and 5 lb of fancy nuts should be used.

28. ***Familiarize***. Let $x =$ the number of gallons of 87-octane gas and $y =$ the number of gallons of 95-octane gas to be used. We arrange the information in a table.

Type of gas	87-octane	95-octane	Mixture
Amount	x	y	10
Octane rating	87	95	93
Mixture	$87x$	$95y$	$10 \cdot 93$, or 930

Translate. The first and last rows of the table give us a system of equations.

$$\begin{aligned} x + \quad y &= 10, &(1) \\ 87x + 95y &= 930 &(2) \end{aligned}$$

Solve. We multiply Equation (1) by -87 and then add.

$$\begin{aligned} -87x - 87y &= -870 \\ 87x + 95y &= 930 \\ \hline 8y &= 60 \\ y &= 7.5 \end{aligned}$$

Now substitute 7.5 for y in one of the original equations and solve for x.

$$x+y=10 \quad (1)$$
$$x+7.5=10$$
$$x=2.5$$

Check. If 2.5 gal of 87-octane gas and 7.5 gal of 95-octane gas are used, then there are $2.5+7.5$, or 10 gal, of gas in the mixture. Also, $87(2.5)+95(7.5)=217.5+712.5=930$, so the answer checks.

State. 2.5 gal of 87-octane gas and 7.5 gal of 95-octane gas should be used.

29. ***Familiarize***. Let x = Jeff's age now and y = his son's age now. In 13 yr their ages will be $x+13$ and $y+13$.

Translate.

Jeff's age is three times his son's age.

$$x = 3 \cdot y$$

In 13 yr,

Jeff's age will be two times his son's age.

$$x+13 = 2 \cdot (y+13)$$

The resulting system is

$$x=3y, \quad (1)$$
$$x+13=2(y+13). \quad (2)$$

Solve. First we substitute $3y$ for x in Equation (2) and solve for y.

$$x+13=2(y+13)$$
$$3y+13=2(y+13)$$
$$3y+13=2y+26$$
$$y+13=26$$
$$y=13$$

Now substitute 13 for y in Equation (1) and find x.

$$x=3y=3\cdot 13=39$$

Check. If Jeff is 39 years old and his son is 13 years old, then Jeff is three times as old as his son. In 13 yr Jeff's age will be $39+13$, or 52, his son's age will be $13+13$, or 26, and $52=2\cdot 26$. The answer checks.

State. Jeff is 39 years old now, and his son is 13 years old.

30. ***Familiarize***. Let x = the measure of the larger angle and y = the measure of the smaller angle. Recall that the sum of the measures of complementary angles is 90°.

Translate.

Measure of larger angle plus measure of smaller angle is 90°.

$$x + y = 90$$

Measure of larger angle minus measure of smaller angle is 26°.

$$x - y = 26$$

The resulting system is

$$x+y=90, \quad (1)$$
$$x-y=26. \quad (2)$$

Solve. We add.

$$x+y=90$$
$$x-y=26$$
$$2x \quad =116$$
$$x=58$$

Substitute 58 for x in one of the original equations and solve for y.

$$x+y=90 \quad (1)$$
$$58+y=90$$
$$y=32$$

Check. $58°+32°=90°$ and $58°-32°=26°$, so the answer checks.

State. The measures of the angles are 58°and 32°.

31. ***Familiarize***. Let x = the measure of the larger angle and y = the measure of the smaller angle. Recall that the sum of the measures of supplementary angles is 180°.

Translate.

Measure of larger angle plus measure of smaller angle is 180°.

$$x + y = 180$$

Measure of larger angle minus measure of smaller angle is 26°.

$$x - y = 26$$

The resulting system is

$$x+y=180, \quad (1)$$
$$x-y=26. \quad (2)$$

Solve. We add.

$$x+y=180$$
$$x-y=26$$
$$2x \quad =206$$
$$x=103$$

Now substitute 103 for x in one of the original equations and solve for y.

$$x+y=180 \quad (1)$$
$$103+y=180$$
$$y=77$$

Check. $103°+77°=180°$ and $103°-77°=26°$; so the answer checks.

State. The measures of the angles are 103°and 77°.

32. ***Familiarize***. Let r = the speed of the airplane in still air, in km/h. Then $r+15$ = the speed with a 15 km/h tail wind and $r-15$ = the speed against a 15 km/h wind. We fill in the table in the text.

	Distance	Speed	Time
With wind	d	$r+15$	4
Against wind	d	$r-15$	5

Translate. We get an equation $d = rt$ from each row of the table.

$$d = (r+15)4, \quad (1)$$
$$d = (r-15)5 \quad (2)$$

Solve. We substitute $(r+15)4$ for d in Equation (2).

$$d = (r-15)5$$
$$(r+15)4 = (r-15)5$$
$$4r + 60 = 5r - 75$$
$$60 = r - 75$$
$$135 = r$$

Check. The plane's speed with the tail wind is $135 + 15$, or 150 km/h. At that speed, in 4 hr it will travel $150 \cdot 4$, or 600 km. The plane's speed against the wind is $135 - 15$, or 120 km/h. At that speed, in 5 hr it will travel $120 \cdot 5$, or 600 km. Since the distances are the same, the answer checks.

State. The speed of the airplane in still air is 135 km/h.

33. ***Familiarize.*** Let $t =$ the number of hours the slower car travels before the second car catches up to it. Then $t-2 =$ the number of hours the faster car travels. We fill in the table in the text.

	Distance	Speed	Time
Slow car	d	55	t
Fast car	d	75	$t-2$

Translate. We get an equation $d = rt$ from each row of the table.

$$d = 55t, \quad (1)$$
$$d = 75(t-2) \quad (2)$$

Solve. We substitute $55t$ for d in Equation (2).

$$d = 75(t-2)$$
$$55t = 75(t-2)$$
$$55t = 75t - 150$$
$$-20t = -150$$
$$t = 7.5$$

Now substitute 7.5 for t in Equation (1) and find d.

$$d = 55t = 55(7.5) = 412.5$$

Check. From the calculation of d above, we see that the slow car travels 412.5 mi in 7.5 hr. The fast car travels $7.5 - 2$, or 5.5 hr. At a speed of 75 mph, it travels $75(5.5)$, or 412.5 mi. Since the distances are the same, the answer checks.

State. The second car catches up to the first car 412.5 mi from Phoenix.

34. $y = x - 2, \quad (1)$

$x - 2y = 6 \quad (2)$

Substitute $x - 2$ for y in Equation (2) and solve for x.

$$x - 2y = 6$$
$$x - 2(x-2) = 6$$
$$x - 2x + 4 = 6$$
$$-x + 4 = 6$$
$$-x = 2$$
$$x = -2$$

Now substitute -2 for x in Equation (1) and find y.

$$y = x - 2 = -2 - 2 = -4$$

The ordered pair $(-2, -4)$ checks in both equations. It is the solution. The y-value of the solution is -4, so answer D is correct.

35. $3x + 2y = 5, \quad (1)$

$x - y = 5 \quad (2)$

Multiply Equation (2) by 2 and then add.

$$\begin{aligned} 3x + 2y &= 5 \\ 2x - 2y &= 10 \\ \hline 5x \phantom{{}+2y} &= 15 \\ x &= 3 \end{aligned}$$

Now substitute 3 for x in one of the original equations and solve for y.

$$x - y = 5 \quad (2)$$
$$3 - y = 5$$
$$-y = 2$$
$$y = -2$$

The ordered pair $(3, -2)$ checks in both equations. It is the solution. The x-value of the solution is 3, so answer A is correct.

36. We substitute 6 for x and 2 for y in each equation.

$$\begin{aligned} 2x - Dy &= 6 & Cx + 4y &= 14 \\ 2 \cdot 6 - D \cdot 2 &= 6 & C \cdot 6 + 4 \cdot 2 &= 14 \\ 12 - 2D &= 6 & 6C + 8 &= 14 \\ -2D &= -6 & 6C &= 6 \\ D &= 3 & C &= 1 \end{aligned}$$

37. $3(x - y) = 4 + x, \quad (1)$

$x = 5y + 2 \quad (2)$

Substitute $5y + 2$ for x in Equation (1) and solve for y.

$$3(x - y) = 4 + x$$
$$3(5y + 2 - y) = 4 + 5y + 2$$
$$3(4y + 2) = 5y + 6$$
$$12y + 6 = 5y + 6$$
$$7y + 6 = 6$$
$$7y = 0$$
$$y = 0$$

Now substitute 0 for y in Equation (2) and find x.

$$x = 5y + 2 = 5 \cdot 0 + 2 = 0 + 2 = 2$$

The solution is $(2, 0)$.

38. ***Familiarize.*** Let $c =$ the compensation agreed upon for 12 months of work and let $h =$ the value of the horse. After 7 months, Stephanie would be owed $\frac{7}{12}$ of the compensation agreed upon for 12 months of work, or $\frac{7}{12}c$.

Translate.

$$\underbrace{\text{Compensation for 12 months}}_{c} \;\underbrace{\text{is}}_{=}\; \underbrace{\text{a horse}}_{h} \;\underbrace{\text{plus}}_{+}\; \underbrace{\$2400.}_{2400}$$

$$\underbrace{\text{Compensation for 7 months}}_{\frac{7}{12}c} \;\underbrace{\text{is}}_{=}\; \underbrace{\text{a horse}}_{h} \;\underbrace{\text{plus}}_{+}\; \underbrace{\$1000.}_{1000}$$

After clearing the fractions we have the following system of equations.

$$c = h + 2400, \quad (1)$$
$$7c = 12h + 12{,}000 \quad (2)$$

Solve. We substitute $h + 2400$ for c in Equation (2) and solve for h.

$$7c = 12h + 12{,}000$$
$$7(h + 2400) = 12h + 12{,}000$$
$$7h + 16{,}800 = 12h + 12{,}000$$
$$16{,}800 = 5h + 12{,}000$$
$$4800 = 5h$$
$$960 = h$$

Check. If the value of the horse is \$960, then the compensation for 12 months of work is \$960 + \$2400, or \$3360; $\frac{7}{12}$ of \$3360 is \$1960. This is the value of the horse, \$960, plus \$1000, so the answer checks.

State. The value of the horse was \$960.

39. The line graphed in red contains the points $(0, 0)$ and $(3, 2)$. We find the slope:

$$m = \frac{2-0}{3-0} = \frac{2}{3}$$

The y-intercept is $(0, 0)$, so the equation of the line is $y = \frac{2}{3}x + 0$, or $y = \frac{2}{3}x$.

The line graphed in blue contains the points $(0, 5)$ and $(3, 2)$. We find the slope:

$$m = \frac{2-5}{3-0} = \frac{-3}{3} = -1$$

The y-intercept is $(0, 5)$, so the equation of the line is $y = -1 \cdot x + 5$, or $y = -x + 5$.

40. The line graphed in red contains the points $(-3, 0)$ and $(0, -3)$. We find the slope:

$$m = \frac{-3-0}{0-(-3)} = \frac{-3}{3} = -1$$

The y-intercept is $(0, -3)$, so the equation of the line is $y = -1 \cdot x - 3$, or $y = -x - 3$, or $x + y = -3$.

The line graphed in blue contains the points $(0, 4)$ and $(4, 0)$. We find the slope:

$$m = \frac{0-4}{4-0} = \frac{-4}{4} = -1$$

The y-intercept is $(0, 4)$, so the equation of the line is $y = -1 \cdot x + 4$, or $y = -x + 4$, or $x + y = 4$.

41. ***Familiarize.*** Let $x =$ the number of rabbits and $y =$ the number of pheasants. Then the rabbits have a total of x heads and $4x$ feet; the pheasants have a total of y heads and $2y$ feet.

Translate.

$$\underbrace{\text{Rabbit heads}}_{x} \;\underbrace{\text{plus}}_{+}\; \underbrace{\text{pheasant heads}}_{y} \;\underbrace{\text{is}}_{=}\; \underbrace{\text{35 heads.}}_{35}$$

$$\underbrace{\text{Rabbit feet}}_{4x} \;\underbrace{\text{plus}}_{+}\; \underbrace{\text{pheasant feet}}_{2y} \;\underbrace{\text{is}}_{=}\; \underbrace{\text{94 feet.}}_{94}$$

The resulting system is

$$x + y = 35, \quad (1)$$
$$4x + 2y = 94. \quad (2)$$

Solve. First we multiply Equation (1) by -2 and then add.

$$-2x - 2y = -70$$
$$4x + 2y = 94$$
$$2x \quad = 24$$
$$x = 12$$

Now substitute 12 for x in one of the original equations and solve for y.

$$x + y = 35 \quad (1)$$
$$12 + y = 35$$
$$y = 23$$

Check. If there are 12 rabbits and 23 pheasants, then there are $12 + 23$, or 35, heads and $4 \cdot 12 + 2 \cdot 23$, or $48 + 46$, or 94, feet. The answer checks.

State. There are 12 rabbits and 23 pheasants.

Chapter 7 Discussion and Writing Exercises

1. If we multiply the first equation by -2 we get the second equation. Thus the graphs of the equations will be the same, and the system of equations has an infinite number of solutions.

2. The multiplication principle might be used to obtain a pair of terms that are opposites. The addition principle is used to eliminate a variable. Once a variable has been eliminated, the multiplication and addition principles are also used to solve for the remaining variable and, after a substitution, are used again to find the variable that was eliminated.

3. Answers will vary.

4. A chart allows us to see the given information and missing information clearly and to see the relationships that yield equations.

Chapter 7 Test

1. We check by substituting alphabetically −2 for x and −1 for y.

$2y - 3x = 4$	
$2(-1) - 3(-2) \ ? \ 4$	
$-2 + 6$	
4	TRUE

$x = 4 + 2y$	
$-2 \ ? \ 4 + 2(-1)$	
	$4 - 2$
	2 FALSE

Although $(-2, -1)$ is a solution of the first equation, it is not a solution of the second equation and thus is not a solution of the system of equations.

2. We graph the equations.

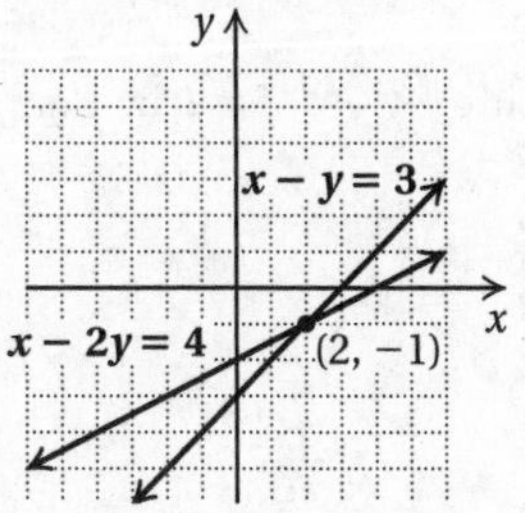

The point of intersection looks as if it has coordinates $(2, -1)$.
Check:

$x - y = 3$	
$2 - (-1) \ ? \ 3$	
$2 + 1$	
3	TRUE

$x - 2y = 4$	
$2 - 2(-1) \ ? \ 4$	
$2 + 2$	
4	TRUE

The solution is $(2, -1)$.

3. $y = 6 - x, \quad (1)$
$2x - 3y = 22 \quad (2)$

We substitute $6 - x$ for y in Equation (2) and solve for x.

$$2x - 3y = 22 \quad (2)$$
$$2x - 3(6 - x) = 22$$
$$2x - 18 + 3x = 22$$
$$5x - 18 = 22$$
$$5x = 40$$
$$x = 8$$

Next we substitute 8 for x in one of the original equations and solve for y.

$$y = 6 - x \quad (1)$$
$$y = 6 - 8$$
$$y = -2$$

The ordered pair $(8, -2)$ checks in both equations. It is the solution.

4. $x + 2y = 5, \quad (1)$
$x + y = 2 \quad (2)$

We solve Equation (1) for x.

$$x + 2y = 5 \quad (1)$$
$$x = -2y + 5 \quad (3)$$

We substitute $-2y + 5$ for x in Equation (2) and solve for y.

$$x + y = 2 \quad (2)$$
$$-2y + 5 + y = 2$$
$$-y + 5 = 2$$
$$-y = -3$$
$$y = 3$$

Now substitute 3 for y in Equation (3) and compute x.

$$x = -2y + 5 = -2 \cdot 3 + 5 = -6 + 5 = -1.$$

The ordered pair $(-1, 3)$ checks in both equations. It is the solution.

5. $y = 5x - 2, \quad (1)$
$y - 2 = x \quad (2)$

Substitute $5x - 2$ for y in Equation (2) and solve for x.

$$y - 2 = x \quad (2)$$
$$5x - 2 - 2 = x$$
$$5x - 4 = x$$
$$-4 = -4x$$
$$1 = x$$

Next we substitute 1 for x in Equation (1) and find y.

$$y = 5x - 2 = 5 \cdot 1 - 2 = 5 - 2 = 3.$$

The ordered pair $(1, 3)$ checks in both equations. It is the solution.

6. $x - y = 6 \quad (1)$
$3x + y = -2 \quad (2)$
$4x \quad = 4$
$x = 1$

Substitute 1 for x in either of the original equations and solve for y.

$$3x + y = -2 \quad (2)$$
$$3 \cdot 1 + y = -2$$
$$3 + y = -2$$
$$y = -5$$

The ordered pair $(1, -5)$ checks in both equations. It is the solution.

7. $\frac{1}{2}x - \frac{1}{3}y = 8,$

$\frac{1}{3}x - \frac{2}{9}y = 1$

First we multiply the first equation by 6 and the second equation by 9 to clear the fractions.

$$6\left(\frac{1}{2}x - \frac{1}{3}y\right) = 6 \cdot 8$$
$$6 \cdot \frac{1}{2}x - 6 \cdot \frac{1}{3}y = 48$$
$$3x - 2y = 48$$

$$9\left(\frac{1}{3}x - \frac{2}{9}y\right) = 9 \cdot 1$$
$$9 \cdot \frac{1}{3}x - 9 \cdot \frac{2}{9}y = 9$$
$$3x - 2y = 9$$

The resulting system is

$3x - 2y = 48,$ (1)

$3x - 2y = 9.$ (2)

We multiply Equation (1) by -1 and then add.

$$\begin{aligned} -3x + 2y &= -48 \\ 3x - 2y &= 9 \\ \hline 0 &= -39 \end{aligned}$$

We obtain a false equation. The system has no solution.

8. $-4x - 9y = 4,$ (1)

$6x + 3y = 1$ (2)

Multiply Equation (2) by 3 and then add.

$$\begin{aligned} -4x - 9y &= 4 \\ 18x + 9y &= 3 \\ \hline 14x &= 7 \\ x &= \frac{1}{2} \end{aligned}$$

Now substitute $\frac{1}{2}$ for x in one of the original equations and solve for y.

$$\begin{aligned} 6x + 3y &= 1 \quad (2) \\ 6 \cdot \frac{1}{2} + 3y &= 1 \\ 3 + 3y &= 1 \\ 3y &= -2 \\ y &= -\frac{2}{3} \end{aligned}$$

The ordered pair $\left(\frac{1}{2}, -\frac{2}{3}\right)$ checks in both equations. It is the solution.

9. $2x + 3y = 13,$ (1)

$3x - 5y = 10$ (2)

Multiply Equation (1) by 5 and Equation (2) by 3 and then add.

$$\begin{aligned} 10x + 15y &= 65 \\ 9x - 15y &= 30 \\ \hline 19x &= 95 \\ x &= 5 \end{aligned}$$

Now substitute 5 for x in one of the original equations and solve for y.

$$\begin{aligned} 2x + 3y &= 13 \quad (1) \\ 2 \cdot 5 + 3y &= 13 \\ 10 + 3y &= 13 \\ 3y &= 3 \\ y &= 1 \end{aligned}$$

The ordered pair $(5, 1)$ checks in both equations. It is the solution.

10. ***Familiarize***. Let l = the length and w = the width, in yd. Recall that the perimeter of a rectangle is given by the formula $P = 2l + 2w$.

Translate.

$$\underbrace{\text{The perimeter}}_{2l + 2w} \ \underbrace{\text{is}}_{=} \ \underbrace{\text{8266 yd.}}_{8266}$$

$$\underbrace{\text{The length}}_{l} \ \underbrace{\text{is}}_{=} \ \underbrace{\text{84 yd}}_{84} \ \underbrace{\text{more than}}_{+} \ \underbrace{\text{the width.}}_{w}$$

The resulting system is

$2l + 2w = 8266,$ (1)

$l = 84 + w.$ (2)

Solve. First we substitute $84 + w$ for l in Equation (1) and solve for w.

$$\begin{aligned} 2l + 2w &= 8266 \\ 2(84 + w) + 2w &= 8266 \\ 168 + 2w + 2w &= 8266 \\ 168 + 4w &= 8266 \\ 4w &= 8098 \\ w &= 2024.5 \end{aligned}$$

Now substitute 2024.5 for w in Equation (2) and find l.

$$l = 84 + w = 84 + 2024.5 = 2108.5$$

Check. If the length is 2108.5 yd and the width is 2024.5 yd, the length is 84 yd more than the width and the perimeter is $2(2108.5) + 2(2024.5)$, or $4217 + 4049$, or 8266 yd. The answer checks.

State. The length is 2108.5 yd and the width is 2024.5 yd.

11. ***Familiarize***. Let a = the amount of solution A and b = the amount of solution B in the mixture, in liters. We organize the information in a table.

Solution	A	B	Mixture
Amount	a	b	60 L
Percent of acid	25%	40%	30%
Amount of acid	$0.25a$	$0.4b$	0.3(60) or 18 L

Translate. The first and third rows of the table give us a system of equations.

$a + b = 60,$

$0.25a + 0.4b = 18$

After clearing decimals, we have

$$a + b = 60, \quad (1)$$
$$25a + 40b = 1800. \quad (2)$$

Solve. First we multiply Equation (1) by -25 and then add.

$$\begin{aligned} -25a - 25b &= -1500 \\ 25a + 40b &= 1800 \\ \hline 15b &= 300 \\ b &= 20 \end{aligned}$$

Now we substitute 20 for b in one of the original equations and solve for a.

$$\begin{aligned} a + b &= 60 \quad (1) \\ a + 20 &= 60 \\ a &= 40 \end{aligned}$$

Check. If 40 L of solution A and 20 L of solution B are used, then there are $40 + 20$, or 60 L, in the mixture. The amount of acid in the mixture is $0.25(40) + 0.4(20)$, or $10 + 8$, or 18 L. The answer checks.

State. 40 L of solution A and 20 L of solution B should be used.

12. ***Familiarize***. Let r = the speed of the motorboat in still water, in km/h. Then the speed of the boat with the current is $r + 8$ and the speed against the current is $r - 8$. We organize the information in a table.

	Distance	Speed	Time
With current	d	$r + 8$	2
Against current	d	$r - 8$	3

Translate. We get an equation from each row of the table.

$$d = (r + 8)2 \quad (1)$$
$$d = (r - 8)3 \quad (2)$$

Solve. We substitute $(r + 8)2$ for d in Equation (2) and solve for r.

$$\begin{aligned} d &= (r - 8)3 \quad (2) \\ (r + 8)2 &= (r - 8)3 \\ 2r + 16 &= 3r - 24 \\ 16 &= r - 24 \\ 40 &= r \end{aligned}$$

Check. When $r = 40$, then $r + 8 = 40 + 8 = 48$ and in 2 hr at this speed the boat travels $48 \cdot 2$, or 96 km. Also, $r - 8 = 40 - 8 = 32$, and in 3 hr at this speed the boat travels $32 \cdot 3$, or 96 km. The distances are the same, so the answer checks.

State. The speed of the motorboat in still water is 40 km/h.

13. ***Familiarize***. Let c = the receipts from concessions and r = the receipts from the rides.

Translate.

$$\underbrace{\text{Receipts from concessions}}_{\downarrow\; c} \;\; \underset{\downarrow\; +}{\text{plus}} \;\; \underbrace{\text{receipts from rides}}_{\downarrow\; r} \;\; \underset{\downarrow\; =}{\text{is}} \;\; \underset{\downarrow\; 4275}{\$4275.}$$

$$\underbrace{\text{Receipts from concessions}}_{\downarrow\; c} \;\; \underset{\downarrow\; =}{\text{were}} \;\; \underset{\downarrow\; 2}{\text{two}} \;\; \underset{\downarrow\; \cdot}{\text{times}} \;\; \underbrace{\text{receipts from rides.}}_{\downarrow\; r}$$

The resulting system is

$$c + r = 4275, \quad (1)$$
$$c = 2r. \quad (2)$$

Solve. First we substitute $2r$ for c in Equation (1) and solve for r.

$$\begin{aligned} c + r &= 4275 \quad (1) \\ 2r + r &= 4275 \\ 3r &= 4275 \\ r &= 1425 \end{aligned}$$

Now we substitute 1425 for r in one of the original equations and find c.

$$c = 2r = 2 \cdot 1425 = 2850$$

Check. If concessions brought in \$2850 and rides brought in \$1425, then the total receipts were \$2850 + \$1425, or \$4275. Also, $2 \cdot \$1425 = \$2850$, so concessions brought in twice as much as rides. The answer checks.

State. Concessions brought in \$2850; rides brought in \$1425.

14. ***Familiarize***. Let x = the number of acres of hay planted and y = the number of acres of oats planted.

Translate.

$$\underbrace{\text{Acres of hay}}_{\downarrow\; x} \;\; \underset{\downarrow\; +}{\text{plus}} \;\; \underbrace{\text{acres of oats}}_{\downarrow\; y} \;\; \underset{\downarrow\; =}{\text{is}} \;\; \underbrace{\text{650 acres.}}_{\downarrow\; 650}$$

$$\underbrace{\text{Acres of hay}}_{\downarrow\; x} \;\; \underset{\downarrow\; =}{\text{is}} \;\; \underset{\downarrow\; 180}{180} \;\; \underbrace{\text{more than}}_{\downarrow\; +} \;\; \underbrace{\text{acres of oats.}}_{\downarrow\; y}$$

The resulting system is

$$x + y = 650, \quad (1)$$
$$x = 180 + y. \quad (2)$$

Solve. First we substitute $180 + y$ for x in Equation (1) and solve for y.

$$\begin{aligned} x + y &= 650 \quad (1) \\ (180 + y) + y &= 650 \\ 180 + 2y &= 650 \\ 2y &= 470 \\ y &= 235 \end{aligned}$$

Now substitute 235 for y in Equation (2) and find x.

$$x = 180 + y = 180 + 235 = 415$$

Check. If 415 acres of hay and 235 acres of oats are planted, a total of $415 + 235$, or 650 acres, is planted. Also, $235 + 180 = 415$, so 180 acres more of hay than of oats are planted. The answer checks.

State. 415 acres of hay and 235 acres of oats should be planted.

15. ***Familiarize.*** Let $x =$ the measure of the larger angle and $y =$ the measure of the smaller angle. Recall that the sum of the measures of supplementary angles is 180°.

Translate.

Measure of larger angle	plus	measure of smaller angle	is 180°.
x	$+$	y	$= 180$

Measure of larger angle	is 45°	more than	twice	measure of smaller angle.
x	$= 45$	$+$	$2\cdot$	y

The resulting system is

$$x + y = 180, \quad (1)$$
$$x = 45 + 2y. \quad (2)$$

Solve. First we substitute $45 + 2y$ for x in Equation (1) and solve for y.

$$x + y = 180 \quad (1)$$
$$(45 + 2y) + y = 180$$
$$45 + 3y = 180$$
$$3y = 135$$
$$y = 45$$

Now substitute 45 for y in Equation (2) and find x.

$$x = 45 + 2y = 45 + 2\cdot 45 = 45 + 90 = 135$$

Check. $135° + 45° = 180°$; and $135° = 45° + 2\cdot 45°$, so the answer checks.

State. The measures of the angles are 135°and 45°.

16. ***Familiarize.*** Let $x =$ the number of gallons of 87-octane gas and $y =$ the number of gallons of 93-octane gas that should be used. We organize the information in a table.

Type of gas	87-octane	93-octane	Mixture
Amount	x	y	12
Octane rating	87	93	91
Mixture	$87x$	$93y$	$12\cdot 91$, or 1092

Translate. The first and last rows of the table give us a system of equations.

$$x + y = 12, \quad (1)$$
$$87x + 93y = 1092 \quad (2)$$

Solve. We multiply Equation (1) by -87 and then add.

$$-87x - 87y = -1044$$
$$87x + 93y = 1092$$
$$6y = 48$$
$$y = 8$$

Then substitute 8 for y in one of the original equations and solve for x.

$$x + y = 12$$
$$x + 8 = 12$$
$$x = 4$$

Check. If 4 gal of 87-octane gas and 8 gal of 93-octane gas are used, then there are $4 + 8$, or 12 gal, of gas in the mixture. Also $87\cdot 4 + 93\cdot 8 = 348 + 744 = 1092$, so the answer checks.

State. 4 gal of 87-octane gas and 8 gal of 93-octane gas should be used.

17. ***Familiarize.*** Let $x =$ the number of minutes used in the \$2.95 plan and $y =$ the number of minutes used in the \$1.95 plan. Expressing 10¢ as \$0.10 and 15¢ as \$0.15, the \$2.95 plan costs $\$2.95 + \$0.10x$ per month and the \$1.95 plan costs $\$1.95 + \$0.15y$ per month.

Translate.

Cost of \$2.95 plan	is the same as	cost of \$1.95 plan.
$2.95 + 0.10x$	$=$	$1.95 + 0.15y$

Minutes used in \$2.95 plan	are the same as	minutes used in \$1.95 plan.
x	$=$	y

After clearing decimals we have the following system of equations.

$$295 + 10x = 195 + 15y, \quad (1)$$
$$x = y \quad (2)$$

Solve. We substitute x for y in Equation (1) and solve for x.

$$295 + 10x = 195 + 15y \quad (1)$$
$$295 + 10x = 195 + 15x$$
$$295 = 195 + 5x$$
$$100 = 5x$$
$$20 = x$$

Substitute 20 for x in Equation (2) to find y.

$$x = y$$
$$20 = y$$

Check. For 20 min the \$2.95 plan costs $\$2.95 + \$0.10(20)$, or \$4.95 per month and the \$1.95 plan costs $\$1.95 + \$0.15(20)$, or \$4.95 per month. The costs are the same, so the answer checks.

State. The two plans cost the same for 20 min.

18. ***Familiarize.*** Let $t =$ the number of hours it will take the SUV to catch up with the car. Then $t + 2 =$ the number of hours the car travels. We organize the information in a table.

	Distance	Speed	Time
Car	d	55	$t + 2$
SUV	d	65	t

Translate. We get an equation $d = rt$ from each row of the table.

$$d = 55(t + 2), \quad (1)$$
$$d = 65t \quad (2)$$

Solve. Substitute $65t$ for d in Equation (1) and solve for t.

$$\begin{aligned} d &= 55(t+2) \quad (1)\\ 65t &= 55(t+2)\\ 65t &= 55t+110\\ 10t &= 110\\ t &= 11 \end{aligned}$$

Check. At a speed of 55 mph for $11+2$, or 13 hr, the car travels $55 \cdot 13$, or 715 mi; at a speed of 65 mph for 11 hr, the SUV travels $65 \cdot 11$, or 715 mi. Since the distances are the same, the answer checks.

State. It will take the SUV 11 hr to catch up to the car.

19. $x - 2y = 4, \quad (1)$
$2x - 3y = 3 \quad (2)$

Multiply Equation (1) by -2 and then add.

$$\begin{aligned} -2x + 4y &= -8\\ 2x - 3y &= 3\\ \hline y &= -5 \end{aligned}$$

Now substitute -5 for y in one of the original equations and solve for x.

$$\begin{aligned} x - 2y &= 4 \quad (1)\\ x - 2(-5) &= 4\\ x + 10 &= 4\\ x &= -6 \end{aligned}$$

The ordered pair $(-6, -5)$ checks in both equations. It is the solution. Since both x and y are negative, answer D is correct.

20. Substitute -2 for x and 3 for y in each equation.

$$\begin{aligned} Cx - 4y &= 7 & 3x + Dy &= 8\\ C(-2) - 4 \cdot 3 &= 7 & 3(-2) + D \cdot 3 &= 8\\ -2C - 12 &= 7 & -6 + 3D &= 8\\ -2C &= 19 & 3D &= 14\\ C &= -\frac{19}{2} & D &= \frac{14}{3} \end{aligned}$$

21. ***Familiarize.*** Let $a =$ the number of people ahead of Lily and $b =$ the number of people behind her. Then in the entire line there are the people ahead of her, the people behind her, and Lily herself, or $a + b + 1$.

Translate.

Number ahead of Lily | is | two | more than | number behind her.
$a \quad = \quad 2 \quad + \quad b$

Number in entire line | is | three | times | number behind her.
$a + b + 1 \quad = \quad 3 \quad \cdot \quad b$

The resulting system is

$$\begin{aligned} a &= 2 + b, & (1)\\ a + b + 1 &= 3b. & (2) \end{aligned}$$

Solve. We substitute $2 + b$ for a in Equation (2) and solve for b.

$$\begin{aligned} a + b + 1 &= 3b \quad (2)\\ (2 + b) + b + 1 &= 3b\\ 2b + 3 &= 3b\\ 3 &= b \end{aligned}$$

Now substitute 3 for b in Equation (1) and find a.

$$a = 2 + b = 2 + 3 = 5$$

Check. If there are 5 people ahead of Lily and 3 people behind her, then the number of people ahead of her is two more than the number behind her. Also, in the entire line there are $5 + 1 + 3$, or 9 people, and $3 \cdot 3 = 9$. The answer checks.

State. There are 5 people ahead of Lily.

22. The line graphed in red contains the points $(-2, 3)$ and $(3, 0)$.

$$m = \frac{3 - 0}{-2 - 3} = \frac{3}{-5} = -\frac{3}{5}$$

Then the equation of the line is $y = -\frac{3}{5}x + b$. To find b we substitute the coordinates of either point for x and y in the equation. We use $(3, 0)$.

$$\begin{aligned} y &= -\frac{3}{5}x + b\\ 0 &= -\frac{3}{5} \cdot 3 + b\\ 0 &= -\frac{9}{5} + b\\ \frac{9}{5} &= b \end{aligned}$$

The equation of the line is $y = -\frac{3}{5}x + \frac{9}{5}$.

The line graphed in blue contains the points $(-2, 3)$ and $(3, 4)$.

$$m = \frac{3 - 4}{-2 - 3} = \frac{-1}{-5} = \frac{1}{5}$$

Then the equation is $y - \frac{1}{5}x + b$. To find b we substitute the coordinates of either point for x and y in the equation. We use $(3, 4)$.

$$\begin{aligned} y &= \frac{1}{5}x + b\\ 4 &= \frac{1}{5} \cdot 3 + b\\ 4 &= \frac{3}{5} + b\\ \frac{17}{5} &= b \end{aligned}$$

The equation of the line is $y = \frac{1}{5}x + \frac{17}{5}$.

23. The line graphed in red is a vertical line, so its equation is of the form $x = a$. The line contains the point $(3, -2)$, so the equation is $x = 3$.

The line graphed in blue is a horizontal line, so its equation is of the form $y = b$. The line contains the point $(3, -2)$, so the equation is $y = -2$.

Cumulative Review Chapters 1 - 7

1. $$\begin{aligned} -2[1.4-(-0.8-1.2)] &= -2[1.4-(-2)] \\ &= -2[1.4+2] \\ &= -2[3.4] \\ &= -6.8 \end{aligned}$$

2. $$\begin{aligned} &(1.3\times 10^8)(2.4\times 10^{-10}) \\ &= (1.3\times 2.4)\times(10^8\times 10^{-10}) \\ &= 3.12\times 10^{-2} \end{aligned}$$

3. $$\begin{aligned} \left(-\frac{1}{6}\right)\div\left(\frac{2}{9}\right) &= -\frac{1}{6}\cdot\frac{9}{2} = -\frac{1\cdot 9}{6\cdot 2} \\ &= -\frac{1\cdot\cancel{3}\cdot 3}{2\cdot\cancel{3}\cdot 2} \\ &= -\frac{3}{4} \end{aligned}$$

4. $$\frac{2^{12}2^{-7}}{2^8} = 2^{12+(-7)-8} = 2^{-3} = \frac{1}{2^3} = \frac{1}{8}$$

5. $$\begin{aligned} \frac{x^2-9}{2x^2-7x+3} &= \frac{(x+3)(x-3)}{(2x-1)(x-3)} \\ &= \frac{x+3}{2x-1}\cdot\frac{x-3}{x-3} \\ &= \frac{x+3}{2x-1} \end{aligned}$$

6. $$\begin{aligned} \frac{t^2-16}{(t+4)^2} &= \frac{(t+4)(t-4)}{(t+4)(t+4)} \\ &= \frac{t+4}{t+4}\cdot\frac{t-4}{t+4} \\ &= \frac{t-4}{t+4} \end{aligned}$$

7. $$\frac{x-\dfrac{x}{x+2}}{\dfrac{2}{x}-\dfrac{1}{x+2}}$$ LCM of the denominators is $x(x+2)$.

$$\begin{aligned} &= \frac{x-\dfrac{x}{x+2}}{\dfrac{2}{x}-\dfrac{1}{x+2}}\cdot\frac{x(x+2)}{x(x+2)} \\ &= \frac{\left(x-\dfrac{x}{x+2}\right)\cdot x(x+2)}{\left(\dfrac{2}{x}-\dfrac{1}{x+2}\right)\cdot x(x+2)} \\ &= \frac{x\cdot x(x+2)-\dfrac{x}{x+2}\cdot x(x+2)}{\dfrac{2}{x}\cdot x(x+2)-\dfrac{1}{x+2}\cdot x(x+2)} \\ &= \frac{x^3+2x^2-x^2}{2x+4-x} = \frac{x^3+x^2}{x+4} \\ &= \frac{x^2(x+1)}{x+4} \end{aligned}$$

8. $(1-3x^2)(2-4x^2) = 2-4x^2-6x^2+12x^4 =$
$2-10x^2+12x^4$

9. $$\begin{aligned} (2a^2b-5ab^2)^2 &= (2a^2b)^2-2\cdot 2a^2b\cdot 5ab^2+(5ab^2)^2 \\ &= 4a^4b^2-20a^3b^3+25a^2b^4 \end{aligned}$$

10. $(3x^2+4y)(3x^2-4y) = (3x^2)^2-(4y)^2 = 9x^4-16y^2$

11. $$\begin{aligned} &-2x^2(x-2x^2+3x^3) \\ &= -2x^2\cdot x-2x^2(-2x^2)-2x^2\cdot 3x^3 \\ &= -2x^3+4x^4-6x^5 \end{aligned}$$

12. $$\begin{array}{rrrr} & 4x^2 & -\ 2x & +\ 1 \\ & & 1 & +\ 2x \\ \hline 8x^3 & -\ 4x^2 & +\ 2x & \\ & 4x^2 & -\ 2x & +\ 1 \\ \hline 8x^3 & & & +\ 1 \end{array}$$

13. $$\left(8-\frac{1}{3}x\right)\left(8+\frac{1}{3}x\right) = 8^2-\left(\frac{1}{3}x\right)^2 = 64-\frac{1}{9}x^2$$

14. $$\begin{aligned} &(-8y^2-y+2)-(y^3-6y^2+y-5) \\ &= -8y^2-y+2-y^3+6y^2-y+5 \\ &= -y^3-2y^2-2y+7 \end{aligned}$$

15. $$\begin{array}{r|l} & x^2 - x - 1 \\ \hline 2x-1 & 2x^3-3x^2-x-1 \\ & \underline{2x^3-x^2} \\ & \quad -2x^2-x \\ & \quad \underline{-2x^2+x} \\ & \qquad -2x-1 \\ & \qquad \underline{-2x+1} \\ & \qquad\quad -2 \end{array}$$

The answer is $x^2-x-1+\dfrac{-2}{2x-1}$.

16. $$\begin{aligned} \frac{7}{5x-25}+\frac{x+7}{5-x} &= \frac{7}{5(x-5)}+\frac{x+7}{5-x} \\ &= \frac{7}{5(x-5)}+\frac{x+7}{5-x}\cdot\frac{-1}{-1} \\ &= \frac{7}{5(x-5)}+\frac{-x-7}{x-5} \\ &= \frac{7}{5(x-5)}+\frac{-x-7}{x-5}\cdot\frac{5}{5} \\ &= \frac{7}{5(x-5)}+\frac{-5x-35}{5(x-5)} \\ &= \frac{7-5x-35}{5(x-5)} \\ &= \frac{-5x-28}{5(x-5)} \end{aligned}$$

17. $\dfrac{2x-1}{x-2}-\dfrac{2x}{2-x}=\dfrac{2x-1}{x-2}-\dfrac{2x}{2-x}\cdot\dfrac{-1}{-1}$

$=\dfrac{2x-1}{x-2}-\dfrac{-2x}{x-2}$

$=\dfrac{2x-1-(-2x)}{x-2}$

$=\dfrac{2x-1+2x}{x-2}$

$=\dfrac{4x-1}{x-2}$

18. $\dfrac{y^2+y}{y^2+y-2}\cdot\dfrac{y+2}{y^2-1}$

$=\dfrac{(y^2+y)(y+2)}{(y^2+y-2)(y^2-1)}$

$=\dfrac{y(y+1)(y+2)}{(y+2)(y-1)(y+1)(y-1)}$

$=\dfrac{(y+1)(y+2)}{(y+1)(y+2)}\cdot\dfrac{y}{(y-1)(y-1)}$

$=\dfrac{y}{(y-1)^2}$

19. $\dfrac{7x+7}{x^2-2x}\div\dfrac{14}{3x-6}=\dfrac{7x+7}{x^2-2x}\cdot\dfrac{3x-6}{14}$

$=\dfrac{7(x+1)(3)(x-2)}{x(x-2)\cdot 2\cdot 7}$

$=\dfrac{7(x-2)}{7(x-2)}\cdot\dfrac{3(x+1)}{2x}$

$=\dfrac{3(x+1)}{2x}$

20. $6x^5-36x^3+9x^2=3x^2(2x^3-12x+3)$

21. $16y^4-81=(4y^2)^2-9^2$

$=(4y^2+9)(4y^2-9)$

$=(4y^2+9)[(2y)^2-3^2]$

$=(4y^2+9)(2y+3)(2y-3)$

22. $3x^2+10x-8$

There is no common factor (other than 1 or −1). We can factor $3x^2$ as $(3x+\quad)(x+\quad)$. We find a pair of factors of −8 that yield a middle term of $10x$. We have $3x^2+10x-8=(3x-2)(x+4)$.

23. $4x^4-12x^2y+9y^2=(2x^2)^2-2\cdot 2x^2\cdot 3y+(3y)^2$

$=(2x^2-3y)^2$

24. $3m^3+6m^2-45m=3m(m^2+2m-15)=3m(m+5)(m-3)$

25. $x^3+x^2-x-1=x^2(x+1)-(x+1)$

$=(x+1)(x^2-1)$

$=(x+1)(x+1)(x-1)$

$=(x+1)^2(x-1)$

26. $3x-4(x+1)=5$

$3x-4x-4=5$

$-x-4=5$

$-x=9$

$-1\cdot(-x)=-1\cdot 9$

$x=-9$

The solution is −9.

27. $x(2x-5)=0$

$x=0$ *or* $2x-5=0$

$x=0$ *or* $2x=5$

$x=0$ *or* $x=\dfrac{5}{2}$

The solutions are 0 and $\dfrac{5}{2}$.

28. $5x+3\geq 6(x-4)+7$

$5x+3\geq 6x-24+7$

$5x+3\geq 6x-17$

$3\geq x-17$

$20\geq x$

The solution set is $\{x|20\geq x\}$, or $\{x|x\leq 20\}$.

29. $1.5x-2.3x=0.4(x-0.9)$

$-0.8x=0.4(x-0.9)$

$-0.8x=0.4x-0.36$

$-1.2x=-0.36$

$x=0.3$

The solution is 0.3.

30. $2x^2=338$

$x^2=169$

$x=13$ *or* $x=-13$

The solutions are 13 and −13.

31. $3x^2+15=14x$

$3x^2-14x+15=0$

$(3x-5)(x-3)=0$

$3x-5=0$ *or* $x-3=0$

$3x=5$ *or* $x=3$

$x=\dfrac{5}{3}$ *or* $x=3$

The solutions are $\dfrac{5}{3}$ and 3.

32.
$$\frac{2}{x}-\frac{3}{x-2}=\frac{1}{x},\ \text{LCM is } x(x-2)$$
$$x(x-2)\left(\frac{2}{x}-\frac{3}{x-2}\right)=x(x-2)\cdot\frac{1}{x}$$
$$x(x-2)\cdot\frac{2}{x}-x(x-2)\cdot\frac{3}{x-2}=x-2$$
$$2x-4-3x=x-2$$
$$-x-4=x-2$$
$$-2x-4=-2$$
$$-2x=2$$
$$x=-1$$
The number -1 checks. It is the solution.

33.
$$1+\frac{3}{x}+\frac{x}{x+1}=\frac{1}{x^2+x}$$
$$1+\frac{3}{x}+\frac{x}{x+1}=\frac{1}{x(x+1)},$$
LCM is $x(x+1)$
$$x(x+1)\left(1+\frac{3}{x}+\frac{x}{x+1}\right)=x(x+1)\cdot\frac{1}{x(x+1)}$$
$$x(x+1)+3(x+1)+x\cdot x=1$$
$$x^2+x+3x+3+x^2=1$$
$$2x^2+4x+3=1$$
$$2x^2+4x+2=0$$
$$2(x^2+2x+1)=0$$
$$2(x+1)(x+1)=0$$
$x+1=0$ *or* $x+1=0$

$x=-1$ *or* $x=-1$

The number -1 does not check. It makes a denominator 0. There is no solution.

34. $y=2x-9,\quad (1)$

$2x+3y=-3\quad (2)$

Substitute $2x-9$ for y in Equation (2) and solve for x.
$$2x+3y=-3$$
$$2x+3(2x-9)=-3$$
$$2x+6x-27=-3$$
$$8x-27=-3$$
$$8x=24$$
$$x=3$$
Now substitute 3 for x in Equation (1) and find y.
$$y=2x-9=2\cdot 3-9=6-9=-3$$
The solution is $(3,-3)$.

35. $6x+3y=-6,\quad (1)$

$-2x+5y=14\quad (2)$

Multiply Equation (2) by 3 and then add.
$$6x+\ \ 3y=-6$$
$$\underline{-6x+15y=42}$$
$$18y=36$$
$$y=2$$
Now substitute 2 for y in one of the original equations and solve for x.
$$6x+3\cdot 2=-6$$
$$6x+6=-6$$
$$6x=-12$$
$$x=-2$$
The solution is $(-2,2)$.

36. $2x=y-2,\quad (1)$

$3y-6x=6\quad (2)$

First we solve Equation (1) for y.
$$2x=y-2$$
$$2x+2=y$$
Now substitute $2x+2$ for y in Equation (2) and solve for x.
$$3(2x+2)-6x=6$$
$$6x+6-6x=6$$
$$6=6$$
We get an equation that is true for all values of x and y. The system of equations has an infinite number of solutions.

37.
$$N=rx-t$$
$$N+t=rx$$
$$\frac{N+t}{r}=x$$

38. ***Familiarize***. Let $p=$ the price of the frame. Then the total amount paid was $p+6\%$ of p, or $p+0.06p$, or $1.06p$.

Translate.

$\underbrace{\text{Total paid}}$ is \$37.10.

$\downarrow \quad \downarrow \quad \downarrow$

$1.06p = 37.10$

Solve.
$$1.06p=37.10$$
$$p=\frac{37.10}{1.06}=35$$
Check. 6% of \$35 is $0.06\times\$35$, or \$2.10, and \$35+\$2.10 = \$37.10, the total amount paid. The answer checks.

State. The price of the frame was \$35.00.

39. ***Familiarize***. Let $t=$ the number of hours required to roof the house, working together. Then in t hr David does $\frac{t}{15}$ of the job and Loren does $\frac{t}{12}$. Together, they do one complete job.

Translate.

$$\frac{t}{15}+\frac{t}{12}=1$$

Solve. The LCM is 60.

$$60\left(\frac{t}{15}+\frac{t}{12}\right)=60\cdot 1$$
$$60\cdot\frac{t}{15}+60\cdot\frac{t}{12}=60$$
$$4t+5t=60$$
$$9t=60$$
$$t=\frac{60}{9}=\frac{20}{3},\text{ or }6\frac{2}{3}$$

Check. In $\frac{20}{3}$ hr David does $\frac{20/3}{15}$, or $\frac{4}{9}$, of the job and Loren does $\frac{20/3}{12}$, or $\frac{5}{9}$, of the job. Since $\frac{4}{9}+\frac{5}{9}=1$ complete job, the answer checks.

State. It would take $\frac{20}{3}$, or $6\frac{2}{3}$ hr, to do the job working together.

40. ***Familiarize.*** Let l = the length of the second leg of the triangle, in inches. Then $l+8$ = the length of the hypotenuse.

Translate. We use the Pythagorean theorem.

$$12^2+l^2=(l+8)^2$$

Solve.

$$12^2+l^2=(l+8)^2$$
$$144+l^2=l^2+16l+64$$
$$144=16l+64$$
$$80=16l$$
$$5=l$$

If $l=5$, then $l+8=5+8=13$.

Check. $12^2+5^2=144+25=169=13^2$. The answer checks.

State. The length of the hypotenuse is 13 in., and the length of the other leg is 5 in.

41. ***Familiarize.*** Let c = the number of defective chips to be expected in a batch of 1800 chips.

Translate. We will use a proportion.

$$\text{Number defective}\longrightarrow\ \frac{5}{120}=\frac{c}{1800}\ \longleftarrow\text{Number defective}$$
$$\text{Number of chips}\longrightarrow\qquad\qquad\qquad\longleftarrow\text{Number of chips}$$

Solve. We multiply both sides by the LCM, 1800.

$$1800\cdot\frac{5}{120}=1800\cdot\frac{c}{1800}$$
$$75=c$$

Check. $\frac{5}{120}=\frac{1}{24}$ and $\frac{75}{1800}=\frac{1}{24}$. The ratios are the same, so the answer checks.

State. 75 defective chips would be expected in a batch of 1800 chips.

42. ***Familiarize.*** Let b = the base of the triangle, in feet. Then $b+5$ = the height.

Translate. We use the formula for the area of a triangle, $A=\frac{1}{2}bh$.

$$18=\frac{1}{2}\cdot b\cdot(b+5)$$

Solve.

$$18=\frac{1}{2}\cdot b\cdot(b+5)$$
$$36=b(b+5)$$
$$36=b^2+5b$$
$$0=b^2+5b-36$$
$$0=(b+9)(b-4)$$
$$b+9=0\quad or\quad b-4=0$$
$$b=-9\quad or\quad b=4$$

Check. The length of the base cannot be negative, so we check only 4. If $b=4$, then $b+5=4+5$, or 9. The area is $\frac{1}{2}\cdot 4\cdot 9=18\text{ ft}^2$. The answer checks.

State. The height is 9 ft, and the base is 4 ft.

43. ***Familiarize and Translate.*** We are told we have inverse variation between the variables h and b. We write an equation of inverse variation.

$$h=\frac{k}{b}$$

Solve. First we substitute 24 for h and 15 for b and find the variation constant, k.

$$h=\frac{k}{b}$$
$$24=\frac{k}{15}$$
$$360=k\qquad\text{Variation constant}$$

Now we find h when b is 5 ft.

$$h=\frac{360}{b}\qquad\text{Equation of variation}$$
$$h=\frac{360}{5}\qquad\text{Substituting 5 for } b$$
$$h=72$$

Check. We could repeat the calculations. We can also observe that when b decreased, h increased as it should in inverse variation. The answer seems reasonable.

State. The height is 72 ft when the base is 5 ft. The variation constant is 360.

44. ***Familiarize.*** Let t = the number of hours the trains have traveled when they are 180 mi apart. We organize the information in a table.

	Distance	Speed	Time
Slower train	Distance of slower train	55	t
Faster train	Distance of faster train	65	t
Total	180		

Translate. The slower train travels $55t$ mi, the faster train travels $65t$ mi, and the total distance traveled is 180 mi. We have

$$55t + 65t = 180.$$

Solve.

$$55t + 65t = 180$$
$$120t = 180$$
$$t = 1.5$$

Check. In 1.5 hr, the slower train travels 55(1.5), or 82.5 mi, and the faster train travels 65(1.5), or 97.5 mi. After 1.5 hr they are 82.5 + 97.5, or 180 mi, apart. The answer checks.

State. In 1.5 hr, the trains will be 180 mi apart.

45. ***Familiarize.*** Let a = the number of milliliters of solution A that should be used and b = the number of milliliters of solution B in the mixture. We organize the information in a table.

	Solution A	Solution B	Mixture
Amount	a	b	100
Percent of salt	10%	40%	22%
Amount of salt	$0.1a$	$0.4b$	0.22(100), or 22 mL

Translate. The first and last rows of the table give us two equations.

$$a + b = 100,$$
$$0.1a + 0.4b = 22$$

After clearing decimals in the second equation, we have

$$a + b = 100, \quad (1)$$
$$a + 4b = 220. \quad (2)$$

Solve. Multiply Equation (1) by -1 and then add.

$$-a - b = -100$$
$$a + 4b = 220$$
$$3b = 120$$
$$b = 40$$

Now substitute 40 for b in one of the original equations and solve for a.

$$a + b = 100 \quad (1)$$
$$a + 40 = 100$$
$$a = 60$$

The solution is $(60, 40)$.

Check. 60 mL + 40 mL is 100 mL; 60 mL of the 10% solution contains 0.1(60), or 6 mL, of salt, and 40 mL of the 40% solution contains 0.4(40), or 16 mL of salt. Then the mixture contains 6 + 16, or 22 mL, of salt. The answer checks.

State. 60 mL of solution A and 40 mL of solution B should be used.

46.

$$y = kx$$
$$2.4 = k \cdot 12$$
$$\frac{2.4}{12} = k$$
$$0.2 = k \quad \text{Variation constant}$$

The equation of variation is $y = 0.2x$.

47. $m = \dfrac{y_2 - y_1}{x_2 - x_1} = \dfrac{3-3}{-1-2} = \dfrac{0}{-3} = 0$

48.

$$2x + 3y = 6$$
$$3y = -2x + 6$$
$$\frac{1}{3} \cdot 3y = \frac{1}{3}(-2x + 6)$$
$$y = -\frac{2}{3}x + 2 \quad \text{Slope-intercept form}$$

The slope is $-\dfrac{2}{3}$; the y-intercept is $(0, 2)$.

49. $m = \dfrac{-4-6}{2-(-5)} = \dfrac{-10}{7} = -\dfrac{10}{7}$

The equation is $y = -\dfrac{10}{7}x + b$. Use one of the given points to find b. We will use $(-5, 6)$.

$$y = -\frac{10}{7}x + b$$
$$6 = -\frac{10}{7}(-5) + b$$
$$6 = \frac{50}{7} + b$$
$$\frac{42}{7} - \frac{50}{7} = b$$
$$-\frac{8}{7} = b$$

The equation is $y = -\dfrac{10}{7}x - \dfrac{8}{7}$.

50. The y-intercept is $(0, -3)$ and the slope is 6. Then the equation is $y = 6x - 3$.

51. Graph: $y = -2$.

This is the equation of a horizontal line with y-intercept $(0, -2)$.

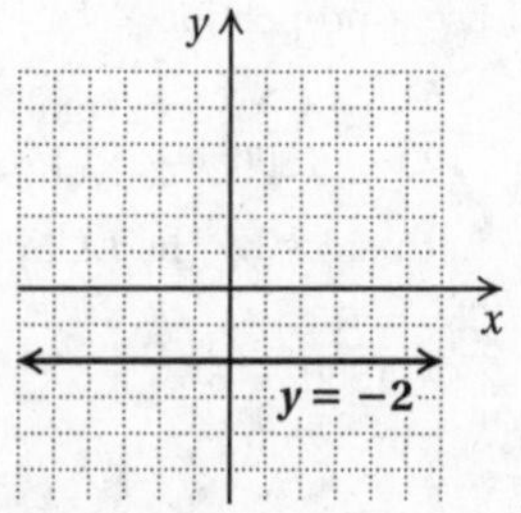

52. Graph: $2x + 5y = 10$

We find the x-intercept.

$$2x + 5 \cdot 0 = 10$$
$$2x = 10$$
$$x = 5$$

The x-intercept is $(5, 0)$.

Now we find the y-intercept.

$$2 \cdot 0 + 5y = 10$$
$$5y = 10$$
$$y = 2$$

The y-intercept is $(0, 2)$.

We find a third point as a check. Let $x = -5$.

$$2(-5) + 5y = 10$$
$$-10 + 5y = 10$$
$$5y = 20$$
$$y = 4$$

A third point is $(-5, 4)$. We graph the intercepts and this point and draw the graph.

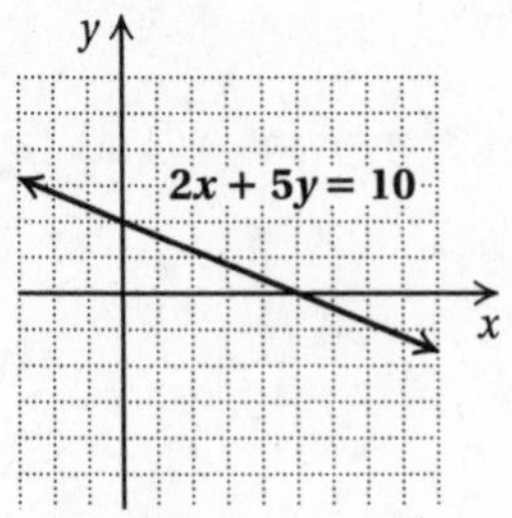

53. Graph: $y \leq 5x$

First we graph the related equation $y = 5x$. We draw a solid line because the inequality symbol is $\leq$. Then we test a point that is not on the line. We try (1,2).

$$\begin{array}{c|c} y & \leq 5x \\ \hline 2 & ?\ 5 \cdot 1 \\ & 5 \quad \text{TRUE} \end{array}$$

We shade the half-plane that contains $(1, 2)$.

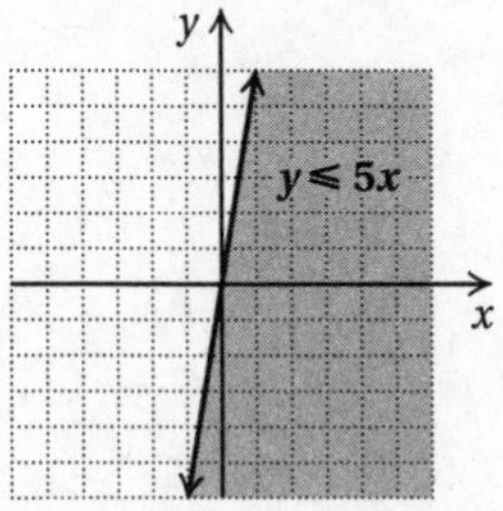

54. Graph on a plane: $5x - 1 < 24$

$$5x - 1 < 24$$
$$5x < 25$$
$$x < 5$$

We graph the related equation $x = 5$. We draw a dashed line because the inequality symbol is $<$. Now we test a point that is not on the line. We try (0,0).

$$\begin{array}{c|c} 5x - 1 & < 24 \\ \hline 5 \cdot 0 - 1 & ?\ 24 \\ -1 & \quad \text{TRUE} \end{array}$$

We shade the half-plane containing (0,0).

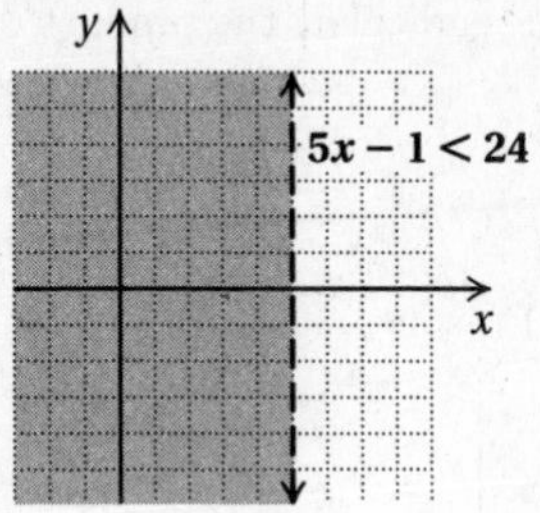

55. We graph the equations.

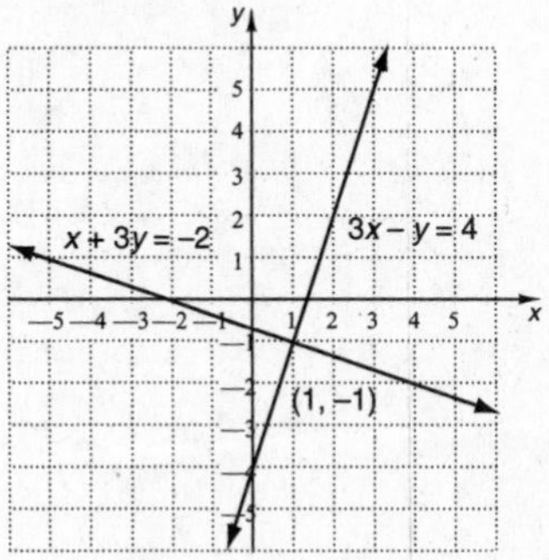

The point of intersection appears to be $(1, -1)$. It checks in both equations, so it is the solution.

56.
$$3(-5) - A \cdot 2 = -7$$
$$-15 - 2A = -7$$
$$-2A = 8$$
$$A = -4$$

$$B(-5) + 4 \cdot 2 = 15$$
$$-5B + 8 = 15$$
$$-5B = 7$$
$$B = -\frac{7}{5}$$

57. $x^2 + 2 < 0$

$x^2 \geq 0$ for any real number x, so $x^2 + 2 \geq 0$. The equation has no solution.

58.
$$\frac{x-5}{x+3} - \frac{x^2-6x+5}{x^2+x-2} \div \frac{x^2+4x+3}{x^2+3x+2}$$
$$= \frac{x-5}{x+3} - \frac{x^2-6x+5}{x^2+x-2} \cdot \frac{x^2+3x+2}{x^2+4x+3}$$
$$= \frac{x-5}{x+3} - \frac{(x-1)(x-5)(x+1)(x+2)}{(x+2)(x-1)(x+1)(x+3)}$$
$$= \frac{x-5}{x+3} - \frac{x-5}{x+3}$$
$$= 0$$

59. We write each equation in slope-intercept form.

$$y - kx = 4 \qquad\qquad 10x - 3y = -12$$
$$y = kx + 4 \qquad\qquad -3y = -10x - 12$$
$$y = \frac{10}{3}x + 4$$

If the lines are perpendicular, the product of their slopes is -1.

$$k \cdot \frac{10}{3} = -1$$

$$k = -\frac{3}{10}$$

Chapter 8

Radical Expressions and Equations

Exercise Set 8.1

RC1. True; see page 604 in the text.

RC3. The radicand is the expression written under the radical, y. The given statement is false.

RC5. True; see page 608 in the text.

1. The square roots of 4 are 2 and -2, because $2^2 = 4$ and $(-2)^2 = 4$.

3. The square roots of 9 are 3 and -3, because $3^2 = 9$ and $(-3)^2 = 9$.

5. The square roots of 100 are 10 and -10, because $10^2 = 100$ and $(-10)^2 = 100$.

7. The square roots of 169 are 13 and -13, because $13^2 = 169$ and $(-13)^2 = 169$.

9. The square roots of 256 are 16 and -16, because $16^2 = 256$ and $(-16)^2 = 256$.

11. $\sqrt{4} = 2$, taking the principal square root.

13. $\sqrt{9} = 3$, so $-\sqrt{9} = -3$.

15. $\sqrt{36} = 6$, so $-\sqrt{36} = -6$.

17. $\sqrt{225} = 15$, so $-\sqrt{225} = -15$.

19. $\sqrt{361} = 19$, taking the principal square root.

21. 2.236

23. 20.785

25. $\sqrt{347.7} \approx 18.647$, so $-\sqrt{347.7} \approx -18.647$.

27. 2.779

29. $-5\sqrt{189 \cdot 6} = -5\sqrt{1134} \approx -5(33.675) = -168.375$

31. a) We substitute 650 for P in the formula.

$W = 118.8\sqrt{650} \approx 3029$

The water flow is about 3029 GPM.

b) We substitute 1500 for P in the formula.

$W = 118.8\sqrt{1500} \approx 4601$

The water flow is about 4601 GPM.

33. Substitute 36 for V in the formula.

$T = 0.144\sqrt{36} \approx 0.864$ sec

35. Substitute 38 for V in the formula.

$T = 0.144\sqrt{38} \approx 0.888$ sec

37. The radicand is the expression under the radical, 200.

39. The radicand is the expression under the radical, x.

41. The radicand is the expression under the radical, $t^2 + 1$.

43. The radicand is the expression under the radical, $\dfrac{3}{x+2}$.

45. No, because the radicand is negative

47. Yes, because the radicand is nonnegative

49. No, because the radicand is negative

51. $\sqrt{c^2} = c$ Since c is assumed to be nonnegative

53. $\sqrt{9x^2} = \sqrt{(3x)^2} = 3x$ Since $3x$ is assumed to be nonnegative

55. $\sqrt{(8p)^2} = 8p$ Since $8p$ is assumed to be nonnegative

57. $\sqrt{(ab)^2} = ab$

59. $\sqrt{(34d)^2} = 34d$

61. $\sqrt{(x+3)^2} = x + 3$

63. $\sqrt{a^2 - 10a + 25} = \sqrt{(a-5)^2} = a - 5$

65. $\sqrt{4a^2 - 20a + 25} = \sqrt{(2a-5)^2} = 2a - 5$

67. $\sqrt{121y^2 - 198y + 81} = \sqrt{(11y-9)^2} = 11y - 9$

69. ***Familiarize***. Let x and y represent the angles. Recall that supplementary angles are angles whose sum is 180°.

Translate. We reword the problem.

$$\underbrace{\text{The sum of two angles}}_{x+y}\ \underbrace{\text{is}}_{=}\ \underbrace{180°}_{180}.$$

$$\underbrace{\text{One angle}}_{x}\ \underbrace{\text{is}}_{=}\ \underbrace{\text{two}}_{2}\ \underbrace{\text{times}}_{\cdot}\ \underbrace{\text{the other}}_{y}\ \underbrace{\text{less}}_{-}\ \underbrace{3°}_{3}.$$

The resulting system is

$x + y = 180,$ (1)

$x = 2y - 3.$ (2)

Solve. We use substitution. We substitute $2y - 3$ for x in Equation (1) and solve for y.

$$\begin{aligned}(2y-3) + y &= 180\\ 3y - 3 &= 180\\ 3y &= 183\\ y &= 61\end{aligned}$$

We substitute 61 for y in Equation (2) to find x.

$x = 2(61) - 3 = 122 - 3 = 119$

Check. $61° + 119° = 180°$, so the angles are supplementary. Also, 3° less than twice 61° is $2 \cdot 61° - 3°$, or $122° - 3°$, or 119°. The numbers check.

State. The angles are 61° and 119°.

71. ***Familiarize***. This problem states that we have direct variation between F and I. Thus, an equation $F = kI$, $k > 0$, applies. As the income increases, the amount spent on food increases.

Translate. We write an equation of variation.

Amount spent on food varies directly as the income.

This translates to $F = kI$.

Solve.

a) First find an equation of variation.

$$F = kI$$
$$10,192 = k \cdot 39,200 \quad \text{Substituting 10,192 for } F \text{ and 39,200 for } I$$
$$\frac{10,192}{39,200} = k$$
$$0.26 = k$$

The equation of variation is $F = 0.26I$.

b) We use the equation to find how much a family spends on food when their income is \$41,000.

$$F = 0.26I$$
$$F = 0.26(\$41,000) \quad \text{Substituting \$41,000 for } I$$
$$F = \$10,660$$

Check. Let us do some reasoning about the answer. The income increased from \$39,200 to \$41,000. Similarly, the amount spend on food increased from \$10,192 to \$10,660. This is what we would expect with direct variation.

State. The amount spent on food is \$10,660.

73.
$$\frac{x^2+10x-11}{x^2-1} \div \frac{x+11}{x+1}$$
$$= \frac{x^2+10x-11}{x^2-1} \cdot \frac{x+1}{x+11}$$
$$= \frac{(x^2+10x-11)(x+1)}{(x^2-1)(x+11)}$$
$$= \frac{(x+11)(x-1)(x+1)}{(x+1)(x-1)(x+11)}$$
$$= 1$$

75. To approximate $\sqrt{3}$, locate 3 on the x-axis, move up vertically to the graph, and then move left horizontally to the y-axis to read the approximation.

$\sqrt{3} \approx 1.7$ (Answers may vary.)

To approximate $\sqrt{5}$, locate 5 on the x-axis, move up vertically to the graph, and then move left horizontally to the y-axis to read the approximation.

$\sqrt{5} \approx 2.2$ (Answers may vary.)

To approximate $\sqrt{7}$, locate 7 on the x-axis, move up vertically to the graph, and then move left horizontally to the y-axis to read the approximation.

$\sqrt{7} \approx 2.6$ (Answers may vary.)

77. If $\sqrt{x^2} = 16$, then $x^2 = 256$ since $\sqrt{256} = 16$. Thus $x = 16$ or $x = -16$.

79. If $t^2 = 49$ then the values of t are the square roots of 49, 7 and -7.

Exercise Set 8.2

RC1. $\sqrt{49w^2} = \sqrt{(7w)^2} = 7w$; the given expression is not in simplest form.

RC3. $\sqrt{121q} = \sqrt{121}\sqrt{q} = 11\sqrt{q}$; the given expression is not in simplest form.

RC5. $t^2 + 1$ has no perfect-square factors, so the given expression is in simplest form.

RC7. 30 has no perfect-square factors, so the given expression is in simplest form.

RC9. $\sqrt{x^3+x^2} = \sqrt{x^2(x+1)} = \sqrt{x^2}\sqrt{x+1} = x\sqrt{x+1}$; the given expression is not in simplest form.

1.
$$\sqrt{12} = \sqrt{4 \cdot 3} \quad \text{4 is a perfect square.}$$
$$= \sqrt{4}\,\sqrt{3} \quad \text{Factoring into a product of radicals}$$
$$= 2\sqrt{3} \quad \text{Taking the square root}$$

3.
$$\sqrt{75} = \sqrt{25 \cdot 3} \quad \text{25 is a perfect square.}$$
$$= \sqrt{25}\,\sqrt{3} \quad \text{Factoring into a product of radicals}$$
$$= 5\sqrt{3} \quad \text{Taking the square root}$$

5.
$$\sqrt{20} = \sqrt{4 \cdot 5} \quad \text{4 is a perfect square.}$$
$$= \sqrt{4}\,\sqrt{5} \quad \text{Factoring into a product of radicals}$$
$$= 2\sqrt{5} \quad \text{Taking the square root}$$

7.
$$\sqrt{600} = \sqrt{100 \cdot 6} \quad \text{100 is a perfect square.}$$
$$= \sqrt{100} \cdot \sqrt{6} \quad \text{Factoring into a product of radicals}$$
$$= 10\sqrt{6} \quad \text{Taking the square root}$$

9.
$$\sqrt{486} = \sqrt{81 \cdot 6} \quad \text{81 is a perfect square.}$$
$$= \sqrt{81} \cdot \sqrt{6} \quad \text{Factoring into a product of radicals}$$
$$= 9\sqrt{6} \quad \text{Taking the square root}$$

11. $\sqrt{9x} = \sqrt{9 \cdot x} = \sqrt{9}\,\sqrt{x} = 3\sqrt{x}$

13. $\sqrt{48x} = \sqrt{16 \cdot 3 \cdot x} = \sqrt{16}\,\sqrt{3x} = 4\sqrt{3x}$

15. $\sqrt{16a} = \sqrt{16 \cdot a} = \sqrt{16}\,\sqrt{a} = 4\sqrt{a}$

17. $\sqrt{64y^2} = \sqrt{64}\,\sqrt{y^2} = 8y$, or

$\sqrt{64y^2} = \sqrt{(8y)^2} = 8y$

19. $\sqrt{13x^2} = \sqrt{13}\,\sqrt{x^2} = \sqrt{13}\cdot x$, or $x\sqrt{13}$

21. $\sqrt{8t^2} = \sqrt{2\cdot 4\cdot t^2} = \sqrt{4}\,\sqrt{t^2}\,\sqrt{2} = 2t\sqrt{2}$

23. $\sqrt{180} = \sqrt{36\cdot 5} = 6\sqrt{5}$

25. $\sqrt{288y} = \sqrt{144\cdot 2\cdot y} = \sqrt{144}\,\sqrt{2y} = 12\sqrt{2y}$

27. $\sqrt{28x^2} = \sqrt{4\cdot 7\cdot x^2} = \sqrt{4}\,\sqrt{x^2}\,\sqrt{7} = 2x\sqrt{7}$

29. $\sqrt{x^2-6x+9} = \sqrt{(x-3)^2} = x-3$

31. $\sqrt{8x^2+8x+2} = \sqrt{2(4x^2+4x+1)} =$
$\sqrt{2(2x+1)^2} = \sqrt{2}\,\sqrt{(2x+1)^2} = \sqrt{2}\,(2x+1)$

33. $\sqrt{36y+12y^2+y^3} = \sqrt{y(36+12y+y^2)} =$
$\sqrt{y(6+y)^2} = \sqrt{y}\,\sqrt{(6+y)^2} = \sqrt{y}\,(6+y)$

35. $\sqrt{t^6} = \sqrt{(t^3)^2} = t^3$

37. $\sqrt{x^{12}} = \sqrt{(x^6)^2} = x^6$

39. $\sqrt{x^5} = \sqrt{x^4\cdot x}$ One factor is a perfect square
$= \sqrt{x^4}\,\sqrt{x}$
$= \sqrt{(x^2)^2}\,\sqrt{x}$
$= x^2\sqrt{x}$

41. $\sqrt{t^{19}} = \sqrt{t^{18}\cdot t} = \sqrt{t^{18}}\,\sqrt{t} = \sqrt{(t^9)^2}\,\sqrt{t} = t^9\sqrt{t}$

43. $\sqrt{(y-2)^8} = \sqrt{[(y-2)^4]^2} = (y-2)^4$

45. $\sqrt{4(x+5)^{10}} = \sqrt{4[(x+5)^5]^2} = \sqrt{4}\,\sqrt{[(x+5)^5]^2} =$
$2(x+5)^5$

47. $\sqrt{36m^3} = \sqrt{36\cdot m^2\cdot m} = \sqrt{36}\,\sqrt{m^2}\,\sqrt{m} = 6m\sqrt{m}$

49. $\sqrt{8a^5} = \sqrt{2\cdot 4\cdot a^4\cdot a} = \sqrt{2\cdot 4\cdot (a^2)^2\cdot a} =$
$\sqrt{4}\,\sqrt{(a^2)^2}\,\sqrt{2a} = 2a^2\sqrt{2a}$

51. $\sqrt{104p^{17}} = \sqrt{4\cdot 26\cdot p^{16}\cdot p} = \sqrt{4\cdot 26\cdot (p^8)^2\cdot p} =$
$\sqrt{4}\,\sqrt{(p^8)^2}\,\sqrt{26p} = 2p^8\sqrt{26p}$

53. $\sqrt{448x^6y^3} = \sqrt{64\cdot 7\cdot x^6\cdot y^2\cdot y} =$
$\sqrt{64\cdot 7\cdot (x^3)^2\cdot y^2\cdot y} =$
$\sqrt{64}\,\sqrt{(x^3)^2}\,\sqrt{y^2}\,\sqrt{7y} = 8x^3y\,\sqrt{7y}$

55. $\sqrt{3}\,\sqrt{18} = \sqrt{3\cdot 18}$ Multiplying
$= \sqrt{3\cdot 3\cdot 6}$ Looking for perfect-square factors or pairs of factors
$= \sqrt{3\cdot 3}\,\sqrt{6}$
$= 3\sqrt{6}$

57. $\sqrt{15}\,\sqrt{6} = \sqrt{15\cdot 6}$ Multiplying
$= \sqrt{5\cdot 3\cdot 3\cdot 2}$ Looking for perfect-square factors or pairs of factors
$= \sqrt{3\cdot 3}\,\sqrt{5\cdot 2}$
$= 3\sqrt{10}$

59. $\sqrt{18}\,\sqrt{14x} = \sqrt{18\cdot 14x} = \sqrt{3\cdot 3\cdot 2\cdot 2\cdot 7\cdot x} =$
$\sqrt{3\cdot 3}\,\sqrt{2\cdot 2}\,\sqrt{7x} = 3\cdot 2\sqrt{7x} = 6\sqrt{7x}$

61. $\sqrt{3x}\,\sqrt{12y} = \sqrt{3x\cdot 12y} = \sqrt{3\cdot x\cdot 3\cdot 4\cdot y} =$
$\sqrt{3\cdot 3\cdot 4\cdot x\cdot y} = \sqrt{3\cdot 3}\,\sqrt{4}\,\sqrt{x\cdot y} = 3\cdot 2\sqrt{xy} = 6\sqrt{xy}$

63. $\sqrt{13}\,\sqrt{13} = \sqrt{13\cdot 13} = 13$

65. $\sqrt{5b}\,\sqrt{15b} = \sqrt{5b\cdot 15b} = \sqrt{5\cdot b\cdot 5\cdot 3\cdot b} =$
$\sqrt{5\cdot 5\cdot b\cdot b\cdot 3} = \sqrt{5\cdot 5}\,\sqrt{b\cdot b}\,\sqrt{3} = 5b\sqrt{3}$

67. $\sqrt{2t}\,\sqrt{2t} = \sqrt{2t\cdot 2t} = 2t$

69. $\sqrt{ab}\,\sqrt{ac} = \sqrt{ab\cdot ac} = \sqrt{a\cdot a\cdot b\cdot c} = \sqrt{a\cdot a}\,\sqrt{b\cdot c} =$
$a\sqrt{bc}$

71. $\sqrt{2x^2y}\,\sqrt{4xy^2} = \sqrt{2x^2y\cdot 4xy^2} = \sqrt{2\cdot x^2\cdot y\cdot 4\cdot x\cdot y^2} =$
$\sqrt{4}\,\sqrt{x^2}\,\sqrt{y^2}\,\sqrt{2xy} = 2xy\sqrt{2xy}$

73. $\sqrt{18}\,\sqrt{18} = \sqrt{18\cdot 18} = 18$

75. $\sqrt{5}\,\sqrt{2x-1} = \sqrt{5(2x-1)} = \sqrt{10x-5}$

77. $\sqrt{x+2}\,\sqrt{x+2} = \sqrt{(x+2)^2} = x+2$

79. $\sqrt{18x^2y^3}\,\sqrt{6xy^4} = \sqrt{18x^2y^3\cdot 6xy^4} =$
$\sqrt{3\cdot 6\cdot x^2\cdot y^2\cdot y\cdot 6\cdot x\cdot y^4} = \sqrt{6\cdot 6\cdot x^2\cdot y^6\cdot 3\cdot x\cdot y} =$
$\sqrt{6\cdot 6}\,\sqrt{x^2}\,\sqrt{y^6}\,\sqrt{3xy} = 6xy^3\sqrt{3xy}$

81. $\sqrt{50x^4y^6}\,\sqrt{10xy} = \sqrt{50x^4y^6\cdot 10xy} =$
$\sqrt{5\cdot 10\cdot x^4\cdot y^6\cdot 10\cdot x\cdot y} = \sqrt{10\cdot 10\cdot x^4\cdot y^6\cdot 5\cdot x\cdot y} =$
$\sqrt{10\cdot 10}\,\sqrt{x^4}\,\sqrt{y^6}\,\sqrt{5xy} = 10x^2y^3\sqrt{5xy}$

83. $\sqrt{99p^4q^3}\sqrt{22p^5q^2} = \sqrt{99p^4q^3\cdot 22p^5q^2} =$
$\sqrt{9\cdot 11\cdot p^4\cdot q^2\cdot q\cdot 2\cdot 11\cdot p^4\cdot p\cdot q^2} =$
$\sqrt{9\cdot 11\cdot 11\cdot p^4\cdot q^2\cdot p^4\cdot q^2\cdot 2\cdot q\cdot p} =$
$3\cdot 11\cdot p^2\cdot q\cdot p^2\cdot q\sqrt{2pq} = 33p^4q^2\sqrt{2pq}$

85. $\sqrt{24a^2b^3c^4}\sqrt{32a^5b^4c^7} = \sqrt{24a^2b^3c^4\cdot 32a^5b^4c^7} =$
$\sqrt{4\cdot 2\cdot 3\cdot a^2\cdot b^2\cdot b\cdot c^4\cdot 16\cdot 2\cdot a^4\cdot a\cdot b^4\cdot c^6\cdot c} =$
$\sqrt{4\cdot 2\cdot 2\cdot 16\cdot a^2\cdot b^2\cdot c^4\cdot a^4\cdot b^4\cdot c^6\cdot 3\cdot b\cdot a\cdot c} =$
$2\cdot 2\cdot 4\cdot a\cdot b\cdot c^2\cdot a^2\cdot b^2\cdot c^3\sqrt{3abc} =$
$16a^3b^3c^5\sqrt{3abc}$

87.

$$\begin{aligned} x - y &= -6 && (1)\\ x + y &= 2 && (2)\\ \hline 2x \quad &= -4 && \text{Adding}\\ x &= -2 \end{aligned}$$

Now we substitute -2 for x in one of the original equations and solve for y.

$$\begin{aligned} x + y &= 2 && \text{Equation (2)}\\ -2 + y &= 2 && \text{Substituting}\\ y &= 4 \end{aligned}$$

Since $(-2, 4)$ checks in both equations, it is the solution.

89. $3x - 2y = 4,\quad (1)$

$2x + 5y = 9\quad (2)$

We will us the elimination method. We multiply on both sides of Equation (1) by 5 and on both sides of Equation (2) by 2. Then we add

$$\begin{aligned} 15x - 10y &= 20 \\ 4x + 10y &= 18 \\ \hline 19x \quad &= 38 \\ x &= 2 \end{aligned}$$

Now we substitute 2 for x in one of the original equations and solve for y.

$$\begin{aligned} 2x + 5y &= 9 \quad \text{Equation (2)} \\ 2 \cdot 2 + 5y &= 9 \\ 4 + 5y &= 9 \\ 5y &= 5 \\ y &= 1 \end{aligned}$$

Since $(2, 1)$ checks, it is the solution.

91. ***Familiarize.*** Let $x =$ the number of liters of 30% solution and $y =$ the number of liters of 50% solution to be used. We organize the information in a table.

Type of solution	30% insecticide	50% insecticide	Mixture
Amount of solution	x	y	200 L
Percent of insecticide	30%	50%	42%
Amount of insecticide in solution	$30\%x$	$50\%y$	$42\% \times 200$, or 84 L

Translate. We get equations from the first and last rows of the table.

$$\begin{aligned} x + y &= 200, \\ 30\%x + 50\%y &= 84, \text{ or} \\ x + y &= 200, \quad (1) \\ 0.3x + 0.5y &= 84 \quad\ \ (2) \end{aligned}$$

Solve. First multiply equation (2) by -2 and add.

$$\begin{aligned} x + y &= 200 \\ -0.6x - y &= -168 \\ \hline 0.4x \quad &= 32 \\ x &= 80 \end{aligned}$$

Now substitute 80 for x in equation (1) and solve for y.

$$\begin{aligned} x + y &= 200 \\ 80x + y &= 200 \\ y &= 120 \end{aligned}$$

Check. We have $80+120 = 200$. Also 30% of 80 is 24, and 50% of 120 is 60. These add to 84, so the answer checks.

State. The mixture should have 80 L of the 30% solution and 120 L of the 50% solution.

93. ***Familiarize.*** We organize the information in a table. The distance traveled downstream is the same as the distance traveled upstream. Let $d =$ this distance, in miles. Let $r =$ the speed at which Greg and Beth paddled. Then their speed downstream was $r + 2$, and their speed upstream was $r - 2$.

	Distance	Speed	Time
Downstream	d	$r+2$	2
Upstream	d	$r-2$	3

Translate. From each row of the table we get an equation $d = rt$.

$$\begin{aligned} d &= (r+2)2, \quad (1) \\ d &= (r-2)3 \quad\ \ (2) \end{aligned}$$

Solve. We solve using substitution. We will substitute $(r+2)2$ for d in equation (2).

$$\begin{aligned} (r+2)2 &= (r-2)3 \quad \text{Substituting} \\ 2r + 4 &= 3r - 6 \\ 4 &= r - 6 \qquad \text{Subtracting } 2r \\ 10 &= r \qquad\quad\ \ \text{Adding 6} \end{aligned}$$

Check. When $r = 10$, then $r + 2 = 10 + 2$, or 12. The distance traveled at 12 mph for 2 hr is $12 \cdot 2$, or 24 mi. When $r = 10$, $r - 2 = 8$. The distance traveled at 8 mph for 3 hr is $8 \cdot 3$, or 24 mi. The distances are the same so the answer checks.

State. Greg and Beth were paddling at a speed of 10 mph.

95. $\sqrt{5x-5} = \sqrt{5(x-1)} = \sqrt{5}\sqrt{x-1}$

97. $\sqrt{x^2-36} = \sqrt{(x+6)(x-6)} = \sqrt{x+6}\sqrt{x-6}$

99. $\sqrt{x^3-2x^2} = \sqrt{x^2(x-2)} = \sqrt{x^2}\sqrt{x-2} = x\sqrt{x-2}$

101. $\sqrt{0.25} = \sqrt{(0.5)^2} = 0.5$

103.
$$\begin{aligned} \sqrt{\sqrt{\sqrt{256}}} &= \sqrt{\sqrt{16}} \qquad (\sqrt{256} = 16) \\ &= \sqrt{4} \qquad\quad\ (\sqrt{16} = 4) \\ &= 2 \end{aligned}$$

105. $\sqrt{18(x-2)}\sqrt{20(x-2)^3} = \sqrt{18(x-2) \cdot 20(x-2)^3} = \sqrt{9 \cdot 2 \cdot 4 \cdot 5(x-2)^4} = 3 \cdot 2(x-2)^2\sqrt{2 \cdot 5} = 6(x-2)^2\sqrt{10}$

107. $\sqrt{2^{109}}\sqrt{x^{306}}\sqrt{x^{11}} = \sqrt{2^{109} \cdot x^{306} \cdot x^{11}} = \sqrt{2^{109}x^{317}} = \sqrt{2^{108} \cdot 2 \cdot x^{316} \cdot x} = 2^{54}x^{158}\sqrt{2x}$

109. $\sqrt{a}(\sqrt{a^3} - 5) = \sqrt{a^4} - 5\sqrt{a} = a^2 - 5\sqrt{a}$

Exercise Set 8.3

RC1. $\sqrt{10} \cdot \sqrt{10} = 10$, so answer (e) is correct.

RC3. $\sqrt{x^5}\sqrt{x} = \sqrt{x^6} = x^3$, so answer (f) is correct.

RC5. $\sqrt{2}\sqrt{2} = 2$, so answer (a) is correct.

1. $\dfrac{\sqrt{18}}{\sqrt{2}} = \sqrt{\dfrac{18}{2}} = \sqrt{9} = 3$

3. $\dfrac{\sqrt{108}}{\sqrt{3}} = \sqrt{\dfrac{108}{3}} = \sqrt{36} = 6$

5. $\dfrac{\sqrt{65}}{\sqrt{13}} = \sqrt{\dfrac{65}{13}} = \sqrt{5}$

7. $\dfrac{\sqrt{3}}{\sqrt{75}} = \sqrt{\dfrac{3}{75}} = \sqrt{\dfrac{1}{25}} = \dfrac{1}{5}$

9. $\dfrac{\sqrt{12}}{\sqrt{75}} = \sqrt{\dfrac{12}{75}} = \sqrt{\dfrac{4}{25}} = \dfrac{2}{5}$

11. $\dfrac{\sqrt{8x}}{\sqrt{2x}} = \sqrt{\dfrac{8x}{2x}} = \sqrt{4} = 2$

13. $\dfrac{\sqrt{63y^3}}{\sqrt{7y}} = \sqrt{\dfrac{63y^3}{7y}} = \sqrt{9y^2} = 3y$

15. $\sqrt{\dfrac{16}{49}} = \dfrac{\sqrt{16}}{\sqrt{49}} = \dfrac{4}{7}$

17. $\sqrt{\dfrac{1}{36}} = \dfrac{\sqrt{1}}{\sqrt{36}} = \dfrac{1}{6}$

19. $-\sqrt{\dfrac{16}{81}} = -\dfrac{\sqrt{16}}{\sqrt{81}} = -\dfrac{4}{9}$

21. $\sqrt{\dfrac{64}{289}} = \dfrac{\sqrt{64}}{\sqrt{289}} = \dfrac{8}{17}$

23. $\sqrt{\dfrac{1690}{1960}} = \sqrt{\dfrac{169 \cdot 10}{196 \cdot 10}} = \sqrt{\dfrac{169}{196} \cdot \dfrac{10}{10}} = \sqrt{\dfrac{169}{196} \cdot 1} =$

$\sqrt{\dfrac{169}{196}} = \dfrac{\sqrt{169}}{\sqrt{196}} = \dfrac{13}{14}$

25. $\sqrt{\dfrac{25}{x^2}} = \dfrac{\sqrt{25}}{\sqrt{x^2}} = \dfrac{5}{x}$

27. $\sqrt{\dfrac{9a^2}{625}} = \dfrac{\sqrt{9a^2}}{\sqrt{625}} = \dfrac{3a}{25}$

29. $\dfrac{\sqrt{50y^{15}}}{\sqrt{2y^{25}}} = \sqrt{\dfrac{50y^{15}}{2y^{25}}} = \sqrt{\dfrac{25}{y^{10}}} = \dfrac{\sqrt{25}}{\sqrt{y^{10}}} = \dfrac{5}{y^5}$

31. $\dfrac{\sqrt{7x^{23}}}{\sqrt{343x^5}} = \sqrt{\dfrac{7x^{23}}{343x^5}} = \sqrt{\dfrac{x^{18}}{49}} = \dfrac{\sqrt{x^{18}}}{\sqrt{49}} = \dfrac{x^9}{7}$

33. $\sqrt{\dfrac{2}{5}} = \sqrt{\dfrac{2}{5} \cdot \dfrac{5}{5}} = \sqrt{\dfrac{10}{25}} = \dfrac{\sqrt{10}}{\sqrt{25}} = \dfrac{\sqrt{10}}{5}$

35. $\sqrt{\dfrac{7}{8}} = \sqrt{\dfrac{7}{8} \cdot \dfrac{2}{2}} = \sqrt{\dfrac{14}{16}} = \dfrac{\sqrt{14}}{\sqrt{16}} = \dfrac{\sqrt{14}}{4}$

37. $\sqrt{\dfrac{1}{12}} = \sqrt{\dfrac{1}{12} \cdot \dfrac{3}{3}} = \sqrt{\dfrac{3}{36}} = \dfrac{\sqrt{3}}{\sqrt{36}} = \dfrac{\sqrt{3}}{6}$

39. $\sqrt{\dfrac{5}{18}} = \sqrt{\dfrac{5}{18} \cdot \dfrac{2}{2}} = \sqrt{\dfrac{10}{36}} = \dfrac{\sqrt{10}}{\sqrt{36}} = \dfrac{\sqrt{10}}{6}$

41. $\dfrac{3}{\sqrt{5}} = \dfrac{3}{\sqrt{5}} \cdot \dfrac{\sqrt{5}}{\sqrt{5}} = \dfrac{3\sqrt{5}}{5}$

43. $\sqrt{\dfrac{8}{3}} = \sqrt{\dfrac{8}{3} \cdot \dfrac{3}{3}} = \sqrt{\dfrac{24}{9}} = \dfrac{\sqrt{4 \cdot 6}}{\sqrt{9}} = \dfrac{\sqrt{4}\,\sqrt{6}}{\sqrt{9}} = \dfrac{2\sqrt{6}}{3}$

45. $\sqrt{\dfrac{3}{x}} = \sqrt{\dfrac{3}{x} \cdot \dfrac{x}{x}} = \sqrt{\dfrac{3x}{x^2}} = \dfrac{\sqrt{3x}}{\sqrt{x^2}} = \dfrac{\sqrt{3x}}{x}$

47. $\sqrt{\dfrac{x}{y}} = \sqrt{\dfrac{x}{y} \cdot \dfrac{y}{y}} = \sqrt{\dfrac{xy}{y^2}} = \dfrac{\sqrt{xy}}{\sqrt{y^2}} = \dfrac{\sqrt{xy}}{y}$

49. $\sqrt{\dfrac{x^2}{20}} = \sqrt{\dfrac{x^2}{20} \cdot \dfrac{5}{5}} = \sqrt{\dfrac{5x^2}{100}} = \dfrac{\sqrt{x^2 \cdot 5}}{\sqrt{100}} = \dfrac{\sqrt{x^2}\,\sqrt{5}}{\sqrt{100}} = \dfrac{x\sqrt{5}}{10}$

51. $\dfrac{1}{\sqrt{3}} = \dfrac{1}{\sqrt{3}} \cdot \dfrac{\sqrt{3}}{\sqrt{3}} = \dfrac{\sqrt{3}}{3}$

53. $\dfrac{\sqrt{9}}{\sqrt{8}} = \dfrac{\sqrt{9}}{\sqrt{8}} \cdot \dfrac{\sqrt{2}}{\sqrt{2}} = \dfrac{\sqrt{9 \cdot 2}}{\sqrt{16}} = \dfrac{3\sqrt{2}}{4}$

55. $\dfrac{\sqrt{11}}{\sqrt{5}} = \dfrac{\sqrt{11}}{\sqrt{5}} \cdot \dfrac{\sqrt{5}}{\sqrt{5}} = \dfrac{\sqrt{55}}{5}$

57. $\dfrac{2}{\sqrt{2}} = \dfrac{2}{\sqrt{2}} \cdot \dfrac{\sqrt{2}}{\sqrt{2}} = \dfrac{2\sqrt{2}}{2} = \sqrt{2}$

59. $\dfrac{\sqrt{5}}{\sqrt{11}} = \dfrac{\sqrt{5}}{\sqrt{11}} \cdot \dfrac{\sqrt{11}}{\sqrt{11}} = \dfrac{\sqrt{55}}{11}$

61. $\dfrac{\sqrt{7}}{\sqrt{12}} = \dfrac{\sqrt{7}}{\sqrt{12}} \cdot \dfrac{\sqrt{3}}{\sqrt{3}} = \dfrac{\sqrt{21}}{\sqrt{36}} = \dfrac{\sqrt{21}}{6}$

63. $\dfrac{\sqrt{48}}{\sqrt{32}} = \sqrt{\dfrac{48}{32}} = \sqrt{\dfrac{3}{2}} = \sqrt{\dfrac{3}{2} \cdot \dfrac{2}{2}} = \sqrt{\dfrac{6}{4}} = \dfrac{\sqrt{6}}{\sqrt{4}} = \dfrac{\sqrt{6}}{2}$

65. $\dfrac{\sqrt{450}}{\sqrt{18}} = \sqrt{\dfrac{450}{18}} = \sqrt{25} = 5$

67. $\dfrac{\sqrt{3}}{\sqrt{x}} = \dfrac{\sqrt{3}}{\sqrt{x}} \cdot \dfrac{\sqrt{x}}{\sqrt{x}} = \dfrac{\sqrt{3x}}{x}$

69. $\dfrac{4y}{\sqrt{5}} = \dfrac{4y}{\sqrt{5}} \cdot \dfrac{\sqrt{5}}{\sqrt{5}} = \dfrac{4y\sqrt{5}}{5}$

71. $\dfrac{\sqrt{a^3}}{\sqrt{8}} = \dfrac{\sqrt{a^3}}{\sqrt{8}} \cdot \dfrac{\sqrt{2}}{\sqrt{2}} = \dfrac{\sqrt{2a^3}}{\sqrt{16}} = \dfrac{\sqrt{a^2 \cdot 2a}}{\sqrt{16}} = \dfrac{a\sqrt{2a}}{4}$

73. $\dfrac{\sqrt{56}}{\sqrt{12x}} = \sqrt{\dfrac{56}{12x}} = \sqrt{\dfrac{14}{3x}} = \sqrt{\dfrac{14}{3x} \cdot \dfrac{3x}{3x}} = \sqrt{\dfrac{42x}{3x \cdot 3x}} =$

$\dfrac{\sqrt{42x}}{3x}$

75. $\dfrac{\sqrt{27c}}{\sqrt{32c^3}} = \sqrt{\dfrac{27c}{32c^3}} = \sqrt{\dfrac{27}{32c^2}} = \sqrt{\dfrac{27}{32c^2} \cdot \dfrac{2}{2}} = \sqrt{\dfrac{54}{64c^2}} =$

$\sqrt{\dfrac{9 \cdot 6}{64c^2}} = \dfrac{3\sqrt{6}}{8c}$

77. $\dfrac{\sqrt{y^5}}{\sqrt{xy^2}} = \sqrt{\dfrac{y^5}{xy^2}} = \sqrt{\dfrac{y^3}{x}} = \sqrt{\dfrac{y^3}{x} \cdot \dfrac{x}{x}} = \sqrt{\dfrac{xy^3}{x^2}} =$

$\sqrt{\dfrac{y^2 \cdot xy}{x^2}} = \dfrac{y\sqrt{xy}}{x}$

79. $\dfrac{\sqrt{45mn^2}}{\sqrt{32m}} = \sqrt{\dfrac{45mn^2}{32m}} = \sqrt{\dfrac{45n^2}{32}} = \sqrt{\dfrac{45n^2}{32} \cdot \dfrac{2}{2}} =$

$\sqrt{\dfrac{90n^2}{64}} = \dfrac{\sqrt{90n^2}}{\sqrt{64}} = \dfrac{\sqrt{9 \cdot n^2 \cdot 10}}{8} = \dfrac{3n\sqrt{10}}{8}$

81.
$$\begin{array}{rll} x + y &= -7 & (1) \\ x - y &= 2 & (2) \\ \hline 2x &= -5 & \text{Adding} \\ x &= -\dfrac{5}{2} & \end{array}$$

Substitute $-\dfrac{5}{2}$ for x in Equation (1) to find y.

$$\begin{array}{rll} x + y &= -7 & \text{Equation (1)} \\ -\dfrac{5}{2} + y &= -7 & \text{Substituting} \\ y &= -\dfrac{9}{2} & \end{array}$$

The ordered pair $\left(-\dfrac{5}{2}, -\dfrac{9}{2}\right)$ checks in both equations. It is the solution.

83. $2x - 3y = 7$ (1)
$2x - 3y = 9$ (2)

We multiply Equation (2) by -1 and add.

$$\begin{array}{r} 2x - 3y = 7 \\ -2x + 3y = -9 \\ \hline 0 = -2 \end{array}$$

We get a false equation. The system of equations has no solution.

85.
$$\begin{aligned} \frac{a^2-25}{6} \div \frac{a+5}{3} &= \frac{a^2-25}{6} \cdot \frac{3}{a+5} \\ &= \frac{(a^2-25)\cdot 3}{6(a+5)} \\ &= \frac{(a+5)(a-5)\cdot 3}{2\cdot 3\cdot (a+5)} \\ &= \frac{\cancel{(a+5)}(a-5)\cdot \cancel{3}}{2\cdot \cancel{3}\cdot \cancel{(a+5)}} \\ &= \frac{a-5}{2} \end{aligned}$$

87. 2 ft: $T \approx 2(3.14)\sqrt{\dfrac{2}{32}} = 6.28\sqrt{\dfrac{1}{16}} = 6.28\left(\dfrac{1}{4}\right) =$
1.57 sec

8 ft: $T \approx 2(3.14)\sqrt{\dfrac{8}{32}} = 6.28\sqrt{\dfrac{1}{4}} = 6.28\left(\dfrac{1}{2}\right) =$
3.14 sec

To find the period for a 10-in. pendulum, we first convert 10 in. to feet.

$10 \text{ in.} = 10 \text{ in.} \times \dfrac{1 \text{ ft}}{12 \text{ in.}} = \dfrac{10}{12} \times \dfrac{\text{in.}}{\text{in.}} \times \text{ ft} = \dfrac{5}{6} \text{ ft}$

Now we use the formula.

$T \approx 2(3.14)\sqrt{\dfrac{5/6}{32}} = 6.28\sqrt{\dfrac{5}{6} \cdot \dfrac{1}{32}} = 6.28\sqrt{\dfrac{5}{192}} \approx 1.01 \text{ sec}$

89. $\sqrt{\dfrac{3x^2y}{a^2x^5}} = \sqrt{\dfrac{3y}{a^2x^3}} = \sqrt{\dfrac{3y}{a^2x^3} \cdot \dfrac{x}{x}} = \sqrt{\dfrac{3xy}{a^2x^4}} =$

$\dfrac{\sqrt{3xy}}{\sqrt{a^2x^4}} = \dfrac{\sqrt{3xy}}{ax^2}$

Chapter 8 Mid-Chapter Review

1. True; see page 604 in the text.

2. For any nonnegative real number A, the principal square root of A^2 is A. The given statement is false.

3. The number 0 has only one square root 0. The given statement is false.

4. True; see page 606 in the text.

5.
$$\begin{aligned} &\sqrt{3x^2 - 48x + 192} \\ &= \sqrt{3(x^2 - 16x + 64)} \\ &= \sqrt{3(x-8)^2} \\ &= \sqrt{3}\sqrt{(x-8)^2} \\ &= \sqrt{3}(x-8) \end{aligned}$$

6.
$$\begin{aligned} &\sqrt{30}\sqrt{40y} \\ &= \sqrt{30 \cdot 40y} \\ &= \sqrt{1200y} \\ &= \sqrt{100 \cdot 12 \cdot y} \\ &= \sqrt{100 \cdot 4 \cdot 3 \cdot y} \\ &= \sqrt{100}\sqrt{4}\sqrt{3y} \\ &= 10 \cdot 2\sqrt{3y} \\ &= 20\sqrt{3y} \end{aligned}$$

7.
$$\begin{aligned} &\sqrt{18ab^2}\sqrt{14a^2b^4} \\ &= \sqrt{18ab^2 \cdot 14a^2b^4} \\ &= \sqrt{2 \cdot 3 \cdot 3 \cdot 2 \cdot 7 \cdot a^3 \cdot b^6} \\ &= \sqrt{2^2 \cdot 3^2 \cdot 7 \cdot a^2 \cdot a \cdot b^6} \\ &= \sqrt{2^2}\sqrt{3^2}\sqrt{a^2}\sqrt{b^6}\sqrt{7a} \\ &= 2 \cdot 3 \cdot a \cdot b^3\sqrt{7a} \\ &= 6ab^3\sqrt{7a} \end{aligned}$$

8. $\sqrt{\dfrac{3y^2}{44}} = \sqrt{\dfrac{3y^2}{2 \cdot 2 \cdot 11}} = \sqrt{\dfrac{3y^2}{2 \cdot 2 \cdot 11} \cdot \dfrac{11}{11}}$

$= \sqrt{\dfrac{33y^2}{2^2 \cdot 11^2}} = \dfrac{y\sqrt{33}}{2 \cdot 11} = \dfrac{y\sqrt{33}}{22}$

9. The square roots of 121 are 11 and -11, because $11^2 = 121$ and $(-11)^2 = 121$.

10. The radicand is the expression under the radical, $\dfrac{x-3}{7}$.

11. a) No, because the radicand is negative.

b) Yes, because the radicand is nonnegative.

12. $\sqrt{128r^7s^6} = \sqrt{64 \cdot 2 \cdot r^6 \cdot r \cdot s^6} = \sqrt{64r^6s^6}\sqrt{2r} =$
$8r^3s^3\sqrt{2r}$

13. $\sqrt{25(x-3)^2} = \sqrt{25}\sqrt{(x-3)^2} = 5(x-3)$

14. $\sqrt{\frac{1}{100}} = \frac{\sqrt{1}}{\sqrt{100}} = \frac{1}{10}$

15. $\sqrt{36} = 6$, so $-\sqrt{36} = -6$.

16. $-\sqrt{\frac{6250}{490}} = -\sqrt{\frac{625 \cdot 10}{49 \cdot 10}} = -\sqrt{\frac{625}{49}} = -\frac{\sqrt{625}}{\sqrt{49}} = -\frac{25}{7}$

17. $\sqrt{225} = 15$, taking the principal square root.

18. $\sqrt{(10y)^2} = 10y$

19. $\sqrt{4x^2 - 4x + 1} = \sqrt{(2x-1)^2} = 2x - 1$

20. $\sqrt{800x} = \sqrt{400 \cdot 2 \cdot x} = \sqrt{400}\sqrt{2x} = 20\sqrt{2x}$

21. $\frac{\sqrt{6}}{\sqrt{96}} = \sqrt{\frac{6}{96}} = \sqrt{\frac{6 \cdot 1}{6 \cdot 16}} = \sqrt{\frac{1}{16}} = \frac{\sqrt{1}}{\sqrt{16}} = \frac{1}{4}$

22. $\sqrt{32q^{11}} = \sqrt{16 \cdot 2 \cdot q^{10} \cdot q} = \sqrt{16q^{10}}\sqrt{2q} = 4q^5\sqrt{2q}$

23. $\sqrt{\frac{81}{x^2}} = \frac{\sqrt{81}}{\sqrt{x^2}} = \frac{9}{x}$

24. $\sqrt{25}\sqrt{25} = \sqrt{25 \cdot 25} = 25$

25. $\frac{\sqrt{18}}{\sqrt{98}} = \sqrt{\frac{18}{98}} = \sqrt{\frac{9 \cdot 2}{49 \cdot 2}} = \sqrt{\frac{9}{49}} = \frac{\sqrt{9}}{\sqrt{49}} = \frac{3}{7}$

26. $\frac{\sqrt{192x}}{\sqrt{3x}} = \sqrt{\frac{192x}{3x}} = \sqrt{\frac{3x \cdot 64}{3x \cdot 1}} = \sqrt{64} = 8$

27. $\sqrt{40c^2d^7}\sqrt{15c^3d^3} = \sqrt{40c^2d^7 \cdot 15c^3d^3} =$
$\sqrt{2 \cdot 2 \cdot 2 \cdot 5 \cdot c^2 \cdot d^6 \cdot d \cdot 3 \cdot 5 \cdot c^2 \cdot c \cdot d^2 \cdot d} =$
$\sqrt{2 \cdot 2 \cdot 5 \cdot 5 \cdot c^2 \cdot d^6 \cdot c^2 \cdot d^2 \cdot d \cdot d \cdot 2 \cdot 3 \cdot c} =$
$\sqrt{2 \cdot 2 \cdot 5 \cdot 5 \cdot c^2 \cdot d^6 \cdot c^2 \cdot d^2 \cdot d \cdot d}\sqrt{6c} =$
$2 \cdot 5 \cdot c \cdot d^3 \cdot c \cdot d \cdot d\sqrt{6c} = 10c^2d^5\sqrt{6c}$

28. $\sqrt{24x^5y^8z^2}\sqrt{60xy^3z} = \sqrt{24x^5y^8z^2 \cdot 60xy^3z} =$
$\sqrt{4 \cdot 2 \cdot 3 \cdot x^4 \cdot x \cdot y^8 \cdot z^2 \cdot 4 \cdot 3 \cdot 5 \cdot x \cdot y^2 \cdot y \cdot z} =$
$\sqrt{4 \cdot 3 \cdot 3 \cdot 4 \cdot x^4 \cdot x \cdot x \cdot y^8 \cdot y^2 \cdot z^2}\sqrt{2 \cdot 5 \cdot y \cdot z} =$
$2 \cdot 3 \cdot 2 \cdot x^2 \cdot x \cdot y^4 \cdot y \cdot z\sqrt{10yz} = 12x^3y^5z\sqrt{10yz}$

29. $\sqrt{2x}\sqrt{30y} = \sqrt{2x \cdot 30y} = \sqrt{2 \cdot x \cdot 2 \cdot 15 \cdot y} =$
$\sqrt{2 \cdot 2 \cdot x \cdot 15 \cdot y} = \sqrt{2 \cdot 2}\sqrt{x \cdot 15 \cdot y} = 2\sqrt{15xy}$

30. $\sqrt{21a}\sqrt{35a} = \sqrt{21a \cdot 35a} = \sqrt{3 \cdot 7 \cdot a \cdot 5 \cdot 7 \cdot a} =$
$\sqrt{7 \cdot 7 \cdot a \cdot a \cdot 3 \cdot 5} = \sqrt{7 \cdot 7 \cdot a \cdot a}\sqrt{3 \cdot 5} =$
$7a\sqrt{15}$

31. $\frac{\sqrt{3y^{29}}}{\sqrt{75y^5}} = \sqrt{\frac{3y^{29}}{75y^5}} = \sqrt{\frac{y^{24}}{25}} = \frac{\sqrt{y^{24}}}{\sqrt{25}} = \frac{y^{12}}{5}$

32. $\frac{x}{\sqrt{3}} = \frac{x}{\sqrt{3}} \cdot \frac{\sqrt{3}}{\sqrt{3}} = \frac{x\sqrt{3}}{3}$

$\sqrt{\frac{3}{x}} = \sqrt{\frac{3}{x} \cdot \frac{x}{x}} = \sqrt{\frac{3x}{x^2}} = \frac{\sqrt{3x}}{x}$

$\frac{3}{\sqrt{x}} = \frac{3}{\sqrt{x}} \cdot \frac{\sqrt{x}}{\sqrt{x}} = \frac{3\sqrt{x}}{x}$

$\frac{3x}{\sqrt{3}} = \frac{3x}{\sqrt{3}} \cdot \frac{\sqrt{3}}{\sqrt{3}} = \frac{3x\sqrt{3}}{3} = x\sqrt{3}$

$\frac{3}{\sqrt{3}} = \frac{3}{\sqrt{3}} \cdot \frac{\sqrt{3}}{\sqrt{3}} = \frac{3\sqrt{3}}{3} = \sqrt{3}$

$\sqrt{\frac{x}{3}} = \sqrt{\frac{x}{3} \cdot \frac{3}{3}} = \sqrt{\frac{3x}{9}} = \frac{\sqrt{3x}}{3}$

33. **The** square root of 10 is the principal, or positive, square root. **A** square root of 10 could refer to either the positive or the negative square root.

34. It is incorrect to take the square roots of the terms in the numerator individually; that is, $\sqrt{a+b}$ and $\sqrt{a} + \sqrt{b}$ are not equivalent. The following is correct:

$$\sqrt{\frac{9+100}{25}} = \frac{\sqrt{9+100}}{\sqrt{25}} = \frac{\sqrt{109}}{5}.$$

35. In general, $\sqrt{a^2 - b^2} \neq \sqrt{a^2} - \sqrt{b^2}$. In this case, let $x = 13$. Then

$\sqrt{x^2 - 25} = \sqrt{13^2 - 25} = \sqrt{169 - 25} = \sqrt{144} = 12$,

but $\sqrt{x^2} - \sqrt{25} = \sqrt{13^2} - \sqrt{25} = 13 - 5 = 8$.

36. 1. If necessary, rewrite the expression as $\sqrt{a}/\sqrt{b}$.

2. Simplify the numerator and denominator, if possible, by taking the square roots of perfect square factors.

3. Multiply by a form of 1 that produces an expression without a radical in the numerator.

Exercise Set 8.4

RC1. The expressions are of the form $a - \sqrt{b}$ and $a + \sqrt{b}$, so they are conjugates.

RC3. The expressions are of the form $\sqrt{a} - \sqrt{b}$ and $\sqrt{b} - \sqrt{a}$, so they are not conjugates.

RC5. The expressions can be written as $3 - \sqrt{5}$ and $3 + \sqrt{5}$. They are of the form $a - \sqrt{b}$ and $a + \sqrt{b}$, so they are conjugates.

RC7. The expressions are reciprocals. They are not conjugates.

RC9. The expressions can be written as $3\sqrt{5} + 5\sqrt{3}$ and $3\sqrt{5} - 5\sqrt{3}$. They are of the form $a\sqrt{b} + c\sqrt{d}$ and $a\sqrt{b} - c\sqrt{d}$, so they are conjugates.

1. $7\sqrt{3} + 9\sqrt{3} = (7+9)\sqrt{3}$
$= 16\sqrt{3}$

3. $7\sqrt{5} - 3\sqrt{5} = (7-3)\sqrt{5}$
$= 4\sqrt{5}$

5. $6\sqrt{x}+7\sqrt{x}=(6+7)\sqrt{x}$
$=13\sqrt{x}$

7. $4\sqrt{d}-13\sqrt{d}=(4-13)\sqrt{d}$
$=-9\sqrt{d}$

9. $5\sqrt{8}+15\sqrt{2}=5\sqrt{4\cdot 2}+15\sqrt{2}$
$=5\cdot 2\sqrt{2}+15\sqrt{2}$
$=10\sqrt{2}+15\sqrt{2}$
$=25\sqrt{2}$

11. $\sqrt{27}-2\sqrt{3}=\sqrt{9\cdot 3}-2\sqrt{3}$
$=3\sqrt{3}-2\sqrt{3}$
$=(3-2)\sqrt{3}$
$=1\sqrt{3}$
$=\sqrt{3}$

13. $\sqrt{45}-\sqrt{20}=\sqrt{9\cdot 5}-\sqrt{4\cdot 5}$
$=3\sqrt{5}-2\sqrt{5}$
$=(3-2)\sqrt{5}$
$=1\sqrt{5}$
$=\sqrt{5}$

15. $\sqrt{72}+\sqrt{98}=\sqrt{36\cdot 2}+\sqrt{49\cdot 2}$
$=6\sqrt{2}+7\sqrt{2}$
$=(6+7)\sqrt{2}$
$=13\sqrt{2}$

17. $2\sqrt{12}+\sqrt{27}-\sqrt{48}=2\sqrt{4\cdot 3}+\sqrt{9\cdot 3}-\sqrt{16\cdot 3}$
$=2\cdot 2\sqrt{3}+3\sqrt{3}-4\sqrt{3}$
$=4\sqrt{3}+3\sqrt{3}-4\sqrt{3}$
$=(4+3-4)\sqrt{3}$
$=3\sqrt{3}$

19. $\sqrt{18}-3\sqrt{8}+\sqrt{50}=\sqrt{9\cdot 2}-3\sqrt{4\cdot 2}+\sqrt{25\cdot 2}$
$=3\sqrt{2}-3\cdot 2\sqrt{2}+5\sqrt{2}$
$=3\sqrt{2}-6\sqrt{2}+5\sqrt{2}$
$=(3-6+5)\sqrt{2}$
$=2\sqrt{2}$

21. $2\sqrt{27}-3\sqrt{48}+3\sqrt{12}=2\sqrt{9\cdot 3}-3\sqrt{16\cdot 3}+3\sqrt{4\cdot 3}$
$=2\cdot 3\sqrt{3}-3\cdot 4\sqrt{3}+3\cdot 2\sqrt{3}$
$=6\sqrt{3}-12\sqrt{3}+6\sqrt{3}$
$=(6-12+6)\sqrt{3}$
$=0\sqrt{3}$
$=0$

23. $\sqrt{4x}+\sqrt{81x^3}=\sqrt{4\cdot x}+\sqrt{81\cdot x^2\cdot x}$
$=2\sqrt{x}+9x\sqrt{x}$
$=(2+9x)\sqrt{x}$

25. $\sqrt{27}-\sqrt{12x^2}=\sqrt{9\cdot 3}-\sqrt{4\cdot 3\cdot x^2}$
$=3\sqrt{3}-2x\sqrt{3}$
$=(3-2x)\sqrt{3}$

27. $\sqrt{8x+8}+\sqrt{2x+2}=\sqrt{4(2x+2)}+\sqrt{2x+2}$
$=2\sqrt{2x+2}+1\sqrt{2x+2}$
$=(2+1)\sqrt{2x+2}$
$=3\sqrt{2x+2}$

29. $\sqrt{x^5-x^2}+\sqrt{9x^3-9}=\sqrt{x^2(x^3-1)}+\sqrt{9(x^3-1)}$
$=x\sqrt{x^3-1}+3\sqrt{x^3-1}$
$=(x+3)\sqrt{x^3-1}$

31. $4a\sqrt{a^2b}+a\sqrt{a^2b^3}-5\sqrt{b^3}$
$=4a\sqrt{a^2\cdot b}+a\sqrt{a^2\cdot b^2\cdot b}-5\sqrt{b^2\cdot b}$
$=4a\cdot a\sqrt{b}+a\cdot a\cdot b\sqrt{b}-5\cdot b\sqrt{b}$
$=4a^2\sqrt{b}+a^2b\sqrt{b}-5b\sqrt{b}$
$=(4a^2+a^2b-5b)\sqrt{b}$

33. $\sqrt{3}-\sqrt{\frac{1}{3}}=\sqrt{3}-\sqrt{\frac{1}{3}\cdot\frac{3}{3}}$
$=\sqrt{3}-\frac{\sqrt{3}}{3}$
$=\left(1-\frac{1}{3}\right)\sqrt{3}$
$=\frac{2}{3}\sqrt{3}$, or $\frac{2\sqrt{3}}{3}$

35. $5\sqrt{2}+3\sqrt{\frac{1}{2}}=5\sqrt{2}+3\sqrt{\frac{1}{2}\cdot\frac{2}{2}}$
$=5\sqrt{2}+\frac{3}{2}\sqrt{2}$
$=\left(5+\frac{3}{2}\right)\sqrt{2}$
$=\frac{13}{2}\sqrt{2}$, or $\frac{13\sqrt{2}}{2}$

37. $\sqrt{\frac{2}{3}}-\sqrt{\frac{1}{6}}=\sqrt{\frac{2}{3}\cdot\frac{3}{3}}-\sqrt{\frac{1}{6}\cdot\frac{6}{6}}$
$=\frac{\sqrt{6}}{3}-\frac{\sqrt{6}}{6}$
$=\left(\frac{1}{3}-\frac{1}{6}\right)\sqrt{6}$
$=\frac{1}{6}\sqrt{6}$, or $\frac{\sqrt{6}}{6}$

39. $\sqrt{3}(\sqrt{5}-1)=\sqrt{3}\,\sqrt{5}-\sqrt{3}\cdot 1$
$=\sqrt{15}-\sqrt{3}$

41. $(2+\sqrt{3})(5-\sqrt{7})$
$=2\cdot 5-2\sqrt{7}+\sqrt{3}\cdot 5-\sqrt{3}\sqrt{7}$ Using FOIL
$=10-2\sqrt{7}+5\sqrt{3}-\sqrt{21}$

43. $(2-\sqrt{5})^2$

$= 2^2 - 2\cdot 2\cdot\sqrt{5} + (\sqrt{5})^2$

Using $(A-B)^2 = A^2 - 2AB + B^2$

$= 4 - 4\sqrt{5} + 5$

$= 9 - 4\sqrt{5}$

45. $(\sqrt{2}+8)(\sqrt{2}-8)$

$= (\sqrt{2})^2 - 8^2$ Using $(A+B)(A-B) = A^2 - B^2$

$= 2 - 64$

$= -62$

47. $(\sqrt{6}-\sqrt{5})(\sqrt{6}+\sqrt{5})$

$= (\sqrt{6})^2 - (\sqrt{5})^2$ Using $(A+B)(A-B) = A^2 - B^2$

$= 6 - 5$

$= 1$

49. $(3\sqrt{5}-2)(\sqrt{5}+1)$

$= 3\sqrt{5}\sqrt{5} + 3\sqrt{5} - 2\sqrt{5} - 2$ Using FOIL

$= 3\cdot 5 + 3\sqrt{5} - 2\sqrt{5} - 2$

$= 15 + \sqrt{5} - 2$

$= 13 + \sqrt{5}$

51. $(\sqrt{x}-\sqrt{y})^2 = (\sqrt{x})^2 - 2\sqrt{x}\sqrt{y} + (\sqrt{y})^2$

Using $(A-B)^2 = A^2 - 2AB + B^2$

$= x - 2\sqrt{xy} + y$

53. We multiply by 1 using the conjugate of $\sqrt{3}-\sqrt{5}$, which is $\sqrt{3}+\sqrt{5}$, as the numerator and denominator.

$\frac{2}{\sqrt{3}-\sqrt{5}} = \frac{2}{\sqrt{3}-\sqrt{5}}\cdot\frac{\sqrt{3}+\sqrt{5}}{\sqrt{3}+\sqrt{5}}$ Multiplying by 1

$= \frac{2(\sqrt{3}+\sqrt{5})}{(\sqrt{3}-\sqrt{5})(\sqrt{3}+\sqrt{5})}$ Multiplying

$= \frac{2\sqrt{3}+2\sqrt{5}}{(\sqrt{3})^2-(\sqrt{5})^2} = \frac{2\sqrt{3}+2\sqrt{5}}{3-5}$

$= \frac{2\sqrt{3}+2\sqrt{5}}{-2} = \frac{2(\sqrt{3}+\sqrt{5})}{-2}$

$= -(\sqrt{3}+\sqrt{5}) = -\sqrt{3}-\sqrt{5}$

55. We multiply by 1 using the conjugate of $\sqrt{3}+\sqrt{2}$, which is $\sqrt{3}-\sqrt{2}$, as the numerator and denominator.

$\frac{\sqrt{3}-\sqrt{2}}{\sqrt{3}+\sqrt{2}} = \frac{\sqrt{3}-\sqrt{2}}{\sqrt{3}+\sqrt{2}}\cdot\frac{\sqrt{3}-\sqrt{2}}{\sqrt{3}-\sqrt{2}}$ Multiplying by 1

$= \frac{(\sqrt{3}-\sqrt{2})^2}{(\sqrt{3}+\sqrt{2})(\sqrt{3}-\sqrt{2})}$

$= \frac{(\sqrt{3})^2 - 2\sqrt{3}\sqrt{2} + (\sqrt{2})^2}{(\sqrt{3})^2-(\sqrt{2})^2}$

$= \frac{3-2\sqrt{6}+2}{3-2} = \frac{5-2\sqrt{6}}{1}$

$= 5 - 2\sqrt{6}$

57. We multiply by 1 using the conjugate of $\sqrt{10}+1$, which is $\sqrt{10}-1$, as the numerator and denominator.

$\frac{4}{\sqrt{10}+1} = \frac{4}{\sqrt{10}+1}\cdot\frac{\sqrt{10}-1}{\sqrt{10}-1}$

$= \frac{4(\sqrt{10}-1)}{(\sqrt{10}+1)(\sqrt{10}-1)}$

$= \frac{4\sqrt{10}-4}{(\sqrt{10})^2-1^2} = \frac{4\sqrt{10}-4}{10-1}$

$= \frac{4\sqrt{10}-4}{9}$

59. We multiply by 1 using the conjugate of $3+\sqrt{7}$, which is $3-\sqrt{7}$, as the numerator and denominator.

$\frac{1-\sqrt{7}}{3+\sqrt{7}} = \frac{1-\sqrt{7}}{3+\sqrt{7}}\cdot\frac{3-\sqrt{7}}{3-\sqrt{7}}$

$= \frac{(1-\sqrt{7})(3-\sqrt{7})}{(3+\sqrt{7})(3-\sqrt{7})}$

$= \frac{3-\sqrt{7}-3\sqrt{7}+\sqrt{7}\sqrt{7}}{3^2-(\sqrt{7})^2}$

$= \frac{3-\sqrt{7}-3\sqrt{7}+7}{9-7} = \frac{10-4\sqrt{7}}{2}$

$= \frac{2(5-2\sqrt{7})}{2} = 5-2\sqrt{7}$

61. We multiply by 1 using the conjugate of $4+\sqrt{x}$, which is $4-\sqrt{x}$, as the numerator and denominator.

$\frac{3}{4+\sqrt{x}} = \frac{3}{4+\sqrt{x}}\cdot\frac{4-\sqrt{x}}{4-\sqrt{x}}$

$= \frac{3(4-\sqrt{x})}{(4+\sqrt{x})(4-\sqrt{x})}$

$= \frac{12-3\sqrt{x}}{4^2-(\sqrt{x})^2}$

$= \frac{12-3\sqrt{x}}{16-x}$

63. We multiply by 1 using the conjugate of $8-\sqrt{x}$, which is $8+\sqrt{x}$, as the numerator and denominator.

$\frac{3+\sqrt{2}}{8-\sqrt{x}} = \frac{3+\sqrt{2}}{8-\sqrt{x}}\cdot\frac{8+\sqrt{x}}{8+\sqrt{x}}$

$= \frac{(3+\sqrt{2})(8+\sqrt{x})}{(8-\sqrt{x})(8+\sqrt{x})}$

$= \frac{3\cdot 8 + 3\cdot\sqrt{x} + \sqrt{2}\cdot 8 + \sqrt{2}\cdot\sqrt{x}}{8^2-(\sqrt{x})^2}$

$= \frac{24+3\sqrt{x}+8\sqrt{2}+\sqrt{2x}}{64-x}$

65. We multiply by 1 using the conjugate of $1+\sqrt{a}$, which is $1-\sqrt{a}$, as the numerator and denominator.

$$\frac{\sqrt{a}-1}{1+\sqrt{a}} = \frac{\sqrt{a}-1}{1+\sqrt{a}}\cdot\frac{1-\sqrt{a}}{1-\sqrt{a}}$$
$$= \frac{(\sqrt{a}-1)(1-\sqrt{a})}{(1+\sqrt{a})(1-\sqrt{a})}$$
$$= \frac{\sqrt{a}\cdot 1-\sqrt{a}\cdot\sqrt{a}-1\cdot 1+1\cdot\sqrt{a}}{1^2-(\sqrt{a})^2}$$
$$= \frac{\sqrt{a}-a-1+\sqrt{a}}{1-a}$$
$$= \frac{2\sqrt{a}-a-1}{1-a}$$

67. We multiply by 1 using the conjugate of $\sqrt{a}-\sqrt{t}$, which is $\sqrt{a}+\sqrt{t}$, as the numerator and denominator.

$$\frac{4+\sqrt{3}}{\sqrt{a}-\sqrt{t}} = \frac{4+\sqrt{3}}{\sqrt{a}-\sqrt{t}}\cdot\frac{\sqrt{a}+\sqrt{t}}{\sqrt{a}+\sqrt{t}}$$
$$= \frac{(4+\sqrt{3})(\sqrt{a}+\sqrt{t})}{(\sqrt{a}-\sqrt{t})(\sqrt{a}+\sqrt{t})}$$
$$= \frac{4\cdot\sqrt{a}+4\cdot\sqrt{t}+\sqrt{3}\cdot\sqrt{a}+\sqrt{3}\cdot\sqrt{t}}{(\sqrt{a})^2-(\sqrt{t})^2}$$
$$= \frac{4\sqrt{a}+4\sqrt{t}+\sqrt{3a}+\sqrt{3t}}{a-t}$$

69.
$$3x+5+2(x-3) = 4-6x$$
$$3x+5+2x-6 = 4-6x$$
$$5x-1 = 4-6x$$
$$11x-1 = 4$$
$$11x = 5$$
$$x = \frac{5}{11}$$

The solution is $\frac{5}{11}$.

71.
$$x^2-5x = 6$$
$$x^2-5x-6 = 0$$
$$(x+1)(x-6) = 0$$
$$x+1=0 \quad or \quad x-6=0$$
$$x=-1 \quad or \quad x=6$$

The solutions are -1 and 6.

73. $$\frac{3}{x^2-9}\cdot\frac{x^2-6x+9}{12} = \frac{\cancel{3}\cancel{(x-3)}(x-3)}{(x+3)\cancel{(x-3)}\cancel{(3)}(4)} = \frac{x-3}{4(x+3)}$$

75. ***Familiarize***. Let $a =$ the highest altitude of the Trail, in feet.

Translate.

Lowest altitude	plus	9990 ft	is	highest altitude
↓	↓	↓	↓	↓
4280	+	9990	=	a

Solve. We add on the left side.

$$4280+9990 = a$$
$$14,270 = a$$

Check. We can repeat the calculation. We can also estimate.

$$4280+9990 \approx 4300+10,000 = 14,300 \approx 14,270$$

The answer seems reasonable.

State. The highest altitude of the Trail is 14,270 ft.

77. Enter $y_1 = \sqrt{9x^3}+\sqrt{x}$ and $y_2 = \sqrt{9x^3+x}$ and look at a table of values. The values of y_1 and y_2 are different, so the given statement is not correct.

79.
$$\frac{3}{5}\sqrt{24}+\frac{2}{5}\sqrt{150}-\sqrt{96}$$
$$= \frac{3}{5}\sqrt{4\cdot 6}+\frac{2}{5}\sqrt{25\cdot 6}-\sqrt{16\cdot 6}$$
$$= \frac{3}{5}\cdot 2\sqrt{6}+\frac{2}{5}\cdot 5\sqrt{6}-4\sqrt{6}$$
$$= \frac{6}{5}\sqrt{6}+2\sqrt{6}-4\sqrt{6}$$
$$= \left(\frac{6}{5}+2-4\right)\sqrt{6}$$
$$= -\frac{4}{5}\sqrt{6}, \text{ or } -\frac{4\sqrt{6}}{5}$$

Exercise Set 8.5

RC1. There is one radical, so the principle of squaring will be used one time.

RC3. There is a single radical on each side of the equation, so the principle of squaring will be used one time.

RC5. There is a radical on the left side of the equation and a constant plus a radical on the right side. When both sides of the equation are squared, a radical will remain on the right side. After it is isolated, both sides of the equation must be squared again, so the principle of squaring will be used two times.

1.
$$\sqrt{x} = 6$$
$$(\sqrt{x})^2 = 6^2 \quad \text{Squaring both sides}$$
$$x = 36 \quad \text{Simplifying}$$

Check: $\sqrt{x} = 6$

$\sqrt{36}$? 6

6 | TRUE

The solution is 36.

3.
$$\sqrt{x} = 4.3$$
$$(\sqrt{x})^2 = (4.3)^2 \quad \text{Squaring both sides}$$
$$x = 18.49 \quad \text{Simplifying}$$

Check: $\sqrt{x} = 4.3$

$\sqrt{18.49}$? 4.3

4.3 | TRUE

The solution is 18.49.

5. $\sqrt{y+4} = 13$

$(\sqrt{y+4})^2 = 13^2$ Squaring both sides

$y+4 = 169$ Simplifying

$y = 165$ Subtracting 4

Check: $\sqrt{y+4} = 13$

$\sqrt{165+4}\ ?\ 13$

$\sqrt{169}$

13 TRUE

The solution is 165.

7. $\sqrt{2x+4} = 25$

$(\sqrt{2x+4})^2 = 25^2$ Squaring both sides

$2x+4 = 625$ Simplifying

$2x = 621$ Subtracting 4

$x = \frac{621}{2}$ Dividing by 2

Check: $\sqrt{2x+4} = 25$

$\sqrt{2\cdot\frac{621}{2}+4}\ ?\ 25$

$\sqrt{621+4}$

$\sqrt{625}$

25 TRUE

The solution is $\frac{621}{2}$.

9. $3+\sqrt{x-1} = 5$

$\sqrt{x-1} = 2$ Subtracting 3

$(\sqrt{x-1})^2 = 2^2$ Squaring both sides

$x-1 = 4$

$x = 5$

Check: $3+\sqrt{x-1} = 5$

$3+\sqrt{5-1}\ ?\ 5$

$3+\sqrt{4}$

$3+2$

5 TRUE

The solution is 5.

11. $6-2\sqrt{3n} = 0$

$6 = 2\sqrt{3n}$ Adding $2\sqrt{3n}$

$6^2 = (2\sqrt{3n})^2$ Squaring both sides

$36 = 4\cdot 3n$

$36 = 12n$

$3 = n$

Check: $6-2\sqrt{3n} = 0$

$6-2\sqrt{3\cdot 3}\ ?\ 0$

$6-2\cdot 3$

$6-6$

0 TRUE

The solution is 3.

13. $\sqrt{5x-7} = \sqrt{x+10}$

$(\sqrt{5x-7})^2 = (\sqrt{x+10})^2$ Squaring both sides

$5x-7 = x+10$

$4x = 17$

$x = \frac{17}{4}$

Check: $\sqrt{5x-7} = \sqrt{x+10}$

$\sqrt{5\cdot\frac{17}{4}-7}\ ?\ \sqrt{\frac{17}{4}+10}$

$\sqrt{\frac{85}{4}-\frac{28}{4}}\ \Big|\ \sqrt{\frac{57}{4}}$

$\sqrt{\frac{57}{4}}$ TRUE

The solution is $\frac{17}{4}$.

15. $\sqrt{x} = -7$

There is no solution. The principal square root of x cannot be negative.

17. $\sqrt{2y+6} = \sqrt{2y-5}$

$(\sqrt{2y+6})^2 = (\sqrt{2y-5})^2$

$2y+6 = 2y-5$

$6 = -5$

The equation $6 = -5$ is false; there is no solution.

19. $x-7 = \sqrt{x-5}$

$(x-7)^2 = (\sqrt{x-5})^2$

$x^2-14x+49 = x-5$

$x^2-15x+54 = 0$

$(x-9)(x-6) = 0$

$x-9 = 0$ or $x-6 = 0$

$x = 9$ or $x = 6$

Check: $x-7 = \sqrt{x-5}$

$9-7\ ?\ \sqrt{9-5}$

$2\ \Big|\ \sqrt{4}$

2 TRUE

$x-7=\sqrt{x-5}$	
$6-7$	$\sqrt{6-5}$
-1	$\sqrt{1}$
	1 FALSE

The number 9 checks, but 6 does not. The solution is 9.

21.
$$\begin{aligned}
x-9&=\sqrt{x-3}\\
(x-9)^2&=(\sqrt{x-3})^2\\
x^2-18x+81&=x-3\\
x^2-19x+84&=0\\
(x-12)(x-7)&=0\\
x-12=0 \quad &\text{or} \quad x-7=0\\
x=12 \quad &\text{or} \quad x=7
\end{aligned}$$

Check:

$x-9=\sqrt{x-3}$	
$12-9\ ?$	$\sqrt{12-3}$
3	$\sqrt{9}$
	3 TRUE

$x-9=\sqrt{x-3}$	
$7-9\ ?$	$\sqrt{7-3}$
-2	$\sqrt{4}$
	2 FALSE

The number 12 checks, but 7 does not. The solution is 12.

23.
$$\begin{aligned}
2\sqrt{x-1}&=x-1\\
(2\sqrt{x-1})^2&=(x-1)^2\\
4(x-1)&=x^2-2x+1\\
4x-4&=x^2-2x+1\\
0&=x^2-6x+5\\
0&=(x-5)(x-1)\\
x-5=0 \quad &\text{or} \quad x-1=0\\
x=5 \quad &\text{or} \quad x=1
\end{aligned}$$

Both numbers check. The solutions are 5 and 1.

25.
$$\begin{aligned}
\sqrt{5x+21}&=x+3\\
(\sqrt{5x+21})^2&=(x+3)^2\\
5x+21&=x^2+6x+9\\
0&=x^2+x-12\\
0&=(x+4)(x-3)\\
x+4=0 \quad &\text{or} \quad x-3=0\\
x=-4 \quad &\text{or} \quad x=3
\end{aligned}$$

Check:

$\sqrt{5x+21}$	$=x+3$
$\sqrt{5(-4)+21}\ ?$	$-4+3$
$\sqrt{1}$	-1
1	FALSE

$\sqrt{5x+21}$	$=x+3$
$\sqrt{5\cdot 3+21}\ ?$	$3+3$
$\sqrt{36}$	6
6	TRUE

The number 3 checks, but -4 does not. The solution is 3.

27.
$$\begin{aligned}
\sqrt{2x-1}+2&=x\\
\sqrt{2x-1}&=x-2 \qquad \text{Isolating the radical}\\
(\sqrt{2x-1})^2&=(x-2)^2\\
2x-1&=x^2-4x+4\\
0&=x^2-6x+5\\
0&=(x-5)(x-1)\\
x-5=0 \quad &\text{or} \quad x-1=0\\
x=5 \quad &\text{or} \quad x=1
\end{aligned}$$

Check:

$\sqrt{2x-1}+2$	$=x$
$\sqrt{2\cdot 5-1}+2\ ?$	5
$\sqrt{10-1}+2$	
$\sqrt{9}+2$	
$3+2$	
5	TRUE

$\sqrt{2x-1}+2$	$=x$
$\sqrt{2\cdot 1-1}+2\ ?$	1
$\sqrt{2-1}+2$	
$\sqrt{1}+2$	
$1+2$	
3	FALSE

The number 5 checks, but 1 does not. The solution is 5.

29.
$$\begin{aligned}
\sqrt{x^2+6}-x+3&=0\\
\sqrt{x^2+6}&=x-3 \qquad \text{Isolating the radical}\\
(\sqrt{x^2+6})^2&=(x-3)^2\\
x^2+6&=x^2-6x+9\\
-3&=-6x \qquad \text{Adding } -x^2 \text{ and } -9\\
\frac{1}{2}&=x
\end{aligned}$$

Check: $\sqrt{x^2+6}-x+3=0$

$$\sqrt{\left(\frac{1}{2}\right)^2+6}-\frac{1}{2}+3 \ ? \ 0$$
$$\sqrt{\frac{25}{4}}-\frac{1}{2}+3$$
$$\frac{5}{2}-\frac{1}{2}+3$$
$$5 \quad \text{FALSE}$$

The number $\frac{1}{2}$ does not check. There is no solution.

31. $\sqrt{x^2-4}-x=6$

$\sqrt{x^2-4}=x+6$ Isolating the radical

$(\sqrt{x^2-4})^2=(x+6)^2$

$x^2-4=x^2+12x+36$

$-40=12x$ Adding $-x^2$ and -36

$-\frac{40}{12}=x$

$-\frac{10}{3}=x$

The number $-\frac{10}{3}$ checks. It is the solution.

33. $\sqrt{(p+6)(p+1)}-2=p+1$

$\sqrt{(p+6)(p+1)}=p+3$ Isolating the radical

$\left(\sqrt{(p+6)(p+1)}\right)^2=(p+3)^2$

$(p+6)(p+1)=p^2+6p+9$

$p^2+7p+6=p^2+6p+9$

$p=3$

The number 3 checks. It is the solution.

35. $\sqrt{4x-10}=\sqrt{2-x}$

$(\sqrt{4x-10})^2=(\sqrt{2-x})^2$

$4x-10=2-x$

$5x=12$ Adding 10 and x

$x=\frac{12}{5}$

Check: $\sqrt{4x-10}=\sqrt{2-x}$

$$\sqrt{4\cdot\frac{12}{5}-10} \ ? \ \sqrt{2-\frac{12}{5}}$$
$$\sqrt{\frac{48}{5}-10} \quad \Big| \quad \sqrt{-\frac{2}{5}}$$

Since $\sqrt{-\frac{2}{5}}$ does not represent a real number, there is no solution that is a real number.

37. $\sqrt{x-5}=5-\sqrt{x}$

$(\sqrt{x-5})^2=(5-\sqrt{x})^2$ Squaring both sides

$x-5=25-10\sqrt{x}+x$

$-30=-10\sqrt{x}$ Isolating the radical

$3=\sqrt{x}$ Dividing by -10

$3^2=(\sqrt{x})^2$ Squaring both sides

$9=x$

The number 9 checks. It is the solution.

39. $\sqrt{y+8}-\sqrt{y}=2$

$\sqrt{y+8}=\sqrt{y}+2$ Isolating one radical

$(\sqrt{y+8})^2=(\sqrt{y}+2)^2$ Squaring both sides

$y+8=y+4\sqrt{y}+4$

$4=4\sqrt{y}$ Isolating the radical

$1=\sqrt{y}$ Dividing by 4

$1^2=(\sqrt{y})^2$

$1=y$

The number 1 checks. It is the solution.

41. $\sqrt{x-4}+\sqrt{x+1}=5$

$\sqrt{x-4}=5-\sqrt{x+1}$ Isolating one radical

$(\sqrt{x-4})^2=(5-\sqrt{x+1})^2$

$x-4=25-10\sqrt{x+1}+x+1$

$-30=-10\sqrt{x+1}$ Isolating the radical

$3=\sqrt{x+1}$ Dividing by -10

$3^2=(\sqrt{x+1})^2$

$9=x+1$

$8=x$

The number 8 checks. It is the solution.

43. $\sqrt{x}-1=\sqrt{x-31}$

$(\sqrt{x}-1)^2=(\sqrt{x-31})^2$

$x-2\sqrt{x}+1=x-31$

$-2\sqrt{x}=-32$

$\sqrt{x}=16$

$(\sqrt{x})^2=16^2$

$x=256$

The number 256 checks. It is the solution.

45. Substitute 400 for d.

$v=3.1\sqrt{d}=3.1\sqrt{400}=3.1(20)=62$ m/sec

47. Substitute 34.1 for v and solve for d.

$v=3.1\sqrt{d}$

$34.1=3.1\sqrt{d}$

$11=\sqrt{d}$

$11^2=(\sqrt{d})^2$

$121=d$

The water is 121 m deep.

49. For 65 mph, substitute 64 for S in the equation and solve for x.

$$S = 2\sqrt{5x}$$
$$65 = 2\sqrt{5x}$$
$$65^2 = (2\sqrt{5x})^2$$
$$4225 = 4 \cdot 5x$$
$$4225 = 20x$$
$$211.25 = x$$

At 65 mph, a car will skid 211.25 ft.

For 75 mph, substitute 75 for S in the equation and solve for x.

$$S = 2\sqrt{5x}$$
$$75 = 2\sqrt{5x}$$
$$75^2 = (2\sqrt{5x})^2$$
$$5625 = 4 \cdot 5x$$
$$5625 = 20x$$
$$\frac{5625}{20} = x$$
$$281.25 = x$$

At 75 mph, a car will skid 281.25 ft.

51.
$$R = \frac{s+t}{2}$$
$$2R = s + t \quad \text{Multiplying by 2}$$
$$2R - t = s \quad \text{Subtracting } t$$

53. $(3x)^3 = [3(-4)]^3 = (-12)^3 = -1728$

55. $2x^2 + 11x + 5 = (2x+1)(x+5)$

57. $9t^2 + 24t + 16 = (3t)^2 + 2 \cdot 3t \cdot 4 + 4^2 = (3t+4)^2$

59.
$$\sqrt{5x^2+5} = 5$$
$$(\sqrt{5x^2+5})^2 = 5^2$$
$$5x^2 + 5 = 25$$
$$5x^2 - 20 = 0$$
$$5(x^2 - 4) = 0$$
$$5(x+2)(x-2) = 0$$
$$x + 2 = 0 \quad or \quad x - 2 = 0$$
$$x = -2 \quad or \quad x = 2$$

Both numbers check, so the solutions are 2 and -2.

61.
$$4 + \sqrt{19-x} = 6 + \sqrt{4-x}$$
$$\sqrt{19-x} = 2 + \sqrt{4-x} \quad \text{Isolating one radical}$$
$$(\sqrt{19-x})^2 = (2 + \sqrt{4-x})^2$$
$$19 - x = 4 + 4\sqrt{4-x} + (4-x)$$
$$19 - x = 4\sqrt{4-x} + 8 - x$$
$$11 = 4\sqrt{4-x}$$
$$11^2 = (4\sqrt{4-x})^2$$
$$121 = 16(4-x)$$
$$121 = 64 - 16x$$
$$57 = -16x$$
$$-\frac{57}{16} = x$$

$-\frac{57}{16}$ checks, so it is the solution.

63.
$$\sqrt{x+3} = \frac{8}{\sqrt{x-9}}$$
$$(\sqrt{x+3})^2 = \left(\frac{8}{\sqrt{x-9}}\right)^2$$
$$x + 3 = \frac{64}{x-9}$$
$$(x-9)(x+3) = 64 \quad \text{Multiplying by } x - 9$$
$$x^2 - 6x - 27 = 64$$
$$x^2 - 6x - 91 = 0$$
$$(x-13)(x+7) = 0$$
$$x - 13 = 0 \quad or \quad x + 7 = 0$$
$$x = 13 \quad or \quad x = -7$$

The number 13 checks, but -7 does not. The solution is 13.

Exercise Set 8.6

RC1. The correct answer is (d).

1.
$$a^2 + b^2 = c^2$$
$$8^2 + 15^2 = c^2 \quad \text{Substituting}$$
$$64 + 225 = c^2$$
$$289 = c^2$$
$$\sqrt{289} = c$$
$$17 = c$$

3.
$$a^2 + b^2 = c^2$$
$$4^2 + 4^2 = c^2 \quad \text{Substituting}$$
$$16 + 16 = c^2$$
$$32 = c^2$$
$$\sqrt{32} = c \quad \text{Exact answer}$$
$$5.657 \approx c \quad \text{Approximation}$$

5. $$\begin{aligned} a^2+b^2 &= c^2 \\ 5^2+b^2 &= 13^2 \\ 25+b^2 &= 169 \\ b^2 &= 144 \\ b &= 12 \end{aligned}$$

7. $$\begin{aligned} a^2+b^2 &= c^2 \\ (4\sqrt{3})^2+b^2 &= 8^2 \\ 16\cdot 3+b^2 &= 64 \\ 48+b^2 &= 64 \\ b^2 &= 16 \\ b &= 4 \end{aligned}$$

9. $$\begin{aligned} a^2+b^2 &= c^2 \\ 10^2+24^2 &= c^2 \\ 100+576 &= c^2 \\ 676 &= c^2 \\ 26 &= c \end{aligned}$$

11. $$\begin{aligned} a^2+b^2 &= c^2 \\ 9^2+b^2 &= 15^2 \\ 81+b^2 &= 225 \\ b^2 &= 144 \\ b &= 12 \end{aligned}$$

13. $$\begin{aligned} a^2+b^2 &= c^2 \\ a^2+1^2 &= (\sqrt{5})^2 \\ a^2+1 &= 5 \\ a^2 &= 4 \\ a &= 2 \end{aligned}$$

15. $$\begin{aligned} a^2+b^2 &= c^2 \\ 1^2+b^2 &= (\sqrt{3})^2 \\ 1+b^2 &= 3 \\ b^2 &= 2 \\ b &= \sqrt{2} \quad \text{Exact answer} \\ b &\approx 1.414 \quad \text{Approximation} \end{aligned}$$

17. $$\begin{aligned} a^2+b^2 &= c^2 \\ a^2+(5\sqrt{3})^2 &= 10^2 \\ a^2+25\cdot 3 &= 100 \\ a^2+75 &= 100 \\ a^2 &= 25 \\ a &= 5 \end{aligned}$$

19. $$\begin{aligned} a^2+b^2 &= c^2 \\ (\sqrt{2})^2+(\sqrt{7})^2 &= c^2 \\ 2+7 &= c^2 \\ 9 &= c^2 \\ 3 &= c \end{aligned}$$

21. Let $d =$ the distance between A and B. From the drawing in the text, we know that $312^2+110^2=d^2$. We solve this equation.

$$\begin{aligned} 312^2+110^2 &= d^2 \\ 97,344+12,100 &= d^2 \\ 109,444 &= d^2 \\ \sqrt{109,144} &= d \\ 330.823 &\approx d \end{aligned}$$

The distance between A and B is about 330.823 ft.

23. We make a drawing, letting h represent the height of the top of the ladder.

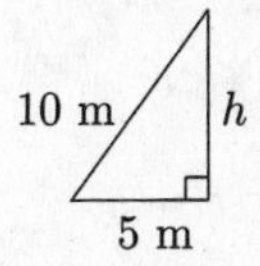

We know that $5^2+h^2=10^2$. We solve this equation.

$$\begin{aligned} 5^2+h^2 &= 10^2 \\ 25+h^2 &= 100 \\ h^2 &= 75 \\ h &= \sqrt{75}\approx 8.660 \end{aligned}$$

The top of the ladder is $\sqrt{75}$ m high, or approximately 8.660 m high.

25. We make a drawing. Let d represent the length of a diagonal of the field.

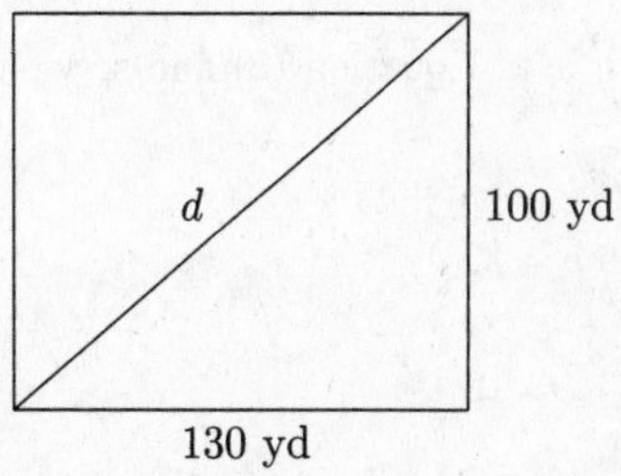

We know that $130^2+100^2=d^2$. We solve this equation.

$$\begin{aligned} 130^2+100^2 &= d^2 \\ 16,900+10,000 &= d^2 \\ 26,900 &= d^2 \\ \sqrt{26,900} &= d \\ 164.012 &\approx d \end{aligned}$$

The length of a diagonal of the field is $\sqrt{26,900}$ yd, or approximately 164.012 yd.

27. $5x+7=8y,$

$3x=8y-4$

$5x-8y=-7 \quad (1)$ Rewriting

$3x-8y=-4 \quad (2)$ the equations

We multiply Equation (2) by -1 and add.

$$\begin{aligned} 5x - 8y &= -7 \\ -3x + 8y &= 4 \\ \hline 2x \quad &= -3 \\ x &= -\frac{3}{2} \end{aligned}$$

Substitute $-\frac{3}{2}$ for x in Equation (1) and solve for y.

$$\begin{aligned} 5x - 8y &= -7 \\ 5\left(-\frac{3}{2}\right) - 8y &= -7 \\ -\frac{15}{2} - 8y &= -7 \\ -8y &= \frac{1}{2} \\ y &= -\frac{1}{16} \end{aligned}$$

The ordered pair $\left(-\frac{3}{2}, -\frac{1}{16}\right)$ checks. It is the solution.

29. $3x - 4y = -11$ (1)

$5x + 6y = 12$ (2)

We multiply Equation (1) by 3 and Equation (2) by 2, and then we add.

$$\begin{aligned} 9x - 12y &= -33 \\ 10x + 12y &= 24 \\ \hline 19x \quad &= -9 \\ x &= -\frac{9}{19} \end{aligned}$$

Substitute $-\frac{9}{19}$ for x in Equation (2) and solve for y.

$$\begin{aligned} 5x + 6y &= 12 \\ 5\left(-\frac{9}{19}\right) + 6y &= 12 \\ -\frac{45}{19} + 6y &= 12 \\ 6y &= \frac{273}{19} && \text{Adding } \frac{45}{19} \\ y &= \frac{273}{6 \cdot 19} && \text{Dividing by 6} \\ y &= \frac{91}{38} && \text{Simplifying} \end{aligned}$$

The ordered pair $\left(-\frac{9}{19}, \frac{91}{38}\right)$ checks. It is the solution.

31. Write the equation in the slope-intercept form.

$$\begin{aligned} 4 - x &= 3y \\ \frac{1}{3}(4 - x) &= y \\ \frac{4}{3} - \frac{1}{3}x &= y, \text{ or} \\ y &= -\frac{1}{3}x + \frac{4}{3} \end{aligned}$$

The slope is $-\frac{1}{3}$.

33.

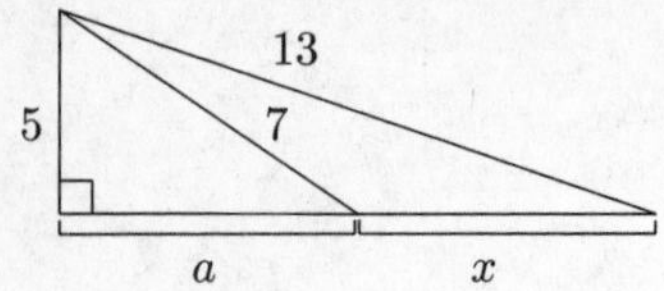

$$\begin{aligned} a^2 + 5^2 &= 7^2 \\ a^2 + 25 &= 49 \\ a^2 &= 24 \\ a &= \sqrt{24}, \text{ or } 2\sqrt{6} \end{aligned}$$

$$\begin{aligned} (a + x)^2 + 5^2 &= 13^2 \\ (2\sqrt{6} + x)^2 + 5^2 &= 13^2 && \text{Substituting } 2\sqrt{6} \text{ for } a \\ (2\sqrt{6} + x)^2 + 25 &= 169 \\ (2\sqrt{6} + x)^2 &= 144 \\ 2\sqrt{6} + x &= 12 && \text{Taking the principal square root} \\ x &= 12 - 2\sqrt{6} \\ x &\approx 7.101 \end{aligned}$$

Chapter 8 Vocabulary Reinforcement

1. The positive square root is called the principal square root.
2. The symbol $\sqrt{}$ is called a radical symbol.
3. The number c is a square root of a if $c^2 = a$.
4. The procedure to find an equivalent expression without a radical in the denominator is called rationalizing the denominator.
5. Like radicals have the same radicands.
6. Expressions such as $\sqrt{6} - \sqrt{a}$ and $\sqrt{6} + \sqrt{a}$ are known as conjugates.
7. In any right triangle, if a and b are the lengths of the legs and c is the length of the hypotenuse, then $a^2 + b^2 = c^2$. The equation $a^2 + b^2 = c^2$ is called the Pythagorean equation.

Chapter 8 Concept Reinforcement

1. True; see page 638 in the text
2. True; see page 616 in the text.
3. True; see page 637 in the text.
4. The statement is false. For example, the equation $2^2 = (-2)^2$ is true, but $2 = -2$ is not true.

Chapter 8 Study Guide

1. The radicand is the expression under the radical, $y^2 - 3$.

2. a) $-\sqrt{-(-3)} = -\sqrt{3}$

Yes; this expression represents a real number, because the radicand is positive.

b) $\sqrt{-200}$

No; the radicand is negative.

3. $\sqrt{1200y^2} = \sqrt{400 \cdot 3 \cdot y^2} = \sqrt{400}\sqrt{y^2}\sqrt{3} = 20y\sqrt{3}$

4. $\sqrt{175a^{12}b^9} = \sqrt{25 \cdot 7 \cdot a^{12} \cdot b^8 \cdot b} =$
$\sqrt{25}\sqrt{a^{12}}\sqrt{b^8}\sqrt{7b} = 5a^6b^4\sqrt{7b}$

5. $\sqrt{8x^3y}\sqrt{12x^4y^3} = \sqrt{8x^3y \cdot 12x^4y^3} =$
$\sqrt{4 \cdot 2 \cdot 4 \cdot 3 \cdot x^2 \cdot x \cdot x^4 \cdot y \cdot y^2 \cdot y} =$
$\sqrt{4}\sqrt{4}\sqrt{x^2}\sqrt{x^4}\sqrt{y \cdot y}\sqrt{y^2}\sqrt{2 \cdot 3 \cdot x} =$
$2 \cdot 2 \cdot x \cdot x^2 \cdot y \cdot y\sqrt{6x} = 4x^3y^2\sqrt{6x}$

6. $\dfrac{\sqrt{15b^7}}{\sqrt{5b^4}} = \sqrt{\dfrac{15b^7}{5b^4}} = \sqrt{3b^3} = \sqrt{3 \cdot b^2 \cdot b} =$
$\sqrt{b^2}\sqrt{3 \cdot b} = b\sqrt{3b}$

7. $\sqrt{\dfrac{50}{162}} = \sqrt{\dfrac{2 \cdot 25}{2 \cdot 81}} = \sqrt{\dfrac{25}{81}} = \dfrac{\sqrt{25}}{\sqrt{81}} = \dfrac{5}{9}$

8. $\dfrac{2a}{\sqrt{50}} = \dfrac{2a}{\sqrt{50}} \cdot \dfrac{\sqrt{2}}{\sqrt{2}} = \dfrac{2a\sqrt{2}}{\sqrt{100}} = \dfrac{2a\sqrt{2}}{10} =$
$\dfrac{2a\sqrt{2}}{2 \cdot 5} = \dfrac{2}{2} \cdot \dfrac{a\sqrt{2}}{5} = \dfrac{a\sqrt{2}}{5}$

9. $\sqrt{x^3 - x^2} + \sqrt{36x - 36} = \sqrt{x^2(x-1)} + \sqrt{36(x-1)} =$
$\sqrt{x^2}\sqrt{x-1} + \sqrt{36}\sqrt{x-1} = x\sqrt{x-1} + 6\sqrt{x-1} =$
$(x+6)\sqrt{x-1}$

10.
$$\begin{aligned}&(\sqrt{13} - \sqrt{2})(\sqrt{13} + 2\sqrt{2})\\ &= \sqrt{13} \cdot \sqrt{13} + \sqrt{13} \cdot 2\sqrt{2} - \sqrt{2} \cdot \sqrt{13} - \sqrt{2} \cdot 2\sqrt{2}\\ &= 13 + 2\sqrt{26} - \sqrt{26} - 4\\ &= 9 + \sqrt{26}\end{aligned}$$

11.
$$\begin{aligned}\frac{5-\sqrt{2}}{9+\sqrt{2}} &= \frac{5-\sqrt{2}}{9+\sqrt{2}} \cdot \frac{9-\sqrt{2}}{9-\sqrt{2}}\\ &= \frac{5 \cdot 9 - 5 \cdot \sqrt{2} - \sqrt{2} \cdot 9 + \sqrt{2} \cdot \sqrt{2}}{9^2 - (\sqrt{2})^2}\\ &= \frac{45 - 5\sqrt{2} - 9\sqrt{2} + 2}{81 - 2}\\ &= \frac{47 - 14\sqrt{2}}{79}\end{aligned}$$

12.
$$\begin{aligned}x - 4 &= \sqrt{x-2}\\ (x-4)^2 &= (\sqrt{x-2})^2\\ x^2 - 8x + 16 &= x - 2\\ x^2 - 9x + 18 &= 0\\ (x-3)(x-6) &= 0\end{aligned}$$
$x - 3 = 0$ *or* $x - 6 = 0$
$x = 3$ *or* $x = 6$

The number 3 does not check, but 6 does. The solution is 6.

13.
$$\begin{aligned}12 - \sqrt{x} &= \sqrt{90 - x}\\ (12 - \sqrt{x})^2 &= (\sqrt{90-x})^2\\ 144 - 24\sqrt{x} + x &= 90 - x\\ -24\sqrt{x} &= -54 - 2x\\ 12\sqrt{x} &= 27 + x \quad \text{Dividing by } -2\\ (12\sqrt{x})^2 &= (27 + x)^2\\ 144x &= 729 + 54x + x^2\\ 0 &= x^2 - 90x + 729\\ 0 &= (x-81)(x-9)\end{aligned}$$
$x - 81 = 0$ *or* $x - 9 = 0$
$x = 81$ *or* $x = 9$

Both numbers check. The solutions are 81 and 9.

14.
$$\begin{aligned}a^2 + b^2 &= c^2\\ a^2 + 32^2 &= 41^2\\ a^2 + 1024 &= 1681\\ a^2 &= 657\\ a &= \sqrt{657} \approx 25.632\end{aligned}$$

Chapter 8 Review Exercises

1. The square roots of 64 are 8 and -8, because $8^2 = 64$ and $(-8)^2 = 64$.

2. The square roots of 400 are 20 and -20, because $20^2 = 400$ and $(-20)^2 = 400$.

3. $\sqrt{36} = 6$, taking the principal square root

4. $\sqrt{169} = 13$, so $-\sqrt{169} = -13$.

5. 1.732

6. 9.950

7. $\sqrt{320.12} \approx 17.892$, so $-\sqrt{320.12} \approx -17.892$.

8. 0.742

9. $\sqrt{\dfrac{47.3}{11.2}} \approx 2.055$, so $-\sqrt{\dfrac{47.3}{11.2}} \approx -2.055$.

10. 394.648

11. The radicand is the expression under the radical, $x^2 + 4$.

12. The radicand is the expression under the radical, x.

13. The radicand is the expression under the radical, $4 - x$.

14. The radicand is the expression under the radical, $\dfrac{2}{y-7}$.

15. Yes, because the radicand is nonnegative

16. No, because the radicand is negative

17. No, because the radicand is negative

18. $\sqrt{(-3)(-27)} = \sqrt{81}$

Yes, because the radicand is positive.

19. $\sqrt{m^2} = m$

20. $\sqrt{(x-4)^2} = x - 4$

21. $\sqrt{(16x)^2} = 16x$

22. $\sqrt{4p^2 - 12p + 9} = \sqrt{(2p-3)^2} = 2p - 3$

23. $\sqrt{48} = \sqrt{16 \cdot 3} = \sqrt{16}\sqrt{3} = 4\sqrt{3}$

24. $\sqrt{32t^2} = \sqrt{16 \cdot 2 \cdot t^2} = \sqrt{16}\sqrt{t^2}\sqrt{2} = 4t\sqrt{2}$

25. $\sqrt{t^2 - 14t + 49} = \sqrt{(t-7)^2} = t - 7$

26. $\sqrt{x^2 + 16x + 64} = \sqrt{(x+8)^2} = x + 8$

27. $\sqrt{x^8} = \sqrt{(x^4)^2} = x^4$

28. $\sqrt{75a^7} = \sqrt{25 \cdot 3 \cdot a^6 \cdot a} = \sqrt{25}\sqrt{a^6}\sqrt{3a} = 5a^3\sqrt{3a}$

29. $\sqrt{3}\sqrt{7} = \sqrt{3 \cdot 7} = \sqrt{21}$

30. $\sqrt{x-3}\sqrt{x+3} = \sqrt{(x-3)(x+3)} = \sqrt{x^2 - 9}$

31. $\sqrt{6}\sqrt{10} = \sqrt{6 \cdot 10} = \sqrt{2 \cdot 3 \cdot 2 \cdot 5} = \sqrt{2 \cdot 2}\sqrt{3 \cdot 5} = 2\sqrt{15}$

32. $\sqrt{5x}\sqrt{8x} = \sqrt{5x \cdot 8x} = \sqrt{5 \cdot x \cdot 4 \cdot 2 \cdot x} =$
$\sqrt{4}\sqrt{x \cdot x}\sqrt{5 \cdot 2} = 2x\sqrt{10}$

33. $\sqrt{5x}\sqrt{10xy^2} = \sqrt{5x \cdot 10xy^2} = \sqrt{5 \cdot x \cdot 2 \cdot 5 \cdot x \cdot y^2} =$
$\sqrt{5 \cdot 5}\sqrt{x \cdot x}\sqrt{y^2}\sqrt{2} = 5xy\sqrt{2}$

34. $\sqrt{20a^3b}\sqrt{5a^2b^2} = \sqrt{20a^3b \cdot 5a^2b^2} =$
$\sqrt{4 \cdot 5 \cdot a^2 \cdot a \cdot b \cdot 5 \cdot a^2 \cdot b^2} =$
$\sqrt{4}\sqrt{5 \cdot 5}\sqrt{a^2}\sqrt{a^2}\sqrt{b^2}\sqrt{ab} = 2 \cdot 5 \cdot a \cdot a \cdot b\sqrt{a \cdot b} =$
$10a^2b\sqrt{ab}$

35. $\sqrt{\dfrac{25}{64}} = \dfrac{\sqrt{25}}{\sqrt{64}} = \dfrac{5}{8}$

36. $\sqrt{\dfrac{49}{t^2}} = \dfrac{\sqrt{49}}{\sqrt{t^2}} = \dfrac{7}{t}$

37. $\dfrac{\sqrt{2c^9}}{\sqrt{32c}} = \sqrt{\dfrac{2c^9}{32c}} = \sqrt{\dfrac{c^8}{16}} = \dfrac{\sqrt{c^8}}{\sqrt{16}} = \dfrac{c^4}{4}$

38. $\sqrt{\dfrac{1}{2}} = \sqrt{\dfrac{1}{2} \cdot \dfrac{2}{2}} = \sqrt{\dfrac{2}{4}} = \dfrac{\sqrt{2}}{\sqrt{4}} = \dfrac{\sqrt{2}}{2}$

39. $\dfrac{\sqrt{x^3}}{\sqrt{15}} = \dfrac{\sqrt{x^3}}{\sqrt{15}} \cdot \dfrac{\sqrt{15}}{\sqrt{15}} = \dfrac{\sqrt{15 \cdot x^2 \cdot x}}{15} = \dfrac{x\sqrt{15x}}{15}$

40. $\sqrt{\dfrac{5}{y}} = \sqrt{\dfrac{5}{y} \cdot \dfrac{y}{y}} = \sqrt{\dfrac{5y}{y^2}} = \dfrac{\sqrt{5y}}{\sqrt{y^2}} = \dfrac{\sqrt{5y}}{y}$

41. $\dfrac{\sqrt{b^9}}{\sqrt{ab^2}} = \sqrt{\dfrac{b^9}{ab^2}} = \sqrt{\dfrac{b^7}{a}} = \sqrt{\dfrac{b^7}{a} \cdot \dfrac{a}{a}} = \sqrt{\dfrac{a \cdot b^6 \cdot b}{a^2}} =$
$\dfrac{\sqrt{a \cdot b^6 \cdot b}}{\sqrt{a^2}} = \dfrac{b^3\sqrt{ab}}{a}$

42. $\dfrac{\sqrt{27}}{\sqrt{45}} = \sqrt{\dfrac{27}{45}} = \sqrt{\dfrac{3}{5}} = \sqrt{\dfrac{3}{5} \cdot \dfrac{5}{5}} = \sqrt{\dfrac{15}{25}} = \dfrac{\sqrt{15}}{\sqrt{25}} = \dfrac{\sqrt{15}}{5}$

43. $\dfrac{\sqrt{45x^2y}}{\sqrt{54y}} = \sqrt{\dfrac{45x^2y}{54y}} = \sqrt{\dfrac{5x^2}{6}} = \sqrt{\dfrac{5x^2}{6} \cdot \dfrac{6}{6}} = \sqrt{\dfrac{30x^2}{36}} =$
$\dfrac{\sqrt{30x^2}}{\sqrt{36}} = \dfrac{x\sqrt{30}}{6}$

44. $10\sqrt{5} + 3\sqrt{5} = (10+3)\sqrt{5} = 13\sqrt{5}$

45.
$$\begin{aligned} \sqrt{80} - \sqrt{45} &= \sqrt{16 \cdot 5} - \sqrt{9 \cdot 5} \\ &= 4\sqrt{5} - 3\sqrt{5} \\ &= (4-3)\sqrt{5} \\ &= \sqrt{5} \end{aligned}$$

46.
$$\begin{aligned} 3\sqrt{2} - 5\sqrt{\frac{1}{2}} &= 3\sqrt{2} - 5\sqrt{\frac{1}{2} \cdot \frac{2}{2}} \\ &= 3\sqrt{2} - 5 \cdot \frac{\sqrt{2}}{2} \\ &= 3\sqrt{2} - \frac{5}{2}\sqrt{2} \\ &= \left(3 - \frac{5}{2}\right)\sqrt{2} \\ &= \frac{1}{2}\sqrt{2}, \text{ or } \frac{\sqrt{2}}{2} \end{aligned}$$

47. $(2+\sqrt{3})^2 = 2^2 + 2 \cdot 2 \cdot \sqrt{3} + (\sqrt{3})^2 = 4 + 4\sqrt{3} + 3 =$
$7 + 4\sqrt{3}$

48. $(2+\sqrt{3})(2-\sqrt{3}) = 2^2 - (\sqrt{3})^2 = 4 - 3 = 1$

49.
$$\begin{aligned} \frac{4}{2+\sqrt{3}} &= \frac{4}{2+\sqrt{3}} \cdot \frac{2-\sqrt{3}}{2-\sqrt{3}} \\ &= \frac{4(2-\sqrt{3})}{(2+\sqrt{3})(2-\sqrt{3})} \\ &= \frac{8-4\sqrt{3}}{2^2 - (\sqrt{3})^2} = \frac{8-4\sqrt{3}}{4-3} \\ &= \frac{8-4\sqrt{3}}{1} = 8 - 4\sqrt{3} \end{aligned}$$

50.
$$\begin{aligned} \sqrt{x-3} &= 7 \\ (\sqrt{x-3})^2 &= 7^2 \\ x - 3 &= 49 \\ x &= 52 \end{aligned}$$

The number 52 checks. It is the solution.

51.
$$\begin{aligned} \sqrt{5x+3} &= \sqrt{2x-1} \\ (\sqrt{5x+3})^2 &= (\sqrt{2x-1})^2 \\ 5x + 3 &= 2x - 1 \\ 3x + 3 &= -1 \\ 3x &= -4 \\ x &= -\frac{4}{3} \end{aligned}$$

The number $-\dfrac{4}{3}$ does not check. There is no solution.

52.
$$1+x=\sqrt{1+5x}$$
$$(1+x)^2=(\sqrt{1+5x})^2$$
$$1+2x+x^2=1+5x$$
$$x^2-3x=0$$
$$x(x-3)=0$$
$$x=0 \quad or \quad x-3=0$$
$$x=0 \quad or \quad x=3$$

Both numbers check. The solutions are 0 and 3.

53.
$$\sqrt{x}=\sqrt{x-5}+1$$
$$(\sqrt{x})^2=(\sqrt{x-5}+1)^2$$
$$x=x-5+2\sqrt{x-5}+1$$
$$x=x-4+2\sqrt{x-5}$$
$$4=2\sqrt{x-5}$$
$$2=\sqrt{x-5}$$
$$2^2=(\sqrt{x-5})^2$$
$$4=x-5$$
$$9=x$$

The number 9 checks. It is the solution.

54. a) We substitute 200 for L in the formula.
$$r=2\sqrt{5\cdot 200}=2\sqrt{1000}\approx 63 \text{ mph}$$

b) We substitute 90 for r and solve for L.
$$90=2\sqrt{5L}$$
$$90^2=(2\sqrt{5L})^2$$
$$8100=4\cdot 5L$$
$$8100=20L$$
$$405 \text{ ft}=L$$

55.
$$a^2+b^2=c^2$$
$$15^2+b^2=25^2$$
$$225+b^2=625$$
$$b^2=400$$
$$b=20$$

56.
$$a^2+b^2=c^2$$
$$1^2+(\sqrt{2})^2=c^2$$
$$1+2=c^2$$
$$3=c^2$$
$$\sqrt{3}=c \quad \text{Exact answer}$$
$$1.732\approx c \quad \text{Approximation}$$

57. First we subtract to find the vertical distance of the descent:
$$30,000 \text{ ft}-20,000 \text{ ft}=10,000 \text{ ft}$$

Let $c=$ the distance the plane travels during the descent. This is labeled "?" in the drawing in the text. We know that $10,000^2+50,000^2=c^2$. We solve this equation.
$$10,000^2+50,000^2=c^2$$
$$100,000,000+2,500,000,000=c^2$$
$$2,600,000,000=c^2$$
$$\sqrt{2,600,00,000} \text{ ft}=c$$
$$50,990 \text{ ft}\approx c$$

58. Let $d=$ the distance each brace reaches vertically. From the drawing in the text we see that $d^2+12^2=15^2$. We solve this equation.
$$d^2+12^2=15^2$$
$$d^2+144=225$$
$$d^2=81$$
$$d=9 \text{ ft}$$

59.
$$x-2=\sqrt{4-9x}$$
$$(x-2)^2=(\sqrt{4-9x})^2$$
$$x^2-4x+4=4-9x$$
$$x^2+5x=0$$
$$x(x+5)=0$$
$$x=0 \quad or \quad x+5=0$$
$$x=0 \quad or \quad x=-5$$

Neither number checks. The equation has no solution. The correct choice is answer B.

60.
$$(2\sqrt{7}+\sqrt{2})(\sqrt{7}-\sqrt{2})$$
$$=2\sqrt{7}\cdot\sqrt{7}-2\sqrt{7}\cdot\sqrt{2}+\sqrt{2}\cdot\sqrt{7}-\sqrt{2}\sqrt{2}$$
$$=2\cdot 7-2\sqrt{14}+\sqrt{14}-2$$
$$=14-\sqrt{14}-2$$
$$=12-\sqrt{14}$$

The correct choice is answer C.

61. After $\frac{1}{2}$ hr, the car traveling east at 50 mph has traveled $\frac{1}{2}\cdot 50$, or 25 mi and the car traveling south at 60 mph has traveled $\frac{1}{2}\cdot 60$, or 30 mi. These distances are the legs of a right triangle. The length of the hypotenuse of the triangle is the distance that separates the cars. Let $d=$ this distance. We know that $25^2+30^2=d^2$. We solve this equation.
$$25^2+30^2=d^2$$
$$625+900=d^2$$
$$1525=d^2$$
$$\sqrt{1525} \text{ mi}=d$$
$$39.051 \text{ mi}\approx d$$

62.
$$A=\sqrt{a^2+b^2}$$
$$A^2=(\sqrt{a^2+b^2})^2$$
$$A^2=a^2+b^2$$
$$A^2-a^2=b^2$$

If $b^2=A^2-a^2$, then $b=\sqrt{A^2-a^2}$ or $b=-\sqrt{A^2-a^2}$, so we have $b=\pm\sqrt{A^2-a^2}$.

63. Using the drawing in the text, let $a =$ the hypotenuse of the triangle with legs 4 and x and let $b =$ the hypotenuse of the triangle with legs 9 and x. Then we have

$$4^2 + x^2 = a^2, \text{ or } 16 + x^2 = a^2$$

and

$$9^2 + x^2 = b^2, \text{ or } 81 + x^2 = b^2.$$

Note that a and b are also legs of the large triangle with hypotenuse $4 + 9$, or 13. Then we have

$$a^2 + b^2 = 13^2, \text{ or } a^2 + b^2 = 169.$$

Adding the two equations containing x, we have

$$\begin{aligned} 16 + x^2 &= a^2 \\ 81 + x^2 &= b^2 \\ \hline 97 + 2x^2 &= a^2 + b^2. \end{aligned}$$

We substitute $97+2x^2$ for a^2+b^2 in the equation pertaining to the large triangle.

$$\begin{aligned} a^2 + b^2 &= 169 \\ 97 + 2x^2 &= 169 \\ 2x^2 &= 72 \\ x^2 &= 36 \\ x &= 6 \end{aligned}$$

Chapter 8 Discussion and Writing Exercises

1. It is important for the signs to differ to ensure that the product of the conjugates will be free of radicals.

2. Since $\sqrt{11 - 2x}$ cannot be negative, the statement $\sqrt{11 - 2x} = -3$ cannot be true for any value of x, including -1.

3. We often use the rules for manipulating exponents "in reverse" when simplifying radical expressions. For example, we might write x^5 as $x^4 \cdot x$ or y^6 and $(y^3)^2$.

4. No; consider the clapboard's height above ground level to be one leg of a right triangle. Then the length of the ladder is the hypotenuse of that triangle. Since the length of the hypotenuse must be greater than the length of a leg, a 28-ft ladder cannot be used to repair a clapboard that is 28 ft above ground level.

5. The square of a number is equal to the square of its opposite. Thus, while squaring both sides of a radical equation allows us to find the solutions of the original equation, this procedure can also introduce numbers that are not solutions of the original equation.

6. a) $\sqrt{5x^2} = \sqrt{5}\sqrt{x^2} = \sqrt{5} \cdot |x| = |x|\sqrt{5}$. The given statement is correct.

b) Let $b = 3$. Then $\sqrt{b^2 - 4} = \sqrt{3^2 - 4} = \sqrt{9 - 4} = \sqrt{5}$, but $b - 2 = 3 - 2 = 1$. The given statement is false.

c) Let $x = 3$. Then $\sqrt{x^2 + 16} = \sqrt{3^2 + 16} = \sqrt{9 + 16} = \sqrt{25} = 5$, but $x + 4 = 3 + 4 = 7$. The given statement is false.

Chapter 8 Test

1. The square roots of 81 are 9 and -9, because $9^2 = 81$ and $(-9)^2 = 81$.

2. $\sqrt{64} = 8$, taking the principal square root

3. $\sqrt{25} = 5$, so $-\sqrt{25} = -5$.

4. $\sqrt{116} \approx 10.770$

5. $\sqrt{87.4} \approx 9.349$, so $-\sqrt{87.4} \approx -9.349$.

6. $4\sqrt{5 \cdot 6} \approx 21.909$

7. The radicand is the expression under the radical, $4 - y^3$.

8. Yes, because the radicand is nonnegative

9. No, because the radicand is negative

10. $\sqrt{a^2} = a$

11. $\sqrt{36y^2} = \sqrt{(6y)^2} = 6y$

12. $\sqrt{5}\sqrt{6} = \sqrt{5 \cdot 6} = \sqrt{30}$

13. $\sqrt{x - 8}\sqrt{x + 8} = \sqrt{(x - 8)(x + 8)} = \sqrt{x^2 - 64}$

14. $\sqrt{27} = \sqrt{9 \cdot 3} = \sqrt{9}\sqrt{3} = 3\sqrt{3}$

15. $\sqrt{25x - 25} = \sqrt{25(x - 1)} = \sqrt{25}\sqrt{x - 1} = 5\sqrt{x - 1}$

16. $\sqrt{t^5} = \sqrt{t^4 \cdot t} = \sqrt{t^4}\sqrt{t} = \sqrt{(t^2)^2}\sqrt{t} = t^2\sqrt{t}$

17. $\sqrt{5}\sqrt{10} = \sqrt{5 \cdot 10} = \sqrt{5 \cdot 2 \cdot 5} = \sqrt{5 \cdot 5}\sqrt{2} = 5\sqrt{2}$

18. $\sqrt{3ab}\sqrt{6ab^3} = \sqrt{3ab \cdot 6ab^3} = \sqrt{3 \cdot a \cdot b \cdot 2 \cdot 3 \cdot a \cdot b^2 \cdot b} = \sqrt{3 \cdot 3}\sqrt{a \cdot a}\sqrt{b \cdot b}\sqrt{b^2}\sqrt{2} = 3 \cdot a \cdot b \cdot b\sqrt{2} = 3ab^2\sqrt{2}$

19. $\sqrt{\dfrac{27}{12}} = \sqrt{\dfrac{9}{4}} = \dfrac{\sqrt{9}}{\sqrt{4}} = \dfrac{3}{2}$

20. $\sqrt{\dfrac{144}{a^2}} = \dfrac{\sqrt{144}}{\sqrt{a^2}} = \dfrac{12}{a}$

21. $\sqrt{\dfrac{2}{5}} = \sqrt{\dfrac{2}{5} \cdot \dfrac{5}{5}} = \sqrt{\dfrac{10}{25}} = \dfrac{\sqrt{10}}{\sqrt{25}} = \dfrac{\sqrt{10}}{5}$

22. $\sqrt{\dfrac{2x}{y}} = \sqrt{\dfrac{2x}{y} \cdot \dfrac{y}{y}} = \sqrt{\dfrac{2xy}{y^2}} = \dfrac{\sqrt{2xy}}{y}$

23. $\dfrac{\sqrt{27}}{\sqrt{32}} = \sqrt{\dfrac{27}{32}} = \sqrt{\dfrac{27}{32} \cdot \dfrac{2}{2}} = \sqrt{\dfrac{54}{64}} = \dfrac{\sqrt{54}}{\sqrt{64}} = \dfrac{\sqrt{9 \cdot 6}}{8} = \dfrac{3\sqrt{6}}{8}$

24. $\dfrac{\sqrt{35x}}{\sqrt{80xy^2}} = \sqrt{\dfrac{35x}{80xy^2}} = \sqrt{\dfrac{7}{16y^2}} = \dfrac{\sqrt{7}}{\sqrt{16y^2}} = \dfrac{\sqrt{7}}{4y}$

25. $3\sqrt{18} - 5\sqrt{18} = (3 - 5)\sqrt{18} = -2\sqrt{18} = -2\sqrt{9 \cdot 2} = -2 \cdot 3\sqrt{2} = -6\sqrt{2}$

26. $\sqrt{5} + \sqrt{\dfrac{1}{5}} = \sqrt{5} + \sqrt{\dfrac{1}{5} \cdot \dfrac{5}{5}} = \sqrt{5} + \sqrt{\dfrac{5}{25}} = \sqrt{5} + \dfrac{\sqrt{5}}{5} = \left(1 + \dfrac{1}{5}\right)\sqrt{5} = \dfrac{6}{5}\sqrt{5}$, or $\dfrac{6\sqrt{5}}{5}$

27. $(4-\sqrt{5})^2 = 4^2 - 2\cdot 4\cdot \sqrt{5} + (\sqrt{5})^2 = 16 - 8\sqrt{5} + 5 =$
$21 - 8\sqrt{5}$

28. $(4-\sqrt{5})(4+\sqrt{5}) = 4^2 - (\sqrt{5})^2 = 16 - 5 = 11$

29. $\dfrac{10}{4-\sqrt{5}} = \dfrac{10}{4-\sqrt{5}}\cdot\dfrac{4+\sqrt{5}}{4+\sqrt{5}} = \dfrac{10(4+\sqrt{5})}{4^2-(\sqrt{5})^2} = \dfrac{40+10\sqrt{5}}{16-5} =$
$\dfrac{40+10\sqrt{5}}{11}$

30.
$$\begin{aligned} a^2+b^2 &= c^2\\ 8^2+4^2 &= c^2\\ 64+16 &= c^2\\ 80 &= c^2\\ \sqrt{80} &= c \quad \text{Exact answer}\\ 8.944 &\approx c \quad \text{Approximation}\end{aligned}$$

31.
$$\begin{aligned} \sqrt{3x}+2 &= 14\\ \sqrt{3x} &= 12\\ (\sqrt{3x})^2 &= 12^2\\ 3x &= 144\\ x &= 48\end{aligned}$$
The number 48 checks. It is the solution.

32.
$$\begin{aligned} \sqrt{6x+13} &= x+3\\ (\sqrt{6x+13})^2 &= (x+3)^2\\ 6x+13 &= x^2+6x+9\\ 0 &= x^2-4\\ 0 &= (x+2)(x-2)\end{aligned}$$
$x+2=0 \quad or \quad x-2=0$
$x=-2 \quad or \quad x=2$
Both numbers check. The solutions are -2 and 2.

33.
$$\begin{aligned} \sqrt{1-x}+1 &= \sqrt{6-x}\\ (\sqrt{1-x}+1)^2 &= (\sqrt{6-x})^2\\ 1-x+2\sqrt{1-x}+1 &= 6-x\\ 2-x+2\sqrt{1-x} &= 6-x\\ 2\sqrt{1-x} &= 4\\ \sqrt{1-x} &= 2\\ (\sqrt{1-x})^2 &= 2^2\\ 1-x &= 4\\ -x &= 3\\ x &= -3\end{aligned}$$
The number -3 checks. It is the solution.

34. a) Substitute 28,000 for h in the formula.
$D = \sqrt{2\cdot 28{,}000} = \sqrt{56{,}000} \approx 237$ mi
b) Substitute 261 for D and solve for h.
$$\begin{aligned} 261 &= \sqrt{2h}\\ 261^2 &= (\sqrt{2h})^2\\ 68{,}121 &= 2h\\ 34{,}060.5 &= h\end{aligned}$$
The airplane is 34,060.5 ft high.

35. Let $d =$ the length of a diagonal, in yd.
$$\begin{aligned} a^2+b^2 &= c^2\\ 60^2+110^2 &= d^2\\ 3600+12{,}100 &= d^2\\ 15{,}700 &= d^2\\ \sqrt{15{,}700}\text{ yd} &= d\\ 125.300\text{ yd} &\approx d\end{aligned}$$

36. $\sqrt{\dfrac{2a}{5b}} = \sqrt{\dfrac{2a}{5b}\cdot\dfrac{5b}{5b}} = \sqrt{\dfrac{10ab}{25b^2}} = \dfrac{\sqrt{10ab}}{\sqrt{25b^2}} = \dfrac{\sqrt{10ab}}{5b}$
Answer A is correct.

37. $\sqrt{\sqrt{\sqrt{625}}} = \sqrt{\sqrt{25}} = \sqrt{5}$

38. $\sqrt{y^{16n}} = \sqrt{(y^{8n})^2} = y^{8n}$

Cumulative Review Chapters 1 - 8

1. $\dfrac{x-6}{2x+1}$
We set the denominator equal to 0 and solve for x.
$$\begin{aligned} 2x+1 &= 0\\ 2x &= -1\\ x &= -\frac{1}{2}\end{aligned}$$
The expression is not defined for the replacement number $-\dfrac{1}{2}$.

2. No, because the radicand is negative

3. $(2+\sqrt{3})(2-\sqrt{3}) = 2^2 - (\sqrt{3})^2 = 4-3 = 1$

4. $\sqrt{196} = 14$, so $-\sqrt{196} = -14$.

5. $\sqrt{3}\sqrt{75} = \sqrt{3\cdot 75} = \sqrt{3\cdot 3\cdot 25} = \sqrt{3\cdot 3}\sqrt{25} = 3\cdot 5 = 15$

6. $(1-\sqrt{2})^2 = 1^2 - 2\cdot 1\cdot\sqrt{2} + (\sqrt{2})^2 = 1-2\sqrt{2}+2 = 3-2\sqrt{2}$

7. $\dfrac{\sqrt{162}}{\sqrt{125}} = \dfrac{\sqrt{81\cdot 2}}{\sqrt{25\cdot 5}} = \dfrac{9\sqrt{2}}{5\sqrt{5}} = \dfrac{9\sqrt{2}}{5\sqrt{5}}\cdot\dfrac{\sqrt{5}}{\sqrt{5}} = \dfrac{9\sqrt{10}}{5\cdot 5} = \dfrac{9\sqrt{10}}{25}$

8.
$$\begin{aligned} 2\sqrt{45}+3\sqrt{20} &= 2\sqrt{9\cdot 5}+3\sqrt{4\cdot 5}\\ &= 2\cdot 3\sqrt{5}+3\cdot 2\sqrt{5}\\ &= 6\sqrt{5}+6\sqrt{5}\\ &= 12\sqrt{5}\end{aligned}$$

9. $(3x^4-2y^5)(3x^4+2y^5) = (3x^4)^2 - (2y^5)^2 = 9x^8 - 4y^{10}$

10. $(x^2+4)^2 = (x^2)^2 + 2\cdot x^2\cdot 4 + 4^2 = x^4 + 8x^2 + 16$

11. $\left(2x+\dfrac{1}{4}\right)\left(4x-\dfrac{1}{2}\right) = 2x\cdot 4x - 2x\cdot\dfrac{1}{2} + \dfrac{1}{4}\cdot 4x - \dfrac{1}{4}\cdot\dfrac{1}{2} =$
$8x^2 - x + x - \dfrac{1}{8} = 8x^2 - \dfrac{1}{8}$

12. $\dfrac{x}{2x-1}-\dfrac{3x+2}{1-2x}=\dfrac{x}{2x-1}-\dfrac{3x+2}{1-2x}\cdot\dfrac{-1}{-1}$

$$=\frac{x}{2x-1}-\frac{-3x-2}{2x-1}$$
$$=\frac{x-(-3x-2)}{2x-1}$$
$$=\frac{x+3x+2}{2x-1}$$
$$=\frac{4x+2}{2x-1}$$
$$=\frac{2(2x+1)}{2x-1}$$

13. $(3x^2-2x^3)-(x^3-2x^2+5)+(3x^2-5x+5)$

$=3x^2-2x^3-x^3+2x^2-5+3x^2-5x+5$

$=-3x^3+8x^2-5x$

14. $\dfrac{2x+2}{3x-9}\cdot\dfrac{x^2-8x+15}{x^2-1}=\dfrac{(2x+2)(x^2-8x+15)}{(3x-9)(x^2-1)}$

$$=\frac{2(x+1)(x-3)(x-5)}{3(x-3)(x+1)(x-1)}$$
$$=\frac{2\cancel{(x+1)}\cancel{(x-3)}(x-5)}{3\cancel{(x-3)}\cancel{(x+1)}(x-1)}$$
$$=\frac{2(x-5)}{3(x-1)}$$

15. $\dfrac{2x^2-2}{2x^2+7x+3}\div\dfrac{4x-4}{2x^2-5x-3}$

$$=\frac{2x^2-2}{2x^2+7x+3}\cdot\frac{2x^2-5x-3}{4x-4}$$
$$=\frac{(2x^2-2)(2x^2-5x-3)}{(2x^2+7x+3)(4x-4)}$$
$$=\frac{2(x+1)(x-1)(2x+1)(x-3)}{(2x+1)(x+3)(4)(x-1)}$$
$$=\frac{\cancel{2}(x+1)\cancel{(x-1)}\cancel{(2x+1)}(x-3)}{\cancel{(2x+1)}(x+3)(\cancel{2}\cdot 2)\cancel{(x-1)}}$$
$$=\frac{(x+1)(x-3)}{2(x+3)}$$

16.

$$\begin{array}{r|l} & 3x^2+4x+9 \\ \hline x-2 & 3x^3-2x^2+x-5 \\ & \underline{3x^3-6x^2} \\ & \qquad 4x^2+x \\ & \qquad \underline{4x^2-8x} \\ & \qquad\qquad 9x-5 \\ & \qquad\qquad \underline{9x-18} \\ & \qquad\qquad\quad 13 \end{array}$$

The answer is $3x^2+4x+9+\dfrac{13}{x-2}$.

17. $\sqrt{2x^2-4x+2}=\sqrt{2(x^2-2x+1)}=\sqrt{2(x-1)^2}=$
$\sqrt{2}\sqrt{(x-1)^2}=\sqrt{2}(x-1)$

18. $x^{-9}\cdot x^{-3}=x^{-9+(-3)}=x^{-12}=\dfrac{1}{x^{12}}$

19. $\sqrt{\dfrac{50}{2x^8}}=\sqrt{\dfrac{25}{x^8}}=\dfrac{\sqrt{25}}{\sqrt{x^8}}=\dfrac{5}{x^4}$

20. $\dfrac{x-\dfrac{1}{x}}{1-\dfrac{x-1}{2x}}=\dfrac{x-\dfrac{1}{x}}{1-\dfrac{x-1}{2x}}\cdot\dfrac{2x}{2x}$

$$=\frac{2x\cdot x-2x\cdot\dfrac{1}{x}}{2x\cdot 1-2x\left(\dfrac{x-1}{2x}\right)}$$
$$=\frac{2x^2-2}{2x-(x-1)}$$
$$=\frac{2x^2-2}{2x-x+1}$$
$$=\frac{2(x^2-1)}{x+1}$$
$$=\frac{2(x+1)(x-1)}{x+1}$$
$$=\frac{2\cancel{(x+1)}(x-1)}{\cancel{(x+1)}\cdot 1}$$
$$=2(x-1)$$

21. $3-12x^8=3(1-4x^8)$

$=3(1^2-(2x^4)^2)$

$=3(1+2x^4)(1-2x^4)$

22. $12t-4t^2-48t^4=4t(3-t-12t^3)$

23. $6x^2-28x+16=2(3x^2-14x+8)=2(3x-2)(x-4)$

24. $4x^3+4x^2-x-1=(4x^3+4x^2)+(-x-1)$

$=4x^2(x+1)-(x+1)$

$=(x+1)(4x^2-1)$

$=(x+1)(2x+1)(2x-1)$

25. $16x^4-56x^2+49=(4x^2)^2-2\cdot 4x^2\cdot 7+7^2=(4x^2-7)^2$

26. $x^2+3x-180=(x+15)(x-12)$

27. $x^2=-17x$

$x^2+17x=0$

$x(x+17)=0$

$x=0$ *or* $x+17=0$

$x=0$ *or* $x=-17$

The solutions are 0 and -17.

28. $-4(x+5)\geq 2(x+5)-3$

$-4x-20\geq 2x+10-3$

$-4x-20\geq 2x+7$

$-6x-20\geq 7$

$-6x\geq 27$

$\dfrac{-6x}{-6}\leq\dfrac{27}{-6}$ Reversing the inequality symbol

$x\leq -\dfrac{9}{2}$

The solution set is $\left\{x\middle|x\leq-\dfrac{9}{2}\right\}$.

29. $\frac{1}{x}+\frac{2}{3}=\frac{1}{4}$, LCM $=12x$

$$12x\left(\frac{1}{x}+\frac{2}{3}\right)=12x\cdot\frac{1}{4}$$
$$12x\cdot\frac{1}{x}+12x\cdot\frac{2}{3}=3x$$
$$12+8x=3x$$
$$12=-5x$$
$$-\frac{12}{5}=x$$

The number $-\frac{12}{5}$ checks. It is the solution.

30. $x^2-30=x$

$$x^2-x-30=0$$
$$(x-6)(x+5)=0$$
$$x-6=0 \quad or \quad x+5=0$$
$$x=6 \quad or \quad x=-5$$

The solutions are 6 and -5.

31. $\sqrt{2x-1}+5=14$

$$\sqrt{2x-1}=9$$
$$(\sqrt{2x-1})^2=9^2$$
$$2x-1=81$$
$$2x=82$$
$$x=41$$

The number 41 checks. It is the solution.

32. $\frac{1}{4}x+\frac{2}{3}x=\frac{2}{3}-\frac{3}{4}x$, LCM is 12

$$12\left(\frac{1}{4}x+\frac{2}{3}x\right)=12\left(\frac{2}{3}-\frac{3}{4}x\right)$$
$$12\cdot\frac{1}{4}x+12\cdot\frac{2}{3}x=12\cdot\frac{2}{3}-12\cdot\frac{3}{4}x$$
$$3x+8x=8-9x$$
$$11x=8-9x$$
$$20x=8$$
$$x=\frac{2}{5}$$

The solution is $\frac{2}{5}$.

33. $\frac{x}{x-1}-\frac{x}{x+1}=\frac{1}{2x-2}$

$$\frac{x}{x-1}-\frac{x}{x+1}=\frac{1}{2(x-1)},$$

LCM is $2(x-1)(x+1)$

$$2(x-1)(x+1)\left(\frac{x}{x-1}-\frac{x}{x+1}\right)=2(x-1)(x+1)\cdot\frac{1}{2(x-1)}$$
$$2(x-1)(x+1)\cdot\frac{x}{x-1}-$$
$$2(x-1)(x+1)\cdot\frac{x}{x+1}=x+1$$
$$2x(x+1)-2x(x-1)=x+1$$
$$2x^2+2x-2x^2+2x=x+1$$
$$4x=x+1$$
$$3x=1$$
$$x=\frac{1}{3}$$

The number $\frac{1}{3}$ checks. It is the solution.

34. $x=y+3,\quad (1)$

$3y-4x=-13\quad (2)$

First substitute $y+3$ for x in Equation (2) and solve for y.

$$3y-4x=-13$$
$$3y-4(y+3)=-13$$
$$3y-4y-12=-13$$
$$-y-12=-13$$
$$-y=-1$$
$$y=1$$

Now substitute 1 for y in Equation (1) and find x.

$$x=y+3=1+3=4$$

The ordered pair $(4,1)$ checks in both equations. It is the solution.

35. $2x-3y=30,$

$5y-2x=-46$

First we rewrite the second equation in the form $Ax+By=C$. Then we add.

$$\begin{array}{rl} 2x-3y=30, & (1)\\ \underline{-2x+5y=-46} & (2)\\ 2y=-16 & \\ y=-8 & \end{array}$$

Now substitute -8 for y in one of the equations and solve for x.

$$2x-3y=30\quad (1)$$
$$2x-3(-8)=30$$
$$2x+24=30$$
$$2x=6$$
$$x=3$$

The ordered pair $(3,-8)$ checks in both equations. It is the solution.

36.

$$\begin{aligned} 4A &= pr + pq \\ 4A &= p(r+q) \\ \frac{4A}{r+q} &= p \end{aligned}$$

37. $3y - 3x > -6$

First we graph the line $3y - 3x = -6$. The intercepts are $(0, -2)$ and $(2, 0)$. We draw a dashed line since the inequality symbol is $>$. Then we test a point that is not on the line. We use $(0, 0)$.

$$\begin{array}{c|l} 3y - 3x > -6 & \\ \hline 3 \cdot 0 - 3 \cdot 0 \ ? \ -6 & \\ 0 - 0 & \\ 0 & \quad \text{TRUE} \end{array}$$

We see that $(0, 0)$ is a solution of the inequality so we shade the half-plane that contains $(0, 0)$.

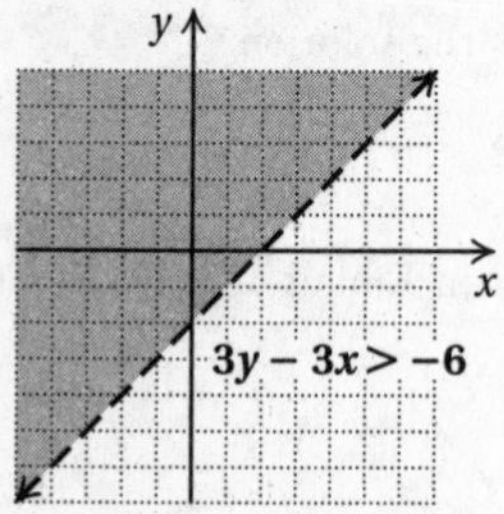

38. $x = 5$, or $x + 0 \cdot y = 5$

Any ordered pair $(5, y)$ is a solution of the equation. The graph is a vertical line with x-intercept $(5, 0)$.

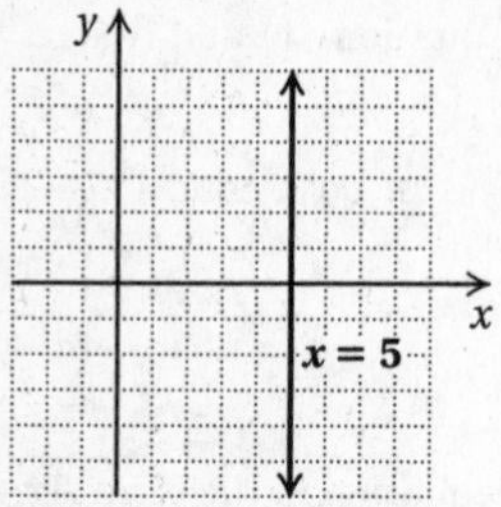

39. $2x - 6y = 12$

First we will find the intercepts. To find the y-intercept we let $x = 0$ and solve for y.

$$\begin{aligned} 2 \cdot 0 - 6y &= 12 \\ -6y &= 12 \\ y &= -2 \end{aligned}$$

The y-intercept is $(0, 2)$.

To find the x-intercept, let $y = 0$ and solve for x.

$$\begin{aligned} 2x - 6 \cdot 0 &= 12 \\ 2x &= 12 \\ x &= 6 \end{aligned}$$

The x-intercept is $(6, 0)$.

We plot the intercepts and draw the graph.

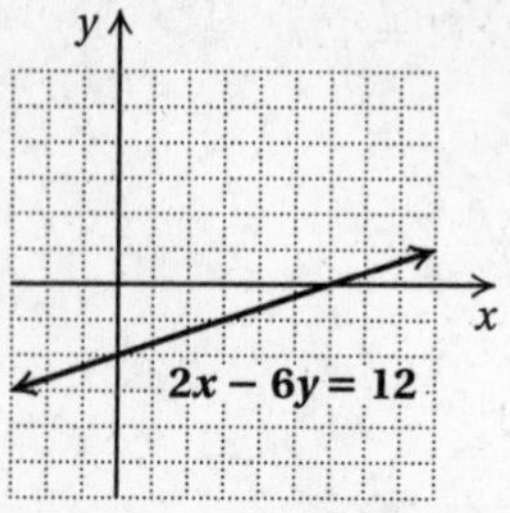

We can find a third point as a check. We let $x = 3$ and solve for y.

$$\begin{aligned} 2 \cdot 3 - 6y &= 12 \\ 6 - 6y &= 12 \\ -6y &= 6 \\ y &= -1 \end{aligned}$$

The point $(3, -1)$ appears to lie on the graph, so the graph is probably correct.

40. First we find the slope.

$$m = \frac{-2-9}{1-5} = \frac{-11}{-4} = \frac{11}{4}$$

Then the equation is $y = \frac{11}{4}x + b$. Using the point $(5, 9)$, we substitute 5 for x and 9 for y and solve for b.

$$\begin{aligned} y &= \frac{11}{4}x + b \\ 9 &= \frac{11}{4} \cdot 5 + b \\ 9 &= \frac{55}{4} + b \\ -\frac{19}{4} &= b \end{aligned}$$

Then the equation is $y = \frac{11}{4}x - \frac{19}{4}$.

41. First we solve for y.

$$\begin{aligned} 5x - 3y &= 9 \\ -3y &= -5x + 9 \\ -\frac{1}{3}(-3y) &= -\frac{1}{3}(-5x + 9) \\ y &= \frac{5}{3}x - 3 \end{aligned}$$

Thus, the slope is $\frac{5}{3}$ and the y-intercept is $(0, -3)$.

42. When $x = -2$, $y = (-2)^3 - 4(-2) - 2 = -8 + 8 - 2 = -2$.
When $x = -1$, $y = (-1)^3 - 4(-1) - 2 = -1 + 4 - 2 = 1$.
When $x = 0$, $y = 0^3 - 4 \cdot 0 - 2 = 0 - 0 - 2 = -2$.
When $x = 1$, $y = 1^3 - 4 \cdot 1 - 2 = 1 - 4 - 2 = -5$.
When $x = 2$, $y = 2^3 - 4 \cdot 2 - 2 = 8 - 8 - 2 = -2$.
We could also estimate these values from the graph.

43. ***Familiarize***. Let h = the cost of one hamburger and m = the cost of one milkshake.

Translate.

$$\underbrace{\text{6 hamburgers}}_{6h} \ \underset{+}{\text{plus}} \ \underbrace{\text{4 milkshakes}}_{4m} \ \underset{=}{\text{cost}} \ \underset{27.70}{\$27.70.}$$

$$\underbrace{\text{3 hamburgers}}_{3h} \text{ plus } \underbrace{\text{1 milkshake}}_{1\cdot m} \text{ cost } \$11.35.$$

$$3h \quad + \quad 1\cdot m \quad = \quad 11.35$$

After clearing decimals we have the following system of equations.

$$60h + 40m = 277, \quad (1)$$
$$300h + 100m = 1135 \quad (2)$$

Solve. We multiply Equation (1) by -5 and then add.

$$-300h - 200m = -1385$$
$$300h + 100m = 1135$$
$$\overline{\quad - 100m = -250}$$
$$m = 2.5$$

Now substitute 2.5 for m in one of the original equations and solve for h.

$$60h + 40m = 277 \quad (1)$$
$$60h + 40(2.5) = 277$$
$$60h + 100 = 277$$
$$60h = 177$$
$$h = 2.95$$

Check. If one hamburger costs \$2.95 and one milkshake costs \$2.50, then 6 hamburgers and 4 milkshakes cost $6(\$2.95) + 4(\$2.50) = \$17.70 + \$10.00 = \$27.70$ and 3 hamburgers and 1 milkshake cost $3(\$2.95) + \$2.50 = \$8.85 + \$2.50 = \$11.35$. The answer checks.

State. A hamburger costs \$2.95 and a milkshake costs \$2.50.

44. ***Familiarize and Translate.*** We have inverse variation between A and d, so an equation $A = k/d$, $k > 0$, applies.

Solve. First we find an equation of variation.

$$A = \frac{k}{d}$$
$$4 = \frac{k}{100}$$
$$400 = k$$

The equation of variation is $A = \dfrac{400}{d}$. Now find A when $d = 1000$.

$$A = \frac{400}{d}$$
$$A = \frac{400}{1000}$$
$$A = 0.4$$

Check. We could repeat the computation. Also, observe that, when the distance from the stage increased, the apparent size of the musicians decreased as we would expect with inverse variation.

State. If you were sitting 1000 ft from the stage, the musicians would appear to be 0.4 ft tall.

45. ***Familiarize.*** Let x = the measure of the first angle. Then $2x$ = the measure of the second angle and $x + 2x - 48$, or $3x - 48$ = the measure of the third angle. Recall that the sum of the measures of the angles of a triangle is 180°.

Translate.

$$\underbrace{\text{The sum of the measures}}_{x + 2x + (3x - 48)} \text{ is } 180°.$$

$$x + 2x + (3x - 48) \quad = \quad 180$$

Solve.

$$x + 2x + (3x - 48) = 180$$
$$6x - 48 = 180$$
$$6x = 228$$
$$x = 38$$

If $x = 38$, then $2x = 2 \cdot 38 = 76$ and $3x - 48 = 3 \cdot 38 - 48 = 66$.

Check. If the angle measures are 38°, 76°, and 66°, then the measure of the second angle is twice the measure of the first angle and the measure of the third angle is 48° less than the sum of the measures of the other two angles. Also, $38° + 76° + 66° = 180°$. The answer checks.

State. The angle measures are 38°, 76°, and 66°.

46. ***Familiarize.*** Let d = the number of defective resistors you would expect to find in a sample of 250.

Translate. We translate to a proportion.

$$\begin{array}{r} \text{Number defective} \rightarrow \\ \text{Sample size} \rightarrow \end{array} \frac{12}{150} = \frac{d}{250} \begin{array}{l} \leftarrow \text{Number defective} \\ \leftarrow \text{Sample size} \end{array}$$

Solve.

$$\frac{12}{150} = \frac{d}{250}$$
$$250 \cdot \frac{12}{150} = 250 \cdot \frac{d}{250}$$
$$20 = d$$

Check.

$$\frac{12}{150} = 0.08 \text{ and } \frac{20}{250} = 0.08.$$

The ratios are the same, so the answer is correct.

State. You would expect to find 20 defective resistors in a sample of 250.

47. ***Familiarize.*** Let w = the width of the rectangle, in meters. Then $w + 3$ = the length. Recall that the area of a rectangle is (length) × (width).

Translate.

$$\underbrace{\text{The area}}_{(w+3)w} \text{ is } 180 \text{ m}^2.$$

$$(w + 3)w \quad = \quad 180$$

Solve.

$$(w + 3)w = 180$$
$$w^2 + 3w = 180$$
$$w^2 + 3w - 180 = 0$$
$$(w + 15)(w - 12) = 0$$
$$w + 15 = 0 \quad \textit{or} \quad w - 12 = 0$$
$$w = -15 \quad \textit{or} \quad w = 12$$

Check. The width cannot be negative, so -15 cannot be a solution. If $w = 12$, then $w + 3 = 12 + 3 = 15$. If the width is 12 m and the length is 15 m, then the length is

3 m more than the width and the area is $15 \cdot 12$, or 180 m^2. The answer checks.

State. The length is 15 m and the width is 12 m.

48. ***Familiarize***. Let $d =$ the number of dimes and $q =$ the number of quarters. Then the value of the dimes is $0.1d$ and the value of the quarters is $0.25q$.

Translate.

$$\underbrace{\text{Value of dimes}}_{\downarrow\ 0.1d} \ \underset{+}{\text{plus}} \ \underbrace{\text{value of quarters}}_{\downarrow\ 0.25q} \ \underset{=}{\text{is}} \ \underset{19}{\$19.00.}$$

$$\underbrace{\text{Number of dimes}}_{\downarrow\ d} \ \underset{+}{\text{plus}} \ \underbrace{\text{number of quarters}}_{\downarrow\ q} \ \underset{=}{\text{is}} \ \underset{115}{115.}$$

After clearing decimals we have the following system of equations.

$$10d + 25q = 1900, \quad (1)$$
$$d + \quad q = \quad 115 \quad (2)$$

Solve. We multiply Equation (2) by -10 and then add.

$$\begin{aligned} 10d + 25q &= 1900 \\ -10d - 10q &= -1150 \\ \hline 15q &= 750 \\ q &= 50 \end{aligned}$$

Now substitute 50 for q in one of the equations and solve for d.

$$\begin{aligned} d + q &= 115 \quad (2) \\ d + 50 &= 115 \\ d &= 65 \end{aligned}$$

Check. If there are 65 dimes and 50 quarters, there are $65 + 50$, or 115, coins in all. The value of the coins is $\$0.1(65) + \$0.25(50) = \$6.50 + \$12.50 = \$19.00$. The answer checks.

State. There are 65 dimes and 50 quarters.

49. ***Familiarize***. Let $x =$ the amount originally invested. Using the formula for simple interest, $I = Prt$, we find that the simple interest for 1 yr is $x \cdot 4.5\% \cdot 1$, or $0.045x$.

Translate.

$$\underbrace{\text{Amount invested}}_{\downarrow\ x} \ \underset{+}{\text{plus}} \ \underset{0.045x}{\text{interest}} \ \underset{=}{\text{is}} \ \underset{2717}{\$2717.}$$

Solve.

$$\begin{aligned} x + 0.045x &= 2717 \\ 1.045x &= 2717 \\ x &= 2600 \end{aligned}$$

Check. 4.5% of \$2600 is 0.045(\$2600), or \$117, and $\$2600 + \$117 = \$2717$. The answer checks.

State. \$2600 was originally invested.

50. ***Familiarize***. Let $r =$ the faster speed. Then $r - 10 =$ the slower speed. We organize the information in a table.

	Distance	Speed	Time
Faster trip	600	r	t_1
Slower trip	600	$r - 10$	t_2

Translate. We use the formula $d = rt$ in each row of the table to get two equations.

$$600 = rt_1 \quad (1)$$
$$600 = (r - 10)t_2 \quad (2)$$

Since $600 = rt_1$, we have $t_1 = \dfrac{600}{r}$. Also, since $600 = (r - 10)t_2$, we have $t_2 = \dfrac{600}{r - 10}$. We know that $t_2 = t_1 + 2$, so we can write the following equation:

$$\frac{600}{r - 10} = \frac{600}{r} + 2.$$

Solve.

$$\begin{aligned} \frac{600}{r - 10} &= \frac{600}{r} + 2 \\ r(r - 10) \cdot \frac{600}{r - 10} &= r(r - 10)\left(\frac{600}{r} + 2\right) \\ 600r &= r(r - 10) \cdot \frac{600}{r} + r(r - 10)(2) \\ 600r &= 600(r - 10) + 2r(r - 10) \\ 600r &= 600r - 6000 + 2r^2 - 20r \\ 0 &= 2r^2 - 20r - 6000 \\ 0 &= 2(r^2 - 10r - 3000) \\ 0 &= r^2 - 10r - 3000 \quad \text{Dividing by 2} \\ 0 &= (r + 50)(r - 60) \end{aligned}$$

$$r + 50 = 0 \quad \text{or} \quad r - 60 = 0$$
$$r = -50 \quad \text{or} \quad r = 60$$

Check. The speed cannot be negative, so -50 cannot be a solution. At a speed of 60 mph it takes 600/60, or 10 hr, to drive 60 miles. At a speed of $60 - 10$, or 50 mph, it takes 600/50, or 12 hr, to drive 600 mi. Since the second time is 2 hr longer than the first, the answer checks.

State. The speed going is 60 mph.

51. ***Familiarize***. Let $x =$ the amount of pure water that should be added to make a 12% salt solution and let $y =$ the total amount of the mixture. We organize the information in a table.

Solution	Salt	Water	Mixture
Amount	200	x	y
Percent of salt	30%	0%	12%
Amount of salt	$30\% \times 200$, or 60	$0\% \times w$, or 0	$12\% \times y$, or $0.12y$

Translate. From the first and last rows of the table we get two equations.

$$200 + x = y, \quad (1)$$
$$60 + 0 = 0.12y, \text{ or } 60 = 0.12y \quad (2)$$

Solve. First we solve Equation (2) for y.

$$60 = 0.12y$$
$$500 = y$$

Substitute 500 for y in Equation (1) and solve for x.

$$200 + x = y$$
$$200 + x = 500$$
$$x = 300$$

Check. If 300 L of pure water is added, the mixture is $200 + 300$ or 500 L and it contains the 60 L of salt from the 200 L of the 30% salt solution. Since $60/500 = 0.12$, or 12%, the mixture is a 12% salt solution. The answer checks.

State. 300 L of pure water should be added.

52. $\sqrt{x} + 1 = y, \quad (1)$

$\sqrt{x} + \sqrt{y} = 5 \quad (2)$

First solve Equation (1) for $\sqrt{x}$.

$$\sqrt{x} + 1 = y$$
$$\sqrt{x} = y - 1 \quad (3)$$

Now substitute $y - 1$ for $\sqrt{x}$ in Equation (2) and solve for y.

$$\sqrt{x} + \sqrt{y} = 5$$
$$y - 1 + \sqrt{y} = 5$$
$$\sqrt{y} = 6 - y$$
$$(\sqrt{y})^2 = (6 - y)^2$$
$$y = 36 - 12y + y^2$$
$$0 = y^2 - 13y + 36$$
$$0 = (y - 4)(y - 9)$$
$$y - 4 = 0 \quad or \quad y - 9 = 0$$
$$y = 4 \quad or \quad y = 9$$

Substitute each value for y in Equation (3) and find x.

For 4:
$$\sqrt{x} = 4 - 1$$
$$\sqrt{x} = 3$$
$$(\sqrt{x})^2 = 3^2$$
$$x = 9$$

For 9:
$$\sqrt{x} = 9 - 1$$
$$\sqrt{x} = 8$$
$$(\sqrt{x})^2 = 8^2$$
$$x = 64$$

The ordered pair $(64, 9)$ does not check, but $(9, 4)$ does. The solution is $(9, 4)$.

Chapter 9

Quadratic Equations

Exercise Set 9.1

RC1. False; the solutions of $x^2 - x - 12 = 0$ are the *first* coordinates of the x-intercepts of the graph of $y = x^2 - x - 12$.

RC3. True; one factor of $85x^2 - 96x$ is x, so the solution of the equation $x = 0$, or 0, is a solution of the equation $85x^2 - 96x = 0$.

1. $x^2 - 3x + 2 = 0$

This equation is already in standard form.

$a = 1,\ b = -3,\ c = 2$

3. $7x^2 = 4x - 3$

$7x^2 - 4x + 3 = 0$ Standard form

$a = 7,\ b = -4,\ c = 3$

5. $5 = -2x^2 + 3x$

$2x^2 - 3x + 5 = 0$ Standard form

$a = 2,\ b = -3,\ c = 5$

7.
$$x^2 + 5x = 0$$
$$x(x + 5) = 0$$
$$x = 0 \text{ or } x + 5 = 0$$
$$x = 0 \text{ or } \quad x = -5$$

The solutions are 0 and -5.

9.
$$3x^2 + 6x = 0$$
$$3x(x + 2) = 0$$
$$3x = 0 \text{ or } x + 2 = 0$$
$$x = 0 \text{ or } \quad x = -2$$

The solutions are 0 and -2.

11.
$$5x^2 = 2x$$
$$5x^2 - 2x = 0$$
$$x(5x - 2) = 0$$
$$x = 0 \text{ or } 5x - 2 = 0$$
$$x = 0 \text{ or } \quad 5x = 2$$
$$x = 0 \text{ or } \quad x = \frac{2}{5}$$

The solutions are 0 and $\frac{2}{5}$.

13.
$$4x^2 + 4x = 0$$
$$4x(x + 1) = 0$$
$$4x = 0 \text{ or } x + 1 = 0$$
$$x = 0 \text{ or } \quad x = -1$$

The solutions are 0 and -1.

15.
$$0 = 10x^2 - 30x$$
$$0 = 10x(x - 3)$$
$$10x = 0 \text{ or } x - 3 = 0$$
$$x = 0 \text{ or } \quad x = 3$$

The solutions are 0 and 3.

17.
$$11x = 55x^2$$
$$0 = 55x^2 - 11x$$
$$0 = 11x(5x - 1)$$
$$11x = 0 \text{ or } 5x - 1 = 0$$
$$x = 0 \text{ or } \quad 5x = 1$$
$$x = 0 \text{ or } \quad x = \frac{1}{5}$$

The solutions are 0 and $\frac{1}{5}$.

19.
$$14t^2 = 3t$$
$$14t^2 - 3t = 0$$
$$t(14t - 3) = 0$$
$$t = 0 \text{ or } 14t - 3 = 0$$
$$t = 0 \text{ or } \quad 14t = 3$$
$$t = 0 \text{ or } \quad t = \frac{3}{14}$$

The solutions are 0 and $\frac{3}{14}$.

21.
$$5y^2 - 3y^2 = 72y + 9y$$
$$2y^2 = 81y$$
$$2y^2 - 81y = 0$$
$$y(2y - 81) = 0$$
$$y = 0 \text{ or } 2y - 81 = 0$$
$$y = 0 \text{ or } \quad 2y = 81$$
$$y = 0 \text{ or } \quad y = \frac{81}{2}$$

The solutions are 0 and $\frac{81}{2}$.

23.
$$x^2 + 8x - 48 = 0$$
$$(x + 12)(x - 4) = 0$$
$$x + 12 = 0 \quad \text{or } x - 4 = 0$$
$$x = -12 \text{ or } \quad x = 4$$

The solutions are -12 and 4.

25.
$$5 + 6x + x^2 = 0$$
$$(5 + x)(1 + x) = 0$$
$$5 + x = 0 \quad \text{or } 1 + x = 0$$
$$x = -5 \text{ or } \quad x = -1$$

The solutions are -5 and -1.

27. $18 = 7p + p^2$
$0 = p^2 + 7p - 18$
$0 = (p+9)(p-2)$
$p+9=0 \quad or \quad p-2=0$
$p=-9 \quad or \quad p=2$
The solutions are -9 and 2.

29. $-15 = -8y + y^2$
$0 = y^2 - 8y + 15$
$0 = (y-5)(y-3)$
$y-5=0 \quad or \quad y-3=0$
$y=5 \quad or \quad y=3$
The solutions are 5 and 3.

31. $x^2 + 10x + 25 = 0$
$(x+5)(x+5) = 0$
$x+5=0 \quad or \quad x+5=0$
$x=-5 \quad or \quad x=-5$
The solution is -5.

33. $r^2 = 8r - 16$
$r^2 - 8r + 16 = 0$
$(r-4)(r-4) = 0$
$r-4=0 \quad or \quad r-4=0$
$r=4 \quad or \quad r=4$
The solution is 4.

35. $6x^2 + x - 2 = 0$
$(3x+2)(2x-1) = 0$
$3x+2=0 \quad or \quad 2x-1=0$
$3x=-2 \quad or \quad 2x=1$
$x=-\frac{2}{3} \quad or \quad x=\frac{1}{2}$
The solutions are $-\frac{2}{3}$ and $\frac{1}{2}$.

37. $3a^2 = 10a + 8$
$3a^2 - 10a - 8 = 0$
$(3a+2)(a-4) = 0$
$3a+2=0 \quad or \quad a-4=0$
$3a=-2 \quad or \quad a=4$
$a=-\frac{2}{3} \quad or \quad a=4$
The solutions are $-\frac{2}{3}$ and 4.

39. $6x^2 - 4x = 10$
$6x^2 - 4x - 10 = 0$
$2(3x^2 - 2x - 5) = 0$
$2(3x-5)(x+1) = 0$
$3x-5=0 \quad or \quad x+1=0$
$3x=5 \quad or \quad x=-1$
$x=\frac{5}{3} \quad or \quad x=-1$
The solutions are $\frac{5}{3}$ and -1.

41. $2t^2 + 12t = -10$
$2t^2 + 12t + 10 = 0$
$2(t^2 + 6t + 5) = 0$
$2(t+5)(t+1) = 0$
$t+5=0 \quad or \quad t+1=0$
$t=-5 \quad or \quad t=-1$
The solutions are -5 and -1.

43. $t(t-5) = 14$
$t^2 - 5t = 14$
$t^2 - 5t - 14 = 0$
$(t+2)(t-7) = 0$
$t+2=0 \quad or \quad t-7=0$
$t=-2 \quad or \quad t=7$
The solutions are -2 and 7.

45. $t(9+t) = 4(2t+5)$
$9t + t^2 = 8t + 20$
$t^2 + t - 20 = 0$
$(t+5)(t-4) = 0$
$t+5=0 \quad or \quad t-4=0$
$t=-5 \quad or \quad t=4$
The solutions are -5 and 4.

47. $16(p-1) = p(p+8)$
$16p - 16 = p^2 + 8p$
$0 = p^2 - 8p + 16$
$0 = (p-4)(p-4)$
$p-4=0 \quad or \quad p-4=0$
$p=4 \quad or \quad p=4$
The solution is 4.

49. $(t-1)(t+3) = t-1$
$t^2 + 2t - 3 = t - 1$
$t^2 + t - 2 = 0$
$(t+2)(t-1) = 0$
$t+2=0 \quad or \quad t-1=0$
$t=-2 \quad or \quad t=1$
The solutions are -2 and 1.

51. $\frac{24}{x-2} + \frac{24}{x+2} = 5$
The LCM is $(x-2)(x+2)$.
$(x-2)(x+2)\left(\frac{24}{x-2} + \frac{24}{x+2}\right) = (x-2)(x+2) \cdot 5$
$(x-2)(x+2) \cdot \frac{24}{x-2} + (x-2)(x+2) \cdot \frac{24}{x+2} = 5(x-2)(x+2)$
$24(x+2) + 24(x-2) = 5(x^2-4)$

$$24x + 48 + 24x - 48 = 5x^2 - 20$$
$$48x = 5x^2 - 20$$
$$0 = 5x^2 - 48x - 20$$
$$0 = (5x + 2)(x - 10)$$

$5x + 2 = 0$ *or* $x - 10 = 0$

$5x = -2$ *or* $x = 10$

$x = -\frac{2}{5}$ *or* $x = 10$

Both numbers check. The solutions are $-\frac{2}{5}$ and 10.

53. $$\frac{1}{x} + \frac{1}{x+6} = \frac{1}{4}$$

The LCM is $4x(x + 6)$.

$$4x(x+6)\left(\frac{1}{x} + \frac{1}{x+6}\right) = 4x(x+6) \cdot \frac{1}{4}$$
$$4x(x+6) \cdot \frac{1}{x} + 4x(x+6) \cdot \frac{1}{x+6} = x(x+6)$$
$$4(x+6) + 4x = x(x+6)$$
$$4x + 24 + 4x = x^2 + 6x$$
$$8x + 24 = x^2 + 6x$$
$$0 = x^2 - 2x - 24$$
$$0 = (x - 6)(x + 4)$$

$x - 6 = 0$ *or* $x + 4 = 0$

$x = 6$ *or* $x = -4$

Both numbers check. The solutions are 6 and -4.

55. $$1 + \frac{12}{x^2 - 4} = \frac{3}{x - 2}$$

The LCM is $(x + 2)(x - 2)$.

$$(x+2)(x-2)\left(1 + \frac{12}{(x+2)(x-2)}\right) = (x+2)(x-2) \cdot \frac{3}{x-2}$$
$$(x+2)(x-2) \cdot 1 + (x+2)(x-2) \cdot \frac{12}{(x+2)(x-2)} = 3(x+2)$$
$$x^2 - 4 + 12 = 3x + 6$$
$$x^2 + 8 = 3x + 6$$
$$x^2 - 3x + 2 = 0$$
$$(x - 2)(x - 1) = 0$$

$x - 2 = 0$ *or* $x - 1 = 0$

$x = 2$ *or* $x = 1$

The number 1 checks, but 2 does not. (It makes the denominators $x^2 - 4$ and $x - 2$ zero.) The solution is 1.

57. $$\frac{r}{r-1} + \frac{2}{r^2 - 1} = \frac{8}{r+1}$$

The LCM is $(r - 1)(r + 1)$.

$$(r-1)(r+1)\left(\frac{r}{r-1} + \frac{2}{(r-1)(r+1)}\right) = (r-1)(r+1) \cdot \frac{8}{r+1}$$
$$(r-1)(r+1) \cdot \frac{r}{r-1} + (r-1)(r+1) \cdot \frac{2}{(r-1)(r+1)} = 8(r-1)$$
$$r(r+1) + 2 = 8(r-1)$$
$$r^2 + r + 2 = 8r - 8$$
$$r^2 - 7r + 10 = 0$$
$$(r - 5)(r - 2) = 0$$

$r - 5 = 0$ *or* $r - 2 = 0$

$r = 5$ *or* $r = 2$

Both numbers check. The solutions are 5 and 2.

59. $$\frac{x-1}{1-x} = -\frac{x+8}{x-8}$$

The LCM is $(1 - x)(x - 8)$.

$$(1-x)(x-8) \cdot \frac{x-1}{1-x} = (1-x)(x-8)\left(-\frac{x+8}{x-8}\right)$$
$$(x-8)(x-1) = -(1-x)(x+8)$$
$$x^2 - 9x + 8 = -(x + 8 - x^2 - 8x)$$
$$x^2 - 9x + 8 = -(-x^2 - 7x + 8)$$
$$x^2 - 9x + 8 = x^2 + 7x - 8$$
$$16 = 16x$$
$$1 = x$$

The number 1 does not check. (It makes the denominator $1 - x$ zero.) There is no solution.

61. $$\frac{5}{y+4} - \frac{3}{y-2} = 4$$

The LCM is $(y + 4)(y - 2)$.

$$(y+4)(y-2)\left(\frac{5}{y+4} - \frac{3}{y-2}\right) = (y+4)(y-2) \cdot 4$$
$$5(y-2) - 3(y+4) = 4(y^2 + 2y - 8)$$
$$5y - 10 - 3y - 12 = 4y^2 + 8y - 32$$
$$2y - 22 = 4y^2 + 8y - 32$$
$$0 = 4y^2 + 6y - 10$$
$$0 = 2(2y^2 + 3y - 5)$$
$$0 = 2(2y + 5)(y - 1)$$

$2y + 5 = 0$ *or* $y - 1 = 0$

$2y = -5$ *or* $y = 1$

$y = -\frac{5}{2}$ *or* $y = 1$

The solutions are $-\frac{5}{2}$ and 1.

63. ***Familiarize.*** We will use the formula

$$d = \frac{n^2 - 3n}{2},$$

where d is the number of diagonals and n is the number of sides.

Translate. We substitute 10 for n.

$$d = \frac{10^2 - 3 \cdot 10}{2}$$

Solve. We do the computation.

$$d = \frac{10^2 - 3 \cdot 10}{2} = \frac{100 - 30}{2} = \frac{70}{2} = 35$$

Check. We can recheck our computation. We can also substitute 35 for d in the original formula and determine whether this yields $n = 10$. Our result checks.

State. A decagon has 35 diagonals.

65. ***Familiarize.*** We will use the formula

$$d = \frac{n^2 - 3n}{2},$$

where d is the number of diagonals and n is the number of sides.

Translate. We substitute 14 for d.

$$14 = \frac{n^2 - 3n}{2}$$

Solve. We solve the equation.

$$\frac{n^2 - 3n}{2} = 14$$

$$n^2 - 3n = 28 \quad \text{Multiplying by 2}$$

$$n^2 - 3n - 28 = 0$$

$$(n-7)(n+4) = 0$$

$$n - 7 = 0 \quad or \quad n + 4 = 0$$

$$n = 7 \quad or \quad n = -4$$

Check. Since the number of sides cannot be negative, -4 cannot be a solution. To check 7, we substitute 7 for n in the original formula and determine if this yields $d = 14$. Our result checks.

State. The polygon has 7 sides.

67. $18t - 9p + 9 = 9(2t - p + 1)$

69. $x^2 + 14x + 49 = x^2 + 2 \cdot x \cdot 7 + 7^2 = (x+7)^2$

71. $t^4 - 81 = (t^2+9)(t^2-9) = (t^2+9)(t+3)(t-3)$

73. $a^2 + ac + a + c = a(a+c) + 1(a+c) = (a+c)(a+1)$

75. $x^2y^2 - 1 = (xy+1)(xy-1)$

77. $3x^4 - 6x^2 - 3 = 3(x^4 - 2x^2 - 1)$

79.

$$4m^2 - (m+1)^2 = 0$$

$$4m^2 - (m^2 + 2m + 1) = 0$$

$$4m^2 - m^2 - 2m - 1 = 0$$

$$3m^2 - 2m - 1 = 0$$

$$(3m+1)(m-1) = 0$$

$$3m + 1 = 0 \quad or \quad m - 1 = 0$$

$$3m = -1 \quad or \quad m = 1$$

$$m = -\frac{1}{3} \quad or \quad m = 1$$

The solutions are $-\frac{1}{3}$ and 1.

81.

$$\sqrt{5}x^2 - x = 0$$

$$x(\sqrt{5}x - 1) = 0$$

$$x = 0 \quad or \quad \sqrt{5}x - 1 = 0$$

$$x = 0 \quad or \quad \sqrt{5}x = 1$$

$$x = 0 \quad or \quad x = \frac{1}{\sqrt{5}}, \text{ or } \frac{\sqrt{5}}{5}$$

The solutions are 0 and $\frac{\sqrt{5}}{5}$.

83. Graph $y_1 = 3x^2 - 7x$ and $y_2 = 20$. Then use the INTERSECT feature to find the first coordinate(s) of the point(s) of intersection. The solutions are 4 and approximately -1.7.

85. Graph $y_1 = 3x^2 + 8x$ and $y_2 = 12x + 15$. Then use the INTERSECT feature to find the first coordinate(s) of the point(s) of intersection. The solutions are 3 and approximately -1.7.

87. Graph $y_1 = (x-2)^2 + 3(x-2)$ and $y_2 = 4$. Then use the INTERSECT feature to find the first coordinate(s) of the point(s) of intersection. The solutions are -2 and 3.

89. Graph $y_1 = 16(x-1)$ and $y_2 = x(x+8)$. Then use the INTERSECT feature to find the first coordinate(s) of the point(s) of intersection. The solution is 4.

Exercise Set 9.2

RC1. False; see Example 2 on page 670 in the text.

RC3. False; we need to add $\left(\frac{1}{2} \cdot 6\right)^2$, or 3^2, or 9.

1. $x^2 = 121$

$$x = 11 \quad or \quad x = -11 \qquad \text{Principle of square roots}$$

The solutions are 11 and -11.

3.

$$5x^2 = 35$$

$$x^2 = 7 \quad \text{Dividing by 5}$$

$$x = \sqrt{7} \quad or \quad x = -\sqrt{7} \qquad \text{Principle of square roots}$$

The solutions are $\sqrt{7}$ and $-\sqrt{7}$.

5.

$$5x^2 = 3$$

$$x^2 = \frac{3}{5}$$

$$x = \sqrt{\frac{3}{5}} \quad or \quad x = -\sqrt{\frac{3}{5}} \qquad \text{Principle of square roots}$$

$$x = \sqrt{\frac{3}{5} \cdot \frac{5}{5}} \quad or \quad x = -\sqrt{\frac{3}{5} \cdot \frac{5}{5}} \qquad \text{Rationalizing denominators}$$

$$x = \frac{\sqrt{15}}{5} \quad or \quad x = -\frac{\sqrt{15}}{5}$$

The solutions are $\frac{\sqrt{15}}{5}$ and $-\frac{\sqrt{15}}{5}$.

7.

$$4x^2 - 25 = 0$$

$$4x^2 = 25$$

$$x^2 = \frac{25}{4}$$

$$x = \frac{5}{2} \quad or \quad x = -\frac{5}{2}$$

The solutions are $\frac{5}{2}$ and $-\frac{5}{2}$.

9. $3x^2 - 49 = 0$
$3x^2 = 49$
$x^2 = \frac{49}{3}$
$x = \frac{7}{\sqrt{3}}$ *or* $x = -\frac{7}{\sqrt{3}}$
$x = \frac{7}{\sqrt{3}} \cdot \frac{\sqrt{3}}{\sqrt{3}}$ *or* $x = -\frac{7}{\sqrt{3}} \cdot \frac{\sqrt{3}}{\sqrt{3}}$
$x = \frac{7\sqrt{3}}{3}$ *or* $x = -\frac{7\sqrt{3}}{3}$

The solutions are $\frac{7\sqrt{3}}{3}$ and $-\frac{7\sqrt{3}}{3}$.

11. $4y^2 - 3 = 9$
$4y^2 = 12$
$y^2 = 3$
$y = \sqrt{3}$ *or* $y = -\sqrt{3}$

The solutions are $\sqrt{3}$ and $-\sqrt{3}$.

13. $49y^2 - 64 = 0$
$49y^2 = 64$
$y^2 = \frac{64}{49}$
$y = \frac{8}{7}$ *or* $y = -\frac{8}{7}$

The solutions are $\frac{8}{7}$ and $-\frac{8}{7}$.

15. $(x+3)^2 = 16$
$x+3 = 4$ *or* $x+3 = -4$ Principle of square roots
$x = 1$ *or* $x = -7$

The solutions are 1 and -7.

17. $(x+3)^2 = 21$
$x+3 = \sqrt{21}$ *or* $x+3 = -\sqrt{21}$ Principle of square roots
$x = -3+\sqrt{21}$ *or* $x = -3-\sqrt{21}$

The solutions are $-3+\sqrt{21}$ and $-3-\sqrt{21}$, or $-3 \pm \sqrt{21}$.

19. $(x+13)^2 = 8$
$x+13 = \sqrt{8}$ *or* $x+13 = -\sqrt{8}$
$x+13 = 2\sqrt{2}$ *or* $x+13 = -2\sqrt{2}$
$x = -13+2\sqrt{2}$ *or* $x = -13-2\sqrt{2}$

The solutions are $-13+2\sqrt{2}$ and $-13-2\sqrt{2}$, or $-13\pm2\sqrt{2}$.

21. $(x-7)^2 = 12$
$x-7 = \sqrt{12}$ *or* $x-7 = -\sqrt{12}$
$x-7 = 2\sqrt{3}$ *or* $x-7 = -2\sqrt{3}$
$x = 7+2\sqrt{3}$ *or* $x = 7-2\sqrt{3}$

The solutions are $7+2\sqrt{3}$ and $7-2\sqrt{3}$, or $7 \pm 2\sqrt{3}$.

23. $(x+9)^2 = 34$
$x+9 = \sqrt{34}$ *or* $x+9 = -\sqrt{34}$
$x = -9+\sqrt{34}$ *or* $x = -9-\sqrt{34}$

The solutions are $-9+\sqrt{34}$ and $-9-\sqrt{34}$, or $-9 \pm \sqrt{34}$.

25. $\left(x+\frac{3}{2}\right)^2 = \frac{7}{2}$
$x+\frac{3}{2} = \sqrt{\frac{7}{2}}$ *or* $x+\frac{3}{2} = -\sqrt{\frac{7}{2}}$
$x = -\frac{3}{2}+\sqrt{\frac{7}{2}}$ *or* $x = -\frac{3}{2}-\sqrt{\frac{7}{2}}$
$x = -\frac{3}{2}+\sqrt{\frac{7}{2}\cdot\frac{2}{2}}$ *or* $x = -\frac{3}{2}-\sqrt{\frac{7}{2}\cdot\frac{2}{2}}$
$x = -\frac{3}{2}+\frac{\sqrt{14}}{2}$ *or* $x = -\frac{3}{2}-\frac{\sqrt{14}}{2}$
$x = \frac{-3+\sqrt{14}}{2}$ *or* $x = \frac{-3-\sqrt{14}}{2}$

The solutions are $\frac{-3\pm\sqrt{14}}{2}$.

27. $x^2 - 6x + 9 = 64$
$(x-3)^2 = 64$ Factoring the left side
$x-3 = 8$ *or* $x-3 = -8$ Principle of square roots
$x = 11$ *or* $x = -5$

The solutions are 11 and -5.

29. $x^2 + 14x + 49 = 64$
$(x+7)^2 = 64$ Factoring the left side
$x+7 = 8$ *or* $x+7 = -8$ Principle of square roots
$x = 1$ *or* $x = -15$

The solutions are 1 and -15.

31. $x^2 - 6x - 16 = 0$
$x^2 - 6x = 16$ Adding 16
$x^2 - 6x + 9 = 16+9$ Adding 9: $\left(\frac{-6}{2}\right)^2 = (-3)^2 = 9$
$(x-3)^2 = 25$
$x-3 = 5$ *or* $x-3 = -5$ Principle of square roots
$x = 8$ *or* $x = -2$

The solutions are 8 and -2.

33. $x^2 + 22x + 21 = 0$
$x^2 + 22x = -21$ Subtracting 21
$x^2 + 22x + 121 = -21+121$ Adding 121: $\left(\frac{22}{2}\right)^2 = 11^2 = 121$
$(x+11)^2 = 100$
$x+11 = 10$ *or* $x+11 = -10$ Principle of square roots
$x = -1$ *or* $x = -21$

The solutions are -1 and -21.

35. $x^2 - 2x - 5 = 0$
$x^2 - 2x = 5$
$x^2 - 2x + 1 = 5+1$ Adding 1: $\left(\frac{-2}{2}\right)^2 = (-1)^2 = 1$

$$(x-1)^2 = 6$$

$$x - 1 = \sqrt{6} \quad or \quad x - 1 = -\sqrt{6}$$
$$x = 1 + \sqrt{6} \quad or \quad x = 1 - \sqrt{6}$$

The solutions are $1 \pm \sqrt{6}$.

37. $x^2 - 22x + 102 = 0$
$$x^2 - 22x = -102$$
$$x^2 - 22x + 121 = -102 + 121 \quad \text{Adding 121: } \left(\frac{-22}{2}\right)^2 = (-11)^2 = 121$$

$$(x-11)^2 = 19$$

$$x - 11 = \sqrt{19} \quad or \quad x - 11 = -\sqrt{19}$$
$$x = 11 + \sqrt{19} \quad or \quad x = 11 - \sqrt{19}$$

The solutions are $11 \pm \sqrt{19}$.

39. $x^2 + 10x - 4 = 0$
$$x^2 + 10x = 4$$
$$x^2 + 10x + 25 = 4 + 25 \quad \text{Adding 25: } \left(\frac{10}{2}\right)^2 = 5^2 = 25$$

$$(x+5)^2 = 29$$

$$x + 5 = \sqrt{29} \quad or \quad x + 5 = -\sqrt{29}$$
$$x = -5 + \sqrt{29} \quad or \quad x = -5 - \sqrt{29}$$

The solutions are $-5 \pm \sqrt{29}$.

41. $x^2 - 7x - 2 = 0$
$$x^2 - 7x = 2$$
$$x^2 - 7x + \frac{49}{4} = 2 + \frac{49}{4} \quad \text{Adding } \frac{49}{4}\text{: } \left(\frac{-7}{2}\right)^2 = \frac{49}{4}$$

$$\left(x - \frac{7}{2}\right)^2 = \frac{8}{4} + \frac{49}{4} = \frac{57}{4}$$

$$x - \frac{7}{2} = \frac{\sqrt{57}}{2} \quad or \quad x - \frac{7}{2} = -\frac{\sqrt{57}}{2}$$
$$x = \frac{7}{2} + \frac{\sqrt{57}}{2} \quad or \quad x = \frac{7}{2} - \frac{\sqrt{57}}{2}$$
$$x = \frac{7+\sqrt{57}}{2} \quad or \quad x = \frac{7-\sqrt{57}}{2}$$

The solutions are $\frac{7 \pm \sqrt{57}}{2}$.

43. $x^2 + 3x - 28 = 0$
$$x^2 + 3x = 28$$
$$x^2 + 3x + \frac{9}{4} = 28 + \frac{9}{4} \quad \text{Adding } \frac{9}{4}: \left(\frac{3}{2}\right)^2 = \frac{9}{4}$$
$$\left(x + \frac{3}{2}\right)^2 = \frac{121}{4}$$

$$x + \frac{3}{2} = \frac{11}{2} \quad or \quad x + \frac{3}{2} = -\frac{11}{2}$$
$$x = \frac{8}{2} \quad or \quad x = -\frac{14}{2}$$
$$x = 4 \quad or \quad x = -7$$

The solutions are 4 and -7.

45. $x^2 + \frac{3}{2}x - \frac{1}{2} = 0$
$$x^2 + \frac{3}{2}x = \frac{1}{2}$$
$$x^2 + \frac{3}{2}x + \frac{9}{16} = \frac{1}{2} + \frac{9}{16} \quad \text{Adding } \frac{9}{16}: \left(\frac{3/2}{2}\right)^2 = \left(\frac{3}{4}\right)^2 = \frac{9}{16}$$

$$\left(x + \frac{3}{4}\right)^2 = \frac{17}{16}$$

$$x + \frac{3}{4} = \frac{\sqrt{17}}{4} \quad or \quad x + \frac{3}{4} = -\frac{\sqrt{17}}{4}$$
$$x = -\frac{3}{4} + \frac{\sqrt{17}}{4} \quad or \quad x = -\frac{3}{4} - \frac{\sqrt{17}}{4}$$
$$x = \frac{-3+\sqrt{17}}{4} \quad or \quad x = \frac{-3-\sqrt{17}}{4}$$

The solutions are $\frac{-3 \pm \sqrt{17}}{4}$.

47. $2x^2 + 3x - 17 = 0$

$$\frac{1}{2}(2x^2 + 3x - 17) = \frac{1}{2} \cdot 0 \quad \text{Multiplying by } \frac{1}{2} \text{ to make the } x^2\text{-coefficient 1}$$

$$x^2 + \frac{3}{2}x - \frac{17}{2} = 0$$
$$x^2 + \frac{3}{2}x = \frac{17}{2}$$
$$x^2 + \frac{3}{2}x + \frac{9}{16} = \frac{17}{2} + \frac{9}{16} \quad \text{Adding } \frac{9}{16}: \left(\frac{3/2}{2}\right)^2 = \left(\frac{3}{4}\right)^2 = \frac{9}{16}$$

$$\left(x + \frac{3}{4}\right)^2 = \frac{145}{16}$$

$$x + \frac{3}{4} = \frac{\sqrt{145}}{4} \quad or \quad x + \frac{3}{4} = -\frac{\sqrt{145}}{4}$$
$$x = \frac{-3+\sqrt{145}}{4} \quad or \quad x = \frac{-3-\sqrt{145}}{4}$$

The solutions are $\frac{-3 \pm \sqrt{145}}{4}$.

49. $3x^2 + 4x - 1 = 0$
$$\frac{1}{3}(3x^2 + 4x - 1) = \frac{1}{3} \cdot 0$$
$$x^2 + \frac{4}{3}x - \frac{1}{3} = 0$$
$$x^2 + \frac{4}{3}x = \frac{1}{3}$$
$$x^2 + \frac{4}{3}x + \frac{4}{9} = \frac{1}{3} + \frac{4}{9}$$
$$\left(x + \frac{2}{3}\right)^2 = \frac{7}{9}$$

$$x + \frac{2}{3} = \frac{\sqrt{7}}{3} \quad or \quad x + \frac{2}{3} = -\frac{\sqrt{7}}{3}$$
$$x = \frac{-2+\sqrt{7}}{3} \quad or \quad x = \frac{-2-\sqrt{7}}{3}$$

The solutions are $\dfrac{-2 \pm \sqrt{7}}{3}$.

51.
$$\begin{aligned} 2x^2 &= 9x+5 \\ 2x^2-9x-5 &= 0 \quad \text{Standard form} \\ \frac{1}{2}(2x^2-9x-5) &= \frac{1}{2}\cdot 0 \\ x^2-\frac{9}{2}x-\frac{5}{2} &= 0 \\ x^2-\frac{9}{2}x &= \frac{5}{2} \\ x^2-\frac{9}{2}x+\frac{81}{16} &= \frac{5}{2}+\frac{81}{16} \\ \left(x-\frac{9}{4}\right)^2 &= \frac{121}{16} \end{aligned}$$

$$x-\frac{9}{4}=\frac{11}{4} \quad or \quad x-\frac{9}{4}=-\frac{11}{4}$$
$$x=\frac{20}{4} \quad or \quad x=-\frac{2}{4}$$
$$x=5 \quad or \quad x=-\frac{1}{2}$$

The solutions are 5 and $-\dfrac{1}{2}$.

53.
$$\begin{aligned} 6x^2+11x &= 10 \\ 6x^2+11x-10 &= 0 \quad \text{Standard form} \\ \frac{1}{6}(6x^2+11x-10) &= \frac{1}{6}\cdot 0 \\ x^2+\frac{11}{6}x-\frac{5}{3} &= 0 \\ x^2+\frac{11}{6}x &= \frac{5}{3} \\ x^2+\frac{11}{6}x+\frac{121}{144} &= \frac{5}{3}+\frac{121}{144} \\ \left(x+\frac{11}{12}\right)^2 &= \frac{361}{144} \end{aligned}$$

$$x+\frac{11}{12}=\frac{19}{12} \quad or \quad x+\frac{11}{12}=-\frac{19}{12}$$
$$x=\frac{8}{12} \quad or \quad x=-\frac{30}{12}$$
$$x=\frac{2}{3} \quad or \quad x=-\frac{5}{2}$$

The solutions are $\dfrac{2}{3}$ and $-\dfrac{5}{2}$.

55. ***Familiarize.*** We will use the formula $s = 16t^2$.

Translate. We substitute 2723 for s.

$2723 = 16t^2$

Solve. We solve the equation.

$2723 = 16t^2$

$\dfrac{2723}{16} = t^2$ Solving for t^2

$170.1875 = t^2$ Dividing

$\sqrt{170.1875} = t$ *or* $-\sqrt{170.1875} = t$ Principle of square roots

$13.0 \approx t$ *or* $-13.0 \approx t$ Using a calculator and rounding to the nearest tenth

Check. The number -13.0 cannot be a solution, because time cannot be negative in this situation. We substitute 13.0 in the original equation.

$s = 16(13.0)^2 = 16(169) = 2704$

This is close. Remember that we approximated a solution. Thus we have a check.

State. It takes about 13.0 sec for an object to fall to the ground from the top of the Burj Khalifa.

57. ***Familiarize.*** We will use the formula $s = 16t^2$.

Translate. We substitute 1353 for s.

$1353 = 16t^2$

Solve. We solve the equation.

$1353 = 16t^2$

$\dfrac{1353}{16} = t^2$ Solving for t^2

$84.5625 = t^2$ Dividing

$\sqrt{84.5625} = t$ *or* $-\sqrt{84.5625} = t$ Principle of square roots

$9.2 \approx t$ *or* $-9.2 \approx t$ Using a calculator and rounding to the nearest tenth

Check. The number -9.2 cannot be a solution, because time cannot be negative in this situation. We substitute 9.2 in the original equation.

$s = 16(9.2)^2 = 16(84.64) = 1354.24$

This is close. Remember that we approximated a solution. Thus we have a check.

State. The fall would take approximately 9.2 sec.

59. $(3x^4 - x^2 + 1) + (4x^3 - x^2) = 3x^4 + 4x^3 - 2x^2 + 1$

61. $(3x^2 + x)^2 = (3x^2)^2 + 2\cdot 3x^2 \cdot x + x^2 = 9x^4 + 6x^3 + x^2$

63. $(12x^4 - 15x^3 - x^2) \div (3x) = \dfrac{12x^4}{3x} - \dfrac{15x^3}{3x} - \dfrac{x^2}{3x} =$
$4x^3 - 5x^2 - \dfrac{x}{3}$

65. $x^2 + bx + 36$

The trinomial is a square if the square of one-half the x-coefficient is equal to 36. Thus we have:

$$\left(\frac{b}{2}\right)^2 = 36$$
$$\frac{b^2}{4} = 36$$
$$b^2 = 144$$

$b = 12$ *or* $b = -12$ Principle of square roots

67. $x^2 + bx + 128$

The trinomial is a square if the square of one-half the x-coefficient is equal to 128. Thus we have:

$$\left(\frac{b}{2}\right)^2 = 128$$
$$\frac{b^2}{4} = 128$$
$$b^2 = 512$$

$b = \sqrt{512}$ *or* $b = -\sqrt{512}$
$b = 16\sqrt{2}$ *or* $b = -16\sqrt{2}$

69. $x^2 + bx + c$

The trinomial is a square if the square of one-half the x-coefficient is equal to c. Thus we have:

$$\left(\frac{b}{2}\right)^2 = c$$
$$\frac{b^2}{4} = c$$
$$b^2 = 4c$$

$b = \sqrt{4c}$ *or* $b = -\sqrt{4c}$
$b = 2\sqrt{c}$ *or* $b = -2\sqrt{c}$

71. $4.82x^2 = 12,000$

$$x^2 = \frac{12,000}{4.82}$$

$x = \sqrt{\frac{12,000}{4.82}}$ *or* $x = -\sqrt{\frac{12,000}{4.82}}$ Principle of square roots

$x \approx 49.896$ *or* $x \approx -49.896$ Using a calculator and rounding

The solutions are approximately 49.896 and −49.896.

73. $\frac{x}{9} = \frac{36}{4x}$, LCM is $36x$

$36x \cdot \frac{x}{9} = 36x \cdot \frac{36}{4x}$ Multiplying by $36x$

$$4x^2 = 324$$
$$x^2 = 81$$

$x = 9$ *or* $x = -9$

Both numbers check. The solutions are 9 and −9.

Exercise Set 9.3

RC1. The given values of a, b, and c yield the equation $x^2 - 7x + 2 = 0$. This is equivalent to $x^2 = 7x - 2$, or $7x - 2 = x^2$, so answer (c) is correct.

RC3. The given values of a, b, and c yield the equation $3x^2 + 7x + 2 = 0$, so answer (a) is correct.

RC5. The given values of a, b, and c yield the equation $3x^2 - 7x + 0 = 0$, or $3x^2 - 7x = 0$. This is equivalent to $3x^2 = 7x$, so answer (e) is correct.

1. $x^2 - 4x = 21$

$x^2 - 4x - 21 = 0$ Standard form

We can factor.

$$x^2 - 4x - 21 = 0$$
$$(x - 7)(x + 3) = 0$$

$x - 7 = 0$ *or* $x + 3 = 0$
$x = 7$ *or* $x = -3$

The solutions are 7 and −3.

3. $x^2 = 6x - 9$

$x^2 - 6x + 9 = 0$ Standard form

We can factor.

$$x^2 - 6x + 9 = 0$$
$$(x - 3)(x - 3) = 0$$

$x - 3 = 0$ *or* $x - 3 = 0$
$x = 3$ *or* $x = 3$

The solution is 3.

5. $3y^2 - 2y - 8 = 0$

We can factor.

$$3y^2 - 2y - 8 = 0$$
$$(3y + 4)(y - 2) = 0$$

$3y + 4 = 0$ *or* $y - 2 = 0$
$3y = -4$ *or* $y = 2$
$y = -\frac{4}{3}$ *or* $y = 2$

The solutions are $-\frac{4}{3}$ and 2.

7. $4x^2 + 4x = 15$

$4x^2 + 4x - 15 = 0$ Standard form

We can factor.

$$4x^2 + 4x - 15 = 0$$
$$(2x - 3)(2x + 5) = 0$$

$2x - 3 = 0$ *or* $2x + 5 = 0$
$2x = 3$ *or* $2x = -5$
$x = \frac{3}{2}$ *or* $x = -\frac{5}{2}$

The solutions are $\frac{3}{2}$ and $-\frac{5}{2}$.

9. $x^2 - 9 = 0$ Difference of squares

$$(x + 3)(x - 3) = 0$$

$x + 3 = 0$ *or* $x - 3 = 0$
$x = -3$ *or* $x = 3$

The solutions are −3 and 3.

11. $x^2 - 2x - 2 = 0$

$a = 1,\ b = -2,\ c = -2$

We use the quadratic formula.

$$x = \frac{-(-2) \pm \sqrt{(-2)^2 - 4 \cdot 1 \cdot (-2)}}{2 \cdot 1}$$
$$x = \frac{2 \pm \sqrt{4 + 8}}{2}$$
$$x = \frac{2 \pm \sqrt{12}}{2} = \frac{2 \pm \sqrt{4 \cdot 3}}{2}$$
$$x = \frac{2 \pm 2\sqrt{3}}{2} = \frac{2(1 \pm \sqrt{3})}{2}$$
$$x = 1 \pm \sqrt{3}$$

The solutions are $1 + \sqrt{3}$ and $1 - \sqrt{3}$, or $1 \pm \sqrt{3}$.

13. $y^2 - 10y + 22 = 0$

$a = 1,\ b = -10,\ c = 22$

We use the quadratic formula.

$$y = \frac{-(-10) \pm \sqrt{(-10)^2 - 4 \cdot 1 \cdot 22}}{2 \cdot 1}$$

$$y = \frac{10 \pm \sqrt{100 - 88}}{2}$$

$$y = \frac{10 \pm \sqrt{12}}{2} = \frac{10 \pm \sqrt{4 \cdot 3}}{2}$$

$$y = \frac{10 \pm 2\sqrt{3}}{2} = \frac{2(5 \pm \sqrt{3})}{2}$$

$$y = 5 \pm \sqrt{3}$$

The solutions are $5 + \sqrt{3}$ and $5 - \sqrt{3}$, or $5 \pm \sqrt{3}$.

15. $x^2 + 4x + 4 = 7$

$x^2 + 4x - 3 = 0$ Adding -7 to get standard form

$a = 1,\ b = 4,\ c = -3$

We use the quadratic formula.

$$x = \frac{-4 \pm \sqrt{4^2 - 4 \cdot 1 \cdot (-3)}}{2 \cdot 1} = \frac{-4 \pm \sqrt{16 + 12}}{2}$$

$$x = \frac{-4 \pm \sqrt{28}}{2} = \frac{-4 \pm \sqrt{4 \cdot 7}}{2}$$

$$x = \frac{-4 \pm 2\sqrt{7}}{2} = \frac{2(-2 \pm \sqrt{7})}{2}$$

$$x = -2 \pm \sqrt{7}$$

The solutions are $-2 + \sqrt{7}$ and $-2 - \sqrt{7}$, or $-2 \pm \sqrt{7}$.

17. $3x^2 + 8x + 2 = 0$

$a = 3,\ b = 8,\ c = 2$

We use the quadratic formula.

$$x = \frac{-8 \pm \sqrt{8^2 - 4 \cdot 3 \cdot 2}}{2 \cdot 3} = \frac{-8 \pm \sqrt{64 - 24}}{6}$$

$$x = \frac{-8 \pm \sqrt{40}}{6} = \frac{-8 \pm \sqrt{4 \cdot 10}}{6}$$

$$x = \frac{-8 \pm 2\sqrt{10}}{6} = \frac{2(-4 \pm \sqrt{10})}{2 \cdot 3}$$

$$x = \frac{-4 \pm \sqrt{10}}{3}$$

The solutions are $\dfrac{-4 + \sqrt{10}}{3}$ and $\dfrac{-4 - \sqrt{10}}{3}$, or $\dfrac{-4 \pm \sqrt{10}}{3}$.

19. $2x^2 - 5x = 1$

$2x^2 - 5x - 1 = 0$ Adding -1 to get standard form

$a = 2,\ b = -5,\ c = -1$

We use the quadratic formula.

$$x = \frac{-(-5) \pm \sqrt{(-5)^2 - 4 \cdot 2 \cdot (-1)}}{2 \cdot 2} = \frac{5 \pm \sqrt{25 + 8}}{4}$$

$$x = \frac{5 \pm \sqrt{33}}{4}$$

The solutions are $\dfrac{5 + \sqrt{33}}{4}$ and $\dfrac{5 - \sqrt{33}}{4}$, or $\dfrac{5 \pm \sqrt{33}}{4}$.

21. $2y^2 - 2y - 1 = 0$

$a = 2,\ b = -2,\ c = -1$

We use the quadratic formula.

$$y = \frac{-(-2) \pm \sqrt{(-2)^2 - 4 \cdot 2 \cdot (-1)}}{2 \cdot 2} = \frac{2 \pm \sqrt{4 + 8}}{4}$$

$$y = \frac{2 \pm \sqrt{12}}{4} = \frac{2 \pm \sqrt{4 \cdot 3}}{4}$$

$$y = \frac{2 \pm 2\sqrt{3}}{4} = \frac{2(1 \pm \sqrt{3})}{2 \cdot 2}$$

$$y = \frac{1 \pm \sqrt{3}}{2}$$

The solutions are $\dfrac{1 + \sqrt{3}}{2}$ and $\dfrac{1 - \sqrt{3}}{2}$, or $\dfrac{1 \pm \sqrt{3}}{2}$.

23. $2t^2 + 6t + 5 = 0$

$a = 2,\ b = 6,\ c = 5$

We use the quadratic formula.

$$t = \frac{-6 \pm \sqrt{6^2 - 4 \cdot 2 \cdot 5}}{2 \cdot 2} = \frac{-6 \pm \sqrt{36 - 40}}{4}$$

$$t = \frac{-6 \pm \sqrt{-4}}{4}$$

Since square roots of negative numbers do not exist as real numbers, there are no real-number solutions.

25. $3x^2 = 5x + 4$

$3x^2 - 5x - 4 = 0$

$a = 3,\ b = -5,\ c = -4$

We use the quadratic formula.

$$x = \frac{-(-5) \pm \sqrt{(-5)^2 - 4 \cdot 3 \cdot (-4)}}{2 \cdot 3} = \frac{5 \pm \sqrt{25 + 48}}{6}$$

$$x = \frac{5 \pm \sqrt{73}}{6}$$

The solutions are $\dfrac{5 + \sqrt{73}}{6}$ and $\dfrac{5 - \sqrt{73}}{6}$, or $\dfrac{5 \pm \sqrt{73}}{6}$.

27. $2y^2 - 6y = 10$

$2y^2 - 6y - 10 = 0$

$y^2 - 3y - 5 = 0$ Multiplying by $\frac{1}{2}$ to simplify

$a = 1,\ b = -3,\ c = -5$

We use the quadratic formula.

$$y = \frac{-(-3) \pm \sqrt{(-3)^2 - 4 \cdot 1 \cdot (-5)}}{2 \cdot 1} = \frac{3 \pm \sqrt{9 + 20}}{2}$$

$$y = \frac{3 \pm \sqrt{29}}{2}$$

The solutions are $\dfrac{3 + \sqrt{29}}{2}$ and $\dfrac{3 - \sqrt{29}}{2}$, or $\dfrac{3 \pm \sqrt{29}}{2}$.

29. $\dfrac{x^2}{x+3} - \dfrac{5}{x+3} = 0$, LCM is $x + 3$

$$(x+3)\left(\frac{x^2}{x+3} - \frac{5}{x+3}\right) = (x+3) \cdot 0$$

$$x^2 - 5 = 0$$

$$x^2 = 5$$

$x=\sqrt{5}$ *or* $x=-\sqrt{5}$ Principle of square roots

Both numbers check. The solutions are $\sqrt{5}$ and $-\sqrt{5}$, or $\pm\sqrt{5}$.

31. $x+2=\dfrac{3}{x+2}$

$(x+2)(x+2)=(x+2)\cdot\dfrac{3}{x+2}$ Clearing the fraction

$x^2+4x+4=3$
$x^2+4x+1=0$

$a=1,\ b=4,\ c=1$

We use the quadratic formula.

$$x=\frac{-4\pm\sqrt{4^2-4\cdot1\cdot1}}{2\cdot1}=\frac{-4\pm\sqrt{16-4}}{2}$$
$$x=\frac{-4\pm\sqrt{12}}{2}=\frac{-4\pm\sqrt{4\cdot3}}{2}$$
$$x=\frac{-4\pm2\sqrt{3}}{2}=\frac{2(-2\pm\sqrt{3})}{2}$$
$$x=-2\pm\sqrt{3}$$

Both numbers check. The solutions are $-2+\sqrt{3}$ and $-2-\sqrt{3}$, or $-2\pm\sqrt{3}$.

33. $\dfrac{1}{x}+\dfrac{1}{x+1}=\dfrac{1}{3}$, LCM is $3x(x+1)$

$$3x(x+1)\left(\frac{1}{x}+\frac{1}{x+1}\right)=3x(x+1)\cdot\frac{1}{3}$$
$$3(x+1)+3x=x(x+1)$$
$$3x+3+3x=x^2+x$$
$$6x+3=x^2+x$$
$$0=x^2-5x-3$$

$a=1,\ b=-5,\ c=-3$

We use the quadratic formula.

$$x=\frac{-(-5)\pm\sqrt{(-5)^2-4\cdot1\cdot(-3)}}{2\cdot1}=\frac{5\pm\sqrt{25+12}}{2}$$
$$x=\frac{5\pm\sqrt{37}}{2}$$

The solutions are $\dfrac{5+\sqrt{37}}{2}$ and $\dfrac{5-\sqrt{37}}{2}$, or $\dfrac{5\pm\sqrt{37}}{2}$.

35. $x^2-4x-7=0$

$a=1,\ b=-4,\ c=-7$

$$x=\frac{-(-4)\pm\sqrt{(-4)^2-4\cdot1\cdot(-7)}}{2\cdot1}$$
$$x=\frac{4\pm\sqrt{16+28}}{2}=\frac{4\pm\sqrt{44}}{2}$$
$$x=\frac{4\pm\sqrt{4\cdot11}}{2}=\frac{4\pm2\sqrt{11}}{2}$$
$$x=\frac{2(2\pm\sqrt{11})}{2}=2\pm\sqrt{11}$$

Using a calculator, we have:

$2+\sqrt{11}\approx5.31662479\approx5.3$, and

$2-\sqrt{11}\approx-1.31662479\approx-1.3$.

The approximate solutions, to the nearest tenth, are 5.3 and -1.3.

37. $y^2-6y-1=0$

$a=1,\ b=-6,\ c=-1$

$$y=\frac{-(-6)\pm\sqrt{(-6)^2-4\cdot1\cdot(-1)}}{2\cdot1}$$
$$y=\frac{6\pm\sqrt{36+4}}{2}=\frac{6\pm\sqrt{40}}{2}$$
$$y=\frac{6\pm\sqrt{4\cdot10}}{2}=\frac{6\pm2\sqrt{10}}{2}$$
$$y=\frac{2(3\pm\sqrt{10})}{2}=3\pm\sqrt{10}$$

Using a calculator, we have:

$3+\sqrt{10}\approx6.16227766\approx6.2$ and

$3-\sqrt{10}\approx-0.1622776602\approx-0.2$.

The approximate solutions, to the nearest tenth, are 6.2 and -0.2.

39. $4x^2+4x=1$

$4x^2+4x-1=0$ Standard form

$a=4,\ b=4,\ c=-1$

$$x=\frac{-4\pm\sqrt{4^2-4\cdot4\cdot(-1)}}{2\cdot4}$$
$$x=\frac{-4\pm\sqrt{16+16}}{8}=\frac{-4\pm\sqrt{32}}{8}$$
$$x=\frac{-4\pm\sqrt{16\cdot2}}{8}=\frac{-4\pm4\sqrt{2}}{8}$$
$$x=\frac{4(-1\pm\sqrt{2})}{4\cdot2}=\frac{-1\pm\sqrt{2}}{2}$$

Using a calculator, we have:

$\dfrac{-1+\sqrt{2}}{2}\approx0.2071067812\approx0.2$ and

$\dfrac{-1-\sqrt{2}}{2}\approx-1.207106781\approx-1.2$.

The approximate solutions, to the nearest tenth, are 0.2 and -1.2.

41. $3x^2-8x+2=0$

$a=3,\ b=-8,\ c=2$

$$x=\frac{-(-8)\pm\sqrt{(-8)^2-4\cdot3\cdot2}}{2\cdot3}$$
$$x=\frac{8\pm\sqrt{64-24}}{6}=\frac{8\pm\sqrt{40}}{6}$$
$$x=\frac{8\pm\sqrt{4\cdot10}}{6}=\frac{8\pm2\sqrt{10}}{6}$$
$$x=\frac{2(4\pm\sqrt{10})}{2\cdot3}=\frac{4\pm\sqrt{10}}{3}$$

Using a calculator, we have:

$\dfrac{4+\sqrt{10}}{3}\approx2.387425887\approx2.4$ and

$\dfrac{4-\sqrt{10}}{3}\approx0.2792407799\approx0.3$.

The approximate solutions, to the nearest tenth, are 2.4 and 0.3.

43.
$$\begin{aligned} 2(x+3)-(x-5) &= 6x \\ 2x+6-x+5 &= 6x \\ x+11 &= 6x \\ 11 &= 5x \\ \frac{11}{5} &= x \end{aligned}$$

The solution is $\frac{11}{5}$.

45.
$$\begin{aligned} \frac{1}{3}y+\frac{1}{2} &= \frac{2}{3}-\frac{5}{6}y \\ 6\left(\frac{1}{3}y+\frac{1}{2}\right) &= 6\left(\frac{2}{3}-\frac{5}{6}y\right) \\ 6\cdot\frac{1}{3}y+6\cdot\frac{1}{2} &= 6\cdot\frac{2}{3}-6\cdot\frac{5}{6}y \\ 2y+3 &= 4-5y \\ 7y+3 &= 4 \\ 7y &= 1 \\ y &= \frac{1}{7} \end{aligned}$$

The solution is $\frac{1}{7}$.

47.
$$\begin{aligned} 3d^2-3 &= 0 \\ 3(d^2-1) &= 0 \\ 3(d+1)(d-1) &= 0 \\ d+1=0 \quad &or \quad d-1=0 \\ d=-1 \quad &or \quad d=1 \end{aligned}$$

The solutions are -1 and 1.

49.
$$\begin{aligned} \sqrt{2x-1} &= x-2 \\ (\sqrt{2x-1})^2 &= (x-2)^2 \\ 2x-1 &= x^2-4x+4 \\ 0 &= x^2-6x+5 \\ 0 &= (x-1)(x-5) \\ x-1=0 \quad &or \quad x-5=0 \\ x=1 \quad &or \quad x=5 \end{aligned}$$

The number 5 checks, but 1 does not. The solution is 5.

51.
$$\begin{aligned} 2n^2-n &= 3 \\ 2n^2-n-3 &= 0 \\ (2n-3)(n+1) &= 0 \\ 2n-3=0 \quad &or \quad n+1=0 \\ 2n=3 \quad &or \quad n=-1 \\ n=\frac{3}{2} \quad &or \quad n=-1 \end{aligned}$$

The solutions are $\frac{3}{2}$ and -1.

53.
$$\begin{aligned} \frac{x}{x-1}+\frac{1}{x} &= 1 \\ x(x-1)\left(\frac{x}{x-1}+\frac{1}{x}\right) &= x(x-1)\cdot 1 \\ x(x-1)\cdot\frac{x}{x-1}+x(x-1)\cdot\frac{1}{x} &= x^2-x \\ x^2+x-1 &= x^2-x \\ x-1 &= -x \\ -1 &= -2x \\ \frac{1}{2} &= x \end{aligned}$$

The number $\frac{1}{2}$ checks. The solution is $\frac{1}{2}$.

55.
$$\begin{aligned} 5x+x(x-7) &= 0 \\ 5x+x^2-7x &= 0 \\ x^2-2x &= 0 \quad \text{We can factor.} \\ x(x-2) &= 0 \\ x=0 \quad &or \quad x-2=0 \\ x=0 \quad &or \quad x=2 \end{aligned}$$

The solutions are 0 and 2.

57.
$$\begin{aligned} 3-x(x-3) &= 4 \\ 3-x^2+3x &= 4 \\ 0 &= x^2-3x+1 \quad \text{Standard form} \end{aligned}$$

$a=1,\ b=-3,\ c=1$

We use the quadratic formula.

$$x=\frac{-(-3)\pm\sqrt{(-3)^2-4\cdot1\cdot1}}{2\cdot1}=\frac{3\pm\sqrt{9-4}}{2}$$

$$x=\frac{3\pm\sqrt{5}}{2}$$

The solutions are $\frac{3+\sqrt{5}}{2}$ and $\frac{3-\sqrt{5}}{2}$, or $\frac{3\pm\sqrt{5}}{2}$.

59.
$$\begin{aligned} (y+4)(y+3) &= 15 \\ y^2+7y+12 &= 15 \\ y^2+7y-3 &= 0 \quad \text{Standard form} \end{aligned}$$

$a=1,\ b=7,\ c=-3$

We use the quadratic formula.

$$y=\frac{-7\pm\sqrt{7^2-4\cdot1\cdot(-3)}}{2\cdot1}=\frac{-7\pm\sqrt{49+12}}{2}$$

$$y=\frac{-7\pm\sqrt{61}}{2}$$

The solutions are $\frac{-7+\sqrt{61}}{2}$ and $\frac{-7-\sqrt{61}}{2}$, or $\frac{-7\pm\sqrt{61}}{2}$.

61.
$$\begin{aligned} x^2+(x+2)^2 &= 7 \\ x^2+x^2+4x+4 &= 7 \\ 2x^2+4x+4 &= 7 \\ 2x^2+4x-3 &= 0 \quad \text{Standard form} \end{aligned}$$

$a=2,\ b=4,\ c=-3$

We use the quadratic formula.

$$x = \frac{-4 \pm \sqrt{4^2 - 4 \cdot 2 \cdot (-3)}}{2 \cdot 2} = \frac{-4 \pm \sqrt{16+24}}{4}$$

$$x = \frac{-4 \pm \sqrt{40}}{4} = \frac{-4 \pm \sqrt{4 \cdot 10}}{4}$$

$$x = \frac{-4 \pm 2\sqrt{10}}{4} = \frac{2(-2 \pm \sqrt{10})}{2 \cdot 2}$$

$$x = \frac{-2 \pm \sqrt{10}}{2}$$

The solutions are $\frac{-2+\sqrt{10}}{2}$ and $\frac{-2-\sqrt{10}}{2}$, or $\frac{-2 \pm \sqrt{10}}{2}$.

63. $x^2 + x = 1$

$x^2 + x - 1 = 0$

Graph $y = x^2 + x - 1$ and observe that the graph intersects the x-axis at two points. This indicates that the equation has two real-number solutions.

65.-71. Left to the student

Chapter 9 Mid-Chapter Review

1. The statement is true. See page 670 in the text.

2. The solutions of $ax^2 + bx + c = 0$ are the first coordinates of the x-intercepts of the graph of $y = ax^2 + bx + c$. The given statement is false.

3. $ax^2 + bx = 0$

$x(ax + b) = 0$

$x = 0 \quad or \quad ax + b = 0$

$x = 0 \quad or \quad ax = -b$

$x = 0 \quad or \quad x = -\frac{b}{a}$

One solution of the equation is 0 and, because $b \neq 0$, then $-\frac{b}{a} \neq 0$, so the other solution is a non-zero number. The given statement is true.

4. $x^2 - 6x - 2 = 0$

$x^2 - 6x = 2$

$x^2 - 6x + 9 = 2 + 9$

$(x-3)^2 = 11$

$x - 3 = \pm\sqrt{11}$

$x = 3 \pm \sqrt{11}$

5. $3x^2 = 8x - 2$

$3x^2 - 8x + 2 = 0$

$a = 3,\ b = -8,\ c = 2$

$$x = \frac{-b \pm \sqrt{b^2 - 4ac}}{2a}$$

$$x = \frac{-(-8) \pm \sqrt{(-8)^2 - 4 \cdot 3 \cdot 2}}{2 \cdot 3}$$

$$x = \frac{8 \pm \sqrt{64 - 24}}{6} = \frac{8 \pm \sqrt{40}}{6} = \frac{8 \pm \sqrt{4 \cdot 10}}{6}$$

$$x = \frac{8 \pm 2\sqrt{10}}{6} = \frac{2(4 \pm \sqrt{10})}{2 \cdot 3} = \frac{4 \pm \sqrt{10}}{3}$$

6. $q^2 - 5q + 10 = 0$ Standard form

$a = 1,\ b = -5,\ c = 10$

7. $6 - x^2 = 14x + 2$

$0 = x^2 + 14x - 4$ Standard form

$a = 1,\ b = 14,\ c = -4$

8. $17z = 3z^2$

$0 = 3z^2 - 17x$ Standard form

$a = 3,\ b = -17,\ c = 0$

9. $16x = 48x^2$

$0 = 48x^2 - 16x$

$0 = 16x(3x - 1)$

$16x = 0 \quad or \quad 3x - 1 = 0$

$x = 0 \quad or \quad 3x = 1$

$x = 0 \quad or \quad x = \frac{1}{3}$

The solutions are 0 and $\frac{1}{3}$.

10. $x(x-3) = 10$

$x^2 - 3x = 10$

$x^2 - 3x - 10 = 0$

$(x-5)(x+2) = 0$

$x - 5 = 0 \quad or \quad x + 2 = 0$

$x = 5 \quad or \quad x = -2$

The solutions are 5 and -2.

11. $20x^2 - 20x = 0$

$20x(x-1) = 0$

$20x = 0 \quad or \quad x - 1 = 0$

$x = 0 \quad or \quad x = 1$

The solutions are 0 and 1.

12. $x^2 = 14x - 49$

$x^2 - 14x + 49 = 0$

$(x-7)(x-7) = 0$

$x - 7 = 0 \quad or \quad x - 7 = 0$

$x = 7 \quad or \quad x = 7$

The solution is 7.

13. $t^2 + 2t = 0$

$t(t+2) = 0$

$$t = 0 \quad or \quad t + 2 = 0$$
$$t = 0 \quad or \quad t = -2$$

The solutions are 0 and -2.

14.
$$18w^2 + 21w = 4$$
$$18w^2 + 21w - 4 = 0$$
$$(3w + 4)(6w - 1) = 0$$
$$3w + 4 = 0 \quad or \quad 6w - 1 = 0$$
$$3w = -4 \quad or \quad 6w = 1$$
$$w = -\frac{4}{3} \quad or \quad w = \frac{1}{6}$$

The solutions are $-\frac{4}{3}$ and $\frac{1}{6}$.

15.
$$9y^2 - 5y^2 = 82y + 6y$$
$$4y^2 = 88y$$
$$4y^2 - 88y = 0$$
$$4y(y - 22) = 0$$
$$4y = 0 \quad or \quad y - 22 = 0$$
$$y = 0 \quad or \quad y = 22$$

The solutions are 0 and 22.

16.
$$2(s - 3) = s(s - 3)$$
$$2s - 6 = s^2 - 3s$$
$$0 = s^2 - 5s + 6$$
$$0 = (s - 3)(s - 2)$$
$$s - 3 = 0 \quad or \quad s - 2 = 0$$
$$s = 3 \quad or \quad s = 2$$

The solutions are 3 and 2.

17.
$$8y^2 - 40y = -7y + 35$$
$$8y^2 - 33y - 35 = 0$$
$$(8y + 7)(y - 5) = 0$$
$$8y + 7 = 0 \quad or \quad y - 5 = 0$$
$$8y = -7 \quad or \quad y = 5$$
$$y = -\frac{7}{8} \quad or \quad y = 5$$

The solutions are $-\frac{7}{8}$ and 5.

18.
$$x^2 + 2x - 3 = 0$$
$$x^2 + 2x = 3$$
$$x^2 + 2x + 1 = 3 + 1$$
$$(x + 1)^2 = 4$$
$$x + 1 = 2 \quad or \quad x + 1 = -2$$
$$x = 1 \quad or \quad x = -3$$

The solutions are 1 and -3.

19.
$$x^2 - 9x + 6 = 0$$
$$x^2 - 9x = -6$$
$$x^2 - 9x + \frac{81}{4} = -6 + \frac{81}{4}$$
$$\left(x - \frac{9}{2}\right)^2 = \frac{57}{4}$$
$$x - \frac{9}{2} = \frac{\sqrt{57}}{2} \quad or \quad x - \frac{9}{2} = -\frac{\sqrt{57}}{2}$$
$$x = \frac{9}{2} + \frac{\sqrt{57}}{2} \quad or \quad x = \frac{9}{2} - \frac{\sqrt{57}}{2}$$
$$x = \frac{9 + \sqrt{57}}{2} \quad or \quad x = \frac{9 - \sqrt{57}}{2}$$

The solutions are $\frac{9 + \sqrt{57}}{2}$ and $\frac{9 - \sqrt{57}}{2}$, or $\frac{9 \pm \sqrt{57}}{2}$.

20.
$$2x^2 = 7x + 8$$
$$2x^2 - 7x = 8$$
$$\frac{1}{2}(2x^2 - 7x) = \frac{1}{2} \cdot 8$$
$$x^2 - \frac{7}{2}x = 4$$
$$x^2 - \frac{7}{2}x + \frac{49}{16} = 4 + \frac{49}{16}$$
$$\left(x - \frac{7}{4}\right)^2 = \frac{113}{16}$$
$$x - \frac{7}{4} = \frac{\sqrt{113}}{4} \quad or \quad x - \frac{7}{4} = -\frac{\sqrt{113}}{4}$$
$$x = \frac{7}{4} + \frac{\sqrt{113}}{4} \quad or \quad x = \frac{7}{4} - \frac{\sqrt{113}}{4}$$
$$x = \frac{7 + \sqrt{113}}{4} \quad or \quad x = \frac{7 - \sqrt{113}}{4}$$

The solutions are $\frac{7 + \sqrt{113}}{4}$ and $\frac{7 - \sqrt{113}}{4}$, or $\frac{7 \pm \sqrt{113}}{4}$.

21.
$$y^2 + 80 = 18y$$
$$y^2 - 18y = -80$$
$$y^2 - 18y + 81 = -80 + 81$$
$$(y - 9)^2 = 1$$
$$y - 9 = 1 \quad or \quad y - 9 = -1$$
$$y = 10 \quad or \quad y = 8$$

The solutions are 10 and 8.

22.
$$t^2 + \frac{3}{2}t - \frac{3}{2} = 0$$
$$t^2 + \frac{3}{2}t = \frac{3}{2}$$
$$t^2 + \frac{3}{2}t + \frac{9}{16} = \frac{3}{2} + \frac{9}{16}$$
$$\left(t + \frac{3}{4}\right)^2 = \frac{33}{16}$$
$$t + \frac{3}{4} = \frac{\sqrt{33}}{4} \quad or \quad t + \frac{3}{4} = -\frac{\sqrt{33}}{4}$$
$$t = -\frac{3}{4} + \frac{\sqrt{33}}{4} \quad or \quad t = -\frac{3}{4} - \frac{\sqrt{33}}{4}$$
$$t = \frac{-3 + \sqrt{33}}{4} \quad or \quad t = \frac{-3 - \sqrt{33}}{4}$$

The solutions are $\frac{-3 + \sqrt{33}}{4}$ and $\frac{-3 - \sqrt{33}}{4}$, or $\frac{-3 \pm \sqrt{33}}{4}$.

23.
$$x + 7 = -3x^2$$
$$3x^2 + x = -7$$
$$\frac{1}{3}(3x^2 + x) = \frac{1}{3}(-7)$$
$$x^2 + \frac{1}{3}x = -\frac{7}{3}$$
$$x^2 + \frac{1}{3}x + \frac{1}{36} = -\frac{7}{3} + \frac{1}{36}$$
$$\left(x + \frac{1}{6}\right)^2 = -\frac{83}{36}$$

Because $-\frac{83}{36}$ has no real-number square root, the equation has no real-number solution.

24. $6x^2 = 384$
$x^2 = 64$
$x = 8$ *or* $x = -8$
The solutions are 8 and -8.

25. $5y^2 + 2y + 3 = 0$
$a = 5,\ b = 2,\ y = 3$
$$y = \frac{-b \pm \sqrt{b^2 - 4ac}}{2a}$$
$$y = \frac{-2 \pm \sqrt{2^2 - 4 \cdot 5 \cdot 3}}{2 \cdot 5} = \frac{-2 \pm \sqrt{4 - 60}}{10}$$
$$y = \frac{-2 \pm \sqrt{-56}}{10}$$
The radicand is negative, so there are no real-number solutions.

26. $6(x - 3)^2 = 12$
$(x - 3)^2 = 2$ Dividing by 6
$x - 3 = \sqrt{2}$ *or* $x - 3 = -\sqrt{2}$
$x = 3 + \sqrt{2}$ *or* $x = 3 - \sqrt{2}$
The solutions are $3 + \sqrt{2}$ and $3 - \sqrt{2}$, or $3 \pm \sqrt{2}$.

27.
$$4x^2 + 4x = 3$$
$$4x^2 + 4x - 3 = 0$$
$$(2x - 1)(2x + 3) = 0$$
$2x - 1 = 0$ *or* $2x + 3 = 0$
$2x = 1$ *or* $2x = -3$
$x = \frac{1}{2}$ *or* $x = -\frac{3}{2}$
The solutions are $\frac{1}{2}$ and $-\frac{3}{2}$.

28. $8y^2 - 5 = 19$
$8y^2 = 24$
$y^2 = 3$ Dividing by 8
$y = \sqrt{3}$ *or* $y = -\sqrt{3}$
The solutions are $\sqrt{3}$ and $-\sqrt{3}$, or $\pm\sqrt{3}$.

29.
$$a^2 = a + 1$$
$$a^2 - a - 1 = 0$$
$a = 1,\ b = -1,\ c = -1$
(Note that a is used in two different ways in this exercise.)
$$a = \frac{-b \pm \sqrt{b^2 - 4ac}}{2a}$$
$$a = \frac{-(-1) \pm \sqrt{(-1)^2 - 4 \cdot 1 \cdot (-1)}}{2 \cdot 1} = \frac{1 \pm \sqrt{1 + 4}}{2}$$
$$a = \frac{1 \pm \sqrt{5}}{2}$$
The solutions are $\frac{1 + \sqrt{5}}{2}$ and $\frac{1 - \sqrt{5}}{2}$, or $\frac{1 \pm \sqrt{5}}{2}$.

30. $(w - 2)^2 = 100$
$w - 2 = 10$ *or* $w - 2 = -10$
$w = 12$ *or* $w = -8$
The solutions are 12 and -8.

31.
$$5m^2 + 2m = -3$$
$$5m^2 + 2m + 3 = 0$$
$a = 5, b = 2, c = 3$
$$m = \frac{-b \pm \sqrt{b^2 - 4ac}}{2a}$$
$$m = \frac{-2 \pm \sqrt{2^2 - 4 \cdot 5 \cdot 3}}{2 \cdot 5} = \frac{-2 \pm \sqrt{4 - 60}}{10}$$
$$m = \frac{-2 \pm \sqrt{-56}}{10}$$
The radicand is negative, so there are no real-number solutions.

32. $\left(y - \frac{1}{2}\right)^2 = \frac{5}{4}$
$y - \frac{1}{2} = \frac{\sqrt{5}}{2}$ *or* $y - \frac{1}{2} = -\frac{\sqrt{5}}{2}$
$y = \frac{1}{2} + \frac{\sqrt{5}}{2}$ *or* $y = \frac{1}{2} - \frac{\sqrt{5}}{2}$
$y = \frac{1 + \sqrt{5}}{2}$ *or* $y = \frac{1 - \sqrt{5}}{2}$
The solutions are $\frac{1 + \sqrt{5}}{2}$ and $\frac{1 - \sqrt{5}}{2}$, or $\frac{1 \pm \sqrt{5}}{2}$.

33. $3x^2 - 75 = 0$
$3x^2 = 75$
$x^2 = 25$
$x = 5$ *or* $x = -5$
The solutions are 5 and -5.

34. $2x^2 - 2x - 5 = 0$

$a = 2,\ b = -2,\ c = -5$

$$x = \frac{-b \pm \sqrt{b^2 - 4ac}}{2a}$$

$$x = \frac{-(-2) \pm \sqrt{(-2)^2 - 4 \cdot 2 \cdot (-5)}}{2 \cdot 2} = \frac{2 \pm \sqrt{4 + 40}}{4}$$

$$x = \frac{2 \pm \sqrt{44}}{4} = \frac{2 \pm \sqrt{4 \cdot 11}}{4} = \frac{2 \pm 2\sqrt{11}}{4}$$

$$x = \frac{2(1 \pm \sqrt{11})}{2 \cdot 2} = \frac{1 \pm \sqrt{11}}{2}$$

The solutions are $\frac{1 + \sqrt{11}}{2}$ and $\frac{1 - \sqrt{11}}{2}$, or $\frac{1 \pm \sqrt{11}}{2}$.

35. $(x + 2)^2 = -5$

Because -5 has no real-number square root, the equation has no real-number solution.

36. $y^2 - y - 8 = 0$

$a = 1,\ b = -1,\ c = -8$

$$y = \frac{-b \pm \sqrt{b^2 - 4ac}}{2a}$$

$$y = \frac{-(-1) \pm \sqrt{(-1)^2 - 4 \cdot 1 \cdot (-8)}}{2 \cdot 1} = \frac{1 \pm \sqrt{1 + 32}}{2}$$

$$y = \frac{1 \pm \sqrt{33}}{2}$$

Using a calculator, we have

$\frac{1 + \sqrt{33}}{2} \approx 3.4$ and

$\frac{1 - \sqrt{33}}{2} \approx -2.4$.

The solutions are approximately 3.4 and -2.4.

37. $2x^2 + 7x + 1 = 0$

$a = 2,\ b = 7,\ c = 1$

$$x = \frac{-b \pm \sqrt{b^2 - 4ac}}{2a}$$

$$x = \frac{-7 \pm \sqrt{7^2 - 4 \cdot 2 \cdot 1}}{2 \cdot 2} = \frac{-7 \pm \sqrt{49 - 8}}{4}$$

$$x = \frac{-7 \pm \sqrt{41}}{4}$$

Using a calculator, we have

$\frac{-7 + \sqrt{41}}{4} \approx -0.1$ and

$\frac{-7 - \sqrt{41}}{4} \approx -3.4$.

The solutions are approximately -0.1 and -3.4.

38. $x^2 - x - 6 = 0$

$(x - 3)(x + 2) = 0$

$x - 3 = 0$ *or* $x + 2 = 0$

$x = 3$ *or* $x = -2$

The equation has two real-number solutions. Answer A is correct.

39. $x^2 = -9$

Because -9 has no real-number square root, the equation has no real-number solutions. Answer B is correct.

40. $x^2 = 31$

$x = \sqrt{31}$ *or* $x = -\sqrt{31}$

The equation has two real-number solutions. Answer A is correct.

41. $x^2 = 0$

$x = 0$

The only solution is 0, so answer C is correct.

42. $x^2 - x + 6 = 0$

$a = 1,\ b = -1,\ c = 6$

$$x = \frac{-b \pm \sqrt{b^2 - 4ac}}{2a}$$

$$x = \frac{-(-1) \pm \sqrt{(-1)^2 - 4 \cdot 1 \cdot 6}}{2 \cdot 1} = \frac{1 \pm \sqrt{1 - 24}}{2}$$

$$x = \frac{1 \pm \sqrt{-23}}{2}$$

Because the radicand is negative, there are no real-number solutions. Answer B is correct.

43. $(x - 3)^2 = 36$

$x - 3 = 6$ *or* $x - 3 = -6$

$x = 9$ *or* $x = -3$

The solutions are 9 and -3. Answer C is correct.

44. $$\frac{-24 \pm \sqrt{720}}{18} = \frac{-24 \pm \sqrt{144 \cdot 5}}{18} = \frac{-24 \pm 12\sqrt{5}}{18} =$$

$$\frac{6(-4 \pm 2\sqrt{5})}{6 \cdot 3} = \frac{-4 \pm 2\sqrt{5}}{3}$$

Answer B is correct.

45. Mark does not recognize that the $\pm$ sign yields two solutions, one in which the radical is added to 3 and the other in which the radical is subtracted from 3.

46. The addition principle should be used at the outset to get 0 on one side of the equation. Since this was not done, the principle of zero products was not used correctly.

47. The first coordinates of the x-intercepts of the graph of $y = (x - 2)(x + 3)$ are the solutions of the equation $(x - 2)(x + 3) = 0$.

48. The quadratic formula would not be the easiest way to solve a quadratic equation when the equation can be solved by factoring or by using the principle of square roots.

49. Answers will vary. Any equation of the form $ax^2 + bx + c = 0$, where $b^2 - 4ac < 0$, will do. Then the graph of the equation $y = ax^2 + bx + c$ will not cross the x-axis.

50. If $x = -5$ or $x = 7$, then $x + 5 = 0$ or $x - 7 = 0$. Thus, the equation $(x + 5)(x - 7) = 0$, or $x^2 - 2x - 35 = 0$, has solutions -5 and 7.

Exercise Set 9.4

RC1. (e)

RC3. (e)

RC5. (d)

1. $q = \dfrac{VQ}{I}$

$I \cdot q = I \cdot \dfrac{VQ}{I}$ Multiplying by I

$Iq = VQ$ Simplifying

$I = \dfrac{VQ}{q}$ Dividing by q

3. $S = \dfrac{kmM}{d^2}$

$d^2 \cdot S = d^2 \cdot \dfrac{kmM}{d^2}$ Multiplying by d^2

$d^2 S = kmM$ Simplifying

$\dfrac{d^2 S}{kM} = m$ Dividing by kM

5. $S = \dfrac{kmM}{d^2}$

$d^2 \cdot S = d^2 \cdot \dfrac{kmM}{d^2}$ Multiplying by d^2

$d^2 S = kmM$ Simplifying

$d^2 = \dfrac{kmM}{S}$ Dividing by S

7. $T = \dfrac{10t}{W^2}$

$W^2 \cdot T = W^2 \cdot \dfrac{10t}{W^2}$ Multiplying by W^2

$W^2 T = 10t$

$W^2 = \dfrac{10t}{T}$ Dividing by T

$W = \sqrt{\dfrac{10t}{T}}$ Principle of square roots. Assume W is nonnegative.

9. $A = at + bt$

$A = t(a+b)$ Factoring

$\dfrac{A}{a+b} = t$ Dividing by $a+b$

11. $y = ax + bx + c$

$y - c = ax + bx$ Subtracting c

$y - c = x(a+b)$ Factoring

$\dfrac{y-c}{a+b} = x$ Dividing by $a+b$

13. $\dfrac{t}{a} + \dfrac{t}{b} = 1$

$ab\left(\dfrac{t}{a} + \dfrac{t}{b}\right) = ab \cdot 1$ Multiplying by ab

$ab \cdot \dfrac{t}{a} + ab \cdot \dfrac{t}{b} = ab$

$bt + at = ab$

$bt = ab - at$ Subtracting at

$bt = a(b-t)$ Factoring

$\dfrac{bt}{b-t} = a$ Dividing by $b-t$

15. $\dfrac{1}{p} + \dfrac{1}{q} = \dfrac{1}{f}$

$pqf\left(\dfrac{1}{p} + \dfrac{1}{q}\right) = pqf \cdot \dfrac{1}{f}$ Multiplying by pqf

$pqf \cdot \dfrac{1}{p} + pqf \cdot \dfrac{1}{q} = pq$

$qf + pf = pq$

$qf = pq - pf$ Subtracting pf

$qf = p(q-f)$ Factoring

$\dfrac{qf}{q-f} = p$ Dividing by $q-f$

17. $A = \dfrac{1}{2}bh$

$2 \cdot A = 2 \cdot \dfrac{1}{2}bh$ Multiplying by 2

$2A = bh$

$\dfrac{2A}{h} = b$ Dividing by h

19. $S = 2\pi r(r+h)$

$S = 2\pi r^2 + 2\pi rh$ Removing parentheses

$S - 2\pi r^2 = 2\pi rh$ Subtracting $2\pi r^2$

$\dfrac{S - 2\pi r^2}{2\pi r} = h$, or Dividing by $2\pi r$

$\dfrac{S}{2\pi r} - r = h$

21. $\dfrac{1}{R} = \dfrac{1}{r_1} + \dfrac{1}{r_2}$

$Rr_1r_2 \cdot \dfrac{1}{R} = Rr_1r_2\left(\dfrac{1}{r_1} + \dfrac{1}{r_2}\right)$ Multiplying by Rr_1r_2

$r_1r_2 = Rr_1r_2 \cdot \dfrac{1}{r_1} + Rr_1r_2 \cdot \dfrac{1}{r_2}$

$r_1r_2 = Rr_2 + Rr_1$

$r_1r_2 = R(r_2 + r_1)$ Factoring

$\dfrac{r_1r_2}{r_2 + r_1} = R$ Dividing by $r_2 + r_1$

23. $P = 17\sqrt{Q}$

$\dfrac{P}{17} = \sqrt{Q}$ Isolating the radical

$\left(\dfrac{P}{17}\right)^2 = (\sqrt{Q})^2$ Principle of squaring

$\dfrac{P^2}{289} = Q$ Simplifying

25.
$$v = \sqrt{\frac{2gE}{m}}$$
$$v^2 = \left(\sqrt{\frac{2gE}{m}}\right)^2 \quad \text{Principle of squaring}$$
$$v^2 = \frac{2gE}{m}$$
$$mv^2 = 2gE \quad \text{Multipying by } m$$
$$\frac{mv^2}{2g} = E \quad \text{Dividing by } 2g$$

27.
$$S = 4\pi r^2$$
$$\frac{S}{4\pi} = r^2 \quad \text{Dividing by } 4\pi$$
$$\sqrt{\frac{S}{4\pi}} = r \quad \text{Principle of square roots. Assume } r \text{ is nonnegative.}$$
$$\sqrt{\frac{1}{4}\cdot\frac{S}{\pi}} = r$$
$$\frac{1}{2}\sqrt{\frac{S}{\pi}} = r$$

29.
$$P = kA^2 + mA$$
$$0 = kA^2 + mA - P \quad \text{Standard form}$$
$$a = k,\ b = m,\ c = -P$$
$$A = \frac{-b \pm \sqrt{b^2 - 4ac}}{2a} \quad \text{Quadratic formula}$$
$$A = \frac{-m \pm \sqrt{m^2 - 4\cdot k\cdot(-P)}}{2\cdot k} \quad \text{Substituting}$$
$$A = \frac{-m + \sqrt{m^2 + 4kP}}{2k} \quad \text{Using the positive root}$$

31.
$$c^2 = a^2 + b^2$$
$$c^2 - b^2 = a^2$$
$$\sqrt{c^2 - b^2} = a \quad \text{Principle of square roots. Assume } a \text{ is nonnegative.}$$

33.
$$s = 16t^2$$
$$\frac{s}{16} = t^2$$
$$\sqrt{\frac{s}{16}} = t \quad \text{Principle of square roots. Assume } t \text{ is nonnegative.}$$
$$\frac{\sqrt{s}}{4} = t$$

35.
$$A = \pi r^2 + 2\pi rh$$
$$0 = \pi r^2 + 2\pi hr - A$$
$$a = \pi,\ b = 2\pi h,\ c = -A$$
$$r = \frac{-b \pm \sqrt{b^2 - 4ac}}{2a}$$
$$r = \frac{-2\pi h \pm \sqrt{(2\pi h)^2 - 4\cdot\pi\cdot(-A)}}{2\cdot\pi}$$
$$r = \frac{-2\pi h + \sqrt{4\pi^2h^2 + 4\pi A}}{2\pi} \quad \text{Using the positive root}$$
$$r = \frac{-2\pi h + \sqrt{4(\pi^2h^2 + \pi A)}}{2\pi}$$
$$r = \frac{-2\pi h + 2\sqrt{\pi^2h^2 + \pi A}}{2\pi}$$
$$r = \frac{2\left(-\pi h + \sqrt{\pi^2h^2 + \pi A}\right)}{2\pi}$$
$$r = \frac{-\pi h + \sqrt{\pi^2h^2 + \pi A}}{\pi}$$

37.
$$F = \frac{Av^2}{400}$$
$$400F = Av^2 \quad \text{Multiplying by 400}$$
$$\frac{400F}{A} = v^2 \quad \text{Dividing by } A$$
$$\sqrt{\frac{400F}{A}} = v \quad \text{Principle of square roots. Assume } v \text{ is nonnegative.}$$
$$\sqrt{400\cdot\frac{F}{A}} = v$$
$$20\sqrt{\frac{F}{a}} = v$$

39.
$$c = \sqrt{a^2 + b^2}$$
$$c^2 = (\sqrt{a^2 + b^2})^2 \quad \text{Principle of squaring}$$
$$c^2 = a^2 + b^2$$
$$c^2 - b^2 = a^2$$
$$\sqrt{c^2 - b^2} = a \quad \text{Principle of square roots. Assume } a \text{ is nonnegative.}$$

41.
$$h = \frac{a}{2}\sqrt{3}$$
$$2h = a\sqrt{3}$$
$$\frac{2h}{\sqrt{3}} = a$$
$$\frac{2h\sqrt{3}}{3} = a \quad \text{Rationalizing the denominator}$$

43.
$$n = aT^2 - 4T + m$$
$$0 = aT^2 - 4T + m - n$$
$$a = a,\ b = -4,\ c = m - n$$

$$T = \frac{-b \pm \sqrt{b^2 - 4ac}}{2a}$$

$$T = \frac{-(-4) \pm \sqrt{(-4)^2 - 4 \cdot a \cdot (m-n)}}{2 \cdot a}$$

$$T = \frac{4 + \sqrt{16 - 4a(m-n)}}{2a} \quad \text{Using the positive root}$$

$$T = \frac{4 + \sqrt{4[4 - a(m-n)]}}{2a}$$

$$T = \frac{4 + 2\sqrt{4 - a(m-n)}}{2a}$$

$$T = \frac{2\left(2 + \sqrt{4 - a(m-n)}\right)}{2 \cdot a}$$

$$T = \frac{2 + \sqrt{4 - a(m-n)}}{a}$$

45.

$$v = 2\sqrt{\frac{2kT}{\pi m}}$$

$$\frac{v}{2} = \sqrt{\frac{2kT}{\pi m}} \quad \text{Isolating the radical}$$

$$\left(\frac{v}{2}\right)^2 = \left(\sqrt{\frac{2kT}{\pi m}}\right)^2 \quad \text{Principle of squaring}$$

$$\frac{v^2}{4} = \frac{2kT}{\pi m}$$

$$\frac{v^2}{4} \cdot \frac{\pi m}{2k} = \frac{2kT}{\pi m} \cdot \frac{\pi m}{2k} \quad \text{Multiplying by } \frac{\pi m}{2k}$$

$$\frac{v^2 \pi m}{8k} = T$$

47. $3x^2 = d^2$

$$x^2 = \frac{d^2}{3} \quad \text{Dividing by 3}$$

$$x = \frac{d}{\sqrt{3}} \quad \text{Principle of square roots. Assume } x \text{ is nonnegative.}$$

$$x = \frac{d}{\sqrt{3}} \cdot \frac{\sqrt{3}}{\sqrt{3}} \quad \text{Rationalizing the denominator}$$

$$x = \frac{d\sqrt{3}}{3}$$

49.

$$N = \frac{n^2 - n}{2}$$

$$2N = n^2 - n \quad \text{Multiplying by 2}$$

$$0 = n^2 - n - 2N \quad \text{Finding standard form}$$

$$a = 1, \ b = -1, \ c = -2N$$

$$n = \frac{-b \pm \sqrt{b^2 - 4ac}}{2a}$$

$$n = \frac{-(-1) \pm \sqrt{(-1)^2 - 4 \cdot 1 \cdot (-2N)}}{2 \cdot 1} \quad \text{Substituting}$$

$$n = \frac{1 + \sqrt{1 + 8N}}{2} \quad \text{Using the positive root}$$

51.

$$S = \frac{a+b}{3b}$$

$$3b \cdot S = 3b \cdot \frac{a+b}{3b}$$

$$3bS = a + b$$

$$3bS - b = a$$

$$b(3S - 1) = a$$

$$b = \frac{a}{3S - 1}$$

53.

$$\frac{A - B}{AB} = Q$$

$$AB \cdot \frac{A - B}{AB} = AB \cdot Q$$

$$A - B = ABQ$$

$$A = ABQ + B$$

$$A = B(AQ + 1)$$

$$\frac{A}{AQ + 1} = B$$

55.

$$S = 180(n - 2)$$

$$S = 180n - 360$$

$$S + 360 = 180n$$

$$\frac{S + 360}{180} = n, \text{ or}$$

$$\frac{S}{180} + 2 = n$$

57.

$$A = P(1 + rt)$$

$$A = P + Prt$$

$$A - P = Prt$$

$$\frac{A - P}{Pr} = t$$

59.

$$\frac{A}{B} = \frac{C}{D}$$

$$BD \cdot \frac{A}{B} = BD \cdot \frac{C}{D}$$

$$AD = BC$$

$$D = \frac{BC}{A}$$

61.

$a^2 + b^2 = c^2$ Pythagorean equation

$4^2 + 7^2 = c^2$ Substituting

$16 + 49 = c^2$

$65 = c^2$

$\sqrt{65} = c$ Exact answer

$8.062 \approx c$ Approximate answer

63.

$a^2 + b^2 = c^2$ Pythagorean equation

$4^2 + 5^2 = c^2$ Substituting

$16 + 25 = c^2$

$41 = c^2$

$\sqrt{41} = c$ Exact answer

$6.403 \approx c$ Approximate answer

65.
$$\begin{aligned} a^2+b^2&=c^2 && \text{Pythagorean equation}\\ 2^2+b^2&=(8\sqrt{17})^2 && \text{Substituting}\\ 4+b^2&=64\cdot 17\\ 4+b^2&=1088\\ b^2&=1084\\ b&=\sqrt{1084} && \text{Exact answer}\\ b&\approx 32.924 && \text{Approximate answer}\end{aligned}$$

67. We make a drawing. Let $l =$ the length of the guy wire.

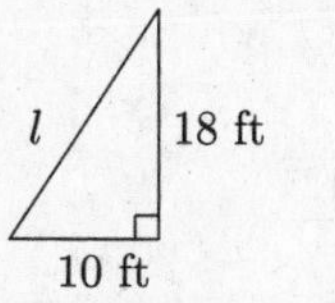

Then we use the Pythagorean equation.

$$\begin{aligned} 10^2+18^2&=l^2\\ 100+324&=l^2\\ 424&=l^2\\ \sqrt{424}&=l && \text{Exact answer}\\ 20.591&\approx l && \text{Approximation}\end{aligned}$$

The length of the guy wire is $\sqrt{424}$ ft ≈ 20.591 ft.

69. $\sqrt{80}=\sqrt{16\cdot 5}=\sqrt{16}\sqrt{5}=4\sqrt{5}$

71. $3\sqrt{t}\cdot\sqrt{t}=3\sqrt{t^2}=3t$

73.
$$\begin{aligned}\sqrt{40}-2\sqrt{10}+\sqrt{90}&=\sqrt{4\cdot 10}-2\sqrt{10}+\sqrt{9\cdot 10}\\ &=\sqrt{4}\sqrt{10}-2\sqrt{10}+\sqrt{9}\sqrt{10}\\ &=2\sqrt{10}-2\sqrt{10}+3\sqrt{10}\\ &=(2-2+3)\sqrt{10}\\ &=3\sqrt{10}\end{aligned}$$

75. a)
$$\begin{aligned} C&=2\pi r\\ \frac{C}{2\pi}&=r\end{aligned}$$

b)
$$\begin{aligned} A&=\pi r^2\\ A&=\pi\cdot\left(\frac{C}{2\pi}\right)^2 && \text{Substituting } \frac{C}{2\pi} \text{ for } r\\ A&=\pi\cdot\frac{C^2}{4\pi^2}\\ A&=\frac{C^2}{4\pi}\end{aligned}$$

c)
$$\begin{aligned} A&=\frac{C^2}{4\pi} && \text{(See part (b).)}\\ A\cdot 4\pi&=C^2\\ \sqrt{4A\pi}&=\sqrt{C^2}\\ \sqrt{4}\sqrt{A\pi}&=C\\ 2\sqrt{A\pi}&=C\end{aligned}$$

Exercise Set 9.5

RC1. True; $A = lw$.

RC3. True

1. ***Familiarize.*** Using the drawing in the text, we have $w =$ the width, in yards, and $3w+1 =$ the length.

Translate. We use the formula for the area of a rectangle.

$$w(3w+1)=80$$

Solve. We solve the equation.

$$\begin{aligned} w(3w+1)&=80\\ 3w^2+w&=80\\ 3w^2+w-80&=0\\ (3w+16)(w-5)&=0\\ 3w+16=0 \quad &or \quad w-5=0\\ 3w=-16 \quad &or \quad w=5\\ w=-\frac{16}{3} \quad &or \quad w=5\end{aligned}$$

Since the width cannot be negative, we consider only 5 as a possible solution.

If $w=5$, then $3w+1=3(5)+1=15+1=16$.

Check. A length of 16 yd is 1 yd longer than three times a width of 5 yd. Also, $16(5)=80$, so the answer checks.

State. The length of the pool is 16 yd, and the width is 5 yd.

3. ***Familiarize.*** Using the drawing in the text, we let s and $s+8$ represent the lengths of the legs, in inches.

Translate. We use the Pythagorean equation.

$$s^2+(s+8)^2=(8\sqrt{13})^2$$

Solve. We solve the equation.

$$\begin{aligned} s^2+(s+8)^2&=(8\sqrt{13})^2\\ s^2+s^2+16s+64&=64\cdot 13\\ 2s^2+16s+64&=832\\ 2s^2+16s-768&=0\\ s^2+8s-384&=0 && \text{Dividing by 2}\\ (s+24)(s-16)&=0\\ s+24=0 \quad &or \quad s-16=0\\ s=-24 \quad &or \quad s=16\end{aligned}$$

Since the length of a leg cannot be negative, we consider only 16 as a possible solution.

If $s=16$, then $s+8=16+8=24$.

Check. The 24-in. side is 8 in. longer than the 16-in. side. Also, $16^2+24^2=256+576=832=(8\sqrt{13})^2$. The answer checks.

State. The lengths of the legs of the square are 16 in. and 24 in.

5. ***Familiarize.*** Let $w =$ the width of the rectangle, in feet. Then $3w =$ the length.

Translate. We use the formula for the area of a rectangle.

$$3w \cdot w = 300$$

Solve. We solve the equation.

$$\begin{aligned} 3w \cdot w &= 300 \\ 3w^2 &= 300 \\ w^2 &= 100 \\ w^2 - 100 &= 0 \\ (w+10)(w-10) &= 0 \end{aligned}$$

$$w + 10 = 0 \quad \text{or} \quad w - 10 = 0$$
$$w = -10 \quad \text{or} \quad w = 10$$

Since the width cannot be negative, we consider only 10 as a possible solution.

If $w = 10$, then $3w = 3 \cdot 10 = 30$.

Check. 30 ft is three times 10 ft. Also, $30 \cdot 10 = 300$ ft^2. The answer checks.

State. The length of the garden is 30 ft, and the width is 10 ft.

7. ***Familiarize.*** We first make a drawing. We let x represent the length. Then $x - 4$ represents the width.

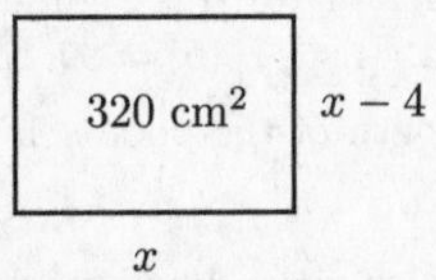

Translate. The area is length × width. Thus, we have two expressions for the area of the rectangle: $x(x-4)$ and 320. This gives us a translation.

$$x(x-4) = 320.$$

Solve. We solve the equation.

$$\begin{aligned} x^2 - 4x &= 320 \\ x^2 - 4x - 320 &= 0 \\ (x-20)(x+16) &= 0 \end{aligned}$$

$$x - 20 = 0 \quad \text{or} \quad x + 16 = 0$$
$$x = 20 \quad \text{or} \quad x = -16$$

Check. Since the length of a side cannot be negative, -16 does not check. But 20 does check. If the length is 20, then the width is $20 - 4$, or 16. The area is 20×16, or 320. This checks.

State. The length is 20 cm, and the width is 16 cm.

9. ***Familiarize.*** We first make a drawing. We let x represent the length of one leg. Then $x + 2$ represents the length of the other leg.

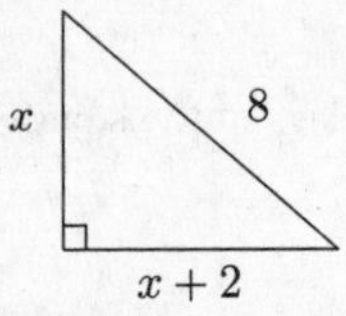

Translate. We use the Pythagorean equation.

$$x^2 + (x+2)^2 = 8^2.$$

Solve. We solve the equation.

$$\begin{aligned} x^2 + x^2 + 4x + 4 &= 64 \\ 2x^2 + 4x + 4 &= 64 \\ 2x^2 + 4x - 60 &= 0 \\ x^2 + 2x - 30 &= 0 \quad \text{Dividing by 2} \end{aligned}$$

$a = 1,\ b = 2,\ c = -30$

$$\begin{aligned} x &= \frac{-2 \pm \sqrt{2^2 - 4 \cdot 1 \cdot (-30)}}{2 \cdot 1} \\ &= \frac{-2 \pm \sqrt{4 + 120}}{2} = \frac{-2 \pm \sqrt{124}}{2} \\ &= \frac{-2 \pm \sqrt{4 \cdot 31}}{2} = \frac{-2 \pm 2\sqrt{31}}{2} \\ &= \frac{2(-1 \pm \sqrt{31})}{2} = -1 \pm \sqrt{31} \end{aligned}$$

Using a calculator we find that $\sqrt{31} \approx 5.568$:

$$-1 + \sqrt{31} \approx -1 + 5.568 \quad \text{or} \quad -1 - \sqrt{31} \approx -1 - 5.568$$
$$\approx 4.6 \quad \text{or} \quad \approx -6.6$$

Check. Since the length of a leg cannot be negative, -6.6 does not check. But 4.6 does check. If the shorter leg is 4.6, then the other leg is $4.6 + 2$, or 6.6. Then $4.6^2 + 6.6^2 = 21.16 + 43.56 = 64.72$ and using a calculator, $\sqrt{64.72} \approx 8.04 \approx 8$. Note that our check is not exact since we are using an approximation.

State. One leg is about 4.6 m, and the other is about 6.6 m long.

11. ***Familiarize.*** We first make a drawing. We let x represent the width and $x + 2$ the length.

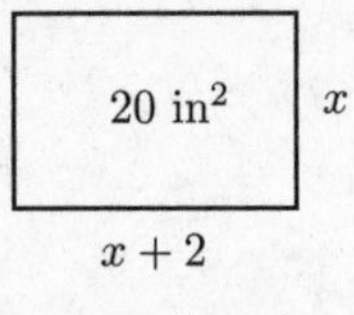

Translate. The area is length × width. We have two expressions for the area of the rectangle: $(x+2)x$ and 20. This gives us a translation.

$$(x+2)x = 20.$$

Solve. We solve the equation.

$$\begin{aligned} x^2 + 2x &= 20 \\ x^2 + 2x - 20 &= 0 \end{aligned}$$

$a = 1,\ b = 2,\ c = -20$

$$\begin{aligned} x &= \frac{-2 \pm \sqrt{2^2 - 4 \cdot 1 \cdot (-20)}}{2 \cdot 1} \\ &= \frac{-2 \pm \sqrt{4 + 80}}{2} = \frac{-2 \pm \sqrt{84}}{2} \\ &= \frac{-2 \pm \sqrt{4 \cdot 21}}{2} = \frac{-2 \pm 2\sqrt{21}}{2} \\ &= \frac{2(-1 \pm \sqrt{21})}{2} = -1 \pm \sqrt{21} \end{aligned}$$

Using a calculator we find that $\sqrt{21} \approx 4.583$:

$$-1+\sqrt{21} \approx -1+4.583 \quad or \quad -1-\sqrt{21} \approx -1-4.583$$
$$\approx 3.6 \quad or \quad \approx -5.6$$

Check. Since the length of a side cannot be negative, -5.6 does not check. But 3.6 does check. If the width is 3.6, then the length is $3.6+2$, or 5.6. The area is $5.6(3.6)$, or $20.16 \approx 20$. This checks.

State. The length is about 5.6 in., and the width is about 3.6 in.

13. ***Familiarize.*** We make a drawing and label it. We let w = the width of the rectangle and $2w$ = the length.

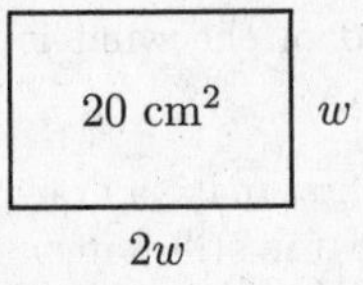

Translate. Recall that area = length × width. Then we have

$$2w \cdot w = 20.$$

Solve. We solve the equation.

$$2w^2 = 20$$
$$w^2 = 10 \quad \text{Dividing by 2}$$

$$w = \sqrt{10} \quad or \quad w = -\sqrt{10} \quad \text{Principle of square roots}$$
$$w \approx 3.2 \quad or \quad w \approx -3.2$$

Check. We know that -3.2 is not a solution of the original problem, because width cannot be negative. When $w \approx 3.2$, then $2w \approx 6.4$ and the area is about $(6.4)(3.2)$, or 20.48. This checks, although the check is not exact since we used an approximation for $\sqrt{10}$.

State. The length is about 6.4 cm, and the width is about 3.2 cm.

15. ***Familiarize.*** Using the drawing in the text, we have x = the width of the frame, $20-2x$ = the width of the picture showing, and $25-2x$ = the length of the picture showing.

Translate. Recall that area = length × width. Then we have

$$(25-2x)(20-2x) = 266.$$

Solve. We solve the equation.

$$500-90x+4x^2 = 266$$
$$4x^2-90x+234 = 0$$
$$2x^2-45x+117 = 0 \quad \text{Dividing by 2}$$
$$(2x-39)(x-3) = 0$$

$$2x-39=0 \quad or \quad x-3=0$$
$$2x=39 \quad or \quad x=3$$
$$x=19.5 \quad or \quad x=3$$

Check. The number 19.5 cannot be a solution, because when $x=19.5$ then $20-2x=-19$, and the width cannot be negative. When $x=3$, then $20-2x = 20-2\cdot 3$, or 14 and $25-2x = 25-2\cdot 3$, or 19 and $19\cdot 14 = 266$. This checks.

State. The width of the frame is 3 cm.

17. ***Familiarize.*** Referring to the drawing in the text, we complete the table.

	d	r	t
Upstream	40	$r-3$	t_1
Downstream	40	$r+3$	t_2
Total Time			14

Translate. Using $t = d/r$ and the rows of the table, we have

$$t_1 = \frac{40}{r-3} \text{ and } t_2 = \frac{40}{r+3}.$$

Since the total time is 14 hr, $t_1+t_2 = 14$, and we have

$$\frac{40}{r-3}+\frac{40}{r+3} = 14.$$

Solve. We solve the equation. We multiply by $(r-3)(r+3)$, the LCM of the denominators.

$$(r-3)(r+3)\left(\frac{40}{r-3}+\frac{40}{r+3}\right) = (r-3)(r+3)\cdot 14$$
$$40(r+3)+40(r-3) = 14(r^2-9)$$
$$40r+120+40r-120 = 14r^2-126$$
$$80r = 14r^2-126$$
$$0 = 14r^2-80r-126$$
$$0 = 7r^2-40r-63$$
$$0 = (7r+9)(r-7)$$

$$7r+9=0 \quad or \quad r-7=0$$
$$7r=-9 \quad or \quad r=7$$
$$r=-\frac{9}{7} \quad or \quad r=7$$

Check. Since speed cannot be negative, $-\frac{9}{7}$ cannot be a solution. If the speed of the boat is 7 km/h, the speed upstream is $7-3$, or 4 km/h, and the speed downstream is $7+3$, or 10 km/h. The time upstream is $\frac{40}{4}$, or 10 hr. The time downstream is $\frac{40}{10}$, or 4 hr. The total time is 14 hr. This checks.

State. The speed of the boat in still water is 7 km/h.

19. ***Familiarize.*** Let r = the speed of the stream. Then the boat's speed upstream is $8-r$ and the speed downstream is $8+r$. We fill in the table in the text.

	d	r	t
Upstream	60	$8-r$	t_1
Downstream	60	$8+r$	t_2
Total Time			16

Translate. Using $t = d/r$ and the rows of the table, we have

$$t_1 = \frac{60}{8-r} \quad \text{and} \quad t_2 = \frac{60}{8+r}.$$

Since the total time is 16 hr, $t_1 + t_2 = 16$, and we have

$$\frac{60}{8-r} + \frac{60}{8+r} = 16.$$

Solve. We first multiply by $(8-r)(8+r)$, the LCM of the denominators.

$$(8-r)(8+r)\left(\frac{60}{8-r} + \frac{60}{8+r}\right) = (8-r)(8+r) \cdot 16$$
$$60(8+r) + 60(8-r) = (64 - r^2)16$$
$$480 + 60r + 480 - 60r = 1024 - 16r^2$$
$$960 = 1024 - 16r^2$$
$$16r^2 - 64 = 0$$
$$16(r^2 - 4) = 0$$
$$16(r+2)(r-2) = 0$$

$r + 2 = 0$ *or* $r - 2 = 0$

$r = -2$ *or* $r = 2$

Check. Since the speed of the stream cannot be negative, -2 cannot be a solution. If the speed of the stream is 2 km/h, the boat's speed upstream is $8 - 2$ or 6 km/h, and the speed downstream is $8 + 2$, or 10 km/h. The time upstream is $\frac{60}{6}$, or 10 hr. The time downstream is $\frac{60}{10}$, or 6 hr. The total time is $10 + 6$, or 16 hr. The answer checks.

State. The speed of the stream is 2 km/h.

21. ***Familiarize***. Let r represent the speed of the wind. Then the speed of the plane flying with the wind is $300 + r$ and the speed against the wind is $300 - r$. We fill in the table in the text.

	d	r	t
With wind	680	$300 + r$	t_1
Against wind	520	$300 - r$	t_2

Translate. Using $t = d/r$ and the rows of the table, we have

$$t_1 = \frac{680}{300+r} \text{ and } t_2 = \frac{520}{300-r}.$$

Since the total time is 4 hr, $t_1 + t_2 = 4$, and we have

$$\frac{680}{300+r} + \frac{520}{300-r} = 4.$$

Solve. We solve the equation. We multiply by $(300 + r)(300 - r)$, the LCM of the denominators.

$$(300+r)(300-r)\left(\frac{680}{300+r} + \frac{520}{300-r}\right) = (300+r)(300-r) \cdot 4$$
$$680(300 - r) + 520(300 + r) = 4(90{,}000 - r^2)$$
$$204{,}000 - 680r + 156{,}000 + 520r = 360{,}000 - 4r^2$$
$$360{,}000 - 160r = 360{,}000 - 4r^2$$
$$4r^2 - 160r = 0$$
$$4r(r - 40) = 0$$

$4r = 0$ *or* $r - 40 = 0$

$r = 0$ *or* $r = 40$

Check. If $r = 0$, then the speed of the wind is 0 km/h. That is, there is no wind. In this case the plane travels 680 km in 680/300, or $2\frac{4}{15}$ hr, and it travels 520 km in 520/300, or $1\frac{11}{15}$ hr. The total time is $2\frac{4}{15} + 1\frac{11}{15}$, or 4 hr, so we have one solution. If the speed of the wind is 40 km/h, then the speed of the airplane with the wind is $300 + 40$, or 340 km/h, and the speed against the wind is $300 - 40$, or 260 km/h. The time with the wind is 680/340, or 2 hr, and the time against the wind is 520/260, or 2 hr. The total time is 2 hr + 2 hr, or 4 hr, so we have a second solution.

State. The speed of the wind is 0 km/h (There is no wind.) or 40 km/h.

23. ***Familiarize.*** We first make a drawing. We let r represent the speed of the boat in still water. Then $r - 4$ is the speed of the boat traveling upstream and $r + 4$ is the speed of the boat traveling downstream.

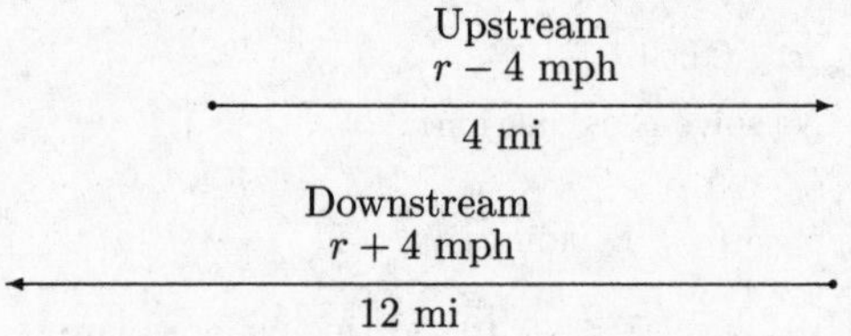

We summarize the information in a table.

	d	r	t
Upstream	4	$r - 4$	t_1
Downstream	12	$r + 4$	t_2
Total Time			2

Translate. Using $t = d/r$ and the rows of the table, we have

$$t_1 = \frac{4}{r-4} \text{ and } t_2 = \frac{12}{r+4}.$$

Since the total time is 2 hr, $t_1 + t_2 = 2$, and we have

$$\frac{4}{r-4} + \frac{12}{r+4} = 2.$$

Solve. We solve the equation. We multiply by $(r - 4)(r + 4)$, the LCM of the denominators.

$$(r-4)(r+4)\left(\frac{4}{r-4} + \frac{12}{r+4}\right) = (r-4)(r+4) \cdot 2$$
$$4(r + 4) + 12(r - 4) = 2(r^2 - 16)$$
$$4r + 16 + 12r - 48 = 2r^2 - 32$$
$$16r - 32 = 2r^2 - 32$$
$$0 = 2r^2 - 16r$$
$$0 = 2r(r - 8)$$

$2r = 0$ *or* $r - 8 = 0$

$r = 0$ *or* $r = 8$

Check. If $r = 0$, then the speed upstream, $0 - 4$, would be negative. Since speed cannot be negative, 0 cannot be a solution. If the speed of the boat is 8 mph, the speed upstream is $8 - 4$, or 4 mph, and the speed downstream is

$8+4$, or 12 mph. The time upstream is $\frac{4}{4}$, or 1 hr. The time downstream is $\frac{12}{12}$, or 1 hr. The total time is 2 hr. This checks.

State. The speed of the boat in still water is 8 mph.

25. ***Familiarize.*** We first make a drawing. We let r represent the speed of the stream. Then $9-r$ represents the speed of the boat traveling upstream and $9+r$ represents the speed of the boat traveling downstream.

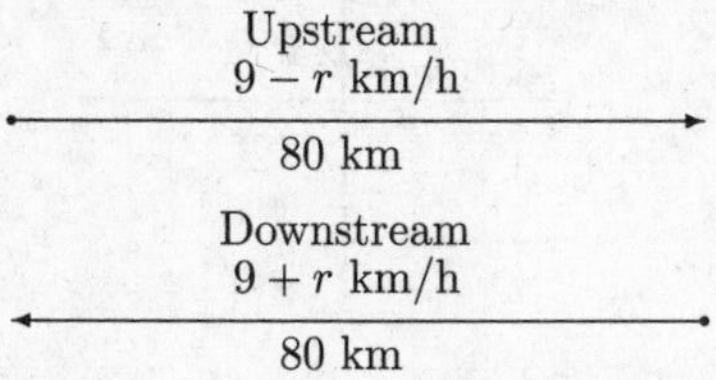

We summarize the information in a table.

	d	r	t
Upstream	80	$9-r$	t_1
Downstream	80	$9+r$	t_2

Translate. Using $t = d/r$ and the rows of the table, we have

$$t_1 = \frac{80}{9-r} \text{ and } t_2 = \frac{80}{9+r}.$$

Since the total time is 18 hr, $t_1 + t_2 = 18$, and we have

$$\frac{80}{9-r} + \frac{80}{9+r} = 18.$$

Solve. We solve the equation. We multiply by $(9-r)(9+r)$, the LCM of the denominators.

$$(9-r)(9+r)\left(\frac{80}{9-r} + \frac{80}{9+r}\right) = (9-r)(9+r)\cdot 18$$
$$80(9+r) + 80(9-r) = 18(81-r^2)$$
$$720 + 80r + 720 - 80r = 1458 - 18r^2$$
$$1440 = 1458 - 18r^2$$
$$18r^2 = 18$$
$$r^2 = 1$$

$r = 1$ *or* $r = -1$ Principle of square roots

Check. Since speed cannot be negative, -1 cannot be a solution. If the speed of the stream is 1 km/h, the speed upstream is $9-1$, or 8 km/h, and the speed downstream is $9+1$, or 10 km/h. The time upstream is $\frac{80}{8}$, or 10 hr. The time downstream is $\frac{80}{10}$, or 8 hr. The total time is 18 hr. This checks.

State. The speed of the stream is 1 km/h.

27. $8x = 4 - y$

To find the y-intercept, we let $x = 0$ and solve for y.

$$8\cdot 0 = 4 - y$$
$$0 = 4 - y$$
$$y = 4$$

The y-intercept is $(0, 4)$.

To find the x-intercept, we let $y = 0$ and solve for x.

$$8x = 4 - 0$$
$$8x = 4$$
$$x = \frac{1}{2}$$

The x-intercept is $\left(\frac{1}{2}, 0\right)$.

29. Graph $y = 3x - 5$

The slope is 3, or $\frac{3}{1}$, and the y-intercept is $(0, -5)$. Start at $(0, -5)$ and move 3 units up and 1 unit right to the point $(1, -2)$. From this point, again move 3 units up and 1 unit right to the point $(2, 1)$. Then draw the line passing through these three points.

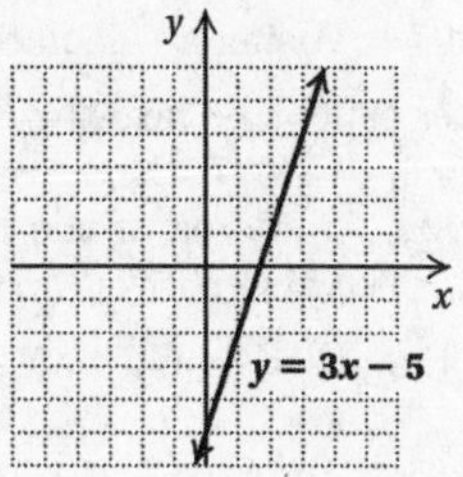

31. Graph $y = 1$.

The graph is a horizontal line with y-intercept $(0, 1)$.

33. Substitute -2 for m and -5 for b in $y = mx + b$.

$$y = mx + b$$
$$y = -2x + (-5)$$
$$y = -2x - 5$$

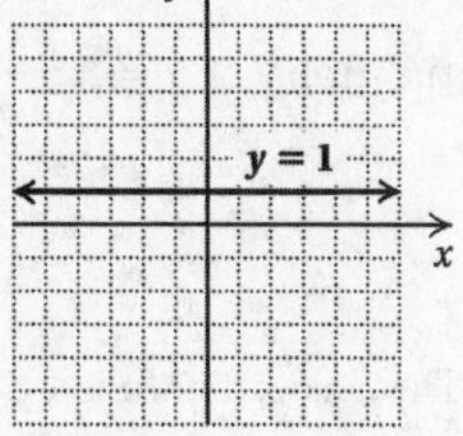

35. First write the second equation in slope-intercept form.

$$3x + 4y = 7$$
$$4y = -3x + 7$$
$$y = -\frac{3}{4}x + \frac{7}{4}$$

The slopes are different, so the lines are not parallel.

37. ***Familiarize.*** The radius of a 12-in. pizza is $\frac{12}{2}$, or 6 in. The radius of a d-in. pizza is $\frac{d}{2}$ in. The area of a circle is πr^2.

Translate.

$$\underbrace{\text{Area of } d\text{-in. pizza}}_{\pi\left(\frac{d}{2}\right)^2} \ \underbrace{\text{is}}_{=} \ \underbrace{\text{Area of 12-in. pizza}}_{\pi \cdot 6^2} \ \underbrace{\text{plus}}_{+} \ \underbrace{\text{Area of 12-in. pizza}}_{\pi \cdot 6^2}$$

Solve. We solve the equation.

$$\frac{d^2}{4}\pi = 36\pi + 36\pi$$
$$\frac{d^2}{4}\pi = 72\pi$$
$$\frac{d^2}{4} = 72 \quad \text{Dividing by } \pi$$
$$d^2 = 288$$
$$d = \sqrt{288} \ \text{ or } \ d = -\sqrt{288}$$
$$d = 12\sqrt{2} \ \text{ or } \ d = -12\sqrt{2}$$
$$d \approx 16.97 \ \text{ or } \ d \approx -16.97 \quad \text{Using a calculator}$$

Check. Since the diameter cannot be negative, -16.97 is not a solution. If $d = 12\sqrt{2}$, or 16.97, then $r = 6\sqrt{2}$ and the area is $\pi(6\sqrt{2})^2$, or 72π. The area of the two 12-in. pizzas is $2 \cdot \pi \cdot 6^2$, or 72π. The value checks.

State. The diameter of the pizza should be $12\sqrt{2}$ in. $\approx$ 16.97 in.

The radius of a 16-in. pizza is $\frac{16}{2}$, or 8 in., so the area is $\pi(8)^2$, or 64π. We found that the area of two 12-in. pizzas is 72π and $72\pi > 64\pi$, so you get more to eat with two 12-in. pizzas than with a 16-in. pizza.

Exercise Set 9.6

RC1. The equation $x^2 - 9x + 8 = 0$ is an example of a quadratic equation.

RC3. The turning point of the graph of $y = x^2 - 9x + 8$ is the vertex.

1. $y = x^2 + 1$

We first find the vertex. The x-coordinate is

$$-\frac{b}{2a} = -\frac{0}{2 \cdot 1} = 0.$$

We substitute into the equation to find the second coordinate of the vertex.

$$y = x^2 + 1 = 0^2 + 1 = 1$$

The vertex is (0,1). This is also the y-intercept. The line of symmetry is $x = 0$, the y-axis.

We choose some x-values on both sides of the vertex and graph the parabola.

When $x = 1$, $y = 1^2 + 1 = 1 + 1 = 2$.
When $x = -1$, $y = (-1)^2 + 1 = 1 + 1 = 2$.
When $x = 2$, $y = 2^2 + 1 = 4 + 1 = 5$.
When $x = -2$, $y = (-2)^2 + 1 = 4 + 1 = 5$.

x	y	
-2	5	
-1	2	
0	1	← Vertex
1	2	
2	5	
3	10	

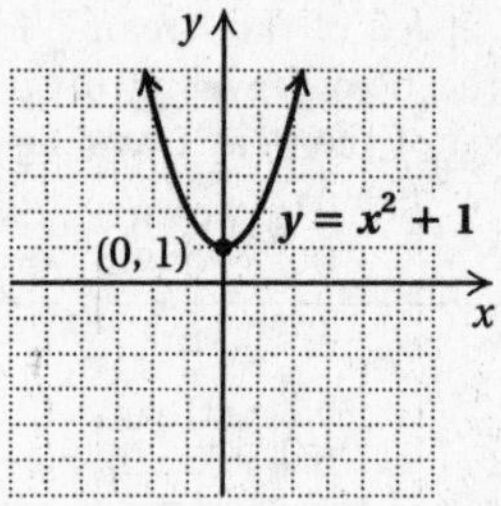

3. $y = -1 \cdot x^2$

Find the vertex. The x-coordinate is

$$-\frac{b}{2a} = -\frac{0}{2(-1)} = 0.$$

The y-coordinate is

$$y = -1 \cdot x^2 = -1 \cdot 0^2 = 0.$$

The vertex is (0,0). This is also the y-intercept. The line of symmetry is $x = 0$, the y-axis.

Choose some x-values on both sides of the vertex and graph the parabola.

When $x = -2$, $y = -1 \cdot (-2)^2 = -1 \cdot 4 = -4$.
When $x = -1$, $y = -1 \cdot (-1)^2 = -1 \cdot 1 = -1$.
When $x = 1$, $y = -1 \cdot 1^2 = -1 \cdot 1 = -1$.
When $x = 2$, $y = -1 \cdot 2^2 = -1 \cdot 4 = -4$.

x	y	
0	0	←Vertex
-2	-4	
-1	-1	
1	-1	
2	-4	

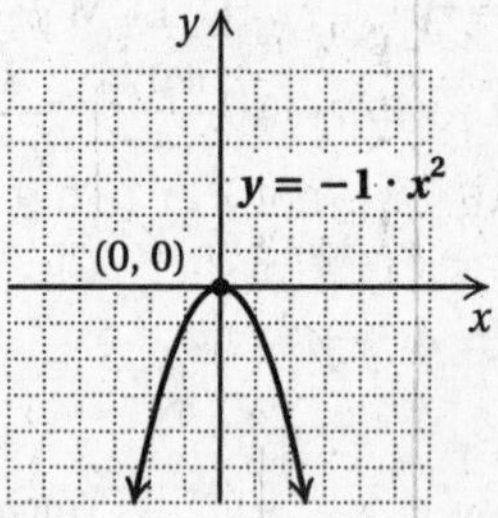

5. $y = -x^2 + 2x$

Find the vertex. The x-coordinate is

$$-\frac{b}{2a} = -\frac{2}{2(-1)} = -(-1) = 1.$$

The y-coordinate is

$$y = -x^2 + 2x = -(1)^2 + 2 \cdot 1 = -1 + 2 = 1.$$

The vertex is (1,1).

We choose some x-values on both sides of the vertex and graph the parabola. We make sure we find y when $x = 0$. This gives us the y-intercept.

x	y	
1	1	←Vertex
0	0	←y-intercept
−1	−3	
2	0	
3	−3	

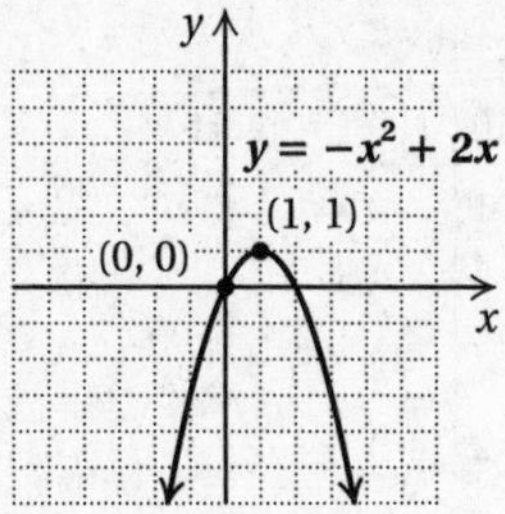

7. $y = 5 - x - x^2$, or $y = -x^2 - x + 5$

Find the vertex. The x-coordinate is

$$-\frac{b}{2a} = -\frac{-1}{2(-1)} = -\frac{1}{2}.$$

The y-coordinate is

$$y = 5 - x - x^2 = 5 - \left(-\frac{1}{2}\right) - \left(-\frac{1}{2}\right)^2 = 5 + \frac{1}{2} - \frac{1}{4} = \frac{21}{4}.$$

The vertex is $\left(-\frac{1}{2}, \frac{21}{4}\right)$.

We choose some x-values on both sides of the vertex and graph the parabola.

x	y	
$-\frac{1}{2}$	$\frac{21}{4}$	←Vertex
0	5	←y-intercept
−1	5	
−2	3	
1	3	

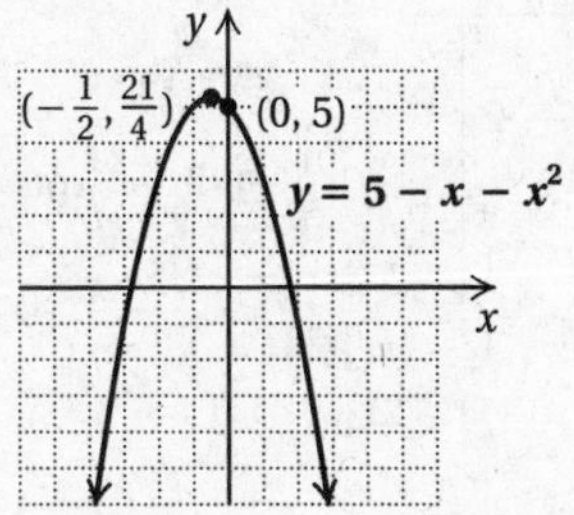

9. $y = x^2 - 2x + 1$

Find the vertex. The x-coordinate is

$$-\frac{b}{2a} = -\frac{-2}{2 \cdot 1} = -(-1) = 1.$$

The y-coordinate is

$$y = x^2 - 2x + 1 = 1^2 - 2 \cdot 1 + 1 = 1 - 2 + 1 = 0.$$

The vertex is $(1, 0)$.

We choose some x-values on both sides of the vertex and graph the parabola.

x	y	
1	0	←Vertex
0	1	←y-intercept
−1	4	
2	1	
3	4	

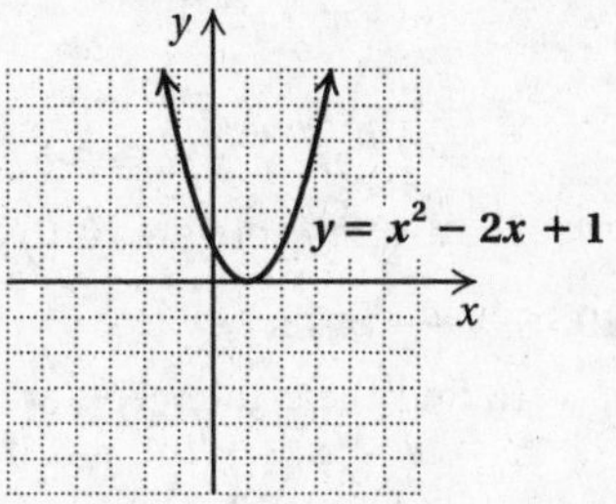

11. $y = -x^2 + 2x + 3$

Find the vertex. The x-coordinate is

$$-\frac{b}{2a} = -\frac{2}{2(-1)} = -(-1) = 1.$$

The y-coordinate is

$$y = -x^2 + 2x + 3 = -(1)^2 + 2 \cdot 1 + 3 = -1 + 2 + 3 = 4.$$

The vertex is $(1, 4)$.

We choose some x-values on both sides of the vertex and graph the parabola.

x	y	
1	4	←Vertex
0	3	←y-intercept
−1	0	
2	3	
3	0	

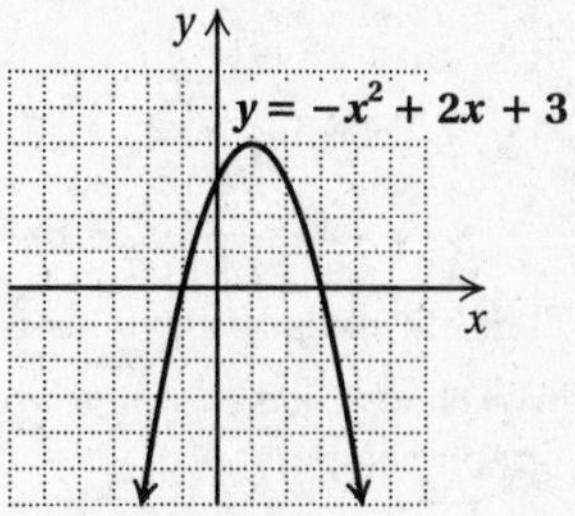

13. $y = -2x^2 - 4x + 1$

Find the vertex. The x-coordinate is

$$-\frac{b}{2a} = -\frac{-4}{2(-2)} = -1.$$

The y-coordinate is

$$y = -2x^2 - 4x + 1 = -2(-1)^2 - 4(-1) + 1 = -2 + 4 + 1 = 3.$$

The vertex is $(-1, 3)$.

We choose some x-values on both sides of the vertex and graph the parabola.

x	y	
−1	3	←Vertex
0	1	←y-intercept
1	−5	
−2	1	
−3	−5	

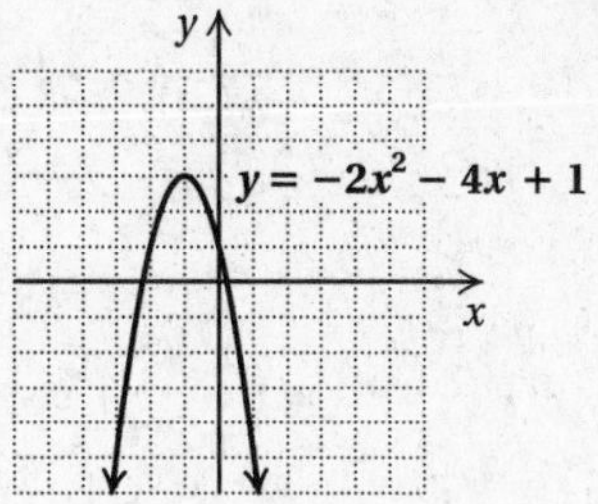

15. $y = 5 - x^2$, or $y = -x^2 + 5$

Find the vertex. The x-coordinate is

$$-\frac{b}{2a} = -\frac{0}{2(-1)} = 0.$$

The y-coordinate is

$$y = 5 - x^2 = 5 - 0^2 = 5.$$

The vertex is $(0, 5)$. This is also the y-intercept.

We choose some x-values on both sides of the vertex and graph the parabola.

x	y	
0	5	←Vertex
−1	4	
−2	1	
1	4	
2	1	

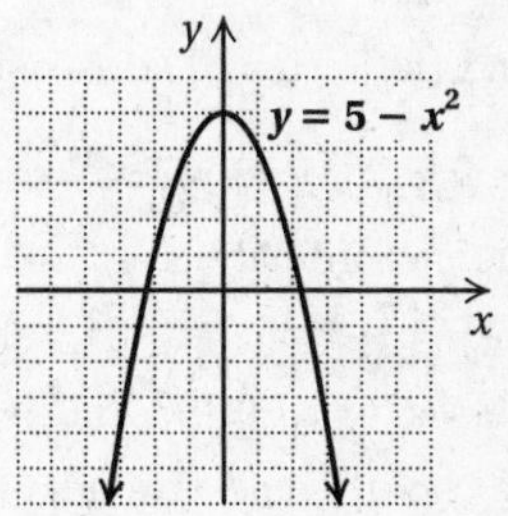

17. $y = \frac{1}{4}x^2$

Find the vertex. The x-coordinate is

$$-\frac{b}{2a} = -\frac{0}{2\left(\frac{1}{4}\right)} = 0.$$

The y-coordinate is

$$y = \frac{1}{4}x^2 = \frac{1}{4}\cdot 0^2 = 0.$$

The vertex is $(0, 0)$. This is also the y-intercept.

We choose some x-values on both sides of the vertex and graph the parabola.

x	y	
0	0	←Vertex
−2	1	
−4	4	
2	1	
4	4	

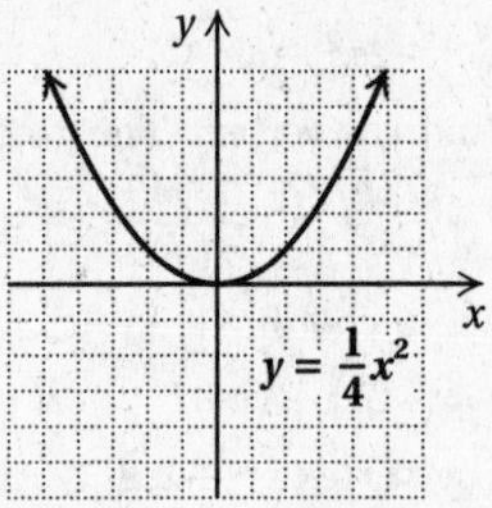

19. $y = -x^2 + x - 1$

Find the vertex. The x-coordinate is

$$-\frac{b}{2a} = -\frac{1}{2(-1)} = -\left(-\frac{1}{2}\right) = \frac{1}{2}.$$

The y-coordinate is

$$y = -x^2 + x - 1 = -\left(\frac{1}{2}\right)^2 + \frac{1}{2} - 1 = -\frac{1}{4} + \frac{1}{2} - 1 = -\frac{3}{4}.$$

The vertex is $\left(\frac{1}{2}, -\frac{3}{4}\right)$.

We choose some x-values on both sides of the vertex and graph the parabola.

x	y	
$\frac{1}{2}$	$-\frac{3}{4}$	←Vertex
0	−1	←y-intercept
−1	−3	
1	−1	
2	−3	

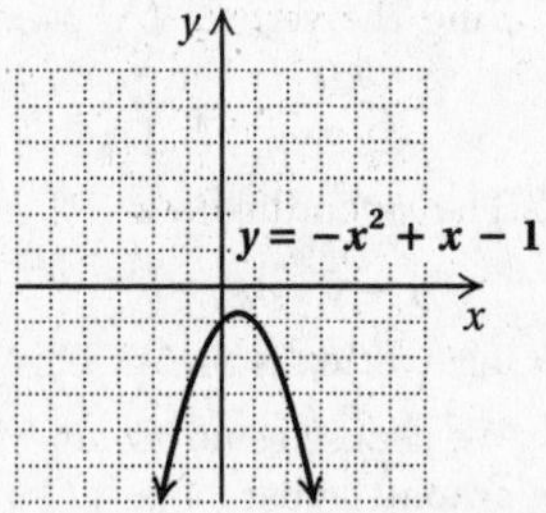

21. $y = -2x^2$

Find the vertex. The x-coordinate is

$$-\frac{b}{2a} = -\frac{0}{2(-2)} = 0.$$

The y-coordinate is

$$y = -2x^2 = -2\cdot 0^2 = 0.$$

The vertex is $(0, 0)$. This is also the y-intercept.

We choose some x-values on both sides of the vertex and graph the parabola.

x	y	
0	0	←Vertex
−1	−2	
−2	−8	
1	−2	
2	−8	

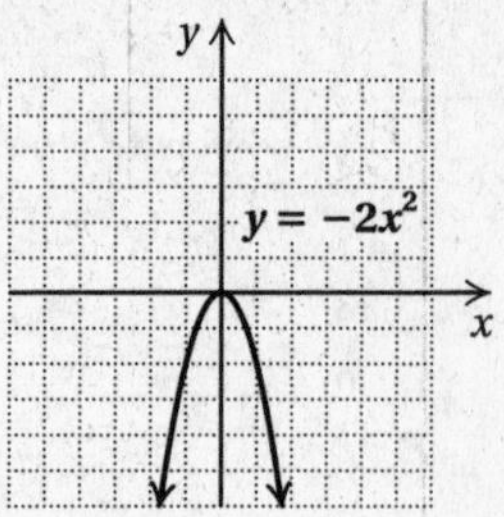

23. $y = x^2 - x - 6$

Find the vertex. The x-coordinate is

$$-\frac{b}{2a} = -\frac{-1}{2\cdot 1} = -\left(-\frac{1}{2}\right) = \frac{1}{2}.$$

The y-coordinate is

$$y = x^2 - x - 6 = \left(\frac{1}{2}\right)^2 - \frac{1}{2} - 6 = \frac{1}{4} - \frac{1}{2} - 6 = -\frac{25}{4}.$$

The vertex is $\left(\frac{1}{2}, -\frac{25}{4}\right)$.

We choose some x-values on both sides of the vertex and graph the parabola.

x	y	
$\frac{1}{2}$	$-\frac{25}{4}$	←Vertex
0	−6	←y-intercept
−1	−4	
1	−6	
2	−4	

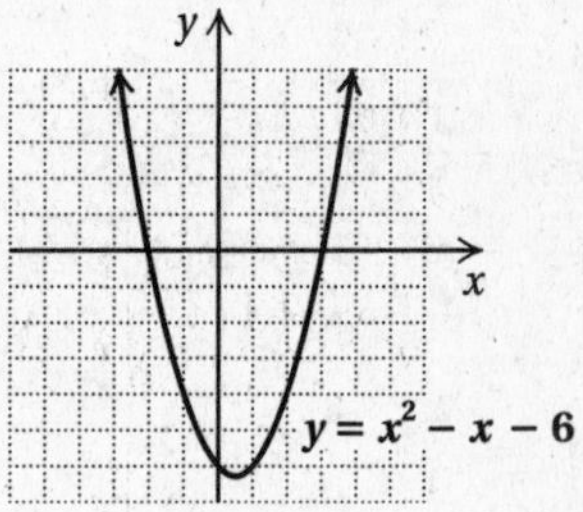

25. $y = x^2 - 2$

To find the x-intercepts we solve the equation $x^2 - 2 = 0$.

$$x^2 - 2 = 0$$
$$x^2 = 2$$

$x = \sqrt{2}$ *or* $x = -\sqrt{2}$ Principle of square roots

The x-intercepts are $(\sqrt{2}, 0)$ and $(-\sqrt{2}, 0)$.

27. $y = x^2 + 5x$

To find the x-intercepts we solve the equation $x^2 + 5x = 0$.

$$x^2 + 5x = 0$$
$$x(x + 5) = 0$$

$x = 0$ *or* $x + 5 = 0$
$x = 0$ *or* $x = -5$

The x-intercepts are $(0, 0)$ and $(-5, 0)$.

29. $y = 8 - x - x^2$

To find the x-intercepts we solve the equation $8 - x - x^2 = 0$.

$$8 - x - x^2 = 0$$
$$x^2 + x - 8 = 0 \quad \text{Standard form}$$

$a = 1,\ b = 1,\ c = -8$

$$x = \frac{-1 \pm \sqrt{1^2 - 4\cdot 1\cdot(-8)}}{2\cdot 1}$$

$$x = \frac{-1 \pm \sqrt{33}}{2}$$

The x-intercepts are $\left(\frac{-1+\sqrt{33}}{2}, 0\right)$ and $\left(\frac{-1-\sqrt{33}}{2}, 0\right)$.

31. $y = x^2 - 6x + 9$

To find the x-intercepts we solve the equation $x^2 - 6x + 9 = 0$.

$$x^2 - 6x + 9 = 0$$
$$(x-3)(x-3) = 0$$

$$x - 3 = 0 \quad or \quad x - 3 = 0$$
$$x = 3 \quad or \quad x = 3$$

The x-intercept is $(3, 0)$.

33. $y = -x^2 - 4x + 1$

To find the x-intercepts we solve the equation $-x^2 - 4x + 1 = 0$.

$$-x^2 - 4x + 1 = 0$$
$$x^2 + 4x - 1 = 0 \quad \text{Standard form}$$

$a = 1,\ b = 4,\ c = -1$

$$x = \frac{-4 \pm \sqrt{4^2 - 4 \cdot 1 \cdot (-1)}}{2 \cdot 1}$$
$$x = \frac{-4 \pm \sqrt{20}}{2} = \frac{-4 \pm \sqrt{4 \cdot 5}}{2} = \frac{-4 \pm 2\sqrt{5}}{2}$$
$$x = \frac{2(-2 \pm \sqrt{5})}{2} = -2 \pm \sqrt{5}$$

The x-intercepts are $(-2 + \sqrt{5}, 0)$ and $(-2 - \sqrt{5}, 0)$.

35. $y = x^2 + 9$

To find the x-intercepts we solve the equation $x^2 + 9 = 0$.

$$x^2 + 9 = 0$$
$$x^2 = -9$$

The negative number -9 has no real-number square roots. Thus there are no x-intercepts.

37.
$$\sqrt{8} + \sqrt{50} + \sqrt{98} + \sqrt{128}$$
$$= \sqrt{4 \cdot 2} + \sqrt{25 \cdot 2} + \sqrt{49 \cdot 2} + \sqrt{64 \cdot 2}$$
$$= 2\sqrt{2} + 5\sqrt{2} + 7\sqrt{2} + 8\sqrt{2}$$
$$= 22\sqrt{2}$$

39.
$$y = \frac{k}{x}$$
$$12.4 = \frac{k}{2.4} \quad \text{Substituting}$$
$$29.76 = k \quad \text{Variation constant}$$

$$y = \frac{29.76}{x} \quad \text{Equation of variation}$$

41.
$$-\frac{1}{5} + \frac{7}{10} - \left(-\frac{4}{15}\right) + \frac{1}{60}$$
$$= -\frac{1}{5} + \frac{7}{10} + \frac{4}{15} + \frac{1}{60}$$
$$= -\frac{1}{5} \cdot \frac{12}{12} + \frac{7}{10} \cdot \frac{6}{6} + \frac{4}{15} \cdot \frac{4}{4} + \frac{1}{60}$$
$$= -\frac{12}{60} + \frac{42}{60} + \frac{16}{60} + \frac{1}{60}$$
$$= \frac{-12 + 42 + 16 + 1}{60}$$
$$= \frac{47}{60}$$

43. a) We substitute 128 for H and solve for t:

$$128 = -16t^2 + 96t$$
$$16t^2 - 96t + 128 = 0$$
$$16(t^2 - 6t + 8) = 0$$
$$16(t-2)(t-4) = 0$$

$$t - 2 = 0 \quad or \quad t - 4 = 0$$
$$t = 2 \quad or \quad t = 4$$

The projectile is 128 ft from the ground 2 sec after launch and again 4 sec after launch. The graph confirms this.

b) We find the first coordinate of the vertex of the function $H = -16t^2 + 96t$:

$$-\frac{b}{2a} = -\frac{96}{2(-16)} = -\frac{96}{-32} = -(-3) = 3$$

The projectile reaches its maximum height 3 sec after launch. The graph confirms this.

c) We substitute 0 for H and solve for t:

$$0 = -16t^2 + 96t$$
$$0 = -16t(t-6)$$

$$-16t = 0 \quad or \quad t - 6 = 0$$
$$t = 0 \quad or \quad t = 6$$

At $t = 0$ sec the projectile has not yet been launched. Thus, we use $t = 6$. The projectile returns to the ground 6 sec after launch. The graph confirms this.

45. $y = x^2 + 2x - 3$

$a = 1,\ b = 2,\ c = -3$

$b^2 - 4ac = 2^2 - 4 \cdot 1 \cdot (-3) = 4 + 12 = 16$

Since $b^2 - 4ac = 16 > 0$, the equation $x^2 + 2x - 3 = 0$ has two real-number solutions.

47. $y = -0.02x^2 + 4.7x - 2300$

$a = -0.02,\ b = 4.7,\ c = -2300$

$b^2 - 4ac = (4.7)^2 - 4(-0.02)(-2300) = 22.09 - 184 = -161.91$

Since $b^2 - 4ac = -161.91 < 0$, the equation $-0.02x^2 + 4.7x - 2300 = 0$ has no real solutions.

Exercise Set 9.7

RC1. A function is a special kind of correspondence between a first set, called the domain, and a second set, called the range.

RC3. We use the vertical-line test to test whether a graph represents a function.

1. Yes; each member of the domain is matched to only one member of the range.

3. Yes; each member of the domain is matched to only one member of the range.

5. No; a member of the domain is matched to more than one member of the range. In fact, each member of the domain is matched to 3 members of the range.

7. Yes; each member of the domain is matched to only one member of the range.

9. This correspondence is a function, because each class member has only one seat number.

11. This correspondence is a function, because each shape has only one number for its area.

13. This correspondence is not a function, because it is reasonable to assume that at least one person has more than one aunt.

The correspondence is a relation, because it is reasonable to assume that each person has at least one aunt.

15. $f(x) = x + 5$

a) $f(4) = 4 + 5 = 9$

b) $f(7) = 7 + 5 = 12$

c) $f(-3) = -3 + 5 = 2$

d) $f(0) = 0 + 5 = 5$

e) $f(2.4) = 2.4 + 5 = 7.4$

f) $f\left(\frac{2}{3}\right) = \frac{2}{3} + 5 = 5\frac{2}{3}$

17. $h(p) = 3p$

a) $h(-7) = 3(-7) = -21$

b) $h(5) = 3 \cdot 5 = 15$

c) $h(14) = 3 \cdot 14 = 42$

d) $h(0) = 3 \cdot 0 = 0$

e) $h\left(\frac{2}{3}\right) = 3 \cdot \frac{2}{3} = \frac{6}{3} = 2$

f) $h(-54.2) = 3(-54.2) = -162.6$

19. $g(s) = 3s + 4$

a) $g(1) = 3 \cdot 1 + 4 = 3 + 4 = 7$

b) $g(-7) = 3(-7) + 4 = -21 + 4 = -17$

c) $g(6.7) = 3(6.7) + 4 = 20.1 + 4 = 24.1$

d) $g(0) = 3 \cdot 0 + 4 = 0 + 4 = 4$

e) $g(-10) = 3(-10) + 4 = -30 + 4 = -26$

f) $g\left(\frac{2}{3}\right) = 3 \cdot \frac{2}{3} + 4 = 2 + 4 = 6$

21. $f(x) = 2x^2 - 3x$

a) $f(0) = 2 \cdot 0^2 - 3 \cdot 0 = 0 - 0 = 0$

b) $f(-1) = 2(-1)^2 - 3(-1) = 2 + 3 = 5$

c) $f(2) = 2 \cdot 2^2 - 3 \cdot 2 = 8 - 6 = 2$

d) $f(10) = 2 \cdot 10^2 - 3 \cdot 10 = 200 - 30 = 170$

e) $f(-5) = 2(-5)^2 - 3(-5) = 50 + 15 = 65$

f) $f(-10) = 2(-10)^2 - 3(-10) = 200 + 30 = 230$

23. $f(x) = |x| + 1$

a) $f(0) = |0| + 1 = 0 + 1 = 1$

b) $f(-2) = |-2| + 1 = 2 + 1 = 3$

c) $f(2) = |2| + 1 = 2 + 1 = 3$

d) $f(-3) = |-3| + 1 = 3 + 1 = 4$

e) $f(-10) = |-10| + 1 = 10 + 1 = 11$

f) $f(22) = |22| + 1 = 22 + 1 = 23$

25. $f(x) = x^3$

a) $f(0) = 0^3 = 0$

b) $f(-1) = (-1)^3 = -1$

c) $f(2) = 2^3 = 8$

d) $f(10) = 10^3 = 1000$

e) $f(-5) = (-5)^3 = -125$

f) $f(-10) = (-10)^3 = -1000$

27. $l(x) = \frac{1700}{x}$

a) $l(50) = \frac{1700}{50} = 34$ years

b) $l(85) = \frac{1700}{85} = 20$ years

29. $F(t) = 22t^2 - 41t + 2.7$

$F(0) = 22 \cdot 0^2 - 41 \cdot 0 + 2.7 = 2.7$ million users

$$\begin{aligned} F(6) &= 22 \cdot 6^2 - 41 \cdot 6 + 2.7 \\ &= 22 \cdot 36 - 41 \cdot 6 + 2.7 \\ &= 792 - 246 + 2.7 \\ &= 548.7 \text{ million users} \end{aligned}$$

$$\begin{aligned} F(8) &= 22 \cdot 8^2 - 41 \cdot 8 + 2.7 \\ &= 22 \cdot 64 - 41 \cdot 8 + 2.7 \\ &= 1408 - 328 + 2.7 \\ &= 1082.7 \text{ million users} \end{aligned}$$

31.

$$\begin{aligned} P(d) &= 1 + \frac{d}{33} \\ P(20) &= 1 + \frac{20}{33} = 1\frac{20}{33} \text{ atm} \\ P(30) &= 1 + \frac{30}{33} = 1\frac{10}{11} \text{ atm} \\ P(100) &= 1 + \frac{100}{33} = 1 + 3\frac{1}{33} = 4\frac{1}{33} \text{ atm} \end{aligned}$$

33. Graph $f(x) = 3x - 1$

Make a list of function values in a table.

When $x = -1$, $f(-1) = 3(-1) - 1 = -3 - 1 = -4$.

When $x = 0$, $f(0) = 3 \cdot 0 - 1 = 0 - 1 = -1$.

When $x = 2$, $f(2) = 3 \cdot 2 - 1 = 6 - 1 = 5$.

x	$f(x)$
-1	-4
0	-1
2	5

Plot these points and connect them.

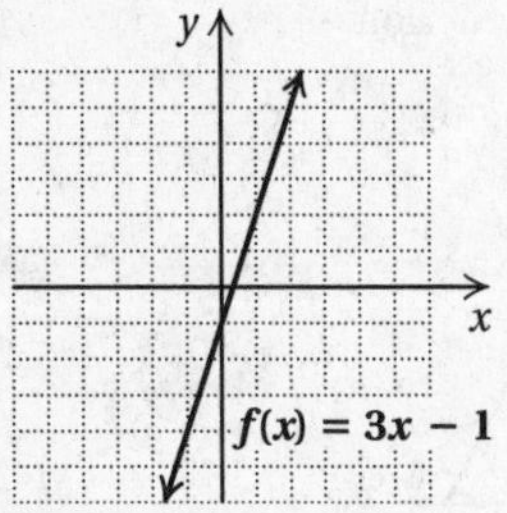

35. Graph $g(x) = -2x + 3$

Make a list of function values in a table.

When $x = -1$, $g(-1) = -2(-1) + 3 = 2 + 3 = 5$.

When $x = 0$, $g(0) = -2 \cdot 0 + 3 = 0 + 3 = 3$.

When $x = 3$, $g(3) = -2 \cdot 3 + 3 = -6 + 3 = -3$.

x	$g(x)$
-1	5
0	3
3	-3

Plot these points and connect them.

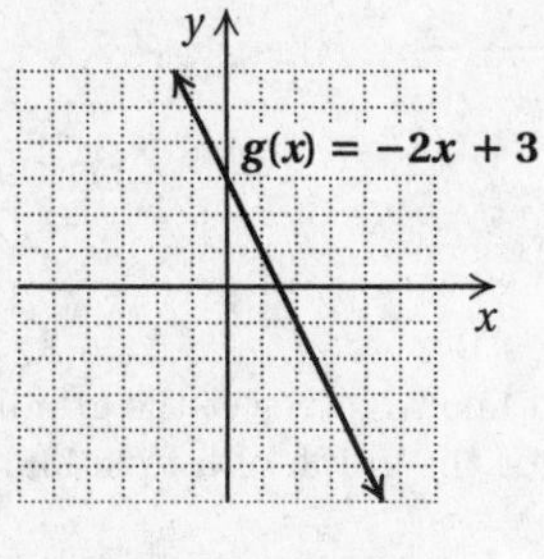

37. Graph $f(x) = \frac{1}{2}x + 1$.

Make a list of function values in a table.

When $x = -2$, $f(-2) = \frac{1}{2}(-2) + 1 = -1 + 1 = 0$.

When $x = 0$, $f(0) = \frac{1}{2} \cdot 0 + 1 = 0 + 1 = 1$.

When $x = 4$, $f(4) = \frac{1}{2} \cdot 4 + 1 = 2 + 1 = 3$.

x	$f(x)$
-2	0
0	1
4	3

Plot these points and connect them.

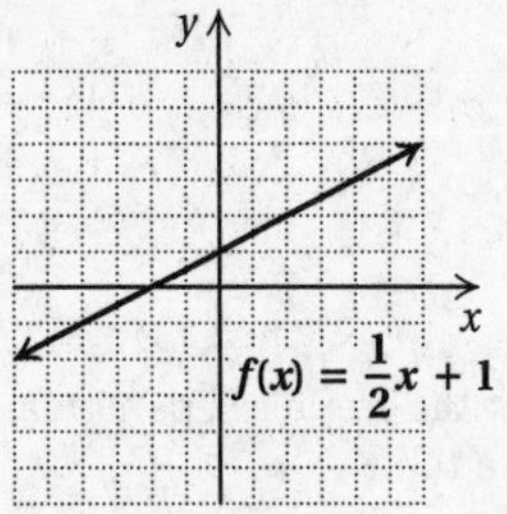

39. Graph $f(x) = 2 - |x|$.

Make a list of function values in a table.

When $x = -4$, $f(-4) = 2 - |-4| = 2 - 4 = -2$.

When $x = 0$, $f(0) = 2 - |0| = 2 - 0 = 2$.

When $x = 3$, $f(3) = 2 - |3| = 2 - 3 = -1$.

x	$f(x)$
-4	-2
0	2
3	-1

Plot these points and connect them.

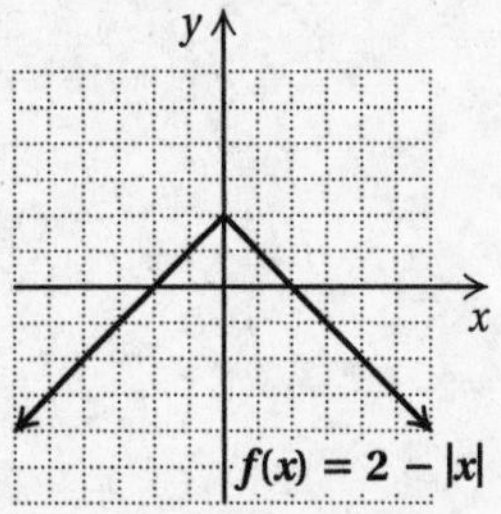

41. Graph $f(x) = x^2$.

Recall from Section 9.6 that the graph is a parabola. Make a list of function values in a table.

When $x = -2$, $f(-2) = (-2)^2 = 4$.

When $x = -1$, $f(-1) = (-1)^2 = 1$.

When $x = 0$, $f(0) = 0^2 = 0$.

When $x = 1$, $f(1) = 1^2 = 1$.

When $x = 2$, $f(2) = 2^2 = 4$.

x	$f(x)$
-2	4
-1	1
0	0
1	1
2	4

Plot these points and connect them.

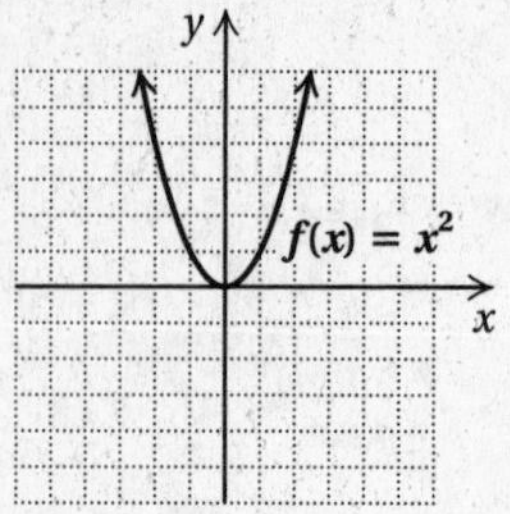

43. Graph $f(x) = x^2 - x - 2$.

Recall from Section 9.6 that the graph is a parabola. Make a list of function values in a table.

When $x = -1$, $f(-1) = (-1)^2 - (-1) - 2 = 1 + 1 - 2 = 0$.

When $x = 0$, $f(0) = 0^2 - 0 - 2 = -2$.

When $x = 1$, $f(1) = 1^2 - 1 - 2 = 1 - 1 - 2 = -2$.

When $x = 2$, $f(2) = 2^2 - 2 - 2 = 4 - 2 - 2 = 0$.

x	$f(x)$
-1	0
0	-2
1	-2
2	0

Plot these points and connect them.

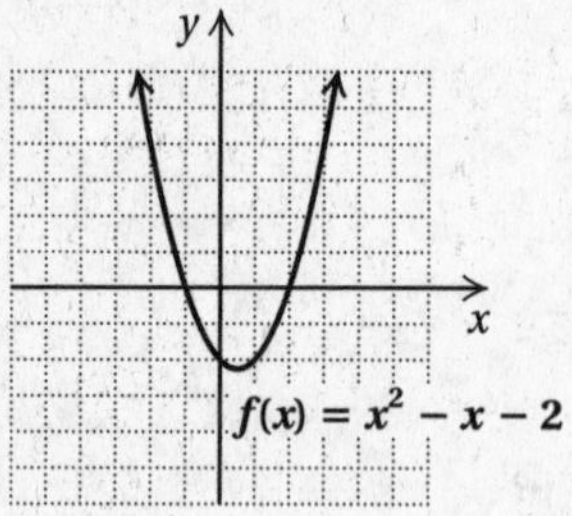

45. We can use the vertical-line test:

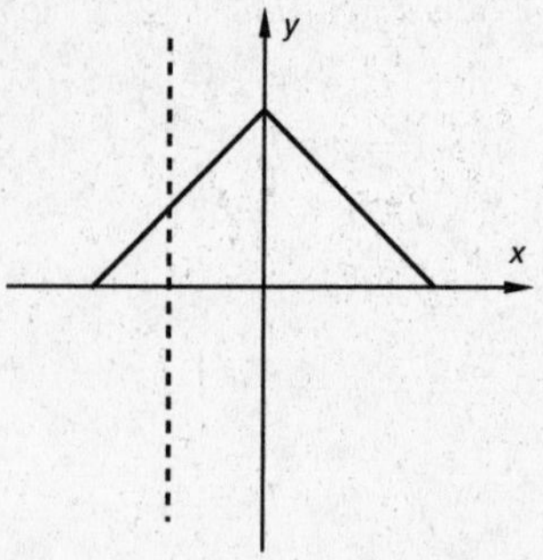

Visualize moving this vertical line across the graph. No vertical line will intersect the graph more than once. Thus, the graph is a graph of a function.

47. We can use the vertical-line test:

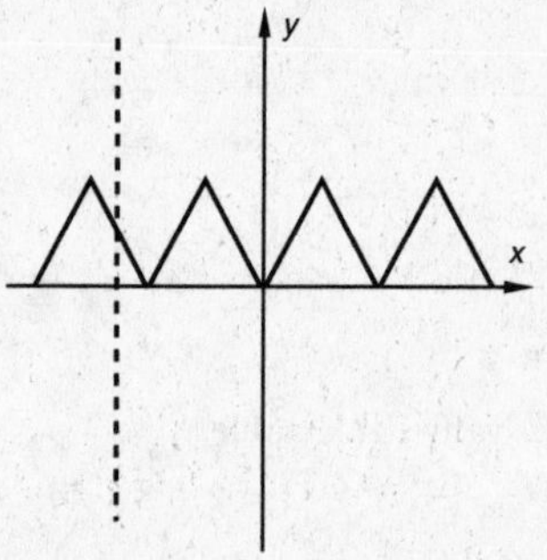

Visualize moving this vertical line across the graph. No vertical line will intersect the graph more than once. Thus, the graph is a graph of a function.

49. We can use the vertical-line test.

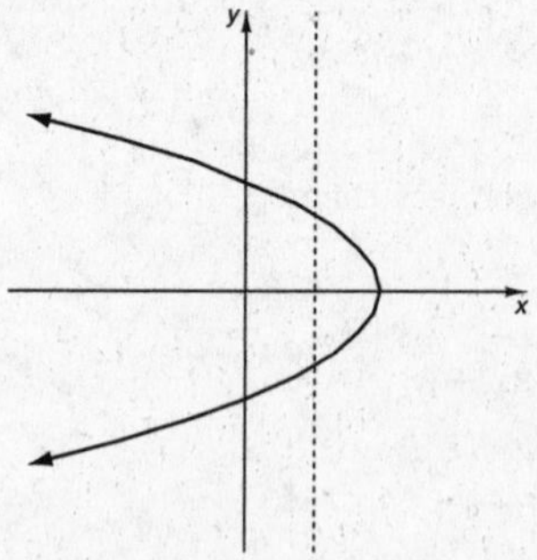

It is possible for a vertical line to intersect the graph more than once. Thus this is not the graph of a function.

51. We can use the vertical-line test.

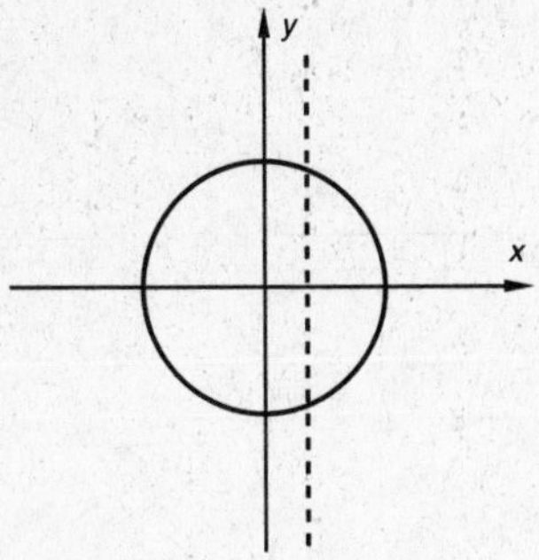

It is possible for a vertical line to intersect the graph more than once. Thus this is not a graph of a function.

53. Locate the point that is directly above 225. Then estimate its second coordinate by moving horizontally from the point to the vertical axis. The rate is about 75 per 10,000 men.

55. $x = 2 - y$, (1)

$3x + y = 5$ (2)

Substitute $2 - y$ for x in equation (2) and solve for y.

$$\begin{aligned} 3x + y &= 5 \\ 3(2 - y) + y &= 5 \\ 6 - 3y + y &= 5 \\ 6 - 2y &= 5 \\ -2y &= -1 \\ y &= \frac{1}{2} \end{aligned}$$

Now substitute $\frac{1}{2}$ for y in equation (1) and find x.

$$x = 2 - y = 2 - \frac{1}{2} = \frac{3}{2}$$

The solution is $\left(\frac{3}{2}, \frac{1}{2}\right)$.

57.
$$\begin{aligned} 2x - 5y &= 7, \quad (1) \\ x + 5y &= 2 \quad (2) \\ \hline 3x \quad &= 9 \\ x &= 3 \end{aligned}$$

Substitute 3 for x in one of the original equations and solve for y.

$$\begin{aligned} x + 5y &= 2 \quad (2) \\ 3 + 5y &= 2 \\ 5y &= -1 \\ y &= -\frac{1}{5} \end{aligned}$$

The solution is $\left(3, -\frac{1}{5}\right)$.

59. Graph $g(x) = x^3$.

Make a list of function values in a table. Then plot the points and connect them.

x	$g(x)$
-2	-8
-1	-1
0	0
1	1
2	8

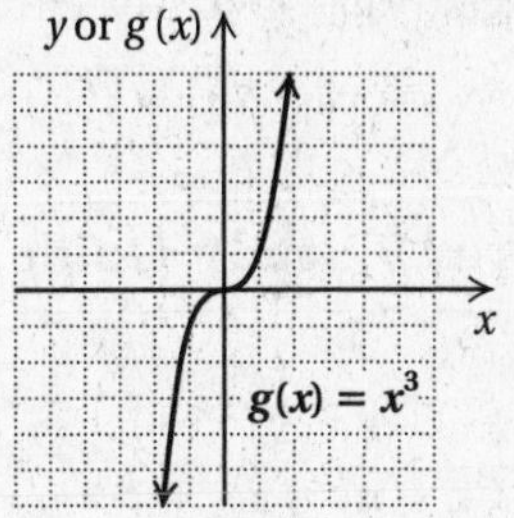

61. Graph $g(x) = |x| + x$.

Make a list of function values in a table. Then plot the points and connect them.

x	$f(x)$
-3	0
-2	0
-1	0
0	0
1	2
2	4
3	6

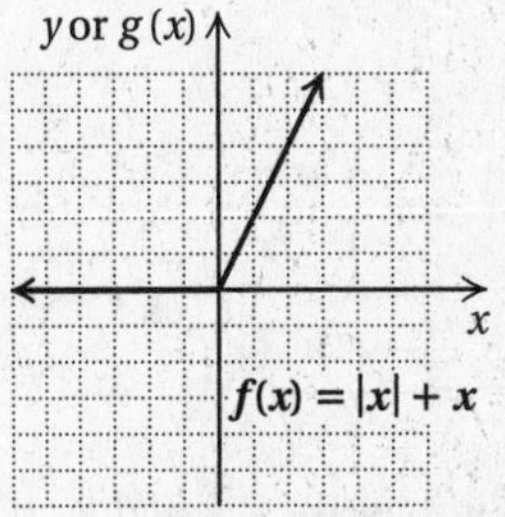

Chapter 9 Vocabulary Reinforcement

1. The equation $ax^2 + bx + c = 0$ is the standard form of a quadratic equation.
2. When we add 25 to $x^2 + 10x$, we are completing the square.
3. The expression $b^2 - 4ac$ is called the discriminant.
4. The turning point of the graph of a quadratic equation is the vertex.
5. The graph of a quadratic equation is called a parabola.
6. The function given by $f(x) = 7x^2 - 3x$ is an example of a quadratic function.
7. A graph is the graph of a function if it passes the vertical-line test.
8. The set of all inputs of a function is called the domain.

Chapter 9 Concept Reinforcement

1. A graph represents a function if it is not possible to draw a vertical line that intersects the graph more than once. The given statement is false.

2. When $x = 0$, $y = a \cdot 0^2 + b \cdot 0 + c = c$, so all graphs of quadratic equations, $y = ax^2 + bx + c$, have the y-intercept $(0, c)$. The given statement is true.

3. If $a < 0$, then the graph opens down and, thus, q is the largest value of y. The given statement is true.

4. True; see page 704 in the text.

Chapter 9 Study Guide

1. $\dfrac{3}{x+2} + \dfrac{1}{x} = \dfrac{5}{4}$

We multiply by the LCM, $4x(x+2)$.

$$4x(x+2)\left(\frac{3}{x+2} + \frac{1}{x}\right) = 4x(x+2) \cdot \frac{5}{4}$$
$$4x(x+2) \cdot \frac{3}{x+2} + 4x(x+2) \cdot \frac{1}{x} = x(x+2) \cdot 5$$
$$4x \cdot 3 + 4(x+2) = 5x^2 + 10x$$
$$12x + 4x + 8 = 5x^2 + 10x$$
$$16x + 8 = 5x^2 + 10x$$
$$0 = 5x^2 - 6x - 8$$
$$0 = (5x+4)(x-2)$$

$5x + 4 = 0$ *or* $x - 2 = 0$

$5x = -4$ *or* $x = 2$

$x = -\dfrac{4}{5}$ *or* $x = 2$

The solutions are $-\dfrac{4}{5}$ and 2.

2. $7x^2 - 3 = 8$

$7x^2 = 11$

$x^2 = \dfrac{11}{7}$

$x = \sqrt{\dfrac{11}{7}}$ *or* $x = -\sqrt{\dfrac{11}{7}}$

$x = \sqrt{\dfrac{11}{7} \cdot \dfrac{7}{7}}$ *or* $x = -\sqrt{\dfrac{11}{7} \cdot \dfrac{7}{7}}$

$x = \dfrac{\sqrt{77}}{7}$ *or* $x = -\dfrac{\sqrt{77}}{7}$

The solutions are $\dfrac{\sqrt{77}}{7}$ and $-\dfrac{\sqrt{77}}{7}$, or $\pm\dfrac{\sqrt{77}}{7}$.

3. $x^2 - 4x + 1 = 0$

$x^2 - 4x = -1$

$x^2 - 4x + 4 = -1 + 4$

$(x-2)^2 = 3$

$x - 2 = \sqrt{3}$ *or* $x - 2 = -\sqrt{3}$

$x = 2 + \sqrt{3}$ *or* $x = 2 - \sqrt{3}$

The solutions are $2 + \sqrt{3}$ and $2 - \sqrt{3}$, or $2 \pm \sqrt{3}$.

4. $4y^2 = 6y + 3$

$4y^2 - 6y - 3 = 0$

$a = 4$, $b = -6$, $c = -3$

$$y = \frac{-b \pm \sqrt{b^2 - 4ac}}{2a}$$
$$y = \frac{-(-6) \pm \sqrt{(-6)^2 - 4 \cdot 4 \cdot (-3)}}{2 \cdot 4} = \frac{6 \pm \sqrt{36 + 48}}{8}$$
$$y = \frac{6 \pm \sqrt{84}}{8} = \frac{6 \pm \sqrt{4 \cdot 21}}{8} = \frac{6 \pm 2\sqrt{21}}{8}$$
$$y = \frac{2(3 \pm \sqrt{21})}{2 \cdot 4} = \frac{3 \pm \sqrt{21}}{4}$$

The solutions are $\dfrac{3 \pm \sqrt{21}}{4}$.

5. $y = x^2 - 4x + 2$.

We first find the vertex. The x-coordinate is

$$-\frac{b}{2a} = -\frac{-4}{2 \cdot 1} = 2.$$

We substitute into the equation to find the second coordinate of the vertex.

$$y = 2^2 - 4 \cdot 2 + 2 = 4 - 8 + 2 = -2$$

The vertex is $(2, -2)$.

We choose some x-values on both sides of the vertex and graph the parabola.

x	y	
-2	14	
-1	7	
0	2	⟵ x-intercept
1	-1	
2	-2	⟵ vertex
3	-1	
4	2	

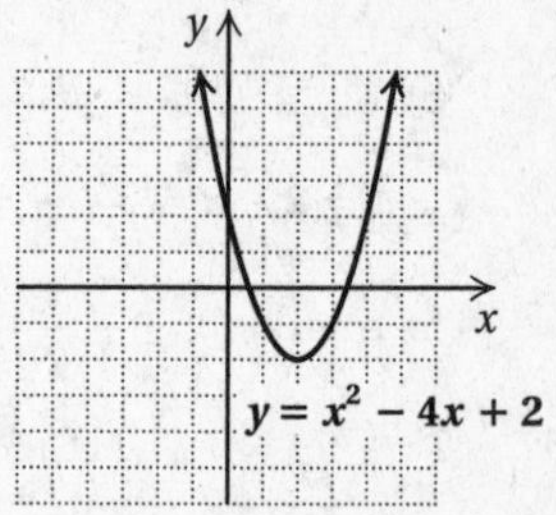

6. a) $h(5) = \dfrac{1}{5} \cdot 5^2 + 5 - 1 = \dfrac{1}{5} \cdot 25 + 5 - 1 = 5 + 5 - 1 = 9$

b) $f(0) = -3 \cdot 0 - 4 = 0 - 4 = -4$

Chapter 9 Review Exercises

1. $8x^2 = 24$

$x^2 = 3$

$x = \sqrt{3}$ *or* $x = -\sqrt{3}$

The solutions are $\sqrt{3}$ and $-\sqrt{3}$.

2. $40 = 5y^2$

$8 = y^2$

$y = \sqrt{8}$ *or* $y = -\sqrt{8}$

$y = 2\sqrt{2}$ *or* $y = -2\sqrt{2}$

The solutions are $2\sqrt{2}$ and $-2\sqrt{2}$.

3. $5x^2 - 8x + 3 = 0$

$(5x-3)(x-1) = 0$

$5x - 3 = 0 \quad or \quad x - 1 = 0$

$5x = 3 \quad or \quad x = 1$

$x = \frac{3}{5} \quad or \quad x = 1$

The solutions are $\frac{3}{5}$ and 1.

4. $3y^2 + 5y = 2$

$3y^2 + 5y - 2 = 0$

$(3y-1)(y+2) = 0$

$3y - 1 = 0 \quad or \quad y + 2 = 0$

$3y = 1 \quad or \quad y = -2$

$y = \frac{1}{3} \quad or \quad y = -2$

The solutions are $\frac{1}{3}$ and -2.

5. $(x+8)^2 = 13$

$x + 8 = \sqrt{13} \quad or \quad x + 8 = -\sqrt{13}$

$x = -8 + \sqrt{13} \quad or \quad x = -8 - \sqrt{13}$

The solutions are $-8 \pm \sqrt{13}$.

6. $9x^2 = 0$

$x^2 = 0$

$x = 0$

The solution is 0.

7. $5t^2 - 7t = 0$

$t(5t - 7) = 0$

$t = 0 \quad or \quad 5t - 7 = 0$

$t = 0 \quad or \quad 5t = 7$

$t = 0 \quad or \quad t = \frac{7}{5}$

The solutions are 0 and $\frac{7}{5}$.

8. $x^2 - 2x - 10 = 0$

$a = 1,\ b = -2,\ c = -10$

$$x = \frac{-(-2) \pm \sqrt{(-2)^2 - 4\cdot 1\cdot(-10)}}{2\cdot 1}$$

$$x = \frac{2 \pm \sqrt{4+40}}{2}$$

$$x = \frac{2 \pm \sqrt{44}}{2} = \frac{2 \pm \sqrt{4\cdot 11}}{2}$$

$$x = \frac{2 \pm 2\sqrt{11}}{2} = \frac{2(1 \pm \sqrt{11})}{2}$$

$x = 1 \pm \sqrt{11}$

The solutions are $1 \pm \sqrt{11}$.

9. $9x^2 - 6x - 9 = 0$

$a = 9,\ b = -6,\ c = -9$

$$x = \frac{-(-6) \pm \sqrt{(-6)^2 - 4\cdot 9\cdot(-9)}}{2\cdot 9}$$

$$x = \frac{6 \pm \sqrt{36+324}}{18}$$

$$x = \frac{6 \pm \sqrt{360}}{18} = \frac{6 \pm \sqrt{36\cdot 10}}{18}$$

$$x = \frac{6 \pm 6\sqrt{10}}{18} = \frac{6(1 \pm \sqrt{10})}{3\cdot 6}$$

$$x = \frac{1 \pm \sqrt{10}}{3}$$

The solutions are $\frac{1 \pm \sqrt{10}}{3}$.

10. $x^2 + 6x = 9$

$x^2 + 6x - 9 = 0$

$a = 1,\ b = 6,\ c = -9$

$$x = \frac{-6 \pm \sqrt{6^2 - 4\cdot 1\cdot(-9)}}{2\cdot 1}$$

$$x = \frac{-6 \pm \sqrt{36+36}}{2}$$

$$x = \frac{-6 \pm \sqrt{72}}{2} = \frac{-6 \pm \sqrt{36\cdot 2}}{2}$$

$$x = \frac{-6 \pm 6\sqrt{2}}{2} = \frac{2(-3 \pm 3\sqrt{2})}{2}$$

$x = -3 \pm 3\sqrt{2}$

The solutions are $-3 \pm 3\sqrt{2}$.

11. $1 + 4x^2 = 8x$

$4x^2 - 8x + 1 = 0$

$a = 4,\ b = -8,\ c = 1$

$$x = \frac{-(-8) \pm \sqrt{(-8)^2 - 4\cdot 4\cdot 1}}{2\cdot 4}$$

$$x = \frac{8 \pm \sqrt{64-16}}{8}$$

$$x = \frac{8 \pm \sqrt{48}}{8} = \frac{8 \pm \sqrt{16\cdot 3}}{8}$$

$$x = \frac{8 \pm 4\sqrt{3}}{8} = \frac{4(2 \pm \sqrt{3})}{2\cdot 4}$$

$$x = \frac{2 \pm \sqrt{3}}{2}$$

The solutions are $\frac{2 \pm \sqrt{3}}{2}$.

12. $6 + 3y = y^2$

$0 = y^2 - 3y - 6$

$a = 1,\ b = -3,\ c = -6$

$$y = \frac{-(-3) \pm \sqrt{(-3)^2 - 4\cdot 1\cdot(-6)}}{2\cdot 1}$$

$$y = \frac{3 \pm \sqrt{9+24}}{2} = \frac{3 \pm \sqrt{33}}{2}$$

The solutions are $\frac{3 \pm \sqrt{33}}{2}$.

13. $3m = 4 + 5m^2$

$0 = 5m^2 - 3m + 4$

$a = 5,\ b = -3,\ c = 4$

$$m = \frac{-(-3) \pm \sqrt{(-3)^2 - 4 \cdot 5 \cdot 4}}{2 \cdot 5}$$

$$m = \frac{3 \pm \sqrt{9 - 80}}{10} = \frac{3 \pm \sqrt{-71}}{10}$$

Since the radicand is negative, there are no real-number solutions.

14. $3x^2 = 4x$

$3x^2 - 4x = 0$

$x(3x - 4) = 0$

$x = 0$ *or* $3x - 4 = 0$

$x = 0$ *or* $3x = 4$

$x = 0$ *or* $x = \frac{4}{3}$

The solutions are 0 and $\frac{4}{3}$.

15. $\frac{15}{x} - \frac{15}{x+2} = 2$, LCM is $x(x+2)$

$$x(x+2)\left(\frac{15}{x} - \frac{15}{x+2}\right) = x(x+2)(2)$$

$$x(x+2) \cdot \frac{15}{x} - x(x+2) \cdot \frac{15}{x+2} = 2x(x+2)$$

$15(x+2) - 15x = 2x^2 + 4x$

$15x + 30 - 15x = 2x^2 + 4x$

$30 = 2x^2 + 4x$

$0 = 2x^2 + 4x - 30$

$0 = x^2 + 2x - 15$ Dividing by 2

$0 = (x+5)(x-3)$

$x + 5 = 0$ *or* $x - 3 = 0$

$x = -5$ *or* $x = 3$

Both numbers check. The solutions are -5 and 3.

16. $x + \frac{1}{x} = 2$, LCM is x

$$x\left(x + \frac{1}{x}\right) = x \cdot 2$$

$$x \cdot x + x \cdot \frac{1}{x} = 2x$$

$x^2 + 1 = 2x$

$x^2 - 2x + 1 = 0$

$(x - 1)^2 = 0$

$x - 1 = 0$

$x = 1$

The number 1 checks. It is the solution.

17. $x^2 - 4x + 2 = 0$

$x^2 - 4x = -2$

$x^2 - 4x + 4 = -2 + 4$ Adding 4: $\left(\frac{-4}{2}\right)^2 = 4$

$(x - 2)^2 = 2$

$x - 2 = \sqrt{2}$ *or* $x - 2 = -\sqrt{2}$

$x = 2 + \sqrt{2}$ *or* $x = 2 - \sqrt{2}$

The solutions are $2 \pm \sqrt{2}$.

18. $3x^2 - 2x - 5 = 0$

$$\frac{1}{3}(3x^2 - 2x - 5) = \frac{1}{3} \cdot 0$$

$$x^2 - \frac{2}{3}x - \frac{5}{3} = 0$$

$$x^2 - \frac{2}{3}x = \frac{5}{3}$$

$$x^2 - \frac{2}{3}x + \frac{1}{9} = \frac{5}{3} + \frac{1}{9} \quad \text{Adding } \frac{1}{9}\text{:}$$

$$\left[\frac{1}{2}\left(-\frac{2}{3}\right)\right]^2 = \left(-\frac{1}{3}\right)^2 = \frac{1}{9}$$

$$\left(x - \frac{1}{3}\right)^2 = \frac{16}{9}$$

$x - \frac{1}{3} = \frac{4}{3}$ *or* $x - \frac{1}{3} = -\frac{4}{3}$

$x = \frac{5}{3}$ *or* $x = -1$

The solutions are $\frac{5}{3}$ and -1.

19. $x^2 - 5x + 2 = 0$

$a = 1,\ b = -5,\ c = 2$

$$y = \frac{-(-5) \pm \sqrt{(-5)^2 - 4 \cdot 1 \cdot 2}}{2 \cdot 1}$$

$$y = \frac{5 \pm \sqrt{25 - 8}}{2} = \frac{5 \pm \sqrt{17}}{2}$$

Using a calculator, we have $\frac{5 + \sqrt{17}}{2} \approx 4.6$ and $\frac{5 - \sqrt{17}}{2} \approx 0.4$.

20. $4y^2 + 8y + 1 = 0$

$a = 4,\ b = 8,\ c = 1$

$$y = \frac{-8 \pm \sqrt{8^2 - 4 \cdot 4 \cdot 1}}{2 \cdot 4}$$

$$y = \frac{-8 \pm \sqrt{64 - 16}}{8} = \frac{-8 \pm \sqrt{48}}{8}$$

Using a calculator, we have $\frac{-8 + \sqrt{48}}{8} \approx -0.1$ and $\frac{-8 - \sqrt{48}}{8} \approx -1.9$.

21.

$$V = \frac{1}{2}\sqrt{1+\frac{T}{L}}$$

$$V^2 = \left(\frac{1}{2}\sqrt{1+\frac{T}{L}}\right)^2$$

$$V^2 = \frac{1}{4}\left(1+\frac{T}{L}\right)$$

$$4 \cdot V^2 = 4 \cdot \frac{1}{4}\left(1+\frac{T}{L}\right)$$

$$4V^2 = 1+\frac{T}{L}$$

$$4V^2 - 1 = \frac{T}{L}$$

$$L(4V^2-1) = L \cdot \frac{T}{L}$$

$$L(4V^2-1) = T$$

22. $y = 2 - x^2$, or $y = -x^2 + 2$

Find the vertex. The x-coordinate is

$$-\frac{b}{2a} = -\frac{0}{2(-1)} = 0.$$

The y-coordinate is

$$y = 2 - x^2 = 2 - 0^2 = 2.$$

The vertex is $(0, 2)$. This is also the y-intercept.

We choose some x-values on both sides of the vertex and graph the parabola.

x	y	
0	2	←Vertex
−1	1	
−2	−2	
1	1	
2	−2	

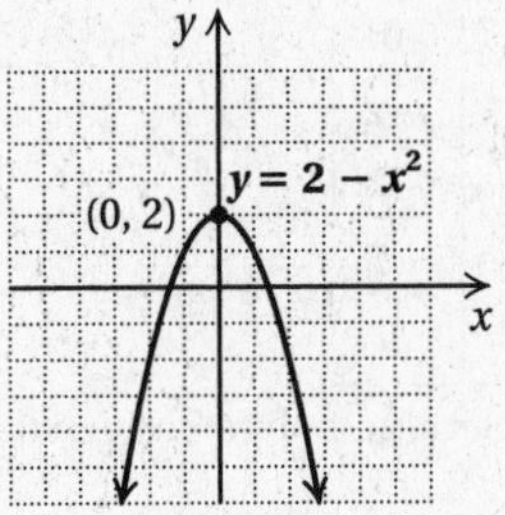

23. $y = x^2 - 4x - 2$

Find the vertex. The x-coordinate is

$$-\frac{b}{2a} = -\frac{-4}{2 \cdot 1} = -(-2) = 2.$$

The y-coordinate is

$$y = 2^2 - 4 \cdot 2 - 2 = 4 - 8 - 2 = -6.$$

The vertex is $(2, -6)$.

We choose some x-values on both sides of the vertex and graph the parabola.

x	y	
2	−6	← Vertex
0	−2	← y-intercept
−1	3	
3	5	
5	3	

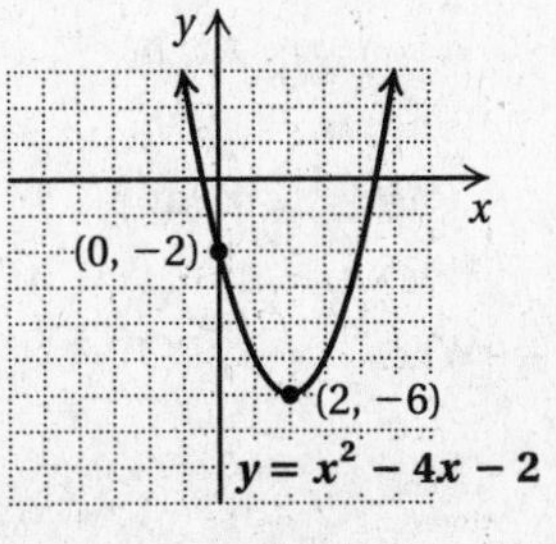

24. $y = 2 - x^2$

To find the x-intercepts we solve the equation $2 - x^2 = 0$.

$$2 - x^2 = 0$$

$$2 = x^2$$

$$x = \sqrt{2} \ \ or \ \ x = -\sqrt{2}$$

The x-intercepts are $(\sqrt{2}, 0)$ and $(-\sqrt{2}, 0)$.

25. $y = x^2 - 4x - 2$

To find the x-intercepts we solve the equation $x^2 - 4x - 2 = 0$.

$$x = \frac{-(-4) \pm \sqrt{(-4)^2 - 4 \cdot 1(-2)}}{2 \cdot 1}$$

$$x = \frac{4 \pm \sqrt{16+8}}{2} = \frac{4 + \sqrt{24}}{2}$$

$$x = \frac{4 \pm \sqrt{4 \cdot 6}}{2} = \frac{4 \pm 2\sqrt{6}}{2}$$

$$x = \frac{2(2 \pm \sqrt{6})}{2} = 2 \pm \sqrt{6}$$

The x-intercepts are $(2 - \sqrt{6}, 0)$ and $(2 + \sqrt{6}, 0)$.

26. ***Familiarize.*** Using the labels on the drawing in the text, we let a and $a + 3$ represent the lengths of the legs, in cm.

Translate. We use the Pythagorean equation.

$$a^2 + (a+3)^2 = 5^2$$

Solve.

$$a^2 + (a+3)^2 = 5^2$$

$$a^2 + a^2 + 6a + 9 = 25$$

$$2a^2 + 6a + 9 = 25$$

$$2a^2 + 6a - 16 = 0$$

$$a^2 + 3a - 8 = 0 \quad \text{Dividing by 2}$$

We use the quadratic formula.

$$a = \frac{-3 \pm \sqrt{3^2 - 4 \cdot 1 \cdot (-8)}}{2 \cdot 1}$$

$$a = \frac{-3 \pm \sqrt{9+32}}{2} = \frac{-3 \pm \sqrt{41}}{2}$$

Using a calculator, we have

$$a = \frac{-3 - \sqrt{41}}{2} \approx -4.7 \text{ and } a = \frac{-3 + \sqrt{41}}{2} \approx 1.7.$$

Check. Since the length of a leg cannot be negative, −4.7 cannot be a solution. If $a \approx 1.7$, then $a + 3 \approx 4.7$ and $(1.7)^2 + (4.7)^4 = 24.98 \approx 25 = 5^2$. The answer checks.

State. The lengths of the legs are about 1.7 cm and 4.7 cm.

27. ***Familiarize.*** Using the labels on the drawing in the text, we let s and $s - 5$ represent the lengths of the legs, in ft.

Translate. We use the Pythagorean equation.

$$s^2 + (s-5)^2 = 25^2$$

Solve.

$$s^2 + (s-5)^2 = 25^2$$
$$s^2 + s^2 - 10s + 25 = 625$$
$$2s^2 - 10s + 25 = 625$$
$$2s^2 - 10s - 600 = 0$$
$$s^2 - 5s - 300 = 0 \quad \text{Dividing by 2}$$
$$(s-20)(s+15) = 0$$
$$s - 20 = 0 \quad or \quad s + 15 = 0$$
$$s = 20 \quad or \quad s = -15$$

Check. Since the length of a leg of the triangle cannot be negative, -15 cannot be a solution. If $s = 20$, then $s - 5 = 20 - 5 = 15$ and $20^2 + 15^2 = 625 = 25^2$. The answer checks.

State. The height of the ramp is 15 ft.

28. ***Familiarize.*** We will use the formula $s = 16t^2$.

Translate. We substitute 1053 for s.

$$1053 = 16t^2$$

Solve.

$$1053 = 16t^2$$
$$\frac{1053}{16} = t^2$$
$$65.8125 = t^2$$
$$t = \sqrt{65.8125} \quad or \quad t = -\sqrt{65.8125}$$
$$t \approx 8.1 \quad or \quad t \approx -8.1$$

Check. Time cannot be negative in this application, so -8.1 cannot be a solution. We check 8.1.

$$16(8.1)^2 = 1049.76 \approx 1053$$

The answer is close, so we have a check. (Remember that we rounded the value of t.)

State. It would take about 8.1 sec for an object to fall to the ground from the top of the Royal Gorge Bridge.

29. $f(x) = 2x - 5$

$f(2) = 2 \cdot 2 - 5 = 4 - 5 = -1$

$f(-1) = 2(-1) - 5 = -2 - 5 = -7$

$f(3.5) = 2(3.5) - 5 = 7 - 5 = 2$

30. $g(x) = |x| - 1$

$g(1) = |1| - 1 = 1 - 1 = 0$

$g(-1) = |-1| - 1 = 1 - 1 = 0$

$g(-20) = |-20| - 1 = 20 - 1 = 19$

31. $C(p) = 15p$

$C(180) = 15 \cdot 180 = 2700$ calories

32. $g(x) = 4 - x$

We find some function values.

When $x = -1$, $g(-1) = 4 - (-1) = 4 + 1 = 5$

When $x = 0$, $g(0) = 4 - 0 = 4$.

When $x = 3$, $g(3) = 4 - 3 = 1$.

x	$g(x)$
-1	5
0	4
3	1

Plot these points and connect them.

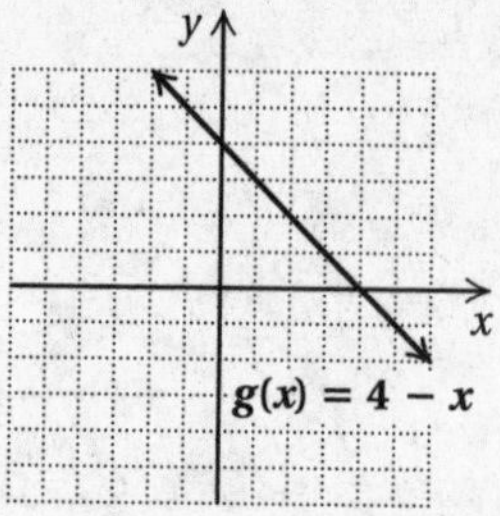

33. $f(x) = x^2 - 3$

Recall from Section 9.6 that the graph is a parabola. We find some function values.

When $x = -3$, $f(-3) = (-3)^2 - 3 = 9 - 3 = 6$.

When $x = -1$, $f(-1) = (-1)^2 - 3 = 1 - 3 = -2$.

When $x = 0$, $f(0) = 0^2 - 3 = 0 - 3 = -3$.

When $x = 1$, $f(1) = 1^2 - 3 = 1 - 3 = -2$.

When $x = 2$, $f(2) = 2^2 - 3 = 4 - 3 = 1$.

x	$f(x)$
-3	6
-1	-2
0	-3
1	-2
2	1

Plot these points and connect them.

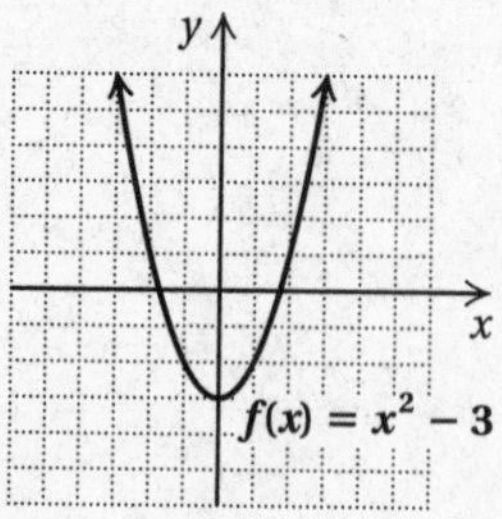

34. $h(x) = |x| - 5$

We find some function values.

When $x = -4$, $h(-4) = |-4| - 5 = 4 - 5 = -1$.

When $x = -2$, $h(-2) = |-2| - 5 = 2 - 5 = -3$.

When $x = 0$, $h(0) = |0| - 5 = 0 - 5 = -5$.

When $x = 1$, $h(1) = |1| - 5 = 1 - 5 = -4$.

When $x = 3$, $h(3) = |3| - 5 = 3 - 5 = -2$.

x	$h(x)$
-4	-1
-2	-3
0	-5
1	-4
3	-2

Plot these points and connect them.

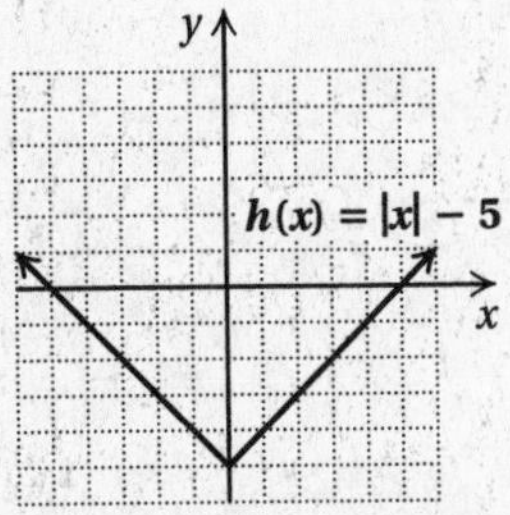

35. $f(x) = x^2 - 2x + 1$

Recall from Section 9.6 that the graph is a parabola.

When $x = -1$, $f(-1) = (-1)^2 - 2(-1) + 1 = 1 + 2 + 1 = 4$.

When $x = 0$, $f(0) = 0^2 - 2 \cdot 0 + 1 = 0 - 0 + 1 = 1$.

When $x = 1$, $f(1) = 1^2 - 2 \cdot 1 + 1 = 1 - 2 + 1 = 0$.

When $x = 2$, $f(2) = 2^2 - 2 \cdot 2 + 1 = 4 - 4 + 1 = 1$.

When $x = 3$, $f(3) = 3^2 - 2 \cdot 3 + 1 = 9 - 6 + 1 = 4$.

x	$f(x)$
-1	4
0	1
1	0
2	1
3	4

Plot these points and connect them.

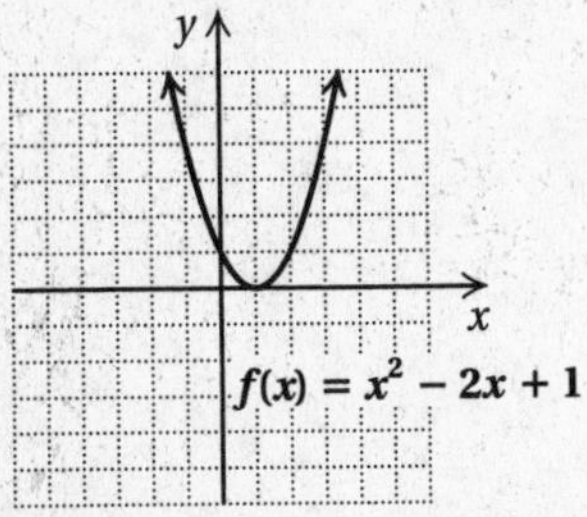

36. It is possible for a vertical line to intersect the graph more than once, so this is not the graph of a function.

37. No vertical line will intersect the graph more than once, so this is the graph of a function.

38. $40x - x^2 = 0$

$x(40 - x) = 0$

$x = 0$ *or* $40 - x = 0$

$x = 0$ *or* $40 = x$

Answer D is correct.

39. $\frac{1}{2}c^2 + c - \frac{1}{2} = 0$

$a = \frac{1}{2}$, $b = 1$, $c = -\frac{1}{2}$

(Note that we are using c in two different ways in this exercise.)

$$c = \frac{-b \pm \sqrt{b^2 - 4ac}}{2a}$$

$$c = \frac{-1 \pm \sqrt{1^2 - 4 \cdot \frac{1}{2} \cdot (-\frac{1}{2})}}{2 \cdot \frac{1}{2}} = \frac{-1 \pm \sqrt{1+1}}{1}$$

$$c = -1 \pm \sqrt{2}$$

Answer A is correct.

40. ***Familiarize.*** Let x = the first integer. Then $x + 1$ = the second integer.

Translate. If the numbers are positive, then $(x+1)^2$ is larger than x^2 and we have:

Square of larger number	minus	square of smaller number	is	63.
↓	↓	↓	↓	↓
$(x+1)^2$	$-$	x^2	$=$	63

If the numbers are negative, then x^2 is larger than $(x+1)^2$ and we have:

Square of smaller number	minus	square of larger number	is	63.
↓	↓	↓	↓	↓
x^2	$-$	$(x+1)^2$	$=$	63

Solve. We solve each equation.

$$\begin{aligned} (x+1)^2 - x^2 &= 63 \\ x^2 + 2x + 1 - x^2 &= 63 \\ 2x + 1 &= 63 \\ 2x &= 62 \\ x &= 31 \end{aligned}$$

If $x = 31$, then $x + 1 = 31 + 1 = 32$.

$$\begin{aligned} x^2 - (x+1)^2 &= 63 \\ x^2 - (x^2 + 2x + 1) &= 63 \\ x^2 - x^2 - 2x - 1 &= 63 \\ -2x - 1 &= 63 \\ -2x &= 64 \\ x &= -32 \end{aligned}$$

If $x = -32$, then $x + 1 = -32 + 1 = -31$.

Check. 31 and 32 are consecutive integers and $32^2 - 31^2 = 1024 - 961 = 63$. Also, -32 and -31 are consecutive integers and $(-32)^2 - (-31)^2 = 1024 - 961 = 63$. Both pairs of numbers check.

State. The integers are 31 and 32 or -32 and -31.

41. ***Familiarize.*** The area of a square with side s is s^2; the area of a circle with radius 5 in. is $\pi \cdot 5^2$, or 25π in^2.

Translate.

Area of square	equals	area of circle.
↓	↓	↓
s^2	$=$	25π

Solve.

$$s^2 = 25\pi$$

$$s = 5\sqrt{\pi} \quad or \quad s = -5\sqrt{\pi}$$

Check. Since the length of a side of the square cannot be negative, $-5\sqrt{\pi}$ cannot be a solution. If the length of a side of the square is $5\sqrt{\pi}$, then the area of the square is $(5\sqrt{\pi})^2 = 25\pi$. Since this is also the area of the circle, $5\sqrt{\pi}$ checks.

State. $s = 5\sqrt{\pi}$ in. ≈ 8.9 in.

42. $x - 4\sqrt{x} - 5 = 0$

Let $u = \sqrt{x}$. Then $u^2 = x$. Substitute u for $\sqrt{x}$ and u^2 for x and solve for u.

$$u^2 - 4u - 5 = 0$$

$$(u+1)(u-5) = 0$$

$$u+1 = 0 \quad or \quad u-5 = 0$$

$$u = -1 \quad or \quad u = 5$$

Now we substitute $\sqrt{x}$ for u and solve for x.

$\sqrt{x} = -1$ has no real-number solutions.

If $u = 5$, then we have:

$$\sqrt{x} = 5$$

$$(\sqrt{x})^2 = 5^2$$

$$x = 25$$

The number 25 checks. It is the solution.

43. The graph of $y = (x+3)^2$ contains the points $(-4, 1)$ and $(-2, 1)$, so the solutions of $(x+3)^2 = 1$ are -4 and -2.

44. The graph of $y = (x+3)^2$ contains the points $(-5, 4)$ and $(-1, 4)$, so the solutions of $(x+3)^2 = 4$ are -5 and -1.

45. The graph of $y = (x+3)^2$ contains the points $(-6, 9)$ and $(0, 9)$, so the solutions of $(x+3)^2 = 9$ are -6 and 0.

46. The graph of $y = (x+3)^2$ contains the point $(-3, 0)$, so the solution of $(x+3)^2 = 0$ is -3.

Chapter 9 Discussion and Writing Exercises

1. The second line should be $x+6 = \sqrt{16}$ *or* $x+6 = -\sqrt{16}$. Then we would have

$$x+6 = 4 \quad or \quad x+6 = -4$$

$$x = -2 \quad or \quad x = -10$$

Both numbers check, so the solutions are -2 and -10.

2. No; since each input has exactly one output, the number of outputs cannot exceed the number of inputs.

3. Find the average, v, of the x-coordinates of the x-intercepts. Then the equation of the line of symmetry is $x = v$. The number v is also the first coordinate of the vertex. Substitute this value for x in the equation of the parabola to find the y-coordinate of the vertex.

4. If $a > 0$ the graph opens up. If $a < 0$, the graph opens down.

5. The solutions will be rational numbers because each is the solution of a linear equation of the form $bx + a = 0$.

Chapter 9 Test

1. $7x^2 = 35$

$$x^2 = 5$$

$$x = \sqrt{5} \quad or \quad x = -\sqrt{5}$$

The solutions are $\sqrt{5}$ and $-\sqrt{5}$.

2. $7x^2 + 8x = 0$

$$x(7x+8) = 0$$

$$x = 0 \quad or \quad 7x+8 = 0$$

$$x = 0 \quad or \quad 7x = -8$$

$$x = 0 \quad or \quad x = -\frac{8}{7}$$

The solutions are 0 and $-\frac{8}{7}$.

3. $48 = t^2 + 2t$

$$0 = t^2 + 2t - 48$$

$$0 = (t+8)(t-6)$$

$$t+8 = 0 \quad or \quad t-6 = 0$$

$$t = -8 \quad or \quad t = 6$$

The solutions are -8 and 6.

4. $3y^2 - 5y = 2$

$$3y^2 - 5y - 2 = 0$$

$$(3y+1)(y-2) = 0$$

$$3y+1 = 0 \quad or \quad y-2 = 0$$

$$3y = -1 \quad or \quad y = 2$$

$$y = -\frac{1}{3} \quad or \quad y = 2$$

The solutions are $-\frac{1}{3}$ and 2.

5. $(x-8)^2 = 13$

$$x - 8 = \sqrt{13} \quad or \quad x - 8 = -\sqrt{13}$$

$$x = 8 + \sqrt{13} \quad or \quad x = 8 - \sqrt{13}$$

The solutions are $8 \pm \sqrt{13}$.

6. $x^2 = x + 3$

$$x^2 - x - 3 = 0$$

$$a = 1,\ b = -1,\ c = -3$$

$$x = \frac{-(-1) \pm \sqrt{(-1)^2 - 4 \cdot 1 \cdot (-3)}}{2 \cdot 1}$$

$$x = \frac{1 \pm \sqrt{1+12}}{2}$$

$$x = \frac{1 \pm \sqrt{13}}{2}$$

The solutions are $\frac{1 \pm \sqrt{13}}{2}$.

7. $m^2 - 3m = 7$

$m^2 - 3m - 7 = 0$

$a = 1,\ b = -3,\ c = -7$

$$m = \frac{-(-3) \pm \sqrt{(-3)^2 - 4 \cdot 1 \cdot (-7)}}{2 \cdot 1}$$

$$m = \frac{3 \pm \sqrt{9+28}}{2}$$

$$m = \frac{3 \pm \sqrt{37}}{2}$$

The solutions are $\dfrac{3 \pm \sqrt{37}}{2}$.

8. $10 = 4x + x^2$

$0 = x^2 + 4x - 10$

$a = 1,\ b = 4,\ c = -10$

$$x = \frac{-4 \pm \sqrt{4^2 - 4 \cdot 1 \cdot (-10)}}{2 \cdot 1}$$

$$x = \frac{-4 \pm \sqrt{16+40}}{2} = \frac{-4 \pm \sqrt{56}}{2}$$

$$x = \frac{-4 \pm \sqrt{4 \cdot 14}}{2} = \frac{-4 \pm 2\sqrt{14}}{2}$$

$$x = \frac{2(-2 \pm \sqrt{14})}{2} = -2 \pm \sqrt{14}$$

The solutions are $-2 \pm \sqrt{14}$.

9. $3x^2 - 7x + 1 = 0$

$a = 3,\ b = -7,\ c = 1$

$$x = \frac{-(-7) \pm \sqrt{(-7)^2 - 4 \cdot 3 \cdot 1}}{2 \cdot 3}$$

$$x = \frac{7 \pm \sqrt{49-12}}{6} = \frac{7 \pm \sqrt{37}}{6}$$

The solutions are $\dfrac{7 \pm \sqrt{37}}{6}$.

10. $x - \dfrac{2}{x} = 1$, LCM is x

$$x\left(x - \frac{2}{x}\right) = x \cdot 1$$

$$x \cdot x - x \cdot \frac{2}{x} = x$$

$$x^2 - 2 = x$$

$$x^2 - x - 2 = 0$$

$$(x-2)(x+1) = 0$$

$x - 2 = 0$ *or* $x + 1 = 0$

$x = 2$ *or* $x = -1$

Both numbers check. The solutions are 2 and -1.

11. $\dfrac{4}{x} - \dfrac{4}{x+2} = 1$, LCM is $x(x+2)$

$$x(x+2)\left(\frac{4}{x} - \frac{4}{x+2}\right) = x(x+2) \cdot 1$$

$$x(x+2) \cdot \frac{4}{x} - x(x+2) \cdot \frac{4}{x+2} = x(x+2)$$

$$4(x+2) - 4x = x^2 + 2x$$

$$4x + 8 - 4x = x^2 + 2x$$

$$8 = x^2 + 2x$$

$$0 = x^2 + 2x - 8$$

$$0 = (x+4)(x-2)$$

$x + 4 = 0$ *or* $x - 2 = 0$

$x = -4$ *or* $x = 2$

Both numbers check. The solutions are -4 and 2.

12. $x^2 - 4x - 10 = 0$

$x^2 - 4x = 10$

$x^2 - 4x + 4 = 10 + 4$ Adding 4: $\left(\dfrac{-4}{2}\right)^2 = (-2)^2 = 4$

$(x-2)^2 = 14$

$x - 2 = \sqrt{14}$ *or* $x - 2 = -\sqrt{14}$

$x = 2 + \sqrt{14}$ *or* $x = 2 - \sqrt{14}$

The solutions are $2 \pm \sqrt{14}$.

13. From Exercise 12 we know that the solutions of the equation are $2 \pm \sqrt{14}$.

Using a calculator, we have

$2 - \sqrt{14} \approx -1.7$ and $2 + \sqrt{14} \approx 5.7$.

14. $d = an^2 + bn$

$0 = an^2 + bn - d$

We will use the quadratic formula with $a = a$, $b = b$, and $c = -d$.

$$n = \frac{-b \pm \sqrt{b^2 - 4 \cdot a \cdot (-d)}}{2 \cdot a}$$

$n = \dfrac{-b + \sqrt{b^2 + 4ad}}{2a}$ Using the positive square root

15. To find the x-intercepts we solve the following equation.

$-x^2 + x + 5 = 0$

$x^2 - x - 5 = 0$ Standard form

$a = 1,\ b = -1,\ c = -5$

$$x = \frac{-(-1) \pm \sqrt{(-1)^2 - 4 \cdot 1 \cdot (-5)}}{2 \cdot 1}$$

$$x = \frac{1 \pm \sqrt{1+20}}{2} = \frac{1 \pm \sqrt{21}}{2}$$

The x-intercepts are $\left(\dfrac{1-\sqrt{21}}{2}, 0\right)$ and $\left(\dfrac{1+\sqrt{21}}{2}, 0\right)$.

16. $y = 4 - x^2$, or $y = -x^2 + 4$

Find the vertex. The x-coordinate is

$$-\frac{b}{2a} = -\frac{0}{2(-1)} = 0.$$

The y-coordinate is

$$y = 4 - x^2 = 4 - 0^2 = 4.$$

The vertex is $(0, 4)$. This is also the y-intercept.

We choose some x-values on both sides of the vertex and graph the parabola.

x	y	
0	4	←Vertex
−1	3	
−2	0	
1	3	
2	0	

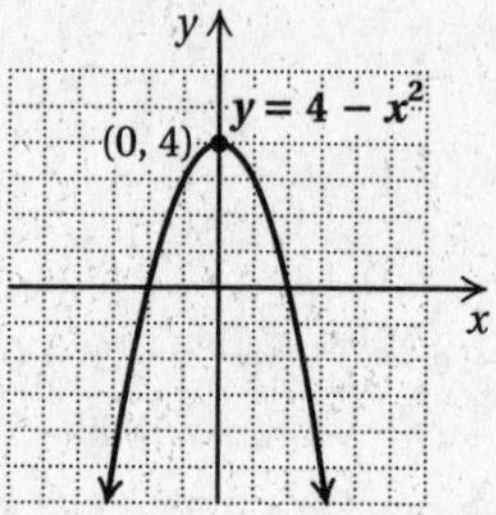

17. $y = -x^2 + x + 5$

Find the vertex. The x-coordinate is

$$-\frac{b}{2a} = -\frac{1}{2(-1)} = -\left(-\frac{1}{2}\right) = \frac{1}{2}.$$

The y-coordinate is

$$y = -\left(\frac{1}{2}\right)^2 + \frac{1}{2} + 5 = -\frac{1}{4} + \frac{1}{2} + 5 = \frac{21}{4}.$$

The vertex is $\left(\frac{1}{2}, \frac{21}{4}\right)$.

We choose some x-values on both sides of the vertex and graph the parabola.

x	y	
$\frac{1}{2}$	$\frac{21}{4}$	← Vertex
0	5	← y-intercept
−2	−1	
−1	3	
2	3	
3	−1	

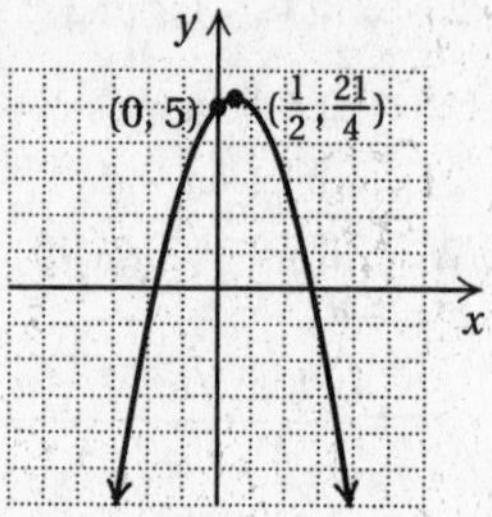

18. $f(x) = \frac{1}{2}x + 1$

$$f(0) = \frac{1}{2} \cdot 0 + 1 = 0 + 1 = 1$$

$$f(1) = \frac{1}{2} \cdot 1 + 1 = \frac{1}{2} + 1 = 1\frac{1}{2}$$

$$f(2) = \frac{1}{2} \cdot 2 + 1 = 1 + 1 = 2$$

19. $g(t) = -2|t| + 3$

$$g(-1) = -2|-1| + 3 = -2 \cdot 1 + 3 = -2 + 3 = 1$$

$$g(0) = -2|0| + 3 = -2 \cdot 0 + 3 = 0 + 3 = 3$$

$$g(3) = -2|3| + 3 = -2 \cdot 3 + 3 = -6 + 3 = -3$$

20. ***Familiarize.*** Using the labels on the drawing in the text, we let l and $l-4$ represent the length and width of the rug, respectively, in meters. Recall that the area of a rectangle is (length) × (width).

Translate.

$$\underbrace{\text{The area}}_{l(l-4)} \;\; \underbrace{\text{is}}_{=} \;\; \underbrace{16.25 \text{ m}^2}_{16.25}.$$

Solve.

$$l(l-4) = 16.25$$
$$l^2 - 4l = 16.25$$
$$l^2 - 4l - 16.25 = 0$$

The factorization of $l^2 - 4l - 16.25$ is not readily apparent so we will use the quadratic formula with $a = 1$, $b = -4$, and $c = -16.25$.

$$l = \frac{-(-4) \pm \sqrt{(-4)^2 - 4 \cdot 1 \cdot (-16.25)}}{2 \cdot 1}$$

$$l = \frac{4 \pm \sqrt{16 + 65}}{2} = \frac{4 \pm \sqrt{81}}{2}$$

$$l = \frac{4 \pm 9}{2}$$

$$l = \frac{4-9}{2} = \frac{-5}{2} = -\frac{5}{2}, \text{ or } -2.5$$

or

$$l = \frac{4+9}{2} = \frac{13}{2} = 6.5$$

Check. Since the length cannot be negative, −2.5 cannot be a solution. If $l = 6.5$, then $l - 4 = 6.5 - 4 = 2.5$ and $6.5(2.5) = 16.25$. Thus, the width is 4 m less than the length and the area is 16.25 m^2. The answer checks.

State. The length of the rug is 6.5 m, and the width is 2.5 m.

21. ***Familiarize.*** Let r = the speed of the boat in still water. Then $r - 2$ = the speed upstream and $r + 2$ = the speed downstream. We organize the information in a table.

	d	r	t
Upstream	44	$r-2$	t_1
Downstream	52	$r+2$	t_2

Translate. Using $t = d/r$ and the rows of the table, we have

$$t_1 = \frac{44}{r-2} \text{ and } t_2 = \frac{52}{r+2}.$$

Since the total time is 4 hr, $t_1 + t_2 = 4$, and we have

$$\frac{44}{r-2} + \frac{52}{r+2} = 4.$$

Solve. We solve the equation. We multiply by $(r-2)(r+2)$, the LCM of the denominators.

$$(r-2)(r+2)\left(\frac{44}{r-2} + \frac{52}{r+2}\right) = (r-2)(r+2) \cdot 4$$
$$44(r+2) + 52(r-2) = 4(r^2 - 4)$$
$$44r + 88 + 52r - 104 = 4r^2 - 16$$
$$96r - 16 = 4r^2 - 16$$
$$0 = 4r^2 - 96r$$
$$0 = 4r(r - 24)$$

$4r = 0$ *or* $r - 24 = 0$

$r = 0$ *or* $r = 24$

Check. The boat cannot travel upstream if its speed in still water is 0 km/h. If the speed of the boat in still water is 24 km/h, then it travels at a speed of 24 − 2, or 22 km/h, upstream and 24 + 2, or 26 km/h, downstream. At 22 km/h the boat travels 44 km in 44/22, or 2 hr. At 26 km/h it travels 52 km in 52/26, or 2 hr. The total time is 2 + 2, or 4 hr, so the answer checks.

State. The speed of the boat in still water is 24 km/h.

22. In 2012, $t = 2012 - 1940 = 72$.

$R(72) = 30.18 - 0.06(72) = 30.18 - 4.32 = 25.86$

We estimate that the record was 25.86 min in 2012.

23. $h(x) = x - 4$

We find some function values.

When $x = -1$, $h(-1) = -1 - 4 = -5$.

When $x = 2$, $h(2) = 2 - 4 = -2$.

When $x = 5$, $h(5) = 5 - 4 = 1$.

x	$h(x)$
−1	−5
2	−2
5	1

Plot these points and connect them.

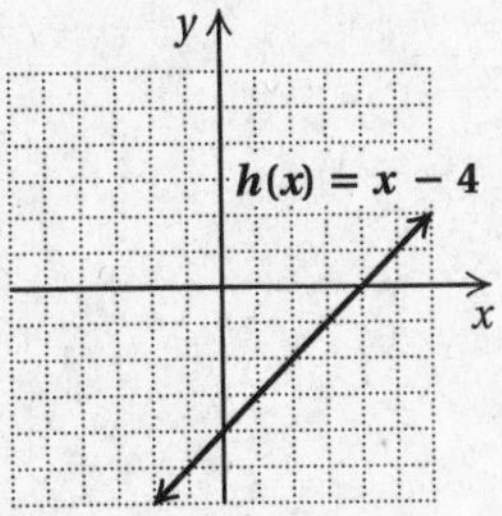

24. $g(x) = x^2 - 4$

Recall from Section 9.6 that the graph is a parabola. We find some function values.

When $x = -3$, $g(-3) = (-3)^2 - 4 = 9 - 4 = 5$.

When $x = -1$, $g(-1) = (-1)^2 - 4 = 1 - 4 = -3$.

When $x = 0$, $g(0) = 0^2 - 4 = 0 - 4 = -4$.

When $x = 2$, $g(2) = 2^2 - 4 = 4 - 4 = 0$.

When $x = 3$, $g(3) = 3^2 - 4 = 9 - 4 = 5$.

x	$g(x)$
−3	5
−1	−3
0	−4
2	0
3	5

Plot these points and connect them.

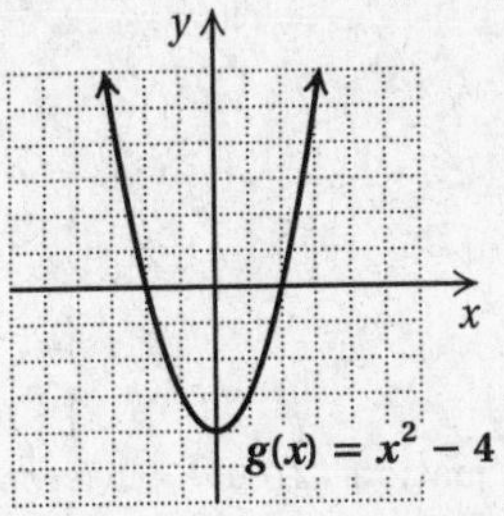

25. No vertical line will intersect the graph more than once, so this is the graph of a function.

26. It is possible for a vertical line to intersect the graph more than once, so this is not the graph of a function.

27. $g(x) = -2x - x^2$

$g(-8) = -2(-8) - (-8)^2 = 16 - 64 = -48$

Answer D is correct.

28. ***Familiarize***. We make a drawing. Let s = the length of a side of the square, in feet. Then $s + 5$ = the length of a diagonal.

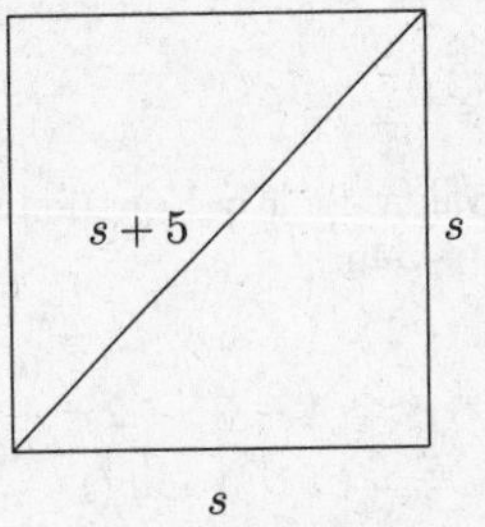

Translate. We use the Pythagorean equation.

$$s^2 + s^2 = (s + 5)^2$$

Solve.

$$s^2 + s^2 = (s + 5)^2$$
$$2s^2 = s^2 + 10s + 25$$
$$s^2 - 10s - 25 = 0$$

We use the quadratic formula with $a = 1$, $b = -10$, and $c = -25$.

$$s = \frac{-(-10) \pm \sqrt{(-10)^2 - 4 \cdot 1 \cdot (-25)}}{2 \cdot 1}$$
$$s = \frac{10 \pm \sqrt{100 + 100}}{2} = \frac{10 \pm \sqrt{200}}{2}$$
$$s = \frac{10 \pm \sqrt{100 \cdot 2}}{2} = \frac{10 \pm 10\sqrt{2}}{2}$$
$$s = \frac{2(5 \pm 5\sqrt{2})}{2} = 5 \pm 5\sqrt{2}$$

Check. Since $5 - 5\sqrt{2}$ is negative, it cannot be the length of a side of the square. If the length of a side is $5+5\sqrt{2}$ ft, then $(5+5\sqrt{2})^2+(5+5\sqrt{2})^2 = 25+50\sqrt{2}+50+25+50\sqrt{2}+50 = 150+100\sqrt{2}$. The length of a diagonal is $5+5\sqrt{2}+5$ ft, or $10 + 5\sqrt{2}$, and $(10 + 5\sqrt{2})^2 = 100 + 100\sqrt{2} + 50 = 150 + 100\sqrt{2}$. Since these lengths satisfy the Pythagorean equation, the answer checks.

State. The length of a side of the square is $5 + 5\sqrt{2}$ ft.

29. $x - y = 2,\quad (1)$
$xy = 4 \quad (2)$

Solve Equation (1) for y.

$$x - y = 2$$
$$-y = -x + 2$$
$$y = x - 2 \quad (3)$$

Substitute $x - 2$ for y in Equation (2) and solve for x.

$$xy = 4$$
$$x(x-2) = 4$$
$$x^2 - 2x = 4$$
$$x^2 - 2x - 4 = 0$$

We use the quadratic formula with $a = 1$, $b = -2$, and $c = -4$.

$$x = \frac{-(-2) \pm \sqrt{(-2)^2 - 4 \cdot 1 \cdot (-4)}}{2 \cdot 1}$$
$$x = \frac{2 \pm \sqrt{4 + 16}}{2} = \frac{2 \pm \sqrt{20}}{2}$$
$$x = \frac{2 \pm \sqrt{4 \cdot 5}}{2} = \frac{2 \pm 2\sqrt{5}}{2}$$
$$x = \frac{2(1 \pm \sqrt{5})}{2} = 1 \pm \sqrt{5}$$

Now substitute these values for x in Equation (3) and solve for the corresponding y-values.

For $x = 1 + \sqrt{5}$: $y = x - 2 = 1 + \sqrt{5} - 2 = -1 + \sqrt{5}$.

For $x = 1 - \sqrt{5}$: $y = x - 2 = 1 - \sqrt{5} - 2 = -1 - \sqrt{5}$.

The solutions are $(1+\sqrt{5}, -1+\sqrt{5})$ and $(1-\sqrt{5}, -1-\sqrt{5})$.

Cumulative Review Chapters 1 - 9

1. x^3 means $x \cdot x \cdot x$.

2. $(x-3)^2 + 5 = (10-3)^2 + 5 = 7^2 + 5 = 49 + 5 = 54$

3. First we divide to find decimal notation for $\frac{3}{11}$.

$$\begin{array}{r} 0.2727 \\ 11\overline{)3.0000} \\ \underline{22} \\ 80 \\ \underline{77} \\ 30 \\ \underline{22} \\ 80 \\ \underline{77} \\ 3 \end{array}$$

We have a repeating decimal.

$\frac{3}{11} = 0.\overline{27}$, so $-\frac{3}{11} = -0.\overline{27}$.

4. $15 = 3 \cdot 5$
$48 = 2 \cdot 2 \cdot 2 \cdot 2 \cdot 3$
$\text{LCM} = 2 \cdot 2 \cdot 2 \cdot 2 \cdot 3 \cdot 5$, or 240

5. This distance of -7 from 0 is 7, so $|-7| = 7$.

6. $-6 + 12 + (-4) + 7 = 6 + (-4) + 7$
$= 2 + 7$
$= 9$

7. $2.8 - (-12.2) = 2.8 + 12.2 = 15$

8. $-\frac{3}{8} \div \frac{5}{2} = -\frac{3}{8} \cdot \frac{2}{5} = -\frac{3 \cdot 2}{8 \cdot 5} = -\frac{3 \cdot \not{2}}{\not{2} \cdot 4 \cdot 5} = -\frac{3}{20}$

9. $13 \cdot 6 \div 3 \cdot 2 \div 13 = 78 \div 3 \cdot 2 \div 13$
$= 26 \cdot 2 \div 13$
$= 52 \div 13$
$= 4$

10. $4m + 9 - (6m + 13) = 4m + 9 - 6m - 13 = -2m - 4$

11.
$$3x = -24$$
$$\frac{3x}{3} = \frac{-24}{3}$$
$$x = -8$$

The solution is -8.

12.
$$3x + 7 = 2x - 5$$
$$3x + 7 - 2x = 2x - 5 - 2x$$
$$x + 7 = -5$$
$$x + 7 - 7 = -5 - 7$$
$$x = -12$$

The solution is -12.

13.
$$3(y-1) - 2(y+2) = 0$$
$$3y - 3 - 2y - 4 = 0$$
$$y - 7 = 0$$
$$y = 7$$

The solution is 7.

14.
$$x^2 - 8x + 15 = 0$$
$$(x-3)(x-5) = 0$$
$$x - 3 = 0 \quad or \quad x - 5 = 0$$
$$x = 3 \quad or \quad x = 5$$

The solutions are 3 and 5.

15. $y - x = 1,\quad (1)$
$y = 3 - x \quad (2)$

Substitute $3 - x$ for y in Equation (1) and solve for x.

$$y - x = 1$$
$$3 - x - x = 1$$
$$3 - 2x = 1$$
$$-2x = -2$$
$$x = 1$$

Substitute 1 for x in one of the original equations and find y. We use Equation (2).

$$y = 3 - x = 3 - 1 = 2$$

The ordered pair $(1, 2)$ checks in both equations. It is the solution.

16. $x + y = 17$, (1)

$x - y = 7$ (2)

$2x = 24$ Adding

$x = 12$

Substitute 12 for x in one of the original equations and solve for y.

$x + y = 17$ (1)

$12 + y = 17$

$y = 5$

The ordered pair $(12, 5)$ checks in both equations. It is the solution.

17. $4x - 3y = 3$, (1)

$3x - 2y = 4$ (2)

First we multiply Equation (1) by 2 and Equation (2) by -3 and then add.

$8x - 6y = 6$

$-9x + 6y = -12$

$-x = -6$

$x = 6$

Substitute 6 for x in one of the original equations and solve for y.

$3x - 2y = 4$ (2)

$3 \cdot 6 - 2y = 4$

$18 - 2y = 4$

$-2y = -14$

$y = 7$

The ordered pair $(6, 7)$ checks in both equations. It is the solution.

18. $x^2 - x - 6 = 0$

$(x - 3)(x + 2) = 0$

$x - 3 = 0$ *or* $x + 2 = 0$

$x = 3$ *or* $x = -2$

The solutions are 3 and -2.

19. $x^2 + 3x = 5$

$x^2 + 3x - 5 = 0$

We will use the quadratic formula with $a = 1$, $b = 3$, and $c = -5$.

$$x = \frac{-3 \pm \sqrt{3^2 - 4 \cdot 1 \cdot (-5)}}{2 \cdot 1}$$

$$x = \frac{-3 \pm \sqrt{9 + 20}}{2} = \frac{-3 \pm \sqrt{29}}{2}$$

20. $3 - x = \sqrt{x^2 - 3}$

$(3 - x)^2 = (\sqrt{x^2 - 3})^2$

$9 - 6x + x^2 = x^2 - 3$

$9 - 6x = -3$

$-6x = -12$

$x = 2$

The number 2 checks. It is the solution.

21. $5 - 9x \le 19 + 5x$

$5 - 14x \le 19$

$-14x \le 14$

$\frac{-14x}{-14} \ge \frac{14}{-14}$ Reversing the inequality symbol

$x \ge -1$

The solution set is $\{x | x \ge -1\}$.

22. $-\frac{7}{8}x + 7 = \frac{3}{8}x - 3$, LCM is 8

$8\left(-\frac{7}{8}x + 7\right) = 8\left(\frac{3}{8}x - 3\right)$

$8\left(-\frac{7}{8}x\right) + 8 \cdot 7 = 8 \cdot \frac{3}{8}x - 8 \cdot 3$

$-7x + 56 = 3x - 24$

$-10x + 56 = -24$

$-10x = -80$

$x = 8$

The solution is 8.

23. $0.6x - 1.8 = 1.2x$

$10(0.6x - 1.8) = 10(1.2x)$ Clearing decimals

$6x - 18 = 12x$

$-18 = 6x$

$-3 = x$

The solution is -3.

24. $-3x > 24$

$\frac{-3x}{-3} < \frac{24}{-3}$ Reversing the inequality symbol

$x < -8$

The solution set is $\{x | x < -8\}$.

25. $23 - 19y - 3y \ge -12$

$23 - 22y \ge -12$

$-22y \ge -35$

$\frac{-22y}{-22} \le \frac{-35}{-22}$ Reversing the inequality symbol

$y \le \frac{35}{22}$

The solution set is $\left\{y \middle| y \le \frac{35}{22}\right\}$.

26. $3y^2 = 30$

$y^2 = 10$

$y = \sqrt{10}$ *or* $y = -\sqrt{10}$

The solutions are $\sqrt{10}$ and $-\sqrt{10}$.

27. $(x - 3)^2 = 6$

$x - 3 = \sqrt{6}$ *or* $x - 3 = -\sqrt{6}$

$x = 3 + \sqrt{6}$ *or* $x = 3 - \sqrt{6}$

The solutions are $3 \pm \sqrt{6}$.

28.
$$\frac{6x-2}{2x-1}=\frac{9x}{3x+1},$$
LCM is $(2x-1)(3x+1)$
$$(2x-1)(3x+1)\cdot\frac{6x-2}{2x-1}=(2x-1)(3x+1)\cdot\frac{9x}{3x+1}$$
$$(3x+1)(6x-2)=9x(2x-1)$$
$$18x^2-2=18x^2-9x$$
$$-2=-9x$$
$$\frac{2}{9}=x$$
The number $\frac{2}{9}$ checks. It is the solution.

29.
$$\frac{2x}{x+1}=2-\frac{5}{2x},\text{ LCM is }2x(x+1)$$
$$2x(x+1)\cdot\frac{2x}{x+1}=2x(x+1)\left(2-\frac{5}{2x}\right)$$
$$2x\cdot 2x=2x(x+1)(2)-2x(x+1)\cdot\frac{5}{2x}$$
$$4x^2=4x(x+1)-5(x+1)$$
$$4x^2=4x^2+4x-5x-5$$
$$4x^2=4x^2-x-5$$
$$0=-x-5$$
$$x=-5$$
The number -5 checks. It is the solution.

30.
$$\frac{2x}{x+3}+\frac{6}{x}+7=\frac{18}{x^2+3x}$$
$$\frac{2x}{x+3}+\frac{6}{x}+7=\frac{18}{x(x+3)},\text{ LCM is }x(x+3)$$
$$x(x+3)\left(\frac{2x}{x+3}+\frac{6}{x}+7\right)=x(x+3)\cdot\frac{18}{x(x+3)}$$
$$2x\cdot x+6(x+3)+7x(x+3)=18$$
$$2x^2+6x+18+7x^2+21x=18$$
$$9x^2+27x+18=18$$
$$9x^2+27x=0$$
$$9x(x+3)=0$$
$9x=0$ *or* $x+3=0$

$x=0$ *or* $x=-3$

Since both 0 and -3 make a denominator 0 in the original equation, there is no solution.

31.
$$\sqrt{x+9}=\sqrt{2x-3}$$
$$(\sqrt{x+9})^2=(\sqrt{2x-3})^2$$
$$x+9=2x-3$$
$$-x+9=-3$$
$$-x=-12$$
$$x=12$$
The number 12 checks. It is the solution.

32.
$$A=\frac{4b}{t}$$
$$t\cdot A=t\cdot\frac{4b}{t}$$
$$At=4b$$
$$\frac{At}{4}=b$$

33.
$$\frac{1}{t}=\frac{1}{m}-\frac{1}{n}$$
$$mnt\cdot\frac{1}{t}=mnt\cdot\left(\frac{1}{m}-\frac{1}{n}\right)$$
$$mn=mnt\cdot\frac{1}{m}-mnt\cdot\frac{1}{n}$$
$$mn=nt-mt$$
$$mn+mt=nt$$
$$m(n+t)=nt$$
$$m=\frac{nt}{n+t}$$

34.
$$r=\sqrt{\frac{A}{\pi}}$$
$$r^2=\left(\sqrt{\frac{A}{\pi}}\right)^2$$
$$r^2=\frac{A}{\pi}$$
$$\pi\cdot r^2=\pi\cdot\frac{A}{\pi}$$
$$\pi r^2=A$$

35. $y=ax^2-bx$

$0=ax^2-bx-y$

We will use the quadratic formula with $a=a$, $b=-b$, and $c=-y$.
$$x=\frac{-(-b)\pm\sqrt{(-b)^2-4\cdot a\cdot(-y)}}{2\cdot a}$$
$$x=\frac{b+\sqrt{b^2+4ay}}{2a}\quad\text{Taking the positive square root}$$

36. $x^{-6}\cdot x^2=x^{-6+2}=x^{-4}=\frac{1}{x^4}$

37. $\frac{y^3}{y^{-4}}=y^{3-(-4)}=y^{3+4}=y^7$

38. $(2y^6)^2=2^2(y^6)^2=4y^{6\cdot 2}=4y^{12}$

39. $2x-3+5x^3-2x^3+7x^3+x=10x^3+3x-3$

40. $(4x^3+3x^2-5)+(3x^3-5x^2+4x-12)=7x^3-2x^2+4x-17$

41. $(6x^2-4x+1)-(-2x^2+7)=6x^2-4x+1+2x^2-7=8x^2-4x-6$

42. $-2y^2(4y^2-3y+1)=-2y^2\cdot 4y^2-(-2y^2)(3y)-2y^2\cdot 1=-8y^4-(-6y^3)-2y^2=-8y^4+6y^3-2y^2$

43.
$$\begin{aligned}(2t-3)(3t^2-4t+2)&=(2t-3)\cdot 3t^2-(2t-3)\cdot 4t+(2t-3)\cdot 2\\&=6t^3-9t^2-(8t^2-12t)+4t-6\\&=6t^3-9t^2-8t^2+12t+4t-6\\&=6t^3-17t^2+16t-6\end{aligned}$$

44. $\left(t-\frac{1}{4}\right)\left(t+\frac{1}{4}\right)=t^2-\left(\frac{1}{4}\right)^2=t^2-\frac{1}{16}$

45. $(3m-2)^2=(3m)^2-2\cdot 3m\cdot 2+2^2=9m^2-12m+4$

46. $(15x^2y^3+10xy^2+5)-(5xy^2-x^2y^2-2)$
$=15x^2y^3+10xy^2+5-5xy^2+x^2y^2+2$
$=15x^2y^3+x^2y^2+5xy^2+7$

47. $(x^2-0.2y)(x^2+0.2y)=(x^2)^2-(0.2y)^2=x^4-0.04y^2$

48. $(3p+4q^2)^2=(3p)^2+2\cdot 3p\cdot 4q^2+(4q^2)^2=$
$9p^2+24pq^2+16q^4$

49. $\frac{4}{2x-6}\cdot\frac{x-3}{x+3}=\frac{4(x-3)}{(2x-6)(x+3)}$
$=\frac{4(x-3)}{2(x-3)(x+3)}$
$=\frac{2\cdot 2\cdot (x-3)}{2(x-3)(x+3)}$
$=\frac{2}{x+3}$

50. $\frac{3a^4}{a^2-1}\div\frac{2a^3}{a^2-2a+1}=\frac{3a^4}{a^2-1}\cdot\frac{a^2-2a+1}{2a^3}$
$=\frac{3a^4(a^2-2a+1)}{(a^2-1)(2a^3)}$
$=\frac{3\cdot a^3\cdot a\cdot(a-1)(a-1)}{(a+1)(a-1)\cdot 2\cdot a^3}$
$=\frac{3\cdot a^3\cdot a(a-1)(a-1)}{(a+1)(a-1)\cdot 2\cdot a^3}$
$=\frac{3a(a-1)}{2(a+1)}$

51. $\frac{3}{3x-1}+\frac{4}{5x}=\frac{3}{3x-1}\cdot\frac{5x}{5x}+\frac{4}{5x}\cdot\frac{3x-1}{3x-1}$
$=\frac{3\cdot 5x}{5x(3x-1)}+\frac{4(3x-1)}{5x(3x-1)}$
$=\frac{15x}{5x(3x-1)}+\frac{12x-4}{5x(3x-1)}$
$=\frac{15x+12x-4}{5x(3x-1)}$
$=\frac{27x-4}{5x(3x-1)}$

52. $\frac{2}{x^2-16}-\frac{x-3}{x^2-9x+20}$
$=\frac{2}{(x+4)(x-4)}-\frac{x-3}{(x-4)(x-5)}$
$=\frac{2}{(x+4)(x-4)}\cdot\frac{x-5}{x-5}-\frac{x-3}{(x-4)(x-5)}\cdot\frac{x+4}{x+4}$
$=\frac{2(x-5)}{(x+4)(x-4)(x-5)}-\frac{(x-3)(x+4)}{(x+4)(x-4)(x-5)}$
$=\frac{2x-10}{(x+4)(x-4)(x-5)}-\frac{x^2+x-12}{(x+4)(x-4)(x-5)}$
$=\frac{2x-10-(x^2+x-12)}{(x+4)(x-4)(x-5)}$
$=\frac{2x-10-x^2-x+12}{(x+4)(x-4)(x-5)}$
$=\frac{-x^2+x+2}{(x+4)(x-4)(x-5)}$, or $\frac{-(x-2)(x+1)}{(x+4)(x-4)(x-5)}$

53. $8x^2-4x=4x\cdot 2x-4x\cdot 1=4x(2x-1)$

54. $25x^2-4=(5x)^2-2^2=(5x+2)(5x-2)$

55. $6y^2-5y-6=(3y+2)(2y-3)$ Using trial and error

56. $m^2-8m+16=m^2-2\cdot m\cdot 4+4^2=(m-4)^2$

57. $x^3-8x^2-5x+40=(x^3-8x^2)+(-5x+40)$
$=x^2(x-8)-5(x-8)$
$=(x-8)(x^2-5)$

58. $3a^4+6a^2-72=3(a^4+2a^2-24)$
$=3(a^2+6)(a^2-4)$
$=3(a^2+6)(a+2)(a-2)$

59. $16x^4-1=(4x^2)^2-1^2$
$=(4x^2+1)(4x^2-1)$
$=(4x^2+1)[(2x)^2-1^2]$
$=(4x^2+1)(2x+1)(2x-1)$

60. $49a^2b^2-4=(7ab)^2-2^2=(7ab+2)(7ab-2)$

61. $9x^2+30xy+25y^2=(3x)^2+2\cdot 3x\cdot 5y+(5y)^2=$
$(3x+5y)^2$

62. $2ac-6ab-3db+dc=(2ac-6ab)+(-3db+dc)$
$=2a(c-3b)+d(-3b+c)$
$=(2a+d)(c-3b)$

63. $15x^2+14xy-8y^2=(5x-2y)(3x+4y)$

64. $\frac{\frac{3}{x}+\frac{1}{2x}}{\frac{1}{3x}-\frac{3}{4x}}=\frac{\frac{3}{x}+\frac{1}{2x}}{\frac{1}{3x}-\frac{3}{4x}}\cdot\frac{12x}{12x}$
$=\frac{\frac{3}{x}\cdot 12x+\frac{1}{2x}\cdot 12x}{\frac{1}{3x}\cdot 12x-\frac{3}{4x}\cdot 12x}$
$=\frac{36+6}{4-9}$
$=\frac{42}{-5}=-\frac{42}{5}$

65. $\sqrt{49} = 7$ Taking the principal square root

66. $\sqrt{625} = 25$, so $-\sqrt{625} = -25$.

67. $\sqrt{64x^2} = \sqrt{(8x)^2} = 8x$

68. $\sqrt{a+b}\sqrt{a-b} = \sqrt{(a+b)(a-b)} = \sqrt{a^2-b^2}$

69. $\sqrt{32ab}\sqrt{6a^4b^2} = \sqrt{32ab \cdot 6a^4b^2} =$
$\sqrt{16 \cdot 2 \cdot a \cdot b \cdot 2 \cdot 3 \cdot a^4 \cdot b^2} = \sqrt{16}\sqrt{2 \cdot 2}\sqrt{a^4}\sqrt{b^2}\sqrt{3ab} =$
$4 \cdot 2 \cdot a^2 \cdot b\sqrt{3ab} = 8a^2b\sqrt{3ab}$

70. $\sqrt{150} = \sqrt{25 \cdot 6} = \sqrt{25}\sqrt{6} = 5\sqrt{6}$

71. $\sqrt{243x^3y^2} = \sqrt{81 \cdot 3 \cdot x^2 \cdot x \cdot y^2} = \sqrt{81}\sqrt{x^2}\sqrt{y^2}\sqrt{3x} =$
$9xy\sqrt{3x}$

72. $\sqrt{\dfrac{100}{81}} = \dfrac{\sqrt{100}}{\sqrt{81}} = \dfrac{10}{9}$

73. $\sqrt{\dfrac{64}{x^2}} = \dfrac{\sqrt{64}}{\sqrt{x^2}} = \dfrac{8}{x}$

74. $4\sqrt{12} + 2\sqrt{12} = 6\sqrt{12} = 6\sqrt{4 \cdot 3} = 6\sqrt{4}\sqrt{3} =$
$6 \cdot 2\sqrt{3} = 12\sqrt{3}$

75. $\dfrac{\sqrt{72}}{\sqrt{45}} = \sqrt{\dfrac{72}{45}} = \sqrt{\dfrac{8}{5}} = \sqrt{\dfrac{8}{5} \cdot \dfrac{5}{5}} = \dfrac{\sqrt{40}}{5} =$
$\dfrac{\sqrt{4 \cdot 10}}{5} = \dfrac{2\sqrt{10}}{5}$

76.
$$\begin{aligned} a^2 + b^2 &= c^2 \\ 9^2 + b^2 &= 41^2 \\ 81 + b^2 &= 1681 \\ b^2 &= 1600 \\ b &= 40 \end{aligned}$$

77. $y = \dfrac{1}{3}x - 2$

We choose some values of x and find the corresponding y-values. We use multiples of 3 to avoid fractions.

When $x = 3$, $y = \dfrac{1}{3}(-3) - 2 = -1 - 2 = -3$.

When $x = 0$, $y = \dfrac{1}{3} \cdot 0 - 2 = 0 - 2 = -2$.

When $x = 3$, $y = \dfrac{1}{3} \cdot 3 - 2 = 1 - 2 = -1$.

x	y
-3	-3
0	-2
3	-1

We plot the points and draw and label the graph.

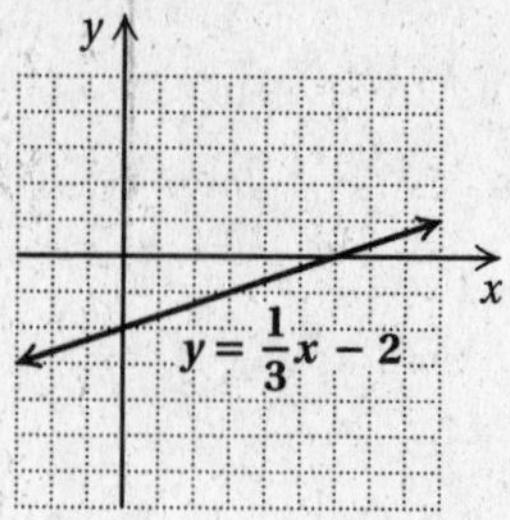

78. $2x + 3y = -6$

First we will find the intercepts. To find the y-intercept let $x = 0$ and solve for y.

$$\begin{aligned} 2 \cdot 0 + 3y &= -6 \\ 3y &= -6 \\ y &= -2 \end{aligned}$$

The y-intercept is $(0, -2)$.

To find the x-intercept let $y = 0$ and solve for x.

$$\begin{aligned} 2x + 3 \cdot 0 &= -6 \\ 2x &= -6 \\ x &= -3 \end{aligned}$$

The x-intercept is $(-3, 0)$.

Plot these points and draw the graph.

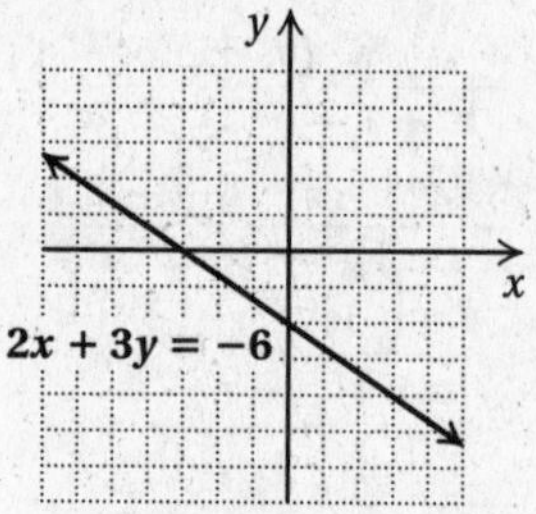

We can use a third point as a check. When $x = 3$ we have:

$$\begin{aligned} 2 \cdot 3 + 3y &= -6 \\ 6 + 3y &= -6 \\ 3y &= -12 \\ y &= -4 \end{aligned}$$

It appears that the point $(3, -4)$ is on the graph, so the graph is probably correct.

79. $y = -3$, or $y = 0 \cdot x - 3$

No matter what value we choose for x, y is always -3. The graph is a horizontal line with y-intercept $(0, -3)$.

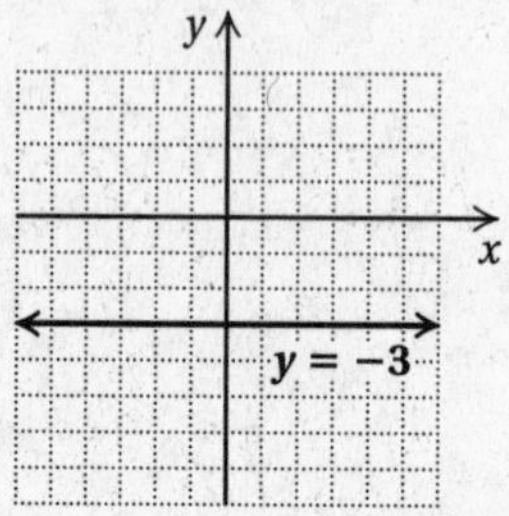

80. $x \geq -3$

First we graph the line $x = -3$. We draw a solid line since the inequality symbol is $\geq$. Then we test a point not on the line. We use $(0,0)$.

$$\begin{array}{c|c} \multicolumn{2}{c}{x \geq -3} \\ \hline 0 \ ? & -3 \ \text{TRUE} \end{array}$$

Since $(0,0)$ is a solution of the inequality we shade the half-plane that contains (0.0).

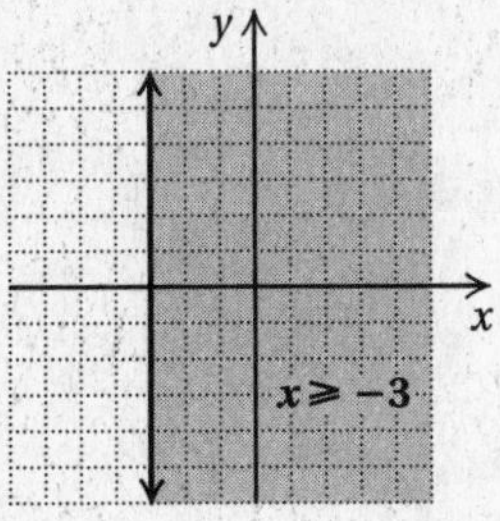

81. $4x - 3y > 12$

First we graph the line $4x - 3y = 12$. The intercepts are $(0, -4)$ and $(3, 0)$. We draw a dashed line since the inequality symbol is $>$. Now we test a point not on the line. We use $(0,0)$.

$$\begin{array}{c|l} \multicolumn{2}{c}{4x - 3y > 12} \\ \hline 4 \cdot 0 - 3 \cdot 0 \ ? & 12 \\ 0 & \quad \text{FALSE} \end{array}$$

Since $(0,0)$ is not a solution we shade the half-plane that does not contain $(0,0)$.

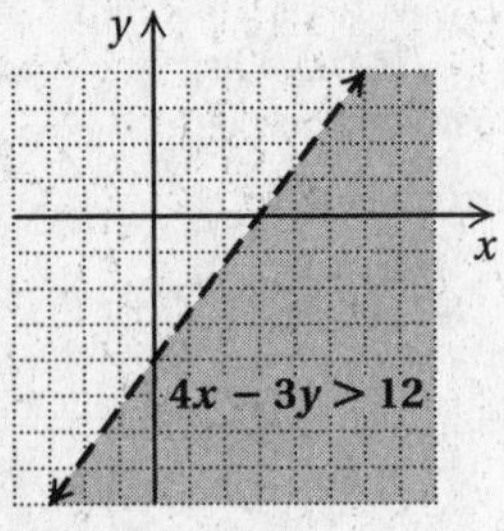

82. $y = x^2 + 2x + 1$

Find the vertex. The x-coordinate is

$$-\frac{b}{2a} = -\frac{2}{2 \cdot 1} = -1.$$

The y-coordinate is

$$y = x^2 + 2x + 1 = (-1)^2 + 2(-1) + 1 = 1 - 2 + 1 = 0.$$

The vertex is $(-1, 0)$.

We choose some x-values on both sides of the vertex and graph the parabola.

x	y	
-1	0	← Vertex
0	1	← y-intercept
-3	4	
-2	1	
1	4	

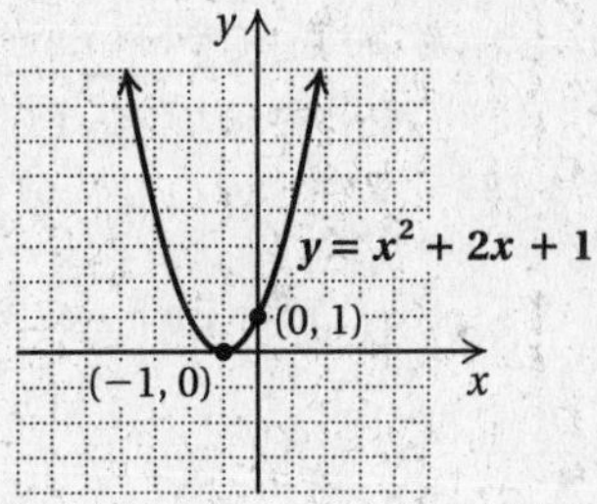

83.

$$9x^2 - 12x - 2 = 0$$

$$\frac{1}{9}(9x^2 - 12x - 2) = \frac{1}{9} \cdot 0$$

$$x^2 - \frac{4}{3}x - \frac{2}{9} = 0$$

$$x^2 - \frac{4}{3}x = \frac{2}{9}$$

$$x^2 - \frac{4}{3}x + \frac{4}{9} = \frac{2}{9} + \frac{4}{9} \quad \text{Adding: } \frac{4}{9}:$$

$$\left[\frac{1}{2}\left(-\frac{4}{3}\right)\right]^2 = \left(-\frac{2}{3}\right)^2 = \frac{4}{9}$$

$$\left(x - \frac{2}{3}\right)^2 = \frac{6}{9}$$

$$x - \frac{2}{3} = \frac{\sqrt{6}}{3} \quad or \quad x - \frac{2}{3} = -\frac{\sqrt{6}}{3}$$

$$x = \frac{2}{3} + \frac{\sqrt{6}}{3} \quad or \quad x = \frac{2}{3} - \frac{\sqrt{6}}{3}$$

$$x = \frac{2 + \sqrt{6}}{3} \quad or \quad x = \frac{2 - \sqrt{6}}{3}$$

The solutions are $\dfrac{2 \pm \sqrt{6}}{3}$.

84.

$$4x^2 = 4x + 1$$

$$4x^2 - 4x - 1 = 0$$

$$a = 4,\ b = -4,\ c = -1$$

$$x = \frac{-(-4) \pm \sqrt{(-4)^2 - 4 \cdot 4 \cdot (-1)}}{2 \cdot 4}$$

$$x = \frac{4 \pm \sqrt{16 + 16}}{8} = \frac{4 \pm \sqrt{32}}{8}$$

Using a calculator, we have

$$\frac{4 - \sqrt{32}}{8} \approx -0.2 \text{ and } \frac{4 + \sqrt{32}}{8} \approx 1.2.$$

85. $\underbrace{\text{What percent}}$ of 52 is 13?

$$\begin{array}{ccccc} \downarrow & \downarrow & \downarrow & \downarrow & \downarrow \\ p & \cdot & 52 & = & 13 \end{array}$$

To solve the equation we divide both sides by 52 and then convert to percent notation.

$$p \cdot 52 = 13$$

$$\frac{p \cdot 52}{52} = \frac{13}{52}$$

$$p = 0.25 = 25\%$$

Thus, 25% of 52 is 13.

86. 12 is 20% of $\underbrace{\text{what number}}$?

$$\begin{array}{ccccc} \downarrow & \downarrow & \downarrow & \downarrow & \downarrow \\ 12 & = & 20\% & \times & b, \text{ or} \\ 12 & = & 0.2 & \times & b \end{array}$$

To solve the equation we divide both sides by 0.2.

$$12 = 0.2 \times b$$
$$\frac{12}{0.2} = \frac{0.2 \times b}{0.2}$$
$$60 = b$$

Thus, 12 is 20% of 60.

87. ***Familiarize***. We will use the work principal, substituting 8 for a and 10 for b.

Translate.

$$\frac{t}{8} + \frac{t}{10} = 1$$

Solve. First we multiply by 40 to clear the fractions.

$$\frac{t}{8} + \frac{t}{10} = 1$$
$$40\left(\frac{t}{8} + \frac{t}{10}\right) = 40 \cdot 1$$
$$40 \cdot \frac{t}{8} + 40 \cdot \frac{t}{10} = 40$$
$$5t + 4t = 40$$
$$9t = 40$$
$$t = \frac{40}{9}, \text{ or } 4\frac{4}{9}$$

Check. In $\frac{40}{9}$ hr crew A does $\frac{40}{9} \cdot \frac{1}{8}$, or $\frac{5}{9}$, of the job and crew B does $\frac{40}{9} \cdot \frac{1}{10}$, or $\frac{4}{9}$, of the job. Together they do $\frac{5}{9} + \frac{4}{9} = 1$ entire job. The answer checks.

State. It would take the crews $4\frac{4}{9}$ hr to do the job working together.

88. ***Familiarize***. Using the labels on the drawing in the text, we let $w =$ the width of the screen, in feet, and $w + 4 =$ the length. Recall that the area of a rectangle is (length) × (width).

Translate. The area is 96 ft^2, so we have

$$(w + 4)w = 96.$$

Solve.

$$(w + 4)w = 96$$
$$w^2 + 4w = 96$$
$$w^2 + 4w - 96 = 0$$
$$(w + 12)(w - 8) = 0$$
$$w + 12 = 0 \quad or \quad w - 8 = 0$$
$$w = -12 \quad or \quad w = 8$$

Check. The width cannot be negative, so -12 cannot be a solution. If $w = 8$, then $w + 4 = 12$. The length is 4 ft more than the width and the area is $12 \cdot 8$, or 96 ft^2. The answer checks.

State. The length of the screen is 12 ft, and the width is 8 ft.

89. ***Familiarize***. Let $r =$ the speed of the stream, in km/h. Then the boat travels upstream at a speed of $8 - r$ and it travels downstream at a speed of $8 + r$. We organize the information in a table.

	d	r	t
Upstream	60	$8 - r$	t_1
Downstream	60	$8 + r$	t_2

Translate. Using $t = d/r$ and the rows of the table we have

$$t_1 = \frac{60}{8 - r} \text{ and } t_2 = \frac{60}{8 + r}.$$

Since the total time is 16 hr, $t_1 + t_2 = 16$, and we have

$$\frac{60}{8 - r} + \frac{60}{8 + r} = 16.$$

Solve. First we multiply by $(8 - r)(8 + r)$, the LCM of the denominators.

$$(8 - r)(8 + r)\left(\frac{60}{8 - r} + \frac{60}{8 + r}\right) = (8 - r)(8 + r) \cdot 16$$
$$60(8 + r) + 60(8 - r) = 16(64 - r^2)$$
$$480 + 60r + 480 - 60r = 1024 - 16r^2$$
$$960 = 1024 - 16r^2$$
$$-64 = -16r^2$$
$$4 = r^2$$
$$r = 2 \quad or \quad r = -2$$

Check. The speed of the current cannot be negative, so -2 cannot be a solution. If the speed of the current is 2 km/h, then the boat travels upstream at $8 - 2$, or 6 km/h. At this speed it will travel 60 km in 60/6, or 10 hr. The boat travels downstream at $8 + 2$, or 10 km/h. At this speed it will travel 60 km in 60/10, or 6 hr. The total time is $10 + 6$, or 16 hr, so the answer checks.

State. The speed of the current is 2 km/h.

90. ***Familiarize***. Let $w =$ the width of the floor, in meters. Then $w + 7 =$ the length.

Translate. We use the Pythagorean equation.

$$a^2 + b^2 = c^2$$
$$w^2 + (w + 7)^2 = 13^2$$

Solve.

$$w^2 + (w + 7)^2 = 13^2$$
$$w^2 + w^2 + 14w + 49 = 169$$
$$2w^2 + 14w + 49 = 169$$
$$2w^2 + 14w - 120 = 0$$
$$w^2 + 7w - 60 = 0 \quad \text{Dividing by 2}$$
$$(w + 12)(w - 5) = 0$$
$$w + 12 = 0 \quad or \quad w - 5 = 0$$
$$w = -12 \quad or \quad w = 5$$

Check. The width cannot be negative, so -12 cannot be a solution. If $w = 5$, then $w + 7 = 12$ and $5^2 + 12^2 = 25 + 144 = 169 = 13^2$. The answer checks.

State. The length of the floor is 12 m.

91. ***Familiarize.*** Let $x =$ the smaller odd integer. Then $x + 2 =$ the larger odd integer.

Translate.

$$\underbrace{\text{The sum of the squares of the integers}}_{x^2 + (x+2)^2} \ \underbrace{\text{is}}_{=} \ \underbrace{74.}_{74}$$

Solve.

$$x^2 + (x+2)^2 = 74$$
$$x^2 + x^2 + 4x + 4 = 74$$
$$2x^2 + 4x + 4 = 74$$
$$2x^2 + 4x - 70 = 0$$
$$x^2 + 2x - 35 = 0 \quad \text{Dividing by 2}$$
$$(x+7)(x-5) = 0$$
$$x + 7 = 0 \quad or \quad x - 5 = 0$$
$$x = -7 \quad or \quad x = 5$$

Check. If $x = -7$, then $x + 2 = -7 + 2 = -5$ and $(-7)^2 + (-5)^2 = 49 + 25 = 74$. If $x = 5$, then $x + 2 = 5 + 2 = 7$ and $5^2 + 7^2 = 25 + 49 = 74$. The answer checks.

State. The numbers are -7 and -5 or 5 and 7.

92. ***Familiarize.*** Let $x =$ the amount of solution A and $y =$ the amount of solution B to be used, in liters. We organize the information in a table.

Solution	A	B	Mixture
Amount	x	y	60
Percent of alcohol	75%	50%	$66\frac{2}{3}\%$
Amount of alcohol	$75\%x$	$50\%y$	$66\frac{2}{3}\% \times 60$, or 40

Translate. The first and last rows of the table give us two equations:

$$x + y = 60 \text{ and}$$
$$75\%x + 50\%y = 40, \text{ or } 0.75x + 0.5y = 40.$$

After clearing decimals we have the following system of equations.

$$x + y = 60, \quad (1)$$
$$75x + 50y = 4000 \quad (2)$$

Solve. First we multiply Equation (1) by -50 and then add.

$$-50x - 50y = -3000$$
$$75x + 50y = 4000$$
$$25x \quad = 1000$$
$$x = 40$$

Substitute 40 for x and one of the equations and solve for y.

$$x + y = 60 \quad (1)$$
$$40 + y = 60$$
$$y = 20$$

Check. If 40 L of solution A and 20 L of solution B are used, the solution consists of $40 + 20$, or 60 L. The amount of alcohol in the solution is $0.75(40) + 0.5(20) = 30 + 10 = 40$ L. The answer checks.

State. 40 L of solution A and 20 L of solution B should be used.

93. ***Familiarize.*** We will use the formula $s = 16t^2$.

Translate. Substitute 315 for s in the formula.

$$s = 16t^2$$
$$315 = 16t^2$$

Solve.

$$315 = 16t^2$$
$$\frac{315}{16} = t^2$$
$$\sqrt{\frac{315}{16}} = t \quad or \quad -\sqrt{\frac{315}{16}} = t$$
$$4.4 \approx t \quad or \quad -4.4 \approx t$$

Check. The time cannot be negative in this application, so -4.4 cannot be a solution. If $t \approx 4.4$, then $16(4.4)^2 = 309.76 \approx 315$. The answer checks. (Remember that we rounded the value of t.)

State. It would take an object about 4.4 sec to fall to the ground from the top of the Elizabeth Tower.

94. ***Familiarize and Translate.*** Let $p =$ the amount of the pay and $h =$ the number of hours worked. We have direct variation between p and h so an equation $p = kh$, $k > 0$, applies.

Solve. First we find an equation of variation.

$$p = kh$$
$$242.52 = k \cdot 43$$
$$5.64 = k$$

The equation of variation is $p = 5.64h$. Note that the variation constant k is the hourly rate at which the student is paid.

Now we find the value of p when $h = 80$.

$$p = 5.64h$$
$$p = 5.64(80) = 451.20$$

Check. We can repeat the computations. Also note that, when the number of hours increased, the pay also increased. This is what we expect with direct variation.

State. The student would earn \$451.20 for working 80 hr. The variation constant is the amount earned per hour.

95. ***Familiarize.*** Let $c =$ the number of cars entering the city each morning.

Translate. We reword and translate to an equation.

$$\underbrace{3654}_{3654} \ \underbrace{\text{is}}_{=} \ \underbrace{\text{three-fifths}}_{\frac{3}{5}} \ \underbrace{\text{of}}_{\cdot} \ \underbrace{\text{what number?}}_{c}$$

Solve.

$$3654 = \frac{3}{5} \cdot c$$

$$\frac{5}{3} \cdot 3654 = c$$

$$6090 = c$$

Check. $\frac{3}{5} \cdot 6090 = \frac{3 \cdot 6090}{5} = \frac{3 \cdot \not{5} \cdot 1218}{\not{5} \cdot 1} = 3654.$

The answer checks.

State. 6090 cars enter the city each morning.

96. ***Familiarize.*** Let r the speed of the plane in still air. Then its speed with the wind is $r + 20$ and the speed against the wind is $r - 20$. We organize the information in a table.

	Distance	Speed	Time
With the wind	d	$r+20$	3
Against the wind	d	$r-20$	4

Translate. Using $d = rt$ in each row of the table we have

$$d = (r + 20)3, \quad (1)$$

$$d = (r - 20)4. \quad (2)$$

Solve. We substitute $(r + 20)3$ for d in Equation (2) and solve for r.

$$(r + 20)3 = (r - 20)4$$

$$3r + 60 = 4r - 80$$

$$60 = r - 80$$

$$140 = r$$

Check. If the speed of the plane in still air is 140 mph, then it travels at $140 + 20$, or 160 mph, with the wind. At this speed, in 3 hr it travels $160 \cdot 3$, or 480 mi. The plane travels at $140 - 20$, or 120 mph, against the wind. At this speed, in 4 hr it travels $120 \cdot 4$, or 480 mi. The distances are the same, so the answer checks.

State. The speed of the plane in still air is 140 mph.

97. ***Familiarize.*** Let $x =$ the number of pounds of nuts and $y =$ the number of pounds of candy to be used. We organize the information in a table.

	Nuts	Candy	Mixture
Cost	\$9.90	\$7.20	\$8.10
Amount	x	y	42
Total cost	$9.9x$	$7.2y$	\$8.10(42), or \$340.20

Translate. The last two rows of the table give us two equations.

$$x + y = 42,$$

$$9.9x + 7.2y = 340.20$$

After clearing decimals we have the following system of equations.

$$x + y = 42, \quad (1)$$

$$99x + 72y = 3402 \quad (2)$$

Solve. First we multiply Equation (1) by -72 and then add.

$$\begin{aligned} -72x - 72y &= -3024 \\ 99x + 72y &= 3402 \\ \hline 27x \qquad &= 378 \\ x &= 14 \end{aligned}$$

Now substitute 14 for x in one of the equations and solve for y.

$$x + y = 42 \quad (1)$$

$$14 + y = 42$$

$$y = 28$$

Check. If 14 lb of nuts and 28 lb of candy are used, then the mixture weighs $14 + 28$, or 42 lb. The total cost of the mixture is \$9.90(14) + \$7.20(28) = \$138.60 + \$201.60 = \$340.20. The answer checks.

State. 14 lb of nuts and 28 lb of candy should be used.

98. The solutions of the equation are the first coordinates of the x-intercepts of the graph. We see that they are -3 and 2.

99. We let $y = 0$ and solve for x.

$$x^2 + 4x + 1 = 0$$

$$a = 1, \; b = 4, \; c = 1$$

$$x = \frac{-4 \pm \sqrt{4^2 - 4 \cdot 1 \cdot 1}}{2 \cdot 1}$$

$$x = \frac{-4 \pm \sqrt{16 - 4}}{2} = \frac{-4 \pm \sqrt{12}}{2}$$

$$x = \frac{-4 \pm \sqrt{4 \cdot 3}}{2} = \frac{-4 \pm 2\sqrt{3}}{2}$$

$$x = \frac{2(-2 \pm \sqrt{3})}{2} = -2 \pm \sqrt{3}$$

The x-intercepts are $(-2 - \sqrt{3}, 0)$ and $(-2 + \sqrt{3}, 0)$.

100. We solve for y to write the equation in the form $y = mx + b$.

$$-6x + 3y = -24$$

$$3y = 6x - 24$$

$$\frac{1}{3} \cdot 3y = \frac{1}{3}(6x - 24)$$

$$y = 2x - 8$$

The slope is 2 and the y-intercept is $(0, -8)$.

101. First we solve each equation for y.

$$y - x = 4 \qquad 3y + x = 8$$

$$y = x + 4 \qquad 3y = -x + 8$$

$$y = -\frac{1}{3}x + \frac{8}{3}$$

The slopes are not the same, so the lines are not parallel. The product of the slopes is $1\left(-\frac{1}{3}\right) = -\frac{1}{3} \neq -1$, so the lines are not perpendicular. Thus, the lines are neither parallel nor perpendicular.

102. $m = \dfrac{y_2 - y_1}{x_2 - x_1} = \dfrac{9-(-6)}{-4-(-5)} = \dfrac{15}{1} = 15$

103.
$$y = kx$$
$$100 = k \cdot 10$$
$$10 = k$$

The equation of variation is $y = 10x$.

When $x = 64$, we have:
$$y = 10x$$
$$y = 10 \cdot 64$$
$$y = 640$$

104.
$$y = \frac{k}{x}$$
$$100 = \frac{k}{10}$$
$$1000 = k$$

The equation of variation is $y = \dfrac{1000}{x}$.

When $x = 125$, we have:
$$y = \frac{1000}{x}$$
$$y = \frac{1000}{125}$$
$$y = 8$$

105. No vertical line will intersect the graph more than once, so this is the graph of a function.

106. It is possible for a vertical line to intersect the graph more than once, so this is not the graph of a function.

107. $f(x) = x^2 + x - 2$

We find some function values. Recall from Section 9.6 that the graph is a parabola.

x	$f(x)$
-3	4
-2	0
0	-2
1	0
2	4

We plot these points and draw the graph.

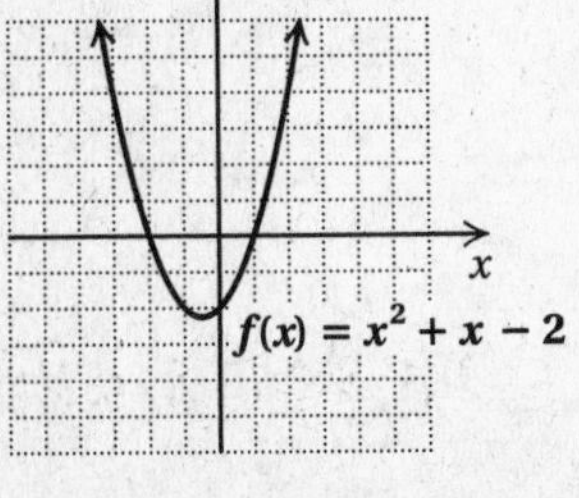

108. $g(x) = |x + 2|$

We find some function values, plot the points, and draw the graph.

x	$f(x)$
-5	3
-3	1
-2	0
0	2
3	5

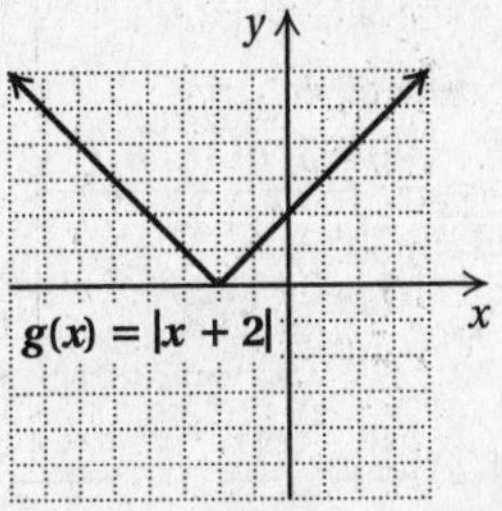

109. $f(x) = 2x^2 + 7x - 4$

$f(0) = 2 \cdot 0^2 + 7 \cdot 0 - 4 = 0 + 0 - 4 = -4$

$f(-4) = 2(-4)^2 + 7(-4) - 4 = 2 \cdot 16 - 28 - 4 = 32 - 28 - 4 = 0$

$f\left(\frac{1}{2}\right) = 2\left(\frac{1}{2}\right)^2 + 7 \cdot \frac{1}{2} - 4 = 2 \cdot \frac{1}{4} + \frac{7}{2} - 4 = \frac{1}{2} + \frac{7}{2} - 4 = 0$

110. ***Familiarize.*** Let r = the speed of the wind, in mph. Then the airplane travels at a speed of $300 - r$ against the wind and at $300 + r$ with the wind. We organize the information in a table.

	Distance	Speed	Time
Against wind	408	$300 - r$	t_1
With wind	492	$300 + r$	t_2

Translate. Using $t = d/r$ in each row of the table we have

$$t_1 = \frac{408}{300-r} \text{ and } t_2 = \frac{492}{300+r}.$$

Since the total time is 3 hr, $t_1 + t_2 = 3$ and we have

$$\frac{408}{300-r} + \frac{492}{300+r} = 3$$

Solve. First we multiply by $(300 - r)(300 + r)$, the LCM of the denominators.

$$(300-r)(300+r)\left(\frac{408}{300-r} + \frac{492}{300+r}\right) = (300-r)(300+r) \cdot 3$$
$$408(300+r) + 492(300-r) = 3(90{,}000 - r^2)$$
$$122{,}400 + 408r + 147{,}600 - 492r = 270{,}000 - 3r^2$$
$$270{,}000 - 84r = 270{,}000 - 3r^2$$
$$0 = 84r - 3r^2$$
$$0 = 3r(28 - r)$$

$3r = 0$ *or* $28 - r = 0$

$r = 0$ *or* $28 = r$

Check. Since we are assuming there is some wind, 0 cannot be a solution. If the speed of the wind is 28 mph, then the plane travels at a speed of $300-28$, or 272 mph, against the wind. At this speed, it travels 408 mi in 408/272, or 1.5 hr. With the wind the plane travels at $300 + 28$, or 328 mph. At this speed, it travels 492 mi in 492/328, or 1.5 hr. The total time is $1.5 + 1.5$, or 3 hr. The answer checks.

State. The speed of the wind is 28 mph. This is between 22 and 29 mph, so answer C is correct.

111. $2x^2 + 6x + 5 = 4$

$2x^2 + 6x + 1 = 0$

$a = 2,\ b = 6,\ c = 1$

$$x = \frac{-6 \pm \sqrt{6^2 - 4 \cdot 2 \cdot 1}}{2 \cdot 2}$$

$$x = \frac{-6 \pm \sqrt{36 - 8}}{4} = \frac{-6 \pm \sqrt{28}}{4}$$

$$x = \frac{-6 \pm \sqrt{4 \cdot 7}}{4} = \frac{-6 \pm 2\sqrt{7}}{4}$$

$$x = \frac{2(-3 \pm \sqrt{7})}{2 \cdot 2} = \frac{-3 \pm \sqrt{7}}{2}$$

Answer D is correct.

112. $S = \dfrac{a+b}{3b}$

$3bS = a + b$ Multiplying by $3b$

$3bS - b = a$

$b(3S - 1) = a$

$b = \dfrac{a}{3S - 1}$

Answer D is correct.

113. $3x - 4y = 12$

We find the intercepts. To find the y-intercept, let $x = 0$ and solve for y.

$3 \cdot 0 - 4y = 12$

$-4y = 12$

$y = -3$

The y-intercept is $(0, -3)$.

To find the x-intercept, let $y = 0$ and solve for x.

$3x - 4 \cdot 0 = 12$

$3x = 12$

$x = 4$

The x-intercept is $(4, 0)$.

Graph C is the only graph with these intercepts, so C is the correct answer.

114. $|x| = 12$

The solutions are the numbers whose distance from 0 is 12. They are -12 and 12.

115. $\sqrt{\sqrt{\sqrt{81}}} = \sqrt{\sqrt{9}} = \sqrt{3}$

116. $x^2 - bx + 225 = x^2 - bx + 15^2$

If the trinomial is a square, then:

$b = 2 \cdot 1 \cdot 15$ *or* $b = -2 \cdot 1 \cdot 15$

$b = 30$ *or* $b = -30$

117. Let y = the second leg of the small right triangle with hypotenuse $\sqrt{2}$ and one leg x.

$x^2 + y^2 = (\sqrt{2})^2$

$x^2 + y^2 = 2$

$y^2 = 2 - x^2$

$y = \sqrt{2 - x^2}$

Let z = the second leg of the small right triangle with hypotenuse 1 and one leg x.

$x^2 + z^2 = 1^2$

$x^2 + z^2 = 1$

$z^2 = 1 - x^2$

$z = \sqrt{1 - x^2}$

Then $y + z$ = the hypotenuse of the large right triangle with legs $\sqrt{2}$ and 1.

$(\sqrt{2})^2 + 1^2 = (y + z)^2$

$2 + 1 = (\sqrt{2 - x^2} + \sqrt{1 - x^2})^2$

$3 = 2 - x^2 + 2\sqrt{2 - x^2}\sqrt{1 - x^2} + 1 - x^2$

$3 = -2x^2 + 3 + 2\sqrt{(2 - x^2)(1 - x^2)}$

$2x^2 = 2\sqrt{2 - 3x^2 + x^4}$

$x^2 = \sqrt{2 - 3x^2 + x^4}$ Dividing by 2

$(x^2)^2 = (\sqrt{2 - 3x^2 + x^4})^2$

$x^4 = 2 - 3x^2 + x^4$

$0 = 2 - 3x^2$

$3x^2 = 2$

$x^2 = \dfrac{2}{3}$

$x = \sqrt{\dfrac{2}{3}}$ Taking the positive square root

$x = \sqrt{\dfrac{2}{3} \cdot \dfrac{3}{3}} = \dfrac{\sqrt{6}}{3}$

118. Since $(x-3)(x+3) = x^2 - 9$, the expressions are equivalent.

119. When $x = 0$, $\dfrac{x+3}{3} = \dfrac{0+3}{3} = 1 \neq 0$, so the expressions are not equivalent.

120. When $x = 1$, $(x+5)^2 = (1+5)^2 = 6^2 = 36$ and $x^2 + 25 = 1^2 + 25 = 1 + 25 = 26$. Since $36 \neq 26$, the expressions are not equivalent.

121. When $x = 3$, $\sqrt{x^2 + 16} = \sqrt{3^2 + 16} = \sqrt{9 + 16} = \sqrt{25} = 5$ and $x + 4 = 3 + 4 = 7$. Since $5 \neq 7$, the expressions are not equivalent.

122. By definition, $\sqrt{x^2} = |x|$ so the expressions are equivalent.